IMPULSE CONTROL and QUASI-VARIATIONAL INEQUALITIES

IMPULSE CONTROL

and QUASI-VARIATIONAL

INEQUALITIES

Alain BENSOUSSAN
Université Paris-Dauphine
& INRIA

Jacques-Louis LIONS
Collège de France
& INRIA

English version produced by
TRANS-INTER-SCIENTIA
P.O. Box 16, Tonbridge, TN 11 8DY, Kent, England

gauthier-villars

Translation of:
"Contrôle impulsionnel et inéquations quasi-variationnelles"

The present English language version was translated by J.M. COLE
and was produced by TRANS-INTER-SCIENTIA

British Library Cataloguing in Publication Data

Bensoussan Alain
Impulsive control and quasi-variational inequalities.
1. Control theory 2. Stochastic processes
3. Calculus of variations 4. Inequalities (Mathematics)
I. Title II. Lions, J.-L. III. Controle impulsionnel et inequations quasi-variationnelles.
English
629.8'312 QA402.3

AMS Subject classifications (1980):
65 L XX

ISBN-2-04-015577-5

Foreword

1. The general goal of this book, which is a sequel to the book *Applications of Variational Inequalities in Stochastic Control*(1), is to establish and study the relations that exist, via Dynamic Programming, between, on the one hand, stochastic control respectively by stopping times and impulse control, and on the other hand Variational(2) and Quasi-Variational(3) Inequalities, with the intention of obtaining constructive methods of solution by numerical methods.

There is a difference between the case of V.I.'s and that of Q.V.I.'s in the following sense: V.I.'s were *primarily* for solving problems in potential theory and unilateral mechanics, whilst Q.V.I.'s were *primarily* introduced (by the authors in 1973) for solving problems in stochastic impulse control, and *then* were used by several authors to solve certain problems in mechanics (or in control of distributed systems).

But if we ignore these purely chronological aspects, the cases of V.I.'s and Q.V.I.'s are conceptually analogous — in one sense that must be specified, V.I.'s are equivalent to control by stopping times, and Q.V.I.'s are equivalent to impulse control.

2. The V.I.'s met in *B.L. Vol.1*, and in this book, appear in the *following formal framework*: $\mathcal{A}$ denotes a second order elliptical or parabolic differential operator on a set $\mathcal{O} \subset \mathbb{R}^n$ or $Q = \mathcal{O} \times]0,T[$; we look for a function u $(=u(x)$ or $u(x,t)$, $x \in \mathcal{O}$, $t \in]0,T[)$ satisfying

$$(1)\qquad \begin{cases} \mathcal{A}u - f \leq 0 \quad , \qquad u - \psi \leq 0, \\ (\mathcal{A}u - f)(u - \psi) = 0 \quad \text{in } \mathcal{O} \text{ (or in } Q) \end{cases}$$

with the addition of suitable boundary conditions (and in the evolutionary case an initial condition(4)). This is what is called a V.I. *with an obstacle*, the obstacle being represented by ψ; this terminology comes from Mechanics, where ψ does *indeed* represent an obstacle! In the cases studied here ψ represents *a cost resulting from the fact of being stopped.*

Other cases are also studied in this book that also correspond to *integro-differential* operators $\mathcal{A}$, or even to the 'abstract' case where $\mathcal{A}$ is the generator of a

(1) Which may be referred to as *B.L. Vol.1.*
(2) Abreviated to V.I..
(3) Abreviated to Q.V.I..
(4) In fact a 'final' condition: $u(x,T)$ is given, since $\mathcal{A}$ is a retrograde parabolic operator.

semi-group(1); the integro-differential case corresponds to the case of diffusions with jumps.

The *formal framework for* Q.V.I.'s *is the following*: $\mathcal{A}$ is given as before, and *in addition there is a non-linear operator*

(2) $$v \to M(v)$$

essentially mapping bounded measurable functions into themselves, and having a certain number of properties, the principal one of which is that of being *increasing*:

(3) $$v_1 \leq v_2 \Rightarrow M(v_1) \leq M(v_2).$$

The *prototype of the operators M encountered* (2) is

(4) $$Mv(x) = k + \inf_{\xi} v(x+\xi), \qquad x \in \mathcal{O}, \quad \xi_i \geq 0, \forall i, x+\xi \in \mathcal{O}.$$

The Q.V.I. problem is then formulated thus: look for a function u ($= u(x)$ $u(x,t)$) such that

(5) $$\begin{cases} \mathcal{A}u - f \leq 0, \quad u - Mu \leq 0, \\ (\mathcal{A}u - f)(u - M(u)) = 0 \text{ in } \mathcal{O} \text{ (or in } Q), \end{cases}$$

(as well as conditions at the limits, and, if $\mathcal{A}$ is evolutionary, initial conditions).

We also add — only because this appears naturally in the theory of impulse control — the condition

(6) $$u \geq 0.$$

Naturally, the structure of (5) is analogous to that of (1), the 'obstacle' ψ here being replaced by an 'implicit obstacle' depending upon the solution sought. The terminology *Quasi*-Variational Inequality has been chosen as a result of this remark.

3. It is now convenient to specify *'what happens at the boundary of* $\mathcal{O}$*'*. In *B.L. Vol.1* we have studied the simplest case for control by stopping times: that in which the process is *stopped* if the trajectory (representing, for example, the state of stocks) arrives at the boundary of the open set $\mathcal{O}$ (which represents, for example, the volume in which stocks have been disposed of). In fact this case is *insufficient* for applications, as much for control by stopping times as for impulse control. We must consider processes *reflected at the boundary*, which leads to conditions of the *Neumann type* instead of *Dirichlet type* on the boundary of $\mathcal{O}$. In this book V.I. problems are studied for Neumann-type of conditions, and Q.V.I. problems are studied for Dirichlet *and* Neumann types of boundary conditions.

---

(1) *Bilateral* Q.V.I.'s are also met, arising in the theory of stochastic games, and which are expressed by:

$$\psi_1 \leq u \leq \psi_2, \qquad \mathcal{A}u = f \text{ if } \psi_1 < u < \psi_2, \ \mathcal{A}u - f < 0 \text{ (resp } > 0) \text{ if}$$
$$u = \psi_2 \text{ (resp. } u = \psi_1).$$

(2) Here we emphasize the fact that this type of non-linear and non-local operator has been introduced for the first time for the purpose of applications to problems of stock management.

4. Let us now give a little detail about the content of the various chapters.

Chapter 1 gives *numerous examples* which occur in the applications and lead to V.I.'s and Q.V.I.'s via dynamic programming (used *formally* in this chapter 1). In particular, numerous variants of the operators M will be found here, of which (4) gives an example.

The reader who is first interested in the motivations and then in the mathematical methods for solving the problems thus formulated can go directly from Chapter 1 to Chapters 4 and 5 (looking at homographic penalisation in Chapter 2 on the way).

Amongst other things, Chapter 1 shows it to be indispensible to consider *reflected processes*, which are therefore presented (as directly as possible) in Chapter 2. Then Chapter 2 adapts, in some way, the results of *B.L. Vol.1* relating to the Dirichlet conditions at the boundary to the case of Neumann boundary conditions.

Chapter 3 studies the case in which $\mathcal{A}$ is an *integro*-differential operator; in the first part of the Chapter the study is carried ahead by purely analytical methods, probabilistic methods being introduced in the second part of the Chapter. Chapter 3 is not indispensible for reading the following chapters, save for that which concerns the interpretation of the penalised equation approximating the Q.V.I.(1).

Chapter 4 studies from an analytical viewpoint the elliptic (stationary) Q.V.I.'s (corresponding to an infinite horizon in the applications), and Chapter 5 — still from an analytical point of view — studies the parabolic Q.V.I.'s of evolution. By way of a supplement we have, in this Chapter, added some remarks — simple and not exhaustive — about second order hyperbolic Q.V.I.'s; this is not pursued very much, although the mathematical problems met seem to present a certain intrinsic interest.

Since the Q.V.I.'s are solved *from an analytical point of view*, we need next to see *how to reconstruct the* (or *an*) *optimal control starting from the solution of the Q.V.I.* (and this does exist!); this is the object of Chapter 6.

5. As has already been indicated from the beginning, a constant concern in this book (as well as in *B.L. Vol.1*) is to arrive at *constructive solutions*. Chapter 6 shows that to the degree that the numerical calculation of the solution of the Q.V.I. is possible, the optimal strategies may be deduced.

A systematic presentation of numerical methods is not given here. The situation here is the following: by using essentially the estimations of Hanouzet-Joly presented in Chapter 4 the convergence of the various iterative procedures is established; numerous cases of actual calculations have been treated (cf. the references in the Bibliography, and Chapter 1).

6. An (interesting and considerable) analytical difficulty met in Q.V.I.'s is the following: even for a regular function u, the function $M(u)$ in general *is not* regular, so that whatever method may be used for solving (5) the V.I. (1) for 'irregular' obstacles ψ has to be considered. The methods used for solving (5) rest upon fixed point theorems: we solve

$$(7) \qquad \mathcal{A}w - f \le 0 \ , \quad w\, M(u) \le 0 \ , \quad (\mathcal{A}w - f)(w - M(u)) = 0$$

(with adequate boundary conditions and initial conditions), and we look for u so that $w = u$. Every *algorithm* allowing us to approximate a fixed point of the mapping $u \to w$ then gives an algorithm for calculating a solution of the Q.V.I..

(1) The case of the delay is brought up in Chapter 1, but it is not treated in this book. We refer to M. ROBIN [3].

7. As is indicated in Chapter 1, we also meet situations leading to problems analogous to those studied here, but with operators $\mathcal{A}$ that are *degenerate* or contain *'small parameters'* which leads to the problems of *singular perturbations in* Q.V.I.'s, a subject that is not embarked upon here.

Table of contents

Chapter 1

Examples of the application of stochastic control and impulse control to management problems

INTRODUCTION

Stochastic control and impulse control have found numerous applications in management problems. That should not be very surprising, because management often encounters the situation of dynamical systems evolving under conditions of uncertainty, and where decisions have to be taken in order to optimise an economic criterion.

The word 'management' must be taken in a very broad sense here, including not just the problems of business, but also of energy, natural resources, information networks, etc., ...

The examples described below are numerous, but they are far from exhausting the range of possibilities. They do not all have the same degree of reality. Certain of them are very close to implementation, others are only theoretically possible applications which will have to be tested. As will be seen, theory as such is rarely applied, and needs to be adapted. Two important difficulties are met almost systematically. First of all, certain components of the system evolve deterministically, others randomly. The mathematical difficulty of partially 'degenerate' problems is then met. *Lastly,* and *above all, there is the difficulty of large systems, generally very big ones.* This difficulty is a *numerical* one, but implies the search for decomposition algorithms, proceeding to approximations, or having recourse to partially heuristic rules.

This gives an idea of the problems met in practice when adapting the general methodology. Another serious difficulty is connected with the identification of models. We shall not speak of it here, for our concern is above all to illustrate the theory of stochastic control, but it is a necessary step.

The majority of examples described here have been developed as much in the area of modelling as in the numerical area, at INRIA in particular about energy, within the confines of an EDF-INRIA collaboration. Therefore we have not sought to be exhaustive in our examples. The literature on the subject is quite vast, and certain fields are not mentioned here, even though they have numerous applications. For certain extra information one can refer to BENSOUSSAN-HURST-NÅSLÜND [1], SETHI-THOMPSON [1], BENSOUSSAN-KLEINDORFER-TAPIERO [1], and to the corresponding references.

In presenting applications we have not sought mathematical rigour, nor gone on to a complete development. A good knowledge of the techniques (or at least of the ideas) of stochastic control and impulsive control is useful, therefore. This is why we shall give in Section 1, to facilitate the reading of this book independently of all other references, an outline of the principal theoretical ideas, limiting ourselves, naturally, to essentials, and leaving out proofs.

1. THE ESSENTIAL IDEAS OF STOCHASTIC CONTROL AND IMPULSE CONTROL

1.1 *Modelling*

Here we shall be interested in dynamical systems. The word 'system' will be applied to any physical or economical phenomenon whose evolution changes with the flow of time. A dynamical system is characterised by its *state* at each instant t. The state of the system represents the set of *quantitative*[1] variables, forming an *exhaustive*[2] description of the system.

Once the state is defined, the task of modelling consists of describing the law of this state's evolution as a function of time. The time may be discrete or continuous. We shall assume that it varies continuously, but everything we shall say goes over to the discrete case.

The horizon (the interval in which time varies) may be finite or infinite. The state variables can be real numbers or integers. In general we shall assume they are real numbers, but we shall say a few words about the case of integers.

The state of the system at the instant t will be denoted y(t). Hence if the state variables are real numbers, and n in number, $y(t) \in R^n$.[3]

The mapping $t \to y(t)$ describes the evolution of the system.

This evolution is provided by a *model*. Naturally, a great variety of them is possible, models can be deterministic, stochastic, delayed, and the mapping y(t) may be continuous or discontinuous.

We shall only consider the deterministic and stochastic cases without delay, leading to a continuous function y(t). In the deterministic case the most frequent model is that in which y(t) is the solution of a differential equation, say,

$$(1.1) \qquad \frac{dy}{dt} = g(y(t),t) \quad , \qquad y(o) = y_o \; .$$

In practice, the model (1.1) expresses the fact that the variation in the state between the instants t and $t + \Delta t$ is given by

$$(1.2) \qquad \Delta y(t) = g(y(t),t)\,\Delta t \; .$$

In the stochastic case the variation $\Delta y(t)$ is random. In fact it is natural to assume that a (small) random number, with zero mean, is added to the right hand side of (1.2). In the absence of additional information, it is simplest to assume that this random number is normally distributed and that two (small) random numbers corresponding to two non-overlapping time intervals are independent. Naturally, the variance of the law for $\Delta y(t)$ (which is a matrix, since Δy is a vector) may depend upon y(t). For this, one generally adopts the model

$$(1.3) \qquad \Delta y(t) = g(y(t),t)\Delta t + \sigma(y(t),t)\Delta w(t)$$

where σ is a matrix and $\Delta w(t)$ is normally distributed with mean 0 and variance $I\Delta t$ (I is the identity matrix). As we want the quantities $\Delta w(t)$ for non-overlapping intervals Δt to be independent variables, it suffices to assume that w(t) is an n-dimensional *Wiener* process, and

---

(1) In practice very many variables are qualitative. It is a question of quantifying them. The precise definition of the state variables is the first step in modelling.
(2) This notion is entirely relative. Descriptions can be more accurate or less so. Everything depends upon the use that one has in view.
(3) Here we shall exclude 'distributed' systems in which the state depends not only upon time, but also upon a spatial variable x.

$$\Delta w(t) = w(t+\Delta t) - w(t) .$$

The properties wanted are then verified. When $\Delta t \to 0$, (1.3) is generally written in the differential form

$$dy(t) = g(y(t),t)dt + \sigma(y(t),t)dw(t) \tag{1.4}$$

and not in the differentiated form, for although the Wiener process is continuous and has bounded variation, it is not differentiable.

Equation (1.4) is called a stochastic differential equation (in Ito's sense).

The process y(t) which is the solution of (1.4) is a continuous Markov process called a *diffusion process* (g is called the *drift of the process* and σ the *diffusion term*).

If σ is not *degenerate* (i.e., $\sigma\sigma^* \geqq \alpha I$, $\alpha > 0$), dy(t), and therefore y(t) can take any value in R^n (y(t) can lie in any region of R^n with a non-zero probability by virtue of the properties of the normal law). This can be troublesome in certain applications where one wants y(t) to remain in the interior of a region $\mathcal{O} \subset R^n$ (for example, if y(t) is a stock one wants $y(t) \geqq 0$). In this case one might prefer, to the model (1.4), a *reflected diffusion* model:

$$dy(t) = g(y(t),t)dt + \sigma(y(t),t)dw(t) + \gamma(y(t),t)d\xi(t) \tag{1.5}$$

where $\gamma(x,t)$ is a vector which, for x on the boundary of $\mathcal{O}$, points to the *interior* of $\mathcal{O}$, and $\xi(t)$ is an increasing scalar process ($d\xi \geqq 0$) such that $\chi_{\mathcal{O}}(y(t))d\xi(t) = 0$, in other words which only increases on the boundary of $\mathcal{O}$. The model (1.5) can be obtained as the limit of a process receiving small impulses on the boundary. In other words, whenever y(t) reaches the boundary, it is made to jump instantaneously in the direction γ by a small amount ε. Since γ points towards the interior of $\mathcal{O}$, such a process remains in the interior of $\mathcal{O}$ for every $\varepsilon > 0$. Nevertheless, it is discontinuous, but at the limit when $\varepsilon \to 0$ one obtains a continuous process, which justifies the interest in the model (1.5).

Naturally, the models described above are models of free evolution, i.e., free from control. We must now describe the way in which control can influence the evolution of the state.

We shall consider two types of control, *continuous control* and *impulse control*.

In continuous control, in general it is assumed that the drift g depends, besides upon the state x and time t, upon a control term v, so $g \equiv g(x,v,t)$. The term σ (and in the case of the model (1.5) γ) may also depend on the control.

The control is a process v(t) whose value can be decided at every instant t as a function of available information at that moment. In general the value v(t) must satisfy certain constraints, $v(t) \in \mathcal{U}_{ad}$ (the admissible set).

If the evolution of y(t) with the course of time can be observed (without, naturally, anticipating the future) it is said that the *observation is complete*. Otherwise is is said that the *observation is partial*.

The model (1.4) is now written

$$dy(t) = g(y(t),v(t),t)dt + \sigma(y(t),t)dw(t) . \tag{1.6}$$

In *impulse control* it is assumed that at certain instants (*impulse instants*) the state undergoes jumps (*impulses*). The impulse instants, the intensity of the impulses, and the number of them are decision variables, the *collection* of which is called an *impulse control*.

Thus, let $\theta^1 \leqq \theta^2 \leqq ..$ be the sequence of impulse instants, and $\xi^1, \xi^2, \ldots$ be the sequence of corresponding intensities of the jumps; the evolution of the state is then described in the following way:

(1.7)
$$\begin{cases} dy = g(y(t),t)dt + \sigma(y(t),t)dw(t) , & \theta_i \leq t < \theta_{i+1} \\ y(\theta_i) = y(\theta_i - 0) + \xi_i \\ y(0) = y_o \end{cases}$$

where $y(\xi_i - 0)$ represents the quantity,

$$y(\theta_i - 0) = y(\theta_{i-1}) + \int_{\theta_{i-1}}^{\theta_i} g(y(t),t)dt + \int_{\theta_{i-1}}^{\theta_i} \sigma(y(t),t)dw(t) .$$

The instants θ_i, ξ_i are random in general, and can be chosen as a function of the available information. Thus the events $\{\theta_i \leq t\}$ depend upon information available up to the instant t, and the value of ξ_i depends upon information available up to θ_i. The θ_i's are stopping times, according to the usual terminology of probability calculus.

Naturally, it is possible to have a *continuous control and an impulsive control at the same time.*

In the case where y(t) has integers as values the variation on the interval Δt cannot be proportional to Δt since they are integers. The law $\Delta y(t) \in \mathcal{Z}$, has to be given, a law dependent upon $y(t), \Delta t$, and on the control.

For example, in the case of a life and death process where the rates of birth and death are controlled, one will have

(1.8)
$$\Delta y(t) = \begin{cases} 1 \text{ with a probability } \lambda(y(t),v(t),t)\Delta t \\ -1 \text{ with a probability } \mu(y(t),v(t),t)\Delta t \\ 0 \text{ with a probability } 1 - (\lambda + \mu)\Delta t . \end{cases}$$

If the process is controlled impulsively, we shall have

(1.9)
$$\begin{cases} \Delta y(t) = 0 , & \theta_i \leq t < \theta_{i+1} \\ y(\theta_i) = y(\theta_{i-1}) + \xi_i \end{cases}$$

where ξ_i has integral values.

Let us indicate a *very important* particular case of impulse control, which is the problem of stopping times. In this problem the evolution of the state, described by (1.6), is not touched, but one decides to interrupt this evolution at an instant θ, which is a stopping time. This instant is the decision variable.

Finally, to close this Section, let us recall that once the nature of model, deterministic, stochastic, real- or integer-valued, with or without constraints, is chosen, it is convenient to *identify* the functions occurring in the analytic description of the model.

We shall not start upon this problem here, but it is a crucial one.

1.2 *The Economic Function*

In continuous control the criterion in general is written (to minimise it, to fix our ideas) in the form

(1.10)
$$J = E\left[\int_0^T e^{-\alpha t} f(y(t),v(t),t)dt + g(y(T))\, e^{-\alpha T}\right]$$

where α is the discount factor (possibly zero) and T may be $+\infty$, T is the horizon, f the integral cost, g the final cost. For an optimal stopping time problem the criterion is written

$$(1.10') \quad J = E\left[\int_0^{T\wedge\theta} e^{-\alpha t} f(y(t),t)dt + g(y(T))\chi_{T<\theta}\, e^{-\alpha T} + \psi(y(\theta),\theta)\chi_{\theta\leq T}\, e^{-\alpha\theta}\right]$$

where ψ represents the cost when the system is stopped before the horizon. In impulse control the criterion is written in the form

$$(1.11) \quad J = E\left[\int_0^{T} e^{-\alpha t} f(y(t),t)dt + \sum_i c(\xi_i)e^{-\alpha\theta_i}\right]$$

where $c(\xi_i)$ is the cost sustained at each impulse. In general we have

$$c(\xi) \geqq k > 0 ,$$

where k represents the minimum fixed cost per impulse. Here there is a very important difference between the situations of continuous and impulse control. In continuous control the cost is 'in a certain sense' proportional to the duration for which control is applied. In impulse control, however, the decision to exercise control implies a minimum fixed cost independent of the level of control. This clearly implies the impossibility of applying such control continuously.

1.3 *Notions of Value and Price*

We now consider a (continuous or impulse) control problem with variable initial conditions, i.e., starting at the instant t in the state x. In other words we consider the evolutions (1.6) or (1.7) for $s \geqq t$, and

$$y(t) = x .$$

To fix our ideas let us take the case of continuous control. We write

$$(1.12) \quad J_{x,t}(v(.)) = E\left[\int_t^{T} e^{-\alpha(s-t)} f(y(s),v(s),s)ds + g(y(T))\, e^{-\alpha(T-t)}\right]$$

and set

$$(1.13) \quad \Phi(x,t) = \operatorname*{Inf}_{v(.)} J_{x,t}(v(.)) .$$

The function $\Phi(x,t)$ has an interesting economic interpretation. To understand it better, it is preferable to go to a maximisation problem. We notice that

$$-\Phi(x,t) = \operatorname*{Sup}_{v(.)} (-J_{x,t}(v(.))) .$$

The function $-J_{x,t}(v(.))$ (which we seek to maximise) is then the *profit* that can be realised by starting from the conditions x,t and by applying the control v(.). Therefore $-\Phi(x,t)$ *represents the profit resulting from the disposition at the instant* t *of the level* x of the state (which could, for example, be a stock of goods, capital, etc., ...).

Hence it is natural to call $-\Phi(x,t)$ the *value of the state* x *at the instant* t. In fact the state x only has a value at the instant t to the extent that having it implies a profit in the future. This is a natural outcome of a purely economic frame of reference.

The vector

(1.14) $$p(x,s) = -\frac{\partial\Phi}{\partial x}(x,s)$$

represents the *prices* vector. In fact, $-\dfrac{\partial\Phi}{\partial x}(x,s)$ is the *marginal revenue*, that is to say the revenue corresponding to having an additional unit of each of the components of the set. According to the law of supply and demand, $p(x,t)$ represents the market price at the instant t when the quantity available is x.

1.4 *Dynamic Programming*

The function $\Phi(x,t)$ which we have just seen has an interesting economic interpretation, is also the solution of an important analytical problem. To obtain it we can carry out a simple (formal) argument based on the optimality principle of dynamic programming introduced by R. BELLMAN [1]. This is not a case of justifying it, but a quick and intuitive way of understanding the origin of the equations or of the sets of inequalities and equations studied in this book.

We begin with the case of continuous control without constraints on the state. Here we systematically place ourselves under the hypothesis of a *complete observation*; the case of a partial observation is very much more complex.

Let us assume that we exercise a control $v(s)$ during the interval $t, t+\delta$ (δ small). At the instant $t+\delta$ the state of the system becomes $y(t+\delta)$, and we know that we can observe it at the instant $t+\delta$.

Let us assume that starting from $t+\delta$ we know the optimal policy that it is suitable to apply, taking into account that at the instant $t+\delta$ the state of the system if $y(t+\delta)$, in other words let us assume known the control $u(s)$, $s \in [t+\delta, T]$, such that

$$\Phi(y(t+\delta),t+\delta) = E\left[\int_{t+\delta}^{T} e^{-\alpha(s-(t+\delta))} f(y(s),u(s),s)ds + g(y(T))\, e^{-\alpha(T-(t+\delta))} \,\Big|\, y(t+\delta)\right]^{(*)}.$$

Applying a control

$$\tilde{v}(s) = \begin{vmatrix} v(s) & \text{on } t,t+\delta, \\ u(s) & \text{for } s > t+\delta, \end{vmatrix}$$

we have

(1.15) $$J_{x,t}(\tilde{v}(.)) = E\left[\int_{t}^{t+\delta} e^{-\alpha(s-t)} f(y(s),v(s),s)ds + e^{-\alpha\delta}\,\Phi(y(t+\delta),t+\delta)\right].$$

The optimality principle tells us that if $v(s)$ is chosen on $t, t+\delta$ so as to minimise the expression (1.15), we then obtain the optimal cost. Another way of saying this is to say that the optimal policy on (t,T) can be split into $u(s)$, s $(t,t+\delta)$ and $u(s)$, s $(t+\delta,T)$; the latter is also the optimal policy for a

---

(1) It must be noted that here it is the conditional expectation that appears, for the value $y(t+\delta)$ is random, and by the definition of $\Phi(x,t+\delta)$ the value of the state at the instant $t+\delta$ has to be known.

problem starting at $t+\delta$ in the state $y(t+\delta)$. Therefore we have

$$(1.16) \qquad \Phi(x,t) = \operatorname*{Inf}_{v(t,t+\delta)} E\left[\int_t^{t+\delta} e^{-\alpha(s-t)} f(y(s),\, v(s),s)ds + e^{-\alpha\delta}\,\Phi(y(t+\delta),t+\delta)\right].$$

We can therefore expand the right hand side of (1.16) up to the first order in δ. Here we note the fact

$$E\,\Phi(y(t+\delta),t+\delta) = \Phi(x,t) + \delta\,\frac{\partial\Phi}{\partial x}.g(x,v,t) + \delta\,\frac{\partial\Phi}{\partial t} + + \delta\sum_{ij}\frac{\partial^2\Phi}{\partial x_i\partial x_j}\,a_{ij}(x,t) + O(\delta)$$

where a_{ij} represents the matrix $\frac{1}{2}\sigma\sigma^*$. This follows from the properties of the Wiener process, i.e.,

$$E(w(t+\delta) - w(t))^2 = \delta\ .$$

Under these conditions we deduce from (1.16), by making δ tend to 0, that

$$(1.17) \qquad -\frac{\partial\Phi}{\partial t} - \sum_{ij} a_{ij}\frac{\partial^2\Phi}{\partial x_i\partial x_j} - \operatorname*{Inf}_{v\in\mathcal{U}_{ad}}\left[\frac{\partial\Phi}{\partial x}\cdot g(x,v,t) + f(x,v,t\right] + \alpha\Phi = 0\ ,$$

which is a non-linear partial differential equation satisfied by Φ. This equation is called the *Dynamic Programming Equation*, or the *Hamilton-Jacobi-Bellman Equation*.

It is necessary to add to (1.17) the initial (or, more exactly, final) condition

$$(1.18) \qquad \Phi(x,T) = g(x)$$

which immediately follows from the definition of Φ.

The problem (1.17),(1.18) is a well set problem (Cauchy problem) whose solution is $\Phi(x,t)$.(1)

When the *state has constraints* we have to specify the criterion which can no longer be reduced to (1.10). There are then two options. The first is to keep the Equation (1.6), but then we take as the horizon not T but $T\wedge\tau$, where τ *represents the first instant* when the process y leaves the constraints' domain. Such a situation arises, as we shall see, in financial problems where τ is interpreted as an instant of insolvency. In this case it is convenient to add to (1.17) and (1.18) the condition

$$(1.19) \qquad \Phi(x,t) = 0 \quad ,\quad x\in\Gamma$$

where Γ is the boundary of $\mathcal{O}$.

The second option is to use a reflected diffusion model (1.5). In this case the cost function can still be given by (1.10). It is then convenient to add to the conditions (1.17),(1.18) by replacing (1.19) with the condition

---

(1) Naturally, this assumes differentiability properties of Φ which are to be specified.

(1.20) $$\frac{\partial\Phi}{\partial x}\cdot\gamma = 0 \quad , \quad x\in\Gamma .$$

This condition is explained as follows. The reflected process can be approximated by a sequence of processes such that on the boundary the process jumps instaneously from ε in the direction γ. In other words

$$y(\theta) = y(\theta - o) + \varepsilon\gamma(y(\theta - o),\theta)$$

if $y(\theta - 0)\in\Gamma$. But then it is clear, for the problem corresponding to the approximation process, and calling $\Phi^{\varepsilon}(x,t)$ the associated Bellman function, that we have

$$\Phi^{\varepsilon}(x,t) = \Phi^{\varepsilon}(x+\varepsilon\gamma(x,t),t) \quad , \quad x\in\Gamma .$$

By making ε tend to 0 we see that the limit function Φ of Φ^{ε} must satisfy (1.20). Let us now go on to the case of *impulsive control*.

The evolution of the system is described by (1.7) (we consider the constraint-free case to simplify matters). The economy function is defined by (1.11). In this case the function $\Phi(x,t)$ is the solution of an analytic problem that cannot be reduced to an equation. Nevertheless, the dynamic programming argument is still valid, but with certain adaptations.

First of all we notice that at the instant t it is permissible to decide to give an immediate impulse. In this case, assuming that at the release of the first decision which transforms the state from x to $x+\xi$ we apply a sequence of optimal decisions, the resultant cost is

$$C(\xi) + \Phi(x+\xi) .$$

Making an optimal choice for ξ, we obtain

$$\operatorname*{Inf}_{\xi}\,[C(\xi) + \Phi(x+\xi)]$$

and by the definition of Φ we can write

(1.21) $$\Phi(x,t) - \operatorname*{Inf}_{\xi}\,[C(\xi) + \Phi(x+\xi,t)] \leq 0 .$$

On the other hand, if an immediate impulse is not given, this means that the system is left to evolve freely, at least during a 'small' time interval $(t,t+\delta)$. In this case the supported cost is given, by an argument similar to that in the proof of (1.17), by

$$\Phi(x,t) + \delta\frac{\partial\Phi}{\partial x}\cdot g(x,t) + \delta\frac{\partial\Phi}{\partial t} + \delta\sum_{ij}\frac{\partial^2\Phi}{\partial x_i\partial x_j}a_{ij}(x,t) + \delta f(x,t) - \alpha\delta\Phi(x,t) + O(\delta) ;$$

this cost must therefore be greater than or equal to $\Phi(x,t)$ by definition. From this it follows that by making δ tend to 0 we have

(1.22) $$-\frac{\partial\Phi}{\partial t} - \frac{\partial\Phi}{\partial x}\cdot g(x,t) - \sum_{ij} a_{ij}\frac{\partial^2\Phi}{\partial x_i\partial x_j} + \alpha\Phi - f \leq 0 .$$

As may be seen, one or the other of the two decisions described above (where an impulse may be given or not) necessarily must be taken, from which it follows that one or the other of the inequalities (1.21),(1.22) is in fact an equation, which can be expressed as the product of the quantities (1.21),(1.22) having to vanish. To this we add the final condition (1.18).

The optimal stopping time problem is simpler. Let us briefly give the result, restricting ourselves to the constrain-free case (the theory will be developed in the general case, cf., Chapter 2).

The evolution of the system is described by (1.6), and the economic function is defined by (1.10'). It is clear that $\Phi(x,t)$ will again satisfy condition (1.22) (corresponding to the case where the system is left to develop freely during a very short period). On the other hand, at the instant t it is possible to stop immediately, in which case we obtain a cost equal to $\Phi(x,t)$, therefore

(1.23) $\varphi(x,t) - \psi(x,t) \leqq 0$

which plays the role of (1.21). We also have that the product of (1.22) and (1.23) is equal to 0.

1.5 *Obtaining the Optimal Policy: Feedback*

We must now specify how to obtain the optimal policy. First let us remark that we must give the value of the optimal control at every instant t as a function of the available information at t. In the case of impulse control we have to decide at every instant whether it is time to give an impulse or not, and if so then at what level.

The information available at the instant t consists of the values of the state and control at the previous instants. We say that we have *Markov control*, or a *feedback* control if the optimal value at every instant depends only on the state at that instant.

This is one of the fundamental results of Stochastic Control Theory, which in the case of complete information says that there exists an optimal control that is Markovian. This is given in the following way for the continuous control case. If a minimum is reached in the bracket in (1.17) considering x,s as parameters, we define a function V(x,s). Then at each instant t the optimal control is

(1.24) $v(t) = V(y(t),t)$

which is certainly Markov.

In the case of impulse control, it is convenient to operate as follows. Write

$$C = \left\{ x,s \,|\, \Phi(x,s) < \operatorname*{Inf}_{\xi} \left[c(\xi) + \Phi(x+\xi,s) \right] \right\}$$

and $\hat{\xi}(x,s)$ is a function with x,s fixed that produces the lower bound of the bracketed expression.

We call C the *continuation set*. Then at every instant t we look to see if the state y(t) is such that $(y(t),t) \in C$. If it is, it is convenient to let the system evolve freely, otherwise an impulse of value $\hat{\xi}(y(t),t)$ must be given at the instant t. From (1.16) and the limited development that follows, we can intuitively understand the feedback (1.24).

Similarly, if $(y(t),t) \in C$ there is no interest in exercising a control at this instant, because we then recover at best $\operatorname*{Inf}_{\xi}[C(\xi) + \Phi(y(t) + \xi,t)]$, when the optimal cost $\Phi(y(t),t)$ is strictly less than the preceding one.

For the stopping time problem the argument is similar, this time defining the continuation set by

$$C = \{x,s \,|\, \Phi(x,s) < \psi(x,s)\} .$$

2. THE MANAGEMENT OF ELECTRICITY PRODUCTION(1)

2.1 *General Description of the Problem*

To provide for the demand for electricity at every instant an electricity company runs a system of hydroelectric and thermal power stations. This system must be managed in such a way as to minimise the running costs, taking shut downs into account. In its entirety the problem is very complex, on account of:

- the very large number of power stations (several hundreds),
- the presence of uncertainties in demand, reserves of water, and the availability of power stations,
- of the very different characteristics of hydroelectric and thermal power stations.

For the thermal group, at every moment it must be decided whether a shut down power station must be started up, or whether one in operation must be shut down. One must also take into account that once started, a power station does not function properly for several hours. Once it is working, the power provided must be decided. Furthermore, the thermal group has several categories of power stations.

For the hydroelectric group the situation is as follows. Valleys are available, each of which has a large dam (at the head) and secondary dams downstream. This is a somewhat simplified view of the situation, we admit. The large dam serves essentially as the stock of water, and the small ones produce a large part of the hydraulic electricity. For each dam we must decide how much water to put through the turbines.

In addition there is a time-scale problem, and a seasonal one also. Demand, as well as water reserves, has a periodicity of the order of one year.

In contrast, the variations in electricity demand during the course of a day change over the order of one hour. It is quite difficult to view such a step in time in the context of one year.

The complete problem is enormous. Therefore we must arrange for simplified models, and even for a battery of overlapping models.

The first thing to do is to work with two distinct time intervals. First of all with an interval of several years, with steps in time of a week, then in an interval of one week with steps of an hour. A model based on an interval of several years is called *long term*, and one based on a week is called *short term*.

The long term model has the object of optimising large dams, that is to say to provide the quantities of water destocked 'on average' each week. It also provides the *value* of the disposable stock of water each week.

To establish the long term model it is necessary to have at our disposal a function for the cost of production of electricity by purely thermal means per unit time (here a week). If we *neglect* at this level the start-up costs of thermal power stations, this function is obtained simply by bringing in the power stations in the order of their increasing running costs, so as to acheive the desired production. This problem would be very much more complex if we wanted to account, at this level, for their start-up costs.

The value of a stock of water, or the quantity of water destocked in a week, then appears as external data for establishing a short term model, over a week, with time increments of one hour. We then need to globally optimise the short term thermal and hydraulic system.

Although we have reduced the horizon, it is still a problem of considerable

---

(1) This Section describes only very partially the works of DELEBECQUE-QUADRAT [1],[2],[3] and COLLETER-DELEBECQUE-FALGARONNE-QUADRAT [1].

magnitude, which is beyond our grasp in general. Here we describe the global solution for a simplified model valid in the cases where the hydroelectric system is concentrated on a small number of large dams.

2.2 *A Simplified Model of Long-Term Management*

We have at our disposal a set of n valleys provided with a dam at the head and stations downstream. Here the essential simplification is the absence of secondary dams. We develop a continuous model with a horizon T. In practice we can take T = several years, and use a discount factor(1).

We write:

(2.1) $y_i(t)$ = disposable stock of water in the valley number i, at the instant t,

(2.2) $v_i(t)$ = quantity of water through the turbines at the instant t in the valley number i per unit term,

(2.3) $\underline{y}_i, \overline{y}_i$ are the minimum and maximum quantities (respectively) stocked in the valley number i,

(2.4) $0 \leq v_i(t) \leq \overline{v}_i$,

where $\overline{v}_i$ denotes the maximum quantity of water that can be put through the turbines per unit time.

The behaviour of the stocks of water is described by the equations

(2.5) $$dy_i(t) = d\eta_i(t) - v_i(t)dt$$

where $d\eta_i(t)$ represents the contributions of water at the instant t in the i-th valley during the time interval dt. We assume the relation

(2.6) $$d\eta_i(t) = g_i(t)dt + \sigma_i(t)dw_i(t)$$

where $g_i(t), \sigma_i(t)$ are deterministic functions, and $w_i(t)$ is a standard Wiener process. The representation (2.6) means that $d\xi_i(t)$ is a Gaussian variable with mean $g_i(t)dt$ and variance $\sigma_i^2(t)dt$. Furthermore, the variables $d\xi_i(t)$ are independent when i and t vary. This is the simplest representation of reality. Naturally, we assume the problem of identifying the functions $g_i(t), \sigma_i(t)$ is solved.

We write

(2.7) $W_i(y_i, v_i, t)$ = amount of electrical energy produced per unit time at the instant t by the stations in valley number i, when the water stock is y_i, and the amount of water run through the turbines is v_i.

Therefore

$W = \sum_i W_i(y_i, v_i, t)$ is the global quantity of electricity produced at the instant t per unit time.

Also, let

(2.8) D(t) = internal demand for electricity per unit time at the instant t.

---

(1) If T = 1 year, it would be necessary to forecast some final conditions.

Here we assume that $D(t)$ is a known deterministic function of the time. We are again given

(2.9) $C_1(W)$ = production cost of a quantity W of hydroelectrically generated electrical energy per unit time,

(2.9') $C_2(D(t) - W)$ = thermal cost per unit time at the instant t.

The preceding function does not simply express a production cost, but results itself from an optimisation problem. In fact we have

$$D(t) - W = P - \Delta$$

where P is the amount of thermal electricity actually produced, and Δ the amount of electricity sold abroad.

When W is fixed, we look for a solution of the problem

$$(2.10) \quad \left| \begin{array}{l} \underset{P}{\mathrm{Min}} \quad C_3(P) - R(P+W-D(t)) \\ \\ P + W - D(t) \geqq 0 \,, \quad P \geqq 0 \end{array} \right.$$

where $C_3(P)$ denotes the production cost of thermal electricity, and $R(\Delta)$ the revenue from the sale abroad of excess electricity. The constraint $P + W - D(t) \geqq 0$ means that the internal demand must be met. Naturally, we can replace this constraint with a penalty. The result of the problem (2.10) yields the function (2.9).

To take the data (2.3) into account we impose upon the controls v_i the additional constraints

$$(2.11) \quad \left| \begin{array}{lll} v_i(t) = \overline{v_i} & \text{if} & y_i(t) \geqq \overline{y_i} \\ v_i(t) = 0 & \text{if} & y_i(t) \leqq \underline{y_i} \end{array} \right.$$

which says that we are now seeking to keep at a maximum the stock of water $y_i(t)$ between $\underline{y_i}$ and $\overline{y_i}$.

If α denotes the discount factor, we have then to solve the problem

$$(2.12) \quad \mathrm{Min}\, E \int_0^T e^{-\alpha t} \left\{ C_1\left(\sum_i W_i(y_i, v_i, t) \right) + C_2\left(D(t) - \sum_i W_i(y_i, v_i, t) \right) \right\} dt$$

$$(2.13) \quad \left| \begin{array}{l} dy_i = g_i(t)dt + \sigma_i(t)dw_i - v_i(t)dt \\ y_i(o) = y_i^o \end{array} \right.$$

$$(2.14) \quad \left| \begin{array}{lll} 0 \leqq v_i(t) \leqq \overline{v_i} & & \\ v_i(t) = \overline{v_i} & \text{if} & y_i(t) \geqq \overline{y_i} \\ v_i(t) = 0 & \text{if} & y_i(t) \leqq \underline{y_i} \,. \end{array} \right.$$

We then consider the function $\Phi(x,t)$, $x \in R^n$, the solution of

$$(2.15) \quad -\frac{\partial \Phi}{\partial t} + \sum_i A_i(t)\Phi - H(\nabla \Phi, x, t) + \alpha \Phi = 0 \qquad \Phi(x,T) = 0$$

where

$$(2.16) \qquad A_i(t)\Phi = -\frac{\partial\Phi}{\partial x_i} g_i(t) - \frac{1}{2}\sigma_i^2(t)\frac{\partial^2\Phi}{\partial x_i^2}$$

$$(2.17) \qquad \left| \begin{array}{l} H(p,x,t) = \displaystyle\inf_{\substack{0 \leq v_i \leq \overline{v_i} \\ v_i = 0 \text{ if } x_i \leq \underline{y_i} \\ v_i = \overline{v_i} \text{ if } x_i \geq \overline{y_i}}} L(p,v,x,t) \end{array} \right.$$

with

$$(2.18) \qquad L(p,v,x,t) = -\sum_i p_i v_i + C_1\left(\sum_i W_i(x_i, v_i, t)\right) + C_2\left(D(t) - \sum_i W_i(x_i, v_i, t)\right)$$

where, naturally,

$$p = (p_1 \ldots, p_n), \qquad v = (v_1, \ldots, v_n).$$

The general theory then gives us that $\Phi(x,0)$ is equal to the function (2.12), and that we obtain an optimal feedback in the following way. Let

$$\hat{v}_i(p,x,t), \quad i = 1 \ldots n$$

give rise to the lower bound of $L(p,v,x,t)$ with fixed p,x,t, for $0 \leq v_i \leq \overline{v_i}$. Then

$$(2.19) \qquad \left| \begin{array}{lll} \hat{v}_i(x,t) = \hat{v}_i(\nabla\Phi(x),x,t) & \text{if} & \underline{y_i} < x_i < \overline{y_i} \\ \hat{v}_i(x,t) = 0 & \text{if} & x_i \leq \underline{y_i} \\ \hat{v}_i(x,t) = \overline{v_i} & \text{if} & x_i \geq \overline{y_i} \end{array} \right.$$

is an optimal feedback.

Theoretically, the problem is completely solved. There is, however, a practical problem arising from n being large (for France $n = 20$). Therefore (2.15) must be solved by decomposition methods.

We can proceed in the following way. In place of working with global feedbacks, as in (2.19), i.e., the feedback depends upon all the variables $x_1, \ldots, x_n$, we work with *local* feedbacks, i.e., $v_i = v_i(x_i,t)$. In other words, the decisions in each valley depend only on the water stock in that valley considered. This is a reasonable approximation.

If $\hat{v}_i(x_i,t)$ denotes an optimal local feedback, and if $\hat{y}_i(t)$ is the process that is the solution of

$$(2.20) \qquad \left| \begin{array}{l} d\hat{y}_i = g_i(t)dt + \sigma_i(t)dw_i(t) - \hat{v}_i(\hat{y}_i,t)dt \\ \hat{y}_i(0) = y_i^o \end{array} \right.$$

and if we write

$$(2.21) \qquad \hat{W}^i(t) = \sum_{j \neq i} W_j\left(\hat{y}_j(t), \hat{v}_j(\hat{y}_j(t),t), t\right)$$

we can then characterise $\hat{v}_i(x_i,t)$ as the solution of a 1-dimensional stochastic control problem. In fact we solve

(2.22)
$$\left|\begin{array}{l}\underset{v_i}{\text{Min}}\ E\displaystyle\int_0^T e^{-\alpha t}\left\{\hat{C}_1^i\left(W_i(y_i,v_i,t),t\right)+\hat{C}_2^i\left(D(t)-W_i(y_i,v_i,t)\right)\right\}dt \\ dy_i = g_i(t)dt + \sigma_i(t)dw_i - v_i(t)dt \\ 0 \leq v_i(t) \leq \overline{v_i} \\ v_i(t) = \overline{v_i} \quad \text{if} \quad y_i(t) \geq \overline{y_i} \\ v_i(t) = 0 \quad \text{if} \quad y_i(t) \leq \underline{y_i}\ .\end{array}\right.$$

In (2.22) the functions $\hat{C}_1^i$, $\hat{C}_2^i$ are defined by

(2.23)
$$\left|\begin{array}{l}\hat{C}_1^i(\theta,t) = E\ C_1(W^i(t)+\theta) \\ \hat{C}_2^i(\theta,t) = E\ C_2(\theta - W^i(t))\ .\end{array}\right.$$

Next we set

(2.24)
$$\hat{L}_i(p_i,v_i,x_i,t) = -p_i v_i + \hat{C}_1^i\left(W_i(x_i,v_i,t),t\right) + \hat{C}_2^i\left(D(t)-W_i(x_i,v_i,t),t\right)$$

(2.25)
$$\hat{H}_i(p_i,x_i,t) = \underset{\substack{0\leq v_i\leq \overline{v_i} \\ v_i = 0 \text{ if } x_i \leq \underline{y_i} \\ v_i = \overline{v_i} \text{ if } x_i \geq \overline{y_i}}}{\text{Inf}} \hat{L}_i(p_i,v_i,x_i,t)$$

and we introduce the functions $\hat{\Phi}_i(x_i,t)$ which are solutions of

(2.26)
$$\left|\begin{array}{l}-\dfrac{\partial\hat{\Phi}_i}{\partial t} + A_i(t)\hat{\Phi}_i - \hat{H}_i\left(\dfrac{\partial\hat{\Phi}_i}{\partial x_i}, x_i, t\right) + \alpha\hat{\Phi}_i = 0 \\ \hat{\Phi}_i(x_i,T) = 0\ .\end{array}\right.$$

We then define $\hat{v}_i(x_i,t)$ as a function of $\hat{\Phi}_i(x,t)$ in the usual way. Clearly (2.26) is a coupled system, since the functions $\hat{C}_1^i, \hat{C}_2^i$, and therefore $\hat{H}_i$, depend on the feedbacks $\hat{v}_j$, for all $j \neq i$, and $\hat{v}_i$ is calculated by starting from the $\hat{\Phi}_i$. But it is then possible to use an iterative algorithm (relaxation). Knowing the feedbacks $\hat{v}_i^k$ at the iteration k, we define

$$\hat{W}^{i,k+1}(t) = \sum_{j<i} W_j\left(\hat{y}_j^{k+1}(t),\ \hat{v}_j^{k+1}(\hat{y}_j^{k+1}(t),t),t\right) + $$
$$+ \sum_{j>i} W_j\left(\hat{y}_j^{k}(t),\ \hat{v}_j^{k}(\hat{y}_j^{k}(t),t),t\right)$$

where $\hat{y}_j^k$ is the solution of (2.20) corresponding to $\hat{v}_j^k$.

We next define $\hat{C}_1^{i,k+1}, \hat{C}_2^{i,k+1}, \hat{L}_i^{k+1}, \hat{H}_i^{k+1}$, and then consider the solution $\hat{\Phi}_i^{k+1}$ of (2.26) with $\hat{H}_i$ replaced by $\hat{H}_i^{k+1}$, and $\hat{v}_i^{k+1}$ is the component in (2.25) reaching the lower bound with $\nabla \hat{\Phi}_i^{k+1}$ in place of p_i.

We can show that $\hat{\Phi}_i^{k+1}(y_i^o,0) \leq \hat{\Phi}_{i-1}^{k+1}(y_{i-1}^o,0)$(1) and $\Phi_1^{k+1}(y_1^o,0) \leq \Phi_n^k(y_n^o,0)$. Therefore the $\Phi_i^k(y_i^o,0)$ converge when $k \to +\infty$, to the same limits.

By restricing ourselves to the class of local feedbacks we have thus got over the difficulty of dimension posed by equation (2.15). But we must calculate the functions (2.23). That assumes we obtain the probability of the process $\hat{y}_i(t)$, and that we do a multiple integral calculation in dimension $n-1$.

Here there is a new difficulty which we can overcome by a new approximation, which is reasonable if n is large enough. In fact, we notice that it is necessary to know only the behaviour of $\hat{W}^i(t)$. Now, by (2.21) $\hat{W}^i(t)$ is the sum of independent random variables.

For n large enough, by the law of large numbers $\hat{W}^i(t)$ behaves in a Gaussian way with mean the sum of the means, and variance the sum of the variances. Therefore in fact it is sufficient to calculate the expectation and variance of

$$W_j\left(\hat{y}_j(t)\ ,\ \hat{v}_j(\hat{y}_j(t),t),t\right) .$$

Knowing the behaviour of $\hat{y}_j(t)$, this reduces to calculating 1-dimensional integrals. As for the behaviour of $\hat{y}_j(t)$, that is obtained by approximating the model (2.20) by a Markov chain, or operating by simulation. For details, refer to DELEBECQUE-QUADRAT [2], QUADRAT [1].

Remark 2.1

In the model presented above we do not have the difficulty of *degeneracy*. This is owed to the fact of water contributions being described by the very simple model (2.6), and therefore disappear at the end of the calculation in the definitive formulation of the problem. This would not be so if $g_i(t),\sigma_i(t)$ depended upon η_i explicitly. □

Remark 2.2

The function $\Phi(x,t)$ represents the value of the stock of water $x = (x_1,\dots,x_n)$ at the instant t. □

2.3 *A More Elaborate Model of Long-Term Management*

Here we consider the optimisation of n valleys, taking the secondary dams into account.

2.3.1 *The General Model*

Consider n valleys, each having a certain number of dams, one barrage at the head (the principal dam serving especially as the stocker of water) and secondary dams down stream. We write

(2.27) $y_{ij}(t)$ = stock of disposable water in the j-th dam of the i-th valley at the instant t,

$$i = 1,\dots,n, \qquad j = 0,1,\dots n_i.$$

The dam $j = 0$ corresponds to the dam at the head of the valley. It will be convenient to write

(2.28) $$y_{i0}(t) = y_i(t).$$

We set

(1) It will be noticed that $\hat{\Phi}_i^{k+1}(y_o^i,0)$ also depends upon y_0^j, $j \neq 1$.

$$(2.29)\quad \begin{cases} v_{ij}(t) = \text{quantity of water through the turbines in the } (i,j)\text{-th dam per unit time,} \\ v_i(t) = v_{i0}(t). \end{cases}$$

We write

$$(2.30)\quad \begin{cases} \eta_i(t) = \text{contributions of water per unit time in the dam at the head of the } i\text{-th valley,} \\ \eta_{ij}(t) = \text{contributions of water per unit time in the } (i,j)\text{-th dam.} \end{cases}$$

For $\eta_i(t), \eta_{ij}(t)$ we use the following models

$$(2.31)\quad d\eta_i(t) = g_i(\eta_i, t)dt + \sigma_i(\eta_i, t)dw_i(t)$$

$$(2.32)\quad \eta_{ij}(t) = f_{ij}(\eta_i(t),t) .$$

Relation (2.31) means that $\eta_i(t)$ behaves as a diffusion, and relation (2.32) means that the contributions in the small dams are linked deterministically to the contributions in the large dams. This is a realistic hypothesis, for the small dams collect water by seepages and from a part of the contributions of the dams at the head.

The 'balance sheet' relations for water stocks allow us to write

$$(2.33)\quad \frac{dy_i}{dt} = \eta_i - v_i$$

$$(2.34)\quad \frac{dy_{ij}}{dt} = \eta_{ij} + v_{i,j-1} - v_{i,j} , \qquad j = 1 \dots n_i .$$

The demand for electricity is going to have periodic characteristics with a period of one week. In fact every working day there are two peaks of demand, and a particular case holds at the weekend. For the demand per unit time we adopt the following model

$$(2.35)\quad d\zeta(t) = \frac{1}{\varepsilon} g(\zeta, \frac{t}{\varepsilon}, t)dt + \frac{1}{\sqrt{\varepsilon}} \sigma(\zeta, \frac{t}{\varepsilon}, t)dw(t)$$

where $g(\zeta,\theta,t),\sigma(\zeta,\theta,t)$ are periodic in θ with period 1 (1 year). If $\varepsilon = \frac{1}{52}$, $g(\zeta, \frac{t}{\varepsilon}, t), \sigma(\zeta, \frac{t}{\varepsilon}, t)$ are periodic with respect to the first variable, with period ε (i.e., 1 week). The model (2.35) allows us to represent the non-negligible fluctuations within a day *taking the year as the unit of time.* In fact, the variation of the instantaneous demand in an interval of Δs weeks (Δs = 3 hours = 1/56 week, for example) corresponds to a variation in an interval of $\Delta t = \varepsilon\Delta s$ years, therefore by (2.35) is approximately a normal law with mean $g(\zeta, \frac{t}{\varepsilon}, t)\Delta s$ and with variance $\sigma^2(\zeta, \frac{t}{\varepsilon}, t)\Delta s$. The model also allows us to take into account the behaviour over some years.

The constraints on the quantities put through turbines are given by

$$(2.36)\quad \begin{cases} 0 \leq v_{ij}(t) \leq \overline{v_{ij}} \\ 0 \leq v_i(t) \leq \overline{v_i} . \end{cases}$$

In addition there are constraints on the storage capacities of the dams. For the large dams we have

$$(2.37) \qquad 0 \leq y_i(t) \leq \overline{y_i} \,.$$

Small dams have a maximum storage capacity very much less than that of large dams, in fact in a ratio of the order ε. To bring this important phenomenon into evidence in the remainder, we write

$$(2.38) \qquad 0 \leq y_{ij}(t) \leq \varepsilon\, \overline{y_{ij}} \,.$$

Contrary to the case (2.11), it is possible to realise the constraints (2.37), (2.38). It suffices to impose upon the controls the constraints

$$(2.39) \qquad \left| \begin{array}{lll} \eta_i - v_i \geq 0 & \text{if} & y_i = 0 \\ \eta_i - v_i \leq 0 & \text{if} & y_i = \overline{y_i} \end{array} \right.$$

$$(2.40) \qquad \left| \begin{array}{lll} \eta_{ij} + v_{i,j-1} - v_{ij} \geq 0 & \text{if} & y_{ij} = 0 \\ \eta_{ij} + v_{i,j-1} - v_{ij} \leq 0 & \text{if} & y_{ij} = \varepsilon\, \bar{y}_{ij} \,. \end{array} \right.$$

The hydroelectrically generated electrical energy is given by

$$(2.41) \qquad W = \sum_i W_i(y_i\,, v_i\,, t) + \sum_{i,j} W_{ij}(y_{ij}\,, v_{ij}\,, t)$$

where W_i, W_{ij} have the same meaning as in the preceding Section (cf., 2.7).

For the thermal system we make the same assumptions as for the preceding Subsection, and therefore we introduce the functions C_1, C_2 (cf. (2.9),(2.9')). Nevertheless, to simplify matters a little we take $C_1 = 0$, $C_2 = C$. The criterion to be minimised is therefore the following

$$(2.42) \qquad J = E\int_0^T C\left(\zeta(t) - W(t)\right) e^{-\alpha t}\, dt$$

where T is a horizon of several years.

It is convenient to consider $\frac{y_{ij}(t)}{\varepsilon}$, instead of $y_{ij}(t)$, as the state variable. For economy of notation we make the transformation $\frac{y_{ij}(t)}{\varepsilon} \to y_{ij}(t)$, therefore the interpretation of $y_{ij}(t)$ is then the following

$$(2.43) \qquad \left| \; y_{ij}(t) = \text{state of the water stock in the } i,j\text{-th dam multiplied by } 52\left(\frac{1}{\varepsilon}\right) . \right.$$

Thus the variables y_i, y_{ij} are now of the same order of magnitude. With this transformation we are led to pose the following stochastic control problem

$$(2.44) \qquad d\eta_i(t) = g_i(\eta_i\,, t)dt + \sigma_i(\eta_i\,, t)dw_i\,,$$

$$(2.45) \qquad \frac{dy_i}{dt} = \eta_i - v_i$$

$$(2.46) \qquad \varepsilon\,\frac{dy_{ij}}{dt} = f_{ij}\left(\eta_i(t), t\right) + v_{i,j-1} - v_{i,j}\,, \quad j = 1 \,..\, n_i$$

(2.47) $$d\zeta = \frac{1}{\epsilon} g(\zeta, \frac{t}{\epsilon}, t)dt + \frac{1}{\sqrt{\epsilon}} \sigma(\zeta, \frac{t}{\epsilon}, t)dw(t)$$

(2.48) $$0 \leqq v_{ij} \leqq \bar{v}_{ij}$$

(2.49) $$\eta_i - v_i \geqq 0 \quad \text{if} \quad y_i = 0, \quad \eta_i - v_i \leqq 0 \quad \text{if} \quad y_i = \bar{y}_i$$

(2.50) $$\left| \begin{array}{ll} f_{ij}(\eta_i, t) + v_{i,j-1} - v_{ij} \geqq 0 & \text{if} \quad y_{ij} = 0 \\ f_{ij}(\eta_i, t) + v_{i,j-1} - v_{ij} \leqq 0 & \text{if} \quad y_{ij} = \bar{y}_{ij} \quad , j = 1 .. n_i \end{array} \right.$$

(2.51) $$\text{Min } J = E \int_0^T C\left(\zeta(t) - \sum_{i=1}^{n} W_i(y_i, v_i, t) - \sum_{i,j=1..n_i} W_{ij}(\epsilon y_{ij}, v_{ij}, t)\right) dt.$$

The state variables are $y_i, \eta_i, y_{ij}, \zeta$; the controls are $v_{ij}(t)$.

Treatment with dynamic programming then leads to introducing a function $\Phi(x_i, \xi_i, x_{ij}, z, t)$, the solution of the following problem

(2.52) $$\left| \begin{array}{l} -\frac{\partial \Phi}{\partial t} + \sum_i A_i \Phi + \frac{1}{\epsilon} A_z^\epsilon \Phi - H^\epsilon\left(\frac{\partial \Phi}{\partial x_i}, \frac{1}{\epsilon}\frac{\partial \Phi}{\partial x_{ij}}, x_i, \xi_i, x_{ij}, z, t\right) \\ \qquad + \alpha\Phi = 0 \\ \Phi(x_i, \xi_i, x_{ij}, z, T) = 0 \end{array} \right.$$ (1)

where

(2.53) $$A_i \Phi = - g_i(\xi_i, t) \frac{\partial \Phi}{\partial \xi_i} - \frac{1}{2} \sigma_i^2(\xi_i, t) \frac{\partial^2 \Phi}{\partial \xi_i^2}$$

(2.54) $$A_z^\epsilon \Phi = - g(z, \frac{t}{\epsilon}, t) \frac{\partial \Phi}{\partial z} - \frac{1}{2} \sigma^2(z, \frac{t}{\epsilon}, t) \frac{\partial^2 \Phi}{\partial z^2}$$

(2.55) $$H^\epsilon(p_i, p_{ij}, x_i, \xi_i, x_{ij}, z, t) =$$

$$= \underset{\substack{0 \leqq v_{ij} \leqq \bar{v}_{ij} \\ v_i \leqq \xi_i \text{ if } x_i = 0 \\ v_i \geqq \xi_i \text{ if } x_i = \bar{y}_i \\ v_{ij} \leqq v_{i,j-1} + f_{ij}(\xi_i, t) \text{ if } x_{ij} = \bar{y}_{ij} \\ v_{ij} \geqq v_{i,j-1} + f_{ij}(\xi_i, t) \text{ if } x_{ij} = 0}}{\text{Inf}} \left[p_i(\xi_i - v_i) + p_{ij}(f_{ij}(\xi_i, t) + v_{i,j-1} - v_{ij}) + C(z - \sum_i W_{ij}(x_i, v_i, t) - \sum_{ij} W_{ij}(\epsilon x_{ij}, v_{ij}, t)) \right]$$

The solving of (2.53) provides the solution of the problem, and $\Phi(x_i, \xi_i, x_{ij}, z, 0)$ represents the optimal cost (2.52).

Remark 2.3

Equation (2.53) presents all the characteristic difficulties of stochastic control problems arising from real problems (degenerate character and very great order, $2n + \sum_{i=1}^{n} n_i + 1$). □

(1) The indices i,j are dummy indices, i.e., all components occur.

2.3.2 *Use of singular perturbation techniques*

The technique of singular perturbations allows us to decouple (2.52) into a problem relating uniquely to the large dams, and to a problem relating to the small dams. The origin of this decoupling can be understood quite simply by using an asymptotic expansion in ε. Denoting the solution of (2.52) as Φ^ε, we write

$$\Phi^\varepsilon(x_i, \xi_i, x_{ij}, z, t) = \Psi(x_i, \xi_i, t) + \varepsilon\chi(x_i, \xi_i, x_{ij}, z, \frac{t}{\varepsilon}, t) \,. \tag{2.56}$$

Substituting in (2.52) and neglecting the terms in ε, we obtain

$$-\frac{\partial\Psi}{\partial t} - \frac{\partial\chi}{\partial\theta} + \sum_i A_i\Psi + A_z(\theta)\chi + \alpha\Psi - H\left(\frac{\partial\Psi}{\partial x_i}, \frac{\partial\chi}{\partial x_{ij}}, x_i, \xi_i, x_{ij}, z, t\right) = 0. \quad (1) \tag{2.57}$$

Moreover, since the function $\frac{t}{\varepsilon}$ appearing in the problem (g, σ, cf., (2.35)) are periodic with period 1, we add the condition

$$\theta \to \chi(x_i, \xi_i, x_{ij}, z, \theta, t) \quad \text{periodic with period 1.}$$

But then we can consider (2.57) as an equation in χ as a function of θ, z, x_{ij}, with the variables x_i, ξ_i, t being parameters. We are thus led to pose the problem

$$(2.58)\quad \left|\begin{array}{l} -\frac{\partial\Psi}{\partial\theta} + A_z(\theta)\chi - H(p_i, \frac{\partial\chi}{\partial x_{ij}}, x_i, \xi_i, x_{ij}, z, t) + \Lambda = 0 \\ \chi \;\; \text{periodic in } \theta \text{ with period 1,} \\ \Lambda \;\; \text{constant, independent of } \theta, z, x_{ij}. \end{array}\right.$$

In (2.58) p_i, x_i, ξ_i, t are parameters and χ, Λ are unknowns. Naturally, $\Lambda = \Lambda(p_i, x_i, \xi_i, t)$. If we assume (2.58) is solved, then Ψ is the solution of

$$(2.59)\quad \left|\begin{array}{l} -\frac{\partial\Psi}{\partial t} + \sum_i A_i\Psi + \alpha\Psi = \Lambda\left(\frac{\partial\Psi}{\partial x_i}, x_i, \xi_i, t\right) \\ \Psi(x_i, \xi_i, T) = 0 \,. \end{array}\right.$$

The equation (2.59) is completely analogous with (2.15), with Λ replacing H.(2)

The decoupling of (2.52) into (2.58),(2.59) brings a simplification into the realm of the dimension, but it is necessary to obtain the function Λ. To do that we begin by interpreting the function Λ. Now (2.58) is an equation of dynamic programming, where p_i, x_i, ξ_i, t are simple parameters. It is interpreted with the help of a stochastic control problem with an ergodic criterion. We introduce the equations of state

$$\frac{dy_{ij}}{d\theta} - f_{ij}(\xi_i, t) + \tilde{v}_{i,j-1}(\theta) - \tilde{v}_{i,j}(\theta) \tag{2.60}$$

$$d\zeta = g(\zeta, \theta, t)d\theta + \sigma(\zeta, \theta, t)dw(\theta) \tag{2.61}$$

$$(2.62)\quad \left|\begin{array}{lll} \tilde{v}_i(\theta) \le \xi_i & \text{if} & x_i = 0 \\ \tilde{v}_i(\theta) \ge \xi_i & \text{if} & x_i = \bar{y}_i \end{array}\right.$$

$$(2.63)\quad \left|\begin{array}{lll} \tilde{v}_{ij}(\theta) \le \tilde{v}_{i,j-1}(\theta) + f_{ij}(\xi_i, t) & \text{if} & y_{ij}(\theta) = 0 \\ \tilde{v}_{ij}(\theta) \ge \tilde{v}_{i,j-1}(\theta) + f_{ij}(\xi_i, t) & \text{if} & y_{ij}(\theta) = \bar{y}_{ij} \end{array}\right.$$

---

(1) H represents the expression (2.55) with εx_{ij} replaced by 0 in W_{ij}.
(2) With the difference, however, that we have the variables x_i, ξ_i instead of x_i alone.

$$(2.64) \quad \underset{S\to\infty}{\text{Min}\ \lim}\ E\frac{1}{S}\int_0^S \Big[\sum p_i(\xi_i - \tilde{v}_i(\theta)) + \sum C(\zeta(\theta) - \sum_i w_i(x_i, \tilde{v}_i(\theta), t) - \sum_{ij} W_{ij}(0, \tilde{v}_{ij}(\theta), t)\Big]\, d\theta .$$

Recall that in (2.60),...,(2.64) x_i,ξ_i,p_i,t are parameters. In addition, the optimal value of the criterion (2.64) corresponds to Λ.

The problem (2.60),...,(2.64) is a problem relating uniquely to the small dams, in which the characteristics (stock and contributions) of the large dams are constants.

As for the problem (2.59), it is not in a very agreeable form (although it only relates to the large dams), for it is not in general a dynamic programming equation. We can, then, make an approximation.

Using the convexity of C and the concavity of W_i,W_{ij}, and setting

$$v_i = \lim_{S\to\infty} E\frac{1}{S}\int_0^S \tilde{v}_i(\theta)d\theta$$

$$v_{ij} = \lim_{S\to\infty} E\frac{1}{S}\int_0^S \tilde{v}_{ij}(\theta)d\theta$$

we deduce from (2.64) that

$$\Lambda \geq \underset{\substack{v_i\\ v_{ij}}}{\text{Min}} \Big[\sum p_i(\xi_i - v_i) + C\Big(\Delta(t) - \sum_{ij} W_{ij}(0, v_{ij}, t) - \sum_i W_i(x_i, v_i, t)\Big)\Big]$$

where

$$\Lambda(t) = \lim_{S\to\infty} E\frac{1}{S}\int_0^S \zeta(\theta, t)d\theta$$

(notice that $\zeta(\theta)$ depends on t as a parameter). Moreover, if we want to keep $Ey_{ij}(S)$ bounded as $S\to+\infty$, by virtue of (2.60) we are led to write

$$(2.65) \quad f_{ij}(\xi_i, t) + v_{i,j-1} - v_{ij} = 0 .$$

The relations (2.65) define the v_{ij}, $i = 1,\dots n$, $j = 1,\dots,n_i$, when ξ_i,v_i,t are known. If we set

$$(2.66) \quad D(t, \xi, v) = \Delta(t) - \sum_{ij} W_{ij}(0, v_{ij}, t)$$

then we finally obtain

$$(2.67) \quad \Lambda \geq \underset{v_i}{\text{Min}}\Big[\sum p_i(\xi_i - v_i) + C\Big(D(t, \xi) - \sum_i W_i(x_i, v_i, t)\Big)\Big] = H^*(p, x, \xi, t).$$

But if we then introduce the equation

$$(2.68) \quad \Big| -\frac{\partial\Psi^*}{\partial t} + \sum_i A_i \Psi^* + \alpha\Psi^* = H^*(\frac{\partial\Psi^*}{\partial x}, x, \xi, t), \qquad \Psi^*(x, \xi, T) = 0$$

we have the relation

(2.69) $\Psi \geqq \Psi^*$.

The advantage of (2.68) over (2.59) is that it is a dynamic programming equation. Thus we are led to *replace* Ψ by Ψ^* and to pose at the large dams problem

$$(2.70)\quad \begin{aligned} &\frac{dy_i}{dt} = \xi_i - v_i \\ &d\xi_i = g_i(\xi_i , t)dt + \sigma_i(\xi_i , t)dw_i \\ &0 \leqq v_i \leqq \bar{v}_i \\ &v_i(t) = \bar{v}_i \quad \text{if} \quad y_i(t) = \bar{y}_i \\ &v_i(t) = 0 \quad \text{if} \quad y_i(t) = 0 \\ &\text{Min } E \int_0^T e^{-\alpha t}\, C\Big(D(t ; \xi , v) - \sum_i W_i(y_i(t) , v_i(t) , t)\Big)\, dt \end{aligned}$$

the dynamic programming equation of which coincides with (2.68).

The problem (2.70) is a control problem for the large dams, analogous to that studied in Section 2.2.

The small dams do not appear at the level of evaluating the demand $D(t;\xi,v)$. The latter appears as the difference between a *mean* real demand $\Delta(t)$ diminished by the energy produced by the small dams, calculated by assuming that the stocks of the small dams are zero, and by fixing the contributions per unit time at each instant t as well as the amounts put through the turbines in the large dams.

We can then completely solve the problem (2.70) by the iterative method in Section 2.2. We obtain the amounts $v_i(t)$ put through the turbines per unit time in the large dams, as well as the behaviour of the function giving the value of water.

In practice we discretise time in the problem (2.70) with a time increment of one week (T = several years). Hence we obtain the 'mean' amounts put through the turbines each week in the large dams.

We can then use the problem (2.60),...,(2.64) for the weekly management of the small dams (and of the large dams, too). We then look for $\tilde{v}_{ij}(\theta)$ (recall that x_i,ξ_i,t are parameters) which is the solution of (2.60),...,(2.64) by taking account of the additional constraints

$$(2.71)\quad \begin{aligned} v_i(t) &= \lim_{S\to\infty} E\,\frac{1}{S}\int_0^S \tilde{v}_i(\theta)d\theta \\ v_{ij}(t) &= \lim_{S\to\infty} E\,\frac{1}{S}\int_0^S \tilde{v}_{ij}(\theta)d\theta \end{aligned}$$

where the $v_{ij}(t)$ are defined by (2.65) assuming the values of ξ_i and v_i are fixed.

Naturally the problem (2.60),...,(2.71),(2.64) is again a high-dimensional problem. Nevertheless, we note that in the criterion (2.64) the state y_{ij} does not appear explicitly. In fact the state only appears by reason of the constraints (2.63). If we neglect the constraints (2.63) we are led to minimise, under the constraints (2.62),(2.71), the functional

$$\lim_{S\to\infty} E\,\frac{1}{S}\int_0^S \Big[C\Big(\zeta(\theta) - \sum_i W_i(x_i , \tilde{v}_i(\theta) , t\Big) - \sum_{ij} W_{ij}\Big(0 , \tilde{v}_{ij}(\theta) , t\Big)\Big]\, d\theta \ .$$

As the function C has a minimum h, we must therefore find $\tilde{v}_i(\theta),\tilde{v}_{ij}(\theta)$ so that we obtain at best the equality

$$(2.72) \qquad \zeta(\theta) - \sum_i W_i(x_i, \tilde{v}_i(\theta), t) - \sum_{ij} W_{ij}(0, \tilde{v}_{ij}(\theta), t) = h$$

under the constraints (2.71). The functions $\tilde{v}_{ij}(\theta)$, $\theta \in [0,1]$ found thus are the amounts put through the turbines in the course of the week in the (i,j)-th dam during the week t. We can now resume the general procedure followed.

The problem (2.70) provides the *mean* amounts put through the turbines each week in the large dams. In this problem the demand to be satisfied depends upon the contributions of water and the amounts turbined in the large dams, for it appears as the difference between an effective *mean* demand in the week, diminished by the *mean* production of the small dams. The latter is the production calculated when the amounts turbined in the small dams are such that the corresponding stocks of water are zero. Therefore they depend (cf. (2.65)) on the amounts turbined and the amounts in the large dams.

Once we know the mean amounts turbined per week in the large and small dams, we look for the allotment of these quantities within the week so that we may realise 'at best' the equality (2.72). Thus we adjust production for the best for the fluctuations in demand within the week, completely respecting the mean constraint.

Meanwhile we may notice that at the weekly level the procedure (2.72),(2.71) for obtaining amounts turbined risks being not very good, because the approximations made in the thermal cost (in particular the cost of starting up no stations) lose their validity.

However, what is very valuable in the practical realm is the optimisation (2.70), i.e., the long-term optimisation.

2.4 *Simplified Model of Short-Term Management of Thermal Power Stations*

The target of short-term management is to give the set of decisions to do with thermal stations (start-up, shut-down, power produced) as well as the set of decisions to do with the hydroelectric stations (amounts turbined) in the large and small dams), all in the horizon of one week.

The link with long-term management can be done in two ways. First of all we can impose constraints on the amounts turbined each hour in a way that makes the mean over one week correspond with the amounts provided by the long-term management model. We can also put into the short-term criterion a final cost corresponding the the 'value' of the water at the end of the week, this being given by the function Φ the solution of (2.52).

The complete treatment of the short term management problem is beyond our grasp in general.

In this Section we give the solution of a short-term management problem of thermal power stations by considering that the hydroelectric system is reduced to a *small* number of dams.

We begin by describing briefly the working of a thermal power station. When a thermal station is set going it does not effectively produce electricity until after a 'warming up' phase, lasting about five hours. We will assume that this delay is fixed, and, in particular, does not depend upon the moment when the station is shut down (in practice this is not actually so). This delay is assimilated by a *fixed cost* for starting up, the corresponding production cost for five hours.

Whilst working, a thermal power station can produce a power p lying between a minimum $\underline{p}$ and a maximum $\overline{p}$. The quantity $\underline{p}$ corresponds to the minimum activity below which the station can be considered as shut down. This means that a new warming up phase is necessary if we want to reactivate the station.

We assume that we have J thermal power stations available, j = 1,...,J. The technical data on each station are

$$(2.73)\quad \left|\begin{array}{l} \text{— unit cost of production } c_j, \\ \text{— start up cost } k_j, \\ \text{— working limits } \underline{p}_j, \overline{p}_j. \end{array}\right.$$

Each station can be in the run up or shut down state. Therefore this gives us 2^J possible states of the system. To simplify this we begin by classifying the stations by the order of their increasing production cost, i.e.,

$$i \geq j \Leftrightarrow c_i \geq c_j .$$

We start up the power stations in the order of increasing number. Therefore there are only J + 1 possible states.

Let m_1, m_2 be two states (i.e., $0 \leq m_1, m_2 \leq J$); we can calculate the cost $K(m_1, m_2)$ of changing from the state m_1 to the state m_2, and the production cost $C(m,p)$ per unit time corresponding to a production p and a state m of the system:

$$(2.74)\quad \left|\begin{array}{l} K(m_1, m_2) = \left|\begin{array}{ll} \displaystyle\sum_{j=m_1+1}^{m_2} k_j , & \text{if } m_2 > m_1 , \\ 0 \text{ otherwise,} & \end{array}\right. \\ C(m, p) = \displaystyle\sum_{j=1}^{h_p} c_j \overline{p}_j + \sum_{j=h_p+1}^{m} c_j \underline{p}_j \\ \text{if there exists } h_p \text{ such that} \\ \displaystyle\sum_{j=1}^{h_p} \overline{p}_j \leq p , \quad \sum_{j=1}^{h_p} \overline{p}_j + \underline{p}_{h_p+1} > p , \end{array}\right.$$

$$(2.74)_1\quad C(m, p) = \sum_{j=1}^{h_p} c_j \overline{p}_j + c_{h_p+1}\left(p - \sum_{j=1}^{h_p} \overline{p}_j\right) + \sum_{j=h_p+2}^{m} c_j \underline{p}_j$$

if there exists h_p such that

$$\sum_{j=1}^{h_p} \overline{p}_j + \underline{p}_{h_p+1} \leq p < \sum_{j=1}^{h_p+1} \overline{p}_j .$$

Notice that for fixed m, $m \geq 1$, the function $C(m,p)$ is defined only for

$$\sum_{j=1}^{m} \underline{p}_j \leq p \leq \sum_{j=1}^{m} \overline{p}_j .$$

If m = 2 the last term on the right hand side of $(2.74)_1$ must be omitted. If $m = 1$ we have

$$C(1,p) = c_1 p \qquad , \quad \underline{p}_1 \leq p \leq \overline{p}_1 .$$

Lastly, if m = 0, then $C(m,p) = 0$, and p is necessarily zero.

The hydroelectric part is described by the model of Section 2.2. Therefore there are n dams, with stocks $y_i(t)$, and amounts turbined $v_i(t)$. The relations (2.13), (2.14) are assumed to be satisfied.

The state variables are the $y_i(t)$, and the state of the system of thermal stations is denoted $M(t)$. The process $M(t)$ has values in $\{0,\dots,J\}$. The control variables are the $v_i(t)$, $p(t)$ the production of the thermal stations, as well as the instants of change of the state $M(t)$ and the corresponding jumps. We write $\tau_1 \leqq \tau_2 \leqq \dots \leqq \tau_\ell \dots$ for the sequence of the instants of jumps of $M(t)$, and $\mu_1,\mu_2,\dots,\mu_\ell,\dots$ for the intensities of the corresponding jumps.

The equations of the model are as follows

$$(2.75)\qquad \begin{aligned} dy_i &= g_i(t)dt + \sigma_i(t)dw_i(t) - v_i(t)dt \\ \frac{dM}{dt} &= \sum_\ell \delta(t-\tau_\ell)\mu_\ell \end{aligned}$$

$$(2.76)\qquad \left|\begin{array}{l} 0 \leqq v_i(t) \leqq \overline{v_i} \\ v_i(t) = \overline{v_i} \quad \text{si} \quad y_i(t) \geqq \overline{y_i} \\ v_i(t) = 0 \quad \text{si} \quad y_i(t) \leqq \underline{y_i} \end{array}\right.$$

$$(2.77)\qquad \left|\begin{array}{l} \mu_\ell \quad \text{entier relatif} \\ -M(\tau_\ell - 0) \leqq \mu_\ell \leqq J - M(\tau_\ell - 0) \end{array}\right.$$

$$(2.78)\qquad \left|\begin{array}{l} \sum_{j=1}^{M(t)} \underline{p_j} \leqq p(t) \leqq \sum_{j=1}^{M(t)} \overline{p_j}\ , \quad \text{si} \quad M(t) \geqq 1 \\ p(t) = 0 \quad \text{si} \quad M(t) = 0\ . \end{array}\right.$$

We now describe the functional to be minimised. Let $D(t)$ be instantaneous demand for electricity, which we assume is deterministic in order to simplify matters. The hydroelectrically originated energy is given by $\sum_i W_i(y_i,v_i,t)$ (cf., (2.7)).

The deficiency in the power network is the quantity

$$\left(D(t) - \sum_i W_i(y_i, v_i, t) - p(t)\right)^+$$

to which is attached a cost

$$\rho\left[\left(D(t) - \sum_i W_i(y_i, v_i, t) - p(t)\right)^+\right] .$$

Hence we are led to the functional

$$(2.79)\qquad \underset{\substack{v_i(.),p(.) \\ \tau_\ell,\mu_\ell}}{\text{Min}}\ E\left[\int_0^T \left\{ C\Big(M(t), p(t)\Big) + \rho\left[\left(D(t) - \sum_i W_i(y_i,v_i,t) - p(t)\right)^+\right]\right\} dt + \sum_\ell K\Big(M(\tau_\ell - 0), M(\tau_\ell)\Big) + \Phi\Big(y(T),T\Big)\right]$$

where, naturally,

$$M(\tau_\ell) = M(\tau_\ell - 0) + \mu_\ell\ .$$

In (2.79) the horizon T is the *week*, and must not be confused with the horizon (of several years) in the long-term problem. The function $\Phi(x,t)$ is the value of the water during the week t, when the stock of water in the dam is represented by the vector x. This function is furnished by the long-term programme (2.15).

As we have said, we could also not introduce this final term, but be given a means constraint on the controls $v_i(t)$.

The dynamic programming method leads to inequalities instead of an equation, as in the preceding Sections (cf., Section 1.4). We look for functions $V_m(x,t)$, $m = 0,\dots,J$, satisfying the following conditions

$$(2.80)\quad \begin{cases} X_m = -\dfrac{\partial V_m}{\partial t} + \sum_i A_i(t)V_m - H_m\left(\dfrac{\partial V_m}{\partial x}, x, t\right) \leq 0 \\ Y_m = V_m(x,t) - \inf\limits_{\substack{\mu \in Z \\ -m \leq \mu \leq J-m}} \left[V_{m+\mu}(x,t) + K(m,m+\mu)\right] \leq 0 \\ X_m Y_m = 0 \\ V_m(x,T) = \Phi(x,T) . \end{cases}$$

In (2.80) the operator $H_m(q,x,t)$ is given by ($q \in R^m$)

$$(2.81)\quad H_m(q,x,t) = \inf_{\substack{0 \leq v_i \leq \bar{v}_i \\ v_i = \bar{v}_i \text{ if } x_i \geq \bar{y}_i \\ v_i = 0 \text{ if } x_i \leq y_i \\ \sum_{j=1}^m \underline{p}_j \leq p \leq \sum_{j=1}^m \bar{p}_j \\ p=0 \text{ if } m=0}} \left[-\sum_i q_i v_i + \right.$$

$$\left. + C(m,p) + \rho\left(\left(D(t) - \sum_i W_i(x_i, v_i, t) - p\right)^+\right)\right] .$$

The operator A_i is the same as in Section 2.2, so

$$A_i(t) = -g_i(t)\frac{\partial}{\partial x_i} - \frac{1}{2}\sigma_i^2(t)\frac{\partial^2}{\partial x_i^2} .$$

Starting from (2.80) we can define the optimal decisions at each instant in the following way. Let t be the present instant, and let $M(t)$, $y(t)$ be the state of the system (the state of the group of stations, the state of the stocks of water in the dams). If $Y_{M(t)}(y(t),t) = 0$ we change the state of the group. Let us choose the number μ that yields the lower bound appearing in the expression for $Y_{M(t)}(y(t),t)$. Then $M(t) + \mu$ is the new state. For this new state $Y_{M(t)}(y(t),t) < 0$, and it remains negative during a time interval $t, t + \delta t$. Then $X_m = 0$ on this interval. We next define the amounts turbined and the thermal production which yield the lower bound in the square brackets of (2.81) when q_i is replaced by $\dfrac{\partial V_m}{\partial x_i}$ and $m = M(t) + \mu$, $x_i = y_i(t)$.

Remark 2.4

C. LEGUAY [1] has treated an example of management of thermal power stations with a single dam and a system of ninety thermal stations.

Remark 2.5

DELEBECQUE and QUADRAT [1],[2] have treated completely the management problem for New Caledonia (the short-term and long-term problems). The problem involves one dam and eight thermal power stations. In addition, the contributions of water are modelled by diffusions with jumps, for they suffer from discontinuities. Application tests are in progress, cf., COLLETER-DELEBECQUE-FALGARONNE-QUADRAT [1].□

Remark 2.6

In Section 3, for the case of management of stocks problem, we shall give an intuitive justification for obtaining quasi-variational inequalities by dynamic programming. The reader can apply a similar argument to understand the origin of the relations (2.80). □

3. STOCK MANAGEMENT

General Approach

Stock management is a typical example of the application of impulse control. Therefore it provides an interesting pedagogic framework for this theory. The theory brings to light optimal decision rules, and characterises them. However, at the practical level we run into a problem of very large dimensionality (as much as of products as of other things, for this number can be very high). It is then necessary to regroup the products into a small number of classes and to use, in addition, heuristic methods for treating the product individually.

3.1 *General Description of the Problems*

The number of goods to be managed is n. The cumulative demand in the interval (s,t) is D(s,t). We will accept the representation

$$D(s,t) = \int_s^t \mu(\theta)d\theta + \int_s^t \sigma(\theta)dw(\theta) \tag{3.1}$$

where $w(\theta)$ is an n-dimensional normalised Wiener process. Other models are possible, in particular those allowing jumps.

Restocking orders are given at (random) instants to be chosen, say

$$\theta^1 \leqq \theta^2 \leqq \dots \leqq \theta^i \leqq \dots$$

At each of these instants we associate an order

$$\xi^1, \xi^2 \dots, \xi^i, \dots$$

which is also a chosen random vector. The set formed by the sequence of instants and quantities ordered makes up an impulse control.

We denote by y(s) the state of the stock at the instant s. The evolution of y(s) is given by

$$dy(s) = -\mu(s)ds - \sigma(s)dw(s) + \sum_i \delta(s-\theta^i)\xi^i \tag{3.2}$$

(cf. (2.75) for an analogous situation).

Let us now give the cost structure. Quite generally, three types of costs are met in stock management: ordering costs, inventory costs, and shortages costs.

In general, when we order a quantity ξ (a vector $\geqq 0$ in R^n), we pay an ordering cost comprising a fixed cost, independent of the quantity ordered, and of a variable cost, dependent upon the quantity ordered. We denote by $C(t;\xi)$ the cost or ordering a quantity $\xi \geqq 0$ at the instant t. Let us give some examples:

$$(3.3)\qquad C(t;\xi) = \begin{cases} 0 \text{ if } \xi = 0 \\ k(t) + \sum_{i=1}^{n} c_i \xi_i , \text{ if } \xi \in \mathcal{U} \\ +\infty \text{ if } \xi \notin \mathcal{U} . \end{cases}$$

In (3.3) $k(t)$ is the fixed cost, $\sum_i c_i \xi_i$ the variable cost, $\mathcal{U}$ is a set of constraints on the quantities ordered.

A natural source of fixed costs lies in transport for delivery of goods ordered. Let us assume that we have lorries available of capacity a, which can be hired for a unit price $k_1(t)$; it is unimportant how the lorries transport the goods. For an order vector $\xi = (\xi_1, \dots, \xi_n)$, the bulk volume which results in a usage charge for the lorry is

$$\sum_{i=1}^{n} \alpha_i \xi_i ,$$

where the α_i are coefficients measuring the bulk of a unit of the i-th goods. The cost of transport is given by

$$(3.4)\qquad C(t;\xi) = \begin{cases} 0 \text{ if } \xi = 0 \\ p k_1(t) \text{ if } (p-1)a < \sum_{i=1}^{n} \alpha_i \xi_i \leqq ap \\ \text{where } p = 1,2, \dots . \end{cases}$$

We often may meet situations where discounts are conceded, starting from a certain level of order, or a certain variety in the order (grouped orders).

We then have the following formulae:

$$(3.5)\qquad C(t;\xi) = \begin{cases} 0 \text{ if } \xi = 0 \\ k(t) + \sum_{i=1}^{n} c_i(t)\xi_i , \text{ if } \sum_{i=1}^{n} c_i(t)\xi_i < b \\ k(t) + (1-\alpha) \sum_{i=1}^{n} c_i(t)\xi_i \text{ if } \sum_{i=1}^{n} c_i(t)\xi_i \geqq b \end{cases}$$

$$(3.6)\qquad C(t;\xi) = \begin{cases} 0 \text{ if } \xi = 0 \\ k(t) + \sum_{i=1}^{n} c_i(t)\xi_i \text{ if } \xi_i \geqq a_i \ \forall\, i \\ k_1(t) + \sum_{i=1}^{n} c_i(t)\xi_i \text{ if at least one of the } \xi_i < a_i , \\ \text{with } k_1(t) > k(t) . \end{cases}$$

If the costs of an order are supported only at the moment of ordering, the costs of stocking and shortages are integral in type, in other words

$$\int_0^T f(y(t), t)dt$$

where $f(y(t),t)$ is the stocking cost per unit time at the instant t if $y(t) \geqq 0$, and the cost of shortages per unit time if $y(t) \leqq 0$.(1)

It often happens that we seek to impose constraints on the stock of products (for example, constraints on the volumes of stocks). Let us assume that we seek to arrange

$$(3.7) \qquad y(t) \in \bar{\mathcal{O}}$$

where $\mathcal{O}$ is a certain open set of R^n (which is bounded, to fix our ideas).

To realise (3.7) we intuitively reject as superfluous the requests that make the level of stocks leave the set of constraints. Naturally we must pay the cost of the request for stock.

We have spoken above of stock shortages and correponding costs. In a certain sense this reduced to considering constraints $y(t) \geqq 0$ on the state of the stock. But in (3.2) we authorise $y(t)$ to be negative, therefore to not satisfy the constraints. This says that the requests are *transferrable* in time. They are even entered if they cannot be satisfied immediately, and therefore imply negative stocks. However, in the case (3.7) the request is not transferrable and must be rejected, and so is lost.

There is therefore a difference (which is expressed naturally in the costs) between stock breakages in the case of a transferrable request and stock breakages in the case of a non-transferrable request, which is definitely lost.

The model used in the case where we impose the constraints (3.7) is a *diffusion* model *with reflection*, so

$$(3.8) \qquad dy = -\mu(s)ds - \sigma(s)dw(s) + \gamma(y(s),s)d\eta(s) + \sum_i \delta(s-\theta^i)\xi^i .$$

where $\gamma(x,s)$ is a vector pointing to the interior of the domain $\mathcal{O}$ and $d\eta(s) \geqq 0$ (i.e., $\eta(s)$ is an increasing scalar process).

The term $\gamma(y(s),s)d\eta(s)$ is interpreted as the *rejected request* at the instant s when the stock is $y(s)$. We show that it is possible to construct uniquely the pair $y(s),\eta(s)$ which is the solution of (3.8) so that (3.7) is satisfied, and so that $d\eta(s) = 0$ if $y(s) \in \mathcal{O}$.

The cost corresponding to the rejected request in an interval ds is expressed by $\varphi(y(s),s)d\eta(s)$. Therefore it is zero on every time interval where the state of the stock satisfies the constraints.

3.2 *Optimisation*

First we consider the constraint-free case (transferrable order). The system evolves according to (3.2) and the cost function is given by

$$(3.9) \qquad J = E\left[\int_t^T f(y(s),s)ds + \sum_i C(\theta_i\,;\xi_i)\right] .$$

We write the lower bound of the criterion (3.9) as $u(x,t)$ when the initial stock at the instant t is x.

The evolution of $u(x,t)$ is governed by a quasi-variational inequality (Q.V.I.). We shall briefly explain how this Q.V.I. is obtained by the dynamic programming argument (cf. Section 1 also).

---

(1) This description is perfectly valid in one dimension. In n dimensions certain components may be positive and others negative. We are always led to an integral type of cost.

To simplify matters we restrict outselves to the case

$$(3.10) \qquad C(t;\xi) = \begin{vmatrix} 0 & \text{if} & \xi = 0 \\ 1 & \text{if} & \xi \neq 0 . \end{vmatrix}$$

In this case we have

$$(3.11) \qquad J = E\left[\int_t^T f(y(s),s)ds + N\right]$$

where N is the number of ordering instants before T. The function u(x,t) is the solution of

$$(3.12) \qquad \begin{vmatrix} -\frac{\partial u}{\partial t} + A(t)u \leqq f \\ u \leqq Mu \qquad , \qquad u(x,T) = 0 \\ (u - Mu)\left(-\frac{\partial u}{\partial t} + A(t)u - f\right) = 0 \end{vmatrix}$$

where we have set

$$(3.13) \qquad \begin{vmatrix} A(t) = +\sum_i \mu_i(t)\frac{\partial}{\partial x_i} - \frac{1}{2}\sum_i \sigma_i^2(t)\frac{\partial^2}{\partial x_i^2} , \\ M\varphi(x) = 1 + \operatorname{Inf}_{\xi \geqq 0} \varphi(x+\xi) . \end{vmatrix}$$

In fact the relation u(x,t) is obvious from the definition of u. Also, at the instant t an immediate order can be made at a level ξ. In this case the system moves instantaneously from t,x to t,x + ξ. Assuming that after this first decision the later decisions are optimal ones, the corresponding cost is 1 + u(x + ξ,t). Hence we necessarily have

$$u(x,t) \leqq 1 + u(x+\xi,t) .$$

Since ξ is arbitrary and $\geqq 0$, we deduce the second relation (3.12) by taking the lower bound in ξ.

Let us now assume that we make no orders at the instant t. Then, we leave the system to evolve freely during a small time interval t,t + δ into the state

$$x - \sum_i \mu_i \delta - \sum_i \sigma_i \, dw_i$$

where

$$dw_i = w_i(t+\delta) - w_i(t) .$$

By assuming that we know the optimal policy after the instant t + δ, the corresponding cost

$$f(x,t)\delta + Eu(x - \sum_i \mu_i \delta - \sum_i \sigma_i \, dw_i \, , \, t + \delta) \sim$$

$$f(x,t)\delta + u(x,t) - \sum_i \mu_i \frac{\partial u}{\partial x_i}\delta + \frac{1}{2}\sum_i \sigma_i^2 \frac{\partial^2 u}{\partial x_i^2}\delta + \frac{\partial u}{\partial t}\delta + O(\delta)$$

which is $\geqq$ u(x,t) by the definition of u.

Dividing by δ and letting δ tend to 0, we obtain the first relation (3.12). The last relation of (3.12) (producing zero) is explained by the fact that one or other of the first two relations of (3.12) has to be satisfied. In fact, at the instant t we necessarily have one or other of the following cases: either an order is made instantaneously, or else the system is left to evolve freely during a very short time interval.

We now move to the case with constraints. This time the evolution of the system is described by (3.8), and the cost function to be minimised is given by

$$(3.14) \qquad J = E\left[\int_t^T f(y(s),s)ds + \sum_i C(\theta_i, \xi_i) + \int_t^T \varphi(y(s),s)d\eta(s)\right].$$

Let us again take the case (3.10), therefore

$$(3.15) \qquad J = E\left[\int_t^T f(y(s),s)ds + N + \int_t^T \varphi(y(s),s)d\eta(s)\right].$$

If $u(x,t)$ is still the lower bound of J with respect to all the impulse controls, then $u(x,t)$ is a function defined when $x \in \bar{\mathcal{O}}$. It satisfies the relations

$$(3.16) \qquad \left|\begin{array}{l} -\dfrac{\partial u}{\partial t} + A(t)u \leq f \\ u \leq Mu \\ (u - Mu)\left(-\dfrac{\partial u}{\partial t} + A(t)u - f\right) = 0 \end{array}\right\} \text{ in } \mathcal{O}$$

$$(3.17) \qquad \left|\begin{array}{l} u \leq Mu \\ 0 \leq \varphi + \gamma.\nabla u \\ (u - Mu)(\varphi + \gamma.\nabla u) = 0 \end{array}\right\} \text{ on } \partial\mathcal{O}$$

$$(3.18) \qquad u(x,T) = 0 \qquad \text{on } \bar{\mathcal{O}}.$$

The relations (3.16) are easily explained by the fact that if we start off from the interior of $\mathcal{O}$, then we remain in $\mathcal{O}$ for a certain time interval. Therefore we have a situation analogous to the constraint-free case. The relations (3.17) need a more intricate justification.

Now we explain how the optimal policy is obtained by starting from the Q.V.I.. We restrict ourselves to the constraint-free case (3.12). Let us introduce the region

$$(3.19) \qquad \left|\begin{array}{l} C = \{x,t \mid u < Mu\} \\ S = \text{complement of } C \text{ in } R^n \times [0,T]. \end{array}\right.$$

We call C the *continuation set* and S the *stopping set*. Right from the start let us assume that $(x,t) \in C$, and it is then of no interest to make an order at this instant, for in this case Mu is reached at best. Now, we know that $u < Mu$, which means that a better policy exists. Hence we must continue (i.e., let the system evolve freely), which justifies the terminology. Continuing step by step, we see that the first ordering instant is that when the system reaches the boundary of C. Let θ^1 be this instant, and $y(\theta^1 - 0)$ the state of the stock at the instant θ^1. In order to know the quantity to order we must know a function $\hat{\xi}(x,t)$ which is equal to the lower bound of $u(x + \xi,t)$ when $\xi \geqq 0$. Under these conditions the optimal order is $\hat{\xi}(y(\theta^1 - 0),\theta^1)$. The new state of the system is then

$$y(\theta^1) = y(\theta^1 - 0) + \hat{\xi}(y(\theta^1 - 0), \theta^1).$$

The point $y(\theta^1),\theta^1$ is in the continuation set. We then begin again the procedure described by starting from x,t, and so on up to the instant T.

3.3 *Some Remarks and Additional Notes*

3.3.1 *The delay problem*

Up to now we have assumed that deliveries were instantaneous. Now let us consider the case of a fixed and deterministic delay h in delivery. If θ^i is an ordering instant, the delivery happens at the instant $\theta^i + h$, so that the stock evolves according to the equations

$$dy(s) = -\mu(s)ds - \sigma(s)dw(s) + \sum_i \delta(s-\theta^i-h)\,\xi^i \,. \tag{3.20}$$

The great difficulty introduced by delays is that theoretically we can make a new order during the period delivery is awaited. Therefore orders during the course of delivery must be taken into account in order to define the optimal policy.

If, however, we forbid the making of an order before the delivery of the preceding order, we can treat the problem with great additional difficulties. The θ^i sequence satisfies the condition

$$\theta^{i+1} \geqq \theta^i + h \,. \tag{3.21}$$

The criterion is the following

$$J = E\left[\int_t^T f(y(s),s)\,ds + N\right]. \tag{3.22}$$

Let u(x,t) be the lower bound of the criterion, and furthermore at the initial instant t there are no orders in the course of delivery. The function u is the solution of the following Q.V.I.

$$\left.\begin{aligned} &-\frac{\partial u}{\partial t} + A(t)u \leqq f \\ &u(t) \leqq M(t)u(t+h) \\ &\Big(u(t) - M(t)u(t+h)\Big)\left(-\frac{\partial u}{\partial t} + A(t)u - f\right) = 0 \end{aligned}\right\}\ t < T-h \tag{3.23}$$

where M(t) is given by the expression

$$M(t)\varphi(x) = 1 + E\int_t^{t+h} f(x - D(t,s),s)\,ds + \inf_{\xi \geqq 0}\left[E\varphi\Big(x + \xi - D(t,t+h)\Big)\right]. \tag{3.24}$$

Recall that D(t,s) is given by (3.1). The notation u(t) in (3.23) represents the function u(x,t) of x. Hence we see that the operator M(t) must be applied to u(t + h) and not to u(t). The origin of (3.24) and of the second relation (3.23) is easily understood. In fact, if we decide to make an order at the instant t, it will only be realised at the instant t + h. But then between t and t + h no order can be made. Hence the system evolves according to the law

$$x - D(t,s) \qquad \text{for} \qquad s \in [t, t+h-0]$$

and at the instant t + h becomes $x - D(t,t+h) + \xi$, since delivery is made at this instant. Furthermore, between T − h and T there can be no orders, therefore

(3.25) $$u(x,t) = E\int_t^T f(x - D(t,s),s)ds \text{ ,for } t\in [T - h,T] \text{ .}$$

Some very much more general cases have been studied by M. ROBIN [1].

First of all we authorise a (fixed) maximum number of orders during the period of a delay in delivery. In this case we obtain systems of Q.V.I.'s. We can also consider the case of random delays, according to a given law.

3.3.2 *Management of hierarchies of stocks*

Here we consider a system consisting of a warehouse and n retail shops. The retail shops serve the external demand and are supplied by the warehouse. The latter does not recieve orders except from the retail shops. To simplify matters we assume that a single product is handled.

We write $D^j(t,s)$ for the order in the interval t,s arriving at shop number j. So we have

(3.26) $$D^j(t,s) = \int_t^s \mu_j(\tau)d\tau + \int_t^s \sigma_j(\tau)dw_j(\tau)$$

and the processes w_j are independent (to simplify matters).

Let $y_j(s)$ be the product stock in retail shop number j, j = 1,...,n, at the instant s. We have the following equation for the evolution

(3.27) $$dy_j(s) = - \mu_j(s)ds - \sigma_j(s)\ dw_j(s) + \sum_i \delta(s - \theta_j^i)\xi_j^i \text{ .}$$

Let $y_0(s)$ be the warehouse stock. Then $y_0(s)$ evolves according to the equation

(3.28) $$dy_0(s) = - \sum_i\sum_j \delta(s - \theta_j^i)\xi_j^i + \sum_i \delta(s - \theta_0^i)\xi_0^i$$

where the θ_0^i, ξ_0^i represent the orders of the warehouse.

Let $C_0(t;\xi),\ldots,C_j(t;\xi)$ be the ordering costs of the warehouse and the retail shops, and $f_0,f_1,\ldots,f_n$ the corresponding costs of stocking and breakages. The functional to be minimised is given by

(3.29) $$J = E\left[\sum_{i=1}^{\infty}\sum_{j=0}^{n} C_j(\theta_j^i ; \xi_j^i) + \sum_{j=0}^{n}\int_0^T f_j(y_j(s),s)ds\right] \text{ .}$$

To simplify, we take

$$C_j(t;\xi) = k_j \quad \text{if } \xi > 0 \text{ ,} \quad 0 \text{ if } \xi = 0 \text{ .}$$

Let $u(x_0,x,t)$ be the lower bound of the criterion where x_0 denotes the initial stock of the warehouse, and $x = (x_1,\ldots,x_n)$ the different stocks in the shops. The function u is the solution of

(3.30) $$u(x_0 \text{ , } x,T) = 0$$

(3.31) $$- \frac{\partial u}{\partial t} + \sum_{j=1}^{n} A_j\ u - \left(\sum_{j=1}^{n} f_j(x_j \text{ , } t) + f_0(x_0 \text{ , } t)\right) \leq 0$$

(3.32) $$u - Mu \leq 0$$

(3.33) the product of (3.31) with (3.32) = 0,

with

$$M\varphi(x_0, \ldots, x_n) = \underset{J \subset \{0,1, \ldots, n\}}{\text{Min}} \left[\sum_{j \in J} k_j + \underset{\xi_j \geq 0}{\text{Inf}}\, \varphi(x' + \xi_J) \right] \tag{3.34}$$

where

$x' = (x_0, x)$,

ξ_J = the vector whose components are zero if their index does not belong to J.

Remark 3.1

The reader may refer to M. GOURSAT-S. MAURIN [1], M. GOURSAT and G. MAAREK [1] for numerous examples and the study of other situations. □

3.3.3 *Management of stocks that deteriorate*

Up to now we have only treated the management of stocks of products whose physical state does not alter with the passage of time. Otherwise the state variable $y(t)$ representing the stock state must be changed into $y(t,x)$ = the quantity in stock at the instant t of a product of quality x. The variable x plays the role of a spatial variable. The evolutionary model of y becomes a partial differential equation. The optimisation problem which then arises is an impulse control problem for a distributed system. We shall not treat this type of problem in this book.

4. PRODUCTION MANAGEMENT

General Approach

We have already seen in Section 2 the problems arising in the management of electricity production. We now give some other examples.

4.1 *Use of Different Production Procedures*

Here we assume that there are two manufacturing systems for the production of n goods (for example, a speeded-up system, and a slow one). The problem consists of adjusting the periods for which we use one system or the other, so as to satisfy demand at the better cost. Let us call the systems 1 and 2. When we use system 1 the evolution of the stocks of the n products is given by

$$dy = g_1(y,t)dt + \sigma_1(y,t)dw(t) \tag{4.1}$$

and when we use system 2 they are given by

$$dy = g_2(y,t)dt + \sigma_2(y,t)dw(t) \tag{4.2}$$

where $g_1, \sigma_1, g_2, \sigma_2$ characterise the systems. Note that equations (4.1),(4.2) result from the modelling of two simultaneous phenomena: on the one hand the production of goods, and the demand for products on the other. The fact that the functions g_i, σ_i depend upon the stock state (the opposite of what happens in the management of stocks) is explained by the fact the we adjust the level of production as a function of the state of the stocks.

The cost structure is as follows. When we change from one system to another we pay a fixed cost: k_1 if we change from system 1 to system 2, and k_2 is we change from system 2 to system 1. Also we encounter a stocking cost and a breakage cost per unit time, which we represent by a function $f(x,t)$.

The problem can be formulated as an impulse control problem. Here a control is a sequence of instants

$$\theta^1 , \theta^2 \dots \theta^i \dots . \tag{4.3}$$

The state of the system is determined by the stock $y(t)$, on the one hand, and on the other by a process $z(t)$ with values in $\{1,2\}$ whose evolution is given by

$$\frac{dz}{dt} = \sum \delta(t - \theta_i)\xi_i \tag{4.4}$$

where

$$\left| \begin{array}{l} \xi_i = +1 \quad \text{if} \quad z(\theta_i - 0) = 1 \\ \xi_i = -1 \quad \text{if} \quad z(\theta_i - 0) = 2 . \end{array} \right.$$

It will be noticed that the ξ_i are not controls. The state of the stock is then described by

$$dy = g_{z(t)}(y(t),t)dt + \sigma_{z(t)}(y(t),t)dw(t) . \tag{4.5}$$

The fixed cost is defined by

$$C(t;\xi) = \left| \begin{array}{ll} k_1 & \text{if} \quad \xi = 1 \\ k_2 & \text{if} \quad \xi = -1 . \end{array} \right. \tag{4.6}$$

The function to be optimised can then be written in the form

$$J = E\left[\int_t^T f(y(s),s)dt + \sum_i C(\theta_i ; \xi_i)\right] \tag{4.7}$$

which is therefore a classical criterion for impulse control. We write the lower bound of J as $u_i(x,t)$, $i = 1,2$, when we start off in the state $y(t) = x$, $z(t) = i$, at the initial instant t.

The functions $u_1(x,t), u_2(x,t)$ are solutions of a system of Q.V.I.'s, say

$$\left| \begin{array}{l} -\dfrac{\partial u_1}{\partial t} + A_1 u_1 \leq f \\ u_1 \leq k_2 + u_2 \\ \left(-\dfrac{\partial u_1}{\partial t} + A_1 u_1 - f\right)\left(u_1 - k_2 - u_2\right) = 0 \end{array} \right. \tag{4.8}$$

$$\left| \begin{array}{l} -\dfrac{\partial u_2}{\partial t} + A_2 u_2 \leq f \\ u_2 \leq k_1 + u_1 \\ \left(-\dfrac{\partial u_2}{\partial t} + A_2 u_2 - f\right)\left(u_2 - k_1 - u_1\right) = 0 \end{array} \right. \tag{4.9}$$

$$u_i(x,T) = 0 . \tag{4.10}$$

In (5.8),(5.9), respectively, we have set

$$(4.11) \qquad A_i = - g_i \cdot \nabla - \frac{1}{2} \operatorname{tr} \sigma_i \sigma_i^* D^2 \quad , \quad i = 1,2$$

where D^2 is the matrix of second derivatives given by

$$\left(D^2\varphi \equiv \frac{\partial^2 \varphi}{\partial x_k \partial x_\ell} \right) .$$

The procedure for changing from one state to the other is easily described by starting from the relations (4.8),(4.9). In fact, let $y(s),z(s)$ be the state of the system at the instant s. Let us assume that $z(s) = 1$, and that $u_1(y(s),s) < k_2 + u_2(y(s),s)$; we then keep to system 1.

If $u_1(y(s),s) = k_2 + u_2(y(s),s)$, however, we then change to system 2. A corresponding argument must be carried out if $z(s) = 2$.

Remark 4.1

The case of m regimes, $m \geqq 2$, can be considered similarly.

4.2 *Whether to Produce or to Buy on the Market*

In managing production we sometimes meet the following problem. It is possible to buy the goods we produce on the open market. Sometimes it may be more advantageous to buy directly on the open market than to produce them ourselves. When we do not buy on the open market the state of stocks evolves according to the equation

$$(4.12) \qquad dy = g(y,v,t)dt + \sigma(y,t)dw(t)$$

where $v(t)$ is a (continuous) control measuring the level of production. If we now denote by

$$\theta^1 , \theta^2 ,\dots, \theta^i ,\dots$$

the sequence of instants of ordering, and by

$$\xi^1 , \xi^2 ,\dots, \xi^i ,\dots$$

the sequence of quantities ordered, we define the cost function to be minimised as

$$(4.13) \qquad J = E\left[\int_t^T f(y,v,s)ds + \sum_i C(\theta_i,\xi_i)\right]$$

where f takes into account production costs, in addition to stocking and breakages costs. Furthermore, $C(t;\xi)$ represents the fixed cost of ordering. For example,

$$C(t;\xi) = \begin{vmatrix} k \text{ if } \xi \neq 0 \quad , \quad \xi \geqq 0 \\ 0 \text{ if } \xi = 0 \; . \end{vmatrix}$$

We write $u(x,t)$ for the lower bound of J when $y(t) = x$. The function u is the solution of

$$(4.14) \qquad \begin{vmatrix} -\dfrac{\partial u}{\partial t} + A(t)u \leqq H(\nabla u,x,t), \qquad u \leqq Mu \\ \left(-\dfrac{\partial u}{\partial t} + A(t)u - H(\nabla u,t)\right)(u - Mu) = 0, \qquad u(x,T) = 0 \end{vmatrix}$$

where

(4.15) $$A(t) = -\frac{1}{2}\,\mathrm{tr}\,\sigma\,\sigma^{*}\,D^2$$

(4.16) $$H(p,x,t) = \operatorname*{Inf}_{v}\{p.g(x,v,t) + f(x,v,t)\}$$

(4.17) $$M\varphi(x) = k + \operatorname*{Inf}_{\xi \geqq 0} \varphi(x+\xi) .$$

4.3 *Production and Management of Raw Materials*

We here consider a continuous production process, which, for simplification, we assume is 'optimised' (i.e., the production level is adapted to the evolution of demand in an optimal way).

Production consumes raw materials. The problem consists of the best management of the stocks of raw materials as a function of the requirements for production. Let $y_1(t)$ be the level of the stocks of finished products at the instant t, and let $y_2(t)$ be the level of raw materials. The process $y_1(t)$ is not controlled and evolves according to the equation

(4.18) $$dy_1 = g(y_1 , t)dt + \sigma(y_1 , t)dw(t) .$$

Now let $(\theta^1,\xi^1;\ldots;\theta^i,\xi^i;\ldots)$ be the sequence of ordering instants and quantities of raw materials ordered. The process $y_2(t)$ evolves according to the equation

(4.19) $$dy_2 = -\gamma\, y_1(t)dt + \sum \delta(t-\theta_i)\xi_i .$$

The functional we are given to optimise is

(4.20) $$J = E\left\{\int_t^T f(y(s),s)ds + \sum_i C(\theta_i ; \xi_i)\right\}$$

where y(s) represents the pair $(y^1(s),y^2(s))$. To simplify, let us take

$$C(t;\xi) = \begin{cases} k & \text{if } \xi \neq 0 \\ 0 & \text{if } \xi = 0 . \end{cases}$$

Let x_1,x_2 be the initial conditions at the instant t for the processes $y_1(s)$, $y_2(s)$. We denote the lower bound of the criterion J by $u(x_1,x_2,t)$. Let A be the operator (at x_1)

$$A = -g.\nabla_{x_1} - \frac{1}{2}\,\mathrm{tr}\,\sigma\sigma^{*}\,D^2_{x_1}$$

then u is the solution of the following Q.V.I.

(4.21) $$\begin{cases} -\dfrac{\partial u}{\partial t} + Au - \gamma\, x_1 \cdot \dfrac{\partial u}{\partial x_2} \leq f \\ u(x_1 , x_2 , t) \leq k + \operatorname*{Inf}_{\xi \geqq 0} u(x_1 , x_2+\xi,t) \\ \left[u-k-\operatorname*{Inf}_{\xi \geqq 0} u(x_1 , x_2+\xi,t)\right]\left[-\dfrac{\partial u}{\partial t} + Au - \gamma x_1 \dfrac{\partial u}{\partial x_2} - f\right] = 0 \\ u(x_1 , x_2 , T) = 0 . \end{cases}$$

The optimal policy is easily defined by starting from the Q.V.I. (4.21).

5. THE PROBLEMS OF MAINTENANCE AND QUALITY CONTROL

5.1 *Case of a Complete Observation*

Consider a machine whose performance is represented by a process y(t) (assumed to be a scalar, to simplify matters). Assume that the value 0 represents the brand new state, and that y(t) represents a deteriorated state (getting worse as y(t) increases). We shall assume that y(t) evolves according to the equation

$$(5.1) \qquad dy = g(y,t)dt + \sigma(y,t)dw(t) .$$

We have taken g, σ to depend upon the state to express the possibility of continuous maintenance of the machine, and assume it to be done in an optimal way (this is to simplify matters; we could add a continuous control in g representing the maintenance effort). The model (5.1) has the inconvenience that, if $\sigma \neq 0$, then y(t) can take all values from $-\infty$ to $+\infty$. Now, only the values of y(t) that are $\geqq 0$ have a physical meaning. However, the data g, σ can be chosen so that $\{y(t) < 0\}$ has a very small probability.

Also, we shall only consider an example here. Different Markov processes from (5.1) would be more adequate in certain cases.

The problem which arises is the following: at which instants must we replace the machine by a new one. Let

$$\theta^1 , \theta^2 ,\dots , \theta^i ,\dots$$

be a sequence of 'replacing' instants. At the instant θ^i the system's state changes from the value $y(\theta^i - 0)$ to the value 0 (the new state). We shall assume that the cost of replacement is equal to k (the purchase price of the brand new machine less the disposal price fetched by the used machine(1)). The controlled process y(t) therefore evolves according to the equation

$$(5.2) \qquad dy = g(y,t)dt + \sigma(y,t)dw(t) - \sum_i \delta(t-\theta_i)y(\theta_i - 0) .$$

The criterion to be minimised is of the form

$$(5.3) \qquad J = E\left[\int_t^T f(y(s),s)ds + kN\right]$$

where N represents the random number of jumps occurring strictly before T. Let u(x,t) be the lower bound of J. Then u is the solution of the Q.V.I.

$$(5.4) \qquad \left|\begin{array}{l} \dfrac{\partial u}{\partial t} + A(t)u \leq f \\ u(x,t) \leqq k + u(0,t) \\ \left(-\dfrac{\partial u}{\partial t} + A(t)u - f\right)\Big(u(x,t) - k - u(0,t)\Big) = 0 \\ u(x,T) = 0 \end{array}\right.$$

In (5.4) we have set

$$(5.5) \qquad A(t) = - g(x,t)\frac{\partial}{\partial x} - \frac{1}{2}\sigma^2(x,t)\frac{\partial^2}{\partial x^2} .$$

---

(1) In general this cost is not a constant, because it depends upon the instant of replacement. To simplify matters we take k to be constant.

The problem of replacing a machine also provides a natural example of the *optimal stopping time problem.* In fact, if instead of a sequence of replacement instants we consider simply finding the best instant to sell off a machine, an unique *instant* θ must be chosen

Let $\psi(x,t)$ then be the replacement cost at the instant t when the quality of service is x (we have taken $\psi = k$ above). We then have the problem

$$dy = g(y,t)dt + \sigma(y,t)dw(t) \ . \tag{5.6}$$

The criterion to be minimised is the following:

$$J = E\left[\int_t^{T\wedge\theta} f(y(s),s)ds + \psi(y(\theta),\theta)\chi_{\theta<T}\right] . \tag{5.7}$$

Let $u(x,t)$ be the lower bound of J. Then u is the solution of the problem

$$\left|\begin{array}{l} -\dfrac{\partial u}{\partial t} + A(t)u \leq f \\ u \leq \psi \\ \left(-\dfrac{\partial u}{\partial t} + A(t)u - f\right)(u-\psi) = 0 \ . \\ u(x,T) = 0 \end{array}\right. \tag{5.8}$$

5.2 *Partial Observation*

Let us now assume that the quality of the machine always evolves according to the model (5.1), but that we no longer can inspect $y(t)$ at every instant. More precisely, we can inspect $y(t)$ at the instant t if we can make a measurement (control of quality). But this measurement implies the cost of an experiment.

Here we assume that the decisions are of two types: either a measurement is made that gives us the value of the quality at the instant considered, or else we pre-emptively decide to replace the machine. The information available bears upon all measurements carried out in the past.

Let

$$t \leq \theta^1 \leq \theta^2 \leq \dots \leq \theta^i \leq \dots$$

be a sequence of decisions. At the instant θ^i two decisions are possible

$$\xi^i = \left|\begin{array}{l} 1 \text{ if we decide to change the machine,} \\ 0 \text{ if we decide to inspect the quality without changing the machine} \end{array}\right. \tag{5.9}$$

The quality of the machine behaves according to the equation

$$dy = g(y,t)dt + \sigma(y,t)dw(t) - \sum_i \delta(t-\theta^i)\, y(\theta^i - 0)\,\xi^i \ . \tag{5.10}$$

The process $y(s)$ is not observable, rather does one have access to the actual values $y(\theta^i)$ at the instants θ^i.

It is convenient to introduce the process

$$(5.11)\qquad z(s) = \begin{vmatrix} x & , & t \leq s < \theta^1 \\ y(\theta^i) & , & \theta^i \leq s < \theta^{i+1} \end{vmatrix}$$

$$(5.12)\qquad \eta(s) = \begin{vmatrix} s - t + h & , & t \leq s < \theta^1 \\ s - \theta^i & , & \theta^i \leq s < \theta^{i+1} . \end{vmatrix}$$

So $z(s)$ then represents the last observation made, and $\eta(s)$ the time elapsed since the last observation. Notice that $z(s)$ evolves according to the equation

$$(5.13)\qquad dz = \sum \delta(s - \theta^i)\Big(\zeta_i(1 - \xi_i) - z(\theta^i - 0)\Big)$$

where

$$(5.14)\qquad \xi^i = y_{z(\theta^{i-1}),\ \theta^{i-1}}(\theta^i)$$

using the notation $y_{xt}(s)$ for the solution of (5.1) with the initial conditions

$$y_{xt}(t) = x.$$

Similarly $\eta(s)$ evolves according to the relation

$$(5.15)\qquad d\eta(s) = ds - \sum \delta(s - \theta^i)\ \eta\ (\theta^i - 0) \quad , \eta(t) = h .$$

The decisions θ^{i+1}, ξ^{i+1} are based on the values of $\theta^j, \xi^j; z(s), \eta(s), s < \theta^{i+1}$ (say, $\theta^j, \xi^j, s(\theta^j), j \leq 1$).

The functional to be minimised is given by the expression

$$(5.16)\qquad J = E\left[\left[\int_t^T f(y(s),s)ds + \sum_i \left(k_o\ \chi_{\xi^i=0} + k_1\ \chi_{\xi^i=1}\right)\right]\right].$$

In order to be able to apply the dynamic programming argument, we have to transform the integral term in (5.16), because the process $y(s)$ is not observable. Now we have

$$(5.17)\qquad E\int_t^T f(y(s),s)ds = E\int_t^{\theta_1} f(y(s),s)ds + \sum_{i=1}^{\infty} E\int_{\theta_i}^{\theta_{i+1}} f(y(s),s)ds .$$

Let us introduce the operator

$$(5.18)\qquad \Phi(t,h)\varphi(x) = E\varphi\Big(y_{x,t-h}(t)\Big).$$

We also write

$$(5.19)\qquad f(t)(x) = f(x,t)$$

$$(5.20)\qquad \psi(x,h,t) = \Phi(t,h)f(t)(x) = E\ f\Big(y_{x,t-h}(t),t\Big) .$$

Now, we have

$$E\int_t^{\theta_1} f(y(s),s)ds = E\int_t^{\theta_1} f(y_{x,t-h}(s),s)ds$$

$$E\int_{\theta^i}^{\theta^{i+1}} f(y(s),s)ds = E\int_{\theta^i}^{\theta^{i+1}} f(y_{y(\theta^i),\theta^i}(s),s)ds \ .$$

Now, by the properties of θ^{i+1}, using the definition of ψ we have

$$E\int_t^{\theta^1} f(y(s),s)ds = E\int_t^{\theta^1} \psi(x,s-t+h,s)ds$$

$$E\int_{\theta^i}^{\theta^{i+1}} f(y(s),s)ds = E\left\{ E\left[\int_{\theta^i}^{\theta^{i+1}} f(y(s),s)ds|\theta^j,\xi^j,j \leq i\right]\right\}$$

$$= E\int_{\theta^i}^{\theta^{i+1}} \psi(y(\theta^i),s-\theta^i,s)ds \ .$$

From this we deduce, therefore, using the definition of $z(s),\eta(s)$, that

$$J = E\left[\int_t^T \psi\left(z(s),\eta(s),s\right)ds + \sum_i\left(k_o\chi_{\xi^i=0} + k_1\chi_{\xi^i=1}\right)\right] . \tag{5.21}$$

We are then led to an impulse control problem with a complete observation, the state of the system being the pair $z(s),\eta(s)$. The state's evolution is given by (5.13),(5.15).

We then call $u(x,h,t)$ the lower bound of the criterion J. The dynamic programming argument leads us to the relations

$$\left|\begin{array}{l} -\dfrac{\partial u}{\partial t} - \dfrac{\partial u}{\partial h} \leq \psi \\ u(x,h,t) \leq \text{Min}\left(k_o + \Phi(t,h)u(0,t)(x) \ , \ k_o + k_1 + u(0,0,t)\right) = Mu \\ \left(-\dfrac{\partial u}{\partial t} - \dfrac{\partial u}{\partial h} - \psi\right)(u - Mu) = 0 \\ u(x,h,T) = 0 \ . \end{array}\right. \tag{5.22}$$

In (5.22) $u(0,t)$ denotes the function $u(x,0,t)$.

We now give the optimal policy starting from the relation (5.22). Let us assume that we start out from a point x,h,t such that $u < Mu$. Let θ^1 be the first instant such that

$$u(x,s-t+h,s) = \text{Min}\left[k_o + \Phi(s,s-t+h)u(0,s)(x), \ k_o+k_1+u(0,0,s)\right] . \tag{5.23}$$

We notice that θ^1 is deterministic. The instant θ^1 is the instant of the first control. Let us assume that the minimum of the two quantities in the right hand member of (5.23) is reached by the first term, then at the instant θ^1 we make an observation. If the minimum is attained by the second term, we change the machine. We are in the state $z(\theta^1),0$ at the instant θ^1. We next look for the first instant $s = \theta^2$ such that

$$u(z(\theta^1),s-\theta^1,s) = \text{Min}\left[k_o + \Phi(s,s-\theta^1)u(0,s)(z(\theta^1)), \ k_o+k_1+u(0,0,s)\right] . \tag{5.24}$$

At the instant θ^2 we decide to make an observation or to replace the machine according as the minimum in the right hand member of (5.24) is realised by the first or the second element (i.e., $\theta^2 = 0$ or 1 according as the element which has the minimum is the first or the second term inside the bracket).

We can see that the pair θ^2,ξ^2 only depends on $\theta^1,z(\theta^1)$. The values of θ^i,ξ^i are defined step by step for all i.

Remark 5.1

The model studied in Section 5.2 presents the principal features of the problems of partial observation (more precisely, of problems where the observation is made at the moment of decision). Nevertheless, we can have more complex situations, in particular, when we consider two types of consecutive decisions (make or not make an observation, then make or not make a change of machine). □

Remark 5.2

We shall refer to M. ROBIN [1],[2], A. FRIEDMAN [1], A. FRIEDMAN and M. ROBIN [1], R. ANDERSON and A FRIEDMAN [1],[2] for other aspects of problems of quality control and their treatment by Q.V.I.'s. □

6. STOCHASTIC CONTROL PROBLEMS IN FINANCE

6.1 *Growth of a Firm in an Uncertain Environment and Considerations of Finance*

Here we present a model studied by A. BENSOUSSAN-J. LESOURNE [1]. We consider a firm which invests in a random environment. We seek to evaluate its optimal requirement for funding. In the absence of all uncertainty it is intuitively clear (and this can be proved) that it is useless to preserve its funds. Indeed, the latter yield nothing, and it is therefore preferable to invest them. In the opposite case, in the presence of uncertainties the funds become of interest as a protection against risks. The model presented allows us to bring this into evidence, and at the same time to define an optimal policy for funding.

We shall write

(6.1) $y(t)$ = running total of investments at the instant t,

(6.2) $m(t)$ = disposable reserves at the instant t,

(6.3) $v(t)$ = investment rate at the instant t,

(6.4) $w(t)$ = rate of dividends at the instant t.

When the level of capital is y, the corresponding level during the interval dt is given by

$$f(y(t))dt + \varepsilon\, dw .$$

Therefore $y(t),m(t)$ behave in accordance with the equations

$$\frac{dy}{dt} = v(t) \tag{6.5}$$

$$dm(t) = f(y(t))dt + \varepsilon\, dw - v(t)dt - w(t)dt . \tag{6.6}$$

In the model above the random quantities affect the unit revenue. We have assumed that the latter is the sum of the deterministic quantity (mean revenue as a function of the level of capital invested) and of a random quantity $\varepsilon dw(t)$, where $w(t)$ is a Wiener process. The random quantity therefore does not depend upon the level of capital invested. Other models where the random quantity would depend upon the invested capital are possible.

We impose the following constraints on the controls v,w

$$\begin{cases} v \geq 0 \ , \quad w \geq 0 \\ v + w \leq f(y) . \end{cases} \tag{6.7}$$

The last constraint in (6.7) says that at every instant the sum of the investment and the dividends cannot exceed the mean available revenue.

We shall assume that the firm is bankrupt when its funding $m(t) = 0$. The firm must then cease trading. Let τ be the moment of bankruptcy

$$\tau = \inf \{s \mid m(s) = 0\} .$$

The criterion to be maximised is given by (the realised profit)

$$J = E \int_0^{\tau} e^{-it} w(t)dt . \tag{6.8}$$

In (6.8) i is the interest rate. We write $u(x,m)$ for the maximum of J with respect to the (adapted) possible controls, when we have the following initial controls

$$y(0) = x \quad , \quad m(0) = m . \tag{6.9}$$

Dynamic programming shows that the function $u(x,m)$ is the solution of

$$\left| \begin{array}{l} iu = \dfrac{\epsilon^2}{2} \dfrac{\partial^2 u}{\partial m^2} + \text{Max}\left(1, \dfrac{\partial u}{\partial x}, \dfrac{\partial u}{\partial m}\right) f(x) \ , \ m > 0 \\ u(x,0) = 0 . \end{array} \right. \tag{6.10}$$

Starting with (6.10) we can bring out the optimal strategy of the firm. Let us assume that at an arbitrary instant t the firm is subject to the conditions

$$y(t) = x \quad , \quad m(t) = m .$$

If

$$1 = \text{Max}\left(1, \frac{\partial u}{\partial x}, \frac{\partial u}{\partial m}\right)$$

then the firm distributes dividends without investing. In this case $v = 0$, $w = f(x)$. If

$$\frac{\partial u}{\partial x} = \text{Max}\left(1, \frac{\partial u}{\partial x}, \frac{\partial u}{\partial m}\right)$$

then the firm makes maximum investment, therefore

$$w = 0 \ , \quad v = \lambda f(x) .$$

Finally, if

$$\frac{\partial u}{\partial m} = \text{Max}\left(1, \frac{\partial u}{\partial x}, \frac{\partial u}{\partial m}\right)$$

then the firm seeks to improve its reserves. It neither invests nor pays dividends, so $v = w = 0$.

It will be noted that the preceding conditions are intuitively clear if we notice that

$\dfrac{\partial u}{\partial x}$ = additional revenue generated when 1 franc extra is added in the state (x,m),

$\frac{\partial u}{\partial x}$ = additional revenue generated when the reserves are increased by 1 franc in the state (x,m),

1 = additional revenue generated when an extra franc is distributed in the state (x,m).

The last statement is quite clear, but allows a better interpretation of the above condition.

Remark 6.1

In A. BENSOUSSAN and J. LESOURNE [1] a study will be found of the regions of the half-plane $x,m \geqq 0$ where it is necessary to invest, distribute, or improve its reserves.

Remark 6.2

In the model above, the firm is not able to borrow. It operates entirely by self-financing. This is clearly a serious limitation, for debt plays a very important role in practice. A more elaborate model, taking account of loans is possible, but technically it is very much more complex. □

6.2 *Portfolio Management*

Here we consider a very simple model of a classical problem of financial management, i.e., the choice between placings and the maintenance of liquidity. We shall write

$$\text{(6.11)}\qquad \begin{aligned} y_j(t) &= \text{sum invested in the j-th placing at the instant } t,\ j = 1,\dots,N, \\ z(t) &= \text{liquid assets at the instant } t. \end{aligned}$$

In the absence of any financial investment we have

$$\text{(6.12)}\qquad dy_j(t) = 0 .$$

On the other hand z(t) is given by

$$\text{(6.13)}\qquad dz = g(z,y,t)dt + \sigma(z,y,t)dw .$$

The model (6.13) takes account, in particular, of revenue (from work and capital) and disbursements (consumption).

Let $\theta^1 \leqq \theta^2 \leqq \dots \leqq \theta^i \leqq \dots$ be a sequence of instruction instants. These are instants when we buy or sell shares. Let ξ_j^i, $j = 1,\dots,N$ be the amount of the share j bought or sold at the i-th instant, with

$$\xi_j^i \geqq 0,\ \text{we buy } \xi_j^i \text{ of the shares j},$$

$$\xi_j^i \leqq 0,\ \text{we sell } \xi_j^i \text{ of the shares j}.$$

The processes $y_j(t), z(t)$ then evolve according to the equations

$$\text{(6.14)}\qquad dy_j = \sum_i \delta(t-\theta^i)\xi_j^i .$$

$$\text{(6.15)}\qquad dz = g(z,y,t)dt + \sigma(z,y,t)dw(t) - \sum_i \sum_j \delta(t-\theta^i)\xi_j^i .$$

The functional to be minimised is given by

(6.16) $$J = E\left[\int_t^T f\left(y(s), z(s)\, s\right), ds + \sum_i k(\theta^i;\xi^i)\right]$$

where

$$y(s) = (y_1(s), \dots, y_N(s))$$
$$\xi^i = (\xi_1^i, \dots, \xi_N^i).$$

The quantity $f(y,z,s)$ represents the cost per unit time at the instant s of holding a share portfolio of level y and a liquid amount z. For example, there may be advantages in holding, and for liquidity a cost in not satisfying the needs of the funding.

The function $k(t;\xi)$ represents the cost of ordering an amount ξ of shares at the instant t. If certain components of ξ are negative then there is a sale cost.

Let $u(x,z,t)$ be the minimum of the function J, when the initial conditions are

$$y(t) = x, \quad z(t) = z.$$

We shall write

$$A_z(x,t) = -g(z,x,t)\frac{\partial}{\partial z} - \frac{1}{2}\sigma^2(z,x,t)\frac{\partial^2}{\partial z^2}.$$

Then the function u is the solution of

(6.17) $$\begin{cases} -\frac{\partial u}{\partial t} + A_z u \leq f \\ u(x,z,t) \leq \operatorname{Inf}_{\xi}\left[k(t;\xi) + u(x+\xi, z-\xi, t)\right] = Mu \\ \left(-\frac{\partial u}{\partial t} + A_z u - f\right)(u - Mu) = 0 \\ u(x,z,T) = 0. \end{cases}$$

From this we deduce the optimal management policy.

6.3 *Dividend Distribution and the Problems of Failure*

We dispose of reserves in a random manner as a function of gains or losses. We seek an optimal policy for distributing dividends, taking into account the risk of failure. When the total of reserves vanishes we are failed, and the game stops. The criterion hinges upon the actual sum of the dividends.

Let $y(s)$ be the reserves held at the instant s $(s \geq t)$. Let $\theta^1 \leq \dots \leq \theta^i \leq \dots$ be the instants of dividend distribution, and τ the instant of failure. At the instant θ^1 we distribute the amount ξ^1. The process $y(s)$ evolves according to the equation

(6.18) $$dy(s) = g(y(s),s)ds + \sigma(y(s),s)dw(s) - \sum_i \delta(s-\theta^i)\xi^i.$$

We seek to minimise the criterion

(6.19) $$J = -E\left\{\sum_i \xi^i \chi_{[t,\tau\wedge T[}(\theta^i)\right\}.$$

The presence of the minus sign allows us to set a minimisation problems, even though we naturally seek to maximise $-J$. Let $u(x,t)$ be the lower bound of the criterion J, where we have $y(t) = x$. The function $u(x,t)$ is the solution of the problem

$$
(6.20)\quad \begin{cases} -\dfrac{\partial u}{\partial t} + A(t)u \leqq f \\ u(x,t) \leqq \inf\limits_{0\leqq \xi \leqq x} \{-\xi + u(x-\xi,t)\} = Mu \,, x > 0 \\ u(0,t) = 0 \\ u(x,T) = 0 \\ \left(-\dfrac{\partial u}{\partial t} + A(t)u - f\right)(u - Mu) = 0 \,. \end{cases}
$$

Remark 6.3

For another approach to the problems of risk and failure, see H. BUHLMANN [1]. □

7. APPLICATIONS IN INFORMATION TECHNOLOGY

Here we treat the management problem of the optimal management of a dual processor computer, studied by G. FAYOLLE and M. ROBIN [1]. This is a question of finding the best balance for the global cost using the possibilities of transferring from one processor to another.

The characteristics of the waiting queues of the processors 1 and 2 are respectively (λ_1,μ_1,N) and (λ_2,μ_2,M). Therefore without a transfer policy the processes $N(t)$ and $M(t)$ representing the number of programs queueing in lines 1 and 2 are Markov processes with values in $\{1,\ldots,N\}$ and $\{1,\ldots,M\}$ respectively, whose infinitesimal generators are given by

$$
A_1\Phi(n) = \chi_{n<N}\, \lambda_1[\Phi(n+1) - \Phi(n)] + \chi_{n>0}\, \mu_1[\Phi(n-1) - \Phi(n)]
$$

$$
A_2\Phi(m) = \chi_{m<M}\, \lambda_2[\Phi(m+1) - \Phi(m)] + \chi_{m>0}\, \mu_2[\Phi(m-1) - \Phi(m)] \,.
$$

If Φ is a function of n,m it will be convenient to set

$$
(7.1)\quad A\Phi(n,m) = \chi_{n<N}\, \lambda_1[\Phi(n+1,n) - \Phi(n,m)] + \chi_{n>0}\, \mu_1[\Phi(n-1,s) - \Phi(n,m)] + \\ + \chi_{m<M}\, \lambda_2[\Phi(n,m+1) - \Phi(n,m)] + \chi_{m>0}\, \mu_2[\Phi(n,m-1) - \Phi(n,m)] \,.
$$

Let $0 \leqq \theta^1 \leqq \theta^2 \leqq \ldots$ be a sequence of impulse instants. At these instants we decide to transfer a program from one queue to another. We shall set

$$
\begin{cases} \xi^i = 1 \text{ if queue 2 is transferred to queue 1,} \\ \xi^i = -1 \text{ if queue 1 is transferred to queue 2.} \end{cases}
$$

Therefore the processes $N(t),M(t)$ evolve according to the relations

$$
(7.2)\quad N(0) = n \,,\quad M(0) = m \,,
$$

(7.3) | conditioned by (7.2), $N(t)$ and $M(t)$ are Markov processes independent of the generators A_1 and A_2 respectively, for $0 \leqq t < \theta^1$, denoted $N_1(t)$ and $M_1(t)$,

$$
(7.4)\quad N(\theta^1) = N_1(\theta^1) + \xi^1 \,,
$$

$$
(7.5)\quad M(\theta^1) = M_1(\theta^1) - \xi^1 \,,
$$

(7.6) | conditioned by (7.4), (7.5), $N(t), M(t)$ are independent Markov processes with generators A_1 and A_2 respectively, for $\theta^1 \leqq t < \theta^2$, denoted $N_1(t)$ and $M_2(t)$,

(7.7) $$N(\theta^2) = N_2(\theta^2) + \xi^2 ,$$

(7.8) $$M(\theta^2) = M_2(\theta^2) - \xi^2 .$$

Thus we define the behaviour of the control process $N(t), M(t)$ for all t.

We now define the functional to be minimised. First there is a cost per unit time linked to the charge for the waiting queues, and next there are transfer costs at each impulse.

Let $f(n,m)$ be the unit cost related to the charge and

$$k(\xi) = \begin{vmatrix} k_1 & \text{if} \quad \xi = 1 \\ k_2 & \text{if} \quad \xi = -1 . \end{vmatrix}$$

The functional J to be minimised is then given by

(7.9) $$J = E\left[\int_0^\infty e^{-\alpha s} f(N(s), M(s))ds + \sum e^{-\alpha\theta^i} k(\xi^i)\right] .$$

We define $u(n,m)$ as being the lower bound of the function J. The function u is the solution of

(7.10) $$\begin{vmatrix} Au + \alpha u \leqq f \\ u \leqq Mu \\ \left(Au + \alpha u - f\right)(u - Mu) = 0 \end{vmatrix}$$

where

(7.11) $$Mu(n,m) = \underset{\xi = \left| \begin{smallmatrix} +1 \\ -1 \end{smallmatrix} \right.}{\text{Min}} [k(\xi) + u(n+\xi, m-\xi)] .$$

Remark 7.1

The model above can be extended to the case of several waiting queues, and to the case where these queues are not characterised by rates of arrival and exponential times of servicing. We refer to M. ROBIN [1] for more general cases. □

8. MEDICAL APPLICATIONS

Here we give an example of the application of impulse control to the problem of therapeutic treatment.

We denote by $x(t)$ a variable characterising the degree of extension of a malady at the instant t. In the absence of treatment the state $x(t)$ evolves according to the equation

(8.1) $$\frac{dx}{dt} = \mu x \quad , \qquad x(0) = x .$$

Consider a sequence of instants $\theta \leqq \theta^1 \leqq \theta^2 \leqq \ldots \leqq \theta^i \leqq \ldots$ at which treatments are carried out. Each treatment is characterised by an intensity ξ^i. After treatments the state x(t) evolves according to the equation

$$(8.2) \qquad \frac{dx}{dt} = \mu x - \sum \xi^i \delta(t - \theta^i) \quad , \quad x(0) = x \; .$$

Meanwhile the treatment may have an injurious influence on the healthy tissues. We then introduce the variable y(t) characterising the degree of deterioration of the organism as a result ot the treatment. The state y(t) evolves according to the equation

$$(8.3) \qquad \frac{dy}{dt} = -\beta y + \frac{1}{\xi_M} \sum_i \xi^i \delta(t - \theta^i) \quad , \quad y(0) = y \; .$$

Let us impose the constraints

$$(8.4) \qquad 0 \leqq \xi^i \leqq \xi_M \; .$$

The aim of the treatment is to keep the state x(t) at the maximum in the situation

$$(8.5) \qquad 0 \leqq x(t) \leqq x_o \; .$$

However, we must also not allow an exaggerated deterioration of the healthy tissues. In other words, y(t) must satisfy

$$(8.6) \qquad 0 \leqq y(t) \leqq y_o \; .$$

Let τ be the first instant when x(t) or y(t), respectively, exceed x_0 or y_0. We want to maximise τ. However, each treatment entails a cost k which also enters into the accounting line. We then pose the problem of minimising the functional

$$(8.7) \qquad J = -\tau + k \sum_i \chi_{\theta^i < \tau} \; .$$

We write u(x,y) for the lower bound of the functional J. This is a solution of the system

$$(8.8) \qquad \left| \begin{array}{l} -\mu \dfrac{\partial u}{\partial x} + \beta \dfrac{\partial u}{\partial y} \leqq -1 \\ u(x,y) \leqq \displaystyle\inf_{0 \leqq \xi \leqq \xi_M} \left[u\left(x - \xi, \; y + \frac{\xi}{\xi_M}\right) + k \right] = Mu \\ \left(-\mu \dfrac{\partial u}{\partial x} + \beta \dfrac{\partial u}{\partial y} + 1\right)(u - Mu) = 0 \\ u(0,y) = u(x,0) = 0 \; . \end{array} \right.$$

We can now deduce the optimal treatment policy from (8.8).

Chapter 2
Optimal stopping time problems for reflected diffusion problems

INTRODUCTION

In this Chapter we consider optimal stopping times and stochastic control for reflected diffusion processes. We restrict ourselves to instantaneous reflections, but we could extend numerous results to the case of processes with diffusion at the boundary.

In Sections 1 and 2 we present the theory of reflected stochastic differential equations in the strong sense, following S. WATANABE [1] in the essentials. Section 3 is devoted to the weak formulation in the form of the sub-martingale problem introduced by D. STROOCK - S.R.S. VARADHAN [1]. In Section 4 we study the Markov properties of reflected diffusions linking it with the probabilistic interpretation of boundary problems with (oblique) Neumann conditions on the boundary.

In Sections 5 and 6 we study the V.I.'s (variational inequalities) with Neumann boundary conditions from the analytic viewpoint, first in the stationary case, then then in the evolutionary case.

In Sections 7 and 8 we give the probabilistic interpretation of these V.I.'s.

In Section 9 we consider the problem of games, and in Section 10 problems with a a non-linear operator, and thus where control and stopping times occur at the same time. In Section 11 we give the interpretation of the V.I. with one-sided conditions on the boundary.

Certain methods and results are very similar to the Dirichlet case treated in the book A. BENSOUSSAN - J.L. LIONS [1]. We give no proofs, but state the results.

1. GENERAL PROPERTIES OF REFLECTED DIFFUSION PROCESSES

Let $(\Omega,\mathcal{C},P)$ be a probability space and $\mathcal{F}^t$ an increasing family of sub-σ-algebras of $\mathcal{C}$. A normalised R^n-valued Wiener process $w(t)$ is given, and $w(t)$ is an $\mathcal{F}^t$-martingale.

We are given functions $g(x,t)$ and $\sigma(x,t)$ of $R^n \times [0,\infty[\to R^n$ and $R^n \times [0,\infty[\to \mathcal{L}(R^n,R^n)$ respectively, satisfying

(1.1) $\quad |g(x,t) - g(x',t)| + |\sigma(x,t) - \sigma(x',t)| \leq C|x-x'|, \ \forall x,x',t$

(1.2) $\quad g,\sigma$ are measurable and bounded.

We are also given

(1.3) $\mathcal{O}$ a bounded open set of R^n whose boundary $\Gamma = \partial\mathcal{O}$ is a C^3-manifold(1)

(1.4) γ a C^2-mapping of $\bar{\mathcal{O}} \to R^n$, such that $\gamma(x).n(x) \geqq \delta > 0$, where $n(x)$ denotes the exterior unit normal at $x \in \Gamma$.

We are further given

(1.5) $x_0, \mathcal{F}_0$ measurable and $\bar{\mathcal{O}}$-valued,

$\xi_0, \mathcal{F}_0$ measurable scalars.

We introduce the following problem: find an R^n-valued process $y(t)$ such that

(1.6) $y(t)$ is adapted and continuous,

(1.7) there exists a scalar process $\xi(t)$ which is continuous, non-decreasing, and adapted such that $y(t)$ has the stochastic differential

$$dy(t) = g(y(t),t)\,dt + \sigma(y(t),t)\,dw(t) - \chi_\Gamma(y(t))\,\gamma(y(t))\,d\xi(t)$$

(1.8) a.s., $y(t) \in \bar{\mathcal{O}} \quad \forall t$,

(1.9) a.s. $$\int_{t_1}^{t_2} \chi_{\mathcal{O}}(y(t))\,d\xi(t) = 0 \qquad \forall\, t_1 \leqq t_2$$

(1.10) $y(0) = x_0, \quad \xi(0) = \xi_0$.

Remark 1.1

It is clear that in the problem (1.6),...,(1.10) the unknown consists of the two processes $y(t), \xi(t)$. □

Remark 1.2

The process $y(t)$ (which will be defined uniquely) is called a diffusion process reflected at the boundary of $\mathcal{O}$. □

Remark 1.3

The preceding formulation is strong. D. STROOCK and S.R.S. VARADHAN [1] have introduced a weak formulation in the form of a sub-martingale problem (cf. Section 3). We also refer to S. WATANABE [1] and N. EL KAROUI [1]. □

Remark 1.4

In (1.7) the term $\chi_\Gamma(y(t))\gamma(y(t))d\xi(t)$ clearly only plays a role on the boundary. To be formal, it is a vector that is always directed towards the interior of $\mathcal{O}$, and which therefore tends to redirect the process to the interior of $\mathcal{O}$. □

We have:

THEOREM 1.1: *Assuming* (1.1),(1.2),(1.3),(1.4),(1.5), *there exists one and only one solution of the problem* (1.6),...,(1.10). □

Remark 1.5

The uniqueness of the solution means that there is only one pair of processes $y(t), \xi(t)$ satisfying (1.6),...,(1.10). □

---

(1) The fact that $\mathcal{O}$ is bounded is not essential, but it will enable us to avoid introducing weighted spaces.

Having undertaken the proof of Theorem 1.1, let us give a formulation of the problem behaving only as a single process, namely y(t), a formulation that is, however, valid only in the following particular case

(1.11) $\mathcal{O}$ convex,

(1.12) $\gamma(x) = n(x)$.

We have:

THEOREM 1.2: *Under the assumption of Theorem 1.1, as well as* (1.11),(1.12), *the process* y(t) *is an unique solution of the stochastic variational inequality defined by the conditions*

(1.13) y(t) *is continuous, adapted,* a.s. $y(t) \in \bar{\mathcal{O}}$ $\forall$ t *and*

$$y(t) - x_0 - \int_0^t g(y(s), s)ds - \int_0^t \sigma(y(s), s)\, dw(s)$$

has bounded variation, and is zero for t = 0,

(1.14) $(z(t) - y(t)).(dy(t) - g(y(t), t) - \sigma(y(t), t)\, dw(t)) \geqq 0$

for every adapted continuous process z(t) *such that* a.s. $z(t) \in \bar{\mathcal{O}}$, $\forall$ t.

Proof: If we set

$$\eta(t) = y(t) - x_0 - \int_0^t g(y(s), s)\, ds - \int_0^t \sigma(y(s), s)\, dw(s)$$

which is thus a process with bounded variation, by (1.13), the formulation (1.14) means that

$$(1.15) \qquad \int_{t_1}^{t_2} (z(t) - y(t)) . d\eta(t) \geqq 0 \qquad \forall\, t_1 \leqq t_2 ,$$

considered as a Stieltjes' integral.

Let us now consider y(t) the solution of (1.6),...,(1.10). By (1.7), we have

$$\eta(t) = - \int_0^t \chi_\Gamma(y(s))\, n(y(s))\, d\xi(s)$$

which is certainly a process with bounded variation. Furthermore,

$$\int_{t_1}^{t_2} (z(t) - y(t)) . d\eta(t) = - \int_{t_1}^{t_2} (z(t) - y(t)) . n(y(t))\, \chi_\Gamma(y(t))\, d\xi(t).$$

But if $y \in \Gamma$, $z \in \bar{\mathcal{O}}$, since $\mathcal{O}$ is convex, we have

$$(y - z).n(y) \geqq 0$$

and thus (1.15) is certainly satisfied.

On the other hand, the solution of (1.13),(1.14) is necessarily unique. In fact if $y_1(t)$,$y_2(t)$ are two solutions, it follows from Ito's Formula that

(1.16) $$\frac{1}{2}|y_2(t)-y_1(t)|^2 = \int_0^t (y_2-y_1)\cdot d(y_2-y_1) + \frac{1}{2}\int_0^t \mathrm{tr}(\sigma(y_2)-\sigma(y_1))(\sigma(y_2)-\sigma(y_1))^* ds .$$

But, from (1.15) written for y_1, y_2, and taking as test process $z(t) = y_2(t)$, then $y_1(t)$, we easily obtain

(1.17) $$\int_0^t (y_2-y_1)\cdot d(y_2-y_1) \leq \int_0^t (y_2-y_1)\cdot(g(y_2)-g(y_1))ds + \int_0^t (y_2-y_1)\cdot(\sigma(y_2)-\sigma(y_1))\, dw .$$

From (1.16),(1.17), by adding, and by passing to the mathematical expectation, we deduce

$$\frac{1}{2}E|y_2(t)-y_1(t)|^2 \leq E\int_0^t (y_2-y_1)\cdot(g(y_2)-g(y_1))\,ds + \frac{1}{2}E\int_0^t \mathrm{tr}(\sigma(y_2)-\sigma(y_1))(\sigma(y_2)-\sigma(y_1))^*\, ds .$$

Using the Lipschitz properties of g,σ and Gronwall's Inequality, we easily obtain

$$E|y_2(t)-y_1(t)|^2 = 0 \quad \forall t .$$

Since y_1, y_2 are two continuous processes, we see that

$$\text{a.s.,} \qquad y_1(t) = y_2(t) \quad \forall\, t,$$

whence the result. □

Remark 1.6

Relation (1.16) uses Ito's formula for processes which are the sum of a process with a stochastic differential and a process with bounded variation. More simply, in the case that interests us it suffices to use the following formula for integrating by parts. If $\xi(t),\eta(t)$ are scalar processes such that

$$d\xi = a(t) + b(t)dw(t),$$

$\eta(t)$ continuous and with bounded variation,

then we have

$$\xi(t_2)\,\eta(t_2) - \xi(t_1)\,\eta(t_1) = \int_{t_1}^{t_2} d\xi\;\eta(t) + \int_{t_1}^{t_2} \xi(t)\, d\eta(t)$$

a relation which is easily proved by discretisation. □

Remark 1.7

We can extend the result of Theorem 1.2 a little, to the case where

$$\gamma(x) = Mn(x),$$

where M is a positive definite symmetric matrix. This reduces simply to providing R^n with a new scalar product. □

Remark 1.8

The formulation in the stochastic V.I. form is very useful in singular perturbation problems (cf. A. BENSOUSSAN - J.L. LIONS [12]). □

2. PROOF OF THEOREM 1.1

The proof of Theorem 1.1 will be done in two steps. Here we treat the case of the half-space, i.e.,

(2.1) $$\mathcal{O} = \{x | x_n > 0\}.$$

We begin with an auxiliary problem. Let us take

(2.2) $$\left| \begin{array}{l} x_o \text{ a random variable with values in } \bar{\mathcal{O}},\ x_o \text{ is } \mathcal{F}^o\text{-measurable and} \\ E|x_o|^4 < \infty, \\ \xi_o, \text{ a scalar R.V., } \mathcal{F}^o\text{-measurable, and } E|\xi_o|^4 < \infty, \end{array} \right.$$

(2.3) $$\left| \begin{array}{l} g(t),\sigma(t) \text{ processes adapted to } \mathcal{F}^t, \text{ with values respectively in } R^n \\ \text{and } \mathcal{L}(R^n,R^n),\ g(t),\sigma(t) \text{ are bounded by a non-random constant.} \end{array} \right.$$

We then look for $y(t)$, *adapted and continuous*, $\xi(t)$ *scalar, adapted, and non-decreasing*, and such that

(2.4) $$\left| \begin{array}{l} dy_i(t) = g_i(t)dt + \sum_{j=1}^{n} \sigma_{ij}(t)\, dw_j(t) \quad , \ i = 1 \,..n-1 \\ dy_n(t) = g_n(t)dt + \sum_{j=1}^{n} \sigma_{nj}(t)\, dw_j(t) + d\xi(t) \end{array} \right.$$

(2.5) $$y_n(t) \geq 0$$

(2.6) $$y_n(t)d\xi(t) = 0$$

(2.7) $$y(o) = x_o, \qquad \xi(o) = \xi_o.$$

We are now going to prove the existence and uniqueness of the pair $y(t),\xi(t)$. Let us show the uniqueness of the solut on. Let $y^1,\xi^1;y^2,\xi^2$ be two solutions. It is clear that $y_i^1 = y_i^2$, for all $i = 1,\dots,n-1$. On the other hand

$$(d\xi^1 - d\xi^2)(\xi^1 - \xi^2) = (d\xi^1 - d\xi^2)(y_n^1 - y_n^2) \leq 0$$

and therefore

$$|\xi^1(t) - \xi^2(t)|^2 \leq 0 \qquad \forall t$$

hence a.s.

$$\xi^1(t) = \xi^2(t) \quad \forall t \;, \quad y^1(t) = y^2(t) \quad \forall t \,.$$

For the existence it is clear that it is sufficient to work in dimension 1. Hence we look for

(2.8) $y(t), \xi(t)$ adapted and continuous, $\xi(t)$ non-decreasing,

(2.9) $dy(t) = g(t) + \sigma(t)dw(t) + d\xi(t)$

(2.10) $y(t) \geqq 0$

(2.11) $y(t)d\xi(t) = 0$

(2.12) $y(o) = x_o, \qquad \xi(o) = \xi_o;$

it is clear that we can take $\xi_o = 0$. Now, write

$$M(t) = \int_o^t g(s)ds + \int_o^t \sigma(s)dw(s) \;. \tag{2.13}$$

The solution of (2.8),(2.12) can then be obtained explicitly. In fact, let

$$\theta = \inf\{t : x_o + M(t) \leqq 0\};$$

then we take

$$\begin{cases} \xi(t) = 0 & \text{for} \quad t \leqq \theta \\ \xi(t) = -\underset{\theta \leqq s \leqq t}{\operatorname{Min}} \; (x_o + M(s)) & \text{for} \quad t > \theta \end{cases} \tag{2.14}$$

and

$$y(t) = x_o + M(t) + \xi(t) \;. \tag{2.15}$$

The pair $(y(t),\xi(t))$ answer our problem. Since the process $M(t)$ is continuous, $\xi(t)$ is continuous, therefore $y(t)$ is also. Furthermore, $y(t)$ and $\xi(t)$ are adapted and $\xi(t)$ is not decreasing. It is clear that (2.9) and (2.10) are satisfied, therefore it remains to prove (2.11). Let t be fixed and such that $y(t) > 0$, then there exists small enough such that

$$\xi(t+\varepsilon) - \xi(t) = 0 \,. \tag{2.16}$$

In fact, if $y(t) > 0$, then

$$x_o + M(t) > \underset{\theta \leqq s \leqq t}{\operatorname{Min}} \; (x_o + M(s)) \;. \tag{2.17}$$

Now,

$$\underset{\theta \leqq s \leqq t+\varepsilon}{\operatorname{Min}} \; (x_o + M(s)) = \begin{cases} \underset{\theta \leqq s \leqq t}{\operatorname{Min}} \; (x_o + M(s)) \\ \text{or} \\ \underset{t \leqq s \leqq t+\varepsilon}{\operatorname{Min}} \; (x_o + M(s)) \,. \end{cases} \tag{2.18}$$

Let us choose $\eta > 0$ such that

$$x_o + M(t) > \underset{\theta \leqq s \leqq t}{\operatorname{Min}} \; (x_o + M(s)) + \eta$$

which is permissible, by (2.17). Since M is continuous, we can choose ε such that

$$\underset{t \leqq s \leqq t+\varepsilon}{\operatorname{Min}} (x_o + M(s)) \geqq x_o + M(t) - \eta$$

$$> \underset{\theta \leqq s \leqq t}{\operatorname{Min}} (x_o + M(s))$$

and therefore, by (2.18), we immediately deduce (2.16). By the Stieltjes' integral construction the property (2.16) implies (2.11).

We can then consider the initial problem in the half-space. We want to find

(2.19) $y(t)$ adapted, and continuous, $\xi(t)$ adapted, continuous, non-decreasing

(2.20) $$E \int_o^T |y(t)|^4 \, dt < \infty$$

(2.21) $$dy_i(t) = g_i(y, t)\, dt + \sum_{j=1}^{n} \sigma_{ij}(y, t)\, dw_j(t), \quad i = 1 \ldots n-1$$

(2.22) $$dy_n(t) = g_n(y, t)\, dt + \sum_{j=1}^{n} \sigma_{nj}(y,t)\, dw_j(t) + d\xi(t)$$

(2.23) $$\text{p.s.} \quad y_n(t) \geqq 0 \quad \forall t$$

(2.24) $$y_n(t) d\xi(t) = 0$$

(2.25) $$y(o) = x_o, \quad \xi(o) = \xi_o.$$

The result is stated in the form:

THEOREM 2.1: *Assume* (1.1),(1.2),(2.1),(2.2). *There exists one and only one solution of the problem* (2.19),...,(2.25).

Proof: First let us show the uniqueness. If y^1, y^2 are two solutions, then we have

$$(y^2 - y^1) \cdot d(y^2 - y^1) \leqq (g(y^2) - g(y^1)) \cdot (y^2 - y^1)\, dt +$$
$$+((\sigma(y^2) - \sigma(y^1))\, dw \cdot (y^2 - y^1)).$$

Using Ito's formula, and then going to the mathematical expectation, and taking into account the Lipschitz properties of g, σ, we obtain

$$E\, |y^2(t) - y^1(t)|^2 \leqq C\, E \int_o^t |y^2(s) - y^1(s)|^2\, ds$$

which implies

$$\text{a.s.} \quad y^1(t) = y^2(t) \quad \forall\, t.$$

To prove the existence we use an iterative procedure as follows. Set

$$y_0(t) = x_o,$$

then having defined $y_p(t)$ we define $y_{p+1}(t)$ by solving the auxiliary problem (2.4),(2.5),(2.6),(2.7) for

$$g(t) = g(y_p(t), t) \quad , \quad \sigma(t) = \sigma(y_p(t), t) \cdot$$

Now,

$$(dy_p - g(y_{p-1}, t)dt - \sigma(y_{p-1}, t)dw(t)).(y_{p+1}(t) - y_p(t)) \geqq 0$$
$$(dy_{p+1} - g(y_p, t)dt - \sigma(y_p, t)\, dw(t)) \,.\, (y_p(t) - y_{p-1}(t)) \geqq 0 \,.$$

By addition we obtain

$$(dy_{p+1} - dy_p).\ (y_{p+1} - y_p) \leqq (g(y_p) - g(y_{p-1}))\,.\,(y_{p+1} - y_p)\, dt +$$
$$+ ((\sigma(y_p) - \sigma(y_{p-1}))\, dw(t))\,.\,(y_{p+1} - y_p)\,.$$

On taking Ito's formula into account, we deduce from this that

$$\frac{1}{2} | y_{p+1}(t) - y_p(t) |^2 \leqq \int_0^t (g(y_p) - g(y_{p-1}))\,.\,(y_{p+1} - y_p)\, ds +$$
$$+ \frac{1}{2} \int_0^t \mathrm{tr}\, (\sigma(y_p) - \sigma(y_{p-1}))\, (\sigma^*(y_p) - \sigma^*(y_{p-1}))\, ds +$$
$$+ \int_0^t (y_{p+1} - y_p).\ (\sigma(y_p) - \sigma(y_{p-1}))\, dw(s)\,.$$

This is squared (the inequality is preserved, because the left hand side is positive). We then obtain

$$\text{(2.26)}\qquad \frac{1}{4} |y_{p+1}(t) - y_p(t)|^4 \leqq 3 \Big[\Big(\int_0^t (g(y_p) - g(y_{p-1}))\,.\,(y_{p+1} - y_p)\, ds\Big)^2$$
$$+ \Big(\int_0^t (y_{p+1} - y_p)\,.\,(\sigma(y_p) - \sigma(y_{p-1}))dw\Big)^2 +$$
$$+ \frac{1}{4}\Big(\int_0^t \mathrm{tr}\, (\sigma(y_p) - \sigma(y_{p-1}))\, (\sigma^*(y_p) - \sigma^*(y_{p-1}))ds\Big)^2 \Big].$$

Taking the mathematical expectation and taking account of the Lipschitz properties once again, we deduce from this that

$$E|y_{p+1}(t) - y_p(t)|^4 \leqq C\ [\,E \int_0^t |y_p(s) - y_{p-1}(s)|^4\, ds + E \int_0^t |y_{p+1}(s) - y_p(s)|^4 ds$$

Using Gronwall's Inequality, it follows from this that

$$\text{(2.27)}\qquad E\, |y_{p+1}(t) - y_p(t)|^4 \leqq C_T\, E \int_0^t |y_p(s) - y_{p-1}(s)|^4\, ds\,.$$

By iterating we obtain

$$E\, |y_{p+1}(t) - y_p(t)|^4 \leqq C_T^p \int_0^t \frac{(t-s)^{p-1}}{(p-1)!}\, E|y_1(s) - x_0|^4\, ds\,.$$

Since

$$y_1 \in C(0,T\,;\, L^4(\Omega, \mathcal{G}, P)),$$

from this we deduce that

$$(2.28) \qquad E\,|y_{p+1}(t) - y_p(t)|^4 \leq C\,C_T^p\,\frac{T^p}{p!} .$$

Furthermore, since

$$E \sup_{0 \leq t \leq T} \left(\int_0^t (y_{p+1} - y_p) \cdot (\sigma(y_p) - \sigma(y_{p-1}))\,dw \right)^2 \leq C\,E \int_0^T dt\,|(\sigma^*(y_p) - \sigma^*(y_{p-1}))\,(y_{p+1} - y_p)|^2$$

it follows from (2.26) that

$$E \sup_{0 \leq t \leq T} |y_{p+1}(t) - y_p(t)|^4 \leq C \left[\int_0^T E\,|y_{p+1} - y_p|^2 |y_p - y_{p-1}|^2 dt + \int_0^T E\,|y_p - y_{p-1}|^4\,dt \right]$$

and by (2.28)

$$\leq C \left[\frac{C_T^p\,T^p}{p\,!} + \frac{C_T^{p-1}\,T^{p-1}}{(p-1)\,!} \right]$$

and for p large enough it is seen that

$$E \sup_{0 \leq t \leq T} |y_{p+1}(t) - y_p(t)|^4 \leq C' \,\frac{C_T^{p-1}\,T^{p-1}}{(p-1)\,!} .$$

From this it follows that

$$\sum_{p=1}^{\infty} P \left\{ \sup_{0 \leq t \leq T} |y_{p+1}(t) - y_p(t)| \geq \frac{1}{p^2} \right\} < \infty .$$

Therefore by the Borel-Cantelli Theorem

$$x_o + \sum_{p=0}^{\infty} (y_{p+1}(t) - y_p(t))$$

converges a.s. uniformly at t to a process y(t). By going to the limit we then obtain

$$y_i(t) = x_{oi} + \int_0^t g_i(y(s),\,s)\,ds + \int_0^t \sum_j \sigma_{ij}(y(s),\,s)\,dw_j(s) ,\quad i = 1..n-1$$

$$y_n(t) \geq y_n(s) + \int_s^t g_n(y(\lambda),\,\lambda)\,d\lambda + \int_s^t \Sigma_j\,\sigma_{nj}\,(y(\lambda),\lambda)\,dw_j(\lambda) .$$

We then set

$$\xi(t) = \xi_0 + y_n(t) - x_{on} - \int_0^t g_n(y(s),s)ds - \int_0^t \sum_j \sigma_{nj}(y(s),s)dw_j(s)$$

and it is then easily verified that $y(t),\xi(t)$ satisfy the relations (2.19), (2.20), (2.21), (2.22), (2.23), (2.25). It remains to verify (2.24). Now, by (2.6) we have

$$y_p(t) \cdot (dy_p - g(y_{p-1})dt - \sigma(y_{p-1})dw(t)) = 0 \tag{2.29}$$

and by Ito's formula we have

$$\frac{1}{2}|y_p(t)|^2 = \frac{1}{2}|x_o|^2 + \int_0^t y_p \cdot dy_p + \frac{1}{2}\int_0^t \operatorname{tr} \sigma(y_{p-1})\sigma^*(y_{p-1})\,ds$$

which, when compared with (2.29), leads to

$$\frac{1}{2}|y_p(t)|^2 = \frac{1}{2}|x_o|^2 + \int_0^t y_p \cdot g(y_{p-1})ds + \int_0^t y_p \cdot \sigma(y_{p-1})\,dw(s) + \frac{1}{2}\int_0^t \operatorname{tr}\sigma(y_{p-1})\sigma^*(y_{p-1})\,ds .$$

Integrating and passing to the limit we obtain

$$\frac{1}{2}|y(t)|^2 = \frac{1}{2}|x_o|^2 + \int_0^t y(s) \cdot g(y(s))ds + \int_0^t y(s) \cdot \sigma(y(s))dw(s) + \frac{1}{2}\int_0^t \operatorname{tr}\sigma(y(s))\,\sigma^*(y(s))\,ds . \tag{2.30}$$

But using Ito's formula starting from the relations (2.21) (2.22), we obtain

$$\int_0^t y_n(s)\,d\xi(s) = 0$$

which certainly implies (2.24). Thus we have completely proved Theorem 2.1. □

For the remainder it will be convenient to consider the case of random initial data. Let τ be a stopping time of $\mathcal{F}^t$. Write

$$\chi_\tau(t) = \begin{cases} 1 \text{ if } t<\tau \\ 0 \text{ if } t\geqq\tau . \end{cases}$$

Given

$$\begin{cases} \eta, \text{ an } \mathcal{F}^\tau\text{-measurable R.V., } E|\eta|^4 < \infty,\ \eta \in R^n,\ \eta_n \geqq 0 \\ \zeta, \text{ an } \mathcal{F}^\tau\text{-measurable scalar R.V., } E\zeta^4 < \infty, \end{cases}$$

we seek

(2.32) $y(t)$ continuous and R^n-valued, $\zeta(t)$ continuous and R-valued

(2.33) $(1-\chi_\tau(t))\,y(t)$, $(1-\chi_\tau(t))\,\xi(t)$ are adapted

(2.34) $$E \int_0^T |y(t)|^4 \, dt < \infty$$

(2.35) $$dy_i(t) = (1-\chi_\tau(t))\,(g_i(y,t)\,dt + \sum_{j=1}^{n} \sigma_{ij}(y,t)dw_j) \quad , \; i = 1..n-1$$

(2.36) $$dy_n(t) = (1-\chi_\tau(t))\,(g_n(y,t)dt + \sum_{j=1}^{n} \sigma_{ij}(y,t)dw_j + d\,\xi(t))$$

(2.37) $$\text{p.s.} \quad y_n(t) \geqq 0 \;\; \forall t \;, \quad y_n(t)\, d\,\xi(t) = 0$$

(2.38) $$y(0) = \eta \;, \quad \xi(t) = \zeta \;, \quad \text{for} \quad t \in [0,\tau].$$

We can state:

THEOREM 2.2: *Assuming* (1.1),(1.2),(2.31), *there exists one and only one solution of the problem* (2.32),(2.33),...,(2.38). □

The proof is very similar to that of Theorem 2.1, and is left to the reader.

Remark 2.1

It is clear that y(t) = η for $t \in [0,\tau]$, $\xi(t) = \zeta$ for $t \in [0,\tau]$, and therefore what actually counts is what has happened after the instant τ. It will be said that y(t) is the reflected diffusion starting from the instant τ to the point η.□

We can then complete the proof of Theorem 1.1. To do this we describe the frontier with the aid of a system of local charts. Let $\mathcal{O}_i$, i = 1,...,L, be a system of local charts. With A we associate a C^3-diffeomorphism of $\mathcal{O}_i$ onto Q = $[-1,+1]^n$. We can choose a diffeomorphism ψ_i which transforms the direction γ at every point of $\Gamma \cap \mathcal{O}_i$ into the direction of the internal normal at the corresponding point of $Q_0 = \{z \mid z \in \overset{o}{Q},\, z_n = 0\}$. It is here that we have used the assumptions about the regularity of γ.

Let us give the principle of the construction of the reflected process. Starting from a point of the open set $\mathcal{O}$ we consider the normal diffusion up to the first instant where it reaches the frontier. At this (random) instant the process is necessarily in a local chart $\mathcal{O}_i$. We can then construct the reflected process stopped at the exit from the local chart. The latter is constructed by returning, via the diffeomorphism γ_i, to a half-space problem, and by using Theorem 2.2. At the exit from the local chart $\mathcal{O}_i$ we are then in another chart, and again being the construction as before.

If we write

$$B_1 = \mathcal{O} \cup \mathcal{O}_1$$
$$B_{i+1} = B_i \cup \mathcal{O}_{i+1}$$
$$B_N = \mathcal{O} \cup \mathcal{O}_1 \ldots \cup \mathcal{O}_N$$

we construct successively the reflected diffusion stopped at T or at the exit from B_1,B_2, etc., ... The uniqueness results step by step from the uniqueness result in the half-space case. □

3. SUB-MARTINGALE PROBLEMS

3.1 *Notation and Statement of the Problem*

We shall need the weak formulation for reflected diffusions, in the form of a sub-martingale problem. This allows us to weaken the assumptions about the coefficients, and in particular will allow us to treat reflection directions that vary with time. We are given functions $g(x,t),\sigma(x,t)$ from $R^n \times [0,\infty[\to R^n$ and $\mathcal{L}(R^n;R^n)$ respectively, such that

(3.1) σ is continuous and bounded

(3.2) g is measurable and bounded

(3.3) $$\left| \begin{array}{l} \gamma : R^n \times [0,\infty[\to R^n, \text{ continuous and bounded, and} \\ \gamma(x,t) \,.\, n(x) \geqq \delta > 0 \;,\; \forall\, x \in \Gamma = \partial \mathcal{O} \;,\; \forall t \,. \end{array} \right.$$

We can assume that the open set $\mathcal{O}$ can be represented in the following way

(3.4) $$\mathcal{O} = \{ x \mid \Phi(x) < 0 \,,\ \text{where } \Phi \in C^2(R^n) \,,\ \Phi \text{ is bounded as well as its first and second derivatives, } \| \frac{\partial \Phi}{\partial x} \| \geqq \beta > 0 \ \text{ sur } \ \Gamma \} \,.$$

We also set

(3.5) $$\left| \begin{array}{l} \Omega_t = C^{\circ}([t,\infty[\,;\bar{\mathcal{O}}) \\ y(s\,;\omega) = \omega(s) \;,\; \text{for} \quad s \geqq t \\ \mathcal{U}_t^s = \sigma(y(\lambda) \mid t \leqq \lambda \leqq s) \,. \end{array} \right.$$

A probability measure P^{xt} on Ω_t is a solution of the sub-martingale problem (with respect to g,σ,γ), starting from x at the instant t if

(3.6) $$P^{xt}\,[y(t) = x] = 1$$

(3.7) $$\left| \begin{array}{l} \varphi(y(s),s) - \displaystyle\int_t^s \frac{\partial \varphi}{\partial x} \,.\, g(y(\lambda),\lambda)\, d\lambda - \frac{1}{2} \int_t^s \operatorname{tr} \frac{\partial^2 \varphi}{\partial x^2}\, \sigma \sigma^*(y(\lambda),\lambda)\, d\lambda \\ \qquad\qquad - \displaystyle\int_t^s \frac{\partial \varphi}{\partial s}\,(y(\lambda),\lambda)\, d\lambda \\ \qquad \text{is a sub-martingale, } \forall \varphi \in C^{2,1} \text{ with compact support in} \\ R^n \times [0,\infty[\quad \text{satisfying} \\ \dfrac{\partial \varphi}{\partial x} \,.\, \gamma \leqq 0 \quad , \quad \forall\, x \in \Gamma \quad , \quad \forall\, t \,. \end{array} \right.$$

3.2 *Some Results on the Sub-Martingale Problem*

Before proving existence, we give some results on the sub-martingale problem. We follow the presentation of D. STROOCK - S.R.S. VARADHAN [1].

LEMMA 3.1: *Let* $\varphi \in C_b^{2,1}(R^n \times [0,\infty[)$, *then*

(3.8) $$X_\varphi(s) = \varphi(y(s),s) + \int_t^s \Big(- \frac{\partial \varphi}{\partial s} + A^{(\lambda)}\varphi\Big)\,(y(\lambda),\lambda)\, d\lambda$$

is a local P^{xt}*-martingale on* $\mathcal{O}$. *If* $\gamma \,.\, \frac{\partial \varphi}{\partial x} \leqq 0$ *on* $\Gamma \times [t,\infty[$, *then* $X_\varphi(s)$ *is a* P^{xt}*-sub-martingale. Here we have written*

(3.9) $$A(t)\,\varphi = - g \,.\, \frac{\partial \varphi}{\partial x} - \frac{1}{2} \operatorname{tr} \frac{\partial^2 \varphi}{\partial x^2}\, \sigma \sigma^* \,.$$

Proof: We begin with the second part. It can always be assumed that

(3.10) $\varphi(x,t) = 0$ for $t \geqq T$.

In fact, if the property is proved under this assumption, consider

$$\beta(t) \in C^1[0,\infty],$$

$$\beta = 1 \text{ for } 0 \leqq t \leqq 1,$$

$$\beta = 0 \text{ for } t \geqq 2,$$

$$0 \leqq \beta \leqq 1 \text{ , and}$$

and

$$\beta_T(t) = \beta(\tfrac{t}{T}) \ .$$

Replace φ by $\varphi\beta_T$, then $X_\varphi(s)$ is a P^{xt}-sub-martingale. By making T tend to $+\infty$ the desired result is obtained. Therefore assume (3.10). Let $\eta(x) \in C_b^2(R^n)$, with

$$\eta = 1 \quad \text{for} \quad |x| \leqq 1, \quad \eta = 0 \quad \text{for} \quad |x| \geqq 2, \quad 0 \leqq \eta \leqq 1 .$$

Set

$$\eta_R(x) = \eta(\tfrac{x}{R}) \ .$$

Also let

$$\bar{\eta}_R(x) \in C_b^2(R^n), \qquad 0 \leqq \bar{\eta}_R \leqq 1 ,$$

$$\bar{\eta}_R(x) = 1 \text{ on the support of } \eta_R \quad (\text{i.e. for } |x| \leqq 2R) \ .$$

Then

$$\varphi_R = \varphi\,\eta_R - \frac{\alpha}{R}\,\bar{\eta}_R\,\Phi$$

satisfies

$$\gamma \cdot \frac{\partial \varphi_R}{\partial x} = \gamma \cdot \frac{\partial \varphi}{\partial x}\,\eta_R + \gamma \cdot \frac{\partial \eta}{dx}\,\frac{\varphi}{R} - \frac{\alpha}{R}\,\bar{\eta}_R\,\gamma \cdot \frac{\partial \Phi}{\partial x} \quad \text{on } \Gamma .$$

Since $\gamma \cdot \frac{\partial \Phi}{\partial \gamma} \geqq \delta\beta$ on Γ, it is clear that we can choose a constant α large enough so that $\gamma \cdot \frac{\partial \varphi_R}{\partial x} \leqq 0$. We can choose $\bar{\eta}_R(x) = \bar{\eta}(\frac{x}{R})$, with $\eta(x) = 1$ for $|x| \leqq 2$, 0 for $|x| \geqq 3$, $0 \leqq \bar{\eta} \leqq 1$, $\bar{\eta} \in C_b^2(R^n)$. It is clear that

$X_{\varphi_R}(x) \to X_\varphi(s)$ a.s. uniformly on every bounded interval.

Since $X_{\varphi_R}(s)$ is a P^{xt}-sub-martingale, we obtain the same result for $X_\varphi(s)$. To say that $X_\varphi(s)$ is a local P^{xt}-martingale on $\mathcal{O}$, means that for all stopping times τ, $\tau \geqq t$, then

(3.11) $X_\varphi(s\wedge\tau') - X_\varphi(s\wedge\tau)$ is a P^{xt}-martingale,

with

$$\tau' = \inf\{s \geqq \tau : y(s) \notin \mathcal{O}\}.$$

Let $\mathcal{O}_k$ be a non-decreasing sequence of open sets such that

$$\bar{\mathcal{O}}_k \subset \mathcal{O}_{k+1} \subset \bar{\mathcal{O}}_{k+1} \subset \mathcal{O}$$

($\bar{\mathcal{O}}_{k+1}$ compact). Let $\eta_k(x) \in \mathcal{D}(R^n)$ be such that $0 \leqq \eta_k \leqq 1$, $\eta_k = 1$ on $\mathcal{O}_k$ and $\eta_k = 0$ on $R^n - \mathcal{O}_{k+1}$.

Set $\varphi_k = \varphi\eta_k$. Since $\gamma \cdot \frac{\partial\varphi_k}{\partial x} = 0$ on Γ, it is clear that $X_{\varphi k}(s)$ is a P^{xt}-martingale.

Let

$$\tau'_k = \inf\{ s \geqq \tau : y(s) \notin \mathcal{O}_k\}.$$

Then

$$X_{\varphi k}(s\wedge\tau'_k) - X_{\varphi k}(s\wedge\tau) \text{ is a } P^{xt}\text{-martingale.}$$

Now,

$$X_{\varphi k}(s\wedge\tau'_k) - X_{\varphi k}(s\wedge\tau) = X_\varphi(s\wedge\tau'_k) - X_\varphi(s\wedge\tau).$$

Hence

$$X_\varphi(s\wedge\tau') - X_\varphi(s\wedge\tau) \quad \text{is a } P^{xt}\text{-martingale.}$$

Now

$$s\wedge\tau'_k \uparrow s\wedge\tau'$$

whence (3.11). □

Let $\varphi \in C_b^{2,1}(R^n\times[0,\infty[)$ be such that $\gamma \cdot \frac{\partial\varphi}{\partial x} \leqq 0$ on Γ. Then $X_\varphi(s)$ is a continuous, locally bounded P^{xt}-sub-martingale. By the Doob-Meyer Theorem (cf. P.A. MEYER [1]) there exists an adapted, increasing process $\xi_\varphi(s)$ such that $\xi_\varphi(t) = 0$, which is integrable (i.e., $E\xi_\varphi(s) < \infty$, for all $s \geqq t$) and

(3.12) $X_\varphi(s) - \xi_\varphi(s)$ is a P^{xt}-martingale.

Now let us assume that φ no longer necessarily satisfies the property $\gamma \cdot \frac{\partial\varphi}{\partial x} \leqq 0$ on Γ. Then for α large enough $\bar\varphi = \varphi - \alpha\Phi$ satisfies $\gamma \cdot \frac{\partial\bar\varphi}{\partial x} \leqq 0$ on Γ. Similarly $-\Phi$ satisfies this property. Since

$$X_\varphi(s) = X_{\bar\varphi}(s) - \alpha X_{-\Phi}(s) = \xi_{\bar\varphi}(s) - \alpha\,\xi_{-\Phi}(s) + P^{xt} \text{ martingale}$$

we see that

(3.13) $X_\varphi(s) - \xi_\varphi(s)$ is a P^{xt}-martingale

with

$$\xi_\varphi(s) = \xi_{\bar{\varphi}}(s) - \alpha\, \xi_{-\Phi}(s) \tag{3.14}$$

and therefore $\xi_\varphi(s)$ is a process that is *adapted* and has *locally bounded variation*,

$$E\,|\xi_\varphi(s)| < \infty \ , \ \forall\, s \geq t, \quad \xi_\varphi(t) = 0 \ .$$

Let us note that there exists at most a single process $\xi_\varphi(s)$ that is adapted and has locally bounded variation, and satisfies $\xi_\varphi(t) = 0$, $E|\xi_\varphi(s)| < \infty$, and (3.13). In fact, if $\xi^1(s),\xi^2(s)$ are two such processes, then $\eta(s) = \xi^1(s) - \xi^2(s)$ is an integrable continuous martingale that has locally bounded variation, and vanishes at t. We can write

$$\eta(s) = \eta^1(s) - \eta^2(s)$$

where η^1,η^2 are adapted, continuous, increasing processes, zero at t. We write

$$\eta^i_T(s) = \eta^i(s \wedge T)$$
$$c^i(s) = \inf\,\{\lambda : \eta^i_T(\lambda) > s\} \ , \ i = 1,2 \ ,$$

for the changes in time associated with the increasing process $\eta^i_T(s)$.

Let ζ be a positive and bounded random variable, and write

$$\zeta(s) = E\,[\zeta \mid \mathcal{F}^s_t] .$$

Then

$$E \int_t^\infty \zeta(s)\, d\,\eta^i_T(s) = E \int_t^\infty \zeta(c^i(s))\, \chi_{\{c_i(s) < \infty\}}\, ds$$

and since $c^i(s)$ is an $\mathcal{F}^s_t$-stopping time,

$$= E \int_t^\infty \zeta\, \chi_{\{c_i(s) < \infty\}}\, ds$$
$$= E \int_t^\infty \zeta\, d\,\eta^i_T(s) = E\,\zeta\,\eta^i(T).$$

As $\eta^i_T(s)$ is continuous, we also have

$$\begin{aligned} E\,\zeta\,\eta^i(T) &= E \int_t^\infty \zeta(s-)\, d\,\eta^i_T(s) \\ &= E \int_t^T \zeta(s-)\, d\eta^i(s) \ . \end{aligned} \tag{3.15}$$

But it is easy to see (by discretising the interval (t,T)), and since $\eta^1 - \eta^2$ is a martingale, that

$$E \int_t^T \zeta(s-)\, d\eta^1(s) = E \int_t^T \zeta(s-)\, d\eta^2(s)$$

and therefore it follows that

$$E\, \zeta\, \eta\, (T) = 0$$

and since ζ is ≥ 0 and arbitrary, we have $\eta(T) = 0$, for all $T \geq t$.

LEMMA 3.2: *Let* $\varphi \in C_b^{2,1}(R^n \times [0,\infty[)$ *and* $\xi_\varphi(s)$ *be uniquely defined by* (3.13). *Then*

$$\int_t^T \chi_{\mathcal{O}}(y(s))\, d\, |\xi_\varphi|\, (s) = 0\ . \tag{3.16}$$

Proof: By (3.14) it suffices to consider the case where $\gamma \cdot \frac{\partial\varphi}{\partial x} \leq 0$ on Γ. In this case $\xi_\varphi(s)$ is an increasing process and it suffices to prove that for every compact set $K \subset \mathcal{O}$, we have

$$\int_t^T \chi_K(y(s))\, d\, \xi_\varphi(s) = 0\ . \tag{3.17}$$

So we then write

$$\tau_0 = \inf\{s \geq t : y(s) \in K\} \wedge T$$

$$\tau_{2n+1} = \inf\{ s \geq \tau_{2n} : y(s) \in \Gamma \} \wedge T$$

$$\tau_{2n} = \inf\{ s \geq \tau_{2n-1} : y(s) \in K \} \wedge T\ .$$

Necessarily, $\tau_{2n}, \tau_{2n+1} \uparrow T$ holds. But then by Lemma 3.1,

$$X_\varphi(\tau_{2n+1} \wedge s) - X_\varphi(\tau_{2n} \wedge s) \text{ is a } P^{xt}\text{-martingale.}$$

Therefore

$$\xi_\varphi(\tau_{2n+1} \wedge s) - \xi_\varphi(\tau_{2n} \wedge s) \text{ is a } P^{xt}\text{-martingale}$$

also, that is continuous and has locally bounded variation. By what we saw further above, it follows from this that

$$\xi_\varphi(\tau_{2n+1} \wedge s) = \xi_\varphi(\tau_{2n} \wedge s) \text{ a.s. } P^{xt}.$$

Therefore

$$\int_{\tau_{2n}}^{\tau_{2n+1}} \chi_K(s(s))\, d\, \xi_\varphi(s) = 0$$

and since

$$\int_{\tau_{2n-1}}^{\tau_{2n}} \chi_K(y(s))\, d\, \xi_\varphi(s) = 0$$

by definition, (3.17) certainly follows from this. □

LEMMA 3.3: *Let* $t \leqq a < b$ *and* $\varphi \in C^{2,1}(R^n \times [0,\infty[)$; *let* U *be a neighbourhood of a point* $x \in \partial \mathcal{O}$, *such that* $\varphi = 0$ *on* $]a,b[\times U$. *Then*

$$(3.18) \qquad \int_a^b \chi_U(y(s))\, d\, |\xi_\varphi|(s) = 0 .$$

holds.

Proof: It suffices to prove the result in the case where ξ_φ is increasing. In this case we take $c < d$ and $a < c$, $d < b$, as well as compact $K \subset U$. It suffices to prove that

$$(3.19) \qquad \int_c^d \chi_K\,(y(s))\, d\, \xi_\varphi(s) = 0 .$$

Set

$$\tau_0 = \inf \{ s \geqq c \ ; \ y(s) \in K \} \wedge d$$

$$\tau_{2n+1} = \inf \{ s \geqq \tau_{2n} \ ; \ y(s) \in \partial U \} \wedge d$$

$$\tau_{2n} = \inf \{ s \geqq \tau_{2n-1} ; \ y(s) \in K \} \wedge d .$$

Again we have

$$\tau_{2n}\,,\ \tau_{2n+1} \uparrow d .$$

And then we have

$$\chi_\varphi(\tau_{2n+1} \wedge s) - \chi_\varphi(\tau_{2n} \wedge s) - (\xi_\varphi(\tau_{2n+1} \wedge s) - \xi_\varphi(\tau_{2n} \wedge s)) =$$
$$\int_{\tau_{2n} \wedge s}^{\tau_{2n+1} \wedge s} (- \frac{\partial \varphi}{\partial s} + A(s)\varphi)\ (y(\lambda),\lambda)\, d\lambda -$$
$$- (\xi_\varphi(\tau_{2n+1} \wedge s) - \xi_\varphi(\tau_{2n} \wedge s))$$
$$= - (\xi_\varphi(\tau_{2n+1} \wedge s) - \xi_\varphi(\tau_{2n} \wedge s))$$

because $\varphi = 0$ on U. But then $\xi_\varphi(\tau_{2n+1} \wedge s) - \xi_\varphi(\tau_{2n} \wedge s)$ is a continuous P^{xt}-martingale with bounded variation that vanishes at t, and is therefore zero,

$$\xi_\varphi(\tau_{2n+1} \wedge s) - \xi_\varphi(\tau_{2n} \wedge s) = 0 .$$

Since $s \in [\tau_{2n-1}, \tau_{2n}]$ for $\chi_K(y(s)) = 0$, we deduce (3.19). □

LEMMA 3.4: *Let* $t \leqq a < b$, $\varphi \in C_b^{2,1}(R^n \times [0,\infty[)$ *and* U *be a neighbourhood of a point* x *of* $\partial \mathcal{O}$ *such that* $\gamma . \frac{\partial \varphi}{\partial x} \leq 0$ *on* $]a,b[\times (U \cap \partial \mathcal{O})$. *Then we have*

$$(3.20) \qquad \int_a^b \chi_U\,(y(s))\, d\, \xi_\varphi(s) \geqq 0 .$$

Proof: Let $a < c < d < b$, and let N be an open set such that $\bar{N} \subset U$. Let $\eta \in C_b^\infty(R^n \times [0,\infty[)$ be such that $0 \leq \eta \leq 1$, $\eta = 0$ on $N \times]c,d[$, $\eta = 1$ in the complement of $U \times]a,b[$. We write

$$\bar{\varphi} = \varphi - \alpha\eta$$

where α is a positive constant yet to be chosen. Then

$$\xi_{\bar{\varphi}} = \xi_\varphi - \alpha\xi_{\eta\Phi}$$

and since $\eta\Phi = 0$ on $N \times]c,d[$, it follows from Lemma 3.3 that

$$\text{(3.21)} \qquad \int_c^d \chi_N(y(s))\, d\, \xi_{\eta\Phi} = 0 \ .$$

Now, for α large enough we have $\gamma . \frac{\partial\bar{\varphi}}{\partial x} \leq 0$ on Γ, therefore $d\xi_{\bar{\varphi}} \geq 0$, hence it follows that

$$0 \leq \int_c^d \chi_N(y(s))\, d\, \xi_{\bar{\varphi}}(s) = \int_c^d \chi_N(y(s))\, d\, \xi_\varphi(s) \ .$$

It then suffices to take an increasing sequence of open sets $\uparrow U$, and an increasing sequence of intervals $\uparrow]a,b[$. Applying Lebesgue's Theorem we obtain (3.20). □

We then have:

THEOREM 3.1: *Under the assumptions* (3.1),...,(3.4), *and if there exists a solution* P^{xt} *of the sub-martingale problem (with respect to* σ, g, v*) (cf.* (3.6),(3.7)*), then with it we can uniquely associate a continuous adapted increasing process* $\xi(s)$ *such that* $E\xi(s) < \infty$, *for all* $s \geq t$, *and*

$$\text{(3.22)} \qquad \int_t^T \chi_{\mathcal{O}}(y(s))\, d\, \xi(s) = 0$$

$$\text{(3.23)} \qquad \left| \begin{array}{l} X_\varphi(s) + \int_t^s \chi_\Gamma(y(\lambda))\, \gamma . \frac{\partial\varphi}{\partial x}(y(\lambda), \lambda)\, d\, \xi(\lambda) \\ \textit{is a } P^{xt}\textit{-martingale } \forall\, \varphi \in C_b^{2,1}(R^n \times [o,\infty[). \end{array} \right.$$

Proof: First we show that $d\xi_\varphi(s)$ is absolutely continuous with respect to $d\xi_\phi(s)$. In fact, if α is a large enough positive constant

$$\bar{\varphi} = \pm\varphi - \alpha\phi$$

satisfies $\gamma . \frac{\partial\bar{\varphi}}{\partial x} \leq 0$ on Γ. Therefore $d\xi_{\bar{\varphi}} \geq 0$. But it then follows from this that

$$d\, \xi_\varphi \geq \alpha\, d\, \xi_\Phi$$

$$-d\, \xi_\varphi \geq \alpha\, d\, \xi_\Phi$$

hence the result. We then set

$$\alpha(s) = \frac{d\, \xi_\varphi(s)}{d\, \xi_\Phi(s)}$$

and will then calculate $\alpha(s)$.

We fix $s_0 \in [t,\infty[$ and $x_0 \in \partial\mathcal{O}$. Now set

$$\beta = \frac{\gamma \cdot \frac{\partial\varphi}{\partial x}(x_0, s_0)}{\gamma \cdot \frac{\partial\Phi}{\partial x}(x_0, s_0)} .$$

For fixed $\varepsilon > 0$ there exists an interval $]a,b[$ and an open set $U > x_0$, such that

$$\beta - \varepsilon \leqq \frac{\gamma \cdot \frac{\partial\varphi}{\partial x}(x, s)}{\gamma \cdot \frac{\partial\Phi}{\partial x}(x, s)} \leqq \beta + \varepsilon$$

if $x \in U$, $s \in\,]a,b[$, and the denominator does not vanish. But then

$$\bar{\varphi} = \pm\, \varphi \mp \Phi\, (\beta \pm \varepsilon)$$

satisfies $\gamma \cdot \frac{\partial\bar{\varphi}}{\partial x} \leqq 0$ on $(U \cap \partial\mathcal{O}) \times\,]a, b[$. By Lemma 3.4 we deduce from this that

$$\int_{a'}^{b'} \chi_{U'}(y(s))\, (d\,\xi_\varphi - (\beta + \varepsilon)\, d\,\xi_\Phi) \geqq 0$$

$$\int_{a'}^{b'} \chi_{U'}(y(s))\, (-d\,\xi_\varphi + (\beta - \varepsilon)\, d\,\xi_\Phi) \geqq 0$$

for

$$]a', b'[\, \subset\,]a, b[\quad , \quad \text{and} \quad U' \subset U \quad , \quad U' \supset x_0 \quad , \quad]a', b'[\, \supset s_0 .$$

But then

$$\int_{a'}^{b'} \chi_{U'}(y(s))\, \alpha(s)\, d\,\xi_\Phi(s) \leqq (\beta - \varepsilon) \int_{a'}^{b'} \chi_{U'}(y(s))\, d\,\xi_\Phi(s)$$
$$\geqq (\beta + \varepsilon) \int_{a'}^{b'} \chi_{U'}(y(s))\, d\,\xi_\Phi(s)$$

(notice that $d\xi_\Phi(s) \leqq 0$). Since $]a',b'[$ is arbitrary in $]a,b[$, from this it follows that

$$\chi_{U'}(y(s))\, (\beta - \varepsilon) \leqq \chi_{U'}(y(s))\, \alpha(s) \leq \chi_{U'}(y(s))\, (\beta + \varepsilon)$$

for almost all $s \in\,]a,b[$ (in the sense of $d\xi_\Phi$) and for all $U' \supset x_0$, $U' \subset U$, a.s. P^{xt}. But $y(s) \in U'$ implies

$$\beta - \varepsilon \leqq \frac{\gamma \cdot \frac{\partial\varphi}{\partial x}(y(s), s)}{\gamma \cdot \frac{\partial\Phi}{\partial x}(y(s), s)} \leqq \beta + \varepsilon$$

hence we again have

$$(3.24)\qquad \chi_U(y(s))\left(\frac{\gamma\cdot\frac{\partial\varphi}{\partial x}(y(s),s)}{\gamma\cdot\frac{\partial\Phi}{\partial x}(y(s),s)} - 2\varepsilon\right) \leq \chi_U(y(s))\,\alpha(s) \leq$$

$$\leq \chi_U(y(s))\left(\frac{\gamma\cdot\frac{\partial\varphi}{\partial x}(y(s),s)}{\gamma\cdot\frac{\partial\Phi}{\partial x}(y(s),s)} + 2\varepsilon\right).$$

Since x_0 is arbitrary on $\partial\mathcal{O}$ and s_0 is arbitrary on $[t,\infty[$, and since

$$1 = \chi_\Gamma(y(s)) \qquad d\xi_\Phi \text{ almost everywhere,}$$

it follows from (3.24) that

$$\frac{\gamma\cdot\frac{\partial\varphi}{\partial x}(y(s),s)}{\gamma\cdot\frac{\partial\Phi}{\partial x}(y(s),s)}\,\chi_\Gamma(y(s)) - 2\varepsilon \leq \alpha(s) \leq$$

$$\leq \frac{\gamma\cdot\frac{\partial\varphi}{\partial x}(y(s),s)}{\gamma\cdot\frac{\partial\Phi}{\partial x}(y(s),s)}\,\chi_\Gamma(y(s)) + 2\varepsilon,$$

$d\xi_\Phi$ almost everywhere.

Therefore

$$(3.25)\qquad \alpha(s) = \frac{\gamma\cdot\frac{\partial\varphi}{\partial x}(y(s),s)}{\gamma\cdot\frac{\partial\Phi}{\partial x}(y(s),s)}\,\chi_\Gamma(y(s)), \quad d\xi_\Phi \text{ a.e..}$$

We then set

$$\xi(s) = -\int_t^s \frac{d\xi_\Phi(\lambda)}{\gamma\cdot\frac{\partial\Phi}{\partial x}(y(\lambda),\lambda)}$$

and it is easy to see that $\xi(s)$ answers the question. It is necessarily unique, for by definition one must have

$$d\xi_\Phi = -\gamma\cdot\frac{\partial\Phi}{\partial x}(y(s),s)\,d\xi(s)$$

whence the result. □

We can then prove:

THEOREM 3.2: *Under the assumptions of Theorem 3.1, and for* P^{xt} *a solution of the sub-martingale problem (with respect to* σ, g, γ*), the process*

$$y(s) - \int_t^s g(y(\lambda),\lambda)\,d\lambda + \int_t^s \gamma(y(\lambda),\lambda)\,d\xi(\lambda)$$

is a continuous P^{xt},μ_t^s*-martingale (with values in* R^n*) that has increasing process*

$$\int_t^s a(y(\lambda),\lambda)\,d\lambda \qquad (a = \frac{1}{2}\sigma\sigma^*)\ .$$

Proof: By Theorem 3.1, for $\varphi \in C_b^{2,1}(R^n \times [0,\infty[\,)$, we have

$$(3.26)\qquad \varphi(y(s),s) + \int_t^s (-\frac{\partial\varphi}{\partial s} + A(\lambda)\varphi)\,(y(\lambda),\lambda)\,d\lambda\ +$$
$$+ \int_t^s \gamma\cdot\frac{\partial\varphi}{\partial x}(y(\lambda),\lambda)\,d\xi(\lambda) \quad \text{is a } P^{xt} \text{ martingale}$$

(Notice, also, that $\zeta_\Gamma = 1\ d\xi$ a.e.).

Let $\theta \in R^n$. Apply (3.26) with

$$\varphi(x,s) = \exp i\theta.x$$

since (3.26) is valid for complex-valued φ (by linearity). From this we deduce that

$$(3.27)\qquad Y(s) = \exp i\,\theta.\,y(s) + \int_t^s \exp i\,\theta.\,y(\lambda)\,[\,(-i\,\theta.\,g + \frac{1}{2}\,a\,\theta.\theta)\,d\lambda$$
$$+\ i\,\theta.\,\gamma\,d\,\xi(\lambda)\,]$$

is a P^{xt}-martingale

Set

$$\tau_k = \inf\,\{s \geqq t \mid \xi(s) \geqq k\,\}$$
$$Y_k(s) = Y(s\wedge\tau_k)\ .$$

Then $Y(s\wedge\tau_k)$ is a uniformly bounded, continuous $P^{xt},\mu_t^{s\wedge\tau_k}$-martingale. Also set

$$(3.28)\qquad Z(s) = \exp\int_t^s [(-i\,\theta.\,g + \frac{1}{2}\,a\,\theta.\theta)\,d\lambda + i\,\theta.\,\gamma\,d\,\xi\,]$$
$$Z_k(s) = Z(s\wedge\tau_k).$$

Then $Z_k(s)$ is continuous, adapted to $\mu_t^{s\wedge\tau_k}$, has bounded variation, and

$$|Z_k|(s) \leqq C_k$$

where $|Z_k|(s)$ represents the total variation of A_k on $[t,s]$. But then, as is easily verified by discretisation,

$$Y_k(s)\,Z_k(s) - \int_t^s Y_k(\lambda)\,d\,Z_k(\lambda) \quad \text{is a } \mu_t^{s\wedge\tau_k}\text{-martingale.}$$

Therefore

$$Y(s \wedge \tau_k)\ Z(s \wedge \tau_k) - \int_t^{s \wedge \tau_k} Y(\lambda)\, dZ(\lambda) \quad \text{is a } \mu_t^{s\wedge\tau_k}\text{-martingale.}$$

But this quantity is equal to

$$\exp i\ \theta .\ y(s \wedge \tau_k)\ \ Z(s \wedge \tau_k).$$

Therefore

$$(3.29) \qquad \exp \left\{ i\ \theta .\ y(s \wedge \tau_k) + \int_t^{s \wedge \tau_k} [(-i\ \theta .\ g + \frac{1}{2}\ a\ \theta . \theta) d\lambda + i \theta . \nu d\xi(\lambda)] \right\}$$

is a $P^{xt}, \mu_t^{s\wedge\tau_k}$-martingale.

We notice that $\tau_k \to \infty$ a.e. as $k \to \infty$. Let us call the quantity defined by (3.29) $X_k(s)$. Then we have

$$(3.30) \qquad \sup_{t \leq s \leq T} |X_k(s)| \leq C_T .$$

We can also write

$$X_k(s) = X(s\wedge\tau_\kappa).$$

For $s_2 \geq s_1$ we have

$$E\left[X(s_2 \wedge \tau_k) - X(s_1 \wedge \tau_k) \mid \mu_t^{s_1 \wedge \tau_k} \right] = 0$$

and thus if $\varphi(x_1,\ldots,x_p)$ is continuous and bounded on $(R^n)^p$, then for

$$t_1 \leq t_2 \ldots \leq t_p \leq s_1$$

we have

$$E\left(X(s_2 \wedge \tau_k) - X(s_1 \wedge \tau_k) \right) \varphi\left(y(t_1 \wedge \tau_k),\ldots,y(t_p \wedge \tau_k) \right) = 0.$$

Passing to the limit and taking (3.30) into account, we obtain

$$E\left(X(s_2) - X(s_1) \right) \varphi\left(y(t_1),\ldots, y(t_p) \right) = 0$$

and therefore

$$E\left[X(s_2) \mid \mu_t^{s_1} \right] = X(s_1) .$$

Thus we have obtained that

(3.31) $$X(s) = \exp\left\{ i\,\theta.y(s) + \int_t^s [-i\,\theta.g(y(\lambda),\lambda) + \frac{1}{2}\,a(y(\lambda),\lambda)\,\theta.\theta\,d\lambda + \right.$$
$$\left. + i\,\theta.\,\gamma(y(\lambda),\lambda)\,d\,\xi\,(\lambda)\,]\right\}$$

is a P^{xt}, μ_t^s-martingale.

Replacing θ by $p\theta$, then differentiating once, then twice with respect to p, and taking p = 0, we obtain that

$$M_\theta(s) = \theta.y(s) - \int_t^s \theta.g(y(\lambda),\lambda)\,d\lambda + \int_t^s \theta.\gamma(y(\lambda),\lambda)\,d\,\xi(\lambda)$$

is a martingale, and

$$M_\theta(s)^2 - \int_t^s a\,\theta.\theta\,d\lambda \quad \text{is a martingale}$$

which completes the proof of the result we wanted. □

COROLLARY 3.1: *There exists a normalised Wiener process which is a* μ_s^t*-martingale such that*

(3.32) $$y(s) = \int_t^s g(y(\lambda),\lambda)d\lambda + \int_t^s \sigma(y(\lambda),\lambda)dw(\lambda) - \int_t^s \chi_\Gamma\,\gamma(y(\lambda),\lambda)d\xi(\lambda) + x$$

where σ *is a symmetric* $n \times n$ *matrix such that* $\frac{\sigma^2}{2} = a$.

Proof: The result is clear from Theorem 3.2 and from the classical result on the representation of continuous martingales whose associated increasing process is absolutely continuous with respect to the Lebesgue measure. This is recalled in Chapter II. □

3.3 *Existence of a Solution*

We begin by treating the case where g = 0. If σ,γ are regular and γ is time-independent, then an obvious solution of the sub-martingale problem is obtained by taking the image measure on $C([o,\infty[;\bar{\mathcal{O}})$ of the reflected diffusion defined in (1.6), ...,(1.10).

LEMMA 3.5: *Assume* (3.1),...,(3.4) *with* g = 0, σ,γ *regular; then there exists a measure* P^{xt} *which is a solution of* (3.6),(3.7).

Proof[(1)]*:* We cannot construct a strong solution, as in Theorem 1.1, since γ varies with time. So we then work by discretisation. Thus, let $k \to 0$, and write

$$\gamma_k^p\,(x) = \gamma(x\,,\,p\,k)\;,\; p \in N.$$

Then construct the sequence of following reflected diffusions

(1) To simplify calculations a little, we shall take 0 as the origin instant.

(3.33)
$$\begin{cases} d y_k^{p+1} = \sigma(y_k^{p+1}, t)\, dw(t) - \chi_\Gamma \gamma_k^{p+1} (y_k^{p+1})\, d\, \xi_k^{p+1} \\ y_k^{p+1}(pk) = y_k^p(pk) \qquad , \quad t \in [pk, (p+1)k] \\ \xi_k^{p+1}(pk) = \xi_k^p(pk) . \end{cases}$$

In (3.33) it is permissible to take w(t) as an a priori given Wiener process, since we are working with strong solutions. Next we write

(3.34)
$$\begin{cases} y_k(t) = y_k^p(t) , \quad t \in [(p-1)k, pk] \\ \xi_k(t) = \xi_k^p(t) \qquad \text{''} \end{cases} .$$

Denote by P^k the measure of $y_k(t)$ on $C^o([o,\infty[;\bar{\mathcal{O}})$.

We begin by verifying that P^k remains in a compact subset of the set of probability measures on $C^o([o,\infty[;\bar{\mathcal{O}})$. We notice that if we set

$$\gamma_k(x,t) = \gamma_k^p(x) , \text{ if } t \in [(p-1)k, pk]$$

then $y_k(t), \xi_k(t)$ satisfy

(3.35)
$$\begin{cases} dy_k(t) = \sigma(y_k, t)\, dw(t) - \chi_\Gamma \gamma_k(y_k, t)\, d\, \xi_k \\ y_k(o) = x \\ \chi_{\mathcal{O}}(y_k(t))\, d\, \xi_k(t) = 0 \\ y_k(t) \in \bar{\mathcal{O}} , \quad y_k(t) \text{ adapted and continuous} \end{cases}$$

$\xi_k(t)$ adapted, continuous, and increasing, $\xi_k(0) = 0$.

We are going to obtain a priori estimates. Consider the function $\Phi(x)$ which defines the open set $\mathcal{O}$. Therefore by Ito's formula

(3.36)
$$\Phi(y_k(t)) - \Phi(y_k(s)) = \int_s^t \frac{\partial \Phi}{\partial x} \cdot \sigma(y_k(\lambda),\lambda)\, dw(\lambda) + $$
$$+ \int_s^t \operatorname{tr} \frac{\partial^2 \Phi}{\partial x^2}\, a\, (y_k(\lambda),\lambda) d\lambda - \int_s^t \gamma_k \cdot \frac{\partial \Phi}{\partial x} (y_k(\lambda),\lambda) d\, \xi_k(\lambda) .$$

and since $\Phi|_\Gamma = 0$, we also have

(3.37)
$$\Phi^2(y_k(t)) - \Phi^2(y_k(s)) = \int_s^t 2\, \Phi \frac{\partial \Phi}{\partial x} \cdot \sigma(y_k,\lambda)\, dw(\lambda) + $$
$$+ \int_s^t \left(a \frac{\partial \Phi}{\partial x} \cdot \frac{\partial \Phi}{\partial x} + \Phi \operatorname{tr} a \frac{\partial^2 \Phi}{\partial x^2} \right) (y_k,\lambda)\, d\lambda .$$

We then use the identity

(3.38)
$$\left(\Phi(y_k(t)) - \Phi(y_k(s))\right)^2 = \Phi^2(y_k(t)) - \Phi^2(y_k(s)) - $$

$$- 2\,\Phi(y_k(s))\left(\Phi(y_k(t)) - \Phi(y_k(s))\right).$$

Since

$$- \Phi(y_k(s)) \geq 0, \quad \gamma_k \cdot \frac{\partial \Phi}{\partial x}(y_k(\lambda),\lambda)\, \chi_\Gamma\,(y_k(\lambda)) \geq 0\,,$$

it follows from (3.36),(3.37),(3.38) that

$$(3.39)\qquad \left(\Phi(y_k(t)) - \Phi(y_k(s)\right)^2 \leq \int_s^t 2\,\Phi\,\frac{\partial \Phi}{\partial x}\cdot \sigma\, dw +$$

$$+ \int_s^t \left(a\,\frac{\partial \Phi}{\partial x}\cdot\frac{\partial \Phi}{\partial x} + \Phi\,\mathrm{tr}\, a\,\frac{\partial^2 \Phi}{\partial x^2}\right) d\lambda\, -$$

$$- 2\,\Phi(y_k(s))\left[\int_s^t \frac{\partial \Phi}{\partial x}(y_k)\cdot\sigma\, dw + \int_s^t \mathrm{tr}\,\frac{\partial^2 \Phi}{\partial x^2}\, a\,(y_k)\, d\lambda\right]$$

From (3.39) we deduce by simple calculations and by using the classical properties of stochastic integrals, that

$$(3.40)\qquad E\left(\Phi(y_k(t)) - \Phi(y_k(s))\right)^6 \leq C\,(t-s)^{3/2}\,.$$

Then returning to (3.36) we deduce

$$E\left(\int_s^t \gamma_k\cdot\frac{\partial \Phi}{\partial x}\,(y_k)\,d\,\xi_k\right)^6 \leq C\,(t-s)^{3/2}\,.$$

Since $\gamma_k\cdot\frac{\partial \Phi}{\partial x} \geq \beta\,\delta$, it also follows from this that

$$(3.41)\qquad E\left(\xi_k(t) - \xi_k(s)\right)^6 \leq C\,(t-s)^{3/2}\,.$$

Then returning to Equation (3.37) we deduce

$$(3.42)\qquad E\,|\,y_k(t) - y_k(s)|^6 \leq C\,(t-s)^{3/2}\,.$$

This situation proves that P^k stays in a compact subset of the set of probability measures on $C^o([o,\infty[;\bar{\mathcal{O}})$.

Now let $\varphi \in C_b^{2,1}(R^n \times [0,\infty[\,)$ be such that

$$\gamma\cdot\frac{\partial \varphi}{\partial x} \leq 0\,.$$

Ito's formula applied to (3.35) gives

$$(3.43)\qquad \varphi(y_k(t)) = \varphi(x) + \int_0^t \left(-\frac{\partial \varphi}{\partial s} + A(s)\,\varphi\right)(y_k(s),\, s)\, ds \qquad \text{(Contd)}$$

(Contd) $$-\int_0^t \gamma_k(y_k(s),s)\,.\,\frac{\partial\varphi}{\partial x}(y_k(s),s)\,d\,\xi_k(s)+\int_0^t \frac{\partial\varphi}{\partial x}(y_k).\sigma(y_k)\,dw.$$

But

$$\int_s^t \gamma_k(y_k,s).\,\frac{\partial\varphi}{\partial x}(y_k,s)\,d\,\xi_k(s)=\int_s^t \gamma(y_k,s).\,\frac{\partial\varphi}{\partial x}(y_k,s)\,d\xi_k\,+$$

$$+\int_s^t (\gamma_k-\gamma)\,(y_k,s).\,\frac{\partial\varphi}{\partial x}(y_k,s)\,d\,\xi_k$$

$$\leq\ \rho_T(k)\,(\xi_k(t)-\xi_k(s))\ ,\quad s\leq t\leq T$$

where $\rho_T(k)\to 0$ as $k\to 0$, and $\rho_T(k)$ is deterministic. On passing to image measures, it follows from (3.43) that

(3.44) $$E^k\left[(\varphi(y(t))-\varphi(y(s)))+\int_s^t\left(-\frac{\partial\varphi}{\partial s}+A(s)\,\varphi\right)(y(\lambda),\lambda)\,d\lambda)\ \psi\right]$$

$$\geq -\,\rho_T(k)\ E(\xi_k(t)-\xi_k(s))\ \|\psi\|,$$

for ψ bounded, μ_0^s-measurable and ≥ 0.

From P^k we can extract a sequence weakly converging to P. By passing to the limit in (3.44), we see that P is a solution of the sub-martingale problem. □

We can then prove:

THEOREM 3.3: *Under the assumptions* (3.1),...,(3.4) *and* $a\geq\alpha I$, $\alpha>0$, *there exists a solution of the sub-martingale problem* (3.6),(3.7).

Proof: We always take $g=0$ and σ,γ satisfying (3.1),(3.3). We take a sequence σ_k,γ_k of regular functions such that

$$\left.\begin{array}{l}\sigma_k\to\sigma\\ \gamma_k\to\gamma\end{array}\right|\quad \text{in}\ \ C^{\circ}\,(R^n\times[\,0,\infty[\,)$$

and

$$\gamma_k\,.\,n\geq\delta'>0\ ,\ \forall\ x\in\Gamma=\partial\mathcal{O}\ ,\ \forall t$$

which is permissible on taking $\delta'<\delta$ and for k large enough.

By Lemma 3.5 there exists a solution P^k of the sub-martingale problem with respect to $(0,\sigma_k,\gamma_k)$, and therefore by Corollary 2.1 there exist (Wiener) processes w_k, and increasing ξ_k such that

$$y(s)=x+\int_t^s \sigma_k(y(\lambda),\lambda)\ dw_k(\lambda)-\int_t^s \chi_\Gamma\ \gamma_k\,(y(\lambda),\lambda)\,d\,\xi_k(\lambda)\,.$$

A calculation identical with that carried out in the course of the proof of Lemma 3.5 shows that we again have

$$(3.45)\quad \begin{cases} E^k\,(\xi_k(s_2)-\xi_k(s_1))^6 \leq C\,(s_2-s_1)^{3/2} \\ E^k\,|\,y_k(s_2)-y_k(s_1)\,|^6 \leq C\,(s_2-s_1)^{3/2} \end{cases}$$

The family P^k stays in a relatively compact subset of the set of probability measures on $C^\circ([t,\infty[\bar{\mathcal{O}})$.

Let $\varphi \in C_b^{2,1}(R^n \times [0,\infty[\,)$, be such that $\gamma \cdot \frac{\partial\varphi}{\partial x} \leq 0$. By Ito's formula

$$\varphi(y(s)) = \varphi(x) + \int_t^s \left(-\frac{\partial\varphi}{\partial s} + A(\lambda)\,\varphi\right)(y(\lambda),\lambda)\,d\lambda - \int_t^s \gamma \cdot \frac{\partial\varphi}{\partial x}\,d\xi_k(\lambda)$$
$$+ \int_t^s \frac{\partial\varphi}{\partial x} \cdot \sigma_k\,dw_k + \int_t^s \operatorname{tr}\,(a-a_k)\,\frac{\partial^2\varphi}{\partial x^2}\,(y(\lambda),\lambda)\,d\lambda$$
$$+ \int_t^s (\gamma-\gamma_k) \cdot \frac{\partial\varphi}{\partial x}\,d\xi_k$$

whence

$$\varphi(y(s_2)) - \varphi(y(s_1)) \geq \int_{s_1}^{s_2} \left(-\frac{\partial\varphi}{\partial s} + A(\lambda)\varphi\right)(y(\lambda),\lambda)\,d\lambda + \int_{s_1}^{s_2} \frac{\partial\varphi}{\partial x} \cdot \sigma_k\,dw_k$$
$$- C_1\,\|a-a_k\| - C_2\,\|\gamma-\gamma_k\|\,(\xi_k(s_2) - \xi_k(s_1)) .$$

Taking the conditional space with respect to $\mu_t^{s_1}$, and taking account of the first estimate (3.45), we easily deduce that every limit point of P^k is a solution.

It now remains to treat the case where $g \neq 0$. To do this we use the Girsanov transformation. In fact let P be a solution of the sub-martingale problem with respect to $(1,\sigma,\gamma)$, then there exists a Wiener process $\tilde{w}(t)$, and an increasing process $\xi(t)$, such that for $\tilde{P}$,

$$dy = \sigma(y(s),s)\,d\tilde{w}(s) - \chi_\Gamma(y(s))\,\gamma\,(y(s),s)\,d\,\xi(s) .$$

We then define on Ω_t a measure $P^{xt} \equiv P$, absolutely continuous with respect to $\tilde{P}$, by the formula

$$(3.46)\quad \left.\frac{dP}{d\tilde{P}}\right|_{\mu_t^s} = \exp\left\{ \int_t^s \sigma^{-1}\,g(y(\lambda),\lambda)\,d\tilde{w}(\lambda) - \int_t^s a^{-1} g.g\,d\lambda \right\} .$$

Next we define the process

$$(3.47)\quad w(s) = \tilde{w}(s) - \int_t^s \sigma^{-1}\,g(y(\lambda),\lambda)\,d\lambda$$

which, by Girsanov's Theorem, is a normalised Wiener process and a μ_t^s-martingale. Hence we see that in the P norm we have

$$dy = g\,(y(s),s)ds + \sigma(y(s),s)\,dw(s) - \chi_\Gamma\,(y(s))\,\gamma\,(y(s),s)\,d\,\xi(s)$$

and since

$$y(s) \in \bar{\mathcal{O}} \quad , \quad \chi_{\mathcal{O}}\,(y(s))\,d\,\xi(s) = 0\,,$$

it is clear that P is a solution of the sub-martingale problem with respect to (g,σ,γ).

4. THE MARKOV PROPERTY AND THE INTERPRETATION OF PROBLEMS WITH LIMITS

4.1 *Problems with Limits*

4.1.1 *Variational formulation*

We are given

(4.1) $\mathcal{O}$ a bounded open set of R^n, $Q = \mathcal{O} \times]0,T[$

(4.2)
$a_{ij}\,(x,t) : R^n \times]0,\infty[\to R$, bounded and measurable, i.j. = 1,...,n

$a_i\,(x,t) : R^n \times]0,\infty[\to R$, bounded and measurable,

$a_o\,(x,t) : R^n \times]0,\infty[\to R$, bounded and measurable,

(4.3) $\sum a_{ij}\,\xi_i\,\xi_j \geq \alpha\,|\xi|^2 \quad \forall\ \xi \in R^n$

We write

$$A\,(t) = -\sum_{i,j} \frac{\partial}{\partial x_i}\,a_{ij}\,\frac{\partial}{\partial x_j} + \sum_i a_i\,\frac{\partial}{\partial x_i}\;. \tag{4.4}$$

On $H^1(\mathcal{O})$ we define the family (indexed by t) of bilinear forms)

$$a(t;u,v) = \sum_{ij} \int_{\mathcal{O}} a_{ij}\,\frac{\partial u}{\partial x_j}\,\frac{\partial v}{\partial x_i}\,dx + \sum_i \int_{\mathcal{O}} a_i\,\frac{\partial u}{\partial x_i}\,v\,dx + \int_{\mathcal{O}} a_o\,u\,v\,dx. \tag{4.5}$$

By (4.3) it is clear that there exists $\lambda \geqq 0$ such that

$$a(t;u,u) + \lambda\,|u|^2 \geqq \beta\,\|u\|^2 \quad \forall\ u \in H^1(\mathcal{O})\ ^{(1)}.$$

Let $f \in L^2(Q)$, $u \in L^2(\mathcal{O})$; it is well known that there exists one and only one solution of the problem

$$-\frac{d}{dt}\,(u(t),v) + a(t;u(t),v) = (f(t),v) \quad \forall\ v \in H^1(\mathcal{O}), \tag{4.6}$$

$$u(T) = \bar{u} \quad \text{a.e.}$$

$$u \in L^2(0,T\,;H^1(\mathcal{O})), \quad \frac{\partial u}{\partial t} \in L^2(0,T;(H^1(\mathcal{O}))')\,.$$

(1) Recall that $|\ |$ denotes the norm in $L^2(\mathcal{O})$, and $\|\ \|$ the norm in $H^1(\mathcal{O})$.

4.1.2 *Regularity*

Numerous regularity results exist under the assumptions made about the coefficients and the open set (cf. for certain details, for example, BENSOUSSAN - LIONS [1]). Here we are content to give the essential results which we shall use constantly in the remainder.

In addition to (4.1),(4.2),(4.3) we assume that

(4.7) $$\frac{\partial a_{ij}}{\partial x_k} \text{ is bounded, } a_{ij} \in C^\circ(R^n \times [0,\infty[\,) ,$$

(4.8) $\mathcal{O}$ is a bounded open set whose boundary Γ is of class C^2 and is unbounded.

Let $f,\bar{u}$

(4.9) $$\left| \begin{array}{ll} f \in L^p(Q) & , \; p \geqq 2 \\ \bar{u} \in W^{2,p}(\mathcal{O}) & , \; \frac{\partial \bar{u}}{\partial x_i}\Big|_\Gamma = 0 \end{array} \right. \quad (1).$$

Then the solution u of (4.6) satisfies

(4.10) $$u \in W^{2,1,p}(Q) \quad (2)$$

(4.11) $$\left| \begin{array}{ll} -\frac{\partial u}{\partial t} + A(t)u + a_0 u = f & \text{a.e. in } Q \\ \frac{\partial u}{\partial \nu_A} = 0 \text{ on } \Gamma, \quad u(x,T) = \bar{u} & \end{array} \right.$$

where

(4.12) $$\frac{\partial}{\partial \nu_A} = \sum_{ij} a_{ij} \frac{\partial u}{\partial x_j} n_i \;.$$

4.1.3 *Oblique derivatives*

The expression (4.12) is called the *conormal* derivative. Let us notice that

$$\frac{\partial}{\partial \nu_A} = \gamma \cdot \frac{\partial u}{\partial x}$$

with

(4.13) $$\gamma_i(x,t) = \sum_j a_{ji} n_j$$

and

$$\gamma \cdot n = \sum_{ij} a_{ji} n_j n_i \geqq \alpha > 0 \;.$$

For the remainder it will be useful to treat the case where (4.13) is not satisfied and where only γ is given in such a way that

---

(1) This last assumption can be weakened.

(2) Recall that $\mathcal{W}^{2,1,p}(Q) = \{ z \in L^p(Q) \mid \frac{\partial z}{\partial t}, \frac{\partial z}{\partial x_i}, \frac{\partial^2 z}{\partial x_i \partial x_j} \in L^p(Q) \}$

(4.14) γ bounded and continuous, $\frac{\partial\gamma}{\partial x}$ bounded, $\gamma . n \geqq \delta > 0$.

We are interested in the problem

(4.15) $$\left| \begin{array}{l} -\frac{\partial u}{\partial t} + A(t)u + a_0 u = f. \\ \gamma . \frac{\partial u}{\partial x} = 0 \quad \text{on} \quad \Gamma \\ u(x,T) = \bar{u} \end{array} \right.$$

Under the assumptions (4.1),(4.2),(4.3),(4.7),(4.8),(4.9),(4.14) there exists one and only one solution of (4.15),(4.10). We can reduce the study of (4.15),(4.10) to the case (4.10),(4.11). Let us now show this. To simplify matters a bit we take a_{ij} to be time-independent.

We begin by considering a vector $b \in C^1(\Gamma)$, with values in R^n and is such that $b . \frac{\partial}{\partial x}$ is *tangent* to Γ. To express this analytically we use a local coordinate system $\sigma_1,\ldots,\sigma_{n-1}$ on Γ and ν normal to Γ, then

(4.16) $$\frac{\partial}{\partial x_j} = \sum_i \alpha_{ji} \frac{\partial}{\partial \sigma_i} + \beta_j \frac{\partial}{\partial \nu} ,$$

and assume that

(4.17) $$\sum b_j \beta_j = 0 .$$

For $u,v \in H^1(\mathcal{O})$ we then set

(4.18) $$b(u,v) = \int_\Gamma \sum_j b_j \frac{\partial u}{\partial x_j} v \, d\Gamma$$

and by (4.17) and (4.16)

$$= \sum_{ij} \int_\Gamma \alpha_{ji} b_j \frac{\partial u}{\partial \sigma_i} v \, d\Gamma.$$

Now

$$\frac{\partial u}{\partial \sigma_i} \in H^{-1/2}(\Gamma) \quad , \quad \left|\frac{\partial u}{\partial \sigma_i}\right|_{H^{-1/2}(\Gamma)} \leqq C \|u\|_{H^1(\mathcal{O})}$$

and thus $b(u,v)$ is continuous on $H^1(\mathcal{O}) \times H^1(\mathcal{O})$. Furthermore,

$$b(u,u) = \frac{1}{2} \sum_{ij} \int_\Gamma \alpha_{ji} b_j \frac{\partial u^2}{\partial \sigma_i} d\Gamma$$
$$= - \frac{1}{2} \sum_{ij} \int_\Gamma \frac{\partial}{\partial \sigma_i} (\alpha_{ji} b_j) u^2 d\Gamma$$

therefore

(4.19) $$|b(u,u)| \leqq C |u|^2_{L^2(\Gamma)} \leqq C |u| \|u\| .$$

Consequently there exists λ such that

$$a(u,u)+b(u,u)+\lambda|u|^2 \geq \alpha_0 \|u\|^2, \quad \forall u \in H^1(\mathcal{O}).$$

We can therefore solve the problem

(4.20)
$$\begin{cases} -\frac{d}{dt}(u(t),v)+a(u,v)+b(u,v)=(f,v), \ \forall v \in H^1(\mathcal{O}), \ \text{a.e.} \\ u(T)=\bar{u} \end{cases}$$

for given f in $L^2(Q)$, $u \in L^2(\mathcal{O})$.

Since, by Green's Formula (at least if $u \in H^2(\mathcal{O})$), we have

$$a(u,v)=b(u,v)=(Au,v)+\int_\Gamma \left(\frac{\partial u}{\partial \nu_A}+\sum_j b_j \frac{\partial u}{\partial x_j}\right) v \, d\Gamma,$$

we have solved (at least formally) with (4.20) the problem

(4.21);
$$\begin{cases} -\frac{\partial u}{\partial t}+Au+a_0 u=f & \text{in } Q \\ \frac{\partial u}{\partial \nu_A}+\sum_j b_j \frac{\partial u}{\partial x_j}=0 & \text{on } \Gamma \\ u(x,T)=\bar{u}. \end{cases}$$

It is now therefore a question of seeing that (4.21) is identical to (4.15), on condition that the b_j are well chosen. We have to identify the b_j in such a way that

$$\sum \gamma_j \frac{\partial u}{\partial x_j}=\left(\frac{\partial u}{\partial \nu_A}+\sum_j b_j \frac{\partial u}{\partial x_j}\right)k$$

$$=\left(\sum_{ij} a_{ij}\frac{\partial u}{\partial x_j} n_i+\sum_j b_j\frac{\partial u}{\partial x_j}\right)k.$$

We then replace $\frac{\partial}{\partial x_j}$ by its expression in (4.16). This yields

$$\sum_j \gamma_j \left(\sum_i \alpha_{ji}\frac{\partial}{\partial \sigma_i}+\beta_j\frac{\partial}{\partial \nu}\right)=k\sum_{hj}(a_{hj}n_h+b_j)\left(\sum_i \alpha_{ji}\frac{\partial}{\partial \sigma_i}+\beta_j\frac{\partial}{\partial \nu}\right).$$

Identifying each side term by term, we obtain the set of relations (also by taking (4.17) into account)

(4.22)
$$\begin{cases} \sum_j \gamma_j \alpha_{ji}=k\sum_j \alpha_{ji}\left(\sum_h a_{h_j} n_h+b_j\right) \\ \sum_j \gamma_j \beta_j = k\sum_j\sum_h a_{hj} n_h \beta_j \\ \sum_j \beta_j b_j = 0. \end{cases}$$

But we notice that $\beta_i = n_i$ and therefore

$$k = \frac{\gamma.n}{an.n} > 0$$

by the assumptions. We can then consider the system at $b_1,\ldots,b_n$ formed by the (n - 1) first relations (4.22) as well as the last. This system has one and only one solution, because

$$\sum \alpha_{ij} n_i = 0 \quad , \quad j = 1 \ldots n - 1.$$

The preceding identification therefore has the essential assumption that $\gamma.n \neq 0$ (and we can always assume $\gamma.n > 0$, by possibly changing γ into $-\gamma$, which does not change (4.15)). Thus we have reduced to the variational formulation of (4.15) to the case (4.7). We can also establish the regularity results identical to those in the case of the conormal derivative.

We now assume (4.3),(4.8) hold, and

$$(4.23) \qquad \left| \begin{array}{l} a_{ij}, a_i \text{ regular} \\ \gamma \text{ regular and } \gamma.n \geqq \delta > 0 \end{array} \right.$$

then if $f \in C^\circ(\bar{Q})$ there exists one and only one solution of

$$(4.24) \qquad \left| \begin{array}{l} -\dfrac{\partial u}{\partial t} + A(t)u = f \\ \gamma.\dfrac{\partial u}{\partial x} = 0 \ , \quad u(x,T) = 0 \\ u \in C^{2,1}(\bar{Q}) \ . \end{array} \right.$$

4.2 *Application to the Sub-Martingale Problem*

We shall now systematically use the assumptions (4.1),(4.2),(4.3),(4.7),(4.8), (4.14),(3.4). We define A(t) by (4.4) and notice that we can always assume

$$(4.25) \qquad a_{ij} = a_{ji}$$

even though we change the definition of a_i. Next we write

$$(4.26) \qquad g_i(n,t) = -a_i(x,t) + \sum_j \frac{\partial a_{ij}}{\partial x_j}$$

so that

$$A(t) = -\sum a_{ij} \frac{\partial^2}{\partial x_i \partial x_j} - \sum g_i \frac{\partial}{\partial x_i} \ .$$

We define σ so that

$$a \equiv a_{ij} = \frac{\sigma^2}{2} \ .$$

By Theorem 3.3 there exists a solution P^{xt} of the sub-martingale problem (3.6), (3.7) with respect to (g,σ,γ).

Let us now further assume that the coefficients satisfy (4.23), i.e., they are regular. For $f \in C^{\circ}(\bar{Q})$ we have existence and uniqueness of the solution of (4.24). But there then exists a normalised Wiener process $w(t)$, and an increasing process $\xi(t)$, such that

$$y(s) = x + \int_t^s g(y(\lambda),\lambda)\, d\lambda + \int_t^s \sigma(y(\lambda),\lambda)\, dw(\lambda) - \int_t^s \chi_\Gamma\, \gamma\, (y(\lambda),\lambda)\, d\xi(\lambda)$$

P^{xt} a.s.. It follows from Ito's formula that

$$(4.27) \qquad u(x,t) = E^{xt} \int_t^T f(y(s),s)\, ds.$$

In particular, let $\varphi \in \mathcal{D}(\mathcal{O})$, and consider the problem

$$(4.28) \qquad \left| \begin{array}{l} -\dfrac{\partial u}{\partial t} + A(t)u = 0 \\ \gamma.\dfrac{\partial u}{\partial x} = 0, \quad u(x,T) = \varphi(x) \\ u \in C^{2,1}(\bar{Q}) \end{array} \right.$$

which has one and only one solution (in fact $u - \varphi$ satisfies a problem of type (4.24)). Under these conditions, we verify that

$$(4.29) \qquad u(x,t) = E^{xt} \varphi(y(T)).$$

Let us now be given $\varphi \in L^2(\mathcal{O})$, and consider the problem

$$(4.30) \qquad \left| \begin{array}{l} -\dfrac{\partial u}{\partial t} + A(t)\, u = 0 \\ \gamma\dfrac{\partial u}{\partial x} = 0 \;, \quad u(x,T) = \varphi(x) \\ \qquad u \in L^2(0,T\,;H^1(\mathcal{O}))\;,\; u \in \mathcal{W}^{2,1,p}(\mathcal{O} \times]0,T-\delta[\,), \forall p \end{array} \right.$$

where $\delta > 0$ is arbitrarily small. We are going to show that (4.30) has an unique solution (its uniqueness is clear). First of all, by the variational formulation there exists a solution $u \in L^2(0,T,H^1(\mathcal{O}))$. We set

$$v_o = (T - t)u$$

and v_o is a solution of

$$\begin{array}{l} -\dfrac{\partial v_o}{\partial t} + A(t)\, v_o = u \\ \gamma.\dfrac{\partial v_o}{\partial x} = 0 \qquad v_o(x\,,\,T) = 0 \end{array}$$

and thus, by what we saw for (4.15), $v \in \mathcal{W}^{2,1,2}(Q)$. From this it follows that

$$v_o \in C(0,T\,;W^{1,2}(\mathcal{O})) \subset C(0,T\,;L^{r_o}(\mathcal{O})) \quad \text{with } \frac{1}{r_o} = \frac{1}{2} - \frac{1}{n}\,,$$

if $n > 2$.

If

$$n \leqq 2, \quad v_o \in C(0,T;L^p(\mathcal{O})) , \ \forall p.$$

We next set

$$v_1 = (T - t)v_o$$

which satisfies

$$- \frac{\partial v_1}{\partial t} + A(t)v_1 = v_o$$
$$\gamma \cdot \frac{\partial v_1}{\partial x} = 0 , \ v_1(x,T) = 0 .$$

Therefore if $n \leqq 2$

$$v_1 \in \mathcal{W}^{2,1,p}(Q) , \ \forall p .$$

If $n > 2$,

$$v_1 \in \mathcal{W}^{2,1,r_o}(Q) .$$

Consequently,

$$v_1 \in C(0,T;W^{2(1-\frac{1}{r_o}),r_o}(\mathcal{O})).$$

If $n \leqq 4$,

$$2(1-\frac{1}{r_o}) \geqq \frac{n}{r_o} \quad \text{and} \quad v_1 \in C(0,T;L^p(\mathcal{O})) , \ \forall p ,$$

whence v_2 defined by

$$- \frac{\partial v_2}{\partial t} + A(t)v_2 = v_1$$
$$\gamma \cdot \frac{\partial v_2}{\partial x} = 0 , \ v_2(x,T) = 0$$

belongs to

$$\mathcal{W}^{2,1,p}(Q) , \ \forall p , \quad (v_2 = (T-t)v_1) .$$

If $n \geqq 5$, then

$$2(1-\frac{1}{r_o}) < \frac{n}{r_o} ,$$

therefore

$$W^{2(1-\frac{1}{r_0}), r_0} \subset L^{r_1} ,$$

with $\frac{1}{r_1} = \frac{1}{2} - \frac{2}{n}$. Therefore

$$v_2 \in \mathcal{W}^{2,1,r_1}(Q) \subset C(0,T ; W^{2(1-\frac{1}{r_1}), r_1}(\mathcal{O})) .$$

if $n \leq 7$,

$$v_2 \in L^p(Q) , \forall p \text{ et } v_3 = (T-t)^4 u \in \mathcal{W}^{2,1,p}(Q).$$

If $n \geq 8$, the procedure is begun again. The number k of iterations necessary is such that

$$k \geq \frac{n^2 - 2n}{4+2n} , \text{ and } (T-t)^{k+2} u \in \mathcal{W}^{2,1,p}(Q) , \forall p .$$

This certainly implies the regularity stated in (4.30). We also have the estimate

(4.31) $$|u(x,t)| \leq C_\delta |\varphi|_{L^2} , \text{ if } t \leq T-\delta , \delta > 0 .$$

Another Proof of (4.30): Another way of proving (4.30) is the following(1): multiply (4.30) by $(T - t)u'$; notice that

$$a(t ; u,v) = a_o(t ; u,v) + a_1(t ; u,v)$$

where a_o is *symmetric* and where

$$|a_1(t ; u,v)| \leq C \|u\| |v| ;$$

this yields

$$- (T-t) |u'(t)|^2 + \frac{(T-t)}{2} \frac{d}{dt} a_o (t ; u,u) - \frac{(T-t)}{2} \dot{a}_o (t ; u,u,) +$$

$$+ (T-t) a_1(t ; u,u') = 0$$

whence

$$- \int_0^T (T-t) |u'|^2 dt + \frac{1}{2} \int_0^T a_o(t ; u,u') dt - \int_0^T (\frac{T-t}{2}) a_o(t ; u,u) dt$$
$$+ \int_0^T (T-t) a_1(t ; u,u') dt = 0$$

from whence it follows that

---

(1) This technique is applied in all the evolutionary cases where there is a regularising effect in the course of time; it was introduced by J.L. LIONS and B. MALGRANGE [1].

$$(T-t)^{1/2}\, u' \in L^2(0,T\,;H)\ ,\ H = L^2(\mathcal{O}).$$

Therefore if $S = T - \delta$, $\delta > 0$, is arbitrary, we have:

$$u' \quad L^2(0,S;H).$$

Differentiate (4.30) with respect to t:

$$- u'' + A(t)u' + A(t)u = 0.$$

Multiply this equation by $(S - t)u'$, and so

$$-\int_0^S \left(\frac{S-t}{2}\right) \frac{d}{dt} |u'|^2 + \int_0^S (S-t)\ a(t,u',u')dt + \int_0^S (S-t)\ \dot{a}(t\,;u,u')\ dt = 0$$

whence

$$-\int_0^S \frac{S}{2}\ |u'|^2\ dt + \int_0^S (S-t)\, a\,(t\,;u',u')\,dt + \int_0^S (S-t)\,\dot{a}\,(t\,;u,u')\ dt = 0$$

whence it follows that(1)

$$(S - t)u' \in L^2(0,T;V).$$

Therefore if $R = T - 2\delta$, we have:

$$u' \in L^2(0,R,V) \quad \text{and} \quad u'' \in L^2(0,R;V')$$

(by the equation) so that u' is continuous and H-valued, and thus A(t)u is continuous and H-valued. Therefore u is continuous and D(A(t))-valued, and so $u(R) \in D(A(R))$. We can then, for example, apply (4.15). □

Thanks to (4.30) and (4.31) we can extend Formula (4.29) to cases where φ does not belong to $\mathcal{D}(\mathcal{O})$. In fact, we can find $\varphi_n \in \mathcal{D}(\mathcal{O})$, such that $\varphi_n \to \varphi$ in $L^2(\mathcal{O})$. If we consider the solution of

$$- \frac{\partial u_n}{\partial t} + A\ (t)\, u_n = 0$$

$$\gamma.\ \frac{\partial u_n}{\partial x} = 0 \quad \text{on } \Gamma,\quad u_n(x,T) = \varphi_n$$

$$u_n \in C^{2,1}\ (\bar{Q})$$

which is permissible by (4.28), and then, by (4.31) $u_n(x,t)$ is a Cauchy sequence for x,t fixed, $t < T$, and thus converges. Since $u_n \to u$ in $L^2(Q)$,

$$u_n(x,t) \to u(x,t) \quad \forall\ x,t < T.$$

(1) We justify the preceding calculation, for example, by an approximation by finite differences, as in J.-L. LIONS, B. MALGRANGE, loc. cit..

Now, by (4.29)

$$u_n(x,t) = E^{xt}\varphi_n(y(T))$$

therefore

(4.32) $$E^{xt}\varphi_n(y(T) \to u(x,t) \qquad \forall\ x,t < T$$

If $\varphi \in C^o(\bar{\mathcal{O}})$, $\varphi|_\Gamma = 0$, we can find a sequence $\varphi_n \in \mathcal{D}(\mathcal{O})$, such that $\varphi_n \to \varphi$ in $C^o(\bar{\mathcal{O}})$. From (4.32) it follows that

(4.33) $$u(x,t) = E^{xt}\varphi(y(T))$$

at least for $\varphi \in C^o(\bar{\mathcal{O}})$, $\varphi|_\Gamma = 0$. We can also find a sequence of functions

$$\eta_k(x) \in C^o(\bar{\mathcal{O}}), \quad \eta_k|_\Gamma = 0\ , \quad 0 \leqq \eta_k \leqq 1$$

and

$$\eta_k(x) \to \chi_{\mathcal{O}}(x) \quad \forall\ x \in \bar{\mathcal{O}}\ .$$

Therefore if $\varphi \in C^o(\bar{\mathcal{O}})$, we can apply (4.33) with $\varphi\eta_k$, and by passing to the limit we find that (4.33) holds with $\varphi\chi_{\mathcal{O}}$ instead of φ, and where $\varphi \in C^o(\bar{\mathcal{O}})$. Taking $\varphi = 1$ we find that $E^{xt}\chi_{\mathcal{O}}(y(T))$ is a solution of (4.30) with $\varphi = \chi_{\mathcal{O}}(x)$. Since $\chi_{\mathcal{O}} = 1$ a.e., it follows from (4.31) that

$$E^{xt}\ \chi_{\mathcal{O}}(y(T)) = 1 \qquad \forall x\ ,\ t < T$$

and therefore

$$E^{xt}\ \chi_\Gamma(y(T)) = 0\ .$$

But then for $\varphi \in C^o(\bar{\mathcal{O}})$,

$$E^{xt}\ \varphi(y(T)) = E^{xt}\ \varphi\,\chi_{\mathcal{O}}\,(y(T))$$

and therefore (4.33) holds with $\varphi \in C^o(\bar{\mathcal{O}})$. But then (4.33) defines a probability measure on $\bar{\mathcal{O}}$,

$$u(x,t) = E^{xt}\varphi(y(T)) = \int_{\bar{\mathcal{O}}} P\ (x,t\ ;\ T,dz)\ \varphi(z)$$

and we easily verify that $P(x,t;T,B)$, $x \in \bar{\mathcal{O}}$, $t \leqq T$, B a Borel of $\bar{\mathcal{O}}$, is a *probability transition function.* We notice that $P(x,t;T,B)$ is the same for every solution of the sub-martingale problem. If we now take fixed t_0, x_0 and P^{x_0,t_0} a corresponding solution of the sub-martingale problem, we have

$$y(s) = x_0 + \int_{t_0}^{s} g\,d\lambda + \int_{t_0}^{s} \sigma\,dw - \int_{t_0}^{s} \chi_\Gamma\ \gamma\,d\,\xi(\lambda)$$

therefore by Ito's formula

$$E^{x_o,t_o}\left[\varphi(y(T)) \mid \mu_{t_o}^{t}\right] = u(y(t),t)$$

at least if $\varphi \in \mathcal{D}(\mathcal{O})$. This formula extends to the case where $\varphi \in C^o(\bar{\mathcal{O}})$, for $t_o \leq t < T$, in the same way as we saw earlier. By the theorem on monotonic classes it is extended to the case where φ is Borel and bounded. We can then apply it with $\varphi = \chi_B$, where B is a Borel of $\bar{\mathcal{O}}$. Therefore

$$P^{x_o t_o}\left[y(T) \in B \mid \mu_{t_o}^{t}\right] = P(y(t),t\,;\,T,B)$$

which proves that $\Omega_{t_o}, \mu_{t_o}^{s}, y(s), P^{x_o,t_o}$ is a Markov process whose transition probability is $P(x,t;T,B)$. This implies, by Dynkin's classical result (DYNKIN [1]) that P^{x_o,t_o} is unique. However, thanks to (4.29) and the regularity of $u(x,t)$, the Markov process is Fellerian. Thus we have proved:

LEMMA 4.1: *Assuming* (4.1),(4.2),(4.3),(4.7),(4.8),(4.14),(3.4) *together with* (4.23), *the solution of the submartingale problem* (3.6),(3.7) *is unique and is a Fellerian Markov process whose transition probability is defined by* (4.29). □

We now consider the general case, i.e., without assuming (4.23). Our goal is to prove that the solution of the sub-martingale problem (3.6),(3.7) is unique. First, we shall need an estimate provided by:

LEMMA 4.2: *Let* P^{xt} *be a solution of the sub-martingale problem* (3.6),(3.7), *then there exists a constant* C *such that*

$$(4.34) \qquad \left| E^{xt}\int_t^T f(y(s),s)ds \right| \leq C\, \|f\|_{L^p(Q)}$$

for $p > 2n + 2$. *The constant* C *depends on* P^{xt} *and* T, *but not on* f.

Proof: Let us first show that this can be reduced to the case where $g = 0$. In fact, with the notations (3.46) we have

$$E^{xt}\int_t^T f(y(s),s)ds = \tilde{E}^{xt}\ Z\int_t^T f(y(s),s)ds$$

where

$$Z = \exp\left\{ \int_t^T \sigma^{-1} g\,(y(\lambda),\lambda)\ d\tilde{w}(\lambda) - \int_t^T a^{-1} g.g\, d\lambda \right\}.$$

Therefore

$$\left|E^{xt}\int_t^T f(y(s),s)ds\right| \leq (\tilde{E}^{xt} Z^2)^{1/2}\left[\tilde{E}^{xt}\left(\int_t^T f(y(s),s)\,ds\right)^2\right]^{1/2}$$

(Contd)

$$\text{(Contd)} \quad \leq C_T \left(\tilde{E}^{xt} \int_t^T f^2(y(s),s)\,ds \right)^{1/2} \leq \left(\| f^2 \|_{L^{P/2}} \right)^{1/2}$$

$$\leq \| f \|_{L^P} \,, \quad \text{for } p > 2n+2$$

provided that we have shown that (4.34) is true in the case $g = 0$ with $p > n + 1$.

Therefore we now assume $g = 0$. We know that for the measure P^{xt} we can construct an increasing process $\xi(t)$ and a Wiener process $w(t)$ such that

$$\text{(4.35)} \quad \left| \begin{array}{l} dy = \sigma\,(y(s),s)\,dw\,(s) - \chi_\Gamma\,(y(s))\,\gamma\,(y(s),s)\,d\xi(s) \\ y(t) = x\,. \end{array} \right.$$

Now consider the boundary problem

$$\text{(4.36)} \quad \left| \begin{array}{l} -\dfrac{\partial u}{\partial t} - \sum\limits_{ij} a_{ij}\,(x,t)\,\dfrac{\partial^2 u}{\partial x_i \partial x_j} = f \\ \gamma \,.\, \dfrac{\partial u}{\partial x} = 0 \quad \text{on} \quad \Gamma \\ u(x,T) = 0 \end{array} \right.$$

which has one and only one solution $u \in \mathcal{W}^{2,1,p}(Q)$. For $p > n + 1$, u, $\frac{\partial u}{\partial x}$ are bounded and continuous functions on $\bar{Q}$. So consider a sequence of $a^k_{ij}; \gamma^k_i$ that are regular and bounded, and such that

$$a^k_{ij} = a^k_{ji} \;, \; \sum a^k_{ij}\,\xi_i\,\xi_j \geq \alpha|\xi|^2 \;, \; \forall\, \xi \in R^n$$

$$\gamma^k .\, n \geq \delta' > 0$$

and

$$a^k \to a \quad \text{in} \quad C^o(R^n \times [0,\infty[)$$

$$\frac{\partial a^k_{ij}}{\partial x_\ell} \to \frac{\partial a_{ij}}{\partial x_\ell} \quad \text{a.e.} \,, \quad \frac{\partial a^k_{ij}}{\partial x_\ell} \quad \text{bounded}$$

$$\gamma^k_i \to \gamma_i \quad \text{in} \quad C^o(R^n \times [0,\infty[\,) \,, \quad \frac{\partial \gamma^k_i}{\partial x_\alpha} \quad \text{bounded.}$$

We next consider the approximate problem of (4.36)

$$\text{(4.37)} \quad \left| \begin{array}{l} -\dfrac{\partial u^k}{\partial t} - \sum\limits_{ij} a^k_{ij}\,\dfrac{\partial^2 u^k}{\partial x_i \partial x_j} = f \\ \gamma^k .\, \dfrac{\partial u^k}{\partial x} = 0 \quad \text{on} \quad \Gamma \\ u^k(x,T) = 0 \end{array} \right.$$

and since we may assume $f \in \mathcal{B}(Q)$, then $u^k \in C^{2,1}(\bar{Q})$ in particular. Furthermore,

(4.38) $\quad u^k \to u$ weakly in $\mathcal{W}^{2,1,p}(Q)$.

Now construct the process $y^k(s)$ which is the solution of

(4.39)
$$\begin{vmatrix} dy^k = \sigma^k(y^k(s),s)\,dw(s) - \chi_\Gamma(y(s))\,\gamma(y(s),s)\,d\xi(s) \\ y^k(t) = x \end{vmatrix}$$

and which is a differential equation in the strong sense, because the second term of the right hand side of (4.39) is given. We easily see from (4.35) and (4.39) that

(4.40)
$$E^{xt} \sup_{t \leq s \leq T} |y^k(s)-y(s)|^2 \leq C \|\sigma^k - \sigma\|_{C^o} .$$

Further, since $u^k \in C^{2,1}(\bar{Q})$, and can be continued outside $\bar{Q}$ in such a way that $u^k \in \mathcal{W}^{2,1,p}(R^n \times [0,T])$ and

$$\|u^k\|_{\mathcal{W}^{2,1,p}(R^n \times]0,T[)} \leq C\|u^k\|_{\mathcal{W}^{2,1,p}(Q)} \leq C\|f\|_{L^p(Q)} .$$

From this we deduce

$$\|u^k\|_{C^o(R^n \times [0,T])} , \|\frac{\partial u^k}{\partial x}\|_{C^o(R^n \times [0,T])} \leq C\|f\|_{L^p(Q)}$$

and by applying Ito's formula

$$- u^k(x,t) = - E^{xt}\int_t^T f(y^k(s),s)ds - E^{xt}\int_t^T \gamma(y(s)).\frac{\partial u^k}{\partial x}(y^k(s))\,\chi_\Gamma(y)\,d\xi(s)$$

and thus we obtain

$$|E^{xt}\int_t^T f(y^k(s),s)\,ds| \leq C\|f\|_{L^p(Q)}$$

and hence the desired result by making k tend to 0. □

Under the assumptions (4.1),(4.2),(4.3),(4.7),(4.8),(4.14),(3.4) let us consider the problem (with a given f in $L^p(Q)$)

(4.41)
$$\begin{vmatrix} -\frac{\partial u}{\partial t} + A(t)u = f \\ \gamma.\frac{\partial u}{\partial x} = 0 \ , \ u(x,T) = 0 \\ u \in \mathcal{W}^{2,1,p}(Q) \end{vmatrix}$$

which has one and only one solution. Naturally, for $p > \frac{n}{2} + 1$, we have $u \in C^o(\bar{Q})$. We then have:

LEMMA 4.3: *Let* P^{xt} *be a solution of the sub-martingale problem* (3.6),(3.7), *then if* $p > \frac{n}{2} + 1$, *and if* u *denotes the solution of* (4.41), *then*

$$(4.42) \qquad u(x,t) = E^{xt} \int_t^T f(y(s),s)\,ds\,.$$

Proof: We can continue u outside $\mathcal{O}$ so that u belong to $\mathcal{W}^{2,1,p}(R^n \times]0,T[)$. Next we can find a sequence $u^k \in C^{2,1}(R^n \times [0,T])$ that is bounded as well as its derivatives, such that

$$u^k \to u \text{ in } \mathcal{W}^{2,1,p}(R^n \times]0,T[).$$

By application of Ito's formula we have

$$(4.43) \qquad -u^k(x,t) = E^{xt} \int_t^T \left(\frac{\partial u^k}{\partial s} - A(s)u^k\right)(y(s),s)ds - E^{xt} \int_t^T \frac{\partial u^k}{\partial x}(y(s)).\ \gamma_\Gamma \chi\, d\xi\,.$$

Let us first assume $f \in \mathcal{D}(Q)$, then

$$\frac{\partial u^k}{\partial s} + A(s)u^k \to f \quad \text{in} \quad L^p(Q^\circ),$$

with p arbitrarily large and

$$u^k,\ \frac{\partial u^k}{\partial x} \to u,\ \frac{\partial u}{\partial x} \quad \text{in} \quad C^o(\bar{Q})\,.$$

Applying Lemma 4.2 we can go to the limit in (4.43), from which we deduce (4.42), at least if $f \in \mathcal{D}(Q)$. The general case is easily deduced by approximation. □

We are then in a situation analogous to the case where the coefficients are regular. A proof analogous to that of Lemma 4.1 allows us to state:

THEOREM 4.1: *Assuming* (4.1),(4.2),(4.3),(4.7),(4.8),(4.14),(3.4), *the solution of the sub-martingale problem* (3.6),(3.7) *is unique and is a Fellerian Markov process.* □

COROLLARY 4.1: *Let ψ be a function defined on $\bar{Q}$, such that*

$$\psi \in C^o(\bar{Q})\,,\ \frac{\partial\psi}{\partial x} \in C^o(\bar{Q})\,,\ \frac{\partial\psi}{\partial t} - A(T)\,\psi \in L^p\,(Q),\ p > \frac{n}{2} + 1\,.$$

Then for every stopping time θ of μ_t^s, we have

$$E^{xt}\psi\,(y(\theta \wedge T),\ \theta \wedge T) = \psi(x,t) + E^{xt} \int_t^{\theta\wedge T} \left(\frac{\partial\psi}{\partial s} - A(s)\psi\right)(y(s),s)ds\ -$$

$$-\ E^{xt} \int_t^{\theta\wedge T} \frac{\partial\psi}{\partial x}\ .\ \gamma\ (y(s),s)\ d\ \xi(s)$$

Proof: This is done by regularisation and use of Lemma 4.3. □

4.3 *A Construction of the Solution of the Sub-Martingale Problem*

In this Section we are going to give an intuitive construction of the sub-martingale problem, which allows us to understand better what a reflected diffusion is. However, we shall need some analytical results.

4.3.1 Analytic results

The assumptions are those of Theorem 4.1. First we recall the following result. Consider the non-homogeneous problem

$$(4.44)\qquad -\frac{\partial u}{\partial t} + A(t)u = f \quad , \quad u\Big|_{\Sigma} = h, \quad u(x,T) = \varphi(x)$$

with the assumptions

$$(4.45)\qquad \left| \begin{array}{l} f \in L^p(Q)\ ,\ p > \frac{n}{2} + 1\ , \quad h \text{ bounded and Borel,} \\ h \in C^0(\bar{\mathcal{O}} \times [0, T-\delta]) \quad \forall \delta > 0, \quad \varphi \text{ bounded and Borel} \end{array} \right.$$

then there exists one and only one solution u satisfying (4.44) such that

$$(4.46)\qquad \left| \begin{array}{l} u \text{ bounded and Borel,} \quad u \in C^0(\bar{\mathcal{O}} \times (0, T-\delta)) \text{ for all } \delta > 0, \\ u \in \mathcal{W}^{2,1,p}_{loc}(Q). \end{array} \right.$$

Furthermore, u is given explicitly by the formula

$$(4.47)\qquad u(x,t) = E\left[\int_t^{T\wedge\tau_{xt}} f(y_{xt}(s),s)ds + h(y_{xt}(\tau_{xt}),\ \tau_{xt})\chi_{\tau_{xt}\leq T} + \right.$$
$$\left. + \varphi(y_{xt}(T))\chi_{T<\tau_{xt}}\right] , \ t<T$$

where $y_{xt}(s)$ denotes the diffusion (g,σ) beginning at the instant t from the point x, constructed on an arbitrary probability space, and τ_{xt} is the first exit instant from $\mathcal{O}$.

We shall note the following useful estimates

$$(4.48)\qquad \|u\|_{L^\infty} \leq C\left(\|f\|_{L^p(Q)} + \|h\|_{L^\infty(Q)} + \|\varphi\|_{L^\infty(\mathcal{O})}\right)$$
$$|u(x,t)| \leq C\left(\|h\|_{L^\infty(Q)} + \|f\|_{L^p(Q)}\right) +$$
$$+ C_\delta\|\varphi\|_{L^p(\mathcal{O})} \quad , \quad 0 \leq t \leq T-\delta\ .$$

Also

$$(4.49)\qquad E\chi_{\tau_{xt}=T} = 0, \quad t < T.$$

Now, $\gamma(x,t)$ is given as in (4.14), and in fact we make a technical assumption of regularity

$$(4.50)\qquad \gamma(x,t) \text{ defined on } R^n \times [0,\infty[\ ,\ \frac{\partial\gamma_i}{\partial x_j}\ ,\ \frac{\partial^2\gamma_i}{\partial x_j\partial x_k} \quad \text{bounded.}$$

Here we are interested in the problem (ε fixed)

$$(4.51)\quad \begin{cases} -\dfrac{\partial u^\varepsilon}{\partial t} + A(t)u^\varepsilon = f \\ u^\varepsilon(x,t) = u^\varepsilon(x-\varepsilon\gamma(x,t),t) + h(x,t) \quad \text{on } \Sigma \\ u^\varepsilon(x,T) = \varphi(x) \end{cases}$$

where f,h,φ satisfy (4.45).

Our aim here is to prove:

THEOREM 4.2: *Taking the assumptions of Theorem 4.1, as well as* (4.50) *and* (4.45), *then there exists one and only one solution of* (4.51) *such that*

$$u^\varepsilon \in L^\infty(Q),\ u^\varepsilon \in C^0(\bar{\mathcal{O}} \times [0,T-\delta])\ \forall \delta > 0,\quad u^\varepsilon \in \mathcal{W}^{2,1,p}_{loc}(Q).$$

If h,φ *are zero, then* $u^\varepsilon \in \mathcal{W}^{2,1,p}(Q)$. *We also have the estimate*

$$(4.52)\quad \|u^\varepsilon\|_{L^\infty(Q)} \leq \left(\|f\|_{L^p(Q)} + \|h\|_{L^\infty(Q)}\right)\frac{C_T}{\varepsilon} + \|\varphi\|_{L^\infty(\mathcal{O})}$$

where C_T *is independent of* ε. □

We begin with:

LEMMA 4.4: *Assuming* (4.50), *then for* ε *small enough, the mapping* $x \to x - \varepsilon\gamma(x,t)$ *is defined and invertible on* R^n, *and maps* $\bar{\mathcal{O}}$ *into a compact subset* $\bar{G}$ *of* $\mathcal{O}$. *If* $\psi \in W^{i,p}_{loc}(\mathcal{O})$, $i = 0,1,2$, *then* $\phi(x - \varepsilon\gamma) \in W^{i,p}(\mathcal{O})$, *the mapping of* $W^{i,p}_{loc}(\mathcal{O})$ *into* $W^{i,p}(\mathcal{O})$ *defined thus is uniformly continuous with respect to* $t \in [0,T]$.

Proof: From the formula

$$\Phi(x-\varepsilon\gamma) = \Phi(x) - \varepsilon\frac{\partial\Phi}{\partial x}\cdot\gamma + \frac{\varepsilon^2}{2}\frac{\partial^2\Phi}{\partial x^2}(\xi)\,\gamma\cdot\gamma$$

and from the fact that $\frac{\partial\Phi}{\partial x}\cdot\gamma \geq \beta\delta$ on Γ, $\|\frac{\partial^2\Phi}{\partial x^2}\|$ bounded, it easily follows that for $\varepsilon \leq \varepsilon_0$, we have $\Phi(x-\varepsilon\gamma) \leq -C\varepsilon$, hence $x \to \Phi(x-\varepsilon\gamma)$: $\bar{\mathcal{O}} \to \bar{G}_\varepsilon$ where $G_\varepsilon = \{z \in \mathcal{O} \mid \Phi(z) < -C\varepsilon\}$. Also, let $z \in \bar{G}_\varepsilon$ be fixed, then the mapping $x \to z + \varepsilon\gamma(x)$ is a contraction if $\varepsilon\|\nabla\gamma\| < 1$, which we can always assume, therefore it has a fixed point which is the solution of

$$x - \varepsilon\gamma = z.$$

Let $\psi \in L^p_{loc}(\mathcal{O})$, then

$$\int_{\mathcal{O}} |\psi(x-\varepsilon\gamma)|^p\,dx \leq \int_{G_2} |\psi(z)|^p \det(I - \varepsilon\frac{\partial\gamma}{\partial x})^{-1}\,dz \leq C\int_{G_\varepsilon} |\psi(z)|^p\,dz < \infty$$

since $\bar{G}_2 \subset \mathcal{O}$. This and the regularity of γ give the result desired. □

Proof of Theorem 4.2: Uniqueness: If $f,h,\varphi = 0$, we first note that $u^\varepsilon \in \mathcal{W}^{2,1,p}(G \times]0,T[)$, for every open set G such that $\bar{G} \subset \mathcal{O}$, and thanks to the boundary conditions $u^\varepsilon \in \mathcal{W}^{2,1,p}(Q)$ for all $p \geq 2$. We then consider the maximum and minimum of u^ε on $\bar{Q}$. By the condition at the boundary they are necessarily reached in $\mathcal{O}$. If the maximum is positive it is reached in Q, which contradicts the maximum principle, therefore the maximum is negative or zero. Similarly the minimum cannot be strictly negative, therefore the function is zero.

Existence: First we consider the case $f \in L^\infty(Q)$, $h,\varphi = 0$. We prove the existence of a solution in $\mathcal{W}^{2,1,p}(Q)$. By linearity we can assume $f \geq 0$. We then consider the following iterative procedure.

$$- \frac{\partial u_o}{\partial t} + A(t)\, u_o = f$$

$$u_o\big|_\Sigma = 0, \quad u_o(x,T) = 0$$

$$- \frac{\partial u_{k+1}}{\partial t} + A(t) u_{k+1} = f$$

$$u_{k+1}\big|_\Sigma = u_k(x - \varepsilon \gamma)\big|_\Sigma, \quad u_{k+1}(x,T) = 0.$$

By induction, and using Lemma 4.4, we show the existence of a sequence $u_k \in \mathcal{W}^{2,1,p}(Q)$. It is easy to verify that

$$0 \leq u_o \leq u_1 \leq \ldots \leq u_k \leq \ldots \leq \|f\|(T - t) < T\|f\| .$$

Since the sequence u_k is bounded in $L^\infty(Q)$, from the equation and $u_k(T) = 0$ we deduce that u_k stays bounded in a bounded set of $\mathcal{W}^{2,1,p}(G \times]0,T[)$ for every open set G such that $\bar{G} \subset \mathcal{O}$. But then $u_k(x - \varepsilon\gamma,t)$ remains in a bounded set of $\mathcal{W}^{2,1,p}(Q)$. By passing to the limit we deduce the result desired. Next we consider the case where $f \in L^\infty(Q)$, $\varphi \in L^\infty(\mathcal{O})$ and $h = 0$, $f,\varphi \geq 0$. Again we define a sequence of the form

$$- \frac{\partial u_o}{\partial t} + A(t)\, u_o = f, \quad u_o\big|_\Sigma = 0, \quad u_0\,(x,T) = \varphi(x)$$

$$- \frac{\partial u_{k+1}}{\partial t} + A(t) u_{k+1} = f, \quad u_{k+1}\big|_\Sigma = u_k(x - \varepsilon\gamma)\big|_\Sigma, \quad u_{k+1}(x,T) = \varphi(x).$$

By recurrence we define a sequence satisfying (4.46) and similarly $u_k \in \mathcal{W}^{2,1,p}(\mathcal{O} \times]0,T - \delta[)$, for all $\delta > 0$. It is not difficult to show by recurrence that

$$0 \leq u_o \leq u_1 \leq \ldots \leq u_k \leq \ldots \leq \|f\|(T-t) + \|\varphi\| .$$

We next verify that u_k remains in a bounded set of $\mathcal{W}^{2,1,p}(\mathcal{O} \times]0,T - \delta[)$ for all $\delta > 0$. By going to the limit we prove the existence of a solution of

(4.53)
$$\begin{cases} -\dfrac{\partial u^\varepsilon}{\partial t} + A(t)u^\varepsilon = f \\ u^\varepsilon\big|_\Sigma = u^\varepsilon(x-\varepsilon\gamma, t)\big|_\Sigma, \quad u(x,T) = \varphi(x) \\ u^\varepsilon \in \mathcal{W}^{2,1,p}(\mathcal{O}\times]0,T-\delta[) \ \forall\, \delta>0 \\ \|u^\varepsilon\|_{L^\infty} \le \|f\|_{L^\infty} T + \|\varphi\|_{L^\infty}. \end{cases}$$

We now move on to the general case. Without loss of generality we can assume $f,\varphi = 0$. In fact, if we introduce the solution of

$$-\frac{\partial w}{\partial t} + A(t)w = f, \quad w\big|_\Sigma = h, \quad w(x,T) = \varphi(x)$$

and if we write $w^\varepsilon = u^\varepsilon - w$, then w^ε is a solution of

$$-\frac{\partial w^\varepsilon}{\partial t} + A(t)w^\varepsilon = 0, \quad w^\varepsilon = w^\varepsilon(x-\varepsilon\gamma) + w(x-\varepsilon\gamma) \quad \text{on } \Sigma, \ w^\varepsilon(x,T) = 0.$$

By (4.46),(4.48), $w(x - \varepsilon\gamma,t)$ has the same properties as h. Therefore we assume $f,\varphi = 0$. Now consider the iterative scheme

(4.54)
$$\begin{cases} -\dfrac{\partial u_0}{\partial t} + A(t)u_0 = 0, \quad u_0\big|_\Sigma = h \quad , u_0(x,T) = 0 \\ -\dfrac{\partial u_{k+1}}{\partial t} + A(t)u_{k+1} = 0, \quad u_{k+1}\big|_\Sigma = u_k(x-\varepsilon\gamma) + h\big|_\Sigma, \ u_{k+1}(x,T) = 0 \end{cases}$$

which defines a sequence satisfying (4.46). By the linearity we can assume $h \geq 0$. We then have no difficulty in showing by induction that

(4.55) $$0 \le u_0 \le u_1 \le \dots \le u_k \le \dots$$

Obtaining an upper bound is more delicate than in the case h = 0. We first consider the case where h = c = constant, and shall use the function $\Phi(x)$ defining the open set in (3.4). Notice that for $x \in \Gamma$

$$\Phi(x) - \Phi(x-\varepsilon\gamma) = \varepsilon\gamma\frac{\partial\Phi}{\partial x} + \frac{\varepsilon^2}{2}\frac{\partial^2\Phi}{\partial x^2}(\xi)\ \gamma.\gamma$$

whence

(4.56) $$\Phi(x) - \Phi(x-\varepsilon\gamma) \geq \varepsilon\delta\beta - \varepsilon^2 M \quad \text{on } \Gamma.$$

We then introduce the solution ζ of

(4.57)
$$\begin{cases} -\dfrac{\partial\zeta}{\partial t} + A(t)\zeta = A(t)\Phi \\ \zeta = \zeta(x-\varepsilon\gamma) \quad \text{on } \Sigma, \ \zeta(x,T) = \Phi(x) \end{cases}$$

and by what we saw for (4.53) it is well defined. Moreover,

(4.58) $$\|\zeta\|_{L^\infty} \leqq T\|A(t)\,\Phi\|_{L^\infty} + \|\Phi\|_{L^\infty}\,.$$

We are going to show by induction that

(4.59) $$\Phi \geqq \zeta + \frac{(\epsilon\delta\beta - \epsilon^2 M)u_k}{C}\,.$$

Let us write $\tilde{\zeta} = \Phi - \zeta$, then $\tilde{\zeta}$ satisfies

(4.60) $$\left|\begin{array}{l} -\dfrac{\partial\tilde{\zeta}}{\partial t} + A(t)\,\tilde{\zeta} = 0 \\ \tilde{\zeta} - \tilde{\zeta}(x-\epsilon\gamma)\Big|_{\Sigma} = \Phi - \Phi(x-\epsilon\gamma)\Big|_{\Sigma} \geqq \epsilon\delta\beta - \epsilon^2 M \\ \tilde{\zeta}(x,T) = 0. \end{array}\right.$$

Starting from the iterative definition of ζ seen above, we easily verify that $\tilde{\zeta} \geqq 0$ (for ϵ small enough $\epsilon\delta\beta - \epsilon^2 M > 0$). But by then starting from the boundary condition (4.60), we see that

$$\tilde{\zeta}\Big|_{\Sigma} \geqq \epsilon\delta\beta - \epsilon^2 M = (\epsilon\delta\beta - \epsilon^2 M)\,\frac{u_0}{C}\,.$$

Since $\tilde{\zeta}$ and u_0 satisfy the same equation and vanish at T, (4.5) follows from it for $k = 0$. Let us assume that (4.59) is proved at step k, and let us show that it is at step $k + 1$. Now, for $x \in \Gamma$ we have

$$\begin{aligned} \tilde{\zeta}\Big|_{\Sigma} &\geqq \tilde{\zeta}\,(x-\epsilon\gamma)\Big|_{\Sigma} + \epsilon\delta\beta - \epsilon^2 M \\ &\geqq \frac{\epsilon\delta\beta - \epsilon^2 M}{C}\,(C + u_k(x-\epsilon\gamma)\Big|_{\Sigma}) \end{aligned}$$

by the induction hypothesis. Hence

$$\tilde{\zeta}\Big|_{\Sigma} \geqq \frac{\epsilon\delta\beta - \epsilon^2 M}{C}\,u_{k+1}\Big|_{\Sigma}\,,$$

and since $\tilde{\zeta}, u_{k+1}$ satisfy the same equation and vanish at T, we deduce that (4.59) holds for $k + 1$. There then follows from (4.59) and (4.58) the estimate (if we recall that $u_k \geqq 0$ and $\Phi \leqq 0$)

$$0 \leqq u_k \leqq \frac{C}{\epsilon\delta\beta - \epsilon^2 M}\,(T\|A(.)\,\Phi\|_{L^\infty} + \|\Phi\|_{L^\infty})\,.$$

h is now no longer assumed to be constant. Since

$$-\|h\|_{L^\infty} \leqq h \leqq \|h\|_{L^\infty}$$

by induction we verify that

$$\|u_k\|_{L^\infty} \leq \|h\|_{L^\infty}(T\|A(.)\Phi\|_{L^\infty} + \|\Phi\|_{L^\infty}) \frac{1}{\varepsilon\delta\beta - \varepsilon^2 M}.$$

The pointwise monotonic convergence of u_k follows from this estimate. Furthermore, u_k remains bounded in $\mathcal{W}^{2,1,p}_{loc}(Q)$, therefore on going to the limit $u \in \mathcal{W}^{2,1,p}_{loc}(Q)$. From this it follows that

$$u(x-\varepsilon\gamma,t) \in \mathcal{W}^{2,1,p}(\mathcal{O}\times]0,T-\delta[) \ \forall \delta>0.$$

By passing to the limit and using (4.46) afresh, we obtain the desired result. Also we have proved the estimate (4.52). □

Let, now,

(4.61) $$f \in L^p(Q),\ p>n+1,\ \varphi \in W^{2,p}(\mathcal{O}),\ \varphi, \nabla\varphi|_\Gamma = 0$$

and let u be the solution of

(4.62) $$\left|\begin{array}{l} -\frac{\partial u}{\partial t} + A(t)u = f, \quad \gamma.\frac{\partial u}{\partial x}\Big|_\Sigma = 0, \quad u(x,T) = \varphi(x) \\ u \in \mathcal{W}^{2,1,p}(Q), \quad p > n+1 \end{array}\right.$$

also let u^ε be the solution of

(4.63) $$\left|\begin{array}{ll} -\frac{\partial u^\varepsilon}{\partial t} + A(t)u^\varepsilon = f & ,u^\varepsilon(x,T) = \varphi(x) \\ u^\varepsilon\Big|_\Sigma = u^\varepsilon(x-\varepsilon\gamma,t)\Big|_\Sigma & ,u^\varepsilon \in \mathcal{W}^{2,1,p}(\mathcal{O}\times]0,T-\delta[)\ \forall\delta>0, \\ u^\varepsilon \in L^\infty(Q) & \end{array}\right.$$

which by Theorem 4.2 is well defined. We then have:

THEOREM 4.3: *Under the assumptions of Theorem 4.1, together with* (4.50) *and* (4.61),

(4.64) $$u^\varepsilon \to u \text{ in } L^\infty(Q).$$

Proof: Since $p > n+1$, $u, \frac{\partial u}{\partial x} \in C^0(\bar{Q})$. Hence we have

$$u(x,t) - u(x-\varepsilon\gamma,t) = -\int_0^1 \varepsilon\gamma(x,t).\frac{\partial u}{\partial x}(x-\varepsilon\gamma\theta,t)\,d\theta$$

and since $\gamma.\frac{\partial u}{\partial x}\Big|_\Sigma = 0$, we see that

(4.65) $$u(x,t) - u(x-\varepsilon\gamma,t)\Big|_\Sigma = \int_0^1 \varepsilon\gamma.\left(\frac{\partial u}{\partial x}(x,t) - \frac{\partial u}{\partial x}(x-\varepsilon\theta\gamma,t)\right)d\theta$$

$$= h_\varepsilon(x,t)\Big|_\Sigma$$

with

$$\|h_\varepsilon\|_{L^\infty(Q)} \leq \varepsilon\rho(\varepsilon)$$

where $\rho(\varepsilon) \to 0$ as $\varepsilon \to 0$ ($\rho(\lambda)$ is a multiplicative constant close to the continuity modulus of $\frac{\partial u}{\partial x}$ on $\bar{Q}$). But then $u^\varepsilon - u$ is the solution of

$$-\frac{\partial}{\partial t}(u^\varepsilon - u) + A(t)(u^\varepsilon - u) = 0$$

$$(u^\varepsilon - u)(x,t) - (u^\varepsilon - u)(x - \varepsilon\gamma, t)\Big|_\Sigma = -h_\varepsilon(x,t)$$

$$(u^\varepsilon - u)(x,T) = 0 .$$

Applying the estimate (4.52), we obtain

$$\|u^\varepsilon - u\|_{L^\infty(Q)} \leq \varepsilon\rho(\varepsilon)\frac{C_T}{\varepsilon}$$

whence the desired result. □

4.3.2 Construction of the approximate process

Let $(\Omega, \mathcal{A}, P, \mathcal{F}^s)$, be an arbitrary probability space on which a normalised Wiener process w(t) is given. We construct the following diffusion process with a jump:

(4.66) $$\begin{cases} dy = g(y,s)\,ds + \sigma(y,s)\,dw(s) - \sum_i \delta(s - \theta_i)\,\xi_i\,, \quad s \geq t \quad (1) \\ y(t) = x \end{cases}$$

where the θ_i, ξ_i are defined by induction by the formulae

(4.67) $$\begin{cases} \theta_0 = t \\ \theta_i = \inf\{s \geq \theta_{i-1} \mid y(s) \notin \mathcal{O}\} \quad i \geq 1 \\ \xi_i = \varepsilon\gamma(y(\theta_i - 0), \theta_i)\ (2)\,. \end{cases}$$

We write the process so defined as $y^\varepsilon(s) = y^\varepsilon_{xt}(s)$. It is clear that $y^\varepsilon(s) \in \bar{\mathcal{O}}$ for all s. The definition of $y^\varepsilon(s)$ is very intuitive. In fact the process evolves according to an ordinary diffusion inside $\mathcal{O}$, then when the process reaches the frontier it jumps instantaneously into the interior of $\mathcal{O}$ in the direction opposite to γ, the size of the jump being ε.

Our objective is to prove that, in terms of measures, the process y^ε converges to the solution of the sub-martingale problem. We begin by proving an important property about the number of jumps of the process $y^\varepsilon(s)$. Let $N^\varepsilon(s_1,s_2)$ be the number of jumps on $[s_1,s_2]$.

LEMMA 4.5: *For* $s_1 \leq s_2 \leq T$ *we have the estimate*

(4.68) $$\varepsilon^2 E[N^2(s_1,s_2) \mid \mathcal{F}^{s_1}] \leq K\varepsilon^2 + C(s_2 - s_1)$$

---

(1) Since g is not Lipschitz, Girsanov's Theorem must be used.
(2) $y(t - o) = x$.

where K *is a constant,* C *is a constant depending only upon bounds on the coefficients* g, σ *as well as on the function* Φ *defining the open set.*

Proof: Notice that

$$\Phi(y(s_2)) - \Phi(y(s_1)) = \sum_{i=1}^{N(s_1,s_2)} \Phi(y(\theta_i)) - \Phi(y(\theta_i-0)) + \int_{s_1}^{s_2} - A\Phi(y(s))ds + \int_{s_1}^{s_2} \frac{\partial\Phi}{\partial x}.\sigma\, dw$$

and

$$\Phi(y(\theta_i)) - \Phi(y(\theta_i-0)) = \Phi(y(\theta_i-0) - \varepsilon\gamma) - \Phi(y(\theta_i-0)) = -\varepsilon\gamma.\frac{\partial\Phi}{\partial x}(y(\theta_i-0)) + \frac{\varepsilon^2}{2}\frac{\partial^2\Phi}{\partial x^2}\gamma.\gamma \leq -\varepsilon\beta\delta + \varepsilon^2 M,$$

therefore

$$\Phi(y(s_2)) - \Phi(y(s_1)) \leq -\varepsilon\beta\delta N + \varepsilon^2 MN + \int_{s_1}^{s_2} - A\,\Phi(y(s))ds + \int_{s_1}^{s_2}\frac{\partial\Phi}{\partial x}.\sigma\, dw.$$

Now apply the argument with Φ^2. We note that $\partial\Phi^2/\partial x$ is zero on Γ, and thus obtain

$$\Phi^2(y(s_2)) - \Phi^2(y(s_1)) \leq \varepsilon^2 M'N + \int_{s_1}^{s_2} - A\,\Phi^2(y(s))ds + \int_{s_1}^{s_2}\frac{\partial\Phi^2}{\partial x}.\sigma\, dw. \tag{4.70}$$

Next we use the identity (3.38), hence taking account of (4.69),(4.70), it follows that

$$\begin{aligned}(4.71)\quad & \Big(\Phi(y(s_2)) - \Phi(y(s_1))\Big)^2 \leq \varepsilon^2 M'N + C'(s_2-s_1) + \int_{s_1}^{s_2}\frac{\partial\Phi^2}{\partial x}.\sigma\, dw - \\ & -2\,\Phi(s_1))\left[-\varepsilon\beta\delta N + \varepsilon^2 MN + C(s_2-s_1) + \int_{s_1}^{s_2}\frac{\partial\Phi}{\partial x}.\sigma\, dw\right] \\ & \leq \varepsilon^2 M''N + C''(s_2-s_1) + \int_{s_1}^{s_2}\frac{\partial\Phi^2}{\partial x}.\sigma\, dw - 2\Phi(y(s_1))\int_{s_1}^{s_2}\frac{\partial\Phi}{\partial x}.\sigma\, dw.\end{aligned}$$

From (4.69),(4.71) we deduce that

$$\varepsilon^2 N^2 \leq \varepsilon^2 M'''N + C'''(s_2-s_1) + C\int_{s_1}^{s_2}\frac{\partial}{\partial x}\Phi^2\sigma\, dw - 2C\,\Phi(y(s_1))\int_{s_1}^{s_2}\frac{\partial\Phi}{\partial x}.\sigma\, dw + C\left(\int_{s_1}^{s_2}\frac{\partial\Phi}{\partial x}.\sigma\, dw\right)^2$$

and so (4.68) follows immediately by passage to the mathematical expectation.

Let φ be a bounded Borel set on $\bar{\mathcal{O}}$, and consider the solution of

$$(4.72)\quad \begin{cases} -\dfrac{\partial u^\varepsilon}{\partial t} + A(t)u^\varepsilon = 0, & u^\varepsilon(x,T) = \varphi(x) \\ u^\varepsilon\big|_\Sigma = u^\varepsilon(x-\varepsilon\gamma,t)\big|_\Sigma, & u^\varepsilon \in \mathcal{W}^{2,1,p}(\mathcal{O}\times]0,T-\delta[),\ \forall\delta>0.\end{cases}$$

We are going to prove:

LEMMA 4.6: *The function* $u^{\varepsilon}(x,t)$ *which is a solution of* (4.72) *is given explicitly by*

(4.73) $$u^{\varepsilon}(x,t) = E\,\varphi(y^{\varepsilon}_{xt}(T)).$$

Proof: By applying Formula (4.47) we have

$$u^{\varepsilon}(x,t) = E\,u^{\varepsilon}(y^{\varepsilon}(\theta_1 - 0), \theta_1)\,\chi_{\theta_1 \leq T} + \\ + E\,\varphi(y^{\varepsilon}(T))\,\chi_{T<\theta_1}$$

and by the condition satisfied by u^{ε} at the boundary, and the definition of $y^{\varepsilon}(\theta_1)$,

$$= E u^{\varepsilon}(y^{\varepsilon}(\theta_1), \theta_1)\,\chi_{\theta_1 \leq T} + E\,\varphi(y^{\varepsilon}(T))\,\chi_{T<\theta_1}, \\ = E u^{\varepsilon}(y^{\varepsilon}(\theta_1 \wedge T), \theta_1 \wedge T).$$

We next show that quite generally

(4.74) $$u^{\varepsilon}(x,t) = E\,u^{\varepsilon}(y^{\varepsilon}(\theta_i \wedge T), \theta_i \wedge T).$$

Now, by Lemma 4.5 the number of jumps in the process y^{ε} on $[0,T]$ is almost surely finite, therefore for i large enough $\theta_i \wedge T = T$ and

$$u^{\varepsilon}(y^{\varepsilon}(\theta_i \wedge T), \theta_i \wedge T) = u^{\varepsilon}(y^{\varepsilon}(T), T) = \varphi(y^{\varepsilon}(T))$$

therefore

$$u^{\varepsilon}(y^{\varepsilon}(\theta_i \wedge T), \theta_i \wedge T) \to \varphi(y^{\varepsilon}(T)) \quad \text{a.s. when } i \to +\infty.$$

By applying Lebesgue's Theorem to (4.74) the desired result is obtained. □

It easily follows from (4.73) that $y^{\varepsilon}(s)$ is a strongly Fellerian Markov process whose transition probability is given by (4.72),(4.73).

4.3.3 Convergence

We now study what happens as $\varepsilon \to 0$.

LEMMA 4.7: *Let* $t \leq s_1 \leq s_2 \leq T$. *We have*

(4.75) $$E[\,|y^{\varepsilon}(s_2) - y^{2}(s_1)|^2 \mid \mathcal{F}^{s_1}\,] \leq C\,(s_2 - s_1) + \varepsilon^2 K$$

where C,K *are constants.*

Proof: We notice that

$$y^{\varepsilon}(s_2) - y^{\varepsilon}(s_1) = -\,\varepsilon \sum_{i=1}^{N} \gamma(y^{\varepsilon}(\theta_i - 0), \theta_i) + \qquad \text{(Contd)}$$

$$\text{(Contd)} \quad + \int_{s_1}^{s_2} g(y^\varepsilon(s),s)ds + \int_{s_1}^{s_2} \sigma(y^\varepsilon(s),s)\,dw(s)$$

whence

$$E[|y^\varepsilon(s_2) - y^\varepsilon(s_1)|^2 |\mathcal{F}^{s_1}] \leq C(s_2-s_1) + \varepsilon^2 E[N^2(s_1,s_2)|\mathcal{F}^{s_1}]$$

which, together with Lemma 4.5, provides the desired result. □

We write $P^\varepsilon = P^\varepsilon_{xt}$ for the image measure of the process $y^\varepsilon_{xt}(s)$ on $D([t,\infty[;\bar{\mathcal{O}})$. For all $T > t$ we write P^ε_T for the image measure of $y^\varepsilon(s)$, $t \leq s < T$, and $y^\varepsilon(T) = \lim_{h \to 0} y^\varepsilon(T - h)$. Then P^ε_T is a probability measure on $D[t,T;\bar{\mathcal{O}}]$. Then we have

LEMMA 4.8: *When* ε *varies, the family* P^ε_T *remains in a relatively compact set of the set of probability measures on* $D[t,T;\bar{\mathcal{O}}]$.

Proof: To simplify the writing, take $t = 0$. First of all, by the calculations done in Lemma 4.7, we have

$$\sup_{0 \leq s \leq T} |y^\varepsilon(s)| \leq C\,[1 + \varepsilon N(0,T) + T + \sup_{0 \leq s \leq T} | \int_0^s \sigma\, dw |\,]$$

and

$$E \sup_{0 \leq s \leq T} |y^\varepsilon(s)| \leq C\,[1 + \varepsilon + \sqrt{T} + T\,]$$

therefore

$$E^\varepsilon_T \sup_{0 \leq s \leq T} |y(s)| \leq C \quad \text{as } \varepsilon \text{ varies.}$$

We next set

$$(4.77) \qquad \omega(\rho) = \sup_{\substack{o \leq t < u < s \leq T \\ |t-s| < \rho}} [\inf(|y(t) - y(u)|, |y(u) - y(s)|)] + $$
$$+ \sup_{0 \leq t \leq \rho} |y(t) - y(0)| + \sup_{T-\rho \leq t \leq T} |y(t) - y(T)| .$$

We are going to show that for all $\eta > 0$

$$(4.78) \qquad \lim_{\rho \to 0} \limsup_{\varepsilon \to 0} P^\varepsilon_T [\omega(\rho) \geq \eta] = 0 .$$

But the result is then a consequence of (4.76),(4.78) and of Theorem 7.2 in PARTHASARATHY [1]. To show (4.78), we set

$$\omega^\varepsilon(\rho) = \sup_{\substack{0 \leq t < u < s \leq T \\ |t - s| < \rho}} [\inf(|y^\varepsilon(t) - y^\varepsilon(u)|, |y^\varepsilon(u) - y^\varepsilon(s)|)] + \qquad \text{(Contd)}$$

(Contd) $$+ \sup_{0 \leq t \leq \rho} |y^{\varepsilon}(t) - y^{\varepsilon}(o)| + \sup_{T-\rho \leq t \leq T} |y^{\varepsilon}(t) - y^{\varepsilon}(T)|.$$

This is a question of showing that we have

$$\lim_{\rho \to 0} \limsup_{\varepsilon \to 0} P[\omega^{\varepsilon}(\rho) \geq \eta] = 0. \tag{4.79}$$

But

$$|y^{\varepsilon}(u) - y^{\varepsilon}(t)| \leq C_T [\varepsilon N(t,u) + u - t + |\int_t^u \sigma \, dw|].$$

From this we deduce that

$$\omega^{\varepsilon}(\rho) \leq X_T^{\varepsilon}(\rho) = C_T [\sup_{\substack{t < s \leq T \\ |t-s| < \rho}} \varepsilon N(t,s) + \rho + \sup_{\substack{0 \leq t < s \leq T \\ |t-s| < \rho}} |\int_t^s \sigma \, dw|].$$

Operating as in Lemma 4.5, we verify that

$$E X_T^{\varepsilon}(\rho) \leq C_T' [\varepsilon + \rho + E \sup_{\substack{0 \leq t < s \leq T \\ |t-s| < p}} |\int_t^s \frac{\partial}{\partial x} \Phi^2 \sigma \, dw| +$$

$$+ E \sup_{\substack{0 \leq t < s \leq T \\ |t-s| < \rho}} |\int_t^s \frac{\partial \Phi}{\partial x} \cdot \sigma \, dw| + E \sup_{\substack{0 \leq t < s \leq T \\ |t-s| < \rho}} |\int_t^s \frac{\partial \Phi}{\partial x} \sigma \, dw|^2 +$$

$$+ E \sup_{\substack{0 \leq t < s \leq T \\ |t-s| < \rho}} |\int_t^s \sigma \, dw|] \leq$$

$$\leq C_T'' [\varepsilon + \rho + \rho^{1/4} + p^{1/2}].$$

Since

$$P[\omega^{\varepsilon}(\rho) \geq \eta] \leq \frac{1}{\eta} E \omega^{\varepsilon}(\rho) \leq \frac{1}{\eta} E X_T^{\varepsilon}(\rho)$$

the result (4.79) is easily deduced. □

LEMMA 4.9: *Every limit point* P_T *of* P_T^{ε} *is concentrated on* $C(t,T;\bar{\mathcal{O}})$.

Proof: Starting from (4.71) we verify that

$$E \varepsilon^6 N^6(s_1, s_2) \leq C_T [\varepsilon^6 + (s_2 - s_1)^{3/2} + (s_2 - s_1)^3]$$

therefore

$$E |y^{\varepsilon}(s_2) - y^{\varepsilon}(s_1)|^6 \leq C_T' [\varepsilon^6 + (s_2 - s_1)^{3/2} + (s_2 - s_1)^3].$$

Consequently, if P_T is a limit point of P_T^{ε}, we have

$$E_T |y(s_2) - y(s_1)|^6 \leq C_T'(s_2-s_1)^{3/2}, \quad t \leq s_1 \leq s_2 \leq T,$$

which implies the desired result. □

We can then state:

THEOREM 4.4: *Under the assumptions of Theorem 4.1, together with* (4.50),(4.61), *the sequence* P_T^ε *converges weakly when* $\varepsilon \to 0$ *towards the restriction on* $[t,T]$ *of the solution of the sub-martingale problem.*

Proof: Starting from the probability interpretation of the limit problem, we verify in the same way as in Theorem 4.3 that

$$E_T^\varepsilon \int_t^T f\,(y(s),s)\,ds \to E^{xt} \int_t^T f(y(s),s)\,ds$$

for f bounded and continuous. This implies the desired result. □

4.4 *Interpretation of Problems with Limits*

We begin by giving some results which are easy, but important, consequences of Corollary 4.1.

Under the assumptions of Theorem 4.1 we consider the problem

$$(4.80) \quad \left| \begin{aligned} & -\frac{\partial u}{\partial t} + A(t)\,u + a_0 u = f \\ & \gamma.\frac{\partial u}{\partial x}\Big|_\Gamma = 0 \\ & u(x,T) = \bar{u}(x) \end{aligned} \right.$$

where

$$f \in L^p(Q), \quad \bar{u} \in W^{2,p}(\mathcal{O}), \quad \frac{\partial \bar{u}}{\partial x}\Big|_\Gamma = 0, \quad p > n+1.$$

There then exists one and only one solution $u \in \mathcal{W}^{2,1,p}(Q)$. In particular, $u, \frac{\partial u}{\partial x} \in C^0(Q)$. Recall that P^{xt} denotes the solution of the sub-martingale problem with respect to the operator A(t), as well as the direction of reflection γ. We then have:

THEOREM 4.5: *Under the assumptions of Theorem 4.1, the solution* u(x,t) *of* (4.80) *is explicitly given by the formula*

$$(4.81) \quad u(x,t) = E^{xt} \int_t^T [f(y(s),s) \exp - \int_t^s a_0(y(\lambda),\lambda)\,d\lambda\,]\,ds + $$
$$+ E^{xt}\,\bar{u}(y(T)) \exp - \int_t^T a_0(y(s),s)\,ds\,.$$

Proof: Notice that from Corollary 4.1 it can be deduced that

$$E^{xt} \psi\,(y(\theta \wedge T),\, \theta \wedge T) \exp - \int_t^T a_0(y(s),s)\,ds = \psi\,(x,t) + \quad \text{(Contd)}$$

(Contd)
$$+E^{xt}\int_t^{\theta\wedge T}\left(\frac{\partial\psi}{\partial s}-A(s)\psi-a_o\psi\right)(y(s),s)\left(\exp-\int_t^s a_o(y(\lambda),\lambda)\,d\lambda\right)ds+$$
$$+E^{xt}\int_t^{\theta\wedge T}\frac{\partial\psi}{\partial x}\cdot\gamma(y(s),s)\left(\exp-\int_t^s a_o(y(\lambda),\lambda)\,d\lambda\right)ds\,.$$

The preceding formula can then be applied with $\psi = u$. The result desired is then immediately obtained. □

Remark 4.1

Under the assumptions of Theorem 4.5, with *coefficients independent of time*, $a_o \geq \beta > 0$), where β is a constant, we can interpret the solution of the stationary problem

$$(4.82)\qquad \begin{cases} Au + a_o u = f \\ \gamma\cdot\dfrac{\partial u}{\partial x}\Big|_\Gamma = 0 \end{cases}$$

where $f \in L^p(\mathcal{O})$ and $p > n/2$. The solution $u \in W^{2,p}(\mathcal{O})$ and is given explicitly by the formula

$$(4.83)\qquad u(x) = E^x\int_0^\infty f(y(t))\left(\exp-\int_0^t a_o(y(s))\,ds\right)dt$$

where P^x is the solution of the sub-martingale problem with respec t to A,γ, starting off at the instant 0 at the point x. □

We shall now also give the interpretation of mixed Dirichlet-Neumann boundary conditions. Let us assume that $\mathcal{O}$ is an annulus the boundaries of which are Γ_o and Γ_1, Γ_o being the boundary of an open set $\mathcal{O}_o$ satisfying the property (3.4)

Consider the problem

$$(4.84)\qquad \begin{cases} Au + a_o u = f \quad \text{dans } \mathcal{O} \\ \gamma\cdot\dfrac{\partial u}{\partial x}\Big|_{\Gamma_o} = 0 \\ u\big|_{\Gamma_1} = 0\ ,\ u\in W^{2,p}(\mathcal{O})\,. \end{cases}$$

With the usual regularity assumptions the problem (4.84) has one and only one solution. Let P^x be the solution of the sub-martingale problem with respect to the pair A,γ, starting off from 0 in the position x, constructed in the open set $\mathcal{O}_o$. We write

$$\tau = \inf\{\,t\geq 0 \mid y(t)\in\Gamma_1\}\,.$$

It is easily verified that we have the formula

$$(4.85)\qquad u(x) = E^x\int_0^\tau f(y(s))\left(\exp-\int_0^t a_o(y(s))\,ds\right)dt.$$

Remark 4.2

The interpretation of a mixed problem in the case where the open set $\mathcal{O}$ is not an annulus is more delicate. The difficulty is as follows: We define a solution by the variational formulation, but this solution is not regular up to the boundary (in general it is not continuous on $\bar{\mathcal{O}}$). When the dimension is small we do, however, have regularity theorems (cf. P. GRISVARD [1]) which allows us still to obtain the interpretation. The general case is open. □

5. STATIONARY VARIATIONAL INEQUALITIES WITH NEUMANN BOUNDARY CONDITIONS

Method of Approach

In A. BENSOUSSAN, J.-L. LIONS [1] there is a systematic study of the V.I.'s (variational inequalities) with Dirichlet boundary conditions in conjunction with optimal stopping time problems.

In this and the following Section we give the analogue for the Neumann boundary condition problem, with the aim of studying optimal stopping time problems for reflected processes. In fact very many results are identical, and result from the abstract framework of the variational formulation. Here we shall be content with bringing out results that show some difference from each other.

5.1 *Formulation of the V.I.'s*

We take $V = H^1(\mathcal{O})$, where $\mathcal{O}$ is a bounded open set of R^n. Functions a_{ij}, a_i, a_o are given on $\mathcal{O}$ such that

$$(5.1)\quad \begin{cases} a_{ij}, a_i, a_o \quad \text{are measurable bounded sets} \\ \sum_{ij} a_{ij}\, \xi_i\, \xi_j \geq \alpha\, |\xi|^2 , \ \forall\, \xi \in R^n \\ a_o \geq \beta > 0. \end{cases}$$

Consider on V the bilinear form defined by (4.5). Let ψ be a measurable function. We shall set

$$(5.2)\qquad K = \{v \mid v \in V,\ v \leq \psi,\ \text{a.e.}\}$$

and always assume that

$$(5.3)\qquad K \neq \Phi.$$

Let $f \in L^2(\mathcal{O})$. We call the problem

$$(5.4)\qquad a(u, v-u) \geq (f, v-u) \qquad \forall \quad v \in K,\ u \in K.$$

a stationary V.I..

A penalised problem(1) is defined in the following way

$$(5.5)\qquad a(u^\epsilon, v) + \frac{1}{\epsilon}((u^\epsilon - \psi)^+, v) = (f,v) \qquad \forall v \in V.$$

We are content with recalling the two theorems on the existence and uniqueness of u (the coercive and non-coercive cases; the proof is identical with that carried out for the Dirichlet case (cf., A. BENSOUSSAN, J.-L. LIONS [1], Chap. 3, Section 1)).

---

(1) Another penalisation process is indicated in Remark 5.2 below.

THEOREM 5.1: *Under the assumptions* (5.1),(5.3) *and*

(5.6) $$a(v,v) \geqq \gamma \|v\|^2 \quad \forall\, v \in V,$$

there exists one and only one solution of the problem (5.4). □

THEOREM 5.2: *Under the assumptions* (5.1),(5.3) *and*

(5.7) $$f \in L^\infty(\mathcal{O})$$

(5.8) $$\psi \in L^\infty(\mathcal{O})$$

there exists one and only one solution of the problem (5.4). □

5.2 *Estimation of the Penalisation Error*

THEOREM 5.3: *Take the assumptions of Theorem 5.1. In addition assume that*

(5.9) $\mathcal{O}$ *is an open set with Lipschitz boundary,*

(5.10) $$\psi \in H^2(\mathcal{O})$$

then the solutions u^ε *and* u *of* (5.5) *and* (5.4) *are related by the estimate*

(5.11) $$\|u^\varepsilon - u\| \leqq C\, \varepsilon^{1/4}.$$

Furthermore, if it is assumed that

(5.12) $$\frac{\partial \psi}{\partial \nu_A}\Big|_\Gamma \geqq 0$$

then the estimate (5.11) *can be improved; more precisely, we have*

(5.11) $$\|u^\varepsilon - u\| \leqq C\, \varepsilon^{1/2}.$$

Proof: Take $v = (u^\varepsilon - \psi)^+$ in (5.5). We obtain

$$a((u^\varepsilon - \psi)^+, (u^\varepsilon - \psi)^+) + \frac{1}{\varepsilon} |(u^\varepsilon - \psi)^+|^2 = (f, (u^\varepsilon - \psi)^+) - a(\psi, (u^\varepsilon - \psi)^+).$$

But by Green's Formula, which is applicable thanks to (5.9) and (5.10), we have

$$a(\psi, (u^\varepsilon - \psi)^+) = \int_{\mathcal{O}} A\psi\, (u^\varepsilon - \psi)^+\, dx + \int_\Gamma \frac{\partial \psi}{\partial \nu_A} (u^\varepsilon - \psi)^+\, d\Gamma.$$

But it will be noticed that we have

$$|(u^\varepsilon - \psi)^+|_{L^2(\Gamma)} \leqq C\, |(u^\varepsilon - \psi)^+|^{1/2}\, \|(u^\varepsilon - \psi)^+\|^{1/2}$$

and therefore we obtain the estimate

$(5.12)_1$ $$\alpha \|(u^\varepsilon - \psi)^+\|^2 + \frac{1}{\varepsilon} |(u^\varepsilon - \psi)^+|^2 \leqq C\Big[\, |(u^\varepsilon - \psi)^+| +$$ (Contd)

(Contd) $$+ |(u^\varepsilon-\psi)^+|^{1/2}\|(u^\varepsilon-\psi)^+\|^{1/2}\Big] \leqq C\,|(u^\varepsilon-\psi)^+|^{1/2}\|(u^\varepsilon-\psi)^+\|^{1/2}$$

from which we deduce

$$\|(u^\varepsilon - \psi)^+\| \leqq \text{constant},$$

$$(5.12)_2 \quad |(u^\varepsilon-\psi)^+|^{3/2} \leqq C\,\varepsilon\,|(u^\varepsilon-\psi)^+|^{1/2} + C\,\varepsilon\,\|(u^\varepsilon-\psi)^+\|^{1/2}, \quad |(u^\varepsilon-\psi)^+| \leqq C\,\varepsilon^{\frac{2}{3}}.$$

From $(5.12)_1$ we then deduce that

$$\|(u^\varepsilon-\psi)^+\| \leqq C\,|(u^\varepsilon-\psi)^+|^{1/3}.$$

Substituting in $(5.12)_2$ we obtain

$$|(u^\varepsilon-\psi)^+|^{3/2} \leqq C\,\varepsilon\,\|(u^\varepsilon-\psi)^+\|^{1/2} \leqq C\,\varepsilon\,|(u^\varepsilon-\psi)^+|^{1/6}$$

therefore

$$|(u^\varepsilon-\psi)^+| \leqq C\,\varepsilon^{3/4}$$

$$\|(u^\varepsilon-\psi)^+\| \leqq C\,\varepsilon^{1/4}.$$

We then finish as in the Dirichlet case (cf. B.L. loc. cit.), and so obtain (5.11). If (5.12) is now satisfied we then have

$$- a(\psi,(u^\varepsilon-\psi)^+) \leqq - \int_{\mathcal{O}} A\psi\,(u^\varepsilon-\psi)^+\,dx$$

thus it follows that

$$(5.12)_3 \quad a((u^\varepsilon-\psi)^+,(u^\varepsilon-\psi)^+) + \frac{1}{\varepsilon}\,|(u^\varepsilon-\psi)^+|^2 \leqq (f-A\psi,\ (u^\varepsilon-\psi)^+)$$

and

$$\|(u^\varepsilon - \psi)^+\| \leqq C\sqrt{\varepsilon},$$

and so we obtain (5.11).

Remark 5.1

As in the Dirichlet case we have the monotonic property, i.e., if $u(f,\psi)$ denotes the solution of the problem (5.4), then

$$\tilde{f} \geqq f,\ \tilde{\psi} \geqq \psi \Rightarrow u(\tilde{f},\tilde{\psi}) \geqq u(f,\psi).$$

As in the Dirichlet case $u_\varepsilon \downarrow u$ (cf. B.L. loc. cit.). □

5.3 *Regularity*

THEOREM 5.4: *Under the assumptions of Theorem 5.3 (including* (5.12)*), we have*

$$(5.14) \quad Au \in L^2(\mathcal{O}).$$

Proof: Starting from (5.5) it is known that u^ε satisfies (at least in the sense of a distribution)

$$A u^\varepsilon + a_o u^\varepsilon + \frac{1}{\varepsilon}(u^\varepsilon - \psi)^+ = f.$$

But by $(5.12)_3$

$$\left|\left(\frac{u^\varepsilon - \psi}{\varepsilon}\right)^+\right| \leq C,$$

hence $|Au^\varepsilon| \leq C$, and so we deduce (5.14). □

COROLLARY 5.1: *Under the assumptions of Theorem 5.4, and if* $a_{ij} \in W^{1,\infty}$, *and if* $\mathcal{O}$ *is a bounded open set of class* C^2 *then the solution* u *of* (5.4) *satisfies*

(5.15) $\quad u \in H^2(\mathcal{O}).$

Proof: A subsequence can be chosen such that

$$\left(\frac{u^\varepsilon - \psi}{\varepsilon}\right)^+ \to h \quad \text{weakly in } L^2(\mathcal{O})$$

and $u^\varepsilon \to u$ weakly in $H^1(\mathcal{O})$. From (5.5) we deduce that u is a solution of the problem:

(5.16) $\quad a(u,v) = (f - h, v) \quad \forall v \in H^1(\mathcal{O}).$

It then suffices to apply the regularity results in the Neumann problem to obtain (5.16). □

We now give the analogous estimate in the L^p spaces, $p > 2$.

We assume that

(5.17) $\quad (p-1)\sum a_{ij}(x)\,\xi_i\,\xi_j + \sum_i a_i(x)\,\xi_i\,\xi_0 + a_0(x)\,\xi_0^2 \geq 0$ a.e. in $\mathcal{O}$.

We have:

THEOREM 5.5: *Taking the assumptions of Theorem 5.4, together with* (5.17) *and*

(5.18) $\quad f \in L^p(\mathcal{O}),\quad \psi \in W^{1,p}(\mathcal{O}),\quad A\psi \in L^p(\mathcal{O}),$

we then have

(5.19) $\quad Au \in L^p(\mathcal{O})$

where u *denotes the solution of* (5.4).

Proof: Thanks to (5.17) we have

$$a(\varphi, (\varphi^+)^{p-1}) \geq 0.$$

Furthermore, we have

$$a(u^\epsilon-\psi,((u^\epsilon-\psi)^+)^{p-1}) + \frac{1}{\epsilon}\|(u^\epsilon-\psi)^+\|^p_{L^p} = (f-A\psi,((u^\epsilon-\psi)^+)^{p-1}) -$$

$$-\int_\Gamma \frac{\partial\psi}{\partial\nu_A}((u^\epsilon-\psi)^+)^{p-1}\,d\Gamma$$

therefore

$$\frac{1}{\epsilon}\|(u^\epsilon-\psi)^+\|^p_{L^p} \leq \|f-A\psi\|_{L^p}\|(u^\epsilon-\psi)^+\|^{p/p'}_{L^p}$$

and thus

$$\|(\frac{u^\epsilon-\psi}{\epsilon})^+\|_{L^p} \leq C.$$

Consequently

$$\|Au^\epsilon\|_{L^p} \leq C,$$

and hence (5.19). □

COROLLARY 5.2: *Under the assumptions of Theorem 5.5, and if* $a_{ij} \in W^{1,\infty}$, *and* $\mathcal{O}$ *is an open set of class* C^2, *then the solution* u *of* (5.4) *satisfies*

(5.20) $\quad u \in W^{2,p}(\mathcal{O})$.

Proof: We again have (5.16) or $h \in L^p(\mathcal{O})$. It then suffices to use the regularity theorems in the Neumann problem. □

THEOREM 5.6: *Take the assumptions of Theorem 5.1. For fixed* f *in* $L^2(\mathcal{O})$, *denote the solution of* (5.4) *by* $u(\psi)$. *Assuming that* ψ *and* $\hat\psi$ *are given in* $L^\infty(\mathcal{O})$, *we have*

(5.21) $\quad \|u(\psi)-u(\hat\psi)\|_{L^\infty(\mathcal{O})} \leq \|\psi-\hat\psi\|_{L^\infty(\mathcal{O})}$.

Proof: This is identical with the Dirichlet case (cf. B.L. loc. cit., Chap. 3). □

COROLLARY 5.3: *Take the assumptions of Theorem 5.1, with*

(5.22) $\quad a_{ij} \in W^{1,\infty}$

(5.23) $\quad f \in L^p(\mathcal{O})$, $p > \frac{n}{2}$ (1)

(5.24) $\quad \psi \in C^0(\bar{\mathcal{O}})$; $\psi = \lim \psi^j$ in $C^0(\bar{\mathcal{O}})$, with $\psi^j \in W^{2,p}$, $\frac{\partial\psi^j}{\partial\nu_A}\Big|_\Gamma \geq 0$

then we have

(5.25) $\quad u \in C^0(\bar{\mathcal{O}})$.

(1) Later on in Lemma 7.1 we shall see that such an approximation is always possible.

Proof: This is a consequence of Theorem 5.6, and is proved identically to the Dirichlet case.

Remark 5.2

A penalisation scheme different from (5.5) can be used, say

$$a(u^\varepsilon,v) + \frac{1}{\varepsilon}((u^\varepsilon-\psi)^+, v) + \frac{1}{\varepsilon}((u^\varepsilon-\psi)^+,v)_\Gamma = (f,v) .$$

By taking $v = (u^\varepsilon - \psi)^+$ we obtain

$$a((u^\varepsilon-\psi)^+) + \frac{1}{\varepsilon} |(u^\varepsilon-\psi)^+|^2 + \frac{1}{\varepsilon} |(u^\varepsilon-\psi)^+|^2_\Gamma = (f-A\psi, (u^\varepsilon-\psi)^+) -$$
$$-\int_\Gamma \frac{\partial\psi}{\partial\nu_A}(u^\varepsilon-\psi)^+ d\Gamma$$

hence

$$\frac{1}{\varepsilon} |(u^\varepsilon-\psi)^+|^2 + \frac{1}{\varepsilon} |(u^\varepsilon-\psi)^+|^2_\Gamma \leq C[|(u^\varepsilon-\psi)^+| + |(u^\varepsilon-\psi)^+|_\Gamma]$$

and therefore

$$\|u^\varepsilon-u\| \leq C\sqrt{\varepsilon} . \quad \square$$

We then obtain, without any great difficulty,

$$|(u^\varepsilon-\psi)^+| + |(u^\varepsilon-\psi)^+|_\Gamma \leq C\,\varepsilon.$$

Now consider the case of games. Two measurable functions ψ_1,ψ_2 are given on $\mathcal{O}$ and we set

(5.26) $K = \{v \mid v \in V, \ \psi_1 \leq v \leq \psi_2\}$.

Assume

(5.27) $K \neq \emptyset$.

Then consider the V.I.

(5.28) find $u \in K$ such that
$a(u,v - u) \geq (f,v - u)$ for all $v \in K$.

A penalised problem corresponding to (5.28) is the following:

(5.29) $a(u^\varepsilon,v) + \frac{1}{\varepsilon}((u^\varepsilon-\psi_2)^+, v) - \frac{1}{\varepsilon}((u^\varepsilon-\psi_1)^-, v) = (f,v), \quad \forall v \in V$.

Naturally, u and u^ε exist and are unique, and u^ε converges to u in V. Our object is to give an estimate of the penalisation error. We have:

THEOREM 5.7: *Assume* (5.1),(5.6),(5.27) *together with* (5.9) *and*

(5.30) $\psi_i \in H^2(\mathcal{O})$ $(i=1,2)$, $\frac{\partial\psi_2}{\partial\nu_A}\Big|_\Gamma \geq 0 , \frac{\partial\psi_1}{\partial\nu_A}\Big|_\Gamma \leq 0$, *then*

(5.31) $\|u^\epsilon - u\| \leq C\,\epsilon^{1/2}$

Proof: From (5.29) we deduce the relations

$$a(u^\epsilon-\psi_2, (u^\epsilon-\psi_2)^+) + \frac{1}{\epsilon}|(u^\epsilon-\psi_2)^+|^2 = (f-A\psi_2, (u^\epsilon-\psi_2)^+) - \int \frac{\partial\psi_2}{\partial\nu_A}(u^\epsilon-\psi_2)^+ \, d\Gamma$$

$$a(u^\epsilon-\psi_1, (u^\epsilon-\psi_1)^-) - \frac{1}{\epsilon}|(u^\epsilon-\psi_1)^-|^2 = (f-A\psi_1, (u^\epsilon-\psi_1)^-) - \int_\Gamma \frac{\partial\psi_1}{\partial\nu_A}(u^\epsilon-\psi_1)^- \, d\Gamma$$

and therefore by our assumptions,

$$|(u^\epsilon-\psi_2)^+| \leq |f-A\psi_2|\,\epsilon$$

$$\|(u^\epsilon-\psi_2)^+\| \leq |f-A\psi_2|\frac{\sqrt{\epsilon}}{\sqrt{\gamma}}$$

$$|(u^\epsilon-\psi_1)^-| \leq |f-A\psi_1|\,\epsilon$$

$$\|(u^\epsilon-\psi_1)^-\| \leq |f-A\psi_1|\frac{\sqrt{\epsilon}}{\sqrt{\gamma}}.$$

The proof is then concluded as in the case of the Dirichlet problem (cf., B.L., loc. cit., Chap. 3). □

Remark 5.3

If the assumptions $\frac{\partial\psi_2}{\partial\nu_A} \geq 0$, $\frac{\partial\psi_1}{\partial\nu_A} \leq 0$, are omitted, then we can prove the estimate:

$$\|u^\epsilon - u\| \leq C\,\epsilon^{1/4}$$

We can also consider the penalised scheme

$$a(u^\epsilon,v) + \frac{1}{\epsilon}((u^\epsilon-\psi_2)^+,v) - \frac{1}{\epsilon}((u^\epsilon-\psi_1)^-,v) + \frac{1}{\epsilon}((u^\epsilon-\psi_2)^+, v)_\Gamma - \frac{1}{\epsilon}((u^\epsilon-\psi_1)^-, v)_\Gamma = (f,v)$$

for which it is easily proved that

$$\|u^\epsilon - u\| \leq C\,\epsilon^{1/2}. \quad \square$$

We now give a Levy-Stampacchia type of estimate. This is inspired by the results of J.L. JOLY [1] and U. MOSCO [3].

THEOREM 5.8: *Take the assumptions of* (5.1), (5.6), *and also*

(5.32) $\psi \in H^1(\mathcal{O}) = V$, $f \in L^2(\mathcal{O})$

(5.33) $a(\psi,v) = \langle L^+, v\rangle - \langle L^-, v\rangle \quad , \forall v \in V$

where

(5.34) $L^+, L^- \in V'$ *are* ≥ 0. (1)

Then if u *is the solution of the V.I.* (5.4), *we have*

(5.35) $$(f,v) \geq a(u,v) \geq (f,v) - \sup_{\substack{0 \leq w \leq v \\ w \in V}} [(f,w) - a(\psi,w)].$$

Proof First notice that the mapping

$$v \to L(v) = \sup_{0 \leq w \leq v^+} [(f,w) - a(\psi,w)] - \sup_{0 \leq w \leq v^-} [(f,w) - a(\psi,w)]$$

is an element of V'. In fact, first consider the case where $v \geq 0$. Then

$$L(v) = \sup_{0 \leq w \leq v} [(f,w) - a(\psi,w)].$$

If $v = v_1 + v_2$, $v_1, v_2 \in V \geq 0$, we notice that every element $w \in V$ such that $0 \leq w \leq v$ can be written in the form

$$w = w_1 + w_2$$

with

$$w_1 = v_1 - (v_1 - w)^+,$$

$$w_2 = (v_1 - w)^-,$$

and

$$w_1, w_2 \in V, \qquad 0 \leq w_1 \leq v_1, \quad 0 \leq w_2 \leq v_2,$$

from which it is easily deduced that

$$L(v_1 + v_2) = L(v_1) + L(v_2);$$

furthermore,

$$\begin{aligned} L(v) &= \sup_{0 \leq w \leq v} [(f,w) - \langle L^+, w\rangle + \langle L^-, w\rangle] \\ &\leq \sup_{0 \leq w \leq v} [(f^+,w) + \langle L^-, w\rangle] \\ &= (f^+,v) + \langle L^-, v\rangle \end{aligned}$$

and therefore

---

(1) i.e. $\langle L^+, v\rangle \geq 0$ if $v \geq 0$.

$$\|L(v)\| \leq M\|v\|, \text{ if } v \geq 0.$$

In the general case we notice that by definition

$$L(v) = L(v^+) - L(v^-).$$

Therefore,

$$L(v_1 + v_2) = L((v_1 + v_2)^+) - L((v_1 + v_2)^-).$$

Since

$$(v_1 + v_2)^+ + v_1^- + v_2^- = v_1^+ + v_2^+ + (v_1 + v_2)^-,$$

we see that

$$\begin{aligned} L(v_1 + v_2) &= L\left((v_1 + v_2)^+ + v_1^- + v_2^-\right) - L(v_1^-) - L(v_2^-) \\ &\qquad\qquad - L\left((v_1 + v_2)^-\right) \\ &= L(v_1^+) + L(v_2^+) - L(v_1^-) - L(v_2^-) \\ &= L(v_1) + L(v_2), \end{aligned}$$

and

$$|L(v)| \leq M(\|v^+\| + \|v^-\|) \leq 2M\|v\|.$$

Hence $L \in V'$.

We then define the V.I.

$$\text{(5.36)} \qquad \begin{aligned} &a(z, v-z) \geq (f, v-z) - L(v-z) \\ &\quad \forall\, v \geq u,\ z \geq u,\quad v \in V,\ z \in V. \end{aligned}$$

The solution z of (5.36) is uniquely defined.

If v is given as belonging to V and is ≥ 0, we immediately deduce from (5.36) that

$$a(z,v) \geq (f,v) - L(v).$$

We are going to show that $z = u$, which will immediately imply the inequalities on the right hand side of (5.35). First of all let us show that

$$z \leq \psi.$$

By assumption (5.34), $(z - \psi)^+ \in V$. Therefore $z - (z - \psi)^+ \in V$, and furthermore

$$z - (z - \psi)^+ \geq u,$$

since

$$z - u \geq (z - \psi)^+,$$

because

$$z - u \geqq z - \psi,$$

and $z - u \geqq 0$. We can take

$$v = z - (z - \psi)^+$$

in (5.36), hence, since

$$v - z = - (z - \psi)^+ \leqq 0,$$

it follows that

$$-a(z,(z-\psi)^+) \geqq -(f,(z-\psi)^+) + \sup_{0 \leqq w \leqq (z-\psi)^+} [(f,w) - a(\psi, w)]$$
$$\geqq -(f,(z-\psi)^+) + (f,(z-\psi)^+) - a(\psi,(z-\psi)^+)$$

hence

$$a(z - \psi,(z - \psi)^+) \leqq 0,$$

and therefore

$$(z - \psi)^+ = 0.$$

Since $z \leqq \psi$, we can take $v = z$ in the V.I. (5.4), therefore it follows that

$$a(u,z - u) \geqq (f,z - u). \tag{5.37}$$

Taking $v = u$ in (5.36) we obtain

$$a(z, u-z) \geqq (f,u- z) + \sup_{0 \leqq w \leqq z-u} [(f,w) - a(\psi,w)] \tag{5.38}$$
$$\geqq (f, u-z)$$

and adding (5.37),(5.38) we obtain

$$a(u - z,z - u) \geqq 0$$

therefore $u = z$. The inequality on the left hand side in (5.35) is immediately obtained from the V.I. (5.4). □

COROLLARY 5.4: *Take the assumptions* (5.1),(5.6),(5.9),(5.10) *and* $f \in L^2(\mathcal{O})$. *Then the solution* u *of the V.I.* (5.4) *satisfies*

$$f \geqq Au \geqq A\psi \wedge f \qquad \text{a.e. in } \mathcal{O}\ ^{(1)} \tag{5.39}$$

$$0 \geqq \frac{\partial u}{\partial \nu_A} \geqq -\left(\frac{\partial \psi}{\partial \nu_A}\right)^- \qquad \text{a.e. on } \Gamma. \tag{5.40}$$

In particular, if $\left.\frac{\partial \psi}{\partial \nu_A}\right|_\Gamma \geqq 0$ *then* $\left.\frac{\partial u}{\partial \nu_A}\right|_\Gamma = 0$.

---

(1) Here we write $A + a_0$ as A in order to abbreviate the written forms.

Proof: We approximate u by the penalty scheme described in Remark 5.2, say

$$Au^{\varepsilon} + \frac{1}{\varepsilon}(u^{\varepsilon}-\psi)^{+} = f$$

$$\frac{\partial u^{\varepsilon}}{\partial \nu_A} + \frac{1}{\varepsilon}(u^{\varepsilon}-\psi)^{+} = 0 .$$

In Remark 5.2 we have seen that under the Corollary's conditions Au^{ε} stays in a bounded set of $L^2(\mathcal{O})$, and $\frac{\partial u^{\varepsilon}}{\partial \nu_A}$ stays in a bounded set of $L^2(\Gamma)$. On passing to the limit we see that

$$(5.41) \qquad Au \in L^2(\mathcal{O}), \quad \frac{\partial u}{\partial \nu_A} \in L^2(\Gamma) .$$

The right hand inequality of (5.35) gives

$$a(u,v) \geqq (f,v) - \sup_{\substack{0 \leqq w \leqq v \\ w \in V}} \left[(f,w) - (A\psi, w) - \left(\frac{\partial \psi}{\partial \nu_A}, w\right)_{L^2(\Gamma)}\right]$$

$$\geqq (f,v) - \sup_{\substack{0 \leqq w \leqq v \\ w \in V}} \left[((f-A\psi)^{+}, w) + \left(\frac{\partial \psi^{-}}{\partial \nu_A}, w\right)_{L^2(\Gamma)}\right]$$

$$= (f,v) - ((f-A\psi)^{+}, v) - \left(\frac{\partial \psi^{-}}{\partial \nu_A}, v\right)_{L^2(\Gamma)}$$

$$= (f \wedge A\psi, v) - \left(\frac{\partial \psi^{-}}{\partial \nu_A}, v\right)_{L^2(\Gamma)} .$$

But thanks to (5.41) Green's Formula is applied, and therefore

$$a(u,v) = (Au, v) + \left(\frac{\partial u}{\partial \nu_A}, v\right)_{L^2(\Gamma)} .$$

Collecting the results together, we obtain

$$(Au - f \wedge A\psi, v)_{L^2(\mathcal{O})} + \left(\frac{\partial u}{\partial \nu_A} + \frac{\partial \psi^{-}}{\partial \nu_A}, v\right)_{L^2(\Gamma)} \geqq 0 .$$

Taking test functions v vanishing on Γ we deduce the inequality (5.39). Taking functions supported in a neighbourhood of Γ and making this support tend to Γ, we deduce the right hand inequality of (5.40). As for the lef-hand inequalities of (5.39),(5.40), they are deduced in the same way as the left hand inequality of (5.35). □

Remark 5.4

The coercivity assumption (5.6) is not, properly speaking, indispensible for obtaining the results of Theorem 5.8 and Corollary 5.4 after the moment when we are assured of the existence and uniqueness of the solution of the V.I. (5.4), for example under the conditions of Theorem 5.2. It is sufficient to work by regularisation of the form a(u,v). □

Remark 5.5

We have already seen in Corollary 5.1 that under the assumptions of Theorem 5.4, therefore in particular if

$$\frac{\partial\psi}{\partial\nu_A}\Big|_\Gamma \geq 0 ,$$

we then have

$$\frac{\partial u}{\partial\nu_A}\Big|_\Gamma = 0 \qquad \Box$$

COROLLARY 5.5: *Under the conditions of Corollary 5.4, and if in addition*

(5.42) $f \in L^p(\mathcal{O}), \quad A\,\psi \wedge f \in L^p(\mathcal{O}), \quad p \geq 2$

then

(5.43) $Au \in L^p\,(\mathcal{O})$.

Furthermore, if

$$\frac{\partial\psi}{\partial\nu_A}\Big|_\Gamma \geq 0 \ \textit{and}\ a_{ij} \in W^{1,\infty} ,$$

then

(5.44) $u \in W^{2,p}\,(\mathcal{O})$.

Proof: This is an immediate consequence of Corollary 5.4 and from the regularity theorems for the Neumann problem. □

Remark 5.6

The Levy-Stampacchia estimate (5.35) can be decoupled into an estimate inside the domain and an estimate on the boundary, as in Corollary 5.4, without making the regularity assumptions of this Corollary. But it is technically complex, for we no longer have available the usual Green's formula. We have to use a Green's formula related to the properties of order. For the details we refer to the works of B. HANOUZET - J.L. JOLY [2]. □

Remark 5.7

We can treat the V.I.'s with oblique derivatives, drawing on what was done in Subsection 4.1 for the equations. □

5.4 *A New Approximation and Its Consequences*

Here we are going to give a *new approximation* of the V.I.'s studied in this Chapter. This approximation(1), in comparison with the earlier ones, displays numerous advantages, which we shall bring into evidence.

In order to avoid a certain number of minus signs, we here consider the problem

(5.45) $Au - f \geq 0,\ u - \psi \geq 0, \quad (Au - f)(u - \psi) = 0 \quad \text{in} \ \ \Omega,$

(5.46) $\dfrac{\partial u}{\partial\nu_A} \geq 0,\ u - \psi \geq 0, \ \dfrac{\partial u}{\partial\nu_A}(u - \psi) = 0 \quad \text{on} \ \ \Gamma.$

First of all we are going to assume that

(1) Obtained in a discussion with M. Brauner; cf., also, M. BRAUNER and B. NICOLENKO [1]. We shall describe this method as *homographic penalisation.*

(5.47) $\left| \begin{array}{l} f \in L^2(\mathcal{O}),\ \psi \in H^2(\mathcal{O}),\ (f - A\psi)^- \in L^\infty(\mathcal{O}), \\ (\frac{\partial\psi}{\partial\nu_A})^+ \in L^\infty(\Gamma) \ . \end{array} \right.$

We choose $g \in L^\infty(\mathcal{O})$, $k \in L^\infty(\Gamma)$ with

(5.48) $\left| \begin{array}{l} g \geqq 0,\ \ k \geqq 0, \\ f + g - A\psi \geqq 0,\ \ k - \frac{\partial\psi}{\partial\nu_A} \geqq 0 \ \text{ on } \ \Gamma \end{array} \right.$

(choices possible by the assumptions (5.47)).

For each $\varepsilon > 0$ we denote by u_ε the solution(1) of

(5.49) $$A u_\varepsilon + g \frac{(u_\varepsilon - \psi)}{u_\varepsilon - \psi + \varepsilon} = f + g \quad \text{in} \quad \mathcal{O}$$

(5.50) $$\frac{\partial u_\varepsilon}{\partial\nu_A} + \frac{k(u_\varepsilon - \psi)}{u_\varepsilon - \psi + \varepsilon} = k \quad \text{on} \quad \Gamma .$$

We prove:

LEMMA 5.1: *The problem* (5.49),(5.50) *admits an unique solution such that* $u_\varepsilon \in H^1(\mathcal{O})$ *and such that*

(5.51) $\psi \leqq u_\varepsilon .$ (2)

Proof: (1): We consider a priori the system

(5.52) $\left| \begin{array}{l} A w + \frac{g(w-\psi)^+}{w - \psi + \varepsilon} = f + g , \\ \frac{\partial w}{\partial\nu_A} + k \frac{(w-\psi)^+}{w - \psi + \varepsilon} = k \quad \text{on} \quad \Gamma . \end{array} \right.$

This system admits an unique solution, by the theory of *monotonic operators*.

(2): Let us show that

(5.53) $w \geqq \psi .$

(From this it will follow that (5.52) coincides with (5.49),(5.50) by taking $u_\varepsilon = w$). In fact, multiplying (5.52) by $v \in H^1(\mathcal{O})$ we obtain

(5.54) $a(w - \psi , v) + (g \frac{(w-\psi)^+}{w-\psi+\varepsilon}, v) + (\frac{k(w-\psi)^+}{w-\psi+\varepsilon}, v)_\Gamma =$ (Contd)

---

(1) We verify here (by Lemma 5.1) that this (non-linear) problem admits a unique solution with (5.38).

(2) We shall see that u_ε is *an approximation* of u; (5.38) shows that this approximation preserves the constraints.

(Contd) $$=(f+g-A\psi,v)+(k-\frac{\partial\psi}{\partial\nu_A},v)_\Gamma .$$

Taking $v = (w-\psi)^-$, we obtain

$$a((w-\psi)^-,(w-\psi)^-)+(f+g-A\psi,(w-\psi)^-)+ + (k-\frac{\partial\psi}{\partial\nu_A},(w-\psi)^-)_\Gamma = 0 ; \tag{5.55}$$

by (5.48) the last two terms are ≥ 0, hence (5.55) implies that $(w-\psi)^- = 0$, whence (5.53). □

We are now going to prove:

THEOREM 5.9: *Assume that* (5.47),(5.48) *hold. Let* u_ε *(resp.* u*) be the solution of* (5.49),(5.50) *(resp. of the V.I.* (5.45),(5.46)*). Then as* $\varepsilon \to 0$ *we have* $u_\varepsilon \to u$ *in* $H^1(\mathcal{O})$ *and, more exactly*

$$\|u_\varepsilon - u\| \leq C\,\varepsilon^{1/2} . \tag{5.56}$$

Remark 5.8

As we have already emphasised, the approximation of u by u_ε *preserves the constraints.*

Proof: We can rewrite (5.49),(5.50) in the form

$$A u_\varepsilon - \frac{\varepsilon g}{u_\varepsilon-\psi+\varepsilon} = f ,$$

$$\frac{\partial u_\varepsilon}{\partial\nu_A} - \frac{\varepsilon k}{u_\varepsilon-\psi+\varepsilon} = 0$$

hence

$$a(u_\varepsilon,v) = (f,v) + (\frac{\varepsilon g}{u_\varepsilon-\psi+\varepsilon},v) + (\frac{\varepsilon k}{u_\varepsilon-\psi+\varepsilon},v)_\Gamma, \forall v\in H^1(\mathcal{O}) . \tag{5.57}$$

Moreover,

$$a(u,v-u) \geq (f,v-u) \quad \forall v\in H^1(\mathcal{O}),\ v\geq\psi . \tag{5.58}$$

Now take $v = u_\varepsilon$ in (5.58), which is permissible by (5.53), and $v = -(u_\varepsilon - u)$ in (5.57), and we obtain

$$-a(u_\varepsilon-u,u_\varepsilon-u) \geq +(\frac{\varepsilon g}{u_\varepsilon-\psi+\varepsilon},u-u_\varepsilon)+(\frac{\varepsilon k}{u_\varepsilon-\psi+\varepsilon},u-u_\varepsilon)_\Gamma$$

$$\geq -(\frac{\varepsilon g}{u_\varepsilon-\psi+\varepsilon},u_\varepsilon-\psi)-(\frac{\varepsilon k}{u_\varepsilon-\psi+\varepsilon},u_\varepsilon-\psi)_\Gamma$$

therefore

(5.59) $$\alpha \| u_\varepsilon - u \|^2 \leq \varepsilon \left(\frac{u_\varepsilon - \psi}{u_\varepsilon - \psi + \varepsilon}, g \right) + \varepsilon \left(\frac{u_\varepsilon - \psi}{u_\varepsilon - \psi + \varepsilon}, k \right)_\Gamma ,$$

where α *is the coercivity constant of* $a(u,v)$. Since

$$0 \leq \frac{u_\varepsilon - \psi}{u_\varepsilon - \psi + \varepsilon} \leq 1 ,$$

we deduce from this that

(5.60) $$\alpha \| u_\varepsilon - u \|^2 \leq \varepsilon \left[\int_{\mathcal{O}} g \, dx + \int_\Gamma k \, d\Gamma \right]$$

whence (5.56). □

Remark 5.9

The better choice of g and k with (5.58) is

(5.61) $$g = (f - A\psi)^- , \; k = \left(\frac{\partial \psi}{\partial \nu_A}\right)^+ ,$$

hence

(5.62) $$\alpha \| u_\varepsilon - u \|^2 \leq \varepsilon \left[\int_{\mathcal{O}} (f - A\psi)^- \, dx + \int_\Gamma \left(\frac{\partial \psi}{\partial \nu_A}\right)^+ d\Gamma \right] . \qquad \square$$

From the approximation introduced in this Section we easily deduce the Levy-Stampacchia result established in Corollary 5.4.

Remark 5.10

Naturally, everything that has just been said is valid for the *Dirichlet boundary conditions*. We then consider (5.49) with the condition $u = 0$ on Γ. The proof which follows is valid and gives a simpler proof of Theorem 1.21 in Vol. 1. □

Proof of the Levy-Stampacchia Inequality

Notice that

$$0 \leq \frac{\varepsilon g}{u_\varepsilon - \psi + \varepsilon} \leq g , \quad 0 \leq \frac{\varepsilon k}{u_\varepsilon - \psi + \varepsilon} \leq k$$

so that

$$f \leq Au_\varepsilon \leq f + g , \quad 0 \leq \frac{\partial u_\varepsilon}{\partial \nu_A} \leq k .$$

Choosing g and k as in (5.61), we obtain

(5.63) $$f \leq Au_\varepsilon \leq \sup (f , A\psi)$$

(5.64) $$0 \leq \frac{\partial u_\varepsilon}{\partial \nu_A} \leq \left(\frac{\partial \psi}{\partial \nu_A}\right)^+ .$$

It follows from (5.63) that Au_ε is bounded in $L^2(\mathcal{O})$, therefore $Au_\varepsilon \to Au$ weakly in

$L^2(\mathcal{O})$ (for example) and therefore (since $u_\varepsilon \to u$) in $H^1(\mathcal{O})$

$$\frac{\partial u_\varepsilon}{\partial \nu_A} \to \frac{\partial u}{\partial \nu_A} \quad \text{in} \quad H^{-1/2}(\Gamma),$$

whence the result

$$(5.65) \qquad \begin{cases} f \geqq A u \leqq \sup (f, A\psi) \\ (\frac{\partial \psi}{\partial \nu_A})^- \leqq \frac{\partial u}{\partial \nu_A} \leqq 0 \end{cases}$$

Remark 5.11

(5.65) will hold whenever $\psi \in H^2(\mathcal{O})$, by approximation and passage to the limit.□

Remark 5.12

It can be proved that u_ε decreases as $\varepsilon \downarrow 0$. □

6. EVOLUTIONARY VARIATIONAL INEQUALITIES WITH NEUMANN BOUNDARY CONDITIONS

Method of Procedure

As in the preceding Section we shall often refer to the results established for the Dirichlet problem, and we shall be content with bringing into evidence the results specific to the Neumann problem.

6.1 *Formulation of the V.I.'s*

Given a bounded open set $\mathcal{O}$ of R^n, we set

$$Q = \mathcal{O} \times]0,T[\ , \quad \Sigma = \Gamma \times]0,T[\ , \quad V = H^1(\mathcal{O})\ , H = L^2(\mathcal{O})\ .$$

Consider the functions a_{ij}, a_i, a_o on Q such that

$$(6.1) \qquad a_{ij}\ , a_i\ , a_o \in L^\infty(Q)$$

$$(6.2) \qquad \sum_{i,j} a_{ij}\, \xi_i\, \xi_j \geqq \alpha\, |\xi|^2\ , \ \forall \xi \in R^n .$$

We shall also have to distinguish the cases

$$(6.3) \qquad a_{ij} \neq a_{ji}, \quad \textit{the non-symmetric case},$$

$$(6.4) \qquad a_{ij} = a_{ji} \quad \textit{the symmetric case}.$$

For u,v $\in$ V we set

$$(6.5) \qquad a(t\,;u,v) = \sum \int_{\mathcal{O}} a_{ij}(x,t)\, \frac{\partial u}{\partial x_j}\, \frac{\partial v}{\partial x_i}\, dx + \sum_i \int_{\mathcal{O}} a_i(x,t)\, \frac{\partial u}{\partial x_i}\, v\, dx + \int_{\mathcal{O}} a_o(x,t)\, u\, v\, dx$$

(6.6) $$A(t) + a_o = -\sum_{ij} a_{ij} \frac{\partial}{\partial x_j} + \sum_i a_i \frac{\partial}{\partial x_i} + a_o .$$

Furthermore, we are given

(6.7) $$f \in L^2 (0,T) ; H)$$

(6.8) $$\psi \in L^2 (0,T) ; H)$$

(6.9) $$\bar{u} \in H.$$

The definition of the strong and weak V.I.'s, such as was given in B.L., loc. cit., Chap. 3, are used again without modification. For the convenience of the reader we shall restate them here.

Strong V.I.

(6.10) $$u \in L^2(0,T;V) , \frac{\partial u}{\partial t} \in L^2(0,T;H)$$

(6.11) $$\left| \begin{array}{l} -(\frac{\partial u}{\partial t}, v-u(t)) + a(t;u(t), v-u(t)) \geqq (f(t), v-u(t)) , \text{ p.p.t} \\ \forall v \in V \text{ tel que } v(x) \leqq \psi(x,t) \text{ a.e. in } Q \end{array} \right.$$

(6.12) $$u(x,t) \leqq \psi(x,t) \text{ a.e. in } Q$$

(6.13) $$u(x,T) = \bar{u}(x).$$

Weak V.I.

We introduce the set

(6.14) $$\mathcal{K} = \{v \mid v \in L^2(0,T;V), \frac{\partial v}{\partial t} \in L^2(0,T;V'), \; v(x,t) \leqq \psi(x,t) \text{ a.e. in } Q\}$$

and assume that

(6.15) $$\mathcal{K} \neq \emptyset.$$

We then look for a u such that

(6.16) $$u \in L^2(0,T;V)$$

(6.17) $$u \leqq \psi \text{ a.e. on } Q$$

(6.18) $$\left| \begin{array}{l} \int_0^T [-(\frac{\partial v}{\partial t}, v-u) + a(t;u,v-u) - (f,v-u)]\, dt + \frac{1}{2} |v(T) - \bar{u}|^2 \geqq 0 \\ \forall v \in \mathcal{K}. \end{array} \right.$$

We define the penalised problem as follows: find u^ε such that

(6.19) $$\left| \begin{array}{l} -(\frac{\partial u^\epsilon}{\partial t}, v) + a(t;u^\epsilon,v) + \frac{1}{\epsilon}((u^\epsilon-\psi)^+, v) = (f,v) , \forall v \in V \\ u^\epsilon \in L^2(0,T;V) , \; u^\epsilon(T) = \bar{u} . \end{array} \right.$$

THEOREM 6.1: *Assuming* (6.1),(6.2),(6.7),(6.9) *and* (6.15), *there exists one and only one solution of* (6.19) *such that*

$$\frac{\partial u^\varepsilon}{\partial t} \in L^2(0,T;V').$$

Proof: This is identical to the Dirichlet case. □

In connection with the existence and uniqueness of the strong V.I., we can state the following theorems.

THEOREM 6.2: *Using the assumptions of Theorem 6.1, together with*

$$(6.20) \qquad \frac{\partial a_{ij}}{\partial t}, \frac{\partial a_i}{\partial t}, \frac{\partial a_o}{\partial t} \in L^\infty(Q)$$

$$(6.21) \qquad \frac{\partial f}{\partial t} \in L^2(0,T;H)$$

$$(6.22) \qquad \psi, \frac{\partial \psi}{\partial t} \in L^2(0,T;V), \quad \frac{\partial^2 \psi}{\partial t} \in L^2(0,T;H)$$

$$(6.23) \qquad \bar{u} \in H^2(\mathcal{O}), \quad \frac{\partial \bar{u}}{\partial \nu_{A(T)}}\Big|_\Gamma = 0, \quad \bar{u} \leqq \psi(T)$$

then the 'strong' problem (6.10),...,(6.13) *has an unique solution such that*

$$(6.24) \qquad \frac{\partial u}{\partial t} \in L^2(0,T;V) \cap L^\infty(0,T;H).$$

Proof: This is analogous to the Dirichlet case. □

In the case of the symmetric principal part we can state:

THEOREM 6.3: *Using the assumptions of Theorem 6.1, together with* (6.4),

$$(6.25) \qquad \frac{\partial a_{ij}}{\partial t} \in L^\infty(Q)$$

$$(6.26) \qquad \psi, \frac{\partial \psi}{\partial t} \in L^2(0,T;V)$$

$$(6.27) \qquad \bar{u} \in V, \quad \bar{u} \leqq \psi(T)$$

then the 'strong' problem (6.10),...,(6.13) *has an unique solution such that*

$$(6.28) \qquad u \in L^\infty(0,T;V).$$

Proof: This is analogous to the Dirichlet case. □

6.2 *Estimation of the Penalisation Error*

THEOREM 6.4: *Taking the assumptions of Theorem 6.2, with the addition of*

$$(6.29) \qquad \psi \in L^2(0,T;H^2(\mathcal{O})); \quad \frac{\partial \psi}{\partial \nu_{A(t)}}\Big|_\Sigma \geqq 0;$$

(6.29)' $\mathcal{O}$ a bounded open set with Lipschitz boundary,

then (if u *denotes the solution of* (6.10),...,(6.13) *and* u^ε *the solution of* (6.19)*) we have*

(6.30) $$\|u-u^{\epsilon}\|_{L^2(0,T;V)} + \|u-u^{\epsilon}\|_{L^{\infty}(0,T;H)} \leq C\sqrt{\epsilon}\,.$$

Proof: We take $v = (u^{\varepsilon} - \psi)^+$ in (6.19) and obtain

$$-\left(\frac{\partial}{\partial t}(u^{\epsilon}-\psi),\ (u^{\epsilon}-\psi)^+\right) + a(t\,;u^{\epsilon}-\psi,\ (u^{\epsilon}-\psi)^+) + \frac{1}{\epsilon}\,|(u^{\epsilon}-\psi)^+|^2 =$$
$$= (f(t),\ (u^{\epsilon}-\psi)^+) + \left(\frac{\partial\psi}{\partial t},\ (u^{\epsilon}-\psi)^+\right) - ((A(t)+a_0)\,\psi,\ (u^{\epsilon}-\psi)^+)$$
$$-\int_{\Gamma}\frac{\partial\psi}{\partial\nu_{A(t)}}(u^{\epsilon}-\psi)^+\,d\Gamma$$

from which we deduce[1]

$$\frac{1}{2}\,|(u^{\epsilon}-\psi)^+(t)|^2 + \alpha\int_t^T \|(u^{\epsilon}-\psi)^+\|^2 ds + \frac{1}{\epsilon}\int_t^T |(u^{\epsilon}-\psi)^+|^2\,ds \leq$$
$$\leq \left\|f + \frac{\partial\psi}{\partial t} - (A(t)+a_0)\,\psi\right\|_{L^2(Q)}\int_t^T |(u^{\epsilon}-\psi)^+|^2 ds + \lambda\int_t^T |(u^{\epsilon}-\psi)^+|^2 ds$$

and therefore

$$\|(u^{\epsilon}-\psi)^+\|_{L^2(Q)} \leq \epsilon\left\|f + \frac{\partial\psi}{\partial t} - (A(t)+a_0)\psi\right\|_{L^2(Q)}\frac{1}{1-\lambda\epsilon}$$
$$\|(u^{\epsilon}-\psi)^+\|_{L^2(0,T;V)} \leq \frac{\sqrt{\epsilon}}{\sqrt{\alpha}}\left\|f + \frac{\partial\psi}{\partial t} - (A(t)+a_0)\,\psi\right\|_{L^2(Q)}\frac{1}{\sqrt{1-\lambda\epsilon}}$$
$$\|(u^{\epsilon}-\psi)^+\|_{L^{\infty}(0,T;H)} \leq \sqrt{2\epsilon}\left\|f + \frac{\partial\psi}{\partial t} - (A(t)+a_0)\,\psi\right\|_{L^2(Q)}\frac{1}{\sqrt{1-\lambda\epsilon}}\,.$$

The proof is completed as in the case of the Dirichlet problem (cf. B.L., loc. cit., Chap. 3). □

Remark 6.1

As in the stationary case, the assumption

$$\frac{\partial\psi}{\partial\nu_{A(t)}}\Big|_{\Sigma} \geq 0$$

is not indispensible for obtaining an estimate of the penalisation error. We then simply obtain the estimate

$$\|u^{\epsilon}-u\|_{L^2(0,T;V)} + \|u^{\epsilon}-u\|_{L^{\infty}(0,T;H)} \leq C\,\epsilon^{1/4}\,.$$

We can also obtain the following penalisation scheme

$$-\left(\frac{\partial u^{\epsilon}}{\partial t},\ v\right) + a(t,u^{\epsilon},v) + \frac{1}{\epsilon}\,((u^{\epsilon}-\psi)^+,v) +$$
$$+\frac{1}{\epsilon}\,((u^{\epsilon}-\psi)^+,v)_{\Gamma} = (f,v)$$

---

(1) We have used the property
$$a(t;u,u) + \lambda|u|^2 \geq \alpha\|u\|^2.$$

and then prove the estimate (6.30). □

6.3 *Maximum Weak Solution*(1)

We are content with restating the result of F. MIGNOT, J.P. PUEL [1],[2] which we proved in the Dirichlet case (in fact the proof is done in a variational framework, and therefore is immediately valid for the Neumann boundary problem).

THEOREM 6.5: *Under the assumptions of Theorem 6.1 there exists a maximum element in the set of weak solutions of* (6.16),(6.17),(6.18). □

We may also recall the important properties of the maximum solution. It is convenient to use the following notation: $u(f,\psi,\bar{u})$ will denote the maximum solution of (6.16),(6.17),(6.18).

THEOREM 6.6: *Take the assumptions of Theorem 6.1. Let* $(f,\psi,\bar{u})$ *and* $(\hat{f},\hat{\psi},\hat{\bar{u}})$ *be two sets of data satisfying the assumptions and which are such that*

$$(6.31)\qquad f \leqq \hat{f}\ ,\quad \psi \leqq \hat{\psi}\ ,\quad \bar{u} \leqq \hat{\bar{u}}$$

then

$$(6.32)\qquad u\,(f,\psi,\bar{u}) \leqq u\,(\hat{f},\ \hat{\psi},\ \hat{\bar{u}}) \qquad \square$$

THEOREM 6.7: *Taking the assumptions of Theorem 6.1 and assuming that we take the case* $a_0 \geqq 0$ *and* $a(t;v,v) \geqq \alpha\|v\|^2$. *Let* $\hat{\psi}$ *be such that* $(f,\hat{\psi},u)$ *satisfies these assumptions also, and that* $\psi - \hat{\psi} \in L^\infty(Q)$. *If we write* $u = u(f,\psi,\bar{u})$ *and* $\hat{u} = u(f,\hat{\psi},\bar{u})$, *then*

$$(6.33)\qquad \|u - \hat{u}\|_{L^\infty(Q)} \leqq \|\psi - \hat{\psi}\|_{L^\infty(Q)} \qquad \square$$

6.4 *Regularity*

THEOREM 6.8: *Take the assumptions of Theorem 6.4. Then the solution of the V.I. satisfies*

$$(6.34)\qquad A(t)u \in L^2(Q).$$

Proof: We have seen that $\left(\frac{u^\varepsilon - \psi}{\varepsilon}\right)^+$ remains in a bounded set of $L^2(Q)$. Therefore by (6.19) we have

$$\left\| -\frac{\partial u^\varepsilon}{\partial t} + A(t)\,u^\varepsilon \right\|_{L^2(Q)} \leqq C$$

from which it follows that $-\frac{\partial u}{\partial t} + A(t)u \in L^2(Q)$. Since we already know that $\frac{\partial u}{\partial t} \in L^2(Q)$, we may deduce (6.34). □

COROLLARY 6.1: *Under the assumptions of Theorem 6.4, and, moreover, that* $\mathcal{O}$ *is a bounded open set of class* C^2, *and*

---

(1) For other aspects (use of the theory of the *parabolic potential* allowing us to 'replace' *measure* by *capacity*), cf. M. PIERRE [1].

(6.35) $\frac{\partial}{\partial x_k} a_{ij} \in L^\infty(Q)$, $a_{ij} \in C^\circ(\bar{Q})$

then we have the additional property

(6.36) $u \in L^2(0,T;H^2(\mathcal{O}))$.

Proof: From the proof of Theorem 6.4 we have

$$-\left(\frac{\partial u}{\partial t}, v\right) + a(t\,;u(t),v) = (f(t)-h(t),v)$$

where $h(t) \in L^2(0,T;H)$ (weak limit of $\left(\frac{u^\varepsilon - \psi}{\varepsilon}\right)^+$). The property (6.36) then follows from the regularity theorems for the parabolic problem with Neumann boundary conditions. □

THEOREM 6.9: *Under the assumptions of Theorem 6.4, and the further assumptions*

(6.37) $f \in L^p(Q)$, $\bar{u} \in L^p(\mathcal{O})$

(6.38) $\psi \in L^p(Q)$, $-\frac{\partial \psi}{\partial t} + A(t)\,\psi \in L^p(Q)$

we then have

(6.39) $-\frac{\partial u}{\partial t} + A(t)\,u \in L^p(Q)$.

COROLLARY 6.2: *Under the assumptions of Corollary 6.1, together with* (6.37), (6.38), *and, furthermore, if*

(6.40) $\bar{u} \in W^{2,p}(\mathcal{O})$; $\frac{\partial \bar{u}}{\partial x_i}\Big|_\Gamma = 0$ (1)

then

(6.41) $u \in L^p(0,T;W^{2,p}(\mathcal{O}))$, $\frac{\partial u}{\partial t} \in L^p(Q)$.

Proof: We show that u(t) is a solution of

(6.42) $$\begin{cases} -\left(\frac{\partial u}{\partial t}, v\right) + a(t\,;u(t),v) = (f(t)-h(t),v) \\ u(T) = \bar{u} \end{cases}$$

with $h(t) \in L^p(Q)$. But then $u(t) - \bar{u} = w(t)$ is a solution of

$$-\left(\frac{\partial w}{\partial t}, v\right) + a(t\,;w(t),v) = (f(t)-h(t),v) - a(t\,;\bar{u},v)$$
$$= (f(t)-h(t)-(A(t)+a_0)\,\bar{u}, v)$$
$$w(T) = 0$$

(1) Which is more than is strictly necessary.

and since

$$f(t) - h(t) - (A(t) + a_0)\, \bar{u} \in L^p(Q),$$

the regularity theorems for the Neumann problem may be applied, hence the result. □

Remark 6.2

For $p > \frac{n}{2} + 1$ the solution u of the strong V.I. is *continuous* in $\bar{Q}$.

For the maximum solution of the weak V.I. we can state the following result (the analogue of Corollary 5.3):

THEOREM 6.10: *Take the assumptions of Theorem 6.1, together with* (6.20),(6 21), (6.23),(6.29),(6.35),(6.37),(6.40), *with* $p > \frac{n}{2} + 1$, *and if*

(6.42) $\psi \in C^\circ(Q)$, $\psi = \lim \psi_j$ *in* $C^\circ(Q)$, *where* ψ_j *satisfies* (6.22),(6.29),(6.38),

then the maximum solution of the weak V.I. belongs to $C^\circ(\bar{Q})$.

Proof: This is a consequence of the property (6.33) and of the regularity results for the strong V.I., and in particular of Remark 6.1. □

6.5 *The V.I. Corresponding to Games*

We are now given two functions $\psi_1(x,t), \psi_2(x,t)$ which are measurable on Q and satisfy

(6.43) $$\psi_1 \leqq \psi_2 \quad \text{a.e.}$$

We define only the concept of strong V.I.. Look for u such that

(6.44) $$u \in L^2(0,T;V), \quad (V = H^1(\mathcal{O})), \quad \frac{\partial u}{\partial t} \in L^2(0,T;H)$$

(6.45) $$\psi_1(x,t) \leqq u(x,t) \leqq \psi_2(x,t) \quad \text{a.e.} \quad \text{in} \quad Q$$

(6.46) $$\left| \begin{array}{l} -\left(\frac{\partial u}{\partial t}, v-u\right) + a(t; u(t), v-u(t)) \geqq (f(t), v-u(t)) \ \text{t. a.e.} \\ \forall\, v \ \text{with} \ v \in V \ \text{et} \ \psi_1 \leqq v \leqq \psi_2 \quad \text{x. a.e.} \end{array} \right.$$

(6.47) $$u(x,T) = \bar{u}\,(x).$$

We define the penalised problem by

(6.48) $$\left| \begin{array}{l} -\left(\frac{\partial u^\epsilon}{\partial t}, v\right) + a(t; u^\epsilon, v) + \frac{1}{\epsilon}(u^\epsilon - \psi_2)^+ - \frac{1}{\epsilon}(u^\epsilon - \psi_1)^- = f \\ u^\epsilon(T) = \bar{u}. \end{array} \right.$$

Under the assumptions (6.1),(6.2),(6.7),(6.9) and

(6.49) $$\mathcal{K} = \{v \mid v \in L^2(0,T;V),\ \frac{\partial v}{\partial t} \in L^2(0,T;V'),$$
$$\psi_1(x,t) \leqq v(x,t) \leqq \psi_2(x,t) \quad \text{a.e.} \quad \text{in} \quad Q\} \quad \text{non-empty}$$

then the solution u^{ε} of (6.48) exists and is unique, with

$$u^{\varepsilon} \in L^2(0,T;V), \quad \frac{\partial u^{\varepsilon}}{\partial t} \in L^2(0,T;V').$$

We then have:

THEOREM 6.11: *Assume* (6.1),(6.2),(6.7),(6.9),(6.49) *and*

$$(6.50) \qquad \psi_i, \frac{\partial \psi_i}{\partial t} \in L^2(0,T;V), \quad \frac{\partial^2 \psi_i}{\partial t^2} \in L^2(Q), \quad i = 1,2$$

$$(6.51) \qquad -\frac{\partial}{\partial t}(\psi_2 - \psi_1) \geqq 0, \quad -\frac{\partial^2}{\partial t^2}(\psi_2 - \psi_1) \geqq 0 \text{ a.e.}$$

$$(6.52) \qquad \bar{u} \in H^2(\mathcal{O}), \quad \frac{\partial \bar{u}}{\partial \nu_A}\Big|_{\Gamma} = 0, \quad \psi_1(x,T) \leqq \bar{u} \leqq \psi_2(x,T) \quad \text{a.e.}$$

Then there exists one and only one solution of the problem (6.44),(6.45),(6.46),

$$(6.53) \qquad \frac{\partial u}{\partial t} \in L^2(0,T;V) \cap L^{\infty}(0,T;H).$$

Proof: This is analogous to the Dirichlet case (cf. B.L., loc. cit., Chap. 3).□

Remark 6.3

We can also consider the case of problems periodic in time, or with infinite horizon, as in the Dirichlet case.□

Remark 6.4

We can obtain an estimate on $-\frac{\partial u}{\partial t} + Au$ analogous to the stationary case (cf. Theorem 5.8).□

Let us recall the concept of weak sub-solution. We shall say that w is a *weak sub-solution* if

$$(6.54) \qquad \left| \begin{array}{l} w \in L^2(0,T;V), \ w \leqq \psi \text{ a.e.} \\ \displaystyle\int_0^T [(w, \frac{\partial \varphi}{\partial t}) + a(t;w,\varphi) - (f,\varphi)]\,dt - (\varphi(T),\bar{u}) \leqq 0 \\ \forall \varphi \in L^2(0,T;V), \ \frac{\partial \varphi}{\partial t} \in L^2(0,T;V'), \ \varphi(0) = 0, \quad \varphi \geqq 0 \end{array} \right.$$

and we have:

THEOREM 6.12: *Under the assumptions of Theorem 6.1 the maximum weak solution of* (6.16),(6.17),(6.18) *is the maximum element of the set of weak sub-solutions.* □

6.6 *Another Approximation*

We briefly indicate here how the methods introduced in Section 5.4 are extended to evolutionary problems. Let us consider the V.I.

$$(6.55) \qquad \frac{\partial u}{\partial t} + Au - f \geqq 0, \quad u - \psi \geqq 0, \quad (\frac{\partial u}{\partial t} + Au - f)(u - \psi) = 0 \qquad \text{in } Q = \Omega \times]0,T[,$$

with

(6.56) $$\frac{\partial u}{\partial \nu_A} \geq 0,\ u - \psi \geq 0,\ \frac{\partial u}{\partial \nu_A}(u-\psi) = 0 \quad \text{on} \quad \Sigma = \Gamma \times]0,T[\ ,$$

(6.57) $$u(x,o) = 0 \quad (\text{e.g.}) \quad (\psi \leq 0 \quad \text{for} \quad t = o)\ .$$

Now introduce g and k — *assumed to exist* — such that

(6.58) $$\begin{cases} g \geq 0 \quad \text{in} \quad Q\ , \quad k \geq 0 \quad \text{on} \quad \Sigma\ , \\ g + f - \frac{\partial \psi}{\partial t} - A\psi \geq 0,\ k - \frac{\partial \psi}{\partial \nu_A} \geq 0\ ,\ g,k \ \text{bounded.} \end{cases}$$

the 'better choices' are

(6.59) $$\left| \begin{array}{l} g = (f - \frac{\partial \psi}{\partial t} - A\ \psi)^-\ , \\ k = (\frac{\partial \psi}{\partial \nu_A})^+ \end{array} \right.$$

and we assume that

$$\left| \begin{array}{l} f \in L^2(Q),\ \frac{\partial \psi}{\partial t} + A\psi \in L^2(Q) \quad \text{(for example) and} \\ (f - \frac{\partial \psi}{\partial t} - A\ \psi)^- \in L^\infty(Q),\ \psi \in L^2(0,T;H^2(\Omega))\ , \\ (\frac{\partial \psi}{\partial \nu_A})^+ \in L^\infty(\Sigma)\ . \end{array} \right.$$

For each $\varepsilon > 0$, we denote by u_ε the solution of

(6.61) $$\left| \begin{array}{l} \frac{\partial u_\varepsilon}{\partial t} + A u_\varepsilon + g\ \frac{u_\varepsilon - \psi}{u_\varepsilon - \psi + \varepsilon} = f + g\ , \\ \frac{\partial u_\varepsilon}{\partial \nu_A} + k\ \frac{u_\varepsilon - \psi}{u_\varepsilon - \psi + \varepsilon} = k \quad \text{on} \quad \Sigma, \\ u_\varepsilon\ (x,o) = 0\ , \quad u_\varepsilon \geq \psi\ ; \end{array} \right.$$

to show that (6.61) has an unique solution, we introduce (as in (5.49)) the 'monotonic' evolutionary equation where $\frac{u_\varepsilon - \psi}{u_\varepsilon - \psi + \varepsilon}$ is replaced by $\frac{(u_\varepsilon - \psi)^+}{u_\varepsilon - \psi + \varepsilon}$, then we we verify that $(u_\varepsilon - \psi)^- = 0$.

Let us assume that we are in a situation where (6.55),(6.56),(6.57) has a *strong solution*. We then have

(6.62) $$\frac{1}{2}\ |u_\varepsilon(t) - u(t)|^2 + \int a(u_\varepsilon - u,\ u_\varepsilon - u)\ d\sigma \leq$$
$$\leq \varepsilon\left[\int_o^t \int_\Omega g(x,\sigma)\ dx\ d\sigma + \int_o^t \int_\Gamma k(x,\sigma)\ d\Gamma\ d\sigma\right]\ .$$

(The proof is an immediate adaptation of the stationary case). From this we deduce, in particular, that (with $H = L^2(\Omega)$, $V = H^1(\Omega)$),

$$(6.63)\quad \left| \begin{array}{l} \| u_\varepsilon - u\|_{L^\infty(0,T;H)} + \| u_\varepsilon - u\|_{L^2(0,T;V)} \leq \\ \leq C\, \varepsilon^{1/2} \left[\int_Q (f - \frac{\partial\psi}{\partial t} - A\psi)^- \, dx\, dt + \int_\Sigma (\frac{\partial\psi}{\partial\nu_A})^+ \, d\Gamma\, dt \right]. \end{array}\right.$$

By successive passages to the limit we deduce that (6.63) is true in all the cases where the second member has a meaning, u then referring to the *minimum solution* (resp., to the maximum solution, if we take the inequalities in the opposite sense in (6.55),(6.56)).

Remark 6.5

We can extend what has just been said to *oblique derivatives* (a remark, moreover, valid also for the stationary cases). □

7. OPTIMAL STOPPING TIME PROBLEMS: STATIONARY CASE

7.1 *Regular Case*

For the convenience of the reader we collect here the notations and assumptions. Hence we have

$$(7.1)\quad A + a_0 = -\sum_{ij} \frac{\partial}{\partial x_i}\, a_{ij} \frac{\partial}{\partial x_j} + \sum_i a_i \frac{\partial}{\partial x_i} + a_0$$

where we assume

$$(7.2)\quad \left| \begin{array}{l} a_{ij},\ a_i,\ a_0 \text{ measurable and bounded on } \mathcal{O};\ a_{ij} \in C^1(\bar{\mathcal{O}}) \\ \sum_{ij} a_{ij}\, \xi_i\, \xi_j \geq \alpha|\xi|^2 \ \forall\, \xi \in R^n,\quad a_0 \geq \beta > 0 \end{array}\right.$$

(7.3) $\mathcal{O}$ a bounded open set of R^n of class C^2, defined by (3.4).

$$(7.4)\quad \begin{array}{l} (p-1) \sum a_{ij}\, \xi_i\, \xi_j + \sum_i a_i\, \xi_i\, \xi_0 + a_0\, \xi_0^2 \geq 0 \quad \text{a.e. } \mathcal{O}, \\ \qquad \forall\, \xi_1 \ldots \xi_n,\ \xi_0 \in R,\quad p > n \end{array}$$

$$(7.5)\quad \left| \begin{array}{l} f \in L^p(\mathcal{O}),\quad \psi \in H^2(\mathcal{O}) \cap W^{1,p}(\mathcal{O}),\quad A\psi \in L^p(\mathcal{O}) \\ \left.\frac{\partial\psi}{\partial\nu_A}\right|_\Gamma \geq 0,\quad p > n. \end{array}\right.$$

For $u,v \in H^1(\mathcal{O})$

$$(7.6)\quad a(u,v) = \sum_{ij} \int_{\mathcal{O}} a_{ij} \frac{\partial u}{\partial x_j} \frac{\partial v}{\partial x_i}\, dx + \sum_i \int_{\mathcal{O}} a_i \frac{\partial u}{\partial x_i} v\, dx + \int_{\mathcal{O}} a_0\, u\, v\, dx$$

and consider the V.I.

$$(7.7)\quad a(u, v - u) \geq (f, v - u),\quad \forall\, v \in H^1(\mathcal{O}),\ v \leq \psi,\ u \leq \psi.$$

We also assume

(7.8) $a(v,v) \geqq \gamma \|v\|^2 \quad \forall\, v \in V$.

Under the assumptions (7.2),(7.3),(7.4),(7.5),(7.8) we can apply Theorem 5.1 as well as Corollary 5.2. We also know that there exists one and only one solution of (7.7), and we further have

$(7.8)_1$ $u \in W^{2,p}(\mathcal{O})$.

We also know (cf. Corollary 5.4) that

(7.9) $a(u,v) = (Au + a_o u, v)$

and therefore by Green's formula

(7.10) $\left.\frac{\partial u}{\partial \nu_A}\right|_{\Gamma} = 0$.

We easily deduce from (7.7) that

(7.11) $Au + a_o u \leqq f$ a.e.

Using (7.10), we can rewrite (7.7) in the form

$$(Au + a_o u - f, v - u) \geqq 0$$

taking $v + \psi$, and taking into account that

(7.12) $u \leqq \psi$

we easily obtain the relation

(7.13) $(Au + a_o u - f)(u - \psi) = 0$ a.e.

and thus u is a solution of (7.8),(7.10),(7.11),(7.12),(7.13).

Our goal is to give the probabilistic interpretation of the function u. □

Introduce $\sigma(x) \in \mathcal{L}(R^n;R^n)$ such that

(7.14) $a(x) = \frac{1}{2}\,\sigma(x)\sigma^*(x)$

and set

(7.15) $g_i(x) = \sum_j \frac{\partial a_{ij}}{\partial x_j} - a_i(x)$

(7.16) $\gamma_i(x) = \sum_j a_{ij}(x)\, n_j(x)$.

We define the solution P^x of the sub-martingale problem with respect to the operator A and the reflection direction γ, starting from the instant 0 at the point x. In other words, we set

$$\Omega = C^o([0,\infty;\bar{\mathcal{O}})), \quad y(t) = \omega(t) , \quad \mu^t = \sigma(y(s), s \leq t) .$$

We define P^x on Ω so that there exists an R^n-valued normalised Wiener process $w(t)$ which is a μ^t-martingale, and a continuous adapted process $\xi(t)$, such that

$$(7.17)\qquad \left| \begin{array}{l} y(t) = x + \int_0^t g(y(s))ds + \int_0^t \sigma(y(s))dw(s) - \int_0^t \chi_\Gamma(y(s))\gamma(y(s))d\xi(s) \\ \chi_{\mathcal{O}}(y(t))d\xi(t) = 0 \quad , \quad \forall\, t \,, \quad \text{a.s.} \end{array} \right.$$

By applying Theorem 4.1 we define P^x uniquely.

For every stopping time θ of μ^t, set

$$(7.18)\qquad J_x(\theta) = E^x \left[\int_0^\theta f\Big(y(t)\Big)\left(\exp - \int_0^t a_o(y(s))ds\right) dt + \right. \\ \left. + \psi\Big(y(\theta)\Big) \exp - \int_0^\theta a_o\Big(y(t)\Big) dt \right] .$$

We also set

$$(7.19)\qquad \hat{\theta} = \inf\{t \geqq 0 \,|u(y(t)) = \psi(y(t))\} \quad (\text{possibly } \hat{\theta} = +\infty) .$$

We can state:

THEOREM 7.1: *Under the assumptions* (7.2),(7.3),(7.4),(7.5),(7.8), *we have*

$$(7.20)\qquad u(x) = \operatorname*{Inf}_\theta J_x(\theta)$$

where u *is a solution of* (7.7). *Furthermore,* $\hat{\theta}$, *defined by* (7.19), *is an optimal stopping time, i.e.,*

$$u(x) = J_x(\hat{\theta}).$$

Proof: Since $u \in W^{2,p}(\mathcal{O})$, with $p > n$, by the Sobolev Theorem we have $u \in C^1(\bar{\mathcal{O}})$. Applying Corollary 4.1 to the function $\phi(x,z) = u(x)z$ and the process $(y(t),z(t))$, where $y(t)$ is the reflected diffusion process that is a solution of (7.17), and where z is the solution of

$$dz = - a_o(y)zdt, \qquad z(0) = 1$$

we obtain

$$(7.21)\qquad E\, u\, y(\theta\wedge T))\exp - \int_0^{\theta\wedge T} a_o(y(s))ds = u(x) + \\ + E\int_0^{\theta\wedge T} - (Au + a_o u)(y(s)\left(\exp - \int_0^t a_o(y(s))ds\right) dt - \\ - E\int_0^{\theta\wedge T} \frac{\partial u}{\partial \nu_A}\Big|_\Gamma (y(t))\left(\exp - \int_0^t a_o(y(s))ds\right) d\xi(t) .$$

By (7.10),(7.11) and (7.12), we deduce from this that

$$(7.22)\qquad u(x) \leq E\int_0^{\theta\wedge T} f(y(t))\left(\exp - \int_0^t a_o(y(s))ds\right)dt + \\ + E\,\psi(y(\theta\wedge T))\exp - \int_0^{\theta\wedge T} a_o(y(t))dt\ .$$

But

$$E\int_0^{\theta\wedge T}|f|\left(\exp - \int_0^t a_o\,ds\right)dt \leq E\int_0^{\theta\wedge T}|f|e^{-\beta t}dt \leq C\,\|f\|_{L^p}$$

(by Theorem 4.2), where C depends on neither θ nor T. We can then make T tend to $+\infty$ in the second member of (7.22), whence we deduce

$$(7.23)\qquad u(x) \leq J_x(\theta).$$

The theorem will be completely proved if we show that $\hat{\theta}$ is optimal. So we introduce the set

$$C_u = \{x\in\mathcal{O} \mid u(x) < \psi(x)\}$$

and χ_{C_u} the characteristic function of C_u. Hence

$$\chi_{C_u}(Au + a_o u - f) = 0 \quad \text{a.e. in } \mathcal{O}.$$

But then by Theorem 4.2 we deduce

$$E\int_0^{\hat{\theta}\wedge T}\chi_{C_u}(Au + a_o u - f)(y(t))\left(\exp - \int_0^t a_o(y(s))ds\right)dt = 0$$

and since for $t < \theta$ we have $\chi_{C_u}(y(t)) = 1$, we obtain

$$(7.24)\qquad E\int_0^{\hat{\theta}\wedge T}(Au + a_o u - f)(y(t))\left(\exp - \int_0^t a_o(y(s))ds\right)dt = 0\ .$$

Applying (7.21) with $\theta = \hat{\theta}$, it follows that

$$E\,u(y(\hat{\theta}\wedge T)\exp - \int_0^{\hat{\theta}\wedge T} a_o(y(t))dt = u(x) - E\int_0^{\hat{\theta}\wedge T} f(y(t))\left(\exp - \int_0^t a_o\right)dt\ .$$

We then make T tend to $+\infty$, and take into account that

$$u(y(\hat{\theta}))\exp - \int_0^{\hat{\theta}} a_o(y(t))dt = \psi(y(\hat{\theta}))\exp - \int_0^{\hat{\theta}} a_o(y(t))dt\ ,$$

which is obvious for $\hat{\theta} < +\infty$ (by the definition of $\hat{\theta}$) and also true for $\theta = +\infty$, for then both quantities are zero. From this we easily deduce (7.20) and the desired result. □

Remark 7.1

The assumption (7.4) is actually of no use. It is sufficient to use Corollaries

5.4 and 5.5, which still lead to $(7.8)_1$. □

Remark 7.2

We can also do without the assumption (7.8) of coercivity, provided we make the assumption

(7.25) $\quad f - A\psi \leqq C.$

In fact we first reduce to the case f = 0 by introducing the solution w of a(w,v) = f(v), and u − w is then the solution of the V.I. corresponding to f = 0 and $\psi \to \psi - w$. Since $f \in L^p(\mathcal{O})$, $w \in W^{2,p}(\mathcal{O})$, and $\left.\frac{\partial w}{\partial \nu_A}\right|_\Gamma = 0$, $\psi - w$ therefore satisfies the assumptions for ψ. As for existence, we can then apply Theorem 5.2. To prove the $u \in W^{2,p}(\mathcal{O})$ we start from the penalised scheme and have to find an estimate for $\left(\frac{u^\varepsilon - w}{\varepsilon}\right)^+$ in $L^p(\mathcal{O})$. Naturally, we can no longer use the arguments carried out in Theorems 5.3 and 5.5, but we can use an argument based upon the Maximum Principle. Let us set $\tilde{u}^\varepsilon = u^\varepsilon - \psi$, then $\tilde{u}^\varepsilon$ satisfies

$$a(\tilde{u}^\varepsilon, v) + \frac{1}{\varepsilon}(\tilde{u}^{\varepsilon^+}, v) = (f - (A + a_0)\psi, v) - \int_\Gamma \frac{\partial\psi}{\partial\nu_A} v \, d\Gamma$$

therefore $\tilde{u}^\varepsilon$ satisfies

$$(7.26) \qquad \left| \begin{array}{l} A\tilde{u}^\varepsilon + a_0\tilde{u}^\varepsilon + \frac{1}{\varepsilon} u^{\varepsilon^+} = f - A\psi - a_0\psi \\ \frac{\partial\tilde{u}^\varepsilon}{\partial\nu_A} = - \frac{\partial\psi}{\partial\nu_A} \quad , \quad \tilde{u}^\varepsilon \in W^{2,p}(\mathcal{O}) \end{array} \right.$$

and therefore we deduce from (7.26) that

$$(7.27) \qquad \left| \begin{array}{l} \varepsilon A\tilde{u}^\varepsilon + \tilde{u}^\varepsilon + \varepsilon a_0\tilde{u}^\varepsilon \leqq \varepsilon(f - A\psi - a_0\psi) \leqq \varepsilon C \\ \frac{\partial\tilde{u}^\varepsilon}{\partial\nu_A} \leqq 0 \end{array} \right.$$

the Maximum Principle allows us to conclude that

(7.28) $\quad \tilde{u}^\varepsilon \leqq \varepsilon C$

and therefore

$$\left(\frac{u^\varepsilon - \psi}{\varepsilon}\right)^+ \leqq C$$

which implies, in particular, that $(u^\varepsilon - \psi)^+/\varepsilon$ remains in a bounded set of $L^p(\mathcal{O})$, for all p. From this we deduce Corollary 5.2. □

Remark 7.3

Theorem 7.1 is therefore true without the assumption (7.4). It has been left in the enunciation because we then do not have to use the results of Theorem 5.8. Further, Theorem 7.1 is still true without the assumptions (7.4) and (7.8), but provided we assume (7.25). □

Remark 7.4

The second condition (7.27) prevents the maximum of $\tilde{u}^\varepsilon$ on $\bar{\mathcal{O}}$ from being on the boundary. Since it is attained, it is necessarily attained in the interior of $\mathcal{O}$; but if x_0 is the maximum, then $A\tilde{u}^\varepsilon(x_0) \geq 0$, hence $\tilde{u}^\varepsilon(x_0) \leq C$, whence (7.28). This argument is perfectly correct if $\tilde{u}^\varepsilon$ is $C^2(\bar{\mathcal{O}})$.

However, we can give a probabilistic proof. In fact, $Au^\varepsilon \in L^p(\mathcal{O})$ and $\tilde{u}^\varepsilon \in C^1(\bar{\mathcal{O}})$, since $\tilde{u}^\varepsilon \in W^{2,p}(\mathcal{O})$ with $p > n$. We can then apply Corollary 4.1 to the process $y(t)$ and to $\psi(x,t) = e^{-t/\varepsilon}\tilde{u}^\varepsilon(x)$. We obtain

$$E^x e^{-\frac{T}{\varepsilon}} \tilde{u}^\varepsilon(y(T)) \exp - \int_0^T a_0\, ds = \tilde{u}^\varepsilon(x) +$$

$$+ E^x \int_0^T \left(-\frac{\tilde{u}^\varepsilon}{\varepsilon} - A\tilde{u}^\varepsilon - a_0\tilde{u}^\varepsilon\right)(y(t)) e^{-\frac{t}{\varepsilon}} \exp - \int_0^t a_0\, ds\, dt$$

$$- E^x \int_0^T \frac{\partial \tilde{u}_\varepsilon}{\partial \nu_A}(y(t))\, e^{-\frac{t}{\varepsilon}} \exp - \int_0^t a_0\, ds\, d\xi(t)$$

and therefore

$$\tilde{u}^\varepsilon(x) \leq C \int_0^T e^{-\frac{t}{\varepsilon}} \exp - \int_0^t a_0\, ds\, dt + E^x e^{-\frac{T}{\varepsilon}} \tilde{u}^\varepsilon(y(T)) \left(\exp - \int_0^T a_0\, ds\right).$$

Making T tend to $+\infty$, we deduce (7.78). □

Remark 7.5

Non-homogeneous problems

Let h be a function defined on Γ such that

$$h = \left.\frac{\partial \Phi}{\partial \nu_A}\right|_\Gamma \leq \frac{\partial \Phi}{\partial \nu_A}, \quad \text{where } \Phi \in W^{2,p}(\mathcal{O}), \ p > n .$$

Then we can prove the existence and uniqueness of u, a solution of

$$(7.29) \qquad \left| \begin{array}{l} Au + a_0 u \leq f , \quad u \leq \psi , \quad (Au + a_0 u - f)(u - \psi) = 0 \\ \left.\dfrac{\partial u}{\partial \nu_A}\right|_\Gamma = h , \quad u \in W^{2,p}(\mathcal{O}) . \end{array} \right.$$

In fact, we are immediately reduced to the homogeneous case by setting $\tilde{u} = u - \Phi$, and therefore $\tilde{u}$ satisfies

$$(7.30) \qquad \left| \begin{array}{l} A\tilde{u} + a_0\tilde{u} \leq f - A\Phi - a_0\Phi , \ \tilde{u} \leq \psi - \Phi , \ (A\tilde{u} + a_0\tilde{u} - f + A\Phi - a_0\Phi)(\tilde{u} - \psi + \Phi) = 0 \\ \left.\dfrac{\partial \tilde{u}}{\partial \nu_A}\right|_\Gamma = 0 , \quad \tilde{u} \in W^{2,p}(\mathcal{O}) \end{array} \right.$$

which defines $\tilde{u}$ uniquely. An argument analogous to that carried out for Theorem 7.1 allows us to state that the solution u(x) of (7.29) is explicitly given by the formula

$$u(x) = \operatorname*{Inf}_{\theta} E^x \left[\int_0^{\theta} f(y(t)) \left(\exp - \int_0^t a_o(y(s))ds \right) dt + \right. \tag{7.31}$$
$$+ \psi(y(\theta)) \exp - \int_0^{\theta} a_o(y(t))dt + \int_0^{\theta} h(y(t)) \left(\exp - \int_0^t a_o(y(s)) \right) d\xi(t) \quad .$$

7.2 *Interpretation of the Penalised Scheme*

Here we are going to give the probabilistic interpretation of the penalised scheme, i.e.,

$$Au^\epsilon + a_o u^\epsilon + \frac{1}{\epsilon} (u^\epsilon - \psi)^+ = f \quad , \quad \left.\frac{\partial u^\epsilon}{\partial \nu_A}\right|_\Gamma = h \quad , \quad u^\epsilon \in W^{2,p}(\mathcal{O}) \ . \tag{7.32}$$

Since $p > n$,

$$u^\epsilon \in C^1(\bar{\mathcal{O}}) \ . \tag{7.33}$$

An admissible control is a process $v(t)$ *adapted* to μ^t, such that

$$0 \leq v(t) \leq 1 \quad \text{a.s.} \quad \forall\ t.$$

We define a functional $J_x^\epsilon(v)$ by the formula

$$J_x^\epsilon(v) = E^x \int_0^\infty \left[f(y(t)) + \frac{1}{\epsilon} \psi(y(t))v(t) \right] \left(\exp - \int_0^t (a_o(y(s)) + \frac{v(s)}{\epsilon}) ds \right) dt \tag{7.34}$$
$$+ E \int_0^\infty h(y(t)) \left(\exp - \int_0^t (a_o(y(s)) + \frac{v(s)}{\epsilon}) \, ds \right) d\xi(t) \ .$$

Let us notice that

$$E^x \int_0^\infty h(y(t)) \left(\exp - \int_0^t (a_o(y(s)) + \frac{v(s)}{\epsilon}) \, ds \right) d\xi(t) = \Phi(x) - \tag{7.35}$$
$$- E^x \int_0^\infty \left((A + a_o)\Phi(y(t)) + \frac{v(t)}{\epsilon}\Phi \right) \left(\exp - \int_0^t (a_o(y(s)) + \frac{v(s)}{\epsilon}) ds \right) dt$$

and therefore the Stieltjes' integral has a meaning, since $A\Phi \in L^p(\mathcal{O})$.

THEOREM 7.2: *Assume* (7.2), (7.3), *and*

$$f \in L^p(\mathcal{O}) \quad , \quad p > n \tag{7.36}$$

$$\psi \in C^o(\bar{\mathcal{O}}) \ . \tag{7.37}$$

Then

$$u^\epsilon(x) = \operatorname*{Inf}_{v(.)} J_x^\epsilon(v) \ . \tag{7.38}$$

Further, there exists an optimal contol given by

$$(7.39)\qquad \hat{v}^{\varepsilon}(t) = \begin{cases} 1 \text{ if } u^{\varepsilon}(y(t)) \geq \psi(y(t)), \\ 0 \text{ otherwise.} \end{cases}$$

Proof: Since the function $u^{\varepsilon}(x)$ is continuous, $\hat{v}^{\varepsilon}(t)$ is an admissible control. Applying Ito's formula (cf. Corollary 4.1), we have

$$(7.40)\qquad E^x u^{\varepsilon}(y(t)) \exp - \int_0^T (a_o + \frac{v}{\varepsilon}) ds = u^{\varepsilon}(x) +$$
$$+ E^x \int_0^T \left[-(A + a_o) u^{\varepsilon}(y(t)) - \frac{v}{\varepsilon} u^{\varepsilon}(y(t)) \right] \exp - \int_0^t (a_o + \frac{v}{\varepsilon}) ds -$$
$$- E^x \int_0^T \frac{\partial u^{\varepsilon}}{\partial \nu_A}(y(t)) \left(\exp - \int_0^t (a_o + \frac{v}{\varepsilon}) ds \right) d\xi(t) .$$

We can make T tend to $+\infty$ in the relation (7.40), and using (7.32) we obtain

$$(7.41)\qquad u^{\varepsilon}(x) = E^x \int_0^{\infty} f(y(t)) \left(\exp - \int_0^t (a_o + \frac{v}{\varepsilon}) ds \right) dt +$$
$$+ E^x \int_0^{\infty} \left[-\frac{1}{\varepsilon} (u^{\varepsilon} - \psi)^+ (y(t)) + \frac{v}{\varepsilon} u^{\varepsilon}(y(t)) \right] \left(\exp - \int_0^t (a_o + \frac{v}{\varepsilon}) ds \right) dt +$$
$$+ E^x \int_0^{\infty} h(y(t)) \left(\exp - \int_0^t (a_o + \frac{v}{\varepsilon}) ds \right) d\xi(t)$$
$$= J_x^{\varepsilon}(v) + E^x \int_0^{\infty} \left[\frac{v}{\varepsilon} (u^{\varepsilon} - \psi)(y(t)) - \frac{1}{\varepsilon} (u^{\varepsilon} - \psi)^+ (y(t)) \right] \left(\exp - \int_0^t (a_o + \frac{v}{\varepsilon}) ds \right) dt.$$

We then finish the proof as in the case of the Dirichlet problem (cf. B.L., Chap. 3, loc. cit.). □

When ψ is regular we can give an estimate of the penalisation error, and this completes Theorem 5.3.

THEOREM 7.3: *Under the assumptions cf Theorem 7.2, and if we further have*

$$(7.42)\qquad \psi \in W^{2,p}(\mathcal{O}) \ , \quad \left. \frac{\partial \psi}{\partial \nu_A} \right|_{\Gamma} \geq h \Big|_{\Gamma}$$

then

$$(7.43)\qquad \|u^{\varepsilon} - u\|_{C^0(\bar{\mathcal{O}})} \leq C\, \varepsilon^{\frac{a}{1+a}}$$

where a *is such that* $\frac{p}{1+a} > \frac{n}{2}$.

Proof: Without loss of generality we assume that $h = 0$, for otherwise we are reduced to a problem identical with $f \to f - A\Phi - a_o\Phi$ and $\psi \to \psi - \Phi$. We then have the relations

$$(7.44)\qquad \psi(x) = E^x \int_0^{\infty} \left((A\psi + a_o\psi)(y(t)) + \frac{v(t)\psi(y(t))}{\varepsilon} \right) \left(\exp - \int_0^t (a_o + \frac{v}{\varepsilon}) ds \right) dt + \text{(Contd)}$$

(Contd) $$+ E^x\int_0^\infty \frac{\partial\psi}{\partial\nu_A}(y(t))\left(\exp - \int_0^t (a_o + \frac{v}{\varepsilon})ds\right)d\xi(t)$$

(7.45) $$\psi(x) = E^x\int_0^\theta \left((A+a_o)\psi(y(t)) + \frac{v(t)\psi(y(t))}{\varepsilon}\right)\left(\exp - \int_0^t (a_o + \frac{v}{\varepsilon})ds\right)dt + $$
$$+ E^x\int_0^\theta \frac{\partial\psi}{\partial\nu_A}(y(t))\left(\exp - \int_0^t (a_o + \frac{v}{\varepsilon})ds\right)d\xi(t) + $$
$$+ E^x\,\psi(y(\theta))\,\exp - \int_0^\theta (a_o + \frac{v}{\varepsilon})ds\ .$$

Consequently we can rewrite the functionals $J_x^\varepsilon(v)$ and $J_x(\theta)$ as follows:

(7.46) $$J_x^\varepsilon(v) = \psi(x) + E^x\int_0^\infty (f - (A+a_o)\psi)(y(t))\left(\exp - \int_0^t (a_o + \frac{v}{\varepsilon})ds\right)dt - $$
$$- E^x\int_0^\infty \frac{\partial\psi}{\partial\nu_A}(y(t))\left(\exp - \int_0^t (a_o + \frac{v}{\varepsilon})ds\right)d\xi(t)\ .$$

(7.47) $$J_x(\theta) = \psi(x) + E^x\int_0^\theta (f - (A+a_o)\psi)(y(t))\left(\exp - \int_0^t a_o\,ds\right)dt - $$
$$- E^x\int_0^\theta \frac{\partial\psi}{\partial\nu_A}(y(t))\left(\exp - \int_0^t a_o\,ds\right)d\xi(t)\ .$$

We also know that

(7.48) $$u^\varepsilon(x) \geqq u(x).$$

We therefore have to prove an estimate in the other direction.

Let θ be an arbitrary stopping time of μ^t, and let $v_\theta(t)$ be a control associated with θ by the formula

(7.49) $$v_\theta(t) = \begin{vmatrix} 1 \text{ if } t < \theta \\ 0 \text{ if } t \geqq \theta. \end{vmatrix}$$

Then we have

$$J_x^\varepsilon(v_\theta) = \psi(x) + E^x\int_0^\theta (f-(A+a_o)\psi)(y(t))\left(\exp - \int_0^t a_o ds\right)dt + $$
$$+ E^x\int_\theta^{+\infty} (f-(A+a_o)\psi)(y(t))\left(\exp - \int_0^t a_o\,ds\right)\exp - \frac{t-\theta}{\varepsilon}\,dt - $$
$$- E^x\int_0^\theta \frac{\partial\psi}{\partial\nu_A}(y(t))\left(\exp - \int_0^t a_o\,ds\right)d\xi(t)$$
$$- E^x\int_\theta^{+\infty} \frac{\partial\psi}{\partial\nu_A}(y(t))\left(\exp - \int_0^t a_o\,ds\right)\left(\exp - \frac{t-\theta}{\varepsilon}\right)d\xi(t) = $$

(Contd)

(Contd) $$= J_x(\theta) + E^x \int_\theta^{+\infty} (f-(A+a_0)\psi) \left(\exp - \int_0^t a_0\, ds\right) \exp - \frac{t-\theta}{\varepsilon}\, dt -$$
$$- E^x \int_\theta^{+\infty} \frac{\partial \psi}{\partial \nu_A} (y(t)) \left(\exp - \int_0^t a_0\, ds\right) \exp - \frac{t-\theta}{\varepsilon}\, d\xi(t)$$

so

(7.50) $$J_x^\varepsilon(v_\theta) \leq J_x(\theta) + E^x \int_\theta^{+\infty} (f - (A+a_0)\psi) \left(\exp - \int_0^t a_0\, ds\right) \exp - \frac{t-\theta}{\varepsilon}\, dt \; .$$

But

$$E^x \int_\theta^{+\infty} |f-A\psi| \exp - \int_0^t a_0\, ds \,\exp - \frac{t-\theta}{\varepsilon} dt \leq E^x \int_\theta^{+\infty} |f-A\psi| \exp - \beta t \exp - \frac{t-\theta}{\varepsilon}\, dt$$

$$\leq \left(E^x \int_\theta^{+\infty} |f-A\psi|^{1+a} \exp - \beta(1+a)t \right)^{1/1+a} \left(E^x \int_\theta^{+\infty} \exp - \left(\frac{t-\theta}{\varepsilon}\right) \frac{1+a}{a} \right)^{\frac{a}{1+a}}$$

$$\leq \; C \|f - A\psi\|_{L^p} \; \varepsilon^{\frac{a}{1+a}}$$

and therefore

$$u^\varepsilon(x) \leq T_x(\theta) + C\, \varepsilon^{\frac{a}{1+a}} .$$

Taking the lower bound at θ we obtain

$$u^\varepsilon(x) \leq u(x) + C\, \varepsilon^{\frac{a}{1+a}}$$

whence the result. □

7.3 *Interpretation of the V.I. in the Case of a Continuous Obstacle*

We are now going to interpret the solution of the V.I. (7.7) in the case where the obstacle ψ is continuous only. We have:

THEOREM 7.4: *Under the assumption of Theorem 7.2 we have*

(7.51) $$u^\varepsilon(x) \to \operatorname*{Inf}_\theta J_x(\theta) \quad \text{in } C^0(\bar{\mathcal{O}}) \; .$$

Hence the solution $u(x)$ *of the V.I.* (7.7) *belongs to* $C^0(\bar{\mathcal{O}})$ *and*

(7.52) $$u(x) = \operatorname*{Inf}_\theta J_x(\theta).$$

If we set

(7.53) $$\left| \begin{array}{l} \hat{\theta}^\varepsilon = \inf \{t \,|\, u^\varepsilon(y(t)) \geq \psi(y(t)\} \\ \theta \;= \inf \{t \,|\, u(y(t)) \;= \psi(y(t)\} \end{array} \right.$$

Then $\hat{\theta}$ *is an optimal stopping time for* $J_x(\theta)$ *and*

(7.54) $\hat{\theta}^\varepsilon \to \hat{\theta}$ a.s.. (1)

Proof: We first assume that we are reduced to the case h = 0. Set

$$w(x) = \operatorname*{Inf}_x J_x(\theta).$$

Notice that

$$\hat{v}^\varepsilon(t) = 1 \quad \text{for} \quad t \in [0, \hat{\theta}^\varepsilon[\quad \text{if} \quad \hat{\theta}^\varepsilon > 0 .$$

Applying Ito's formula we have

$$u^\varepsilon(x) = E^x \int_0^{\hat{\theta}^\varepsilon} (A + a_0) u^\varepsilon(y(t)) \exp - \int_0^t a_0 \, ds + E^x u^\varepsilon(y(\theta^\varepsilon)) \exp - \int_0^{\hat{\theta}^\varepsilon} a_0 \, dt .$$

But

$$Au^\varepsilon + a_0 u^\varepsilon = f - \frac{1}{\varepsilon}(u^\varepsilon - \psi)^+$$

and for $t < \hat{\theta}^\varepsilon$,

$$u^\varepsilon(y(t)) < \psi(y(t)) .$$

Furthermore,

$$u^\varepsilon(y(\hat{\theta}^\varepsilon)) \exp - \int_0^{\hat{\theta}^\varepsilon} a_0 \, dt = \psi(y(\hat{\theta}^\varepsilon)) \exp - \int_0^{\hat{\theta}^\varepsilon} a_0 \, dt$$

whether $\hat{\theta}^\varepsilon$ is finite or infinite. From this we deduce that

$$u^\varepsilon(x) = J_x(\theta)$$

and therefore

(7.55) $u^\varepsilon(x) \geqq w(x)$.

We now seek to establish an inequality in the opposite direction, analogous to that proved in Theorem 7.3. However, the situation is more complex here, because ψ is not regular. We evaluate the difference

(7.56)
$$J_x^\varepsilon(v_\theta) - J_x(\theta) = E^x \int_\theta^{+\infty} \left(f + \frac{\psi}{\varepsilon}\right) \left(\exp - \int_0^t a_0 \, ds\right) \exp - \frac{t-\theta}{\varepsilon} \, dt - $$
$$- E\psi(y(\theta)) \exp - \int_0^\theta a_0 \, dt .$$

Let $\delta > 0$. Then

---

(1) These stopping times may possibly be equal to $+\infty$.

(7.57)
$$E^x\int_\theta^{+\infty}\frac{\psi}{\epsilon}(y(t))\ \exp -\int_0^t a_o\,ds\ \ \exp -\frac{t-\theta}{\epsilon}dt - E^x\psi(y(\theta))\exp -\int_0^\theta a_o\,dt =$$
$$= E^x\int_\theta^{\theta+\delta}\frac{\psi}{\epsilon}\left(\exp -\int_0^t a_o ds\right)\exp -\frac{t-\theta}{\epsilon}dt + E^x\int_{\theta+\delta}^{+\infty}\frac{\psi}{\epsilon}\exp -\int_0^t a_o ds\ \exp -\frac{t-\theta}{\epsilon}dt -$$
$$- E^x\psi(y(\theta))\exp -\int_0^\theta a_o\,dt\ .$$

Now

$$E^x\int_{\theta+\delta}^{+\infty}\left|\frac{\psi}{\epsilon}\right|\exp -\int_0^t a_o ds\ \exp -\frac{t-\theta}{\epsilon}dt \leq \frac{1}{\beta}\exp -\frac{\delta}{\epsilon}\ ||\psi||_{C^o(\bar{\mathcal{O}})}$$

and

(7.59)
$$E^x\int_\theta^{\theta+\delta}\frac{\psi}{\epsilon}\left(\exp -\int_0^t a_o\,ds\right)\exp -\frac{t-\theta}{\epsilon}dt - E^x\psi(y(\theta))\exp -\int_0^\theta a_o\,dt =$$
$$E^x\int_0^{\theta+\delta}\frac{\psi(y(s))-\psi(y(\theta))}{\epsilon}\quad \exp -\int_0^t a_o\,ds\ \exp -\frac{t-\theta}{\epsilon}\ +$$
$$+ E^x\int_0^{\theta+\delta}\frac{\psi(y(t))}{\epsilon}\left(\exp -\int_0^t a_o\,ds - \exp -\int_0^\theta a_o ds\right)\exp -\frac{t-\theta}{\epsilon}\,dt\ +$$
$$+ E^x\ \psi(y(\theta))\ \exp -\int_0^\theta a_o\,ds\left(\int_\theta^{\theta+\delta}\frac{1}{\epsilon}\exp -\frac{t-\theta}{\epsilon}\,dt - 1\right) = I + II + III\ .$$

But for $t \in [\theta,\theta+\delta]$

$$\left|\exp -\int_0^t a_o\,ds - \exp -\int_0^\theta a_o\,ds\right| \leq \left(\exp -\int_0^\theta a_o\,ds\right)\delta M$$

where $M \geq a_o$, therefore

(7.60) $\quad |II| \leq \delta||\psi||\ ||a_o||$

(7.61) $\quad |III| \leq ||\psi||\ \exp -\dfrac{\delta}{\epsilon}\ .$

It remains to estimate $|I|$. First, as ψ is continuous on $\bar{\mathcal{O}}$, we introduce the continuity modulus

(7.62)
$$\rho(\lambda) = \sup_{\substack{\xi_1,\xi_2\in\bar{\mathcal{O}}\\ |\xi_1-\xi_2|\leq\lambda}} |\psi(\xi_1)-\psi(\xi_2)|\ .$$

Notice that

(7.63)
$$|I| \leq E^x \sup_{\theta\leq s\leq\theta+\delta}|\psi(y(s))-\psi(y(\theta))|\exp -\beta\theta$$

But if we refer back to the proof of Lemma 3.5 (in particular to (3.39)), we see that

$$E^x\sup_{\theta\leq t\leq\theta+\delta}\left(\Phi(y(t))-\Phi(y(\theta))\right)^2 \leq C\left(\delta+\delta^{1/2}\right)$$

and thus

$$E^x \sup_{\theta \leq t \leq \theta+\delta} \left(\int_\theta^t \gamma(y) . \frac{\partial \Phi}{\partial x}(y) d\xi(s) \right)^2 \leq C_1 \left(\delta^{1/2} + \delta + \delta^2 \right)$$

so also

$$E^x \sup_{\theta \leq t \leq \theta+\delta} \left(\xi(t) - \xi(\theta) \right)^2 \leq C_2 \left(\delta^{1/2} + \delta + \delta^2 \right)$$

hence, by the equation for y

$$E^x \sup_{\theta \leq t \leq \theta+\delta} |y(t) - y(\theta)|^2 \leq C_3 \left(\delta^{1/2} + \delta + \delta^2 \right) . \tag{7.64}$$

But then we have

$$|I| \leq E^x \sup_{\theta \leq t \leq \theta+\delta} |\psi(y(t)) - \psi(y(\theta))| \tag{7.65}$$

$$|I| \leq \rho(h) + \frac{2\|\psi\|}{h} C \delta^{1/4} \tag{7.66}$$

Lastly, we have (cf. Theorem 7.3)

$$E \int_0^{+\infty} |f| \left(\exp - \int_0^t a_0 \, ds \right) \exp - \frac{t-\theta}{\varepsilon} \, dt \leq C \varepsilon^{\frac{a}{1+a}} . \tag{7.67}$$

Using the estimates (7.60), (7.61), (7.66), (7.67) and making ε tend to 0, then δ to 0, then h to 0, in this order, we obtain

$$\sup_x |J_x^\varepsilon(v_\theta) - J_x(\theta)| \to 0 , \quad \text{when } \varepsilon \to 0$$

therefore

$$J_x^\varepsilon(v_\theta) \leq J_x(\theta) + \varphi(\varepsilon)$$

where φ tends to 0 with ε. From this it follows that

$$u^\varepsilon(x) \leq w(x) + \varphi(\varepsilon)$$

and thus

$$\sup_x |u^\varepsilon(x) - w(x)| \to 0 \quad \text{when } \varepsilon \to 0. \tag{7.68}$$

Since we also know that $u^\varepsilon \to u$, for example weakly in $H^1(\mathcal{O})$, we have proved the results (7.51) and (7.52).

The proof that $\hat{\theta}$ is an optimal stopping time is completely analogous to the Dirichlet case, once the uniform convergence of u^ε to u is known. Hence we shall not do it again here (cf. B.L., Chap. 3, loc. cit.); similarly for the property (7.54). □

Remark 7.7

Theorems 7.2 and 7.4 are valid under the assumptions $f \in L^p(\mathcal{O})$, $p > \frac{n}{2}$, which are weaker than (7.36). In fact, let us first notice that if $f \in L^p(\mathcal{O})$, $p > \frac{n}{2}$, then the solution of the problem

$$Aw + a_o w = f \ , \quad \left.\frac{\partial w}{\partial \nu_A}\right|_\Gamma = 0$$

belongs to $W^{2,p}(\mathcal{O})$ and thus $w \in C^o(\bar{\mathcal{O}})$. But then, if we set

$$u^\varepsilon(x) = u^\varepsilon(x) - w(x)$$

we see that $\tilde{u}^\varepsilon$ satisfies

$$A\tilde{u}^\varepsilon + a_o \tilde{u}^\varepsilon + \frac{1}{\varepsilon}(\tilde{u}^\varepsilon - (\psi - w))^+ = 0 \quad \left.\frac{\partial \tilde{u}^\varepsilon}{\partial \nu_A}\right|_\Gamma = 0$$

and $(\tilde{u}^\varepsilon - \psi + w)^+ \in L^\infty(\mathcal{O})$, therefore $\tilde{u}^\varepsilon \in W^{2,p}(\mathcal{O})$, for all $p > 1$. From this it follows that Theorem 7.2 applies to u^ε, in other words we have

$$\tilde{u}^\varepsilon(x) = \operatorname*{Inf}_v E^x \int_0^\infty \frac{1}{\varepsilon} v(t) \Big(\psi(y(t)) - w(y(t))\Big)\left(\exp - \int_0^t (a_o + \frac{v}{\varepsilon}) ds\right) dt \ . \tag{7.69}$$

But the probabilistic interpretation of the linear Neumann problem allows us to state that

$$w(x) = E^x \int_0^\infty \left(f(y(t)) + \frac{v(t)}{\varepsilon} w(y(t))\right)\left(\exp - \int_0^t (a_o + \frac{v}{\varepsilon}) ds\right) dt \tag{7.70}$$

and therefore

$$\begin{aligned}\tilde{u}^\varepsilon(x) = \operatorname*{Inf}_v E^x \Bigg[&\int_0^\infty f(y(t)) \left(\exp - \int_0^t (a_o + \frac{v}{\varepsilon}) ds\right) dt + \\ &+ \int_0^\infty \frac{1}{\varepsilon} v(t)\psi(y(t)) \left(\exp - \int_0^t (a_o + \frac{v}{\varepsilon}) ds\right) dt - w(x)\Bigg]\end{aligned}$$

hence we again have the property (7.38). We can, moreover, easily extend the proof of Theorem 7.4. □

Remark 7.8

In the non-homogeneous case we need to assume that $h = \left.\frac{\partial \Phi}{\partial \nu_A}\right|_\Gamma$ with $\Phi \in W^{2,p}(\mathcal{O})$, $p > n$ (and not $p > \frac{n}{2}$). The reason is that we need to be certain that $\Phi \in C^1(\bar{\mathcal{O}})$, so that we can write formula (7.35). □

Remark 7.9

In Corollary 5.3 we have stated the continuity property of the solution u of the V.I. in the case of an obstacle that is continuous only, by approximating the obstacle ψ by a sequence of obstacles $\psi^j \in W^{2,p}(\mathcal{O})$, $\left.\frac{\partial \psi^j}{\partial \nu_A}\right|_\Gamma \geqq 0$, and $\psi^j \to \psi$ in $C^o(\bar{\mathcal{O}})$.

Through the following Lemma we are going to prove probabilistically that such an approximation is legitimate. □

LEMMA 7.1: *Assume that* $\mathcal{O}$ *is a bounded open set of* R^n*, defined by* (3.4), *and that the coefficients* a_{ij} *satisfy* (7.2). *Let* $\psi \in C^o(\bar{\mathcal{O}})$ *and* ψ^ε *be a solution of*

$$(7.71)\quad \left| \begin{array}{l} -\varepsilon^2 \sum_{ij} a_{ij} \dfrac{\partial^2\psi^\varepsilon}{\partial x_i \partial x_j} + \psi^\varepsilon = \psi \quad \text{in} \quad \mathcal{O}, \quad \left.\dfrac{\partial\psi^\varepsilon}{\partial\nu_A}\right|_\Gamma = 0 \\ \psi^\varepsilon \in W^{2,p}(\mathcal{O}), \quad \forall\, p > 1 . \end{array} \right.$$

Then

$$(7.72)\quad \psi^\varepsilon \to \psi \quad \text{in } C^o(\bar{\mathcal{O}}).$$

Proof: We can always assume that $a_{ij} = a_{ji}$. Define γ by (7.16) and $\sigma(x)$ by (7.14). On an arbitrary probability space construct the reflected diffusion $y_x^\varepsilon(t)$ which is the solution of

$$(7.73)\quad \left| \begin{array}{l} dy^\varepsilon = \varepsilon\sigma(y^\varepsilon)dw(t) - \chi_\Gamma(y^\varepsilon)\gamma(y^\varepsilon)d\xi^\varepsilon(t) \text{ (1)} \\ y^\varepsilon(0) = x . \end{array} \right.$$

We then have

$$(7.74)\quad \psi^\varepsilon(x) = E\int_0^\infty \psi(y_x^\varepsilon(t))e^{-t}\,dt .$$

Therefore

$$(7.75)\quad \psi^\varepsilon(x) - \psi(x) = E\int_0^\infty \left(\psi(y_x^\varepsilon(t)) - \psi(x)\right) e^{-t}\,dt .$$

We then retrace certain points in the proof of Lemma 3.5:

$$(7.76)\quad \Phi(y_x^\varepsilon(t)) - \Phi(x) = \varepsilon\int_0^t \frac{\partial\Phi}{\partial x}(y^\varepsilon(s)).\sigma\, dw(s) + \varepsilon^2\int_0^t \mathrm{tr}\frac{\partial^2\Phi}{\partial x^2}\, a(y^\varepsilon)ds - \\ - \int_0^t \gamma(y^\varepsilon).\frac{\partial\Phi}{\partial x}(y^\varepsilon)d\xi^\varepsilon(s)$$

$$(7.77)\quad \Phi^2(y_x^\varepsilon(t)) - \Phi^2(x) = \varepsilon\int_0^t 2\Phi\frac{\partial\Phi}{\partial x}.\sigma\, dw + 2\varepsilon^2\int_0^t \left(a\,\frac{\partial\Phi}{\partial x}.\frac{\partial\Phi}{\partial x} + \Phi\,\mathrm{tr}\, a\,\frac{\partial^2\Phi}{\partial x^2}\right) ds$$

from which we deduce

$$(7.78)\quad \left(\Phi(y_x^\varepsilon(t)) - \Phi(x)\right)^2 \leq \varepsilon\int_0^t 2\Phi\frac{\partial\Phi}{\partial x}.\sigma\, dw + 2\varepsilon^2\int_0^t \left(a\,\frac{\partial\Phi}{\partial x}.\frac{\partial\Phi}{\partial x} + \Phi\,\mathrm{tr}\, a\,\frac{\partial^2\Phi}{\partial x^2}\right) ds$$

(Contd)

(1) To shorten things a little, we do not write all the relations satisfied by y^ε, ξ^ε (cf. (1.6),...,(1.10)).

(Contd) $$- 2\Phi(x)\left[\varepsilon\int_0^t \frac{\partial\Phi}{\partial x}\cdot\sigma\, dw + \varepsilon^2\int_0^t \operatorname{tr}\frac{\partial^2\Phi}{\partial x^2}\, a\, ds\right]$$

hence

$$E\left(\Phi(y_x^\varepsilon(t)) - \Phi(x)\right)^2 \leq C\,(\varepsilon\sqrt{t} + \varepsilon^2 t\,) \,. \tag{7.79}$$

Returning to (7.76), it follows that

$$E\left(\int_0^t \gamma(y^\varepsilon)\cdot\frac{\partial\Phi}{\partial x}(y^\varepsilon)d\xi^\varepsilon(s)\right)^2 \leq C\,(\varepsilon\sqrt{t} + \varepsilon^2 t + \varepsilon^4 t^2)$$

so

$$E\,\xi^\varepsilon(t)^2 \leq C\,(\varepsilon\sqrt{t} + \varepsilon^2 t + \varepsilon^4 t^2)$$

and therefore using (7.73) this yields

$$E\left|y_x^\varepsilon(t) - x\right|^2 \leq C\,(\varepsilon\sqrt{t} + \varepsilon^2 t + \varepsilon^4 t^2) \,. \tag{7.80}$$

Let $\rho(h)$ be the modulus of continuity of the function ψ on $\bar{\mathcal{O}}$. Then we have

$$E\left|\psi(y_x^\varepsilon(t)) - \psi(x)\right| \leq \rho(h) + 2\,\frac{\|\psi\|}{h^2}\,C(\varepsilon\sqrt{t} + \varepsilon^2 t + \varepsilon^4 t^2) \,.$$

We can take $h = \varepsilon^{1/4}$, and thus obtain

$$|\psi^\varepsilon(x) - \psi(x)| \leq \rho(\varepsilon^{1/4}) + 2\|\psi\|C\left(\varepsilon^{1/2}\int_0^\infty \sqrt{t}e^{-t}dt + \varepsilon^{3/2}\int_0^\infty te^{-t}\,dt + \right.$$
$$\left. + \varepsilon^{7/2}\int_0^\infty t^2e^{-t}\,dt\right)$$

whence the result (7.72). □

Remark 7.10

Variational techniques easily give $\psi^\varepsilon \to \psi$ in $L^2(\mathcal{O})$. □

Remark 7.11

The estimate (5.21) is quite obvious in the formulae (7.52) and (7.17). □

7.4 *A Supplement on Regularity*

We have seen that the solution of the V.I. with a continuous obstacle is continuous. We are going to prove that under certain additional conditions the solution is Lipschitz continuous when the obstacle is Lipschitz continuous.

We have:

THEOREM 7.5: *Under the assumptions of Theorem 7.2, together with*

(7.81) $a_{ij} = \delta_{ij}$, a_i, a_o *Lipschitz continuous*, $a_o \geqq \beta > 0$ *where* β *large enough*

(7.82) $f \in L^p(\mathcal{O})$, $p > n$

(7.83) ψ *Lipschitz continuous*

(7.84) $\mathcal{O}$ *convex (i.e.,* Φ *is a convex mapping),*

the solution u *of the V.I.* (7.7) *then belongs to* $W^{1,\infty}(\mathcal{O})$.

Proof: Without loss of generality, we assume $f = 0$. This is the same as changing ψ into $\psi - w$, where $w \in W^{2,p}(\mathcal{O})$, hence $w \in C^1(\bar{\mathcal{O}})$, since $p > n$.

To simplify the proof a little we assume that $a_o = \beta$.

Thanks to the assumption of regularity of the coefficients, and by noticing that $g_i = a_i$, we can construct the reflected diffusion (in the strong sense), so

(7.85) $$\begin{cases} dy_x = g(y_x(t))dt + \sqrt{2}\, dw(t) - n\chi_\Gamma(y_x(t))d\xi(t) \\ y_x(0) = x\ . \end{cases}$$

We then know that the solution u of the V.I. is interpreted in the form

(7.86) $$u(x) = \operatorname*{Inf}_\theta \psi(y_x(\theta))e^{-\beta\theta}\ .$$

Further, since $\mathcal{O}$ is convex, and as the direction of reflection is the normal pointing into the interior of $\mathcal{O}$, we can use the characterisation of the reflected process given by Theorem 1.2(1). Therefore (cf. (1.13),(1.14)) we have

(7.87) $$\begin{cases} (z(t) - y_x(t)) \cdot (dy_x(t) - g(y_x(t))dt - \sqrt{2}\, dw(t)) \geq 0 \\ \forall\ z(t) \text{ an adapted continuous process such that a.x. } z(t) \in \bar{\mathcal{O}}\ \forall\ t. \end{cases}$$

By Ito's formula we have

(7.88) $$\frac{1}{2}|y_x(t) - y_{x'}(t)|^2 e^{-2\gamma t} = \frac{1}{2}|x - x'|^2 - \beta\int_0^t e^{-2\gamma s}|y_x(s) - y_{x'}(s)|^2 ds + \int_0^t (y_x(s) - y_{x'}(s)) \cdot (dy_x(s) - dy_{x'}(s))e^{-2\gamma s}\ .$$

But taking $z(t) = y_{x'}(t)$ in (7.87), we have

(7.89) $$(y_{x'}(t) - y_x(t)) \cdot (dy_x(t) - g(y_x(t))dt - \sqrt{2}\, dw(t)) \geq 0\ .$$

Similarly, we can write

(7.90) $$(y_x(t) - y_{x'}(t)) \cdot (dy_{x'}(t) - g(y_{x'}(t))dt - \sqrt{2}\, dw(t)) \geq 0\ .$$

By adding (7.89),(7.90) we obtain

$$(y_{x'}(t) - y_x(t)) \cdot (dy_x(t) - dy_{x'}(t)) - (y_{x'}(t) - y_x(t)) \cdot \Big(g(y_x(t)) - g(y_{x'}(t))\Big)dt \geq 0$$

and thus (7.88) becomes

(1) This is what serves as the convexity hypothesis, which is quite useless.

$$\frac{1}{2}|y_x(t) - y_{x'}(t)|^2 e^{-2\gamma t} \leq \frac{1}{2}|x-x'|^2 - \gamma\int_0^t e^{-2\gamma s}|y_x(s) - y_{x'}(s)|^2 ds$$
$$+\int_0^t \Big(g(y_x(s)) - g(y_{x'}(s))\Big) \cdot (y_x(s) - y_{x'}(s))e^{-2\gamma s}\, ds \ .$$

Therefore for $\gamma \geq \|\frac{\partial g}{\partial x}\|$, we have

$$|y_x(t) - y_{x'}(t)|e^{-\gamma t} \leq |x-x'| \ . \tag{7.91}$$

Consequently if $\beta \geqslant \gamma$ we also have

$$|y_x(t) - y_{x'}(t)|e^{-\beta t} \leq |x-x'| \ .$$

Since ψ is Lipschitz continuous, it immediately results that

$$|\psi(y_x(\theta))e^{-\beta\theta} - \psi(y_{x'}(\theta))e^{-\beta\theta}| \leq C|x-x'|$$

hence

$$|u(x) - u(x')| \leq C|x-x'| \ . \quad \square$$

7.5 *The Elliptic Regularisation Method*

In A. BENSOUSSAN - J.L. LIONS [1], Chap.s 2 and 3 we have studied the method of elliptic regularisation which consists of approximating a parabolic problem by a sequence of elliptic problems. We are now going to study this method from the the probabilistic point of view.

Let us first consider the case of a linear problem. Let us recall the method in the analytic plane. Let $\mathcal{O}$ be a bounded open set of R^n, $Q = \mathcal{O} \times]0,T[$, $\Sigma = \Gamma \times]0,T[$. Let the coefficients a_{ij}, a_i, a_o satisfy (6.1),(6.2). We shall use the notations (6.5),(6.6). Let $f \in L^2(Q)$. Now consider the problem (with Dirichlet boundary condition)

$$-\frac{\partial u}{\partial t} + A(t)u + a_o u = f \ , \quad u|_{\Sigma} = 0 \ , \quad u(x,T) = 0 \ .^{(1)} \tag{7.92}$$

Naturally, there is one and only one solution of (7.92) such that $u \in L^2(0,T;V)$, $\frac{\partial u}{\partial t} \in L^2(0,T;V')$. We approximate the problem (7.92) by a sequence of elliptic problems

$$\left|\begin{array}{l} -\eta \dfrac{\partial^2 u^\eta}{\partial t^2} - \dfrac{\partial u^\eta}{\partial t} + (A(t) + a_o)u^\eta = f \\ u^\eta(T) = 0 \ , \quad \dfrac{\partial u^\eta}{\partial t}(0) = 0 \end{array}\right. \tag{7.93}$$

(Contd)

(1) The initial condition of zero is not indispensible, but allows us to simplify the presentation a little.

(Contd) $\left| \; u^{\eta} \in L^2(0,T;H_0^1(\mathcal{O})) \; , \; \frac{\partial u^{\eta}}{\partial t} \in L^2(Q) \; .\right.$ (1)

We know that

(7.94) $\quad u^{\eta} \to u$ weakly in $L^2(0,T;H_0^1(\mathcal{O}))$, $\frac{\partial u^{\eta}}{\partial t} \to \frac{\partial u}{\partial t}$ in $L^2(0,T;V')$.

We assume that

(7.95) $\quad \mathcal{O}$ is an open set of class C^2 satisfying (3.4)

(7.96) $\quad a_{ij} \in C^2(\bar{\mathcal{O}})$, $a_i \in C^1(\bar{\mathcal{O}})$, $a_0 = 0$ (2) ; $\sum a_{ij}\xi_i\xi_j \geq \alpha|\xi|^2$

(7.97) $\quad f \in C^0(\bar{Q})$.

We can always assume $a_{ij} = a_{ji}$; define $\sigma(x)$ by

(7.98) $$\frac{\sigma\sigma^*}{2} = a \quad ; \quad g_i = -a_i + \sum_j \frac{\partial a_{ij}}{\partial x_j} \; .$$

We then construct on an arbitrary probability space the diffusion process $y(s) = y_{xt}(s)$, the solution of

(7.99) $$\left| \begin{array}{l} dy(s) = g(y(s),s)ds + \sigma(y(s)s)ds \; . \\ y(t) = x \; . \end{array}\right.$$

Now write $\tau = \tau_{xt}$ for the exit time from

(7.100) $$\tau_{xt} = \inf \{s \geq t \mid y(s) \notin \mathcal{O}\} \; .$$

It is well known that the solution u of (7.92) can be represented by the formula

(7.101) $$u(x,t) = E \int_t^{\tau_{xt} \wedge T} f(y_{xt}(s),s)ds \; .$$

For the remainder it will be convenient to introduce the process

(7.102) $$z_{xt}(\lambda) = y_{xt}(\lambda + t) \qquad \lambda \geq 0 \; .$$

By (7.100) we have

(7.103) $$\tau_{xt} = t + \tilde{\tau}_{xt}$$

where

(7.104) $$\tilde{\tau}_{xt} = \inf \{\lambda \geq 0 \mid z_{xt}(\lambda) \notin \mathcal{O}\} \; .$$

By (7.99) we have

---

(1) This is a mixed problem (Dirichlet and Neumann boundary conditions).
(2) This latter assumption is entirely directed a simplifying things a little.

$$y(\lambda + t) = x + \int_t^{\lambda + t} g(y(\theta),\theta)\, d\theta + \int_t^{\lambda + t} \sigma(y(\theta),\theta)\, dw(\theta)$$

so

$$z(\lambda) = x + \int_0^{\lambda} g\left(z(\mu), t+\mu\right) d\mu + \int_0^{\lambda} \sigma\left(z(\mu), t+\mu\right) d\tilde{w}(\mu)$$

where

$$\tilde{w}(\mu) = w(t + \mu) - w(t).$$

If we set $\theta(\lambda) = \lambda + t$ we see that the pair $z(\lambda),\theta(\lambda)$ is the solution of

$$\text{(7.105)} \quad \begin{vmatrix} dz = g(z,\theta)d\lambda + \sigma(z,\theta)d\tilde{w}(\lambda) \\ d\theta = d\lambda \\ z(0) = x, \quad \theta(0) = t. \end{vmatrix}$$

Furthermore, we can rewrite (7.101) as follows:

$$\text{(7.106)} \quad u(x,t) = E\int_0^{\tilde{\tau}\wedge(T-t)} f\left(z(\lambda),\theta(\lambda)\right) d\lambda .$$

We now seek to interpret the problem (7.93). To do this we construct the (one-dimensional) reflected process

$$\text{(7.107)} \quad \begin{vmatrix} d\theta^{\eta} = d\lambda + \sqrt{2\eta}\, dw_1(\lambda) + \chi_{\theta^{\eta}=0}\, d\xi^{\eta}(\lambda) \\ \theta^{\eta}(0) = t. \end{vmatrix}$$

In (7.107) $w_1(\lambda)$ is a Wiener process independent of $\tilde{w}(\lambda)$, and $\xi^{\eta}(\lambda)$ is an increasing process, $\xi^{\eta}(0) = 0$, and

$$\text{(7.108)} \quad \chi_{\theta^{\eta}>0}\, d\xi^{\eta}(\lambda) = 0 .$$

We next construct the process $z^{\eta}(\lambda)$ the solution of

$$\text{(7.109)} \quad dz^{\eta} = g(z^{\eta},\theta^{\eta})d\lambda + \sigma(z^{\eta},\theta^{\eta})d\tilde{w}(\lambda), \quad z^{\eta}(0) = x .$$

Notice that by virtue of the assumptions about a_{ij} and a_i, we have $g_i, \sigma_{ij} \in C^1(\bar{Q})$, and we can always assume them to be Lipschitz continuous on R^{n+1} and bounded. The construction of $z^{\eta}(\lambda)$ in the strong sense therefore poses no problem.

On the other hand $u^{\eta} \in W^{2,p}(Q)$, for all p, since u^{η} is the restriction to Q of a Dirichlet problem on $\mathcal{O}\times\,]-T,T[$. We can therefore affirm that

$$\text{(7.110)} \quad u^{\eta}(x,t) = E\int_0^{\tau^{\eta}\wedge\nu^{\eta}} f\left(z^{\eta}(\lambda), \theta^{\eta}(\lambda)\right) d\lambda ,$$

where

$$(7.111)\qquad \left|\begin{array}{l} \tau^{\eta} = \inf\{\lambda \geqq 0 \mid z^{\eta}(\lambda) \notin \mathcal{O}\} \\ \nu^{\eta} = \inf\{\lambda \geqq 0 \mid \theta^{\eta}(\lambda) = T\} . \end{array}\right.$$

Everything therefore reduces to proving that when $\eta \to 0$ the second member of (7.110) converges to the second member of (7.106). Note that the integral (7.110) has a meaning, since $\tau^{\eta} \to +\infty$ a.s.. Therefore there is no need for an adaptation coefficient.

We are going to prove:

THEOREM 7.6: *Assume* (7.95),(7.96),(7.97). *Then*

$$(7.112)\qquad u^{\eta}(x,t) \to u(x,t) \quad \textit{in } C^{o}(\bar{Q}).$$

Proof: First of all we have

$$d(\theta^{\eta} - \theta) = \sqrt{2\eta}\, dw_1(\lambda) + \chi_{\theta^{\eta}=0}\, d\xi^{\eta}$$

therefore

$$(\theta^{\eta} - \theta)d(\theta^{\eta} - \theta) = \sqrt{2\eta}(\theta^{\eta} - \theta)dw_1(\lambda) + (\theta^{\eta} - \theta)\chi_{\theta^{\eta}=0}\, d\xi^{\eta}$$

hence

$$(\theta^{\eta} - \theta)d(\theta^{\eta} - \theta) \leqq \sqrt{2\eta}(\theta^{\eta} - \theta)dw_1(\lambda) .$$

From this it follows that

$$\frac{1}{2}|\theta^{\eta}(s) - \theta(s)|^2 = \int_0^s \left(\theta^{\eta}(s) - \theta(s)\right) d(\theta^{\eta} - \theta) + \eta s$$
$$\leqq \sqrt{2\eta} \int_0^s (\theta^{\eta} - \theta)dw_1(\lambda) + \eta s .$$

Consequently

$$(7.113)\qquad E \sup_{0 \leqq s \leqq S} |\theta^{\eta}(s) - \theta(s)|^2 \leqq (2\sqrt{2} + 1)\eta S .$$

Next we have

$$d(z^{\eta} - z) = \left(g(z^{\eta},\theta^{\eta}) - g(z,\theta)\right) d\lambda + \left(\sigma(z^{\eta},\theta^{\eta}) - \sigma(z,\theta)\right) d\tilde{w}(\lambda)$$
$$\frac{1}{2}|z^{\eta}(s) - z(s)|^2 = (z^{\eta} - z).\left(g(z^{\eta},\theta^{\eta}) - g(z,\theta)\right) ds +$$
$$+ (z^{\eta} - z).\left(\sigma(z^{\eta},\theta^{\eta}) - \sigma(z,\theta)\right) d\tilde{w}(s) + \frac{1}{2}\operatorname{tr}\left(\sigma(z^{\eta},\theta^{\eta}) - \sigma(z,\theta)\right)$$
$$\times \left(\sigma(z^{\eta},\theta^{\eta}) - \sigma(z,\theta)\right)^{*} ds .$$

From this we deduce

$$E|z^\eta(s) - z(s)|^2 \leq C\left[E\int_0^s |z^\eta(\lambda) - z(\lambda)|^2 ds + \eta s\right] .$$

Therefore by Gronwall's Inequality

(7.114) $$E|z^\eta(s) - z(s)|^2 \leq \eta\, C_S \quad , \quad \forall\, s \leq S .$$

Also

(7.115) $$E \sup_{0 \leq s \leq S} |z^\eta(s) - z(s)|^2 \leq D_S \sqrt{\eta} .$$

Consider also the solution of the differential equation

(7.116) $$-\eta \frac{d^2 v^\eta}{dt^2} - \frac{dv^\eta}{dt} = 1 \quad , \qquad v^\eta(T) = 0 \ , \ \frac{dv^\eta}{dt}(0) = 0 .$$

It is easy to verify that

(7.117) $$\begin{aligned} v^\eta(t) &= E\nu^\eta \\ &= T - t + \eta\left(e^{-\frac{T}{\eta}} - e^{-\frac{t}{\eta}}\right) \leq T . \end{aligned}$$

We are now going to show that

(7.118) $\forall$ S fixed, $E|\tau^\eta \wedge \nu^\eta \wedge S - \tilde{\tau} \wedge (T-t) \wedge S| \to 0$. uniformly at x,t.

In fact, the sequence

$$\sup_{x,t} E|\tau^\eta \wedge \nu^\eta \wedge S - \tilde{\tau} \wedge (T-t) \wedge S|$$

is bounded. If we consider a convergent subsequence, thanks to (7.113),(7.114) we can always assume that

(7.119) $$\left|\begin{aligned} &\sup_{0 \leq s \leq S} |\theta^\eta(s) - \theta^\eta(s)| \to 0 \\ &\sup_{0 \leq s \leq S} |z^\eta(s) - \theta^\eta(s)| \to 0 . \end{aligned}\right.$$

uniformly at x,t. If we write

$$\Lambda = \inf\, \{\lambda \geq \tilde{\tau} \,|\, z(\lambda) \in \complement\mathring{\mathcal{O}}\}$$

then we verify as in B.L., Chap. 3, Lemma 3.2, that

$$\Lambda = \tilde{\tau} \quad \text{a.s.}$$

and also

$$\tau^\eta \wedge \nu^\eta \wedge S \to \tilde{\tau} \wedge (T-t) \wedge S \text{ a.s.}$$

uniformly at (x,t). Formula (7.118) follows from this.

Now let us take S > T. Then

$$u^\eta(x,t) = E\left(\int_0^{\tau^\eta \wedge \nu^\eta \wedge S} f(z^\eta,\theta^\eta)d\lambda\right)\chi_{\nu^\eta \leq S} +$$

$$+ E\left(\int_0^{\tau^\eta \wedge \nu^\eta} f\, d\lambda\right)\chi_{\nu^\eta > S} = I^\eta + II^\eta .$$

And then

$$|I^\eta - u| \leq \|f\|E|\tau^\eta \wedge \nu^\eta \wedge S - \tilde{\tau} \wedge (T-t) \wedge S| + \tag{7.120}$$

$$+ E\int_0^S |f(z^\eta,\theta^\eta) - f(z,\theta)|d\lambda + T\|f\|P(\nu^\eta > S) .$$

Moreover,

$$|II^\eta| \leq \|f\|E\,\nu^\eta\,\chi_{\nu^\eta > S}$$

But using the function $v^\eta(t)$ we have

$$T\,E\,\chi_{\nu^\eta > S} = v^\eta(x)\,E\,\chi_{\nu^\eta > S} + E\,\nu^\eta\,\chi_{\nu^\eta > S}$$

from which we easily deduce

$$|II^\eta\| \leq \frac{C}{S} . \tag{7.121}$$

Since

$$E|f(z^\eta,\theta^\eta) - f(z,\theta)|(\lambda) \to 0 , \quad \text{uniformly at } x,t \text{ and } \lambda \in [0,S]$$

from (7.120) and (7.121) we deduce that

$$u^\eta \to u, \text{ uniformly at } x,t.$$

whence the result. □

We now go on to the case of inequalities and consider an obstacle ψ, satisfying the property

$$\psi \in C^0(\bar{Q}) , \quad \psi/_\Sigma \geq 0 , \quad \psi(x,T) \geq 0 . \tag{7.122}$$

The penalised regularised problem is defined by

$$\left|\begin{array}{l} -\eta \dfrac{\partial^2 u^{\epsilon\eta}}{\partial t^2} - \dfrac{\partial u^{\epsilon\eta}}{\partial t} + A(t)u^{\epsilon\eta} + \dfrac{1}{\epsilon}(u^{\epsilon\eta}-\psi)^+ = f \\ u^{\epsilon\eta}(T) = 0 , \quad \dfrac{\partial u^{\epsilon\eta}}{\partial t}(0) = 0 , \quad u^{\epsilon\eta} \in W^{1,p}(\bar{Q}) , \quad \forall\, p > 1 . \end{array}\right.$$

We can interpret the function $u^{\varepsilon\eta}(x,t)$ as:

$$(7.124)\quad u^{\varepsilon\eta}(x,t) = \operatorname*{Inf}_{v(.)} \int_0^{\tau^\eta\wedge\nu^\eta} \left[f(z^\eta(\lambda),\theta^\eta(\lambda)) + \frac{1}{\varepsilon}\psi(z^\eta(\lambda),\theta^\eta(\lambda))v(\lambda)\right] \times \times \left(\exp - \int_0^\lambda \frac{v(\mu)}{\varepsilon}\, d\mu\right) d\lambda$$
$$= \operatorname*{Inf}_{v} J^{\varepsilon\eta}_{xt}(v)\ .$$

Similarly, we consider the penalised problem

$$(7.125)\quad \left|\begin{array}{l} -\dfrac{\partial u^\varepsilon}{\partial t} + A(t)u^\varepsilon + \dfrac{1}{\varepsilon}(u^\varepsilon - \psi)^+ = f \\ u^\varepsilon(T) = 0\ ,\quad u^\varepsilon \in L^p\left(0,T;W^{2,p}(\mathcal{O}) \cap W_0^{1,p}(\mathcal{O})\right)\ ,\ \dfrac{\partial u^\varepsilon}{\partial t} \in L^p(Q)\ . \end{array}\right.$$

We have

$$(7.126)\quad u^\varepsilon(x,t) = \operatorname*{Inf}_{v} \int_0^{\tilde\tau\wedge(T-t)} \left[f(z(\lambda),\theta(\lambda)) + + \frac{1}{\varepsilon}\psi\,(z(\lambda),\theta(\lambda))v(\lambda)\right]\left(\exp - \int_0^\lambda \frac{v(\mu)d\mu}{\varepsilon}\right) d\lambda$$
$$= \operatorname*{Inf}_{v} J^{\varepsilon}_{xt}(v)\ .$$

Now consider the difference

$$J^{\varepsilon\eta}_{xt}(v) - J^\varepsilon_{xt}(v) = E\int_{\tilde\tau\wedge(T-t)}^{\tau^\eta\wedge\nu^\eta} \left[f(z^\eta,\theta^\eta) + \frac{1}{\varepsilon}\psi(z^\eta,\theta^\eta)v(\lambda)\right]\left(\exp - \int_0^\lambda \frac{v}{\varepsilon}\right)d\lambda\ +$$
$$+ E\int_0^{\tilde\tau\wedge(T-t)} \left[f(z^\eta,\theta^\eta) + \frac{1}{\varepsilon}\psi(z^\eta,\theta^\eta)v(\lambda) - f(z,\theta) - \frac{1}{\varepsilon}\psi(z,\theta)v\right]\left(\exp - \int_0^\lambda \frac{v}{\varepsilon}\right)d\lambda$$
$$= I + II$$

$$|I| \leq \|f\|\ E|\tau^\eta \wedge \nu^\eta - \tilde\tau\wedge(T-t)| + \frac{\|\psi\|}{\varepsilon}\ E|\tau^\eta\wedge\nu^\eta - \tilde\tau\wedge(T-t)|$$
$$|II| \leq E \sup_{0\leq s\leq T} |f(z^\eta,\theta^\eta) - f(z,\theta)| + E \sup_{0\leq S\leq T} |\psi(z^\eta,\theta^\eta) - \psi(z,\theta)|\ .$$

Since these estimates are independent of v, we can deduce identical estimates for $|u^{\varepsilon\eta}(x,t) - u^\varepsilon(x,t)|$. Thus we easily obtain:

THEOREM 7.7: *Under the assumptions of Theorem 7.6 and* (7.122) *in* $C^0(\bar Q)$ *we have*

$$(7.127)\quad u^{\varepsilon\eta}(x,t) \to u^\varepsilon(x,t) \quad \textit{when } \eta \to 0,\ \varepsilon \textit{ fixed.}\ \square$$

Now consider the weak maximum solution of the evolutionary V.I.

$$(7.128)\quad \left|\begin{array}{l} u \in L^2(0,T;H_0^1(\mathcal{O}))\ ,\quad u \leq \psi \quad \text{p.p.} \\ -\int_0^T (\frac{\partial v}{\partial t}, v-u)dt + \int_0^T a(t;u,v-u)dt + \frac{1}{2}|(v(T)|^2 \geq \int_0^T (f,v-u)dt \\ \forall\ v \in L^2(0,T;H_0^1)\ ,\ \frac{\partial v}{\partial t} \in L^2(0,T;H^{-1})\ ,\ v \leq \psi \quad \text{a.e.} \end{array}\right.$$

This maximum weak solution is denoted u. Its existence has been proved by F. MIGNOT - J.P. PUEL [1]. In B.L., Chap. 3. Section 4.4, it is proved that u(x,t) has a probabilistic interpretation.

By using the process $z(\lambda)$, in preference to $y(s)$, we can write

$$(7.129)\quad u(x,t) = \inf_{\zeta} J_{xt}(\zeta)$$

with

$$(7.130)\quad J_{xt}(\zeta) = E\left[\int_0^{\tilde{\tau}\wedge (T-t)\wedge \zeta} f(z(\lambda),\theta(\lambda))d\lambda + \psi(z(\zeta),\theta(\zeta))\chi_{\zeta<\tilde{\tau}\wedge (T-t)}\right].$$

We now define the regularised V.I. by

$$(7.131)\quad \left|\begin{array}{l} u^{\eta} \in L^2(0,T;H^1(\mathcal{O}))\ ,\ \frac{\partial u^{\eta}}{\partial t} \in L^2(0,T;L^2(\mathcal{O}))\ ,\ u^{\eta}(T) = 0\ , \\ u^{\eta} \leq \psi \quad \text{a.e.} \\ \int_0^T \left[\eta\left(\frac{\partial u^{\eta}}{\partial t}, \frac{\partial v}{\partial t} - \frac{\partial u^{\eta}}{\partial t}\right) - \left(\frac{\partial u^{\eta}}{\partial t}, v-u^{\eta}\right) + a\left(t;u^{\eta},v-u^{\eta}\right) - \left(f,v-u^{\eta}\right)\right] dt \geq 0 \\ \forall\ v \in L^2(0,T;H_0^1(\mathcal{O}))\ ,\ \frac{\partial v}{\partial t} \in L^2(0,T;L^2(\mathcal{O})), v(T) = 0\ ,\ v \leq \psi \quad \text{a.e}\ . \end{array}\right.$$

We can then interpret $u^{\eta}(x,t)$ by

$$(7.132)\quad u^{\eta}(x,t) = \inf_{\zeta} J_{xt}^{\eta}(\zeta)$$

with

$$(7.133)\quad J_{xt}^{\eta}(\zeta) = E\left[\int_0^{\tau^{\eta}\wedge \nu^{\eta}\wedge \zeta} f(z^{\eta}(\lambda),\theta^{\eta}(\lambda))d\lambda + \psi(z^{\eta}(\zeta),\theta^{\eta}(\zeta))\chi_{\zeta<\tau^{\eta}\wedge \nu^{\eta}}\right].$$

We can then state:

THEOREM 7.8: *Under the assumptions of Theorem 7.6 and* (7.122), *together with*

$$(7.134)\quad a(t;v,v) \geq \alpha \|v\|^2 \quad \forall\ v \in H_0^1(\mathcal{O})$$

we then have

$$(7.135)\quad u^{\eta}(x,t) \to u(x,t) \quad \textit{in } C^0(\bar{Q}).$$

Proof: The assumption (7.134) allows us to prove the existence and uniqueness of u^η, the solution of (7.131), as well as the existence of the maximum solution u. We cannot use the result obtained in the preceding theorem on the penalised scheme, because the estimate obtained there was not independent of ε. We must proceed differently. We introduce w^η the solution of

$$-\eta \frac{\partial^2 w^\eta}{\partial t^2} - \frac{dw^\eta}{\partial t} + (A(t)+a_o)w^\eta = f \quad , \quad w^\eta(T) = 0 \quad , \quad \frac{\partial w^\eta}{\partial t}(0) = 0$$

and w is the solution of

$$-\frac{\partial w}{\partial t} + A(t)w = f \quad , \quad w(T) = 0 \quad , \quad \frac{\partial w}{\partial t}(0) = 0 \, .$$

By Theorem 7.6 we have

(7.136) $\quad w^\eta \to w \quad \text{in} \quad C^o(\bar{Q})$.

We can then rewrite things,

$$(7.137) \quad \left| \begin{array}{l} J^\eta_{xt}(\zeta) = w^\eta(x,t) + E\,(\psi - w^\eta)(z^\eta(\zeta),\theta^\eta(\zeta))\chi_{\zeta<\tau^\eta\wedge\nu^\eta} \\ \\ \text{and} \\ J_{xt}(\zeta) = w(x,t) + E\,(\psi - w)\,(z(\zeta),\,\theta(\zeta))\,\chi_{\zeta<\tilde{\tau}\wedge(T-t)} \, . \end{array} \right.$$

It is clear that

$$u^\eta(x,t) - w^\eta(x,t) \leq 0$$

and therefore

$$(7.138) \quad u^\eta(x,t) - w^\eta(x,t) = \operatorname*{Inf}_\zeta E(\psi - w^\eta)\ \chi_\sigma\ (z^\eta(\zeta\wedge\tau^\eta\wedge\nu^\eta),\theta^\eta(\zeta,\tau^\eta,\nu^\eta))$$

$$(7.139) \quad u(x,t) - w(x,t) = \operatorname*{Inf}_\zeta E(\psi - w)\ \ \chi_\sigma\ (z(\zeta\wedge\tilde{\tau}\wedge(T-t)),\theta(\zeta,\tilde{\tau}\wedge(T-t)) \, .$$

But then

$$(7.140) \quad \left| E\left[(\psi-w^\eta)\,\chi_\sigma\,(z^\eta(\zeta\wedge\tau^\eta\wedge\nu^\eta),\theta^\eta(\zeta,\tau^\eta,\nu^\eta) - (\psi-w)\,\chi_\sigma\,(z(\zeta\wedge\tilde{\tau}\wedge(T-t))\,, \right.\right. \\ \left.\left. \theta(\zeta,\tilde{\tau}\wedge T-t))\right]\right| \leqq$$

$$\left| E\left[(\psi - w^\eta)\ \chi_\sigma(z^\eta(\zeta\wedge\tau^\eta\wedge\nu^\eta\wedge S),\theta^\eta(\zeta\wedge\tau^\eta\wedge\nu^\eta\wedge S)) - \right.\right. \\ \left.\left. -(\psi-w)\ \chi_\sigma\ (z(\zeta\wedge\tilde{\tau}\wedge T-t),\theta(\zeta\wedge\tilde{\tau}\wedge T-t))\right]\right| + \frac{C}{S} = X^\eta + \frac{C}{S} \, .$$

Then

$$X^{\eta} \leq \|w^{\eta} - w\|_{C^{0}(\bar{Q})} + \Big| E\Big[(\psi - w)\, \chi_{\sigma} \,|z^{\eta}(\zeta \wedge \tau^{\eta} \wedge \nu^{\eta} \wedge S),$$

$$\theta^{\eta}(\zeta \wedge \tau^{\eta} \wedge \nu^{\eta} \wedge S)) - (\psi - w)\, \chi_{\sigma}\,(z(\zeta \wedge \tau^{\eta} \wedge \nu^{\eta} \wedge S), \theta(\zeta \wedge \tau^{\eta} \wedge \nu^{\eta} \wedge S))\Big|$$

$$+ E\Big| (\psi - w)\, \chi_{\sigma}\,(z(\zeta \wedge \tau^{\eta} \wedge \nu^{\eta} \wedge S), \theta(\zeta \wedge \tau^{\eta} \wedge \nu^{\eta} \wedge S)) -$$

$$- (\psi - w)\, \chi_{\sigma}\,(z(\zeta \wedge \tilde{\tau} \wedge T-t), \theta(\zeta \wedge \tilde{\tau} \wedge T-t))\Big| .$$

But taking account of (7.113) and (7.115) as well as

$$E|z(\zeta \wedge \tau^{\eta} \wedge \nu^{\eta} \wedge S) - z(\zeta \wedge \tilde{\tau} \wedge T-t)|^2 \leq C\, E|\tau^{\eta} \wedge \nu^{\eta} \wedge S - \tilde{\tau} \wedge (T-t)|$$

$$E|\theta(\zeta \wedge \tau^{\eta} \wedge \nu^{\eta} \wedge S) - \theta(\zeta \wedge \tilde{\tau} \wedge T-t)|^2 \leq \quad ''$$

from which, taking into account that ψ, w are bounded, we easily deduce that $X^{\eta} \to 0$. Since the preceding estimates are uniform in ζ and also in x,t, we certainly obtain the result desired. □

8. OPTIMAL STOPPING TIME PROBLEMS: EVOLUTIONARY CASE

Method of Approach

We seek to interpret the evolutionary V.I.'s studied in Section 6.

8.1 *Regular Case*

We assume the conditions of Corollary 6.2, therefore there exists a function $u(x,t)$ on $\bar{Q}$ that is unique and such that

$$(8.1) \quad \begin{cases} u \in L^p(0,T;W^{2,p}(\mathcal{O})), \ \dfrac{\partial u}{\partial t} \in L^p(Q) \\ -\dfrac{\partial u}{\partial t} + (A(t) + a_0)u \leq f \quad , \quad u \leq \psi \quad \text{a.e.} \\ (u - \psi)\left(-\dfrac{\partial u}{\partial t} + (A(t) + a_0)u - f\right) = 0 \quad \text{a.e.} \\ \dfrac{\partial u}{\partial \nu_{A(t)}}\Big|_{\Sigma} = 0 \quad , \quad u(x,T) = \bar{u}(x) . \end{cases}$$

We define σ by $a = \dfrac{\sigma\sigma^*}{2}$ and g by

$$g_i = a_i + \sum_j \frac{\partial a_{ij}}{\partial x_j} ,$$

then

$$\gamma_i(x,t) = \sum_j a_{ij}(x,t)\, u_j(x) .$$

We assume that the open set $\mathcal{O}$ satisfies (7.3). Under these conditions the functions a_{ij}, g_i, γ_i, and the open set $\mathcal{O}$ satisfy the assumptions of Theorem 4.1. We then consider the probability measure P^{xt}, the solution of the sub-martingale problem (3.6),(3.7). For every stopping time θ of μ^s we define the functional

(8.2) $$J_{xt}(\theta) = E^{xt}\left[\int_t^{\theta\wedge T} f(y(s),s)\left(\exp -\int_t^s a_o(y(\lambda),\lambda)d\lambda\right) ds + \right.$$

$$+ \psi(y(\theta),\theta)\chi_{\theta<T} \exp -\int_t^{\theta} a_o(y(s),s)ds +$$

$$\left. + \bar{u}(y(T))\,\chi_{\theta\geqq T} \exp -\int_t^T a_o(y(s),s)ds\right] .$$

We then have:

THEOREM 8.1: *If we assume the conditions of Corollary 6.2, together with* (7.3), *with* $p > \frac{n}{2} + 1$, *then the solution* u *of* (8.1) *is given by*

(8.3) $$u(x,t) = \operatorname*{Inf}_{\theta} J_{xt}(\theta)$$

Furthermore, there exists an optimal stopping time $\hat{\theta}$ *defined by*

(8.4) $$\hat{\theta} = \inf\{s \geqq t \mid u(y(s),s) = \psi(y(s),s)\} .$$

Proof: This is done as for the stationary case (cf. Theorem 7.1). □

8.2 *Non-Regular Case*

Consider the penalised problem

(8.5) $$\left|\begin{array}{l} -\frac{\partial u^\epsilon}{\partial t} + (A(t)+a_o)u^\epsilon + \frac{1}{\epsilon}(u^\epsilon-\psi)^+ = f \quad , \quad \frac{\partial u^\epsilon}{\partial \nu_{A(t)}}\Big|_{\Sigma} = 0 \ , \\ u^\epsilon(x,T) = \bar{u}(x) \ , \ u^\epsilon \in L^p(0,T;W^{2,p}(\mathcal{O})) \ , \ \frac{\partial u^\epsilon}{\partial t} \in L^p(Q) \ . \end{array}\right.$$

Consider, further, the functional

(8.6) $$J^\epsilon_{xt}(v) = E^{xt}\left[\int_t^T \left(f(y(s),s) + \frac{1}{\epsilon}\psi(y(s),s)v(s)\right)\left(\exp -\int_t^s (a_o + \frac{v}{\epsilon})d\lambda\right) + \right.$$

$$\left. + \bar{u}(y(T)) \exp -\int_t^T (a_o + \frac{v}{\epsilon})ds\right] .$$

THEOREM 8.2: *The coefficients are assumed to satisfy* (6.1),(6.2),(6.35). *Assume* (7.3) *also, and*

(8.7) $$f \in L^p(Q) \ , \ p > \frac{n}{2} + 1 \ ; \ \bar{u} \in W^{2,p}(\mathcal{O}) \ , \ \frac{\partial \bar{u}}{\partial x}\Big|_{\Gamma} = 0$$

(8.8) $$\psi \in C^o(\bar{Q}) \ , \ \bar{u}(x) \leq \psi(x,T) .$$

Then

(8.9) $$u^\epsilon(x,t) = \operatorname*{Inf}_{v(.)} J^\epsilon_{xt}(v) .$$

Moreover, there then exists an optimal control defined by

(8.10) $$\hat{v}^{\varepsilon}(s) = \begin{cases} 1 & \text{if } u^{\varepsilon}(y(s),s) \geqq \psi(y(s),s) \\ 0 & \text{otherwise.} \end{cases}$$

Proof: This is analogous to the stationary cases (cf. Theorem 7.2). □

Lastly, we consider the maximum weak solution of the evolutionary V.I., i.e., the maximum element of the set of functions satisfying (6.16),(6.17),(6.18).

We have:

THEOREM 8.3: *Take the assumptions of Theorem 8.2. Then*

(8.11) $$u(x,t) = \operatorname*{Inf}_{\theta} J_{xt}(\theta)$$

where u *is the maximum weak solution of the evolutionary V.I., and* $J_{xt}(\theta)$ *is defined by* (8.2). *Further,* $u \in C^{o}(\bar{Q})$, *and*

(8.12) $$u^{\varepsilon} \to u \quad \text{in } C^{o}(\bar{Q}).$$

Finally, there exists an optimal stopping time, defined by

(8.13) $$\hat{\theta} = \operatorname{Inf}\{s \geqq t \mid u(y(s),s) = \psi(y(s),s)\} .$$

Proof: This is analogous to the stationary case. □

9. GAMES PROBLEMS

9.1 *Stationary Case*

We assume that the coefficients satisfy

(9.1) $$\left| \begin{array}{l} a_{ij} = a_{ji} \;, \; \sum a_{ij}\xi_i\xi_j \geqq \alpha|\xi|^2 \;, \; a_{ij} \in W^{1,\infty}(\mathcal{O}) \\ a_i, a_o \text{ bounded, measurable, and } a_o \geqslant \beta > 0. \end{array} \right.$$

(9.2) $$f \in L^p(\mathcal{O}),\ p > \frac{n}{2}\ ,\ \psi_1, \psi_2 \in W^{2,p}(\mathcal{O}),\ p > n\ ,\ \psi_1 \leqq \psi_2\ ,\ \left.\frac{\partial \psi_2}{\partial \nu_A}\right|_{\Gamma} \geqq 0, \left.\frac{\partial \psi_1}{\partial \nu_A}\right|_{\Gamma} \leqq 0$$

Remark 9.1

We did not make the assumption of coercivity. □

We now consider the measure $P^x = P^{x,o}$ which is the solution of the sub-martingale problem (cf. Section 7.1). Let θ_1, θ_2 be two stopping times of μ^t, and define the functional

(9.3) $$J_x(\theta_1, \theta_2) = E^x\left[\int_0^{\theta_1 \wedge \theta_2} f(y(t)) \left(\exp - \int_0^t a_o(y(s))ds\right) dt + \right.$$

$$+ \psi_1(y(\theta_1))\chi_{\theta_1 \leqq \theta_2} \exp - \int_0^{\theta_1} a_o(y(s))ds + \qquad \text{(Contd)}$$

(Contd) $$+ \psi_2(y(\theta_2))\chi_{\theta_2<\theta_1} \exp - \int_0^{\theta_1} a_o(y(s))ds \Big] .$$

We introduce the following problems

(9.4)
$$\left|\begin{array}{l} u \in W^{2,p}(\mathcal{O}) \quad , \quad \psi_1 \leq u \leq \psi_2 \quad , \quad \left.\dfrac{\partial u}{\partial \nu_A}\right|_\Gamma = 0 \\ Au + a_o u = f \quad \text{if} \quad \psi_1 < u < \psi_2 \\ Au + a_o u \geq f \quad \text{if} \quad u = \psi_1 \\ Au + a_o u \leq f \quad \text{if} \quad u = \psi_2 \end{array}\right.$$

(9.5)
$$\left|\begin{array}{l} Au^\epsilon + a_o u^\epsilon + \dfrac{1}{\epsilon}(u^\epsilon - \psi_2)^+ - \dfrac{1}{\epsilon}(u^\epsilon - \psi_1)^- = f \\ \left.\dfrac{\partial u^\epsilon}{\partial \nu_A}\right|_\Gamma = 0 \ , \quad u^\epsilon \in W^{2,p}(\mathcal{O}) \ . \end{array}\right.$$

Then if $v_1(t), v_2(t)$ are processes adapted to μ^t such that

(9.6) $$v_i(t) \in [0,1], \quad \forall\, t,$$

the functional

(9.7)
$$J_x^\epsilon(v_1, v_2) = E^x \int_0^\infty \Big[f(y(t)) + \frac{1}{\epsilon}\psi_1(y(t))v_1(t) + $$
$$+ \frac{1}{\epsilon}\psi_2(y(t))v_2(t)\Big] \left(\exp - \int_0^t (a_o(y(s)) + \frac{1}{\epsilon} v_1(s) + \frac{1}{\epsilon} v_2(s))\, ds \right) dt \ .$$

Lastly we shall make the assumptions

(9.8) $$f - A\psi_2 \leq C \qquad f - A\psi_1 \geq -C \ , \qquad C \geq 0 \ .$$

We are now going to prove the following Theorems:

THEOREM 9.1: *Under the assumptions* (9.1),(9.2),(9.8), *there exists one and only one solution of* (9.5)

(9.9) $$u^\epsilon(x) = \operatorname*{Inf}_{v_2} \operatorname*{Sup}_{v_1} J_x^\epsilon(v_1, v_2) = \operatorname*{Sup}_{v_1} \operatorname*{Inf}_{v_2} J_x^\epsilon(v_1, v_2) \qquad \square$$

THEOREM 9.2: *Under the assumptions* (9.1),(9.2),(9.8), *there exists one and only one solution of* (9.4)

(9.10) $$u(x) = \operatorname*{Inf}_{\theta_2} \operatorname*{Sup}_{\theta_1} J_x(\theta_1, \theta_2) = \operatorname*{Sup}_{\theta_1} \operatorname*{Inf}_{\theta_2} J_x(\theta_1, \theta_2) \qquad \square$$

The controls that realise the saddle point in (9.9) are defined by

$$(9.11)\quad \left|\begin{array}{l} \hat{v}_{1\varepsilon}(t) = \begin{cases} 1 & \text{if } u^\varepsilon(y(t)) \leq \psi_1(y(t)) \\ 0 & \text{otherwise} \end{cases} \\ \hat{v}_{2\varepsilon}(t) = \begin{cases} 1 & \text{if } u^\varepsilon(y(t)) \geq \psi_2(y(t)) \\ 0 & \text{otherwise} \end{cases} \end{array}\right.$$

The stopping times realising the saddle point in (9.10) are defined by

$$(9.12)\quad \left|\begin{array}{l} \hat{\theta}_1 = \inf\{s \geq 0 \mid u(y(s)) = \psi_1(y(s))\} \\ \hat{\theta}_2 = \inf\{s \geq 0 \mid u(y(s)) = \psi_2(y(s))\} . \end{array}\right.$$

Proof of Theorem 9.1: For $\varphi \in C^0(\bar{\mathcal{O}})$, we define $S(\varphi) = z$ as the solution of

$$(9.13)\quad \left|\begin{array}{l} Az^\varepsilon + a_0 z^\varepsilon + \frac{2}{\varepsilon} z^\varepsilon = f - \frac{1}{\varepsilon}(\varphi - \psi_2)^+ + \frac{1}{\varepsilon}(\varphi - \psi_1)^- + \frac{2}{\varepsilon}\varphi \\ \left.\frac{\partial z^\varepsilon}{\partial \nu_A}\right|_\Gamma = 0 , \qquad z^\varepsilon \in W^{2,p}(\mathcal{O}) . \end{array}\right.$$

The mapping S is therefore defined from $C^0(\bar{\mathcal{O}})$ into itself. As in the Dirichlet case we easily show that

$$(9.14)\quad \|z_1 - z_2\|_{C^0(\bar{\mathcal{O}})} \leq \frac{1}{\frac{1+\varepsilon\beta}{2}} \|\varphi_1 - \varphi_2\|_{C^0(\bar{\mathcal{O}})}$$

and thus S is a contraction. It is clear that every fixed point of S is a solution of (9.5) and conversely; hence we have the existence and the uniqueness of the solution of (9.5).

Next we prove that

$$(9.15)\quad J_\varepsilon^x(\hat{v}_1, \hat{v}_{2\varepsilon}) \leq J_\varepsilon^x(\hat{v}_{1\varepsilon}, \hat{v}_{2\varepsilon}) = u^\varepsilon(x) \leq J_\varepsilon^x(\hat{v}_{1\varepsilon}, v_2) \qquad \forall\, v_1, v_2$$

which evidently suffices to prove (9.9). We first reduce to the case f = 0, by considering the solution of

$$(9.16)\quad Aw + a_0 w = f , \quad \left.\frac{\partial w}{\partial \nu_A}\right|_\Gamma = 0 , \quad w \in W^{2,p}(\mathcal{O})$$

and then

$$(9.17)\quad w(x) = E^x \int_0^\infty f(y(s)) \exp - \int_0^s a_0(y(\lambda))d\lambda .$$

Everything reduces to changing ψ_1,ψ_2 into $\psi_1 - w$ and $\psi_2 - w$, which satisfy the same assumptions as ψ_1,ψ_2. The interest of this is that the function $u^\varepsilon - w \in W^{2,p}(\mathcal{O})$ for all p > 1. Therefore $u^\varepsilon - w \in C^1(\bar{\mathcal{O}})$, and we can apply the formula (8.14). In fact, with the arguments already seen, for arbitrary controls $v_1(t)$, $v_2(t)$ we obtain

$$(9.18)\qquad u^\varepsilon(x) - w(x) = E^x\int_0^\infty \left[(A+a_o)(u^\varepsilon - w)\exp -\int_0^t\left(a_o + \frac{1}{\varepsilon}(v_1+v_2)\right)ds\right.$$
$$\left. + \frac{1}{\varepsilon}(v_1+v_2)(u^\varepsilon - w)\exp -\int_0^t\left(a_o+\frac{1}{\varepsilon}(v_1+v_2)\right)ds\right] dt$$

hence we have

$$(9.19)\qquad u^\varepsilon(x) = E^x\int_0^\infty\left(Au^\varepsilon + a_o u^\varepsilon + \frac{1}{\varepsilon}(v_1+v_2)u^\varepsilon\right)\left(\exp -\int_0^t (a_o + \frac{1}{\varepsilon}(v_1+v_2))\, ds\right)dt .$$

Using equation (9.5) and arguing as for the Dirichlet case, we easily deduce the property (9.15), and hence the result desired. □

Remark 9.2

As to the regularity of ψ_1,ψ_2, only the fact that $\psi_1,\psi_2 \in L^\infty(\mathcal{O})$ has been used in the proof of Theorem 9.1. The assumptions (9.2) only arise when $\varepsilon \to 0$. □

Proof of Theorem 9.2: Since $\psi_1,\psi_2 \in W^{2,p}(\mathcal{O})$, $p > n$, we have $\psi_1,\psi_2 \in C^0(\bar{\mathcal{O}})$. Therefore we have the formula

$$(9.20)\qquad \psi_i(x) = E^x\int_0^\infty\left[(A+a_o)\psi_i(y(t)) + \frac{1}{\varepsilon}\psi_i(y(t))(v_1(t)+v_2(t))\right]\left(\exp -\int_0^t (a_o +\right.$$
$$\left. + \frac{v_1+v_2}{\varepsilon})ds\right)dt + E^x\int_0^\infty \frac{\partial\psi_i}{\partial\nu_A}\left(\exp -\int_0^t (a_o + \frac{1}{\varepsilon}(v_1+v_2))ds\right)d\xi(t)$$

from which we deduce

$$(9.21)\qquad J_x^\varepsilon(v_1,v_2) - \psi_1(x) = E^x\left\{\int_0^\infty\left[(f-(A+a_o)\psi_1)(y(t)) + \frac{1}{\varepsilon}(\psi_2-\psi_1)(y(t))v_2(t)\right]\times\right.$$
$$\left.\times\left(\exp -\int_0^t (a_o+\frac{1}{\varepsilon}(v_1+v_2))ds\right)dt - \int_0^\infty\frac{\partial\psi_1}{\partial\nu_A}\left(\exp -\int_0^t (a_o+\frac{1}{\varepsilon}(v_1+v_2))ds\right)d\xi(t)\right\}$$

$$(9.22)\qquad J_x^\varepsilon(v_1,v_2) - \psi_2(x) = E^x\left\{\int_0^\infty\left[(f-(A+a_o)\psi_2) + \frac{1}{\varepsilon}(\psi_1-\psi_2)v_1(t)\right]\left(\exp -\int_0^t (a_o+\frac{1}{\varepsilon}(v_1+v_2))ds\right)dt\right.$$
$$\left. - \int_0^\infty\frac{\partial\psi_2}{\partial\nu_A}\left(\exp -\int_0^t (a_o+\frac{1}{\varepsilon}(v_1+v_2))ds\right)d\xi(t)\right\} .$$

Since

$$\psi_1 \leqq \psi_2 \quad\text{and}\quad \left.\frac{\partial\psi_2}{\partial\nu_A}\right|_\Gamma \geqq 0 ,$$

we deduce from (9.22) that

$$J_x^\varepsilon(v_1,v_2) - \psi_2(x) \leqq E^x\int_0^\infty (f-(A+a_o)\psi_2)\left(\exp -\int_0^t (a_o+\frac{1}{\varepsilon}(v_1+v_2))ds\right)dt$$

and using (9.8)

$$J_x^\varepsilon(v_1,v_2) - \psi_2(x) \leq C\, E^x \int_0^\infty \left(\exp - \int_0^t (a_o + \frac{1}{\varepsilon}(v_1 + v_2))ds \right) dt \, . \tag{9.23}$$

Therefore

$$\underset{v_2}{\text{Inf}}\ J_x^\varepsilon(v_1,v_2) - \psi_2(x) \leq C\, E^x \int_0^\infty \exp - \int_0^t (a_o + \frac{v_1}{\varepsilon})ds\ \exp - \frac{t}{\varepsilon}\, dt \leq C\varepsilon$$

and thus

$$\frac{(u^\varepsilon(x) - \psi_2(x))^+}{\varepsilon} \leq C \, .$$

An analogous proof, this time using $\left.\frac{\partial \psi 1}{\partial \nu_A}\right|_\Gamma \leqslant 0$, shows that

$$\frac{(u^\varepsilon(x) - \psi_1(x))^-}{\varepsilon} \leq C \, .$$

We then deduce from (9.5) that u^ε remains in a bounded set of $W^{2,p}(\mathcal{O})$. It is then easy to deduce from (9.5) that we can pick a subsequence, denoted again by u^ε, weakly converging in $W^{2,p}(\mathcal{O})$ to a function u the solution of (9.4).

Notice also that $u^\varepsilon - w$ stays in a bounded set of $W^{2,p}(\mathcal{O})$, for all $p > 1$. Therefore $u - w \in W^{2,p}(\mathcal{O})$, for all $p > 1$.

We can then write the formula (where θ^1, θ^2 are arbitrary stopping times)

$$u(x) - w(x) = E^x \Big[\int_0^{\theta^1 \wedge \theta^2} (Au - Aw + a_o(u - w)(y(t)) \left(\exp - \int_0^t a_o(y(s))ds \right) dt + \\ + (u(y(\theta^1 \wedge \theta^2)) - w(y(\theta^1 \wedge \theta^2))) \left(\exp - \int_0^{\theta^1 \wedge \theta^2} a_o(y(s))ds \right) \Big]$$

therefore

$$u(x) = E^x \Big[\int_0^{\theta^1 \wedge \theta^2} (Au + a_o u)(y(t)) \left(\exp - \int_0^t a_o(y(s))ds \right) dt + \\ + u(y(\theta^1 \wedge \theta^2)) \exp - \int_0^{\theta^1 \wedge \theta^2} a_o(y(t))dt \Big] . \tag{9.24}$$

Using the relations (9.4), we easily deduce (as in the Dirichlet case) that

$$J_x(\theta_1, \hat{\theta}_2) \leq J_x(\hat{\theta}_1, \hat{\theta}_2) \leq J_x(\hat{\theta}_1, \theta_2) \tag{9.25}$$

and hence (9.10), which completes the proof of Theorem 9.2. □

We now give a result on the uniform convergence of u^ε to u:

THEOREM 9.3: *Assuming* (9.1),(9.2),(9.8), *then*

(9.26) $$\|u^\varepsilon - u\|_{C^0(\bar{\mathcal{O}})} \to 0 \text{ with } \varepsilon .$$

Proof: We again use formulae (9.20), this time on the functionals $J^x(\theta_1,\theta_2)$. First

$$J_x(\theta_1,\theta_2) = E^x\Big[\int_0^{\theta_1\wedge\theta_2} f(y(t))\Big(\exp - \int_0^t a_o(y(s))ds\Big)dt + (\psi_1-\psi_2)(y(\theta_1))\,\chi_{\theta_1\leqq\theta_2} \exp - \int_0^{\theta_1} a_o(y(t))dt + \psi_2(y(\theta_1\wedge\theta_2)) \exp - \int_0^{\theta_1\wedge\theta_2} a_o(y(t))dt\Big]$$

and using (9.20), we then have

(9.27) $$J_x(\theta_1,\theta_2) - \psi_2(x) = E^x\Big[\int_0^{\theta_1\wedge\theta_2} (f-(A+a_o)\psi_2)(y(t))\Big(\exp - \int_0^t a_o(y(s))ds\Big)dt + (\psi_1-\psi_2)(y(\theta_1))\,\chi_{\theta_1\leqq\theta_2} \exp - \int_0^{\theta_1} a_o(y(t))dt - \int_0^{\theta_1\wedge\theta_2} \frac{\partial\psi_2}{\partial\nu_A}(y(t))\Big(\exp - \int_0^t a_o(y(s))ds\Big)d\xi(t)\Big] .$$

Similarly

(9.28) $$J_x(\theta_1,\theta_2) - \psi_1(x) = E^x\Big[\int_0^{\theta_1\wedge\theta_2} (f-(A+a_o)\psi_1)(y(t))\Big(\exp - \int_0^t a_o(y(s))ds\Big)dt + (\psi_2-\psi_1)(y(\theta_2))\,\chi_{\theta_2<\theta_1} \exp - \int_0^{\theta_2} a_o(y(s))ds - \int_0^{\theta_1\wedge\theta_2} \frac{\partial\psi_1}{\partial\nu_A}(y(t))\Big(\exp - \int_0^t a_o(y(s))ds\Big)d\xi(t)\Big] .$$

We now define the controls

(9.29) $$\left|\begin{array}{l} \hat{v}_1(t) = \begin{cases} 0 & \text{if } t < \hat{\theta}_1 \\ 1 & \text{if } t \geqq \hat{\theta}_1 \end{cases} \\ \hat{v}_2(t) = \begin{cases} 0 & \text{if } t < \hat{\theta}_2 \\ 1 & \text{if } t \geqq \hat{\theta}_2 \end{cases} \end{array}\right.$$

and then the stopping times

$$(9.30)\quad \left|\begin{array}{lll} \hat\theta_{1\varepsilon} & = & \inf\{t \geq 0 \mid u^\varepsilon(y(t)) \leq \psi_1(y(t))\} \\ \hat\theta_{2\varepsilon} & = & \inf\{t \geq 0 \mid u^\varepsilon(y(t)) \geq \psi_2(y(t))\}\ . \end{array}\right.$$

Notice that

$$(9.31)\quad \left|\begin{array}{llll} \hat v_{1\varepsilon}(t) = 0 & \text{if} & t < \hat\theta_{1\varepsilon} \\ \hat v_{2\varepsilon}(t) = 0 & \text{if} & t < \hat\theta_{2\varepsilon}\ . \end{array}\right.$$

Further, we modify the controls $\hat v_{1\varepsilon}$ and $\hat v_{2\varepsilon}$ in the following way:

$$(9.32)\quad \hat v^\delta_{1\varepsilon}(t) = \left|\begin{array}{lll} \hat v_{1\varepsilon}(t) & \text{if} & t < \hat\theta_{1\varepsilon} \\ 1 & \text{if} & \hat\theta_{1\varepsilon} \leq t < \hat\theta_{1\varepsilon} + \delta \\ \hat v_{1\varepsilon}(t) & \text{if} & t < \hat\theta_{1\varepsilon} + \delta \end{array}\right.$$

and a similar modification for $\hat v_{2\varepsilon}(t)$.

From (3.19) and equation (9.5), we also have the relation

$$(9.33)\quad u^\varepsilon(x) = J^\varepsilon_x(v_1,v_2) + \frac{1}{\varepsilon} E^x \int_0^{+\infty} \left(\exp - \int_0^t (a_0 + \frac{1}{\varepsilon}(v_1+v_2))ds\right)\Big\{v_1(t)(u^\varepsilon - \psi_1) + \\ + (u^\varepsilon - \psi_1)^- + v_2(t)(u^\varepsilon - \psi_2) - (u^\varepsilon - \psi_2)^+\Big\}dt\ .$$

Therefore

$$(9.34)\quad J_x(\hat v^\delta_{1\varepsilon}, v_2) \geq u^\varepsilon(x) - \frac{1}{\varepsilon} E^x \int_0^\infty \left(\exp - \int_0^t (a_0 + \frac{1}{\varepsilon}(\hat v^\delta_{1\varepsilon} + v_2))ds\right)\Big\{\hat v^\delta_{1\varepsilon}(u^\varepsilon - \psi_1) + \\ + (u^\varepsilon - \psi_1)^-\Big\}dt = u^2(x) - E^x X\ .$$

Now,

$$X = \frac{1}{\varepsilon}\int_{\hat\theta_{1\varepsilon}}^{\hat\theta_{1\varepsilon}+\delta} \left(\exp - \int_0^t (a_0 + \frac{1}{\varepsilon}v_2)ds\right) \exp - \frac{t - \hat\theta_{1\varepsilon}}{\varepsilon}\,(u^\varepsilon - \psi_1)^+(y(t))dt$$

$$\leq \operatorname*{Sup}_{\hat\theta_{1\varepsilon} \leq t \leq \hat\theta_{1\varepsilon}+\delta} |(u^\varepsilon - \psi_1)(y(t))|\ .$$

But

$$E^x \operatorname*{Sup}_{\hat\theta_{1\varepsilon} \leq t \leq \hat\theta_{1\varepsilon}+\delta} |y(t) - y(\hat\theta_{1\varepsilon})|^2 \leq C\,\delta^{1/2}$$

and $(u^\varepsilon - \psi_1)(y(\hat\theta_{1\varepsilon})) = 0$. Since $\|u^\varepsilon\|_{W^{2,p}(\mathcal{O})} \leqslant C$ also, we have

$$|u^\varepsilon(x) - u^\varepsilon(x')| \leq C\,|x - x'|^\alpha \quad \forall\; x, x' \in \bar{\mathcal{O}}\;,$$

where C, α are independent of ε.

Also let $\rho_1(\lambda)$ be the modulus of continuity of the function ψ_1. From the preceding considerations we easily obtain

$$E^x X \leq C\left[\lambda^\alpha + \rho_1(\lambda) + \frac{\delta^{1/2}}{\lambda^2}\right] \quad \forall\; \lambda\;.$$

Now choose $\lambda = \delta^{1/8}$. Hence we obtain

$$J_x^\varepsilon(\hat{v}_{1\varepsilon}^\delta, v_2) \geq u^\varepsilon(x) - C\left[\delta^{\alpha/8} + \rho_1(\delta^{1/8}) + \delta^{1/4}\right]. \tag{9.35}$$

Define $\hat{v}_{2\varepsilon}^\delta$ similarly to $\hat{v}_{1\varepsilon}^\delta$ and then

$$J_x^\varepsilon(v_1, v_{2\varepsilon}^\delta) \leq u^\varepsilon(x) + C\left[\delta^{\alpha/8} + \rho_2(\delta^{1/8}) + \delta^{1/4}\right]. \tag{9.36}$$

Let us define the controls

$$\tilde{v}_{1\varepsilon}(t) = \left|\begin{array}{ll} 0 & \text{for } t < \hat{\theta}_1 \text{ ou } \hat{\theta}_{2\varepsilon} < \hat{\theta}_1 \\ 1 & \text{for } \hat{\theta}_1 \leq t \leq \hat{\theta}_{2\varepsilon} \\ 0 & \text{for } t > \hat{\theta}_{2\varepsilon} \end{array}\right. \quad \Big|\text{if } \hat{\theta}_{2\varepsilon} \geq \hat{\theta}_1 \tag{9.37}$$

$$\tilde{v}_{2\varepsilon}(t) = \left|\begin{array}{ll} 0 & \text{for } t < \hat{\theta}_2 \text{ ou } \hat{\theta}_{1\varepsilon} \leq \hat{\theta}_2 \\ 1 & \text{for } \hat{\theta}_2 \leq t \leq \hat{\theta}_{1\varepsilon} \\ 0 & \text{for } t > \hat{\theta}_{1\varepsilon} \end{array}\right. \quad \Big|\text{if } \hat{\theta}_{1\varepsilon} > \hat{\theta}_2. \tag{9.38}$$

Now calculate the difference

$$J_x^\varepsilon(\tilde{v}_{1\varepsilon}, \hat{v}_{2\varepsilon}^\delta) - J_x(\hat{\theta}_1, \hat{\theta}_{2\varepsilon})\;.$$

For $t < \hat{\theta}_1 \wedge \hat{\theta}_{2\varepsilon}$, we have $\tilde{v}_{1\varepsilon}(t) = 0$ and $\tilde{v}_{2\varepsilon}(t) = 0$.

For $t \geq \hat{\theta}_1 \wedge \hat{\theta}_{2\varepsilon}$, then if $\hat{\theta}_{2\varepsilon} < \hat{\theta}_1$, $\tilde{v}_{1\varepsilon}(t) = 0$ and $\tilde{v}_{2\varepsilon}(t) = 1$ at least on $[\hat{\theta}_{2\varepsilon}, \hat{\theta}_{2\delta} + \delta]$, therefore $\tilde{v}_{1\varepsilon}(t) + \hat{v}_{2\varepsilon}(t) \leqslant 1$ and $= 1$ on $[\hat{\theta}_{2\varepsilon}, \hat{\theta}_{2\varepsilon} + \delta]$; if in the opposite case $\hat{\theta}_{2\varepsilon} \geqslant \hat{\theta}_1$, then $\tilde{v}_{1\varepsilon}(t) + \tilde{v}_{2\varepsilon}^\delta(t) = 1$ on $[\hat{\theta}_1, \hat{\theta}_{2\varepsilon} + \delta]$ and $\tilde{v}_{1\varepsilon}(t) + \hat{v}_{2\varepsilon}^\delta(t) \leqslant 1$.

Therefore in all cases $\tilde{v}_{1\varepsilon}(t) + \hat{v}_{2\varepsilon}^\delta(t) = 1$ for $t \in [\hat{\theta}_1 \wedge \hat{\theta}_{2\varepsilon}, \hat{\theta}_{2\varepsilon} + \delta]$ and $\tilde{v}_{1\varepsilon}(t) + \hat{v}_{2\varepsilon}^\delta(t) \leqslant 1$. Then

$$J_x^\varepsilon(\tilde{v}_{1\varepsilon}, \hat{v}_{2\varepsilon}^\varepsilon) - J_x(\hat{\theta}_1, \hat{\theta}_{2\varepsilon}) = E^x \int_{\hat{\theta}_1 \wedge \hat{\theta}_{2\varepsilon}}^{+\infty} dt\, f(y(t)) \exp -\int_0^t a_0\, ds \exp -\int_{\hat{\theta}_1 \wedge \hat{\theta}_{2\varepsilon}}^t \frac{1}{\varepsilon}(\tilde{v}_{1\varepsilon} + \hat{v}_{2\varepsilon}^\delta)\, d \tag{9.39}$$

(Contd)

(Contd)
$$+E^x\Bigg[\int_{\hat\theta_1\wedge\hat\theta_{2\varepsilon}}^{+\infty}\frac{1}{\varepsilon}\left(\tilde v_{1\varepsilon}\psi_1+\tilde v_{2\varepsilon}^{\delta}\psi^2\right)\left(\exp-\int_0^t a_o\, d_s\right)\left(\exp-\int_{\hat\theta_1\wedge\hat\theta_{2\varepsilon}}^t\frac{1}{\varepsilon}(\tilde v_{1\varepsilon}+\hat v_{2\varepsilon}^{\delta})ds\right)dt$$
$$-\psi_1\left(y(\hat\theta_1)\right)\chi_{\hat\theta_1\leqq\hat\theta_{2\varepsilon}}\exp-\int_0^{\theta_1}a_o\,ds\ -$$
$$-\psi_2\left(y(\hat\theta_{2\varepsilon})\right)\chi_{\hat\theta_{2\varepsilon}<\hat\theta_1}\exp-\int_0^{\hat\theta_{2\varepsilon}}a_o\,ds\Bigg]=E^xI+E^xII\ .$$

For $\hat\theta_{2\varepsilon}<\hat\theta_1$

$$II=\int_{\hat\theta_{2\varepsilon}}^{+\infty}\frac{1}{\varepsilon}\hat v_{2\varepsilon}^{\delta}\psi_2\left(\exp-\int_0^t a_o\,ds\right)\left(\exp-\frac{1}{\varepsilon}\int_{\hat\theta_{2\varepsilon}}^t\hat v_{2\varepsilon}^{\delta}\,ds\right)dt\ -$$
$$-\psi_2(y(\hat\theta_{2\varepsilon}))\exp-\int_0^{\theta_{2\varepsilon}}a_o\,ds=$$
$$=\int_{\hat\theta_{2\varepsilon}}^{\hat\theta_{2\varepsilon}+\delta}\frac{1}{\varepsilon}\psi_2\exp-\int_0^t a_o\,ds\ \exp-\frac{t-\hat\theta_{2\varepsilon}}{\varepsilon}dt-\psi_2(y(\hat\theta_{2\varepsilon}))\exp-\int_0^{\hat\theta_{2\varepsilon}}a_o\,ds$$
$$+\exp-\frac{\delta}{\varepsilon}\int_{\hat\theta_{2\varepsilon+\delta}}^{+\infty}\frac{1}{\varepsilon}\hat v_{2\varepsilon}\psi_2\left(\exp-\int_0^t a_o\,ds\right)\left(\exp-\frac{1}{\varepsilon}\int_{\hat\theta_{2\varepsilon+\delta}}^t\hat v_{2\varepsilon}\,ds\right)dt$$

and thus

$$|II|\leqq C\left[\exp-\frac{\delta}{\varepsilon}+\delta+\sup_{\hat\theta_{2\varepsilon}\leqq t\leqq\hat\theta_{2\varepsilon}+\delta}|\psi_2(y(t))-\psi_2(y(\hat\theta_{2\varepsilon}))|\right].\tag{9.40}$$

For $\hat\theta_{2\varepsilon}>\hat\theta_1$ (since $\psi_2\geqslant\psi_1$),

$$II\geqq\int_{\theta_1}^{+\infty}\frac{1}{\varepsilon}(\hat v_{1\varepsilon}+\hat v_{2\varepsilon}^{\delta})\psi_1\left(\exp-\int_0^t a_o\,ds\right)\left(\exp-\int_{\hat\theta_1}^t\frac{1}{\varepsilon}(\hat v_{1\varepsilon}+\hat v_{2\varepsilon}^{\delta})ds\right)dt\ -$$
$$-\psi_1(y(\hat\theta_1))\exp-\int_0^{\hat\theta_1}a_o\,ds=$$
$$=\int_{\theta_1}^{\hat\theta_1+\delta}\frac{1}{\varepsilon}\psi_1\left(\exp-\int_0^t a_o\,ds\right)\left(\exp-\frac{t-\hat\theta_1}{\varepsilon}\right)dt-\psi_1(y(\hat\theta_1))\exp-\int_0^{\hat\theta_1}a_o\,ds$$
$$+\exp-\frac{\delta+\hat\theta_{2\varepsilon}-\hat\theta_1}{\varepsilon}\int_{\hat\theta_{2\varepsilon}+\delta}^{+\infty}dt\,\frac{1}{\varepsilon}(\tilde v_{1\varepsilon}+\hat v_{2\varepsilon}^{\delta})\psi_1\left(\exp-\int_0^t a_o\,ds\right)\left(\exp-\int_{\theta_{2\varepsilon+\delta}}^t\frac{1}{\varepsilon}(\tilde v_{1\varepsilon}+\hat v_{2\varepsilon}^{\delta})\right)-$$

(Contd)

(Contd) $$+\exp -\frac{\delta}{\varepsilon}\int_{\hat\theta_1+\delta}^{\hat\theta_{2\varepsilon}+\delta}\frac{1}{\varepsilon}\psi_1\left(\exp-\int_0^t a_0\,ds\right)\exp-\frac{t-(\hat\theta_1+\delta)}{\varepsilon}\,dt$$

and thus

$$II \geqq -C\left[\exp-\frac{\delta}{\varepsilon}+\delta+\sup_{\hat\theta_1\leqq t\leqq\hat\theta_1+\delta}|\psi_1(y(t))-\psi_1(y(\hat\theta_1))|\right].$$

Finally, we obtain

$$E^x\, II \geqq -C\left[\exp-\frac{\delta}{\varepsilon}+\delta+E^x\sup_{\hat\theta_1\leqq t\leqq\hat\theta_1+\delta}|\psi_1(y(t))-\psi_1(y(\hat\theta_1))| + \right. \tag{9.41}$$

$$\left. + E^x\sup_{\hat\theta_{2\varepsilon}\leqq t\leqq\hat\theta_{2\varepsilon}+\delta}|\psi_2(y(t))-\psi_2(y(\hat\theta_{2\varepsilon}))|\right] \geqq$$

$$\geqq -C\left[\exp-\frac{\delta}{\varepsilon}+\delta+\rho_1(\delta^{1/8})+\rho_2(\delta^{1/8})+\delta^{1/4}\right].$$

Moreover,

$$|I| \leqq \int_{\hat\theta_1\wedge\hat\theta_{2\varepsilon}}^{\hat\theta_{2\varepsilon}+\delta}|f(y(t))|\left(\exp-\int_0^t a_0\,ds\right)\left(\exp-\frac{t-\hat\theta_1\wedge\hat\theta_{2\varepsilon}}{\varepsilon}\right)dt +$$

$$+\exp-\frac{\delta+\hat\theta_{2\varepsilon}-\hat\theta_1\wedge\hat\theta_{2\varepsilon}}{\varepsilon}\int_{\hat\theta_{2\varepsilon}+\delta}^{+\infty}|f|\left(\exp-\int_0^t a_0\,ds\right)\,dt$$

and therefore

$$E^x|I| \leqq C\left[\varepsilon^{\frac{a}{1+a}}+\exp-\frac{\delta}{\varepsilon}\right]. \tag{9.42}$$

But

$$-J_x(\hat\theta_1,\hat\theta_{2\varepsilon}) \leqq -\inf_{\theta_2} J_x(\hat\theta_1,\theta_2) = -u(x).$$

Therefore

$$J_x^\varepsilon(\tilde v_{1\varepsilon},\hat v_{2\varepsilon}^\delta) - J_x(\hat\theta_1,\hat\theta_{2\varepsilon}) \leqq J_x^\varepsilon(\tilde v_{1\varepsilon},\hat v_{2\varepsilon}^\delta) - u(x)$$

and by (9.36)

$$\leqq u^\varepsilon(x)-u(x)+C\left[\delta^{\alpha/8}+\rho_2(\delta^{1/8})+\delta^{1/4}\right].$$

But from the estimates (9.34),(9.35), we see that

$$u^{\varepsilon}(x) - u(x) \geq - C\left[\varepsilon^{\frac{a}{1+a}} + \delta^{\alpha/8} + \exp - \frac{\delta}{\varepsilon} + \delta + \rho_1(\delta^{1/8}) + \rho_2(\delta^{1/8}) + \delta^{1/4}\right] .$$

Arguing with $\tilde{v}_{2\varepsilon}, \tilde{v}_{1\varepsilon}, \hat{\theta}_2, \hat{\theta}_{1\varepsilon}$, we also obtain an estimate in the opposite sense, hence the result. □

We now no longer assume (9.2) and replace it by

(9.43) $\quad f \in L^p(\mathcal{O})\ ,\ p > \frac{n}{2}\ ,\quad \psi_1, \psi_2 \in C^0(\bar{\mathcal{O}})\ ,\quad \psi_1 \leq \psi_2\ .$

Naturally we assume (9.8) no more.

Now approximate ψ_1, ψ_2 by the procedure of Lemma 7.1, and thus construct a sequence ψ_1^N, ψ_2^N such that

(9.44)
$$\left|\begin{array}{l} \psi_1^N\ ,\ \psi_2^N \in W^{2,p}(\mathcal{O})\quad \forall\ p > 1\quad (y \text{ including } W^{2,\infty}(\mathcal{O})) \\ \left.\dfrac{\partial \psi_1^N}{\partial \nu_A}\right|_{\Gamma} = \left.\dfrac{\partial \psi_2^N}{\partial \nu_A}\right|_{\Gamma} = 0 \\ \psi_1^N \leq \psi_2^N\ . \end{array}\right.$$

The last property of (9.44) arises from $\psi_1 \leqslant \psi_2$ and from the maximum principle. We now seek to interpret the V.I. (5.28). Then we have:

THEOREM 9.4: *Assume* (9.1) *and* (9.43) *together with* (5.6). *Then the solution* $u(x)$ *of the V.I.* (5.28) *belongs to* $C^0(\bar{\mathcal{O}})$. *It is given explicitly by* (9.10). *Moreover, the functional* $J_x(\theta_1, \theta_2)$ *has a saddle point given by* (9.12).

Proof: This is analogous to the Dirichlet case. Consider the functions $u^N(x)$, $u^{N,\varepsilon}(x)$ corresponding to ψ_1^N, ψ_2^N instead of ψ_1, ψ_2. We then have Theorem 9.3. The important point is the uniform convergence of $u^{N,\varepsilon}$ to u^{ε} as $N \to +\infty$ with respect to ε. For details, cf. B.L., Chap. Subsection 5.1.5, loc. cit.. □

9.2 *Evolutionary Case*

We shall be content with stating the results, which are very much that analogues of those obtained in the stationary case. Assume that the coefficients satisfy the conditions

(9.45) $\quad a_{ij}\ ,\ a_i\ ,\ a_o \in L^{\infty}(Q)$

(9.46) $\quad \Sigma a_{ij}\xi_i\xi_j \geq \alpha|\xi|^2\ ,\quad \forall\ \xi \in R^n\ ,\quad a_{ij} = a_{ji}$

(9.47) $\quad \dfrac{\partial a_{ij}}{\partial t}\ ,\ \dfrac{\partial a_i}{\partial t}\ ,\ \dfrac{\partial a_o}{\partial t} \in L^{\infty}(Q)$

(9.48) $\quad \dfrac{\partial a_{ij}}{\partial x_k} \in L^{\infty}(Q)\ .$

Also, let

(9.49)
$$\left|\begin{array}{l} f\in L^p(Q)\,,\ p>\frac{n}{2}+1\,,\quad \psi_1,\psi_2\,,\quad \psi_i\in \mathcal{W}^{2,1,p}(Q)\,,\\ p>n+2\,,\quad \left.\frac{\partial\psi_2}{\partial\nu_{A(t)}}\right|_\Sigma \geqq 0\,,\quad \left.\frac{\partial\psi_1}{\partial\nu_{A(t)}}\right|_\Sigma \leqq 0\,,\quad \psi_1\leqq\psi_2 \end{array}\right.$$

(9.50)
$$\bar u\in W^{2,p}(\mathcal{O})\,,\quad p>\frac{n}{2}+1\,,\quad \left.\frac{\partial\bar u}{\partial x}\right|_\Gamma = 0\,,\quad \psi_1(x,T)\leq\bar u(x)\leq\psi_2(x,T)$$

(9.51)
$$f+\frac{\partial\psi_2}{\partial t}-A(t)\psi_2\leqq C\,,\quad f+\frac{\partial\psi_1}{\partial t}-A(t)\psi_1\geqq -C\,.$$

Define the measure $P^{x,t}$ as the solution of the sub-martingale problem. Then consider the functionals

(9.52)
$$\begin{aligned} J_{xt}(\theta_1,\theta_2) = E^{xt}\Bigg[\bigg[\int_t^{\theta_1\wedge\theta_2\wedge T} & f(y(s),s)\left(\exp-\int_t^s a_o(y(\lambda),\lambda)d\lambda\right)ds\ +\\ &+\psi_1(y(\theta_1),\theta_1)\chi_{\theta_1\leqq\theta_2,\theta_1<T}\exp-\int_t^{\theta_1} a_o(y(s))ds\ +\\ &+\psi_2(y(\theta_2),\theta_2)\chi_{\theta_2<\theta_1\wedge T}\ \exp-\int_t^{\theta_2} a_o(y(s),s)ds\ +\\ &+\bar u(y(T))\,\chi_{T\leqq\theta_1\wedge\theta_2}\ \exp-\int_t^T a_o(y(s),s)ds\Bigg]\,, \end{aligned}$$

(9.53)
$$\begin{aligned} J^\epsilon_{xt}(v_1,v_2) = E^{xt}\Bigg[\bigg[\int_t^T & \left(f+\frac{1}{\epsilon}v_1\psi_1+\frac{1}{\epsilon}v_2\psi_2\right)(y(s),s)\ \times\\ &\times\exp-\int_t^s\left(a_o+\frac{1}{\epsilon}v_1+\frac{1}{\epsilon}v_2\right)d\lambda+\bar u(y(T))\exp-\int_t^T\left(a_o+\frac{1}{\epsilon}v_1+\frac{1}{\epsilon}v_2\right)ds\Bigg]\,. \end{aligned}$$

Consider also the problem

(9.54)
$$\left|\begin{array}{l} u\in\mathcal{W}^{2,1,p}(Q)\,,\quad \psi_1\leqq u\leqq\psi_2\,,\quad u(x,T)=\bar u(x)\\ \left.\frac{\partial u}{\partial\nu_{A(t)}}\right|_\Sigma = 0\\ -\frac{\partial u}{\partial t}+(A(t)+a_o)u = f \quad \text{if}\quad \psi_1<u<\psi_2\\ -\frac{\partial u}{\partial t}+(A(t)+a_o)u \geqq f \quad \text{if}\quad u=\psi_1\\ -\frac{\partial u}{\partial t}+(A(t)+a_o)u \leqq f \quad \text{if}\quad u=\psi_2 \end{array}\right.$$

(9.55)
$$\left|\begin{array}{l} u^\epsilon\in\mathcal{W}^{2,1,p}(Q)\,,\quad u^\epsilon(x,T)=\bar u(x)\,,\quad \left.\frac{\partial u^\epsilon}{\partial\nu_{A(t)}}\right|_\Sigma = 0\\ -\frac{\partial u^\epsilon}{\partial t}+(A(t)+a_o)u^\epsilon+\frac{1}{\epsilon}(u^\epsilon-\psi_2)^+-\frac{1}{\epsilon}(u^\epsilon-\psi_1)^- = f. \end{array}\right.$$

Then we have the following Theorems:

THEOREM 9.5: *Assuming* (9.45) *to* (9.51), *there then exists one and only one solution of the problems* (9.54) *and* (9.55). *Moreover,*

(9.56) $$u(x,t) = \operatorname*{Inf}_{\theta_2} \sup_{\theta_1} J_{xt}(\theta_1, \theta_2) = \sup_{\theta_1} \operatorname*{Inf}_{\theta_2} J_{xt}(\theta_1, \theta_2)$$

(9.57) $$u^\varepsilon(x,t) = \operatorname*{Inf}_{v_2} \sup_{v_1} J^\varepsilon_{xt}(v_1, v_2) = \sup_{v_1} \operatorname*{Inf}_{v_2} J^\varepsilon_{xt}(v_1, v_2) .$$

Also

(9.58) $u^\varepsilon \to u$ *in* $C^o(\bar{Q})$.

Proof: Analgous to the stationary case. □

We now consider the case with continuous obstacles only. We replace (9.49) by

(9.59) $$\begin{cases} f \in L^p(Q) \ , \quad p > \frac{n}{2} + 1 \ , \quad \psi_1, \psi_2 \in C^o(\bar{Q}) \ , \quad \psi_1 \leqq \psi_2 \ , \\ \psi_1, \psi_2, \dfrac{\partial \psi_1}{\partial t}, \dfrac{\partial \psi_2}{\partial t} \in L^2(0,T;H^1(\mathcal{O})) \ , \quad \dfrac{\partial^2 \psi_1}{\partial t^2} \in L^2(Q) \\ -\dfrac{\partial}{\partial t}(\psi_2 - \psi_1) \geqq 0 \ , \quad -\dfrac{\partial^2}{\partial t^2}(\psi_2 - \psi_1) \geqq 0 \end{cases}$$

and can then consider the evolutionary V.I. (9.51),(9.52),(9.53),(9.54). We have:

THEOREM 9.6: *Assuming* (9.45),(9.46),(9.47),(9.48),(9.50),(9.59), *the solution* u *of the evolutionary V.I.* (9.51),(9.52),(9.53),(9.54) *is given explicitly by* (9.56). *Furthermore,* $u \in C^o(Q)$, *and the functional* $J_{xt}(\theta_1,\theta_2)$ *has a saddle point* $\hat{\theta}_1, \hat{\theta}_2$ *given by*

(9.60) $$\begin{cases} \hat{\theta}_1 = \inf \{ s \in [t,T] \mid u(y(s),s) = \psi_1(y(s),s) \} \\ \hat{\theta}_2 = \inf \{ s \in [t,T] \mid u(y(s),s) = \psi_2(y(s),s) \}. \end{cases}$$

Proof: Analogous to the stationary case. □

Remark 9.3

The separation principle which we have given for the Dirichlet problem, cf. B.L., Chap. 3, Subsections 4.12 and 5.2.3, extends to the Neumann problem. The proof, which is very similar, is left to the reader. □

9.3 *Camoo and Homographic Penalisation*

We return to the stationary problem of Section 9.2, but with assumptions about the data f and ψ_i that are a little different. Let, therefore, there be an operator $A + a_o$; we shall write A instead of $A + a_o$, and denote by a(u,v) the *bilinear form* on $H^1(\mathcal{O})$ associated with A. Let there be ψ_1 and ψ_2 with

(9.61) $$\begin{cases} \psi_i \in H^2(\mathcal{O}) \ , \quad i = 1,2 \quad \text{and} \\ \dfrac{\partial \psi_1}{\partial \nu_A} \leqq C \ , \quad \dfrac{\partial \psi_2}{\partial \nu_A} \geqq -C \quad \text{a.e. on} \quad \Gamma \end{cases}$$

(where the C's denote the various constants)

and let there be f with

(9.62) $$f \in L^2(\mathcal{O}) \ , \quad f - A\psi_1 \geqq -C \ , \quad f - A\psi_2 \leqq C \quad \text{p.p.} \quad \text{in} \quad \mathcal{O} \ .$$

In addition we shall assume that ψ_1 is *uniformly below* ψ_2 in the sense that

(9.63) $$\psi_1 \leqq \psi_2 \ , \quad \frac{1}{\psi_2 - \psi_1} \in L^\infty(\mathcal{O}) \ .$$

We introduce

(9.64) $$K = \{v \mid v \in H^1(\mathcal{O}) \ , \quad \psi_1 \leqq v \leqq \psi_2 \quad \text{a.e.} \quad \text{in} \quad \mathcal{O}\} \ .$$

The problem to be solved is

(9.65) $$a(u,v-u) \geqq (f,v-u) \quad \forall \ v \in K \ , \quad u \in K \ .$$

We know that it has an unique solution; our aim is to give an approximation of it by 'homographic penalisation', an approximation which, as we shall see, *preserves constraints*. □

Notations

For $\varepsilon > 0$ set:

(9.66) $$R_\varepsilon = \frac{u_\varepsilon - \psi_1}{\varepsilon + |u_\varepsilon - \psi_1|} \ , \quad S_\varepsilon = \frac{u_\varepsilon - \psi_2}{\varepsilon + |u_\varepsilon - \psi_2|} \ .$$

The approximated problem

We seek $u_\varepsilon \in H^1(\mathcal{O})$ such that

(9.67) $$\left| \begin{array}{lll} Au_\varepsilon + g_\varepsilon(R_\varepsilon + S_\varepsilon) = f & \text{in} & \mathcal{O} \ , \\ \dfrac{\partial u_\varepsilon}{\partial \nu_A} + k_\varepsilon(R_\varepsilon + S_\varepsilon) = 0 & \text{on} & \Gamma \end{array} \right.$$

where g_ε and k_ε are functions in $L^\infty(\mathcal{O})$ and $L^\infty(\Gamma)$, $\geqslant 0$, which are *to be determined*. We are going to choose g_ε and k_ε in the following way:

(9.68) $$\left| \begin{array}{l} g_\varepsilon = \dfrac{\varepsilon + \psi_2 - \psi_1}{\psi_2 - \psi_1} \max\left((f - A\psi_1)^-, \ (f - A\psi_2)^+ \right) , \\ k_\varepsilon = \dfrac{\varepsilon + \psi_2 - \psi_1}{\psi_2 - \psi_1} \max\left(\left(\dfrac{\partial \psi_1}{\partial \nu_A}\right)^+ , \ \left(\dfrac{\partial \psi_2}{\partial \nu_A}\right)^- \right) . \end{array} \right.$$

(Every bounded function greater than g_ε or k_ε given by (9.61) would be acceptable; in a certain sense the choice (9.61) is optimal). □

Remark 9.4

The approximation (9.60) is adapted from BRAUNER and NICOLAENKO [1] (who considered the analogous problem, but for the Dirichlet conditions at the boundary). □

In the remainder we are going to prove the following results:

THEOREM 9.7: *The problem* (9.60) *has an unique solution, and it satisfies:*

(9.69) $\psi_1 \leqq u_\varepsilon \leqq \psi_2$

(9.70) $u_\varepsilon \in H^{3/2}(\mathcal{O})$.

THEOREM 9.8: *When* $\varepsilon \to 0$, *then* $u_\varepsilon \to u$ *in* $H^1(\mathcal{O})$, *where* u *is the solution of* (9.65). *The error estimate is*

(9.71) $$a(u_\varepsilon - u) \leqq 2\varepsilon \int_{\mathcal{O}} g_\varepsilon \, dx + 2\varepsilon \int_\Gamma k_\varepsilon \, d\Gamma ,$$

where $a(v) = a(v,v)$. *Therefore, in particular,*

(9.72) $$\|u_\varepsilon - u\|_{H^1(\mathcal{O})} \leqq C\,\varepsilon^{1/2} .$$

THEOREM 9.9: (Levy-Stampacchia Inequality):

(9.73) $$f - \max\left((f - A\psi_1)^-, \ (f - A\psi_2)^+\right) \leqq Au_\varepsilon \leqq f + \max\left((f - A\psi_1)^-, \ (f - \psi_2)^+\right)$$

(and the analogous inequality for Au*): Also*

(9.74) $$-\max\left(\left(\frac{\partial\psi_1}{\partial\nu_A}\right)^+, \left(\frac{\partial\psi_2}{\partial\nu_A}\right)^-\right) \leqq \frac{\partial u_\varepsilon}{\partial\nu_A} \leqq \max\left(\left(\frac{\partial\psi_1}{\partial\nu_A}\right)^+, \left(\frac{\partial\psi_2}{\partial\nu_A}\right)^-\right)$$

(and the analogous inequality for $\frac{\partial u}{\partial\nu_A}$ *).*

Remark 9.5

If, as in (9.2), we assume that $\frac{\partial\psi_1}{\partial\nu_A} \leqq 0$, $\frac{\partial\psi_2}{\partial\nu_A} \geqq 0$, then (9.74) gives

(9.75) $$\frac{\partial u_\varepsilon}{\partial\nu_A} = \frac{\partial u}{\partial\nu_A} = 0 . \quad \square$$

Proof of Theorem 9.7: If u_ε is the solution of (9.67), then we are going to prove that

(9.76) $u_\varepsilon \leqslant \psi_2$.

We would do this in the same way as for $u_\varepsilon \geqslant \psi_1$. That then gives a priori estimates allowing us to prove the Theorem. We are therefore limited to establishing (9.76).

Let us rewrite (9.67) in the form:

(9.77) $$\left| \ A(u_\varepsilon - \psi_2) + g_\varepsilon[R_\varepsilon + S_\varepsilon] = f - A\psi_2 , \right| \ \frac{\partial(u_\varepsilon - \psi_2)}{\partial\nu_A} + k_\varepsilon[R_\varepsilon + S_\varepsilon] = -\frac{\partial\psi_2}{\partial\nu_A}$$

hence we deduce

$$(9.78)\quad \left| \begin{aligned} & a(u_\varepsilon - \psi_2, v) + (g_\varepsilon(R_\varepsilon + S_\varepsilon), v) + (k_\varepsilon(R_\varepsilon + S_\varepsilon), v)_\Gamma = \\ & \quad = (f - A\psi_2, v) - \left(\frac{\partial \psi_2}{\partial \nu_A}, v\right)_\Gamma \quad , \quad \forall\, v \in H^1(\mathcal{O}) \,. \end{aligned} \right.$$

We then take

$$v = (u_\varepsilon - \psi_2)^+$$

in (9.78) and deduce

$$(9.79)\quad a((u_\varepsilon - \psi_2)^+) + (X_\varepsilon, (u_\varepsilon - \psi_2)^+) + (Y_\varepsilon, (u_\varepsilon - \psi_2)^+)_\Gamma = 0$$

where

$$(9.80)\quad \left| \begin{aligned} & X_\varepsilon = g_\varepsilon(R_\varepsilon + S_\varepsilon) - (f - A\psi_2) \,, \\ & Y_\varepsilon = k_\varepsilon(R_\varepsilon + S_\varepsilon) + \frac{\partial \psi_2}{\partial \nu_A} \,. \end{aligned} \right.$$

We are going to show that

$$(9.81)\quad \text{if } u_\varepsilon \geqslant \psi_2 \text{ then } X_\varepsilon \geqslant 0,\ Y_\varepsilon \geqslant 0.$$

From this it will follow that

$$(X_\varepsilon, (u_\varepsilon - \psi_2)^+) \geqq 0 \,,\ (Y_\varepsilon, (u_\varepsilon - \psi_2)^+)_\Gamma \geqq 0$$

so that (9.79) implies $(u_\varepsilon - \psi_2)^+ = 0$, and hence (9.76).

Now if $\xi_\varepsilon \geqslant \psi_2$,

$$R_\varepsilon + S_\varepsilon = \frac{u_\varepsilon - \psi_1}{\varepsilon + u_\varepsilon - \psi_1} + \frac{u_\varepsilon - \psi_2}{\varepsilon + u_\varepsilon - \psi_2}$$

and since

$$\inf_{\lambda \geqq \psi_2} \left[\frac{\lambda - \psi_1}{\varepsilon + \lambda - \psi_1} + \frac{\lambda - \psi_2}{\varepsilon + \lambda - \psi_2} \right] \geqq \frac{\psi_2 - \psi_1}{\varepsilon + \psi_2 - \psi_1}$$

so that (9.74) holds if

$$g_\varepsilon \frac{\psi_2 - \psi_1}{\varepsilon + \psi_2 - \psi_1} \geqq f - A\psi_2 \,,$$

$$k_\varepsilon \frac{\psi_2 - \psi_1}{\varepsilon + \psi_2 - \psi_1} \geqq - \frac{\partial \psi_2}{\partial \nu_A}$$

and certainly

$$g_\varepsilon \frac{\psi_2 - \psi_1}{\varepsilon + \psi_2 - \psi_1} \geq 0 \ , \quad k_\varepsilon \frac{\psi_2 - \psi_1}{\varepsilon + \psi_2 - \psi_1} \geq 0 \ ;$$

hence (9.81) holds if

$$(9.82) \qquad g_\varepsilon \geq \frac{\varepsilon + \psi_2 - \psi_1}{\psi_2 - \psi_1} (f - A\psi_2)^+ , \quad k_\varepsilon \geq \frac{\varepsilon + \psi_2 - \psi_1}{\psi_2 - \psi_1} \left(\frac{\partial \psi_2}{\partial \nu_A} \right)^- .$$

In the same way we would show that $u_\varepsilon \geqslant \psi_1$ if

$$(9.83) \qquad g_\varepsilon \geq \frac{\varepsilon + \psi_2 - \psi_1}{\psi_2 - \psi_1} (f - A\psi_1)^- , \quad k_\varepsilon \geq \frac{\varepsilon + \psi_2 - \psi_1}{\psi_2 - \psi_1} \left(\frac{\partial \psi_1}{\partial \nu_A} \right)^+ ,$$

whence the 'optimal' choice (9.68). □

Proof of Theorem 9.8: The variational formulation of (9.68) is

$$(9.84) \qquad a(u_\varepsilon, v) + (g_\varepsilon (R_\varepsilon + S_\varepsilon), v) + (k_\varepsilon (R_\varepsilon + S_\varepsilon), v)_\Gamma = (f, v) \quad \forall\, v \in H^1(\mathcal{O}) \ .$$

By Theorem 9.7 we can take $v = u$ in (9.65), and we take $v = u - u_\varepsilon$ in (9.84). This yields

$$(9.85) \qquad a(u_\varepsilon - u) \leq (g_\varepsilon (R_\varepsilon + S_\varepsilon), u - u_\varepsilon) + (k_\varepsilon (R_\varepsilon + S_\varepsilon), u - u_\varepsilon)_\Gamma \ .$$

Since we know that $\psi_1 \leqslant u_\varepsilon \leqslant \psi_2$, we have

$$(9.86) \qquad R_\varepsilon + S_\varepsilon = \varepsilon \frac{u_\varepsilon - \psi_1 + u_\varepsilon - \psi_2}{(\varepsilon + u_\varepsilon - \psi_1)(\varepsilon + \psi_2 - u_\varepsilon)} = \varepsilon\, T_\varepsilon$$

therefore

$$(9.87) \qquad a(u_\varepsilon - u) \leq \varepsilon (g_\varepsilon, T_\varepsilon (u - u_\varepsilon)) + \varepsilon (R_\varepsilon, T_\varepsilon (u - u_\varepsilon))_\Gamma \ .$$

We shall therefore have (9.71) if we show that

$$(9.88) \qquad T_\varepsilon (u - u_\varepsilon) \leqslant 2.$$

Now,

$$T_\varepsilon (u - u_\varepsilon) = \frac{(u_\varepsilon - \psi_1)(u - u_\varepsilon)}{(\varepsilon + u_\varepsilon - \psi_1)(\varepsilon + \psi_2 - u_\varepsilon)} - \frac{(\psi_2 - u_\varepsilon)(u - u_\varepsilon)}{(\varepsilon + u_\varepsilon - \psi_1)(\varepsilon + \psi_2 - u_\varepsilon)}$$

hence

$$T_\varepsilon (u - u_\varepsilon) \leq 2 \frac{(u_\varepsilon - \psi_1)(\psi_2 - u_\varepsilon)}{\varepsilon + u_\varepsilon - \psi_1)(\varepsilon + \psi_2 - u_\varepsilon)} \leq 2 \ . \qquad \square$$

Proof of Theorem 9.9: Now,

$$R_\varepsilon + S_\varepsilon = \frac{u_\varepsilon - \psi_1}{\varepsilon + u_\varepsilon - \psi_1} + \frac{u_\varepsilon - \psi_2}{\varepsilon + \psi_2 - u_\varepsilon} \ , \quad \psi_1 \leq u_\varepsilon \leq \psi_2$$

therefore

$$\frac{\psi_1 - \psi_2}{\epsilon + \psi_2 - \psi_1} \leq R_\epsilon + S_\epsilon \leq \frac{\psi_2 - \psi_1}{\epsilon + \psi_2 - \psi_1} \tag{9.89}$$

and therefore the result follows. □

Remark 9.6

We shall obtain analogous results for parabolic evolutionary problems. □

10. NON-LINEAR PROBLEMS

10.1 *Formulation of the V.I.'s*

Assume that $\mathcal{O}$ is a bounded open set of R^n, and let a_{ij}, a_i, a_o satisfy

$$a_{ij}, a_i, a_o \in L^\infty(Q) \tag{10.1}$$

$$\Sigma a_{ij}\, \xi_i \xi_j \geq \alpha|\xi|^2 \quad \forall\ \xi \in R^n \tag{10.2}$$

$$\frac{\partial a_{ij}}{\partial t}\ ,\ \frac{\partial a_i}{\partial t}\ ,\ \frac{\partial a_o}{\partial t} \in L^\infty(Q)\ . \tag{10.3}$$

Consider the measurable Hamiltonian $H(x,t,u,p)$ of $Q \times R \times R^n$ which satisfies the properties

$$|H(x,t,u,p)| \leq C(h(x,t) + |p|)\ ,\quad h \in L^2(Q) \tag{10.4}$$

$$\left|\frac{\partial H}{\partial t}\right| \leq C(h + |u| + |p|) \tag{10.5}$$

$$\left|\frac{\partial H}{\partial u}\right| + \left|\frac{\partial H}{\partial p}\right| \leq C\ . \tag{10.6}$$

Given two functions ψ_1, ψ_2 with

$$\psi_1 \leqslant \psi_2 \tag{10.7}$$

we seek $u(x,t)$ such that

$$\psi_1(x,t) \leq u(x,t) \leq \psi_2(x,t)\ . \tag{10.8}$$

We set

$$\mathcal{P}u = -\frac{\partial u}{\partial t} + A(t)\,u + a_o u + H(x,t,u,Du)$$

and then,

$$\left|\begin{array}{lll} \text{if} & \psi_1 < u < \psi_2\ , & \mathcal{P}u = f \\ \text{if} & u = \psi_2\ , & \mathcal{P}u \leq f \\ \text{if} & u = \psi_1\ , & \mathcal{P}u \geq f \end{array}\right. \tag{10.9}$$

$$\left.\frac{\partial u}{\partial \nu_{A(t)}}\right|_\Sigma = 0\ ,\quad u(x,t) = \bar{u}(x)\ . \tag{10.10}$$

The formulation as a V.I. is

$$(10.11)\quad \left| \begin{array}{ll} u \in L^2(0,T;V) \;, & V = H^1(\mathcal{O}) \\ \frac{\partial u}{\partial t} \in L^2(0,T;H) \;, & H = L^2(\mathcal{O}) \end{array} \right.$$

$$(10.12)\quad \left| \begin{array}{l} -(\frac{\partial u}{\partial t}, v-u) + a(t;u,v-u) + (H(x,t,u,Du),v-u) \geqq (f,v-u) \\ \forall\, v \quad \text{with} \quad \psi_1 \leqq v \leqq \psi_2 \end{array} \right.$$

with (10.8),(10.10).

The formulation as a weak V.I. is as follows:

$$(10.13)\quad u \in L^2(0,T;V), \quad \text{with} \quad (10.8)$$

$$(10.14)\quad \int_0^T \left[-(\frac{\partial v}{\partial t}, v-u) + a(t;u,v-u) + (H(x,t,u,Du),v-u) - (f,v-u) \right] dt + \\ + \frac{1}{2}\,|v(T) - \bar{u}|^2 \geqq 0 \;, \quad \forall\, v \in \mathcal{K}$$

$$(10.15)\quad \mathcal{K} = \left\{ v \,\middle|\, v \in L^2(0,T;V) \;, \frac{\partial v}{\partial t} \in L^2(Q) \;, \quad \psi_1 \leqq v \leqq \psi_2 \right\} .$$

10.2 *Existence Results*

THEOREM 10.1: *Assume* (10.1),(10.2),(10.3),(10.4),(10.5),(10.6),(10.7), *and*

$$(10.16)\quad \left| \begin{array}{l} \psi_i \;, \frac{\partial \psi_i}{\partial t} \in L^2(0,T;V) \;, \frac{\partial^2 \psi_i}{\partial t^2} \in L^2(Q) \\ \frac{\partial}{\partial t}(\psi_1 - \psi_2) \geqq 0 \;, \frac{\partial^2}{\partial t^2}(\psi_1 - \psi_2) \geqq 0 \end{array} \right.$$

$$(10.17)\quad \mathcal{K} \quad \textit{is non-empty}$$

$$(10.18)\quad f \;, \frac{\partial f}{\partial t} \in L^2(Q)$$

$$(10.19)\quad \left| \begin{array}{l} \bar{u} \in V \;, \quad A(T)\bar{u} \in H \;, \quad \left.\frac{\partial \bar{u}}{\partial x}\right|_{\Gamma} = 0 \\ \psi_1(x,T) \leqq \bar{u} \leqq \psi_2(x,T) \;. \end{array} \right.$$

Under these conditions there exists one and only one function u *that is the solution of* (10.11),(10.12),(10.8),(10.10). *Also*

$$(10.20)\quad \frac{\partial u}{\partial t} \in L^2(0,T;V) \cap L^\infty(0,T;H) \;.$$

Proof: This is identical to the Dirichlet case (cf. B.L., Chap. 4, Section 1, loc. cit.). □

The penalised problem is defined by

$$(10.21)\quad \left| \begin{array}{l} -\frac{\partial u^\varepsilon}{\partial t} + A(t)u^\varepsilon + a_0 u^\varepsilon + H(x,t,u^\varepsilon,Du^\varepsilon) + \frac{1}{\varepsilon}(u^\varepsilon - \psi_2)^+ - \frac{1}{\varepsilon}(u^\varepsilon - \psi_1)^- = f \\ u^\varepsilon \in L^2(0,T;V) \quad , \quad u^\varepsilon(x,T) = \bar{u}(x) \;. \end{array} \right.$$

We now give a result for the estimate of the penalisation error $u^\varepsilon - u$:

THEOREM 10.2: *Under the assumptions of Theorem 10.1, and*

(10.22) $$\psi_i \in L^2(0,T;H^2(\mathcal{O})) \ , \quad \frac{\partial \psi_1}{\partial \nu_{A(t)}} \leq 0 \ , \quad \frac{\partial \psi_2}{\partial \nu_{A(t)}} \geq 0$$

(10.23) *$\mathcal{O}$ an open set with Lipschitz continuous boundary,*

we then have the estimate

(10.24) $$\|u^\varepsilon - u\|_{L^2(0,T;V)} + \|u^\varepsilon - u\|_{L^\infty(0,T;H)} \leq C\sqrt{\varepsilon} .$$

Proof: Since we can change the bilinear form a(t;u,v) into a(t;u,v) + λ(u,v), without loss of generality we can therefore assume that

(10.25) $$\left| \begin{array}{l} a(t;u-v, u-v) + (H(x,t,u,Du) - H(x,t,v,Dv), u-v) \geq \gamma\|v\|^2 \\ \forall\, v \in V \ , \quad \gamma > 0 \ . \end{array} \right.$$

From (10.21) we deduce that

$$-\left(\frac{\partial u^\varepsilon}{\partial t}, (u^\varepsilon - \psi_2)^+\right) + a\left(t;u^\varepsilon, (u^\varepsilon - \psi_2)^+\right) + \left(H(x,t,u^\varepsilon, Du^\varepsilon), (u^\varepsilon - \psi_2)^+\right) + \\ + \frac{1}{\varepsilon}|(u^\varepsilon - \psi_2)^+|^2 = \left(f, (u^\varepsilon - \psi_2)^+\right)$$

and so

$$-\left(\frac{\partial}{\partial t}(u^\varepsilon - \psi_2), (u^\varepsilon - \psi_2)^+\right) + a\left(t;u^\varepsilon - \psi_2, (u^\varepsilon - \psi_2)^+\right) + \frac{1}{\varepsilon}|(u^\varepsilon - \psi_2)^+|^2 = \\ = \left(f + \frac{\partial \psi_2}{\partial t} - (A(t) + a_0)\psi_2, (u^\varepsilon - \psi_2)^+\right) - \int_\Gamma \frac{\partial \psi_2}{\partial \nu_{A(t)}} (u^\varepsilon - \psi_2)^+ \, d\Gamma - \\ - \left(H(u^\varepsilon, Du^\varepsilon), (u^\varepsilon - \psi_2)^+\right) .$$

Now $\|u^\varepsilon\|_V \leqslant C$, therefore by integrating and using $\dfrac{\partial \psi_2}{\partial \nu_A} \geqslant 0$ and $\bar{u} \leqslant \psi_2(x,t)$, we obtain

$$\frac{1}{2}|(u^\varepsilon - \psi_2)^+(t)|^2 + \alpha \int_t^T \|(u^\varepsilon - \psi_2)^+\|^2 ds + \frac{1}{\varepsilon}\int_t^T |(u^\varepsilon - \psi_2)^+|^2 ds \leq C\left(\int_0^T |(u^\varepsilon - \psi_2)^+|^2 dt\right)^{1/2}$$

whence

(10.26) $$\left| \begin{array}{l} |(u^\varepsilon - \psi_2)^+|_{L^2(Q)} \leq C\varepsilon \\ \|(u^\varepsilon - \psi_2)^+\|_{L^\infty(0,T;H)} \leq C\sqrt{\varepsilon} \ , \quad \|(u^\varepsilon - \psi_2)^+\|_{L^2(0,T;V)} \leq C\sqrt{\varepsilon} \ . \end{array} \right.$$

Similarly,

$$(10.27)\qquad \left|\begin{array}{l} |(u^\varepsilon-\psi_1)^-|_{L^2(Q)} \leq C\varepsilon \\ \|(u^\varepsilon-\psi_1)^-|_{L^\infty(0,T;H)} \leq C\sqrt{\varepsilon}\ , \quad \|(u^\varepsilon-\psi_1)^-\|_{L^2(0,T;V)} \leq C\sqrt{\varepsilon}\ . \end{array}\right.$$

We then set

$$(10.28)\qquad r^\varepsilon = u^\varepsilon - (u^\varepsilon-\psi_2)^+ + (u^\varepsilon-\psi_1)^-$$

and notice that

$$(10.29)\qquad \psi_1 \leqslant r^\varepsilon \leqslant \psi_2.$$

To prove (10.24), taking (10.27) into account it suffices us to show that

$$(10.30)\qquad \|r^\varepsilon - u\|_{L^2(0,T;V)} \leq C\sqrt{\varepsilon}\ , \quad \|r^\varepsilon - u\|_{L^\infty(0,T;H)} \leq C\sqrt{\varepsilon}\ .$$

Thanks to (10.29) we can take $v = r^\varepsilon$ in (10.12). We then deduce from this, with (10.21), that

$$(10.31)\qquad -\left(\frac{\partial}{\partial t}(u^\varepsilon-u), r^\varepsilon-u\right) + a(t;u^\varepsilon-u, r^\varepsilon-u) + \left(H(u^\varepsilon,Du^\varepsilon) - H(u,Du), r^\varepsilon-u\right) \leq 0$$

since

$$\left((u^\varepsilon-\psi_2)^+, r^\varepsilon-u\right) \geqq 0\ , \left((u^\varepsilon-\psi_1)^-, r^\varepsilon-u\right) \leq 0\ .$$

From (10.31) we deduce that

$$(10.32)\qquad -\left(\frac{\partial}{\partial t}(r^\varepsilon-u), r^\varepsilon-u\right) + a(t;r^\varepsilon-u, r^\varepsilon-u) + \left(H(r^\varepsilon,Dr^\varepsilon) - H(u,Du), r^\varepsilon-u\right) \leq$$
$$\leq -\left(\frac{\partial}{\partial t}(r^\varepsilon-u^\varepsilon), r^\varepsilon-u\right) + a(t;r^\varepsilon-u^\varepsilon, r^\varepsilon-u) + \left(H(r^\varepsilon,Dr^\varepsilon) - H(u^\varepsilon,Du^\varepsilon), r^\varepsilon-u\right).$$

Now,

$$(10.33)\qquad -\int_t^T \left(\frac{\partial}{\partial t}(r^\varepsilon-u^\varepsilon), r^\varepsilon-u\right) ds = \frac{1}{2}\left(r^\varepsilon(t)-u^\varepsilon(t), r^\varepsilon(t)-u(t)\right) +$$
$$+ \int_t^T \left(r^\varepsilon-u^\varepsilon, \frac{\partial}{\partial t}(r^\varepsilon-u)\right) ds$$

and

$$(10.34)\qquad \int_t^T \left(r^\varepsilon-u^\varepsilon, \frac{\partial}{\partial t}(r^\varepsilon-u)\right) ds = \int_t^T \left(-(u^\varepsilon-\psi_2)^+ + (u^\varepsilon-\psi_2)^-, \frac{\partial}{\partial t}(u^\varepsilon-u) - \right.$$
$$\left. - \frac{\partial}{\partial t}(u^\varepsilon-\psi_2)^+ + \frac{\partial}{\partial t}(u^\varepsilon-\psi_1)^-\right) dt =$$
$$\int_t^T \left(-(u^\varepsilon-\psi_2)^+, \frac{\partial}{\partial t}(\psi_2-u)\right) ds + \int_t^T \left((u^\varepsilon-\psi_1)^-, \frac{\partial}{\partial t}(\psi_1-u)\right) ds \leq C\varepsilon$$

so, from (10.32)

$$\frac{1}{2}|r^{\epsilon}(t)-u(t)|^2 + \chi\int_t^T \|r^{\epsilon}-u\|^2 ds \leq \frac{1}{2}|r^{\epsilon}(t)-u^{\epsilon}(t)|\ |r^{\epsilon}(t)-u(t)| + $$
$$+ C\int_t^T \|r^{\epsilon}-u^{\epsilon}\|\ \|r^{\epsilon}-u\| ds .$$

Using (10.28) and the estimates (10.26),(10.27), we easily obtain (10.30), and hence the result. □

10.3 *Regularity*

THEROEME 10.3: *Assume the conditions of Theorem 10.2. In addition assume that the function* h *occurs in* (10.4) *satisfies* $h \in L^p(Q)$. *Also assume*

(10.35) $f \in L^p(Q)$

(10.36) $\psi_i,\ D\psi_i \in L^p(Q)\ ,\ A(t)\psi_i \in L^p(Q)\ ,\quad i = 1,2\ .$

Then

(10.37) $A(t)u \in L^p(Q)$.

Proof: This is analogous to the Dirichlet case. □

COROLLARY 10.1: *Under the assumptions of Theorem 10.2, and if, furthermore,*

(10.38) $\frac{\partial}{\partial x_k} a_{ij} \in L^\infty(Q)$

and $\mathcal{O}$ *is a bounded open set of class* C^2, *then*

(10.39) $u \in L^p(0,T;W^{2,p}(\mathcal{O}))$. □

THEOREM 10.4: *Using the framework of Theorem 10.1, let there be another set of data* $(\hat{f},\hat{\psi}_1,\hat{\psi}_2,\hat{u})$ *satisfying the same assumptions as* $(f,\psi_1,\psi_2,\bar{u})$. *Denote by the solution corresponding to* $(\hat{f},\hat{\psi}_1,\hat{\psi}_2,\hat{u})$. *If*

$$f \leqslant \hat{f},\quad \hat{\psi}_1 \leqslant \hat{\psi}_1,\quad \psi_2 \leqslant \hat{\psi}_2,\quad \bar{u} \leqslant \hat{u}$$

then

$$u \leqslant \hat{u}.$$

Proof: This is analogous to the Dirichlet case. □

10.4 *Maximum Weak Sloution*

Let us give the existence result for the maximum weak solution of (10.13), (10.14):

THEOREM 10.5: *Assume* (10.1),(10.2), *and* (10.4),(10.6). *Assume that* $\psi_1 = -\infty$, $\psi_2 = \psi$ *and that* $\mathcal{K}$ *is non-empty.*

If $f \in L^2(0,T;H)$, $\bar{u} \in H$, *then there exists a maximum solution of the weak V.I.* (10.13),(10.14).

Proof: Analogous to the Dirichlet case. □

10.5 *Interpretation of the Hamilton-Jacobi Equation*

Let us have

(10.40) $\mathcal{V}_1 \mathcal{V}_2$ compact sets of R^p, R^q respectively

(10.41) $\mathcal{O}$ a bounded open set of R^n satisfying (3.4); $Q = \mathcal{O} \times]0,T[$

$$(10.42)\quad \left| \begin{array}{l} f;c : \bar{Q} \times \mathcal{V}_1 \times \mathcal{V}_2 \to R\,, \quad g : \bar{Q} \times \mathcal{V}_1 \times \mathcal{V}_2 \to R^n\,, \text{ measurable with} \\ |f|\,,\ |\frac{\partial f}{\partial t}| \leq h(x,t) \in L^p(Q),\ p > n+2\,;\ c,\ \frac{\partial c}{\partial t},\ g,\ \frac{\partial g}{\partial t} \quad \text{bounded.} \end{array} \right.$$

For $u \in R$, $p \in R^n$ we define

$$(10.43)\quad L(x,t,u\,;\,p\,;\,v_1,v_2) = f(x,t,v_1,v_2) + u\,c(x,\,t,v_1,v_2) + p\,.\,g(x,t,v_1,v_2)\,.$$

Assume that

$$(10.44)\quad \underset{v_2 \in \mathcal{V}_2}{\text{Inf}}\ \underset{v_1 \in \mathcal{V}_1}{\text{Sup}}\ L(x,t,u,p;v_1,v_2) = \underset{v \in \mathcal{V}_1}{\text{Sup}}\ \underset{v_2 \in \mathcal{V}_2}{\text{Inf}}\ L(x,t,u;p;v_1,v_2) = H(x,t,u;p)\,.$$

It is easy to show that the function H (the *Hamiltonian*) satisfies the assumptions (10.4),(10.5),(10.6).

Let the coefficients a_{ij} be such that

$$(10.45)\quad \left| \begin{array}{l} a_{ij} = a_{ji}\,,\ \sum a_{ij}\,\xi_i \xi_j \geq \alpha |\xi|^2 \\ a_{ij} \quad \text{bounded, continuous,}\ \frac{\partial a_{ij}}{\partial x_k} \in L^\infty(Q)\,. \end{array} \right.$$

Let

$$(10.46)\quad \bar{u} \in W^{2,p}(\mathcal{O})\,,\ \left.\frac{\partial \bar{u}}{\partial \nu}\right|_\Gamma = 0\,,\quad p > n+2\,.$$

The Hamilton-Jacobi equation is the quasi-linear equation

$$(10.47)\quad \left| \begin{array}{l} -\frac{\partial u}{\partial t} - \sum a_{ij} \frac{\partial^2 u}{\partial x_i \partial x_j} - H(x,t,u,Du) = 0 \quad \text{in} \quad Q \\ \left.\frac{\partial u}{\partial \nu_{A(t)}}\right|_\Sigma = 0\,,\quad u(x,T) = \bar{u}(x)\,. \end{array} \right.$$

Remark 10.1

We can rewrite (10.47) in the form of a divergence. We are then under the conditions of Theorem 10.3 and Corollary 10.1, obviously with $\psi_1 = -\infty$, $\psi_2 = +\infty$, with the exception, however, of the assumptions of regularity in time for the a_{ij}. But we shall see that we can lighten these assumptions. Thus there is the existence of a solution of (10.47), and $u \in \mathcal{W}^{2,1,p}(Q)$.□

We can construct the measure $\tilde{P}^{xt}$ which is the solution of the sub-martingale problem for a derivative equal to nought. Moreover, let there be two controls $v_1(s), v_2(s)$, processes adapted to μ_t^s, and for which $v_1(s) \in \mathcal{V}_1, v_2(s) \in \mathcal{V}_2$ for all s. We then define the measure $P^{x,t}_{v_1,v_2}$ by

$$(10.48) \quad \left.\frac{dP^{x,t}_{v_1,v_2}}{d\tilde{P}^{x,t}}\right|_{\mu_t^T} = \exp\left\{\int_t^T \sigma^{-1} g_{v_1,v_2}(s).dw(s) - \frac{1}{2}\int_t^T (\sigma\sigma^*)^{-1} g_{v_1,v_2}.g_{v_1,v_2}\, ds\right\}$$

where we have set

$$g_{v_1,v_2}(s) = g(y(s),s,v_1(s),v_2(s)) .$$

Similarly we shall use the notations $c_{v_1,v_2}(s)$, $f_{v_1,v_2}(s)$. We then define the functional

$$(10.49) \quad J_{xt}(v_1,v_2) = E^{x,t}_{v_1,v_2}\left[\int_t^{+T} f_{v_1,v_2}(s) \exp\int_t^s c_{v_1,v_2}(\lambda)d\lambda + \right.$$
$$\left. + \bar{u}(y(T)) \exp\int_t^T c_{v_1,v_2}(\lambda)d\lambda\right] .$$

We shall also assume that

(10.50) | f,c,g are uniformly continuous at x,t with respect to v_1, v_2 and continuous in v_1, v_2 for fixed x,t.

Under the assumptions (10.44),(10.50), there exists a saddle point $\hat{v}_1(x,t,u,p)$, $\hat{v}_2(x,t,u,p)$ of (10.14), which is a Borel function of x,t,u,p

$$H(x,t,u,p) = L(x,t,u,p;\hat{v}_1 , \hat{v}_2) = L(x,t,u,p;\hat{v}_1 , \hat{v}_2) .$$

If $u \in \mathcal{W}^{2,1,p}(Q)$, with $p > n + 2$, then $u, Du \in C^0(\bar{Q})$. We can then define

$$\hat{v}_1(x,t) = \hat{v}_1(x,t,u(x,t), Du(x,t))$$
$$\hat{v}_2(x,t) = \hat{v}_2(x,t,u(x,t), Du(x,t))$$

which are Borel functions of x,t. We can define, then,

$$(10.51) \quad \begin{cases} \hat{v}_1(s) = \hat{v}_1(y(s),s) \\ \hat{v}_2(s) = \hat{v}_2(y(s),s) . \end{cases}$$

THEOREM 10.6: *Under the assumptions* (10.40),(10.41),(10.42),(10.45),(10.46), (10.50), *the solution* $u \in \mathcal{W}^{2,1,p}(Q)$, $p > n + 2$, *of* (10.47) *is necessarily unique.*

It is given explicitly by

(10.52) $$u(x,t) = \operatorname*{Inf}_{v_2} \sup_{v_1} J_{xt}(v_1, v_2) = \sup_{v_1} \operatorname*{Inf}_{v_2} J_{xt}(v_1, v_2) .$$

Moreover, $\hat{v}_1, \hat{v}_2$ *defined by* (10.51) *form a saddle point of* J_{xt}.

Proof: Analogous to the Dirichlet case. □

Remark 10.2

Since f is bounded, by (10.50), and since $c, \bar{u}$ are bounded, it follows that $J_{xt}(v_1,v_2)$ is bounded by a constant depending only upon the bounds on f,c, and T. We can then prove the existence of a solution $\mathcal{W}^{2,1,p}(Q)$ of (10.47) by regularising coefficients a_{ij} in t and using the techniques of Theorems 10.1,10.2,10.3. This solution remains bounded in $L^\infty(Q)$ when the regularisation parameter tends to 0, and even in $\mathcal{W}^{2,1,p}(Q)$, hence the existence. The details are identical to the Dirichlet case.□

10.6 *The Hamilton-Jacobi Inequality*

Now let ψ_1, ψ_2 be such that

(10.53) $$\psi_i \in \mathcal{W}^{2,1,p}(Q) \ , \ p > n+2 \ , \ \psi_1 \leq \psi_2 \ , \ \frac{\partial \psi_1}{\partial \nu_{A(t)}}\bigg|_\Sigma \geq 0 \ , \ \frac{\partial \psi_2}{\partial \nu_{A(t)}}\bigg|_\Sigma \geq 0$$

(10.54) $$\frac{\partial \psi_2}{\partial t} - A(t)\psi_2 \leq C \ , \ \frac{\partial \psi_1}{\partial t} - A(t)\psi_1 \geq -C .$$

We define the functional

(10.55) $$J_{xt}(v_1,v_2;\theta_1,\theta_2) = E^{xt}_{v_1,v_2} \left[\int_t^{T\wedge\theta_1\wedge\theta_2} f_{v_1,v_2}(s) \exp \int_t^s c_{v_1,v_2}(\lambda)d\lambda \ + \right.$$
$$+ \psi_1(y(\theta_1),\theta_1)\, \chi_{\theta_1 \leq \theta_2, \theta_1 < T} \exp \int_t^{\theta_1} c_{v_1,v_2}(\lambda)d\lambda \ +$$
$$+ \psi_2(y(\theta_2),\theta_2)\, \chi_{\theta_2 < \theta_1 \wedge T} \exp \int_t^{\theta_2} c_{v_1,v_2}(\lambda)d\lambda \ +$$
$$\left. + \bar{u}(y(T))\chi_{T \leq \theta_1 \wedge \theta_2} \exp \int_t^T c_{v_1,v_2}(\lambda)d\lambda \right] .$$

and then consider the problem

(10.56) $$\left| \begin{array}{l} u \in \mathcal{W}^{2,1,p}(Q) \ , \quad \psi_1 \leq u \leq \psi_2 \\ \dfrac{\partial u}{\partial \nu_{A(t)}}\bigg|_\Sigma = 0 \ , \quad u(x,T) = \bar{u}(x) \end{array} \right.$$

$$(10.57)\quad \begin{cases} -\dfrac{\partial u}{\partial t} - \sum a_{ij}\dfrac{\partial^2 u}{\partial x_i \partial x_j} - H(x,t,u,Du) = 0 & \text{if} \quad \psi_1 < u < \psi_2 \\ -\dfrac{\partial u}{\partial t} - \sum a_{ij}\dfrac{\partial^2 u}{\partial x_i \partial x_j} - H(x,t,u,Du) \geqq 0 & \text{if} \quad u = \psi_1 \\ -\dfrac{\partial u}{\partial t} - \sum a_{ij}\dfrac{\partial^2 u}{\partial x_i \partial x_j} - H(x,t,u,Du) \leqq 0 & \text{if} \quad u = \psi_2 \,. \end{cases}$$

We then have:

THEOREM 10.7: *Under the assumptions of Theorem 10.6 and* (10.53),(10.54), *there exists one and only one solution of* (10.56),(10.57). *It is given explicitly by the formula*

$$(10.58)\quad u(x,t) = \operatorname*{Inf}_{\theta_2}\ \operatorname*{Sup}_{\theta_1}\ \operatorname*{Inf}_{v_2}\ \operatorname*{Sup}_{v_1}\ J_{xt}(v_1,v_2;\theta_1,\theta_2) = \operatorname*{Sup}_{\theta_1}\ \operatorname*{Inf}_{\theta_2}\ \operatorname*{Sup}_{v_1}\ \operatorname*{Inf}_{v_2}\ J_{xt}(v_1,v_2;\theta_1,\theta_2)\,.$$

Moreover, if

$$\hat{\theta}_1 = \inf\{s \geqq t \mid u(y(s),s) = \psi_1(y(s),s)\}$$
$$\hat{\theta}_2 = \inf\{s \geqq t \mid u(y(s),s) = \psi_2(y(s),s)\}\ ,$$

then

$\hat{\theta}_1,\ \hat{v}_1(t)\ ;\ \hat{\theta}_2,\ \hat{v}_2(t)$ *is a saddle point of* J_{xt}.

Proof: This is analogous to the Dirichlet case. □

We can also introduce the penalised scheme

$$(10.59)\quad \begin{cases} -\dfrac{\partial u^\varepsilon}{\partial t} - \sum a_{ij}\dfrac{\partial^2 u^\varepsilon}{\partial x_i \partial x_j} - H(x,t,u^\varepsilon,Du^\varepsilon) + \dfrac{1}{\varepsilon}(u^\varepsilon - \psi_2)^+ - \dfrac{1}{\varepsilon}(u^\varepsilon - \psi_1)^- = 0 \\ u^\varepsilon \in \mathcal{W}^{2,1,p}(Q)\ , \quad \left.\dfrac{\partial u^\varepsilon}{\partial \nu_{A(t)}}\right|_{\Sigma} = 0\ , \quad u^\varepsilon(x,T) = \bar{u}(x) \end{cases}$$

and

$$(10.60)\quad u^\varepsilon \to u \quad \text{weakly in } \mathcal{W}^{2,1,p}(Q).$$

We can then use this result to interpret the solution of the V.I. in the case of obstacles that are continuous only. Thus we have:

THEOREM 10.8: *Assume* (10.1),(10.2),(10.3),(10.7),(10.16),(10.17),(10.18),(10.19), (10.38),(10.40),(10.41),(10.42),(10.50) *and* $\psi_i \in C^0(\bar{Q})$, $a_0 = 0$, $a_i = \sum_j \dfrac{\partial a_{ij}}{\partial x_j}$. *Then the solution* u *of* (10.11),(10.12),(10.8),(10.10) *belongs to* $C^0(\bar{Q})$. *It is given explicitly by* (10.58).

Proof: This is analogous to the Dirichlet case. □

Remark 10.3

We do not have $Du \in C^0(Q)$. But all the same we can define $\hat{v}_1(x,t), \hat{v}_2(x,t)$ by taking a Borel extension of Du, then substituting in (10.51). Then $\hat{v}_1(t), \hat{\theta}_1; \hat{v}_2(t)$, $\hat{\theta}_2$ still define a saddle point whatever representative of Du is chosen. This is the same situation as in the Dirichlet case. □

In the case of the control we have:

THEOREM 10.9: *Take the assumptions of Theorem 10.5, with* $a_{ij} \in C^0(\bar{Q})$, $\frac{\partial a_{ij}}{\partial x_k} \in L^\infty(Q)$, *and* (10.50)[1], (8.7),(8.8),(10.41). *The maximum solution of the weak V.I.* (10.13),(10.14) *is given explicitly by*

$$(10.61) \qquad u(x,t) = \operatorname*{Inf}_{\theta,v} J_{xt}(v,\theta).$$

Proof: This is identical to the Dirichlet case. □

Remark 10.4

In the case of Theorem 10.9, it is not known if the lower bound is reached, contrary to the linear case. This is owed to the fact that we must prove that

$$u^\varepsilon \to u \quad \text{weakly in} \quad L^2(0,T;V)$$

$$H(x,t,u^\varepsilon,Du^\varepsilon) \to H(x,t,u,Du) \quad \text{weakly in} \quad L^2(Q).$$

Although in the case of Theorem 10.8 we have, further,

$$(10.62) \qquad u^\varepsilon \to u \quad \text{strongly in} \quad L^2(0,T;V),$$

therefore

$$H(x,t,u^\varepsilon,Du^\varepsilon) \to H(x,t,u,Du) \quad \text{strongly in} \quad L^2(Q).$$

Let us prove the property (10.62). Naturally we know that

$$(10.63) \qquad u^\varepsilon \to u \quad \text{weakly in} \quad L^2(0,T;V),$$

$$H(x,t,u^\varepsilon,Du^\varepsilon) \to H(x,t,u,Du) \quad \text{weakly in} \quad L^2(Q).$$

Now, from (10.21) we deduce

$$(10.64) \qquad -(\frac{\partial}{\partial t}(u^\varepsilon - u),v) + a(t;u^\varepsilon - u,v) + (H(u^\varepsilon,Du^\varepsilon),v) - (H(u,Du),v) + \\ + \frac{1}{\varepsilon}((u^\varepsilon - \psi_2)^+,v) - (\frac{1}{\varepsilon}(u^\varepsilon - \psi_1)^-,v) = (f,v) + \\ + (\frac{\partial u}{\partial t},v) - a(t;u,v) - (H(u,Du),v) .$$

In (10.64) let us take $v = u^\varepsilon - u$. Using this yields

$$((u^\varepsilon - \psi_2)^+, u^\varepsilon - u) = ((u^\varepsilon - \psi_2)^+, u^\varepsilon - \psi_2 + \psi_2 - u) \geq 0 \qquad \text{(Contd)}$$

(1) Applied with one single control.

(Contd) $((u^\epsilon-\psi_1)^-, u^\epsilon - u) = ((u^\epsilon-\psi_1)^-, u^\epsilon-\psi_1+\psi_1-u) \leq 0$

$$-\left(\frac{\partial}{\partial t}(u^\epsilon-u), u^\epsilon-u\right) + a(t;u^\epsilon-u,u^\epsilon-u) + (H(u^\epsilon,Du^\epsilon) - H(u,Du), u^\epsilon-u) \leq$$
$$\leq \left(\frac{\partial u}{\partial t}, u^\epsilon-u\right) - a(t;u,u^\epsilon-u) - (H(u,Du),u^\epsilon-u) .$$

Integrating from t to T, using (10.63) and the coercivity (10.25), we easily obtain (10.62). □

11. VARIATIONAL INEQUALITY WITH A ONE-SIDED CONDITION ON THE DOMAIN BOUNDARY

11.1 *Statement of the Problem*

Let $\mathcal{O}$ be a bounded open set of R^n defined by (3.4). Let the coefficients be a_{ij}, a_i, a_o such that

(11.1)
$$\left|\begin{array}{l} a_{ij},\ a_i,\ a_o \in L^\infty(\mathcal{O}) , \quad \dfrac{\partial a_{ij}}{\partial x_k} \in L^\infty(\mathcal{O}) , \\ a_{ij} = a_{ji} \quad ; \ \sum a_{ij}\xi_i\xi_j \geq \alpha|\xi|^2 \quad \forall \xi \in R^n \\ a_o \geq \beta > 0 . \end{array}\right.$$

As usual, we define

(11.2) $$A + a_o - \sum_{ij} \frac{\partial}{\partial x_i} a_{ij} \frac{\partial}{\partial x_j} + \sum_i a_i \frac{\partial}{\partial x_i} + a_o$$

(11.3) $$a(u,v) = \int_{\mathcal{O}} \left(\sum_{ij} a_{ij} \frac{\partial u}{\partial x_j} \frac{\partial v}{\partial x_i} + \sum_i a_i \frac{\partial u}{\partial x_i} v + a_o u v \right) dx .$$

Assume, also,

(11.4) $$a(u,u) \geq \gamma \|u\|^2 \quad \forall u \in V = H_0^1(\mathcal{O}) .$$

Now, we are interested in the V.I.

(11.5) $$a(u,v-u) \geq (f,v-u) , \quad u|_\Gamma \leq 0 , \qquad \forall v \in H^1(\mathcal{O}) , \quad v|_\Gamma \leq 0 ,$$

with

(11.6) $$f \in L^2(\mathcal{O}).$$

Naturally, the solution u of (11.5) exists and is unique.

A penalised problem is the following.

(11.7)
$$\left|\begin{array}{ll} Au^\epsilon + a_o u^\epsilon = f & \text{a.e. in } \mathcal{O} \\ \dfrac{\partial u^\epsilon}{\partial \nu_A} + \dfrac{1}{\epsilon}(u^\epsilon)^+ = 0 & \text{on } \Gamma . \end{array}\right.$$

The variational formulation of (11.7) is given by

(11.8) $$a(u^\varepsilon,v) + \frac{1}{\varepsilon}\int_\Gamma (u^\varepsilon)^+ v \, d\Gamma = \int_{\mathcal{O}} f \, v dx , \quad \forall \, v \in H^1(\mathcal{O}) .$$

Note that the existence and uniqueness of the solution of (11.7) results from the theory of monotonic equations, for if we define a non-linear operator $\mathcal{A}:V \to V'$ by the formula

$$< \mathcal{A}(u),v > = a(u,v) + \frac{1}{\varepsilon}\int_\Gamma u^+ \, v d\Gamma$$

then

$$< \mathcal{A}(u_1) - \mathcal{A}(u_2),u_1-u_2 > = a(u_1-u_2,u_1-u_2) + \frac{1}{\varepsilon}\int_\Gamma (u_1^+-u_2^+)(u_1-u_2)d\Gamma$$
$$\geq \alpha\|u_1-u_2\|^2 .$$

We easily verify that

(11.9) $$u^\varepsilon \to u \quad \text{in } V.$$

Our aim is to interpret u^ε and u, and to give some extra results on regularity.

11.2 *Study of the Penalised Problem*

First of all we are going to prove the following regularity result:

THEOREM 11.1: *Assume* $f \in L^p(\mathcal{O})$ *together with* (11.1),(11.8); *then the solution* u^ε *of* (11.6) *belongs to* $W^{2,p}(\mathcal{O})$.

Proof: We know that $u_\varepsilon \in H^1(\mathcal{O})$, therefore $u_\varepsilon|_\Gamma \in H^{1/2}(\Gamma)$, and thus $u_\varepsilon^+|_\Gamma \in H^{1/2}(\Gamma)$, therefore $\frac{\partial u_\varepsilon}{\partial \nu_A} \in H^{1/2}(\Gamma)$, and consequently

(11.10) $$u_\varepsilon \in H^2(\mathcal{O}).$$

Then

$$u_\varepsilon \in W^{1,p_1}(\mathcal{O}) , \quad \frac{1}{p_1} = \frac{1}{2} - \frac{1}{n} \quad \text{if } n \geq 3,$$

and belongs to $W^{1,p_1}(\mathcal{O})$ for all finite p_1 if $n = 2$. Consequently $u_\varepsilon^+|_\Gamma$ belongs to $W^{1-1/p_1,p_1}(\Gamma)$, and therefore

(11.11) $$u_\varepsilon \in W^{2,\overline{p_1}}(\mathcal{O}) , \quad \overline{p_1} = \inf(p_1 , p) \quad \text{if} \quad n \geq 3 \quad (\text{and } \overline{p_1} = p \quad \text{if} \quad n = 2) .$$

We then have

$$u_\varepsilon \in W^{1,p_2}(\mathcal{O}) \ , \ \frac{1}{p_2} = \frac{1}{p_1} - \frac{1}{n} \ \text{si} \ \frac{1}{p_1} - \frac{1}{n} > 0$$

or

$$p_2 \ \text{is arbitrary and finite if} \ \frac{1}{p_1} - \frac{1}{n} \leqslant 0,$$

and then proceed step by step. □

COROLLARY 11.1: *Assume* $p > n$. *Then* $u^\varepsilon \in C^1(\bar{\mathcal{O}})$. □

We now construct the reflected diffusion at the exit from $\mathcal{O}$, corresponding to the coefficients (11.1). Set

$$a = \frac{\sigma\sigma^*}{2} \ , \quad g_i = - a_i + \sum_j \frac{\partial a_{ij}}{\partial x_j}$$

and on an arbitrary $(\Omega, \mathcal{a}, \mathcal{F}^t)$ we construct a measure P^x, a Wiener process $w(t)$, a process $y(t)$ the solution of

(11.12) $$\begin{cases} dy = g(y)dt + \sigma(y)dw - \chi_\Gamma \, \gamma(y(t))d\xi(t) \\ y(0) = x \ , \end{cases}$$

(11.13) $$\chi_{\mathcal{O}}(y(t))d\xi(t) = 0$$

(11.14) $$\begin{cases} y(t) \ \text{an adapted continuous process,} \quad y(t) \in \bar{\mathcal{O}} \quad \forall t \\ \xi(t) \ \text{a continuous increasing process,} \quad \xi(0) = 0. \end{cases}$$

In (11.12) we took

(11.15) $$\gamma_i(x) = \sum_j a_{ij}(x) \, n_j(x) \ .$$

A scalar adapted process $v(t)$ such that $0 \leqslant v(t) \leqslant 1$ is called an *admissible control*. Now introduce the functional

(11.16) $$J_x^\varepsilon(v) = E^x \int_0^{+\infty} f(y(t))(\exp - \frac{1}{\varepsilon} \int_0^t v(s)d\xi(s))(\exp - \int_0^t a_0 ds)dt$$

and we can then state:

THEOREM 11.2: *Under the assumptions of Theorem 11.1 the solution* u^ε *of* (11.6) *is given explicitly by*

(11.17) $$u^\varepsilon(x) = \operatorname*{Inf}_v J_x^\varepsilon(v).$$

Furthermore, there exists an optimal control defined by

(11.18) $$\hat{v}(t) = \begin{cases} 0 & \textit{if } y(t) \in \mathcal{O} \textit{ where } y(t) \in \Gamma_\varepsilon^- = \{x | u^\varepsilon(x)|_\Gamma < 0 \\ 1 & \textit{otherwise.} \end{cases}$$

Proof: Since $u^\varepsilon \in W^{2,p}(\mathcal{O}) \cap C^1(\bar{\mathcal{O}})$, we can apply Ito's formula to the function $u^\varepsilon(x)z$ and to both processes $y(t), z(t)$, where $z(t)$ is defined by

$$dz = a_o(y)z\,dt - \frac{1}{\varepsilon} v(t) z\,d\xi(t), \quad z(0) = 1 .$$

Then we obtain

$$Eu^\varepsilon(y(T)) \exp - \frac{1}{\varepsilon}\int_0^T v(s)d\xi(s) - \int_0^T a_o(y(s))ds = u^\varepsilon(x) +$$

$$+ E\int_0^T (-A+a_o)u^\varepsilon(y(s))dt - \frac{1}{\varepsilon} v(t)u^\varepsilon(y(t))d\xi(t))\exp - \int_0^t (\frac{1}{\varepsilon} v d\xi + a_o ds) +$$

$$+ E\int_0^T - \frac{\partial u^\varepsilon}{\partial \nu_A}(y(t)) \; (\exp - \int_0^t (\frac{1}{\varepsilon} v d\xi + a_o ds))d\xi(t) .$$

Letting T tend to $+\infty$, and taking into account equation (11.6), we obtain

$$u^\varepsilon(x) = E\int_0^\infty f(y(t)) \; (\exp - \int_0^t (\frac{1}{\varepsilon} v d\xi + a_o ds))dt + \tag{11.19}$$

$$+ E\int_0^\infty (\frac{\partial u^\varepsilon}{\partial \nu_A} + \frac{1}{\varepsilon} v(t)\; u^\varepsilon(y(t)) \; (\exp - \int_0^t (\frac{1}{\varepsilon} v d\xi + a_o ds))d\xi(t) .$$

We easily deduce from (11.19) that

$$u^\varepsilon(x) = J_x^\varepsilon(\hat{v}^\varepsilon)$$

$$u^\varepsilon(x) \leq J_x^\varepsilon(v) , \quad \forall\, v$$

whence the result (11.17). □

11.3 *Interpretation of the V.I.*

Let θ be a stopping time of $\mathcal{F}^t$. Introduce the functional

$$J_x(\theta) = E^x \int_0^\theta f(y(t))(\exp - \int_0^t a_o(y(s))ds)dt . \tag{11.20}$$

We are now going to prove:

THEOREM 11.3: *Under the assumptions of Theorem 11.1, the solution* u *of the V.I.* (11.5) *satisfies*

$$u \in C^o(\bar{\mathcal{O}}), \tag{11.21}$$

$$u(x) = \inf_{\{\theta | y(\theta) \in \Gamma \;,\; if \;\; \theta < \infty\}} J_x(\theta) \tag{11.22}$$

$$u^\varepsilon \to u \;\; in \;\; C^o(\bar{\mathcal{O}}). \tag{11.23}$$

In addition, there exists an optimal stopping time $\hat{\theta}$ *defined as follows: let*

$\Gamma^- = \{x \in \Gamma | u(x) = 0\}$, *then*

(11.24) $\hat{\theta} = \inf\{t \leqslant 0 | y(t) \in \Gamma^-\}$

satisfies

(11.25) $u(x) = J_x(\hat{\theta})$.

Proof: Let us set

$$w(x) = \inf_{\{\theta | y(\theta) \in \Gamma,\ \text{if}\ \theta < \infty\}} J_x(\theta) .$$

We are going to prove that

(11.26) $u^\varepsilon \to w$ in $C^0(\bar{\mathcal{O}})$.

Since $u^\varepsilon \to u$ in $H^1(\mathcal{O})$, the properties (11.21),(11.22), and (11.23) will certainly follow.

Set

$$\hat{\theta}^\varepsilon = \inf\{t \geqq 0 | \ y(t) \in \Gamma_\varepsilon^+\}$$

where

$$\Gamma_\varepsilon^+ = \{x \in \Gamma | \ u^\varepsilon(x) \geqq 0\} .$$

Since Γ_ε^+ is closed, and since

$$\hat{\theta}^\varepsilon = \inf\{t \geqq 0 \ | \ y(t) \notin R^n - \Gamma_\varepsilon^+\}$$

$\hat{\theta}^\varepsilon$ is the first exit time from $R^n - \Gamma_\varepsilon^+$, which is an open set, therefore $\hat{\theta}^\varepsilon$ is a stopping time. Furthermore, $\hat{\theta}^\varepsilon$ is admissible for $J_x(\theta)$, since $y(\hat{\theta}^\varepsilon) \in \Gamma_\varepsilon^+$, because $y(t)$ is a continuous process(1). On the other hand

$$\hat{v}^\varepsilon(t) = 0, \quad \text{for} \quad t < \hat{\theta}^\varepsilon$$

and if $\hat{\theta}^\varepsilon < +\infty$, then $y(\hat{\theta}^\varepsilon) \in \Gamma_\varepsilon^+$, and thus $u^\varepsilon(y(\hat{\theta}^\varepsilon)) = 0$.

Since we have (a formula analogous to (11.19))

(11.27)
$$u^\varepsilon(x) = E\int_0^{\hat{\theta}^\varepsilon} f(y(t))(\exp - \int_0^t (\frac{1}{\varepsilon}\hat{v}^\varepsilon(s)d\xi + a_0 ds))dt +$$
$$+ E\int_0^{\hat{\theta}^\varepsilon}\left(\frac{\partial u^\varepsilon}{\partial \nu_A} + \frac{1}{\varepsilon}\hat{v}^\varepsilon(t)u^\varepsilon(y(t))\right)\left(\exp - \int_0^t (\frac{1}{\varepsilon}\hat{v}\, d\xi + a_0 ds)\right) d\xi(t)\Big) +$$

(Contd)

(1) Naturally $\hat{\theta}^\varepsilon$ can be equal to $+\infty$. It suffices to interpret what has been said in a suitable way.

(Contd) $$+ u^\varepsilon(y(\hat{\theta}^\varepsilon)) \exp - \int_0^{\hat{\theta}^\varepsilon} (\frac{1}{\varepsilon} \hat{v}^\varepsilon(s) d\xi(s) + a_0 ds)$$

we see easily that

(11.28) $$u^\varepsilon(x) = J_x(\theta^\varepsilon)$$

and therefore

$$u^\varepsilon(x) \geqslant w(x).$$

Now we look for an estimate in the opposite sense. Let θ be a stopping time such that $y(\theta) \in \Gamma$. We define a control

$$v_\theta(t) = \begin{vmatrix} 0 & \text{for } t < \theta \\ 1 & \text{for } t \geqslant \theta. \end{vmatrix}$$

Therefore

(11.29) $$J_x^\varepsilon(v_\theta) - J_x(\theta) = - E\int_\theta^{+\infty} f(y(t))(\exp - \frac{1}{\varepsilon}(\xi(t)-\xi(\theta))\left(\exp - \int_0^t a_0 ds\right) ds$$

$$\leq \left(E\int_\theta^{+\infty} |f|^{1+a}(\exp - (1+a)\beta t) dt\right)^{\frac{1}{1+a}} \times \left(E\int_\theta^{+\infty} [\exp - \frac{1+a}{\varepsilon a}(\xi(t)-\xi(\theta))] dt\right)^{\frac{a}{1+a}}$$

and for

$$p > \frac{n}{2(1 + a)}$$

we obtain

(11.30) $$J_x^\varepsilon(v_\theta) - J_x(\theta) \leq C\left(E\int_\theta^{+\infty} [\exp - \frac{1+a}{\varepsilon a}(\xi(t) - \xi(\theta))] dt\right)^{\frac{a}{1+a}}.$$

We now define

$$\tau(\lambda) = \inf \{t \geqq 0 | \xi(t) - \xi(\theta) \geqq \lambda\}.$$

Since $\xi(t)$ is continuous,

$$\xi(\tau(\lambda)) - \xi(\theta) = \lambda.$$

and

$$\tau(\xi(t) - \xi(\theta)) = t.$$

Furthermore,

$$\xi(t) \to +\infty \quad \text{as} \quad t \to +\infty, \text{ a.s..}$$

In fact, unless we start from a certain (random) instant S, $\xi(t) = \xi(S)$, and therefore for $t > S$, y behaves like a non-reflected diffusion; but then it is impossible that $y(t) \in \bar{\mathcal{O}}$ for all $t > S$, hence a contradiction. Consequently we have

$$\tau(\lambda) = +\infty \Rightarrow \lambda = +\infty.$$

In the integral on the right hand side of (11.30) we make the change of variables $t = \tau(\lambda)$. This gives

$$\tau(\lambda) = 0 \Rightarrow \lambda = 0 \quad \text{and} \quad \tau(\lambda) = +\infty \Rightarrow \lambda = +\infty$$

hence

$$J_x^\epsilon(v_\theta) - J_x(\theta) \le C\left(E\int_0^{+\infty} (\exp - \frac{1+a}{\epsilon a}\lambda)\, d\tau(\lambda)\right)^{\frac{a}{1+a}} . \tag{11.31}$$

Now consider the Dirichlet problem

$$\left| \begin{aligned} & -\sum_{ij} \frac{\partial}{\partial x_i} a_{ij} \frac{\partial v}{\partial x_j} + \sum_i a_i \frac{\partial v}{\partial x_i} = 1 \\ & v\big|_\Gamma = 0 . \end{aligned} \right. \tag{11.32}$$

Next set

$$\varphi(x,t) = v(x) + t.$$

Then

$$E[v(y(\tau(\lambda)\wedge T)) + \tau(\lambda)\wedge T] = E[v(y(\theta\wedge T)) + \theta\wedge T] - E\int_{\theta\wedge T}^{\tau(\lambda)\wedge T} \frac{\partial v}{\partial \nu_A}(y(s))d\xi(s)$$

and we see that we can make T tend to $+\infty$, whence

$$E[v(y(\tau(\lambda)) + (\tau(\lambda) - \theta)] = Ev(y(\theta)) - E\int_\theta^{\tau(\lambda)} \frac{\partial v}{\partial \nu_A}(y(s))d\xi(s) .$$

Since $\xi(t) - \xi(\theta) < \lambda$ for $t < \xi(\lambda)$ and $\xi(\tau(\lambda)) - \xi(\theta) = \lambda$, we necessarily have $y(\tau(\lambda)) \in \Gamma$. Since $y(\theta) \in \Gamma$ also, and then using (11.32), we obtain

$$\begin{aligned} E(\tau(\lambda) - \theta) &= - E\int_\theta^{\tau(\lambda)} \frac{\partial v}{\partial \nu_A}(y(s))d\xi(s) \\ &= - E\int_0^\lambda \frac{\partial v}{\partial \nu_A}(y(\tau(\mu)))d\mu \end{aligned}$$

whence it follows that

$$\frac{d}{d\lambda} E\tau(\lambda) = - E\frac{\partial v}{\partial \nu_A}(y(\tau(\lambda))) .$$

Since $v \in W^{2,p}(\mathcal{O})$ for all $p > 1$, in particular $v \in C^1(\bar{\mathcal{O}})$, therefore

$$\left\|\frac{\partial v}{\partial \nu_A}\right\|_{L^\infty(\Gamma)} \leq C ,$$

whence

$$\left|\frac{d}{d\lambda} E\tau(\lambda)\right| \leqslant C.$$

Since the expectation and differentiation operators can be interchanged in the integral on the right hand side of (11.31), we obtain

$$J_x^\varepsilon(v_\theta) - J_x(\theta) \leq C\left(\int_0^\infty (\exp - \frac{1+a}{\varepsilon a} \lambda) d\lambda\right)^{\frac{a}{1+a}} \leq C \varepsilon^{\frac{a}{1+a}}$$

hence

$$u^\varepsilon(x) - v(x) \leq C \varepsilon^{\frac{a}{1+a}} .$$

Therefore we have proved (11.21),(11.22),(11.23).

We now define

$$\Gamma_\delta^- = \{x \in \Gamma | u(x) \geq - \frac{\delta}{2}\}$$

and

$$\hat{\theta}_\delta = \inf \{t \geq 0 | y(t) \in \Gamma_\delta^-\} .$$

Let $\delta_1 < \delta_2$; then $\hat{\theta}_{\delta_2} < \hat{\theta}_{\delta_1}$. In fact, for $t < \hat{\theta}_{\delta_2}$, we have $y(t) \in \mathcal{O}$ or $y(t) \in \Gamma$, and $u(y(t)) < -\frac{\delta_2}{2} < -\frac{\delta_1}{2}$, and thus $t < \hat{\theta}_{\delta_1}$. Similarly, $\hat{\theta}_\delta < \hat{\theta}$. Consequently, as $\delta \downarrow 0$, $\hat{\theta}_\delta \uparrow$ and $\hat{\theta}_\delta < \hat{\theta}$. Therefore $\hat{\theta}_\delta \uparrow \Lambda \leqslant \hat{\theta}$. But if $\hat{\theta} < +\infty$, then

$$u(y(\hat{\theta}_\delta)) = -\frac{\delta}{2}$$

thus by going to the limit

$$u(y(\Lambda)) = 0,$$

so $\Lambda \geqslant \hat{\theta}$. Therefore $\Lambda = \hat{\theta}$ and $\theta_\delta \uparrow \hat{\theta}$ when $\delta \downarrow 0$.

If $\hat{\theta} = +\infty$, then $y(t) \in \mathcal{O}$ or $y(t) \in \Gamma$ and $u(y(t)) < 0$ for all t, and $\hat{\theta}_\delta \uparrow +\infty$, for otherwise we would have $\hat{\theta}_\delta \uparrow \Lambda$ and $u(y(\Lambda)) = 0$, hence a contradiction. Further, we have

$$\|u^\varepsilon - u\|_{C^0(\bar{\mathcal{O}})} \leqslant C(\varepsilon).$$

Therefore, for

$t < \hat\theta_\delta$, we have $y(t) \in \mathcal{O}$, or if $y(t) \in \Gamma$, $u(y(t)) < -\frac{\delta}{2}$, thus

$u^\varepsilon(y(t)) < \frac{\delta}{2} + C(\varepsilon)$;

but then for $\varepsilon \leqslant \varepsilon_\delta$ we have $-\frac{\delta}{2} + C(\varepsilon) < 0$, and thus $t < \hat\theta^\varepsilon$. Consequently $\hat\theta_\delta \leqslant \hat\theta^\varepsilon$. We can then write (the analogue of (11.27))

$$u^\varepsilon(x) = E\int_0^{\hat\theta_\delta} f(y(t))\left(\exp - \int_0^t a_0\, ds\right) dt + E\, u^\varepsilon(y(\hat\theta_\delta))\exp - \int_0^{\hat\theta_\delta} a_0\, dt\,. \tag{11.33}$$

Then make $\varepsilon \to 0$; taking into account the uniform convergence of u^ε to u, we obtain

$$u(x) = E\left[\int_0^{\hat\theta_\delta} f(y(t))\left(\exp - \int_0^t a_0\, ds\right) dt + u(y(\hat\theta_\delta))\exp - \int_0^{\hat\theta_\delta} a_0\, dt\right].$$

Next, make δ tend to 0. Taking note that $u(y(\hat\theta)) = 0$, we obtain

$$u(x) = J_x(\hat\theta)$$

hence the result. □

Remark 11.1

Thus we see the difference between V.I.'s on the boundary and V.I.'s with an obstacle. The process is the same in both cases (reflected diffusion). For the problem with an obstacle we can stop the process anywhere on $\bar{\mathcal{O}}$. For the V.I. on the boundary we stop the process on the boundary only.□

11.4 *Interpretation of the Second Penalisation Scheme for Problems with an Obstacle*

We now return to the penalisation scheme considered in Remark 5.2; let

$$a(u^\varepsilon, v) + \frac{1}{\varepsilon}((u^\varepsilon - \psi)^+, v) + \frac{1}{\varepsilon}((u^\varepsilon - \psi)^+, v)_\Gamma = (f, v)\,. \tag{11.34}$$

We assume that

$$\psi \in W^{1,p}(\mathcal{O})\,,\quad f \in L^p(\mathcal{O})\,,\quad p > n\,, \tag{11.35}$$

as well as (11.1). Then (cf. Theorem 11.1) $u^\varepsilon \in W^{2,p}(\mathcal{O})$, and (11.34) is interpreted as

$$\left|\begin{aligned} &Au^\varepsilon + a_0 u^\varepsilon + \frac{1}{\varepsilon}(u^\varepsilon - \psi)^+ = f \quad \text{in } \mathcal{O} \\ &\frac{\partial u^\varepsilon}{\partial \nu_A} + \frac{1}{\varepsilon}(u^\varepsilon - \psi)^+ = 0 \quad \text{on } \Gamma\,. \end{aligned}\right. \tag{11.36}$$

Consider the reflected diffusion process defined by (11.12), and the admissible controls v(t), and introduce the functional

$$(11.37)\qquad J_x^\epsilon(v) = E^x\int_0^\infty (f(y(t)) + \frac{1}{\epsilon}\psi(y(t))v(t))\exp - \left[\int_0^t a_0 ds + \frac{1}{\epsilon}\int_0^t v ds + \frac{1}{\epsilon}\int_0^t v d\xi\right] dt$$

$$+ E^x\int_0^\infty \frac{1}{\epsilon}\psi(y(t))v(t)\left(\exp - \left[\int_0^t a_0 ds + \frac{1}{\epsilon}\int_0^t v ds + \frac{1}{\epsilon}\int_0^t v d\xi\right]\right) d\xi(t) .$$

We then have:

THEOREM 11.4: *Under the assumptions* (11.1),(4.30),(11.8) *and* (11.35), *the solution* $u^\epsilon(x)$ *of* (11.36) *is given explicitly by*

$$(11.38)\qquad u^\epsilon(x) = \operatorname*{Inf}_v J_x^\epsilon(v).$$

Furthermore, there exists an optimal control defined by

$$(11.39)\qquad v^\epsilon(t) = \begin{cases} 1 & \text{if } u^\epsilon(y(t)) \geqslant \psi(y(t)) \\ 0 & \text{otherwise.}\end{cases}$$

Proof: We use the approach of the proof of Theorem 11.2. We know that $u^\epsilon \in W^{2,p}(\mathcal{O}) \cap C^1(\bar{\mathcal{O}})$, and we define $z(t)$ by

$$dz = - a_0(y)z\,dt - \frac{1}{\epsilon}v(t)z\,dt - \frac{1}{\epsilon}v(t)z\,d\xi(t) \quad , \quad z(0) = 1 .$$

We then apply Ito's formula to the function $u^\epsilon(x)z$ and to the pair of processes $(y(t),z(t))$, and obtain

$$Eu^\epsilon(y(T))\exp - \frac{1}{\epsilon}\int_0^T v(s)d\xi(s) - \int_0^T a_0(y(s))ds - \frac{1}{\epsilon}\int_0^T v(s)ds = u^\epsilon(x) +$$

$$+ E\int_0^T -(A+a_0)u^\epsilon(y(t))dt - \frac{1}{\epsilon}v(t)u^\epsilon(y(t))d\xi(t) - \frac{1}{\epsilon}v(t)u^\epsilon(y(t))dt \quad \times$$

$$\times \left(\exp - \int_0^t (\frac{1}{\epsilon}v d\xi + a_0 ds + \frac{1}{\epsilon}v ds)\right) dt +$$

$$+ E\int_0^T -\frac{\partial u^\epsilon}{\partial \nu_A}(y(t))\left(\exp - \int_0^t (\frac{1}{\epsilon}v\,d\xi + a_0 ds + \frac{1}{\epsilon}v\,ds)\right) d\xi(t) .$$

Making T tend to $+\infty$, and taking into account the relations (11.36), we deduce from this that

$$(11.40)\qquad u^\epsilon(x) = E\int_0^\infty f(y(t))\left(\exp - \int_0^t (\frac{1}{\epsilon}v\,d\xi + a_0 ds + \frac{1}{\epsilon}v\,ds)\right) dt +$$

$$+ E\int_0^\infty (-\frac{1}{\epsilon}(u^\epsilon - \psi)^+(y(t)) + \frac{1}{\epsilon}v(t)u^\epsilon(y(t)))\left(\exp - \int_0^t (\frac{1}{\epsilon}v d\xi + a_0 ds + \frac{1}{\epsilon}v d s)\right) dt$$

$$+ E\int_0^\infty (-\frac{1}{\epsilon}(u^\epsilon - \psi)^+(y(t)) + \frac{1}{\epsilon}v(t)u^\epsilon(y(t)))\left(\exp - \int_0^t \frac{1}{\epsilon}v d\xi + a_0 ds + \frac{1}{\epsilon}v ds\right) d\xi(t)$$

and from this we easily deduce the result (11.38). □

We know that $u^\varepsilon \to u$, where u is the solution of the V.I.

$$a(u,v-u) \geq (f,v-u) \quad \forall\, v\in H^1(\mathcal{O}) \;,\; v \leq \psi \;,\; u \leq \psi \;. \tag{11.41}$$

We are now going to give an estimate for the penalisation error.

THEOREM 11.5: *Take the assumptions of Theorem 11.4, and*

$$\psi \in W^{2,p}(\mathcal{O}), \qquad p > n. \tag{11.42}$$

Then we have the estimate

$$\|u^\varepsilon - u\|_{C^0(\bar{\mathcal{O}})} \leq C\, \varepsilon^{\frac{a}{1+a}} \quad \textit{where} \quad \frac{p}{1+a} > \frac{n}{2} \;. \tag{11.43}$$

Proof: We have the following relation (valid for every admissible control)

$$\begin{aligned}\psi(x) = {} & E\int_0^\infty \left((A+a_0)\psi(y(t)) + \frac{v(t)\psi(y(t))}{\varepsilon}\right)\left(\exp - \int_0^t (a_0 + \frac{v}{\varepsilon})ds - \int_0^t \frac{v d\xi}{\varepsilon}\right) dt \\ & + E\int_0^\infty \left(\frac{\partial\psi}{\partial\nu_A}(y(t)) + \frac{v(t)\psi(y(t))}{\varepsilon}\right)\left(\exp - \int_0^t (a_0 + \frac{v}{\varepsilon})ds - \int_0^t \frac{v d\xi}{\varepsilon}\right) d\xi(t) \;.\end{aligned} \tag{11.44}$$

Further, we also have

$$\psi(x) = E\int_0^\theta (A+a_0)\psi(y(t)) \left(\exp - \int_0^t a_0\, ds\right) dt + E\psi(y(0)) \exp - \int_0^\theta a_0\, ds \;. \tag{11.45}$$

From (11.37) and (11.44) we deduce that

$$\begin{aligned} J_x^\varepsilon(v) - \psi(x) = {} & E\int_0^\infty (f - A\psi)(y(t)) \exp - \left[\int_0^t a_0\, ds + \frac{1}{\varepsilon}\int_0^t v ds + \frac{1}{\varepsilon}\int_0^t v d\xi\right] dt \,- \\ & - E\int_0^\infty \frac{\partial\psi}{\partial\nu_A}(y(t)) \left(\exp - \int_0^t (a_0 + \frac{v}{\varepsilon})ds - \int_0^t \frac{v d\xi}{\varepsilon}\right) d\xi(t) \;.\end{aligned} \tag{11.46}$$

On the other hand

$$\begin{aligned} J_x(\theta) - \psi(x) = {} & E\int_0^\theta (f - A\psi)(y(t)) \left(\exp - \int_0^t a_0\, ds\right) dt \,- \\ & - E\int_0^\theta \frac{\partial\psi}{\partial\nu_A}(y(t)) \left(\exp - \int_0^t (a_0\, ds\right) d\xi(t) \;.\end{aligned} \tag{11.47}$$

Moreover, it is easy to verify that $u^\varepsilon \geqslant u$. We must therefore obtain an estimate in the reverse sense. Let θ be a stopping time of $\mathcal{F}^t$ and a control $v_\theta(t)$ defined by

$$v_\theta(t) = \begin{vmatrix} 1 & \text{if } t < \theta \\ 0 & \text{if } t \geqslant \theta. \end{vmatrix}$$

Then

$$J_x^\varepsilon(v_\theta) = \psi(x) + E\int_0^\theta (f-(A+a_0)\psi)(y(t))\left(\exp-\int_0^t a_0\,ds\right)dt +$$

$$+ E\int_\theta^{+\infty} (f-(A+a_0)\psi)(y(t))\left(\exp-\int_0^t a_0\,ds\right)\left(\exp-\frac{t-\theta}{\varepsilon}\right)\left(\exp-\frac{\xi(t)-\xi(\theta)}{\varepsilon}\right)dt$$

$$- E\int_0^\theta \frac{\partial\psi}{\partial\nu_A}(y(t))\left(\exp-\int_0^t a_0\,ds\right)d\xi(t) -$$

$$- E\int_\theta^{+\infty} \frac{\partial\psi}{\partial\nu_A}(y(t))\left(\exp-\int_0^t a_0\,ds\right)\left(\exp-\frac{t-\theta}{\varepsilon}\right)\left(\exp-\frac{\xi(t)-\xi(\theta)}{\varepsilon}\right)d\xi(t)$$

$$= J_x(\theta) + E\int_\theta^{+\infty} (f-(A+a_0)\psi)\left(\exp-\int_0^t a_0\,ds\right)\exp-\frac{t-\theta}{\varepsilon}\exp-\frac{\xi(t)-\xi(\theta)}{\varepsilon}\,dt$$

$$- E\int_\theta^{+\infty} \frac{\partial\psi}{\partial\nu_A}(y(t)\left(\exp-\int_0^t a_0\,ds\right)\exp-\frac{t-\theta}{\varepsilon}\left(\exp-\frac{\xi(t)-\xi(\theta)}{\varepsilon}\right)d\xi(t) .$$

Moreover (cf. the proof of Theorem 7.3)

$$E\int_\theta^{+\infty} |f-A\psi|\exp-\int_0^t a_0\,ds\exp-\frac{t-\theta}{\varepsilon}\,dt \le C\,\|f-A\psi\|_{L^p}\,\varepsilon^{\frac{a}{1+a}} ,$$

and

$$E\int_\theta^{+\infty}\left|\frac{\partial\psi}{\partial\nu_A}(y(t))\right|\left(\exp-\int_0^t a_0\,ds\right)\exp-\frac{t-\theta}{\varepsilon}\exp-\frac{\xi(t)-\xi(\theta)}{\varepsilon}d\xi(t) \le$$

$$C\,E\int_\theta^{+\infty}\exp-\frac{\xi(t)-\xi(\theta)}{\varepsilon}\,d\xi(t) = -\varepsilon\,C\,E\int_\theta^{+\infty} d\ \exp-\frac{\xi(t)-\xi(\theta)}{\theta} = C\varepsilon$$

hence the result (11.43). □

We now consider the case where the obstacle is not regular. Therefore we assume that

(11.48) $\quad \psi \in C^0(\bar{\mathcal{O}})$

Naturally, u^ε defined by the variational formulation (11.34) has a meaning only in $H^1(\mathcal{O})$. On the other hand the functional $J_x(v)$, defined by (11.37), still has a meaning. The equality (11.38) is still valid. To see this it suffices to remark that if ψ_1,ψ_2 satisfy (11.35), and if u_1,u_2 are the corresponding solutions of (11.36) we then have

(11.49) $\quad \|u_1^\varepsilon - u_2^\varepsilon\|_{C^0(\bar{\mathcal{O}})} \le \|\psi_1 - \psi_2\|_{C^0(\bar{\mathcal{O}})}$.

The relation (11.49) is almost immediate if we use the probabilistic interpretation (11.38). It can also be proved using the maximum principle. Starting from the estimate (11.49) we extend the equality (11.38) without difficulty to the case where ψ is continuous only. We also know that under the assumption (11.48) the probabilistic interpretation of u remains valid (cf. (7.52)).

We can then verify that $u^\varepsilon \to u$ in $C^0(\bar{\mathcal{O}})$ using the result for ψ regular, and working by approximation with the estimate (11.49). We can also obtain an estimate of the penalisation error as in Theorem 7.4.

11.5 *Antoher Approximation of the One-Sided V.I. on the Boundary*

Here we give an approximation of the V.I. (11.4), apart from that given in (11.6). To do this, let

$$(11.50) \quad m(x) \in C^0(\bar{\mathcal{O}}), \quad m(x) > 0 \quad \text{on } \mathcal{O}, \quad m(x) = 0 \quad \text{on } \Gamma.$$

Set

$$(11.51) \quad K^\varepsilon = \{v \mid v \in H^1(\mathcal{O}) \ , \ v \leq \frac{1}{\varepsilon} m \quad \text{a.e.} \quad \text{in} \quad \mathcal{O}\} \ .$$

We now define u^ε by (no confusion is possible with (11.6))

$$(11.52) \quad \left| \begin{array}{l} u_\varepsilon \in K^\varepsilon \\ a(u^\varepsilon, v-u^\varepsilon) \geqq (f, v-u^\varepsilon) \quad \forall\, v \in K^\varepsilon \ . \end{array} \right.$$

We have:

THEOREM 11.6: *Assuming* (11.50),(11.1)[1],(11.5),(11.8), *then*

$$(11.53) \quad u^\varepsilon \to u \quad \textit{weakly in } H^1(\mathcal{O}),$$

where u *is the solution of* (11.4).

Proof: Let $v \in \hat{K}^0$, where

$$\hat{K}^0 = \{v \in C^1(\bar{\mathcal{O}}) \mid v \leq 0 \quad \text{on} \quad \Gamma\} \ .$$

Note that $\hat{K}^0 \subset K^0$, where

$$\hat{K}^0 = \{v \in H^1(\mathcal{O}) \mid v \leqq 0 \quad \text{on} \quad \Gamma\} \ .$$

Let $\lambda > 0$, then there exists ε_0 such that for $\varepsilon \leqslant \varepsilon_0$

$$v_0 = v - \lambda \leq \frac{1}{\varepsilon} m \quad \forall\, x \in \bar{\mathcal{O}} \ .$$

Consequently, by assuming $\varepsilon \leqslant \varepsilon_0$, there exists $v_0 \in K^\varepsilon$, for all ε. It then follows from (11.52) that

$$(11.54) \quad a(u^\varepsilon, u^\varepsilon) \leq a(u^\varepsilon, v_0) - (f, v_0 - u^\varepsilon)$$

and therefore u^ε stays bounded in $H^1(\mathcal{O})$. But (11.54) means

$$a(u^\varepsilon, v-\lambda - u^\varepsilon) \geqq (f, v-\lambda - u^\varepsilon) \ .$$

(1) Without the Lipschitz assumption on the coefficients.

On passing to the limit we obtain

(11.55) $a(u,v-\lambda-u) \geqq (f,v-\lambda-u)$, $\forall\ v \in \hat{K}^o$, $\forall\ \lambda > 0$.

Notice that $u \in K^o$, and making λ tend to 0, it follows from (11.55) that

$$a(u,v-u) \geqq (f,v-u) \ , \quad \forall\ v \in \hat{K}^o \ .$$

Since $\hat{K}^o$ is dense in K^o, we easily obtain that u is a solution of (11.4). □

Let us now give a probabilistic proof of (11.53) by taking the assumptions of Theorem 11.1, and even simplify a little by assuming $f \in L^\infty(\mathcal{O})$. We know that u(x) is given explicitly by (11.22). As for $u^\varepsilon(x)$, it is clearly given by the formula

(11.56) $u^\varepsilon(x) = \underset{\theta}{\text{Inf}}\ J_x^\varepsilon(\theta)$

where $J_x^\varepsilon(\theta)$ is given by (no confusion is possible with (11.16))

(11.57) $J_x^\varepsilon(\theta) = E\left\{\int_0^\theta f(y(t)) \left(\exp -\int_0^t a_o(y(s))ds\right) dt + \frac{1}{\varepsilon} m(y(\theta)) \exp -\int_0^\theta a_o(y(s))ds\right\}$.

Since $J_x^\varepsilon(\theta) = J_x^\varepsilon(\theta)$ for θ such that if $\theta < \infty$, then $y(\theta) \in \Gamma$, from which it follows that

(11.58) $u^\varepsilon(x) \leqslant u(x)$.

We now need to obtain an estimate in the opposite sense. Let θ^ε be such that

(11.59) $J_x^\varepsilon(\theta^\varepsilon) = \underset{\theta}{\text{Inf}}\ J_x^\varepsilon(\theta) = u^\varepsilon(x)$

and let

$$\hat{\theta}^\varepsilon = \inf\{t \geqq \theta^\varepsilon | y(t) \notin \mathcal{O}\} \ , \quad \hat{\theta}^\varepsilon = +\infty \ , \text{ if } \ \theta^\varepsilon = +\infty \ .$$

Since $y(\hat{\theta}^\varepsilon) \in \Gamma$, if $\hat{\theta}^\varepsilon < \infty$ we have

$$m(y(\hat{\theta}^\varepsilon)) \exp -\int_0^{\hat{\theta}^\varepsilon} a_o(y(s))ds = 0 \ .$$

Therefore

$$J_x^\varepsilon(\theta^\varepsilon) - J_x^\varepsilon(\hat{\theta}^\varepsilon) = E\,\frac{1}{\varepsilon} m(y(\theta^\varepsilon)) \exp -\int_0^{\theta^\varepsilon} a_o(y(t))dt +$$
$$+ E\int_{\theta^\varepsilon}^{\hat{\theta}^\varepsilon} f(y(t)) \left(\exp -\int_0^t a_o(y(s))ds\right) dt \ .$$

Hence,

(11.60) $$J_x^\varepsilon(\hat\theta^\varepsilon) \leq J_x^\varepsilon(\theta^\varepsilon) + E\int_{\theta^\varepsilon}^{\hat\theta^\varepsilon} f(y(t)) \left(\exp - \int_0^t a_0(y(s))ds\right) dt \ .$$

But

$$J_x^\varepsilon(\hat\theta^\varepsilon) = J_x(\hat\theta^\varepsilon) \text{ and } \hat\theta^\varepsilon \text{ is admissible for the problem (11.22).}$$

Therefore

$$u(x) \leqslant J_x^\varepsilon(\hat\theta^\varepsilon).$$

We then deduce from (11.60) that

(11.61) $$u(x) \leq u^\varepsilon(x) + E\int_{\theta^\varepsilon}^{\hat\theta^\varepsilon} f(y(t)) \left(\exp - \int_0^t a_0(y(s))ds\right) dt \ .$$

Since f is bounded, by (11.61) and (11.58) we again have

(11.62) $$|u^\varepsilon(x) - u(x)| \leq C\ E\left(\exp - \int_0^{\theta^\varepsilon} a_0(y(t))dt - \exp - \int_0^{\hat\theta^\varepsilon} a_0(y(t))dt\right) .$$

It remains to estimate the quantity on the right hand side of (11.62). To do this we introduce

(11.63) $$\mathcal{O}_\delta = \{x\in\mathcal{O} |\ d(x,\Gamma) > \delta\} \ .$$

Then

(11.64) $$E\left(\exp - \int_0^{\theta^\varepsilon} a_0(y(t))dt - \exp - \int_0^{\hat\theta^\varepsilon} a_0(y(t))dt\right) =$$

$$= E\left[\left(\exp - \int_0^{\theta^\varepsilon} a_0(y(t))dt - \exp - \int_0^{\hat\theta^\varepsilon} a_0(y(t))dt\right) \chi_{y(\theta^\varepsilon)\in\mathcal{O}_\delta}\ \chi_{\theta^\varepsilon<\infty}\right]$$

$$+ E\left[\left(\exp - \int_0^{\theta^\varepsilon} a_0(y(t))dt - \exp - \int_0^{\hat\theta^\varepsilon} a_0(y(t))dt\right) \chi_{y(\theta^\varepsilon)\notin\mathcal{O}_\delta}\ \chi_{\theta^\varepsilon<\infty}\right.$$

$$= I + II \ .$$

First we have

(11.65) $$I \leq E\left(\exp - \int_0^{\theta^\varepsilon} a_0(y(t))dt\right) \chi_{y(\theta^\varepsilon)\in\mathcal{O}_\delta}\ \chi_{\theta^\varepsilon<\infty} \ .$$

But if

$$\rho(\delta) = \sup_{x\in\bar{\mathcal{O}}_\delta} m(x) > 0 \ ,$$

we deduce from (11.57) that

$$u^{\varepsilon}(x) = J_x^{\varepsilon}(\theta^{\varepsilon}) \geqq E\Big[\int_0^{\theta^{\varepsilon}} f\left(\exp - \int_0^t a_o\, ds\right) dt + \\ + \frac{1}{\varepsilon}\rho(\delta)\left(\exp - \int_0^{\theta^{\varepsilon}} a_o\, dt\right)\chi_{y(\theta^{\varepsilon})\in \mathcal{O}_\delta}\,\chi_{\theta^{\varepsilon}<\infty}\Big]$$

from which, since $u^{\varepsilon}(x)$ is bounded, we deduce

$$E\left(\exp - \int_0^{\theta^{\varepsilon}} a_o(y(t))dt\right)\chi_{y(\theta^{\varepsilon})\in \mathcal{O}_\delta}\,\chi_{\theta^{\varepsilon}<\infty} < \frac{C\varepsilon}{\rho(\delta)}$$

so

(11.66) $$I \leqslant \frac{C\varepsilon}{\rho(\delta)}\,.$$

We now introduce the function ψ the solution of

(11.67) $$(A + a_o)\psi = 0, \quad \psi|_{\Gamma} = 1.$$

Note that

$$II = E\Big[\Big(\exp - \int_0^{\theta^{\varepsilon}} a_o(y(t))dt\ \chi_{y(\theta^{\varepsilon})\notin \mathcal{O}_\delta}\,\chi_{\theta^{\varepsilon}<\infty}(1 - \\ - E\Big\{\exp - \int_{\theta^{\varepsilon}}^{\hat\theta^{\varepsilon}} a_o(y(t))dt/\mathcal{F}^{\theta^{\varepsilon}}\Big\}\Big)\Big]\,.$$

But between θ^{ε} and $\hat\theta^{\varepsilon}$, $y(t)$ coincides with a diffusion in $\mathcal{O}$, i.e.,

$$y(\hat\theta^{\varepsilon}) = y(\theta^{\varepsilon}) + \int_{\theta^{\varepsilon}}^{\hat\theta^{\varepsilon}} g(y(t))dt + \int_{\theta^{\varepsilon}}^{\hat\theta^{\varepsilon}} \sigma(y(t))dw(t)$$

and therefore

$$E\left[\exp - \int_{\theta^{\varepsilon}}^{\hat\theta^{\varepsilon}} a_o(y(t))dt \,/\, \mathcal{F}^{\theta^{\varepsilon}}\right] = \psi(y(\theta^{\varepsilon}))\,.$$

But then

$$II = E\left[\left(\exp - \int_0^{\theta^{\varepsilon}} a_o(y(t))dt\right)\chi_{y(\theta^{\varepsilon})\notin \mathcal{O}_\delta}\,\chi_{\theta^{\varepsilon}<\infty}\left(1 - \psi(y(\theta^{\varepsilon}))\right)\right]\,.$$

Since $y(\theta^{\varepsilon}) \notin \mathcal{O}_\delta$, there exists a point ξ of Γ such that $|y(\theta^{\varepsilon}) - \xi| \leqslant \delta$. Since ψ is continuous on $\bar{\mathcal{O}}$, there exists a function $\varpi(\lambda)$ and $\lambda \in R^+$ such that $\varpi(\lambda) \to 0$

when $\lambda \to 0$, and $|\psi(x) - \psi(x')| \leqslant \omega(\lambda)$ if $|x - x'| \leqslant \lambda$. Since $\psi(\xi) = 1$, we have

$$|1 - \psi(y(\theta^\varepsilon))| \leq \varpi(\delta) \text{ , si } y(\theta^\varepsilon) \notin \mathcal{O}_\delta$$

and thus

$$II \leqslant \omega(\delta).$$

Consequently we have obtained

$$|u^\varepsilon(x) - u(x)| \leq C\left[\frac{\varepsilon}{\rho(\delta)} + \varpi(\delta)\right]$$

hence

$$\limsup_{\varepsilon \to 0} |u^\varepsilon(x) - u(x)| \leq C\,\varpi(\delta)$$

and making δ tend to 0, we obtain

(11.68) $u^\varepsilon \to u$ uniformly on $\bar{\mathcal{O}}$. □

11.6 *Another Approximation Procedure*

Here we indicate how to adapt the approximation given in Section 5.4.

Consider the problem

(11.69) $Au = f$ in $\mathcal{O}$,

(11.70) $u - \psi \geq 0 \text{ , } \dfrac{\partial u}{\partial \nu_A} \geq 0 \text{ , } (u - \psi)\dfrac{\partial u}{\partial \nu_A} = 0$ on Γ .

We introduce Ψ by

(11.71) $A\Psi \leqslant f$ in $\mathcal{O}$

(11.72) $\Psi = \psi$ on Γ.

To fix our ideas let us assume for the moment that ψ is quite regular, as well as all the data for Ψ to be regular (clearly there is an infinity of possible choices for Ψ; later we shall indicate 'the best choice'!) and that

(11.73) $g = \left(\dfrac{\partial \Psi}{\partial \nu_A}\right)^+ \in L^\infty(\Gamma)$.

We then consider the solution u_ε of

(11.74) $Au_\varepsilon = f$ in $\mathcal{O}$,

(11.75) $\dfrac{\partial u_\varepsilon}{\partial \nu_A} + g\,\dfrac{u_\varepsilon - \Psi}{u_\varepsilon - \Psi + \varepsilon} = g$ on Γ,

(11.76) $u_\varepsilon \geqslant \Psi$.

To show that there certainly exists an unique u_ε satisfying the conditions, we consider the 'monotonic' equation:

(11.77) $Aw = f$ in $\mathcal{O}$,

(11.78) $$\frac{\partial w}{\partial \nu_A} + g\,\frac{(w-\psi)^+}{w-\psi+\varepsilon} = g \quad \text{on } \Gamma,$$

which defines w uniquely. Let us show that $w \geq \psi$; (next take $u_\varepsilon = w$). To do this we note that we can write (11.77),(11.78) in the form

$$A(w - \Psi) = f - A\Psi \geq 0,$$

$$\frac{\partial(w-\Psi)}{\partial \nu_A} + g\,\frac{(w-\psi)^+}{w-\Psi+\varepsilon} = g - \frac{\partial \Psi}{\partial \nu_A} \geq 0 \quad \text{sur } \Gamma$$

therefore $w - \Psi \geq 0$ (multiply, for example, by $(w - \Psi)^-$).

We then have the following results:

THEOREM 11.7: *Denote by* Ψ_0 *the solution of*

(11.79) $A\Psi_0 = f$ *in* $\mathcal{O}$, $\Psi_0 = \psi$ *on* Γ.

Then if u_ε *(resp.* u*) is the solution of* (11.74),(11.75),(11.76) *(with* Ψ_0*) (resp. of* (11.69),(11.70))

(11.80) $$a(v_\varepsilon - u,\, u_\varepsilon - u) \leq \varepsilon \int_\Gamma \left(\frac{\partial \Psi_0}{\partial \nu_A}\right)^+ d\Gamma\,,$$

and

(11.81) $$0 \leq \frac{\partial u}{\partial \nu_A} \leq \left(\frac{\partial \Psi_0}{\partial \nu_A}\right)^+ .$$

Remark 11.2

The inequality (11.81) is an inequality of the 'Levy-Stampacchia' type.

Remark 11.3

The choice of $\Psi = \Psi_0$ is optimal, as will be seen below.

Remark 11.4

The inequalities (11.80),(11.81) are valid whenever $\left(\frac{\partial \Psi_0}{\partial \nu}\right)^+ \in L^1(\Gamma)$.

Remark 11.5

Naturally, we deduce from (11.81) a regularity result:

$$\text{if } \psi \in W^{1,p}(\Gamma)\,,\ f \in L^p(\mathcal{O})\,,\ a_{ij} \in W^{1,\infty}(\mathcal{O})\,,\ \text{then } \left(\frac{\partial \Psi_0}{\partial \nu_A}\right)^+ \in L^p(\Gamma)$$

and therefore

$$u \in W^{1+1/p,p}(\mathcal{O}).$$

By (11.75) we have

$$(11.83)\qquad \frac{\partial u_\varepsilon}{\partial \nu} = \frac{g\varepsilon}{u_\varepsilon - \psi + \varepsilon} \leqq g\left(=\left(\frac{\partial \Psi_o}{\partial \nu_A}\right)^+\right)$$

so that

$$(11.84)\qquad u_\varepsilon \to u \quad \text{weakly in } W^{1+1/p,p}(\mathcal{O}). \text{ (1)}$$

Proof of Theorem 11.7: (1): Take Ψ defined by (11.71),(11.72); at the end of the proof we shall show that the choice of $\Psi = \Psi_o$ is optimal.

(2): Use (11.75) in the form (11.83); the variational formulation is then

$$(11.85)\qquad a(u_\varepsilon, v) = (f,v) + \left(\frac{\varepsilon g}{u_\varepsilon - \psi + \varepsilon}, v\right)_\Gamma ;$$

and the exit V.I. is written

$$(11.86)\qquad a(u, v-u) \geqq (f, v-u) \quad \forall\, v \in H^1(\Omega)\,,\ v \geqq \psi \quad \text{on} \quad \Gamma\,.$$

Taking $v = u_\varepsilon$ in (11.86), which is permissible, and taking $v = -(u_\varepsilon - u)$ in (11.85), we deduce

$$-\,a(u_\varepsilon - u,\ u_\varepsilon - u) \geqq \left(\frac{\varepsilon g}{u_\varepsilon - \psi + \varepsilon}\ ,\ u - \psi - (u_\varepsilon - \psi)\right)_\Gamma \geqq -\left(\frac{\varepsilon g}{u_\varepsilon - \psi + \varepsilon}\ ,\ u_\varepsilon - \psi\right)_\Gamma$$

hence (11.80), with Ψ instead of Ψ_o.

(3): Note now that, with (11.72) fixed, $\frac{\partial \Psi}{\partial \nu}$ (and thus also $(\frac{\partial \Psi}{\partial \nu})^+$) *decrease* if $A\Psi$ increases, so that the best inequality (11.80) is obtained for $\Psi = \Psi_o$.

(4): From (11.83) we deduce that

$$(11.87)\qquad 0 \leqslant \frac{\partial u_\varepsilon}{\partial \nu_A} \leqslant g \quad \text{on } \Gamma.$$

By (11.80) and $Au_\varepsilon = f$ we know that $\frac{\partial u_\varepsilon}{\partial \nu_A} \to \frac{\partial u}{\partial \nu_A}$ in $H^{-1/2}(\Gamma)$ (for example), so that (11.87) yields (11.81) in the limit. □

12. COMMENTS

Reflected diffusion processes are related to the concept of stochastic V.I.'s. Theorem 1.2 is one step in this direction.

This result has been extended since the publication of the French version by P.L. LIONS, J.L. MENALDI, A.S. SNITZMAN [1] and J.L. MENALDI [4]. Formulation as a stochastic V.I. is very useful for proving uniqueness results, as well as a priori estimation results used in singular perturbation problems (cf. Remark 1.7). For this line, we indicate the works of ANDERSON-OREY [1], A. BENSOUSSAN - J.L. LIONS [12], C. HOLLAND [1].

(1) Very probably there is strong convergence.

In the presentation of the sub-martingale problem we studied existence by the regularisation of the coefficients and using the theory of reflected diffusion equations (in the strong formulation). This method seems the most natural, although numerous other methods are possible. We gave one of them in Section 4 which had the advantage of being very intuitive. We can also think of penalisation techniques for the domain, especially when the stochastic V.I. formulation is possible (i.e., a convex domain). In S.V.(1) they begin by treating the non-instantaneous case, the instantaneous case bein obtained as the limit of it.

We studied uniqueness coupled with Markov properties using as much as possible the properties of limit problems. In fact it is more usual first to prove the uniqueness of the sub-martingale problem, which leads very rapidly to the Markov property. The reader is referred to S.V. for the complete theory.

The approximation process studied in Section 4.3, so it seems, is here studied for the first time. However, it was inspired by the process used by S.V. in the non-instantaneous case. It corresponds to an approximation of the Neumann problem by a sequence of Dirichlet problems. From this starting point, by controlling the direction of reflection we can obtain limit problems with a non-linear condition on the boundary of the domain.

The estimates of the error between the solution of the V.I. and the solution of the penalised problem given in Sections 5 and 6 are original. The process of Section 5.4 was inspired by the works of BRAUNER and NICOLAENKO [1]. The assumption about the sign of the conormal derivative of the obstacle allows the estimates to be improved. It seems to play a role (at least a technique) for proving the $W^{2,p}$-regularity of the solution of the V.I.. However, let us indicate that if the obstacle is continuous the solution is continuous, and (at least in certain cases) if the obstacle is Lipschitz continuous the solution of the V.I. is Lipschitz continuous, regardless of the sign of the conormal derivative of the obstacle.

The results of Section 11 appear for the first time here.

The V.I. on the boundary is an example of an equation with non-linear boundary conditions. It is interesting also to establish that two penalised schemes are possible for approximating the obstacle problem with Neumann boundary conditions. They both have a probabilistic interpretation.

Another regularisation procedure is indicated for the first time here. It has a probabilistic interpretation, but it seems to us to be too complicated to be able to be useful.

---

(1) D. STROOCK - S.R.S. VARADHAN [1].

Chapter 3
Stopping times and stochastic control related to diffusions

INTRODUCTION

The object of this Chapter is to study the problems of stopping times and stochastic control of diffusion processes with jumps. From the analytic point of view it is a question of considering (elliptic or parabolic) V.I.'s corresponding to an operator of the form $A - B$, where A is a second order differential operator

(1) $$A = -\sum \frac{\partial}{\partial x_i} a_{ij} \frac{\partial}{\partial x_j} + \sum a_i \frac{\partial}{\partial x_i} + a_0 \; ,$$

and B is an integro-differential operator of the form

(2) $$B\varphi(x) = \int_{R^n} \left[\varphi(x+z) - \varphi(x) - z.D\varphi(x)\, \chi_{|z|\leq 1}\right] M(x,dz).$$

In (2), for fixed x $M(x,dz)$ is a (unbounded) positive measure on $R^n - \{0\}$. Our presentation here is mostly given under the assumption

(3) $$M(x,dz) = c_0(x,z)m(dz) \quad 0 \leq c_0 \leq 1,$$

with m an unbounded positive measure on $R^n - \{0\}$, with a singularity at the origin, and satisfying

(4) $$\int_{\{|z|\leq 1\}} |z|^2 m(dz) + \int_{\{|z|>1\}} |z| m(dz) < \infty \; .$$

However, we distinguish two cases according as the problem is set in an open set or in the whole space. This distinction is important by reason of the non-local character of the operator B, and thus it is indispensible for defing the values of the solution in the whole space, even if the problem is set in an open set.

The case where $M(x,R^n - \{0\})$ is bounded is simple and is the object of a seperate treatment, for the condition (3) is then no longer necessary.

Since we are interested in control problems we study directly the Hamilton-Jacobi-Bellman type of equations associated with $A - B$, where the control occurs in the first order terms of the diffusion as well as in the function c_0.

Sections 1 to 4 make no reference to probability theory:

In Section 1 we study (non-linear) elliptic problems in the whole space. In Section 2 we study (non-linear) elliptic problems in a bounded open set. Boundary conditions of Dirichlet or Neumann type are given.

Section 3 is devoted to (non-linear) evolutionary problems in the whole space or in an open set. In Section 4 we study variational inequalities (V.I.'s) and we shall restrict ourselves to the parabolic case for the sake of brevity. We consider directly the case of two obstacles (games). Also we distinguish the case of the problem in the whole space or an open set (the Dirichlet condition only). Another distinction is the assumptions about the regularity of the obstacles.

With the exception of a brief summary of the essential results about martingales that are frequently used in the remainder, we give complete proofs. The reader who has a proper grounding in analysis should not have any major difficulties. Naturally, we have developed just what was necessary for us for the interpretation of analytic problems. A martingale problem corresponds to the operator A - B, the solution of which is a diffusion process with a jump. The basic works are those of STROOCK [1], KOMATSU [1], GIKHMAN - SKOROKHOD [1], LEPELTIER - MARCHAL [1]. We give the existence and uniqueness of the solution of the martingale problem in a framework that corresponds exactly with the assumptions made for the solution of analytic problems, but which is not the most general (the most general treatment will be found in the papers by STROOCK and by LEPELTIER - MARCHAL).

We have used analytic results as much as possible, in particular for obtaining L^p estimates (of Krylov type). We refer to the Comments at the end of the Chapter for a statement of how our presentation stands in relation to those given in the articles cited.

Section 5 is devoted to stochastic integral and differential calculus adapted to the processes that are of interest to us. There also we have not sought the greatest generality, in particular in the Ito form. We refer to MEYER [1], DOLEANS - DADE - MEYER [1] for the most general Ito formulae. We have followed the presentation of GIKHMAN - SKOROHOD [1], but in a simpler framework, which seems better adapted to our object.

Section 6 is devoted to the approach using diffusions with a jump, by the method of stochastic differential equations. This is the theory of GIKHMAN - SKOROHOD.

Section 7 is devoted to the martingale problem corresponding to the operator A - B, which we study under different assumptions (existence and uniqueness), essentially (3),(4), or bounded measure $M(x,R^n - \{0\})$.

Section 8 is devoted to the interpretation of linear and non-linear (Hamilton-Jacobi-Bellman) equations in R^n or in a bounded open set with Dirichlet boundary conditions. We restrict ourselves to evolutionary problems for brevity.

Section 9 is devoted to the interpretation of V.I.'s with two obstacles, and we restrict ourselves to the V.I.'s with a linear operator, in order not to make the exposition heavy.

1. PARTIAL DIFFERENTIAL EQUATIONS WITH INTEGRO-DIFFERENTIAL OPERATOR: ELLIPTIC CASE IN THE WHOLE SPACE

1.1 *Assumptions: Notations*

Let us take some functions $a_{ij}(x), a_i(x), a_0(x)$, i.j. = 1,...,n satisfying the relations

$$(1.1)\quad \left|\begin{array}{l} a_{ij} = a_{ji} \in W^{1,\infty}(R^n),\ a_i \in L^\infty(R^n),\ a_0 \in L^\infty(R^n) \\ \sum a_{ij}\xi_i\xi_j \geq \alpha|\xi|^2,\quad \forall \xi \in R^n,\ \alpha > 0\ ;\ a_0(x) \geq \beta > 0\ . \end{array}\right.$$

We define the second order differential operator

(1.2) $$A\varphi = -\sum_{ij} \frac{\partial}{\partial x_i} a_{ij} \frac{\partial \varphi}{\partial x_j} + \sum_i a_i \frac{\partial \varphi}{\partial x_i} + a_0 \varphi .$$

Next we give a positive measure on R^n, not necessarily bounded, with a singularity at 0, denoted M(dz), and satisfying

(1.3) $$\int_{\{|z|\leqslant 1\}} |z|^2 m(dz) + \int_{\{|z|>1\}} |z| m(dz) < \infty .$$

Let $c_0(x,z)$ be a real function defined on $R^n \times R^n$ such that

(1.4) c_0 measurable, $0 \leqslant c_0 \leqslant 1$.

We define the integro-differential operator

(1.5) $$B\varphi(x) = \int_{R^n} \left[\varphi(x+z) - \varphi(x) - z \cdot \nabla\varphi(x)\, \chi_{\{|z|\leqslant 1\}}\right] c_0(x,z) m(dz)$$

which is well defined, at least for $\varphi \in \mathcal{D}(R^n)$. It will be convenient to set

(1.6) $$\tilde{B}\varphi(x) = \int_{R^n} \left[\varphi(x+z) - \varphi(x) - z \cdot \nabla\varphi(x)\right] \chi_{\{|z|\leqslant 1\}}\, c_0(x,z) m(dz)$$

and

(1.7) $$K\varphi(x) = \int_{R^n} \left[\varphi(x+z) - \varphi(x)\right] \chi_{\{|z|>1\}}\, c_0(x,z) m(dz)$$

so that

(1.8) $B = \tilde{B} + K$.

We shall use weighted Sobolev spaces. Let us set

(1.9) $\beta_\mu(x) = \exp - \mu[1+|x|^2]^{\frac{1}{2}}, \ \mu \geqslant 0$.

We write $L^{p,\mu}(R^n)$ for the space of functions u such that $u\beta_\mu \in L^p(R^n)$, $W^{1,p,\mu}(R^n)$ for space of functions u such that $u\beta_\mu, \frac{\partial u}{\partial x_i}\beta_\mu \in L^p(R^n)$, for all i, and $W^{2,p,\mu}(R^n)$ for the space of functions u such that $u\beta_\mu, \frac{\partial u}{\partial x_i}\beta_\mu, \frac{\partial^2 u}{\partial x_i \partial x_j}\beta_\mu \in L^p(R^n)$, for all i,j.

Let $\mathcal{V}_1$ and $\mathcal{V}_2$ be two measurable spaces. Consider functions

(1.10) $$\left|\begin{array}{l} f : R^n \times \mathcal{V}_1 \times \mathcal{V}_2 \to R, \ g : R^n \times \mathcal{V}_1 \times \mathcal{V}_2 \to R^n, \\ \tilde{a}_1 : R^n \times \mathcal{V}_1 \times \mathcal{V}_2 \to R, \ c_1 : R^n \times \mathcal{V}_1 \times \mathcal{V}_2 \times R^n \to R \\ \text{bounded and measurable, } \sup \tilde{a}_1^- < \beta \end{array}\right.$$

(1.11) $1 + c_1(x,v_1,v_2,z) \geqslant 0$

(1.11') $|c_1(x,v_1,v_2,z)| \leq c|z|$.

Define the non-linear operators

(1.12) $$H(\varphi)(x) = \inf_{v_1} \sup_{v_2} \{f(x,v_1,v_2) + \nabla\varphi(x)\cdot g(x,v_1,v_2) - \varphi(x)\tilde{a}_1(x,v_1,v_2) + \int_{R^n} \left[\varphi(x+z) - \varphi(x) - z\cdot\nabla\varphi \, \chi_{\{|z|\leq 1\}}\right] c_0(x,z)\, c_1(x,v_1,v_2,z) m(dz)\}$$

(1.13) $$\tilde{H}(\varphi)(x) = \inf_{v_1} \sup_{v_2} \{f(x,v_1,v_2) + \nabla\varphi(x)\cdot g(x,v_1,v_2) - \varphi(x)\tilde{a}_1(x,v_1,v_2) + \int_{R^n} \left[\varphi(x+z) - \varphi(x) - z\cdot\nabla\varphi(x)\right] \chi_{\{|z|\leq 1\}} c_0(x,z)\, c_1(x,v_1,v_2,z) m(dz)\}.$$

In the remainder we shall expound results valid in Sobolev spaces with weights $\beta\mu$. Certain results will be valid in the case $\mu = 0$, but that then supposes a modification of the assumption (1.10), which becomes

(1.10)' the same statement as (1.10) and in addition

$$|f(x,v_1,v_2)| \leq \bar{f}(x), \text{ où } \bar{f} \in L^2(R^n) \cap L^\infty(R^n).$$

The necessity of modifying (1.10) into (1.10)' arises from the fact $L^\infty \subset L^{p,\mu}$ with $\mu > 0$, but $L^\infty \not\subset L^p$. Note that $\bar{f} \in L^p$, for all $p > 2$.

Before presenting the equations which we want to solve, we give the fundamental properties of the operators $B, \tilde{B}, H, \tilde{H}$.

1.2 *Fundamental Properties of Operators*

LEMMA 1.1: *The operator* $\tilde{B} \in \mathcal{L}(W^{2,p,\mu}(R^n); L^{p,\mu}(R^n))$, $p \geq 2$, $u \geq 0$, *and*

(1.14) $$|\tilde{B}\varphi|_{L^{p,\mu}} \leq C\left[\sigma(r)\, \|\varphi\|_{W^{2,p,\mu}} + \tau(r)\, |\varphi|_{L^{p,\mu}}\right]$$

where $\sigma(r) \to 0$, *as* $r \to 0$, *and where* $\tau(r)$ *is arbitrary and unbounded as* $r \to 0$, *and where* C *is a constant independant of* φ *and* p. *The operator* B *satisfies the same property only in the case* $\mu = 0$.

LEMMA 1.2: *The operator* $\tilde{H}$ *maps* $W^{1,p,\mu}(R^n)$ *into* $L^{p,\mu}(R^n)$. *If* $\psi \in W^{1,p,\mu} \cap W^{1,q,\mu}$ *and* $\varphi \in W^{1,p,\mu}$ *then* $H(\varphi + \psi) - \tilde{H}(\varphi) \in L^{p,\mu} \cap L^{q,\mu}$ *and*

(1.15) $$|\tilde{H}(\varphi+\psi) - \tilde{H}(\varphi)|_{L^{q,\mu}} \leq C\|\psi\|_{W^{1,q,\mu}},$$

where C *depends neither on* φ *nor on* ψ, *nor on* p,q. *The operator* H *satisfies the same property only in the case* $\mu = 0$.

Proof of Lemma 1.1: Let $\varphi \in \mathcal{D}(R^n)$. We can therefore calculate $\tilde{B}\varphi(x)$, and easily show that

(1.16) $$\tilde{B}\varphi(x) = \int_0^1 d\theta \int_0^\theta d\theta' \sum \frac{\partial^2\varphi}{\partial x_j \partial x_k}(x+\theta' z) z_j z_k\, \chi_{|z|\leq 1}\, c_0\, m(dz)$$

hence

$$\left(\int_{R^n} (\beta\mu)^p |\tilde{B}\varphi|^p dx\right)^{\frac{1}{p}} \leqslant$$

$$\leqslant \sum\left(\int_{R^n} dx(\beta\mu)^p\left[\int_0^1 d\theta\int_0^\theta d\theta'\int_{R^n}\left|\frac{\partial^2\varphi}{\partial x_j\partial x_k}(x+\theta' z)\right| |z_j||z_k|\ \chi_{|z|\leqslant 1} m(dz)\right]^p\right)^{\frac{1}{p}}$$

$$\leqslant C\ \sum\left(\int_{R^n} dx(\beta\mu)^p\int_0^1 d\theta\int_0^\theta d\theta'\int_{R^n}\left|\frac{\partial^2\varphi}{\partial x_j\partial x_k}(x+\theta' z)\right|^p |z|^2\ \chi_{|z|\leqslant 1} m(dz)\right)^{\frac{1}{p}} .$$

Now we have

$$\text{(1.17)}\qquad \exp - p\mu[1+|x|^2]^{\frac{1}{2}} \leqslant \exp \mu p(1+|z|) \exp - \mu p(1+|x+\theta' z|^2)^{\frac{1}{2}}$$

hence

$$|\tilde{B}\varphi|_{L^{p,\mu}} \leqslant C\ \sum\left(\int_{R^n} dx\int_0^1 d\theta\int_0^\theta d\theta'\int_{R^n} \beta\mu^p(x+\theta' z)\left|\frac{\partial^2\varphi}{\partial x_j\partial x_k}(x+\theta' z)\right|^p |z|^2\chi_{|z|\leqslant 1} m(dz)\right)^{\frac{1}{p}}$$

$$\leqslant C\|\varphi\|_{W^{2,p,\mu}}$$

whence we deduce that $\tilde{B} \in \mathcal{L}(W^{2,p,\mu};L^{p,\mu})$. Now let $r < 1$. Then

$$\tilde{B}\varphi(x) = \int_{R^n} [\varphi(x+z) - \varphi(x) - z\cdot\nabla\varphi(x)]\chi_{|z|\leqslant r}\ c_0\ m(dz) +$$

$$+ \int_{R^n} [\varphi(x+z) - \varphi(x) - z\cdot\nabla\varphi(x)]\chi_{r<|z|\leqslant 1}\ c_0\ m(dz),$$

$$= \int_0^1 d\theta\int_0^\theta d\theta'\int_{R^n} \sum \frac{\partial^2\varphi}{\partial x_j\partial x_k}(x+\theta' z)z_j z_k\ \chi_{|z|\leqslant r}\ c_0\ m(dz) +$$

$$+ \int_0^1 d\theta\int_{R^n} \sum \frac{\partial\varphi}{\partial x_j}(x+\theta z)z_j\ \chi_{r<|z|\leqslant 1}\ c_0\ m(dz) -$$

$$- \int_{R^n} \sum \frac{\partial\varphi}{\partial x_j}(x)z_j\quad \chi_{r<|z|\leqslant 1}\ c_0\ m(dz)$$

$$= I + II + III.$$

If we do a more exact calculation than that above, we see that

$$|I|_{L^{p,\mu}} \leqslant C\ \sum\left(\int_{R^n} dx(\beta\mu)^p\int_0^1 d\theta\int_0^\theta d\theta'\int_{R^n}\left|\frac{\partial^2\varphi}{\partial x_j\partial x_k}(x+\theta' z)\right|^p |z|^2\ \chi_{|z|\leqslant r}\ m(dz)\ \times\right.$$

$$\left.\times\left(\int_0^1 d\theta\int_0^\theta d\theta'\int_{R^n} |z|^2\ \chi_{|z|\leqslant r}\ m(dz)\right)^{\frac{p}{q}}\right)^{\frac{1}{p}}$$

$$\leqslant C\|\varphi\|_{W^{2,p,\mu}} \int_{R^n} |z|^2\ \chi_{|z|\leqslant r}\ m(dz).$$

Then

$$|II|_{L^{p,\mu}} \leq \sum\left(\left(\int_{R^n} dx(\beta\mu)^p \int_0^1 d\theta \int_{R^n} \left|\frac{\partial\varphi}{\partial x_j}(x+\theta z)\right|^p |z| \;\chi_{r<|z|\leq 1}\; m(dz) \;\times\right.\right.$$

$$\left.\left.\times \left(\int_0^1 d\theta \int_{R^n} |z| \;\chi_{r<|z|\leq 1}\; m(dz)\right)^{\frac{p}{q}}\right)\right)^{\frac{1}{q}}$$

$$\leq C\|\varphi\|_{W^{1,p,\mu}} \int_{R^n} |z| \;\chi_{r<|z|\leq 1}\; m(dz)$$

and an identical estimate for $|III|_{L^{p,\mu}}$. Lastly, taking into account that

$$\|\varphi\|_{W^{1,p,\mu}} \leq \frac{\varepsilon}{2}\|\varphi\|_{W^{2,p,\mu}} + \frac{1}{2\varepsilon}|\varphi|_{L^{p,\mu}}$$

for arbitrary ε, and collecting together the results we obtain (1.14). We show that an analogous calculation is valid with B instead of $\tilde{B}$ in the case where $\mu = 0$ but not in the case $\mu > 0$, for then in (1.17) we cannot estimate $\exp(\mu p(1 + |z|))$ by a constant, since $|z|$ is no longer bounded by 1. □

Proof of Lemma 1.2: We have, for example

$$\text{(1.18)} \qquad |\tilde{H}(\varphi+\psi)(x) - \tilde{H}(\varphi)(x)| \leq C\Big[|\nabla\psi(x)| + |\psi(x)| + $$

$$+ \int_0^1 d\theta \int_{R^n} |\nabla\psi(x+\theta z)| \; |z|^2 \;\chi_{|z|\leq 1}\; m(dz)\Big]$$

hence working out as in Lemma 1.1, we deduce the result desired. □

For B and H, however, we can give the following results:

LEMMA 1.3: *The operator* $B \in \mathcal{L}(W^{2,p,\mu} \cap L^\infty; L^{p,\mu})$, $p \geq 2$, $\mu > 0$ *and*

$$\text{(1.19)} \qquad |B\varphi|_{L^{p,\mu}} \leq C\left[\sigma(r)\|\varphi\|_{W^{2,p,\mu}} + \tau(r)\|\varphi\|_{L^\infty}\right]$$

where $\sigma(r) \to 0$ *as* $r \to 0$.

LEMMA 1.4: *The operator* H *maps* $W^{1,p,\mu} \cap L^\infty$ *into* $L^{p,\mu}$, $p \geq 2$, $\mu > 0$

$$\text{(1.20)} \qquad |H(\varphi+\psi) - H(\varphi)|_{L^{p,\mu}} \leq C\left[\|\psi\|_{W^{1,p,\mu}} + \|\psi\|_{L^\infty}\right]$$

for $\psi \in W^{1,p,\mu} \cap W^{1,q,\mu} \cap L^\infty$ *and* $\varphi \in W^{1,q,\mu} \cap L^\infty$, $q,p \geq 2$, *where the constant* C *is independent of* φ,ψ.

Proof of Lemma 1.3: This follows immediately from (1.8), from Lemma 1.1, and from the fact $K \in \mathcal{L}(L^\infty(R^n); L^\infty(R^n))$. □

Proof of Lemma 1.4: We verify that we can pointwise replace (1.18) by

$$\text{(1.21)} \qquad |H(\varphi+\psi)(x) - H(\varphi)(x)| \leq C\Big[|\nabla\psi(x)| + |\psi(x)| +$$

$$+\int_0^1 d\theta\int_{R^n} |\nabla\psi(x+\theta z)|\ |z|^2\ \chi_{|z|\leq 1}\ m(dz) +$$
$$+\int_{R^n} |\psi(x+z) - \psi(x)|\ \chi_{|z|>1}\ |z|m(dz)\Big]$$

from which we easily deduce (1.20). □

1.3 *Existence and Uniqueness of Results with* $\tilde{B}$ *and* $\tilde{H}$.

Our first objective is to prove:

THEOREM 1.1: *Assume* (1.1),(1.3),(1.4),(1.10),(1.11),(1.11)'. *Let* $h\in L^\infty$. *Then there exists one and only one solution of*

(1.22) $$Au - \tilde{B}u - \tilde{H}(u) = h,\ u\in W^{2,p,\mu}(R^n)\cap L^\infty(R^n),\quad \forall\mu>0,\ p\geq 2\ .$$

We begin by recalling the following classical results about the diffusion operator A. Let p,μ be fixed, then there exists $\lambda_0 \geq 0$ such that the equation

(1.23) $$A\psi + \lambda_0\psi = \varphi$$

has one and only one solution $\psi\in W^{2,p,\mu}$ for a given φ in $L^{p,\mu}$, where λ_0 is independent of φ. For $\lambda \geq \lambda_0$, the solution of the equation

(1.24) $$A\psi_\lambda + \lambda\psi_\lambda = \varphi,\ \psi_\lambda\in W^{2,p,\mu}$$

exists and is defined uniquely for a given φ in $L^{p,\mu}$. We then have the estimates

(1.25) $$|\psi_\lambda|_{L^{p,\mu}} \leq \frac{|\varphi|_{L^{p,\mu}}}{\lambda - \lambda_0}$$

(1.26) $$\|\psi_\lambda\|_{W^{2,p,\mu}} \leq C|\varphi|_{L^{p,\mu}}$$

(1.27) $$\|\psi_\lambda\|_{W^{1,p,\mu}} \leq C\,\frac{|\varphi|_{L^{p,\mu}}}{\sqrt{\lambda-\lambda_0}}$$

where the constant C does not depend on λ for $\lambda\in[\lambda_0',\infty[,\lambda_0' > \lambda_0$.

LEMMA 1.5: *Under the assumptions of Theorem 1.1, but with* h *assumed only to be in* $L^{p,\mu}$, $p\geq 2$, $\mu\geq 0$ (*if* $\mu = 0$ (1.10)' *is taken instead of* (1.10)), *then for* $\lambda \geq \lambda_1$ *large enough but fixed and independent of* h, *the problem*

$$Au_\lambda - \tilde{B}u_\lambda - \tilde{H}(u_\lambda) + \lambda u_\lambda = h,\ u_\lambda\in W^{2,p,\mu}(R^n)$$

has one and only one solution.

Proof: For fixed $z\in W^{2,p,\mu}$ we can solve the problem

$$Aw + \lambda w = h + \tilde{B}z + \tilde{H}(z),\ w\in W^{2,p,\mu}$$

since $\tilde{B}z + \tilde{H}(z)\in L^{p,\mu}$. Thus we have defined a mapping $T_\lambda : z \to w$ of $W^{2,p,\mu}$ into itself. We are going to prove that it is possible to choose λ_1 large enough, but fixed in such a way that T_λ is a contraction for $\lambda \geqslant \lambda_1$. In fact, let $z_1, z_2 \in W^{2,p,\mu}$ and $w_1 = T_\lambda z_1, w_2 = T_\lambda z_2$, then

$$A(w_1-w_2) + \lambda(w_1-w_2) = \tilde{B}(z_1-z_2) + \tilde{H}(z_1) - \tilde{H}(z_2). \tag{1.29}$$

Using (1.26) and Lemmas 1.1 and 1.2, we see that

$$\|w_1-w_2\|_{W^{2,p,\mu}} \leqslant C\left[\sigma(r)\,\|z_1-z_2\|_{W^{2,p,\mu}} + \tau'(r)\,\|z_1-z_2\|_{W^{1,p,\mu}}\right]$$

$$\leqslant C\left[\left(\sigma(r) + \frac{\tau'(r)\varepsilon}{2}\right)\|z_1-z_2\|_{W^{2,p,\mu}} + \frac{\tau'(r)}{2\varepsilon}\left|z_1-z_2\right|_{L^{p,\mu}}\right].$$

Using (1.25) instead of (1.26), we see that $(\lambda-\lambda_0)|w_1-w_2|_{L^{p,\mu}}$ satisfies the same estimate where the constant C is independent of λ. We can then provide $W^{2,p,\mu}$ with the equivalent norm

$$|||\ \ |||_{W^{2,p,\mu}} = \|\ \ \|_{W^{2,p,\mu}} + (\lambda-\lambda_0)\ |\ \ |_{L^{p,\mu}}.$$

First we choose r and ε small enough so that

$$C\left(\sigma(r) + \frac{\tau'(r)\varepsilon}{2}\right) < \frac{k}{2}, \quad k \text{ fixed } < 1,$$

then λ_1 large enough in order that $C\frac{\tau'(r)}{2\varepsilon} < (\lambda_1-\lambda_0)\frac{k}{2}$. From this it follows for $\lambda \geqslant \lambda_1$ that

$$|||T_\lambda z_1 - T_\lambda z_2|||_{W^{2,p,\mu}} \leqslant k\,|||z_1-z_2|||_{W^{2,p,\mu}}.$$

Thus T_λ is a contraction, hence the result desired. □

Remark 1.1

By the proof of Lemma 1.5 we see that the only properties (1.14),(.115) of $\tilde{B}$ and $\tilde{H}$ are brought in for obtaining the existence and uniqueness of the solution of (1.28). The explicit form of $\tilde{B}$ and $\tilde{H}$ is not brought in, however, it will be used for eliminating λ.

Let $\varepsilon < 1$, and set

$$\tilde{m}_\varepsilon(dz) = m(dz)\,\chi_{|z|>\varepsilon}, \quad m_\varepsilon(dz) = m(dz)\,\chi_{\varepsilon<|z|\leqslant 1/\varepsilon}$$

and denote by $\tilde{B}_\varepsilon$, B_ε, $\tilde{H}_\varepsilon$, H_ε, the operators defined identically to $\tilde{B}$, B, $\tilde{H}$, H with $m_\varepsilon(dz)$ replacing M(dz), for $\tilde{B}_\varepsilon$, $\tilde{H}_\varepsilon$, and $m_\varepsilon(dz)$ replacing m(dz) for B_ε, H_ε.

We easily verify that the operators so defined satisfy the properties of Lemmas 1.1 to 1.4 with constants independent of ε. Further, $\tilde{m}_\varepsilon(dz)$ and $m_\varepsilon(dz)$ are bounded measures on R^n, and the measures $\tilde{m}_\varepsilon(R^n)$ and $m_\varepsilon(R^n)$ tend to $+\infty$ as $\varepsilon \to 0$. It is not not difficult to establish the following convergence properties:

$$\left|\tilde{B}_\varepsilon\psi-\tilde{B}\psi\right|_{L^{p,\mu}} \leqslant C\,\|\psi\|_{W^{2,p,\mu}} \int_{R^n} |z|^2\,\chi_{|z|\leqslant\varepsilon}\,m(dz) \tag{1.30}$$

$$\left|\tilde{H}_\varepsilon(\psi)-\tilde{H}(\psi)\right|_{L^{p,\mu}} \leqslant C\,\|\psi\|_{W^{1,p,\mu}} \int_{R^n} |z|^2\,\chi_{|z|\leqslant\varepsilon}\,m(dz) \tag{1.31}$$

and

$$(1.32)\qquad |(B_\varepsilon - B)\psi|_{L^{p,\mu}} \leq C\Big[\|\psi\|_{W^{2,p,\mu}} \int |z|^2 \chi_{|z|\leq\varepsilon}\, m(dz) + \|\psi\|_{L^\infty} \int \chi_{|z|>1/\varepsilon}\, m(dz)\Big]$$

$$(1.33)\qquad |H_\varepsilon(\psi) - H(\psi)|_{L^{p,\mu}} \leq C\Big[\|\psi\|_{W^{1,p,\mu}} \int_{R^n} |z|^2 \chi_{|z|\leq\varepsilon}\, m(dz) + \|\psi\|_{L^\infty} \int \chi_{|z|>1/\varepsilon}\, m(dz)\Big] .$$

Let ψ be given in $W^{2,p,\mu}$ and $\varphi \in L^\infty$. Fixing $\mu > 0$, we can solve the problem $(\lambda \geq \lambda_1)$

$$(1.34)\qquad A\zeta - \tilde{B}\zeta - \tilde{H}(\psi+\zeta) + \tilde{H}(\psi) + \lambda\zeta = \varphi, \quad \zeta \in W^{2,p,\mu} .$$

Exept in the case $\psi \in W^{2,\infty}$ the existence and the uniqueness of the solution ζ of (1.34) is not an immediate consquence of Lemma 1.5. Nevertheless, the proof carried out may be applied to the problem (1.34) in an identical way.

LEMMA 1.6: *The solution* ζ *of* (1.34) *belongs to* L^∞. *Furthermore, we have to estimate*

$$(1.35)\qquad \|\zeta\|_{L^\infty} \leq \frac{\|\varphi\|_{L^\infty}}{\lambda+\beta-\beta_1}$$

where $\beta_1 = \sup \tilde{a}_1^-$.

Proof: Write ζ_ε for the solution

$$(1.36)\qquad A\zeta_\varepsilon - \tilde{B}_\varepsilon\zeta_\varepsilon - \tilde{H}_\varepsilon(\psi+\zeta_\varepsilon) + \tilde{H}_\varepsilon(\psi) + \lambda\zeta_\varepsilon = \varphi,\ \zeta_\varepsilon \in W^{2,p,\mu}.$$

Then we are going to show that

$$(1.37)\qquad \zeta_\varepsilon \to \zeta \text{ weakly in } W^{2,p,\mu}.$$

We begin by establishing a priori estimates. Taking into account the uniformness (with respect to ε) of the estimates for $\tilde{B}_\varepsilon$, $\tilde{H}_\varepsilon$, we have

$$\|\zeta_\varepsilon\|_{W^{2,p,\mu}} \leq C\Big[\sigma(r)\, \|\zeta_\varepsilon\|_{W^{2,p,\mu}} + \tau'(r)\, \|\zeta_\varepsilon\|_{W^{1,p,\mu}} + \|\varphi\|_{L^{p,\mu}}\Big]$$

and $(\lambda - \lambda_0)|\zeta_\varepsilon|_{L^{p,\mu}}$ satisfies the same majorisation. Majorising

$$\|\zeta_2\|_{W^{1,p,\mu}} \text{ by } \frac{\varepsilon}{2}\|\zeta_\varepsilon\|_{W^{2,p,\mu}} + \frac{1}{2\varepsilon}|\zeta_\varepsilon|_{L^{p,\mu}} ,$$

we obtain

$$\begin{aligned}\|\zeta_\varepsilon\|_{W^{2,p,\mu}} + (\lambda-\lambda_0)|\zeta_\varepsilon|_{L^{p,\mu}} &\leq 2C\Big[\Big(\sigma(r) + \frac{\tau'(r)\varepsilon}{2}\Big)\|\zeta_\varepsilon\|_{W^{2,p,\mu}} \\ &\quad + \frac{\tau'(r)}{2\varepsilon}|\zeta_\varepsilon|_{L^{p,\mu}} + \|\varphi\|_{L^{p,\mu}}\Big] \\ &\leq k\Big[\|\zeta_\varepsilon\|_{W^{2,p,\mu}} + (\lambda-\lambda_0)|\zeta_\varepsilon|_{L^{p,\mu}}\Big] + 2C\|\varphi\|_{L^{p,\mu}}\end{aligned}$$

with $k < 1$, by the choice of λ, hence ζ_ε stays in a bounded set of $W^{2,p,\mu}$. We can then take a subsequence, again denoted ζ_ε such that

$$\zeta_\varepsilon \to \zeta \text{ weakly in } W^{2,p,\mu}. \tag{1.38}$$

By (1.30) $\tilde{B}_\varepsilon\zeta_\varepsilon - \tilde{B}\zeta_\varepsilon \to 0$ in $L^{p,\mu}$. Since $\tilde{B} \in \mathcal{L}(W^{2,p,\mu}; L^{p,\mu})$, $\tilde{B}\zeta_\varepsilon \to \tilde{B}\zeta$ weakly in $L^{p,\mu}$. Consequently

$$\tilde{B}_\varepsilon\zeta_\varepsilon \to \tilde{B}\zeta \text{ weakly in } L^{p,\mu}. \tag{1.39}$$

Further, since the injection of $W^{2,p,\mu}$ into $W^{1,p,\nu}$ is compact for $\nu > \mu$, $\zeta_\varepsilon \to \zeta$ strongly in $W^{1,p,\nu}$. By Lemma 1.2

$$\left|\tilde{H}_\varepsilon(\psi+\zeta_\varepsilon) - \tilde{H}_\varepsilon(\psi+\zeta)\right|_{L^{p,\nu}} \leqslant C \left\|\zeta_\varepsilon - \zeta\right\|_{W^{1,p,\nu}} \longrightarrow 0$$

and by (1.31), $\tilde{H}_\varepsilon(\psi+\zeta) \to \tilde{H}(\psi+\zeta)$ strongly in $L^{p,\nu}$. But as $\tilde{H}_\varepsilon(\psi+\zeta_\varepsilon)$ stays bounded in $L^{p,\mu}$, we can, if necessary by taking a second subsequence, affirm that

$$\tilde{H}_\varepsilon(\psi+\zeta_\varepsilon) \to \tilde{H}(\psi+\zeta) \text{ weakly in } L^{p,\mu}.$$

Collecting the results together, it is clear that ζ is the solution of (1.34). Because of the uniqueness of the limit (1.38) holds for the whole sequence ζ_ε.

To obtain (1.35) it is clear that it suffices to prove the same estimate with ζ_ε. Let γ_ε be a constant which will be specified later. Set

$$\tilde{\zeta} = \tilde{\zeta}_\varepsilon = \zeta_\varepsilon - \gamma_\varepsilon.$$

Hence by (1.36) we have

$$A\tilde{\zeta} + \lambda\tilde{\zeta} + (a_0+\lambda)\gamma_\varepsilon = \varphi + (\tilde{B}_\varepsilon+\tilde{H}_\varepsilon)(\psi+\tilde{\zeta}+\gamma_\varepsilon) - (\tilde{B}_\varepsilon+\tilde{H}_\varepsilon)(\psi) = \tag{1.40}$$

$$= \varphi + \inf\sup\{f + \nabla\psi\cdot g - \tilde{a}_1\psi +$$

$$+ \int_{R^n} [\psi(x+z) - \psi(x) - z\cdot\nabla\psi]\, \chi_{|z|\leqslant 1}\, c_0(1+c_1)\, \tilde{m}_\varepsilon(dz) +$$

$$+ \nabla\zeta\cdot g - \tilde{\zeta}a_1 + \int_{R^n} [\tilde{\zeta}(x+z) - \tilde{\zeta}(x) - z\cdot\nabla\tilde{\zeta}(x)]\, \chi_{|z|\leqslant 1}\, c_0(1+c_1)\, \tilde{m}_\varepsilon(dz)$$

$$- \gamma_\varepsilon a_1\} - \inf\sup\{f + \nabla\psi\cdot g - \tilde{a}_1\psi +$$

$$+ \int_{R^n} [\psi(x+z) - \psi(x) - z\cdot\nabla\psi]\, \chi_{|z|\leqslant 1}\, c_0(1+c_1)\, \tilde{m}_\varepsilon(dz)\} \leqslant$$

$$\leqslant \|\varphi\| + \gamma_\varepsilon\beta_1 + C[|\nabla\tilde{\zeta}(x)| + |\tilde{\zeta}(x)|]\,\left(1 + \tilde{m}_\varepsilon(R^n)\right) +$$

$$+ C_1 \int_{R^n} \tilde{\zeta}^+(x+z)\, \chi_{|z|\leqslant 1}\, \tilde{m}_\varepsilon(dz).$$

We take γ_ε in the form

$$\gamma_\varepsilon = \frac{\|\varphi\|_\infty + \gamma'_\varepsilon \|\zeta_\varepsilon\|_\infty}{\lambda + \beta - \beta_1 + \gamma'_\varepsilon}$$

where γ'_ε still remains to be specified. It follows from (1.40) that

$$A\tilde{\zeta} + (\lambda + a_0 + \gamma'_\varepsilon)\tilde{\zeta} \leqslant C[|\nabla\tilde{\zeta}(x)| + |\tilde{\zeta}(x)|]\left(1 + \tilde{m}_\varepsilon(R^n)\right) + \tag{1.41}$$

$$+ C_1 \int_{R^n} \tilde{\zeta}^+(x+z)\, \chi_{|z|\leqslant 1}\, \tilde{m}_\varepsilon(dz).$$

Now multiply by $\tilde{\zeta}^+(x)\beta_\mu^2(x)$ and integrate over x. Noting the relation (the choice of λ_0)

$$\int_{R^n} A\tilde{\zeta}\, \tilde{\zeta}^+ \beta_\mu^2(x) \geqslant \delta \|\tilde{\zeta}^+\|^2_{W^{1,2,\mu}} - \lambda_0 |\tilde{\zeta}^+|^2_{L^{2,\mu}}, \quad \delta > 0$$

then

$$\delta \|\tilde{\zeta}^+\|^2_{W^{1,2,\mu}} - \lambda_0 |\tilde{\zeta}^+|^2_{L^{2,\mu}} + (\lambda + \gamma'_\varepsilon) |\tilde{\zeta}^+|^2_{L^{2,\mu}} \leqslant$$

$$\leqslant C_\varepsilon \left[\|\tilde{\zeta}^+\|_{W^{1,2,\mu}} |\tilde{\zeta}^+|_{L^{2,\mu}} + |\tilde{\zeta}^+|^2_{L^{2,\mu}} \right]$$

and we can always choose γ'_ε large enough so as to obtain $|\tilde{\zeta}^+|_{L^{2,\mu}} = 0$. We have therefore proved that $\zeta_\varepsilon \leqslant \gamma_\varepsilon$. Using a symmetric argument we show that $\zeta_\varepsilon \geqslant -\gamma_\varepsilon$. Therefore

$$\|\zeta_\varepsilon\|_\infty \leqslant \frac{\|\varphi\|_\infty + \gamma'_\varepsilon \|\zeta_\varepsilon\|_\infty}{\lambda + \beta - \beta_1 + \gamma'_\varepsilon}$$

whence it follows that ζ_ε satisfies (1.35).

Proof of Theorem 1.1: Let $z \in L^\infty$, then by Lemma 1.5 we can solve the problem

$$Aw - \tilde{B}w - \tilde{H}(w) + \lambda w = h + \lambda z, \quad w \in W^{2,p,\mu}$$

$$\forall p \geqslant 2 \text{ et } \mu > 0 \qquad (\text{since } h + \lambda z \in L^{p,\mu} \quad \forall p \geqslant 2,\ \mu > 0).$$

By Lemma 1.6 $w \in L^\infty$, and, more precisely,

$$\|w\|_{L^\infty} \leqslant \frac{\|h + \lambda z\|_{L^\infty} + \sup f}{\lambda + \beta - \beta_1}. \tag{1.42}$$

In fact it suffices to apply Lemma 1.6 with $\psi = 0$, $\varphi = h + \lambda\zeta + \tilde{H}(0)$, and to notice that $\|\tilde{H}(0)\|_{L^\infty} \leqslant \sup f$. We have thus defined a mapping $z \to w = T_\lambda z$ of L^∞ into itself. Let us show that T_λ is a contraction. Let $z_1, z_2 \in L^\infty$ and $w_1 = T_\lambda z_1$, $w_2 = T_\lambda z_2$. Then

$$A(w_1 - w_2) - \tilde{B}(w_1 - w_2) - \tilde{H}(w_1) + \tilde{H}(w_2) + \lambda(w_1 - w_2) = \lambda(z_1 - z_2).$$

Hence $w_1 - w_2$ is the solution of (1.34) with $\psi = w_2$ and $\varphi = \lambda(z_1 - z_2)$. Applying Lemma 1.6 again we obtain

$$\| w_1 - w_2 \|_{L^\infty} \leqslant \frac{\lambda \| z_1 - z_2 \|_{L^\infty}}{\lambda + \beta - \beta_1}$$

and since $\beta > \beta_1$, we have proved that T_λ is a contraction. From this follows the existence and uniqueness of the solution of (1.22).

We now give a variant of Theorem 1.1, corresponding to the case $h \in L^p(R^n)$, $p \geqslant 2$ (but $h \notin L^\infty$). We have:

THEOREM 1.2: *Assume* (1.1),(1.3),(1.4),(1.10)',(1.11),(1.11)'. *Let* $h \in L^p(R^n)$. *Then there exists one and only one solution of*

$$\text{(1.43)} \qquad Au - \tilde{B}u - \tilde{H}(u) = h, \; u \in W^{2,p,\mu}(R^n), \quad \forall \mu > 0.$$

Remark 1.2

It would be normal to expect $u \in W^{2,p}(R^n)$ instead of $u \in W^{2,p,\mu}$, with $\mu > 0$. We shall ignore wether this is true.

For the proof of Theorem 1.2 it will be useful to remark that if we consider the equation (1.34) with $\psi \in W^{2,p,\mu}$ and $\varphi \in L^{p,\mu} \cap L^{p,\mu}$, where $\mu \geqslant 0$, p and q are fixed then $2 \leqslant p \leqslant q$, then ζ exists and is uniquely defined in the space $W^{2,p,\mu} \cap W^{2,p,\mu}$. In fact, thanks to the property (1.15) we can apply the fixed point argument used in the proof of Lemma 1.5, both in $W^{2,p,\mu}$ and in $W^{2,p,\mu}$. Naturally, λ_1 *depends* upon the values of μ, p, q chosen.

Proof of Theorem 1.2: Let $j = [n/2p]$. We fix $\mu > 0$ and $\lambda \geqslant \lambda_1$ so that we can solve (1.34) with the data of $\psi \in W^{2,p,\mu}$ and $\varphi \in L^{p,\mu} \cap L^{p_i,\mu}$, where

$$\frac{1}{p_i} = \frac{1}{p} - \frac{2i}{n}, \qquad i = 1,\ldots,j,$$

as well as with the data $\psi \in W^{2,p}$ and $\varphi \in L^p \cap L^{p_i}$.

We consider a sequence of functions defined in the following way:

$$\text{(1.44)} \qquad Au_0 - \tilde{B}u_0 - \tilde{H}(u_0) + \lambda u_0 = h, \; u_0 \in W^{2,p}$$

and as u_k is defined in $W^{2,p}$, u_{k+1} is the solution of

$$\text{(1.45)} \qquad Au_{k+1} - \tilde{B}u_{k+1} - \tilde{H}(u_{k+1}) + \lambda u_{k+1} = h + \lambda u_k, \; u_{k+1} \in W^{2,p} .$$

We can consider that $u_{k+1} - u_k$ is the solution of

$$\text{(1.46)} \qquad A(u_{k+1}-u_k) - \tilde{B}(u_{k+1}-u_k) - \tilde{H}(u_{k+1}-u_k+u_k) + \tilde{H}(u_k) + \\ + \lambda(u_{k+1}-u_k) = \lambda(u_k-u_{k-1}), \; k \geqslant 0$$

by assuming that $u_{-1} = 0$. That is to say, $u_{k+1} - u_k$ is the solution of (1.34) with $\psi = u_k$, and $\varphi = \lambda(u_k - u_{k-1})$. By the preceding we can deduce other regularity properties of $u_{k+1} - u_k$. In fact, since $u_0 \in W^{2,p}$ we also have $u_0 \in L^{p_1}$, with $\frac{1}{p_1} = \frac{1}{p} - \frac{2}{n}$. Hence $u_1 - u_0 \in W^{2,p} \cap W^{2,p_1}$. But then $u_1 - u_0 \in L^{p_2}$, with $\frac{1}{p_2} = \frac{1}{p_1} - \frac{2}{n}$, hence $u_2 - u_1 \in W^{2,p} \cap W^{2,p_2}$.

By induction we prove $u_k - u_{k-1} \in W^{2,p} \cap W^{2,p_k}$, with k such that $\frac{1}{p} - \frac{2k}{n} > 0$, i.e., $k \leq j$, (at least if $\frac{n}{2p} > j$). But then $u_j - u_{j-1} \in W^{2,p_j}$ where $p_j > \frac{n}{2}$, thus $u_j - u_{j-1}$ is uniformly continuous on R^n. Since it belongs to L^p, $p \geq 2$, we necessarily have $u_j - u_{j-1} \in L^\infty$. We have assumed $j \geq 1$. If $j = 0$, then $0 < \frac{n}{2p} < 1$, therefore u_0 is uniformly continuous and bounded. Consequently, for $k \geq j$, $u_k - u_{k-1} \in L^\infty$.

If $\frac{n}{2p} = j \geq 1$, then $u_j - u_{j-1} \in W^{2,n/2}$ therefore in particular $W^{1,n/2}$, hence $u_j - u_{j-1} \in L^n$. But then $u_{j+1} - u_j \in W^{2,n}$. Since $u_{j+1} - u_j \in W^{2,p}$, then also $u_{j+1} - u_j \in W^{2,\theta p + (1-\theta)n}$. But

$$\theta p + (1-\theta)n = \theta\frac{n}{2j} + (1-\theta)n = n(1 - \theta(1 - \frac{1}{2j})) = p'.$$

But then $u_{j+1} - u_j \in W^{2,p'}$ with

$$2 - \frac{n}{p'} = 2 - \frac{1}{1 - \theta\left(1 - \frac{1}{2j}\right)} = \frac{1 - 2\theta\left(1 - \frac{1}{2j}\right)}{1 - \theta\left(1 - \frac{1}{2j}\right)} = \alpha$$

and we can choose θ small enough so that $\alpha \in]0,1[$.

From this it follows that $u_{j+1} - u_j$ is Hölderian with exponent α, and hence L^∞, since $u_{j+1} - uj \in L^p$. Thus in all cases for $k \geq j+1$ we obtain $u_k - u_{k-1} \in L^\infty$. By Lemma 1.6, we then have

$$\|u_{k+1} - u_k\|_{L^\infty} \leq \rho \|u_k - u_{k-1}\|_{L^\infty}, \quad 0 < \rho < 1, \quad k \geq j+1 \tag{1.47}$$

where

$$\rho = \frac{\lambda}{\lambda + \beta - \beta_1}.$$

Therefore $u_k - u_j$ is a Cauchy sequence in L^∞. By (1.45) we see that u_k remains bounded in $W^{2,p,\mu}$ for arbitrary $\mu > 0$. Therefore we can pass to the limit in (1.45) in a way similar to that carried out in Lemma 1.6. It is clear that $u_k \to u$, the solution of (1.43).

Let us now show the uniqueness. Let u_1 and u_2 be two solutions of (1.43). We see that $u_1 - u_2$ is a solution of

$$A(u_1 - u_2) - \tilde{B}(u_1 - u_2) - \tilde{H}(u_2 + u_1 - u_2) + \tilde{H}(u_2) = 0,$$

hence a solution of (1.34) with $\psi = u_2$ and $\varphi = 0$. From this it follows that $u_1 - u_2 \in L^\infty$, and by the estimation (1.35) $u_1 - u_2 = 0$, since $\varphi = 0$. □

Remark 1.3

The argument for Theorem 1.2 was inspired by P.L. LIONS [1]. □

1.4 *Existence Results with* B *and* H

We commence by stating the equivalent of Theorem 1.1:

THEOREM 1.3: *Under the assumptions of* 1.1, *let* $h \in L^\infty$. *Then there exists one and only one solution of*

$$\text{(1.48)} \qquad Au - Bu - H(u) = h, \; u \in W^{2,p,\mu} \cap L^\infty, \quad \forall p \geq 2, \; \mu \geq 0.$$

LEMMA 1.7: *Fix* $p \geq 2$ *and* $\mu > 0$. λ_2 *can be found large enough such that the equation*

$$\text{(1.49)} \qquad Au_\lambda - Bu_\lambda - H(u_\lambda) + \lambda u_\lambda = h, \; u_\lambda \in W^{2,p,\mu} \cap L^\infty$$

has one and only one solution for $\lambda \geq \lambda_2$.

Proof: Let $w \in L^\infty$. Define η as the solution of

$$\text{(1.50)} \qquad A\eta - \tilde{B}\eta - \inf\sup\Big\{ f + \nabla\eta\cdot g - \eta\tilde{a}_1 + \\ + \int \big(\eta(x+z) - \eta(x) - z\cdot\nabla\eta\big) c_0 c_1 \, \chi_{|z|\leq 1} \, m(dz) + \\ + \int w(x+z) c_0(c_1+1) \, \chi_{|z|>1} \, m(dz) - \eta \int c_0(c_1+1) \, \chi_{|z|>1} \, m(dz) \Big\} + \\ + \lambda\eta = h.$$

We notice that equation (1.50) is none other than (1.28) if we make the following transformations

$$f \to f + \int w(x+z) c_0(1+c_1) \, \chi_{|z|>1} \, m(dz)$$
$$\tilde{a}_1 \to a_2 + \int c_0(1+c_1) \, \chi_{|z|>1} \, m(dz) \, .$$

Let us fix p, μ arbitrarily. By Lemma 1.5, we can find λ_1 large enough so that (1.50) has a solution in $W^{2,p,\mu}$ for $\lambda \geq \lambda_1$. The choice of λ_1 depends neither on w nor on h. Further, by Lemma 1.6, $\eta \in L^\infty$. We have thus defined a mapping $L^\infty \to L^\infty$, $T_\lambda w = \eta$.

Let $w_1, w_2 \in L^\infty$ and $\eta_1 = T_\lambda w_1$, $\eta_2 = T_\lambda w_2$. By subtraction, using (1.50), we establish that $\eta_1 - \eta_2$ is a solution of type (1.34) corresponding to

$$\psi = \eta_2$$

$$\varphi = \inf\sup\Big\{ f + \nabla\eta_2\cdot g - \eta_2\tilde{a}_1 + \\ + \int \big(\eta_2(x+z) - \eta_2(x) - z\cdot\nabla\eta_2\big) c_0 c_1 \, \chi_{|z|\leq 1} \, m(dz) - \\ \eta_2 \int c_0(1+c_1) \, \chi_{|z|>1} \, m(dz) + \int w_2(x+z) \, c_0(1+c_1) \, \chi_{|z|>1} m(dz) \Big\}$$

$$- \inf \sup\{f + \nabla\eta_2\cdot g - \eta_2\tilde{a}_1 + \int\left(\eta_2(x+z) - \eta_2(x) - z\cdot\nabla\eta_2\right) c_0c_1 \chi_{|z|\leq 1} m(dz) + \eta_2 \int c_0(1+c_1) \chi_{|z|\leq 1} m(dz) + \int w_1(x+z) c_0(1+c_1) \chi_{|z|>1} m(dz)\}.$$

We establish $\varphi \in L^\infty$ and

$$\|\varphi\|_\infty \leq \|w_1-w_2\|_\infty \sup\int c_0(1+c_1) \chi_{|z|>1} m(dz).$$

Since $\tilde{a}_1$ has been increased, $\tilde{a}_1^-$ can only have been decreased, therefore

$$\left(a_1 + \int c_0(1+c_1) \chi_{|z|>1} m(dz)\right)^- \leq \beta_1 .$$

Applying Lemma 1.6 then gives us

$$\|T_\lambda w_1 - T_\lambda w_2\|_{L^\infty} \leq \frac{\|w_1-w_2\| \sup \int c_0(1+c_1) \chi_{|z|>1} m(dz)}{\lambda + \beta - \beta_1} .$$

Thus for $\lambda \geq \lambda_2$, T_λ is a contraction. The fixed points of T_λ are solutions of (1.49), and so we obtain the result desired. □

In a way similar to that for Lemmm 1.7 we can solve the equation analogous to (1.34), so

(1.51) $$A\zeta - B\zeta - H(\psi+\zeta) + H(\psi) + \lambda\zeta = \varphi, \quad \zeta \in W^{2,p,\mu} \cap L^\infty$$

for $\psi \in W^{2,p,\mu} \cap L^\infty$ and $\varphi \in L^\infty$.

LEMMA 1.8: *The solution of* (1.51) *satisfies the estimate*

(1.52) $$\|\zeta\|_{L^\infty} \leq \frac{\|\varphi\|_{L^\infty}}{\lambda + \beta - \beta_1} .$$

Proof: Consider an aproximation procedure similar to that of Lemma 1.6 (but not identical). Then we solve

(1.53) $$A\zeta_\varepsilon - B_\varepsilon\zeta_\varepsilon - H_\varepsilon(\psi+\zeta_\varepsilon) + H_\varepsilon(\psi) + \lambda\zeta_\varepsilon = \varphi, \quad \zeta_\varepsilon \in W^{2,p,\mu} \cap L^\infty.$$

We begin by proving the estimate

(1.54) $$\|\zeta_\varepsilon\|_{L^\infty} \leq \frac{\|\varphi\|_{L^\infty}}{\lambda + \beta - \beta_1} .$$

Contrary to Lemma 1.6, this step is essential for obtaining a priori estimates allowing passage to the limits.

To within some evident variations, the proof of (1.54) is very similar to that used in the proof of Lemma 1.6, and it is not repeated here. From (1.54) and the estimates (1.19),(1.20), which are satisfied by B_ε and H_ε, we deduce that ζ_ε remains in a bounded set of $W^{2,p,\mu}$.

We then pick a sub-sequence, again denoted ζ_ε, which weakly converges to ζ in $W^{3,p,\mu}$. Using the properties (1.20),(1.32),(1.33) in place of (1.15),(1.30), (1.31), as in Lemma 1.6, we see that we can pass to the limit in (1.53), and ζ is the solution of (1.51). From this we then deduce the desired result (1.52).□

Proof of Theorem 1.3: Let $z \in L^\infty$, then we solve the problem

$$(1.55) \qquad Aw - Bw - H(w) + \lambda w = h + \lambda z, \quad w \in W^{2,p,\mu} \cap L^\infty.$$

Thus we define a mapping $T_\lambda : L^\infty \to L^\infty$, $\tau_\lambda z = w$, and, using Lemma 1.8, we again show that this mapping is a contraction, hence the result desired.□

Remark 1.4

The solution u of (1.48) satisfies the estimate

$$(1.56) \qquad \|u\|_{L^\infty} \leqslant \frac{\|h\|_{L^\infty} + \sup |f|}{\beta - \beta_1}.$$

In fact u is the solution of (1.51) with $\psi = 0$, $\varphi = h + \lambda u + H(0)$. Applying (1.52) we obtain

$$\|u\|_{L^\infty} \leqslant \frac{\|h\|_{L^\infty} + \lambda \|u\|_{L^\infty} + \|H(0)\|_{L^\infty}}{\lambda + \beta - \beta_1}$$

and since $\|H(0)\|_{L^\infty} \leqslant \sup|f|$, we obtain (1.56). Naturally an identical result is true for the solution of (1.22).□

We can also state the analogue of Theorem 1.2:

THEOREM 1.4: *Take the assumptions of Theorem 1.2. Let* $h \in L^p(R^n)$, $p \geqslant 2$. *Then there exists one and only one solution of*

$$(1.57) \qquad Au - Bu - H(u) = h, \qquad u \in W^{2,p,\mu}, \quad \forall \mu > 0.$$

Proof: In the case $\mu = 0$ it is simple to establish that B and H have properties identical with $\tilde{B}$ and $\tilde{H}$ (cf. Lemmas 1.1 and 1.2). Therefore Lemma 1.5 is true with B,H in the case $\mu = 0$. Further, the Lemma 1.8 is also valid for L^∞ data. We then finish with a proof similar to that of Theorem 1.2, with $\tilde{B}$ and $\tilde{H}$ replacing B,H, and so deduce the desired result. □

Remark 1.5

By arguments similar to those of Lemmas 1.6 and 1.8 we would show that if the data f and h were positive, then the solutions of (1.22) and 1.48) would be positive.

2. PARTIAL DIFFERENTIAL EQUATIONS WITH INTEGRO-DIFFERENTIAL OPERATOR: ELLIPTIC CASE IN A BOUNDED OPEN SET

General Approach

Here we are interested in the Dirichlet problem for the operator A - B - H. There is then an important difference from the R^n case, owing to the non-local character of the operators B and H. In fact, if we look for the solution u of

$$Au - Bu - H(u) = f \quad \text{in } \mathcal{O}, \qquad u|_\Gamma = 0$$

there is an ambiguity, because the written forms of Bu,H(u) require the definition of u in the whole space, and not in $\mathcal{O}$ alone. The natural way of doing this is what will become useful to us in the probabilistic interpretation, and consists of continuing u by 0 outside $\mathcal{O}$. However, contrary to the case for the whole space, the function u continued by 0 could not have second derivatives in $L^p(R^n)$, and therefore the techniques of the R^n case will not be applicable. To get over these difficulties we shall be led to make certain additional assumptions.

We can also define u outside of $\mathcal{O}$ with the aid of a continuation operator $P \in \mathcal{L}(W^{2,p}(\mathcal{O});W^{2,p}(R^n))$ such that $P\varphi = \varphi$ on $\mathcal{O}$. We shall come back to this method at the end of this Section to indicate how to use it in a way that is compatible with the probabilistic interpretation.

2.1 *Assumptions: Notations*

Let us take

(2.1) $\mathcal{O}$ a bounded open set of R^n, whose boundary Γ is of class C^2.

Let B be such that

(2.2) $$B \in \mathcal{L}\left(W_0^{1,p}(\mathcal{O})\ ;\ W^{-1,p}(\mathcal{O})\right) \quad \forall p \geq 2,\ p < \infty$$

and for all r > 0, B admits the decomposition

$$B = B_r^1 + B_r^2,$$

where

$$B_r^1 \in \mathcal{L}\left(W_0^{1,p}\ ;\ W^{-1,p}(\mathcal{O})\right)\ ,\ \|B_r^1\| \leq O(r)$$

(where $O(r) \to 0$ with r)

$$B_r^2 \in \mathcal{L}(W_0^{1,p}\ ;\ L^p).$$

Let us also take a non-linear operator H such that

(2.3) $$H : W_0^{1,p}(\mathcal{O}) \to L^p(\mathcal{O}), \quad \forall p \geq 2,\ p < \infty$$

$$H(0) \in L^\infty \text{ and if } \varphi \in W_0^{1,p},\ \psi \in W_0^{1,q},\ q \geq p,$$

$$H(\varphi+\psi) - H(\varphi) \in L^q \text{ and } \left|H(\varphi+\psi) - H(\varphi)\right|_{L^q} \leq C\|\psi\|_{W_0^{1,q}}$$

where C depends neither on φ, nor on ψ. We also assume the property

(2.4) $$< B\psi + H(\varphi+\psi) - H(\varphi),\ (\psi-k)^+ > \leq C|(\psi-k)^+|\ \|(\psi-k)^+\| + k\beta_1 \int_{\mathcal{O}} (\psi-k)^+ dx,$$

where $\beta_1 > 0$, for all $\varphi,\psi \in H_0^1(\mathcal{O})$, $k \geq 0$; C independent of φ,ψ,k.

Before going ahead, we are going to verify the consistency of the abstract assumptions (2.2),(2.3),(2.4) for the operators B and H defined by (1.5) and (1.12).

It will be assumed, once and for all, that a function belonging to $W_0^{1,p}(\mathcal{O})$ is continued by 0 outside $\mathcal{O}$. The function continued thus belongs to $W^{1,p}(R^n)$.

LEMMA 2.1: *Assume* (1.10),(1.11) *and that*

$$\int_{R^n} |z| m(dz) < \infty. \tag{2.5}$$

Then B *and* H *defined by* (1.5),(1.12) *satisfy* (2.2),(2.3),(2.4).

Proof: Let us show that $B \in \mathcal{L}(W_0^1{}^{,p}(\mathcal{O}); L^p(\mathcal{O}))$, so that (2.2) will be satisfied with $B_r^1 = 0$ and $B_r^2 = B$. Let $\varphi \in \mathcal{D}(\mathcal{O})$, then

$$B\varphi(x) = \int_{R^n} c_0 [\varphi(x+z) - \varphi(x)] m(dz) - \nabla\varphi(x) \int_{R^n} c_0 z \chi_{|z|\leqslant 1} m(dz). \tag{2.6}$$

$$\int_{R^n} c_0 [\varphi(x+z) - \varphi(x)] m(dz) = \int_0^1 d\theta \int_{R^n} c_0 \nabla\varphi(x+\theta z).z \, m(dz)$$

Therefore, thanks to (2.5)

$$\int_{\mathcal{O}} \left(\int_{R^n} c_0 |\varphi(x+z) - \varphi(x)| m(dz) \right)^p dx = \int_{\mathcal{O}} \left(\int_0^1 d\theta \int_{R^n} c_0 |\nabla\varphi(x+\theta z)| |z| m(dz) \right)^p dx$$

$$\leqslant \int_{R^n} \left(\int_0^1 d\theta \int_{R^n} |\nabla\varphi(x+\theta z)| \, |z| m(dz) \right)^p dx$$

$$\leqslant C \|\varphi\|_{W^{1,p}(R^n)} = C \|\varphi\|_{W_0^{1,p}(R^n)},$$

whence the result. Further, by (1.12)

$$|H(\varphi+\psi)(x) - H(\varphi)(x)| \leqslant C \Big[|\nabla\psi(x)| + |\psi(x)| + \tag{2.7}$$

$$+ \int_{R^n} |\psi(x+z) - \psi(x)| m(dz) + |\nabla\psi(x)| \int_{R^n} |z| \, \chi_{|z|\leqslant 1} m(dz) \Big]$$

whence easily follows (2.3), thanks to (2.5).

To prove (2.4) we shall need to introduce the operators B_ε and H_ε defined in the same way as B and H by replacing the measure in (dz) by $m_\varepsilon(dz) = \chi_{|z|>\varepsilon} m(dz)$ (there is no use in introducing the measure $\chi_{\varepsilon<|z|\leqslant 1/\varepsilon} m(dz)$, because we do not use weighted spaces). We easily verify that

$$|B_\varepsilon\varphi - B\varphi|_{L^p(\mathcal{O})} \leqslant C \|\varphi\|_{W_0^{1,p}} \int_{R^n} |z| \, \chi_{|z|\leqslant\varepsilon} m(dz) \tag{2.8}$$

$$|H_\varepsilon(\varphi) - H(\varphi)|_{L^p(\mathcal{O})} \leqslant C \|\varphi\|_{W_0^{1,p}} \int_{R^n} |z| \, \chi_{|z|\leqslant\varepsilon} m(dz) \tag{2.9}$$

where C does not depend on ε. Consequently it suffices us to prove (2.4) with B_ε

and H_ε and $\varphi,\psi \in \mathcal{D}(\mathcal{O})$. Now, if $\beta_1 = \sup a_1$, then

$$(B_\varepsilon+H_\varepsilon)(\varphi+\psi)(x) - (B_\varepsilon+H_\varepsilon)\varphi(x) \leqslant k\beta_1$$

$$+ C\Big[|\nabla(\psi-k)(x)| + |(\psi-k)(x)| +$$

$$+ \int_{R^n} |(\psi-k)^+(x+z) - (\psi-k)(x)|\ \chi_{|z|>\varepsilon}\ m(dz) +$$

$$+ |\nabla(\psi-k)(x)| \int \chi_{\varepsilon\leqslant|z|\leqslant 1}\ |z|m(dz)\Big]$$

and therefore

$$(\psi-k)^+(x)\Big[(B_\varepsilon+H_\varepsilon)(\varphi+\psi)(x) - (B_\varepsilon+H_\varepsilon)\varphi(x)\Big] \leqslant$$

$$k\beta_1(\psi-k)^+(x) + C\Big[(\psi-k)^+(x)\ |\nabla(\psi-k)^+(x)| +$$

$$+ |(\psi-k)^+(x)|^2 + (\psi-k)^+(x) \int_{R^n} |(\psi-k)^+(x+z) - (\psi-k)^+(x)\ |m_\varepsilon(dz)$$

$$+ (\psi-k)^+(x)\ |\nabla(\psi-k)^+(x)| \int_{R^n} \chi_{\varepsilon\leqslant|z|\leqslant 1}\ |z|m(dz)\Big] \ ;$$

integrating with respect to x over R^n, from which we easily deduce (2.4), with constants independent of . □

LEMMA 2.2: *Assume* (1.3),(1.4),(1.10),(1.11),(1.11)', *together with the additional assumption*

$$\left|\frac{\partial c_0}{\partial x_k}(x,z)\right| \leqslant \text{constant}. \tag{2.10}$$

Then B *and* H *defined by* (1.5),(1.12) *satisfy* (2.2),(2.3),(2.4).

Proof: Write the operator B in the form

$$B = B_r^1 + B_r^2$$

where

$$B_r^1\varphi(x) = \int_{R^n}\Big(\varphi(x+z) - \varphi(x) - z - \nabla\varphi(x)\Big)\ \chi_{|z|\leqslant r}\ c_0\ m(dz)$$

$$B_r^2\varphi(x) = -\nabla\varphi(x)\int_{R^n} z\ \chi_{r<|z|\leqslant 1}\ c_0\ m(dz) +$$

$$+ \int_{R^n}\Big(\varphi(x+z) - \varphi(x)\Big)\ \chi_{r<|z|}\ c_0\ m(dz).$$

then

$$B_r^1\varphi(x) = \int_0^1 d\theta\int_0^\theta d\theta'\int_{R^n}\sum_{j,k}\frac{\partial^2\varphi}{\partial x_j\partial x_k}(x+\theta' z)\; z_j z_k\;\chi_{|z|\leq r}\; c_0\; m(dz).$$

Let $\varphi,\psi\in\mathcal{D}(R^n)$. Then

$$\int_{R^n} B_r^1\varphi(x)\psi(x)dx = \int_0^1 d\theta\int_0^\theta d\theta'\int_{R^n}dx\int_{R^n}m(dz)\Big[-\sum z_j z_k\frac{\partial\varphi}{\partial x_k}(x+\theta' z)$$
$$\psi(x)\frac{\partial c_0}{\partial x_j}(x,z) + \frac{\partial\psi}{\partial x_j}(x)c_0(x,z)\Big]\chi_{|z|\leq r}$$

whence it follows that

$$\Big|\int_{R^n} B_r^1\varphi(x)\psi(x)dx\Big| \leq C\|\varphi\|_{W^{1,p}(R^n)}\|\psi\|_{W^{1,q}(R^n)}\int|z|^2\;\chi_{|z|\leq r}\;m(dz)$$

with $\frac{1}{p}+\frac{1}{q} = 1$. Thus $B_r^1\in\mathcal{L}(W_0^{1,p}(\mathcal{O});\; W^{-1,p}(\mathcal{O}))$, and

$$\|B_r^1\| \leq C\int|z|^2\;\chi_{|z|\leq r}\;m(dz) = C\sigma(r).$$

Moreover, we easily verify that $B_r^2\in\mathcal{L}(W_0^{1,p};\; L^p)$. Hence we have (2.2).

In order to estimate H this time we use the assumption (1.11)'. It is then necessary to replace (2.7) by (1.21). We easily deduce (2.3) from this.

As for (2.4), we begin by proving it with B_ε and H_ε. For $\varphi,\psi\in\mathcal{D}(\mathcal{O})$, we have

$$(B_\varepsilon+H_\varepsilon)(\varphi+\psi)(x) - (B_\varepsilon+H_\varepsilon)\varphi(x) \leq k\beta_1 +$$
$$+ C\Big[|\nabla(\psi-k)(x)| + |(\psi-k)(x)| +$$
$$+\int_{R^n}|(\psi-k)^+(x+z) - (\psi-k)(x)|\;|z|\;\chi_{|z|>\varepsilon}m(dz) +$$
$$+ |\nabla(\psi-k)(x)|\int_{R^n}|z|^2\;\chi_{\varepsilon\leq|z|\leq 1}\;m(dz)\Big] +$$
$$+\int_{R^n}\Big[(\psi-k)^+(x+z) - (\psi-k)(x) - z\cdot\nabla(\psi-k)\;\chi_{|z|\leq 1}\Big]c_0\;m_\varepsilon(dz).$$

Therefore

$$(\psi-k)^+(x)\Big[(B_\varepsilon+H_\varepsilon)(\varphi+\psi)(x) - (B_\varepsilon+H_\varepsilon)\varphi(x)\Big] \leq k\beta_1(\psi-k)^+(x)$$
$$+ C\Big[|\nabla(\psi-k)(x)| + |(\psi-k)(x)|\Big](\psi-k)^+(x) +$$
$$+ (\psi-k)^+(x)\int_{R^n}|(\psi-k)^+(x+z) - (\psi-k)^+(x)|\;|z|m_\varepsilon(dz) +$$

(Contd)

$$+ (\psi-k)^+(x) \int_{R^n} c_0 \Big[(\psi-k)^+(x+z) - (\psi-k)^+(x) - z \cdot \nabla(\psi-k)^+(x) \chi_{|z|\leqslant 1}\Big] m_\varepsilon(dz).$$

So

$$(2.11) \qquad \int_{R^n} dx(\psi-k)^+(x)\Big[(B_\varepsilon+H_\varepsilon)(\varphi+\psi)(x) - (B_\varepsilon+H_\varepsilon)\varphi(x)\Big] \leqslant k\beta_1 \int_\theta (\psi-k)^+dx + C|(\psi-k)^+| \; \|(\psi-k)^+\| + \langle B_\varepsilon(\psi-k)^+, (\psi-k)^+\rangle .$$

The desired result will then be a consquence of

$$(2.12) \qquad \langle B_\varepsilon\varphi,\varphi\rangle \leqslant C|\varphi| \; \|\varphi\| \; \forall \in H_0^1(\mathcal{O}) ,$$

Where C does not depend on ε.

In fact, let $\varphi \in \mathcal{D}(R^n)$. We notice first that

$$\varphi(x)\cdot B_\varepsilon\varphi(x) \leqslant \frac{1}{2} B_\varepsilon\varphi^2(x)$$

for

$$\varphi(x)\Big(\varphi(x+z) - \varphi(x) - z\cdot\nabla\varphi(x)\Big) \leqslant \frac{1}{2}\Big[\varphi^2(x+z) - \varphi^2(x) - z\cdot\nabla\varphi^2(x)\Big].$$

Therefore

$$\langle B_\varepsilon\varphi,\varphi\rangle \leqslant \frac{1}{2}\int_{R^n} B_\varepsilon\varphi^2(x),$$

and

$$\frac{1}{2}\int_{R^n} B_\varepsilon\varphi^2 dx = -\int_0^1 d\theta \int_0^\theta d\theta' \int_{R^n} dx \int_{R^n} m_\varepsilon(dz) \sum z_j z_k \varphi \frac{\partial\varphi}{\partial x_k}(x+\theta' z) \frac{\partial c_0}{\partial x_j} \chi_{|z|\leqslant 1}$$
$$+ \int_0^1 d\theta \int_{R^n} dx \int_{R^n} m_\varepsilon(dz) \, c_0 \sum \varphi \frac{\partial\varphi}{\partial x_j}(x+\theta z) \, \chi_{|z|>1}$$

whence

$$\left|\frac{1}{2}\int_{R^n} B_\varepsilon\varphi^2 dx\right| \leqslant C|\varphi| \; \|\varphi\| .$$

We have therefore proved (2.4) with B_ε and H_ε, and with constants independent of ε. To pass to the limit in ε, we establish, by using (1.11)', that the property (2.9) stays satisfied. Furthermore, for $\varepsilon < 1$, we have

$$B_\varepsilon = B_\varepsilon^2$$

hence

$$B - B_\varepsilon = B_\varepsilon^1 \text{ et } \|B_\varepsilon^1\|_{\mathcal{L}(H_0^1;H^{-1})} \to 0.$$

We can therefore pass to the limit in (2.4) written with B_ε and H_ε, hence the result desired. □

Remark 2.1

In the course of the proof of Lemma 2.2, we proved a little more than the property (2.2). In fact we showed that

$$B_r^1 \in \mathcal{L}\left(W_0^{1,p}(\mathcal{O});\ W^{-1,p}(\mathcal{O})\right)$$

$$\|B_r^1\| \leqslant C\sigma(r), \quad \sigma(r) \to 0 \text{ with } r,$$

for all $p \geqslant 2$, $p < \infty$, for all $\mathcal{O}$ bounded or not, and $C, \sigma(r)$ depending on neither p, nor $\mathcal{O}$.

Remark 2.2

There is no gain in introducing the operators $\tilde{B}$ and $\tilde{H}$ here (although it is possible). In fact, since we do not consider weighted spaces, B and $\tilde{B}$, H and $\tilde{H}$ have exactly the same properties (cf, Lemma 1.1 and 1.2).

2.2 *Existence Results*

Our object is to prove:

THEOREM 2.1: *Assume* (1.1),(2.1),(2.2),(2.3),(2.4) *and* $\beta > \beta_1$. *Let* $h \in L^p(\mathcal{O})$, $p \geqslant 2$, *then there exists one and only one solution of*

$$\text{(2.13)} \qquad Au - Bu - H(u) = h, \ u \quad W_0^{1,p}(\mathcal{O})$$

We begin with:

LEMMA 2.3: *We can find a large enough fixed* λ_1 *such that for* $\lambda \geqslant \lambda_1$ *the problem*

$$\text{(2.14)} \qquad Au_\lambda - Bu_\lambda - H(u_\lambda) + \lambda u_\lambda = h, \ u_\lambda \in W^{1,p}(\mathcal{O})$$

has one and only one solution for all $h \in L^p(\mathcal{O})$, $p \geqslant 2$.

Proof: Let $\varphi_1, \varphi_2 \in H_0^1(\mathcal{O})$. Then

$$\begin{aligned} <(A-B-H)(\varphi_1) - (A-B-H)(\varphi_2), \varphi_1-\varphi_2> &\leqslant \delta\|\varphi_1-\varphi_2\|^2 - \\ - \lambda_0|\varphi_1-\varphi_2|^2 - C\Big[\sigma(r)\|\varphi_1-\varphi_2\|^2 &+ \tau(r)|\varphi_1-\varphi_2|\ \|\varphi_1-\varphi_2\| \\ + |\varphi_1-\varphi_2|\ \|\varphi_1-\varphi_2\|\Big], \end{aligned}$$

where $\lambda > 0$, $\lambda_0 \geqslant 0$. Choosing r small enough, we see that we can find $\lambda_1 \geqslant 0$ such that

$$\begin{aligned} <(A-B-H)(\varphi_1) - (A-B-H)(\varphi_2), \varphi_1-\varphi_2> &\geqslant \frac{\delta}{2}\|\varphi_1-\varphi_2\|^2 - \\ - \lambda_1|\varphi_1-\varphi_2|^2, \quad \varphi_1, \varphi_2 \in H_0^1(\mathcal{O}). \end{aligned}$$

Therefore for $\lambda \geq \lambda_1$, $(A - B - H) + \lambda$ is a monotone operator on $H_0^1(\mathcal{O})$. From this the existence and uniqueness of the solution of (2.14) in $H_0^1(\mathcal{O})$ follow.

Next, notice that if we consider the equation

$$(2.15) \qquad Az - B_r^1 z = g, \; g \in W^{-1,p}(\mathcal{O}), \; p \geq 2,$$

then the solution of (2.15) exists and is defined uniquely in $W_0^{1,p}(\mathcal{O})$, at least if r is small enough. In fact, for fixed $w \in W_0^{1,p}(\mathcal{O})$ we consider the solution of

$$(2.16) \qquad A\zeta = g + B_r^1 w, \; \zeta \in W_0^{1,p}(\mathcal{O})$$

which exists and is defined uniquely, by the classical results on elliptic diffusion equations. We further have, if $w_1, w_2 \in W_0^{1,p}(\mathcal{O})$ and ζ_1, ζ_2 are the corresponding solutions of (2.16)

$$\|\zeta_1 - \zeta_2\|_{W_0^{1,p}} \leq C \, \|B_r^1(w_1 - w_2)\|_{W^{-1,p}}$$
$$\leq C\sigma(r) \, \|w_1 - w_2\|_{W_0^{1,p}}$$

and for r small enough the mapping $w \to \zeta$ is a contraction, hence the result (2.15). From this we then deduce that the solution of (2.14) belongs to $W_0^{1,p}(\mathcal{O})$ and not to $H_0^1(\mathcal{O})$ only. In fact we write

$$Au - B_r^1 u = h + B_r^2 u - \lambda u = \varphi.$$

Since $u \in H_0^1(\mathcal{O})$, then $H(u) + B_r^2 u \in L^2(\mathcal{O})$. But then $\varphi \in L^2(\mathcal{O}) \subset W^{-1,p}(\mathcal{O})$, with $\frac{1}{q_1} = \frac{1}{2} - \frac{1}{n}$ (if $n > 2$). Therefore, by (2.14), $u \in W_0^{1,q_1}(\mathcal{O})$. But then $H(u) + B_r^2 u \in L^{q_1}(\mathcal{O})$ hence $\varphi \in L^{q_1}(\mathcal{O})$. If $q_1 \geq p$, we have the result desired, otherwise $\varphi \in W^{-1,q_2}(\mathcal{O})$, with $\frac{1}{q_2} = \frac{1}{2} - \frac{2}{n}$, hence $u \in W_0^{1,q_2}(\mathcal{O})$. Proceeding step by step we obtain

$$u \in W^{1,q_k}(\mathcal{O}), \; \frac{1}{q_k} = \frac{1}{2} - \frac{k}{n},$$

as long as $k < \frac{n}{2} < k+1$ and $q_k < p$. Let $k_0 = [\frac{n}{2}]$. If $k_0 < \frac{n}{2}$ and $q_{k_0} < p$, then $n < q_{k_0} < p$. But then

$$\varphi \in L^{q_{k_0}} \subset L^n \subset W^{-1,r}$$

for all $r \geq 1$, hence $u \in W_0^{1,r}$ whence the result. If $p \leq q_{k_0}$, the preceding argument shows that $\varphi \in L^p$, whence the result. If $k_0 = \frac{n}{2}$ then $q_{k_0-1} = n$, and if $p \geq n$, then $\varphi \in L^n$, hence the result again. Similarly if $p < n$. □

Let $\psi \in H_0^1(\mathcal{O})$ and $\varphi \in L^\infty(\mathcal{O})$; similarly as in Lemma 2.3, we prove the existence and uniqueness of the solution ζ of

$$(2.17) \qquad A\zeta - B\zeta - H(\zeta+\psi) + H(\psi) + \lambda\zeta = \varphi, \ \zeta \in W_0^{1,p}(\mathcal{O}), \ \forall p \geq 2.$$

We then have:

LEMMA 2.4: *The solution of* (2.17) *satisfies the estimate*

$$(2.18) \qquad \|\zeta\|_{L^\infty} \leq \frac{\|\varphi\|_{L^\infty}}{\lambda + \beta - \beta_1} .$$

Proof: We know that $\zeta \in L^\infty$, since $\zeta \in W_0^{1,p}(\mathcal{O})$ for all $p \geq 2$, $p < \infty$. But this does not give us, a priori, an estimate of the type (2.18). For that we shall need to use the assumption (2.4) which has not been used up to now. Let k be a constant which will be determined later. Set

$$\tilde{\zeta} = \zeta - k.$$

(2.17) may then be written

$$A\tilde{\zeta} - B\zeta + \lambda\tilde{\zeta} - H(\zeta+\psi) + H(\psi) = \varphi - \lambda k - a_0 k.$$

If we multiply by $\tilde{\zeta}^+$ and integrate over x, this yields

$$<A\tilde{\zeta}^+,\tilde{\zeta}^+> + \lambda|\tilde{\zeta}^+|^2 \leq \left(\|\varphi\| - \lambda k - \beta k\right)\int \tilde{\zeta}^+ dx + $$
$$+ C|\tilde{\zeta}^+|\,\|\tilde{\zeta}^+\| + k\beta_1 \int \tilde{\zeta}^+ dx$$

and choosing

$$k = \frac{\|\varphi\|}{\lambda + \beta - \beta_1}$$

yields

$$<A\tilde{\zeta}^+,\tilde{\zeta}^+> + \lambda|\tilde{\zeta}^+|^2 \leq C|\tilde{\zeta}^+|\ \|\tilde{\zeta}^+\|$$

which implies, at least if λ is large enough, that $\tilde{\zeta}^+ = 0$, and hence $\zeta \leq k$. Let us now show that $\zeta \geq -k$. If we set $\zeta' = -\zeta$, notice that (2.17) may be written

$$A\zeta' - B\zeta' - H(\zeta'+\psi-\zeta') + H(\psi-\zeta') + \lambda\zeta' = -\varphi$$

and thus if $\zeta' = \zeta' - k$. This yields

$$A\tilde{\zeta}' - B\zeta' - H(\zeta'+\psi-\zeta') + H(\psi-\zeta') + \lambda\tilde{\zeta}' = -\varphi - (a_0+\lambda)k$$

so by multiplying by $\tilde{\zeta}'^+$ and integrating, we deduce $\zeta' \leq k$. We have therefore proved the property (2.18). □

Proof of Theorem 2.1: Consider a sequence $u_k \in W_0^{1,p}(\mathcal{O})$ defined as follows:

$$Au_0 - Bu_0 - H(u_0) + \lambda u_0 = h,$$

$$Au_{k+1} - Bu_{k+1} - H(u_{k+1}) + \lambda u_{k+1} = h + \lambda u_k.$$

Notice that $u_{k+1} - u_k$ is a solution of

$$A(u_{k+1} - u_k) - B(u_{k+1} - u_k) - H(u_{k+1} - u_k + u_k) + H(u_k) =$$

$$+ \lambda(u_{k+1} - u_k) = \lambda(u_k - u_{k-1}),\ k \geqslant 0$$

on assuming that $u_{-1} = 0$. The functions $u_{k+1} - u_k$ have regularity properties increasing with k. Since $u_0 \in W_0^{1,p}$, we have $u_0 \in L^{p_1}$, with $\frac{1}{p_1} = \frac{1}{p} - \frac{1}{n}$. Therefore

$$u_1 - u_0 \in W_0^{1,p_1} \subset L^{p_2},\ \frac{1}{p_2} = \frac{1}{p} - \frac{2}{n}.$$

By induction

$$u_k - u_{k-1} \in W_0^{1,p_k},\ \text{with } \frac{1}{p_k} = \frac{1}{p} - \frac{k}{n},$$

as long as $k < \frac{n}{p}$. Let $j = [\frac{n}{p}]$, if $j < \frac{n}{p}$, then $u_j - u_{j-1} \in W_0^{1,p_j}$ where $p_j > n$, hence $u_j - u_{j-1} \in C^0(\bar{\mathcal{O}})$. If $j = 0$, then $p > n$, i.e. $u_0 \in C^0(\bar{\mathcal{O}})$. If $\frac{n}{p} = j$, then $u_{j-1} - u_{j-2} \in W_0^{1,n} \subset L^q$, for all q, hence, $u_j - u_{j-1} \in W_0^{1,q}$ $\forall\, q \geqslant 2$. Thus for $k \geqslant j$, $u_k - u_{k-1} \in L^\infty(\mathcal{O})$. By Lemma 2.4, we have

$$\|u_{k+1} - u_k\|_{L^\infty} \leqslant \rho \|u_k - u_{k-1}\|_{L^\infty},$$

where

$$= \frac{\lambda}{\lambda + \beta - \beta_1}.$$

We see thus, as in Theorem 2.1, that u_k converges to a solution of (2.13). Uniqueness is proved as in Theorem 2.1, thanks to the estimate (2.18). □

2.3 *The Case of Bounded Measures*

We are given here a family of measures $M(x,v_1,v_2;dz)$ on $\mathbb{R}^n$ such that

(2.19) $\quad 0 \leqslant M(x,v_1,v_2;dz),\ x,v_1,v_2 \to M(x,v_1,v_2;A)$

is measurable for all fixed Borel sets A of $\mathbb{R}^n$,

$$\sup_{x,v_1,v_2} \int_{\mathbb{R}^n} M(x,v_1,v_2;dz) < \infty \ .$$

Let there also be given

$$(2.19)_2 \quad \left| \begin{array}{l} f : R^n \times \mathcal{V}_1 \times \mathcal{V}_2 \to R, \ g : R^n \times \mathcal{V}_1 \times \mathcal{V}_2 \to R^n \\ \tilde{a}_1 : R^n \times \mathcal{V}_1 \times \mathcal{V}_2 \to R \text{ , measurable and bounded} \\ \sup \tilde{a}_1^- = \beta_1 < \beta \ . \end{array} \right.$$

We write

$$(2.20) \quad H(\varphi) = \inf_{v_1} \sup_{v_2} \{ f(x,v_1,v_2) + \nabla\varphi(x) \cdot g(x,v_1,v_2) - \\ - \varphi(x)\tilde{a}_1(x,v_1,v_2) + \int_{R^n} \left[\varphi(x+z) - \varphi(x) - z \cdot \nabla\varphi \, \chi_{|z|\leq 1} \right] M(x,v_1,v_2;dz) \}.$$

It is clear that $H : W_0^{1,p}(\mathcal{O}) \times L^\infty(\mathcal{O}) \to L^p(\mathcal{O})$, for all $p \geq 2$.

Our objective is to prove:

THEOREM 2.2: *Assume* (1.1), $(2.19)_1$, $(2.19)_2$, *then the problem*

$$(2.21) \quad Au_\lambda - H(u_\lambda) + \lambda u_\lambda = h, \ u_\lambda \in W^{2,p}(\mathcal{O}) \cap W_0^{1,p}(\mathcal{O})$$

has one and only one solution for all $h \in L^p(\mathcal{O})$, $p \geq n/2$. □

LEMMA 2.5: λ_1 *can be found large enough such that for* $\lambda \geq \lambda_1$ *the equation*

$$(2.22) \quad Au - H(u) = h, \ u \in W^{2,p}(\mathcal{O}) \cap W_0^{1,p}(\mathcal{O}),$$

has one and only one solution, $h \in L^p(\mathcal{O})$, $p \geq n/2$.

Proof: Let $w \in L^\infty(\mathcal{O})$, and consider the solution ζ of

$$(2.23) \quad A\zeta + \lambda\zeta - \inf \sup \{ f + \nabla\zeta \cdot g - \zeta\tilde{a}_1 + \\ + \int_{R^n} \left[w(x+z) - \zeta(x) - z \cdot \nabla\zeta(x) \, \chi_{|z|\leq 1} \right] M(x,v_1,v_2;dz) \} = h \\ \zeta \in W^{2,p}(\mathcal{O}) \cap W_0^{1,p}(\mathcal{O}), \ (\zeta \in L^\infty, \text{ car } p > \tfrac{n}{2}).$$

Equation (2.23) is in fact a Hamilton-Jacobi equation for the operator A, where we have changed f and $\tilde{a}_1$ into

$$f + \int_{R^n} w(x+z) M(x,v_1,v_2;dz)$$

$$\tilde{a}_1 + \int_{R^n} M(x,v_1,v_2,dz) \geq \tilde{a}_1 .$$

If $w_1, w_2 \in L^\infty$ and ζ_1, ζ_2 denote the corresponding solutions of (2.23), we have the estimate

$$\|\zeta_1-\zeta_2\|_{L^\infty} \leq \frac{\|w_1-w_2\|_{L^\infty} \sup \int_{R^n} M(x,v_1,v_2;dz)}{\lambda+\beta-\beta_1}$$

and thus for $\lambda \geq \lambda_1$, large enough, the mapping so defined is a contraction on L^∞, hence the result. □

Similarly we prove the existence and the uniqueness of the solution of

$$A\zeta - H(\psi+\zeta) + H(\psi) + \lambda\zeta = \varphi, \quad \zeta \in W^{2,p}(\mathcal{O}) \cap W_0^{1,p}(\mathcal{O}) \tag{2.24}$$

for $\varphi \in L^\infty(\mathcal{O})$ and $\psi \in H_0^1 \cap L^\infty$. We see that $\zeta \in W^{2,p} \cap W_0^{1,p}$ for all $p \geq 2$, in particular $\zeta \in L^\infty$.

LEMMA 2.6: *The solution* ζ *of* (2.24) *satisfies the estimate*

$$\|\zeta\|_{L^\infty} \leq \frac{\|\varphi\|_{L^\infty}}{\lambda+\beta-\beta_1}. \tag{2.25}$$

Proof: For $w \in L^\infty$ let χ be the solution of

$$\begin{aligned}
&A\chi + \lambda\chi - \inf\sup\Big\{ f + \nabla\chi\cdot g - \chi\tilde{a}_1 + \\
&+ \int_{R^n}\Big[w(x+z) - \chi(x) - z\cdot\nabla\chi(x)\,\chi_{|z|\leq 1}\Big]M(x,v_1,v_2;dz) \\
&+ \nabla\psi\cdot g - \psi\tilde{a}_1 + \int_{R^n}\Big[\psi(x+z) - \psi(x) - z\cdot\nabla\psi\,\chi_{|z|\leq 1}\Big]M(x,v_1,v_2;dz)\Big\} \\
&+ \inf\sup\Big\{ f + \nabla\psi\cdot g - \psi a_1 + \int_{R^n}\Big[\psi(x+z) - \psi(x) - z\cdot\nabla\psi\,\chi_{|z|\leq 1}\Big]M(dz)\Big\} \\
&= \varphi.
\end{aligned}$$

Let $k = \dfrac{\|\varphi\|}{\lambda+\beta-\beta_1}$, and assume $\|w\| \leq k$. Let $\tilde{\chi} = \chi - k$, then

$$\begin{aligned}
&A\tilde{\chi} + \lambda\tilde{\chi} + (a_0+\lambda)k - \inf\sup\Big\{ f + \nabla\tilde{\chi}\cdot g - \tilde{\chi}\tilde{a}_1 - k\tilde{a}_1 + \\
&\qquad + \int_{R^n}\Big[w(x+z) - k - \tilde{\chi} - z\cdot\nabla\tilde{\chi}\,\chi_{|z|\leq 1}\Big]M(dz) + \\
&\qquad + \nabla\psi\cdot g - \psi\tilde{a}_1 + \int_{R^n}\Big[\psi(x+z) - \psi(x) - z\cdot\nabla\psi\,\chi_{|z|\leq 1}\Big]M(dz)\Big\} + \\
&\qquad + \inf\sup\Big\{ f + \nabla\psi\cdot g - \psi\tilde{a}_1 + \int_{R^n}\Big[\psi(x+z) - \psi(x) - z\cdot\nabla\psi\,\chi_{|z|\leq 1}\Big]M(dz)\Big\} = \\
&\qquad = \varphi
\end{aligned}$$

whence

$$A\tilde{\chi} + \lambda\tilde{\chi} + (\beta-\beta_1+\lambda)k \leqslant \|\varphi\| + C[|\nabla\tilde{\chi}| + |\tilde{\chi}|]$$

and by the choice of

$$A\tilde{\chi} + \lambda\tilde{\chi} \leqslant C[|\nabla\tilde{\chi}| + |\tilde{\chi}|] .$$

Multiplying by $\tilde{\chi}^+$ and integrating over x, we deduce that if λ is large enough, $\tilde{\chi}^+ = 0$, hence $\chi \leqslant k$.

Similarly we prove that $\chi \geqslant -k$. Now ζ is a fixed point of the mapping $w \to \chi$. To establish (2.25) it suffices to define an iterative procedure $\chi^h \to \chi^{h+1}$ with $\chi^0 = 0$. Then $\chi^h \to \zeta$ in L^∞ and $\|\chi^h\| \leqslant k$, by induction. Hence the result desired. □

We can then give the proof of Theorem 2.2. For $w \in L^\infty$ we consider the solution of

$$Az - H(z) + \lambda z = h + \lambda w, \quad z \in W^{2,p} \cap W_0^{1,p}(\mathcal{O}) \tag{2.26}$$

and since $p > n/2$, $z \in L^\infty$. Let $w_1, w_2 \in L^\infty$ and z_1, z_2 be the corresponding solutions of (2.26). Then $z_1 - z_2$ is the solution of

$$A(z_1 - z_2) - H(z_2 + z_1 - z_2) + H(z_2) + \lambda(z_1 - z_2) = \lambda(w_1 - w_2)$$

and by Lemma 2.6,

$$\|z_1 - z_2\|_{L^\infty} \leqslant \frac{\lambda\|w_1 - w_2\|}{\lambda + \beta - \beta_1}$$

therefore the mapping $w \to z$ is a contraction, whence the result desired. □

2.4 *Use of a Continuation Operator*

In this section we shall see how it is possible to use the continuation operator technique for solving certain problems at the boundaries for the operator A-B-H. Let $\mathcal{O}$ be an open set satisfying (2.1), then there exist operators P (called continuations) such that

$$P \in \mathcal{L}(W^{i,p}(\mathcal{O}); W^{i,p}(R^n)), \quad i = 0,1,2, \quad \forall p \geqslant 2 \text{ et } P\varphi = \varphi \text{ on } \mathcal{O}. \tag{2.27}$$

In the remainder we are given such an operator P.

Let $\gamma(x)$ be such that

(2.28) $\gamma : R^n \to R^n$, C^1 and bounded, as well as its partial derivatives

$\gamma(x) \cdot n(x) \geqslant \delta > 0$, $\forall x \in \Gamma = \partial\mathcal{O}$ and $n(x)$ the unit normal at x

pointing towards the exterior of $\mathcal{O}$.

We take the assumptions (1.1),(1.3),(1.4),(1.10),(1.11),(1.11)' and let $h \in L^p(\mathcal{O})$, $p \geqslant 2$; our objective is to solve the problem

$$Au - BPu - H(Pu) = h, \quad \gamma \cdot \frac{\partial u}{\partial x}\Big|_\Gamma = 0, \quad u \in W^{2,p}(\mathcal{O}). \tag{2.29}$$

Naturally, the solution of (2.29) depends a priori upon the choice of operator P. There is, however, one case where it does not so depend (a case which will have a probalistic interpretation), namely when

(2.30) $$c_0(x,z) = 0 \text{ for } x + z \notin \mathcal{O}$$

THEOREM 2.3: *Under the assumptions* (1.1),(1.3),(1.4),(1.10),(1.11),(1.11)' *and* (2.1),(2.28), *for* $h \in L^p(\mathcal{O})$ *there then exists one and only one solution of* (2.29). □

LEMMA 2.7: *Under the assumptions of Theorem 2.3, but without assuming* (2.30), *there then exists a fixed* λ_1 *large enough such that for* $\lambda \geqslant \lambda_1$, *the equation*

(2.31) $$Au_\lambda - BPu_\lambda - H(Pu_\lambda) + \lambda u_\lambda = h, \quad \gamma\cdot\frac{\partial u_\lambda}{\partial x}\Big|_\Gamma = 0, \; u_\lambda \in W^{2,p}(\mathcal{O})$$

has one and only one solution for all $h \in L^p(\mathcal{O})$.

Proof: We know (cf. Chapter 2, Section 4.1.3) that we have the Green's formula

(2.32) $$(Au,v) + \frac{1}{k}\int_\Gamma \gamma\cdot\frac{\partial u}{\partial x} v \, d\Gamma = a(u,v) + b(u,v)$$

for $u \in H^2(\mathcal{O})$, $v \in H^1(\mathcal{O})$, where

$$a(u,v) = \sum \int a_{ij} \frac{\partial u}{\partial x_j} \frac{\partial x}{\partial x_j} dx + \sum \int a_i \frac{\partial u}{\partial x_i} v \, dx + \int a_0 \, u \, v \, dx$$

$$b(u,v) = \sum \int_\Gamma b_j \frac{\partial u}{\partial x_j} v \, d\Gamma = \sum_{\substack{i=1 \dots n-1 \\ j=1 \dots n}} \int_\Gamma \alpha_{ji} \, b_j \frac{\partial u}{\partial \sigma_i} v \, d\Gamma$$

where $\frac{\partial u}{\partial \sigma_i}$ is the derivative tangential on Γ, $k = \frac{\gamma\cdot n}{an.n}$, $(a \equiv a_{ij})$, $\sum \alpha_{ij} n_i = 0$, $j = 1$, and the b_j are defined uniquely from the γ_j and α_{ij}. Further, the problem

(2.33) $$A\psi_\lambda + \lambda\psi_\lambda = \varphi, \quad \gamma\cdot\frac{\partial\psi}{\partial x}\Big|_\Gamma = 0, \quad \psi \in W^{2,p}(\mathcal{O})$$

has one and only one solution for $\lambda \geqslant \lambda_0$, for all $\varphi \in L^p(\mathcal{O})$, and

(2.34) $$\|\psi_\lambda\|_{W^{2,p}} \leqslant C|\varphi|_{L^p}$$

(2.35) $$(\lambda-\lambda_0)|\psi_\lambda|_{L^p} \leqslant C|\varphi|_{L^p} .$$

Let $z \in W^{2,p}(\mathcal{O})$, then we can solve the problem

$$Aw + \lambda w = h + BPz + H(Pz), \; \gamma\cdot\frac{\partial w}{\partial x}\Big|_\Gamma = 0, \quad w \in W^{2,p}$$

since $BPz + H(Pz) \in L^p(R^n)$. We can then work as in Lemma 1.5, to prove that the mapping $z \to w$ is a contraction, at least for $\lambda \geqslant \lambda_1$.

Similarly we let $\psi \in W^{2,p}(\mathcal{O})$ and $\varphi \in L^\infty(\mathcal{O})$ and solve the problem

(2.36)
$$\left|\begin{array}{l} A\zeta - BP\zeta - H\big(P(\zeta+\psi)\big) + H(P\psi) + \lambda\zeta = \varphi, \\ \gamma\cdot\frac{\partial\zeta}{\partial x}\Big|_\Gamma = 0, \quad \zeta\in W^{2,p}(\mathcal{O}). \end{array}\right.$$

We have:

LEMMA 2.8: *Under the assumptions of Theorem 2.3 (including* (2.30)*), the solution of* (2.36) *satisfies the estimate*

(2.37)
$$\|\zeta\|_{L^\infty} \leqslant \frac{\|\varphi\|_{L^\infty}}{\lambda+\beta-\beta_1}.$$

Proof: Consider the operators B and H, where m(dz) is replaced by $m_\varepsilon(dz) = m(dz)\chi_{|z|>\varepsilon}$. Let ζ_ε be the solution of

(2.38)
$$\left|\begin{array}{l} A\zeta_\varepsilon - B_\varepsilon P\zeta_\varepsilon - H_\varepsilon\big(P(\zeta_\varepsilon+\psi)\big) + H_\varepsilon(P\psi) + \lambda\zeta_\varepsilon = \varphi, \\ \gamma\cdot\frac{\partial\zeta_\varepsilon}{\partial x}\Big|_\Gamma = 0, \quad \zeta_\varepsilon\in W^{2,p}(\mathcal{O}), \end{array}\right.$$

as in Lemma 1.6, we prove that $\zeta_\varepsilon \to \zeta$ weakly in $W^{2,p}(\mathcal{O})$. Therefore it suffices to prove (2.37) for ζ_ε. We work as in Lemma 1.6, noting that thanks to (2.30) only the values of the functions inside $\mathcal{O}$ are used. Thus we obtain (cf. (1.40), (1.41))

(2.39)
$$\begin{aligned} & A\tilde\zeta + (\lambda+\gamma_\varepsilon)\tilde\zeta \leqslant C\big[|\nabla\tilde\zeta(x)| + |\tilde\zeta(x)|\big]\,\big(1+m_\varepsilon(R^n)\big) + \\ & \quad + C_1\int_{\{z|z+x\in\mathcal{O}\}} \tilde\zeta^+(x+z)\, m_\varepsilon(dz) \\ & \gamma\cdot\frac{\partial\tilde\zeta}{\partial x}\Big|_\Gamma = 0. \end{aligned}$$

By (2.32),

$$\begin{aligned} \int_{\mathcal{O}} A\tilde\zeta\,\tilde\zeta^+ dx &= a(\tilde\zeta,\tilde\zeta^+) + b(\tilde\zeta,\tilde\zeta^+) \\ &= a(\tilde\zeta^+,\tilde\zeta^+) + b(\tilde\zeta^+,\tilde\zeta^+) \\ &\geqslant \delta\|\tilde\zeta^+\|^2 - \lambda_0|\tilde\zeta^+|^2 \end{aligned}$$

and for $x\in\mathcal{O}$

$$\tilde\zeta^+(x)\int_{\{z|z+x\in\mathcal{O}\}} \tilde\zeta^+(x+z)\, m_\varepsilon(dz) = P\tilde\zeta^+(x)\int_{\{z|z+x\in\mathcal{O}\}} P\tilde\zeta^+(x+z)\, m_\varepsilon(dz)$$

therefore

$$\begin{aligned} \int_{\mathcal{O}} dx\,\tilde\zeta^+(x)\int_{\{z|z+x\in\mathcal{O}\}} \tilde\zeta^+(x+z)\, m_\varepsilon(dz) &\leqslant \int_{R^n} dx\,|P\tilde\zeta^+(x)|\int_{R^n} |P\tilde\zeta^+(x+z)|\, m_\varepsilon(dz) \\ &\leqslant m_\varepsilon(R^n)\,|P\zeta^+|^2_{L^2(R^n)} \leqslant m_\varepsilon(R^n)\,|\tilde\zeta^+|^2_{L^2(\mathcal{O})}. \end{aligned}$$

Multiply (2.39) by $\tilde{\zeta}^+$ and integrate over $\mathcal{O}$; using the preceding, and arguing as in Lemma 1.6, we obtain $\zeta^+ = 0$. The proof is completed as in Lemma 1.6. □

Proof of Theorem 2.3: We notice that by arguments already used the solution of (2.36) belongs to $W^{2,q}(\mathcal{O})$ for all $q \geq 2$. We then use the iteration procedure already seen in Theorem 1.2, with some obvious variations.

Remark 2.3

Naturally, we can replace the Neumann type of condition at the boundary in (2.29) by the Dirichlet boundary condition $u|_\Gamma = 0$. Theorem 2.3 applies. The solution u thus found coincides with that of (2.13) provided that the assumptions are compatible. However, it does not seem to be possible to obtain the analogue of Theorem 2.1 in the Neumann case.

3. PARTIAL DIFFERENTIAL EQUATIONS WITH INTEGRO-DIFFERENTIAL OPERATOR: PARABOLIC CASE

Method of Approach

We consider the parabolic version of the problems treated in Sections 1 and 2. In fact, certain proofs are simpler, because we can always replace A by $A + \lambda$, where λ is fixed and arbitrarily large. We do not give all the details in order not to weigh down our progress.

3.1 *Assumptions and Notations*

Let $a_{ij}(x,t)$, $a_i(x,t)$, $a_0(x,t)$ be defined on $R^n \times]0,T[$, and satisfy

$$(3.1)\qquad \left| \begin{array}{l} a_{ij},\ \dfrac{\partial a_{ij}}{\partial x_k},\ a_i,\ a_0 \in L^\infty(R^n \times]0,T[) \\ \sum a_{ij}\,\xi_i\xi_j \geq \alpha|\xi|^2, \quad \forall \xi \in R^n,\ \alpha > 0\ . \end{array} \right.$$

We define A(t) associated with the coefficients (3.1) by (1.2). Let $c_0(x,z,t)$ be defined on $R^n \times R^n \times]0,T[$ such that

$$(3.2)\qquad c_0 \text{ is measurable, } 0 \leq c_0 \leq 1$$

and B(t) and $\tilde{B}(t)$ associated with c_0 are defined by (1.5), (1.6).

We write $\mathfrak{b}^{2,1,p,\mu}(R^n \times]0,T[)$ for the space of functions $u \in L^p(0,T;W^{2,p,\mu}(R^n))$ such that $\frac{\partial u}{\partial t} \in L^p(0,T;L^{p,\mu}(R^n))$. We replace (1.10),(1.11),(1.11)' by

$$(3.3)\qquad \left| \begin{array}{l} f : R^n \times [0,T] \times \mathcal{V}_1 \times \mathcal{V}_2 \to R, \quad g : R^n \times [0,T] \times \mathcal{V}_1 \times \mathcal{V}_2 \to R^n, \\ \tilde{a}_1 : R^n \times [0,T] \times \mathcal{V}_1 \times \mathcal{V}_2 \to R, \quad c_1 : R^n \times [0,T] \times \mathcal{V}_1 \times \mathcal{V}_2 \times R^n \to R \\ \text{measurable and bounded} \end{array} \right.$$

$$(3.4)\qquad 1 + c_1(x,t,v_1,v_2,z) \geq 0$$

$$(3.4)'\qquad |c_1(x,t,v_1,v_2,z)| \leq C|z|.$$

For the case $\mu = 0$, we are led to replace the assumption (3.3) by

(3.3)' $\left| \begin{array}{l} |f(x,t,v_1,v_2)| \leqslant \bar{f}(x,t), \quad \bar{f} \in L^2(0,T;L^2(R^n)) \cap L^\infty, \\ \text{the other properties of (3.3) remain valid.} \end{array} \right.$

We define $H(t)$ and $\tilde{H}(t)$ by formulae (1.12) and (1.13). The operators $B(t),\tilde{B}(t)$, $H(t),\tilde{H}(t)$, satisfy for almost all t, the properties given in Section 1.2, where *the constants do not depend upon* t, on $]0,T[$.

3.2 *The Case in the Whole Space*

The analogue of (1.25),(1.26),(1.27) is the following: Let $\varphi \in L^p(0,T;L^{p,\mu})$, $p \geqslant 2$, $\mu \geqslant 0$; for $\lambda \geqslant 0$ we solve

(3.5) $$\left| \begin{array}{l} -\dfrac{\partial \psi_\lambda}{\partial t} + A(t)\psi_\lambda + \lambda\psi_\lambda = \varphi, \quad \psi_\lambda(T) = 0 \\ \psi_\lambda \in \mathfrak{W}^{2,1,p,\mu} \quad . \end{array} \right.$$

Then for $\lambda > 0$

(3.6) $$|\psi_\lambda|_{0,0,p,\mu} \leqslant \frac{C}{\lambda} |\varphi|_{\mathfrak{W}^{0,0,\mu,\mu}}$$

(3.7) $$\|\psi_\lambda\|_{1,0,p,\mu} \leqslant \frac{C}{\sqrt{\lambda}} |\varphi|_{L^{p,\mu}}$$

(3.8) $$\|\psi_\lambda\|_{2,1,p,\mu} \leqslant C\, |\varphi|_{L^{p,\mu}}$$

where the constants depend neither on φ, nor $\lambda \geqslant \lambda_0 > 0$. We then have:

THEOREM 3.1: *Take the assupmtions* (3.1),(3.2),(3.3),(3.4)' *((*3.3)' *is used instead of* (3.3) *if* $\mu = 0$*), then for* $h \in L^p(0,T;L^{p,\mu}(R^n))$ *and* $\bar{u}$ $W^{2,p,\mu}(R^n)$, *the problem*

(3.9) $$\left| \begin{array}{l} -\dfrac{\partial u}{\partial t} + A(t)u - \tilde{B}(t)u - \tilde{H}(t)(u) = h \\ u(T) = \bar{u}, \ u \in \mathfrak{W}^{2,1,p,\mu} \end{array} \right.$$

has one and only one solution.

Proof: By making the change of unknown $u = e^{-\lambda t}v$, we reduce to the problem

$$\left| \begin{array}{l} -\dfrac{\partial v}{\partial t} + A(t)v - \tilde{B}(t)v - e^{\lambda t}\,\tilde{H}(t)\,(ve^{-\lambda t}) + \lambda v = he^{\lambda t} \\ v(T) = e^{\lambda T}\,\bar{u}. \end{array} \right.$$

By changing f into $fe^{\lambda t}$ and $\bar{u}$ into $\bar{u}e^{\lambda t}$, which obviously does not change their properties, we can therefore consider that the operator is $A(t) + \lambda$ instead of $A(t)$. We then procede as in Lemma 1.5, using (3.6),(3.7),(3.8) to reduce to a contraction in $\mathfrak{W}^{2,1,p,\mu}$.

We can now state an equivalent result with B instead of $\tilde{B}$, and H instead of $\tilde{H}$ in the case $\mu = 0$:

THEOREM 3.2: *Let us take the assumptions of Theorem 3.1, with* (3.3)' *instead of* (3.3). *Then for* $h \in L^p(0,T;L^p(R^n))$, $\bar{u} \in W^{2,p}(R^n)$, $2 \leqslant p < \infty$, *the problem*

(3.10) $$\left| \begin{array}{l} -\dfrac{\partial u}{\partial t} + A(t)u - B(t)u - H(t)(u) = h \\ u(T) = \bar{u},\ u \in \mathfrak{W}^{2,1,p,o} \end{array} \right.$$

has one and only one solution.

When the data $h,\bar{u}$ are bounded in addition, the solution u of (3.9) or (3.10) is then bounded. This does not follow from Theorems 3.1,3.2, and requires a maximum principle analogous to that of Section 1. Consider two constants $\beta_1,\beta_2 \geq 0$ such that

(3.11) $$(a_0 + \tilde{a}_1)^- \leq \beta_1,\ (a_0 + \tilde{a}_1)^+ \geq \beta_2.$$

Let $\psi \in \mathfrak{W}^{2,1,p,\mu}$, $\mu > 0$, $2 \leq p < \infty$, $\varphi \in L^\infty(R^n \times]0,T[)$, $\chi \in W^{2,p,\mu} \cap L^\infty$; then we can solve the equation

(3.12) $$\left| \begin{array}{l} -\dfrac{\partial \zeta}{\partial t} + A(t)\zeta - \tilde{B}(t)\zeta - \tilde{H}(t)\big(\psi(t)+\zeta\big) + \tilde{H}(t)\big(\psi(t)\big) = \varphi \\ \zeta(T) = \chi,\quad \zeta \in \mathfrak{W}^{2,1,p,\mu}. \end{array} \right.$$

We then have:

LEMMA 3.1: *The solution* ζ *of* (3.12) *satisfies the estimate*

(3.13) $$\|\zeta(t)\|_{L^\infty(R^n)} \leq \|\varphi\|_{L^\infty(R^n\times]0,T[)} \frac{\big(1 - \exp - (\beta_2-\beta_1)(T-t)\big)}{\beta_2 - \beta_1} + \|\chi\|_{L^\infty(R^n)} \exp - (\beta_2-\beta_1)(T-t),\quad t \in [0,T].$$

Proof: As in the proof of Lemma 1.6, we show that we can reduce to the operators $\tilde{B}_\varepsilon(t)$ and $\tilde{H}_\varepsilon(t)$. Let ζ_ε be the solution of (3.12) with $\tilde{B}_\varepsilon(t)$ instead of $B_\varepsilon(t)$ and $\tilde{H}_\varepsilon(t)$ instead of $H_\varepsilon(t)$. We then set

$$\tilde{\zeta}_\varepsilon = \zeta_\varepsilon - \|\varphi\| \frac{\big(1 - \exp - (\beta_2-\beta_1)(T-t)\big)}{\beta_2 - \beta_1} - \|\chi\| \exp - (\beta_2-\beta_1)(T-t).$$

Next we work as in Lemma 1.6. In fact matters are simpler, for we end up with a Gronwall inequality

$$|\tilde{\zeta}_\varepsilon^+(t)|^2_{L^{2,\mu}} \leq \lambda_\varepsilon \int_t^T |\tilde{\zeta}_\varepsilon^+(s)|^2_{L^{2,\mu}} ds \quad \forall t \leq T$$

from which we conclude $\tilde{\zeta}_\varepsilon^+(t) = 0$. (3.13) is then easily deduced. □

We can then establish the analogue of Theorem 3.1:

THEOREM 3.3: *Take the assumptions of Theorem 3.1. Let* $h \in L^\infty$, $\bar{u} \in W^{2,p,\mu} \cap L^\infty$, $p \geq 2$, $\mu > 0$. *Then there exists one and only one solution of*

(3.14) $$\left| \begin{array}{l} -\dfrac{\partial u}{\partial t} + A(t)u - B(t)u - H(t)(u) = h, \\ u(T) = \bar{u},\quad u \in \mathfrak{W}^{2,1,p,\mu} \cap L^\infty. \end{array} \right.$$

Proof: This is very analogous to that of Lemma 1.7, using, in particular, the result of Lemma 3.1. □

If we consider equation (3.12) with B(t),H(t) instead of B(t),H(t), and $\psi \in \mathfrak{W}^{2,1,p,\mu} \cap L^\infty$, we again have the same result as in Lemma 3.1.

Remark 3.1

When $\beta_1 = \beta_2$ the expression on the right hand side of (3.13) must be replaced by $T-t$. □

Remark 3.2

From Lemma 3.1 (cf. also remarks 1.4 amd 1.5) we deduce that the solution u of (3.14) is positive if the data h, f, $\bar{u}$ are positive and if, further,

$$(3.15)\qquad \|u(t)\|_{L^\infty(R^n)} \leq \left(\|h\|_{L^\infty} + \sup_{v_1,v_2} \|f\|_{L^\infty}\right) \frac{\left(1 - \exp - (\beta_2-\beta_1)(T-t)\right)}{\beta_2 - \beta_1}$$

$$+ \|\bar{u}\|_{L^\infty} \exp - (\beta_2-\beta_1)(T-t) \quad . \square$$

3.3 *The Case in a Bounded Open Set*

In a way analogous to what we saw in Section 2 we are led to make one or another additional assumptions (with respect to those of Section 1.1), either (2.5) or

$$(3.16)\qquad \left|\frac{\partial C_0}{\partial x_k}(x,z,t)\right| \leq \text{Constant}.$$

The operators B(t),H(t) then satisfy the properties (2.2),(2.3),(2.4) for almost all t, with constants independent of t.

Our objective is to prove:

THEOREM 3.4: *Assume* (1.3),(2.1),(3.1),(3.2(,(3.3),(3.4),(3.4)' *and* (2.5) (3.16). *Then for*

$$h \in L^p(Q) \;\; (Q = \mathcal{O} \times]0,T[), \;\; \bar{u} \in W_0^{1,p}(\mathcal{O}) \;, \;\; 2 \leq p < \infty,$$

there exists one and only one solution of the problem

$$(3.17)\qquad \left|\begin{array}{l} -\dfrac{\partial u}{\partial t} + A(t)u - B(t)u - H(t)\big(u(t)\big) = h \\ u(T) = \bar{u};\; u \in L^p\big(0,T;W_0^{1,p}(\mathcal{O})\big),\; \dfrac{\partial u}{\partial t} \in L^p\big(0,T;W^{-1,p}(\mathcal{O})\big) \end{array}\right. .$$

Proof: The case where (2.5) is satisfied is actually much simpler. Then we have existence and uniqueness of a solution of (3.17), and it further satisfies

$$u \in L^p\big(0,T;W^{2,p}(\mathcal{O})\big),\; \frac{\partial u}{\partial t} \in L^p(Q).$$

In this case, moreover, B(t) is not seperated from H(t). The interesting case is therefore that where (3.16) is satisfied. The situation is then the parabolic analogue of Lemma 2.3, for we can always replace A(t) by A(t) + λ. We cannot, however, adopt the proof of Lemma 2.3, owing to the dependence on t. We then proceed in a slightly different way, and begin by solving the equation

$$(3.18)\qquad \left|\begin{array}{l} -\frac{\partial\psi}{\partial t} + A(t)\psi - H(t)\big(\psi(t)\big) = \varphi \\ \psi(T) = \bar{u}, \quad \psi\in L^p\big(0,T;W_0^{1,p}(\mathcal{O})\big),\ \frac{\partial\psi}{\partial t}\in L^p\big(0,T;W^{-1,p}(\mathcal{O})\big) \end{array}\right.$$

where φ is given in $L^p(0,T;W^{-1,p}(\mathcal{O}))$. For this we define ζ to be the unique solution of

$$(3.19)\qquad \left|\begin{array}{l} -\frac{\partial\zeta}{\partial t} + A(t)\zeta = \varphi \\ \zeta(T) = \bar{u}, \quad \zeta\in L^p\big(0,T;W_0^{1,p}(\mathcal{O})\big),\ \frac{\partial\zeta}{\partial t}\in L^p\big(0,T,W^{-1,p}(\mathcal{O})\big). \end{array}\right.$$

If we set

$$\theta = \psi - \zeta$$

we reduce (3.18) to

$$(3.20)\qquad \left|\begin{array}{l} -\frac{\partial\theta}{\partial t} + A(t)\theta - H(t)\big(\zeta(t)+\theta(t)\big) = 0, \ \theta(T) = 0 \\ \theta\in L^p\big(0,T;W_0^{1,p}(\mathcal{O})\big),\ \frac{\partial\theta}{\partial t}\in L^p\big(0,T;W^{-1,p}(\mathcal{O})\big). \end{array}\right.$$

We then easily verify that (3.20) has one and only one solution. The solution of (3.20) even belongs to $L^p(0,T;W^{2,p}(\mathcal{O}))$, $\frac{\partial\theta}{\partial t}\in L^p(Q)$. So we have solved (3.18), and we further have

$$(3.21)\qquad \|\psi\|_{L^p(0,T;W_0^{1,p})} + \left\|\frac{\partial\psi}{\partial t}\right\|_{L^p(0,T;W^{-1,p})} \leq C\left(\|\varphi\|_{L^p(0,T;W^{-1,p})} + \|\bar{u}\|_{W_0^{1,p}} + 1\right).$$

Next we solve the equation

$$(3.22)\qquad \left|\begin{array}{l} -\frac{\partial z}{\partial t} + A(t)z - B_r^1(t)z - H(t)(z) = \varphi \\ z(T) = \bar{u}, \quad z\in L^p\big(0,T,W_0^{1,p}(\mathcal{O})\big),\ \frac{\partial z}{\partial t}\in L^p\big(0,T;W^{-1,p}(\mathcal{O})\big) \end{array}\right.$$

which is possible if r is fixed and small enough (and independent of the data φ, $\bar{u}$), as in Lemma 2.3. Using next a solution u of (3.17) in $L^2(0,T;H_0^1(\mathcal{O}))$ we can conclude the proof by a regularity argument similar to that of Lemma 2.3. □

Remark 3.3

We may state results of maximum principle type, as in Lemma 2.4, modified as in Lemma 3.1, since a parabolic equation is involved. □

3.4 *Bounded Measures*

The situation is analogous to that of Section 2.3. We are given a family of measures $((x,t,v_1,v_2;dz)$ on R^n satisfying

$$(3.23)\qquad |M \geq 0,\ x,t,v_1,v_2 \to M(x,t,v_1,v_2;A) \text{ is measurable, } \forall A \text{ fixed Borel sets of } R^n;$$

(Contd)

$$\left| \sup_{x,t,v_1,v_2} \int_{R^n} M(x,t,v_1,v_2;dz) < \infty. \right.$$

We are also given

$$(3.24) \quad \left| \begin{array}{l} f : R^n \times [0,T] \times \mathcal{V}_1 \times \mathcal{V}_2 \to R, \quad g : R^n \times [0,T] \times \mathcal{V}_1 \times \mathcal{V}_2 \to R^n, \\ a_1 : R^n \times [0,T] \times \mathcal{V}_1 \times \mathcal{V}_2 \to R, \text{ measurable and bounded.} \end{array} \right.$$

We define $H(t)(\varphi)$ as in (2.20). We then have:

THEOREM 3.5: *Assume* (2.1),(3.1),(3.23),(3.24). *Let*

$$h \in L^p(Q) \ , \ \bar{u} \in W^{2,p}(\mathcal{O}) \cap W_0^{1,p}(\mathcal{O}) , \ p > \frac{n}{2} + 1.$$

Then the problem

$$(3.25) \quad \left| \begin{array}{l} - \frac{\partial u}{\partial t} + A(t)u - H(t)\big(u(t)\big) = h, \\ u(T) = \bar{u}, \ u \in L^p\big(0,T;W^{2,p}(\mathcal{O}) \cap W_0^{1,p}(\mathcal{O})\big), \ \frac{\partial u}{\partial t} \in L^p(Q) \end{array} \right.$$

has one and only one solution.

Proof: This is similar to that of Lemma 2.5. □

3.5 *Penalisation of the Domain*

In this section we show that we can approximate the solution of (3.17) by a sequence of more regular functions defined in the whole space, which are solutions of problems of the type (3.10). We shall take the assumptions of Theorem 3.4 (in fact (3.16) instead of (2.5), otherwise the approximation is of no interest). We shall assume (3.3)'. We shall continue h by 0 outside the open set $\mathcal{O}$.

Let $q(x) = 1 - \chi_{\bar{\mathcal{O}}}(x)$ be the characteristic function of the complement of $\bar{\mathcal{O}}$. We then consider the problem (for $\eta > 0$)

$$(3.26) \quad \left| \begin{array}{l} - \frac{\partial u_\eta}{\partial t} + A(t)u_\eta - B(t)u_\eta - H(t)\big(u_\eta(t)\big) + \frac{1}{\eta} q u_\eta = h \\ u_\eta(T) = \bar{u}, \quad u_\eta \in \mathbb{W}^{2,1,p,o}. \end{array} \right.$$

the solution u_η of (3.26) exists and is defined uniquely, by Theorem 3.2. We then have

THEOREM 3.6: *Take the assumptions of Theorem* (3.4) *and* (3.3)'. *Then the solution* u_η *of* (3.26) *converges to* u *(solution of* (3.17) *continued by* 0) *in* $L^2(0,T;H^1(R^n))$ *and* $C^0(0,T;L^2(R^n))$.

Proof: From (3.26) we deduce the energy equation

$$(3.27) \quad - \frac{d}{dt}|u_\eta(t)|^2 + \langle A(t)u_\eta,u_\eta \rangle - \langle B(t)u_\eta,u_\eta \rangle$$
$$- \big(H(t)\big(u_\eta(t)\big),u_\eta(t)\big) + \frac{1}{\eta}\int_{R^n - \mathcal{O}} u_\eta^2 dx = \big(h(t),u_\eta(t)\big).$$

The properties of the operators B(t),H(t) (cf. Remark 2.1 and Lemma 1.2) allow us to write

$$-\frac{d}{dt}|u_\eta(t)|^2 + \alpha\|u_\eta(t)\|^2 - C\sigma(r)\|u_\eta(t)\|^2 -$$
$$C\tau(r)\|u_\eta(t)\|\ |u_\eta(t)| - C|u_\eta(t)|\ \left(1+\|u_\eta(t)\|\right) +$$
$$+\frac{1}{\eta}\int_{R^n-\mathcal{O}} u_\eta^2 dx \leq |h(t)|\ |u_\eta(t)|$$

where $\sigma(r) \to 0$ as $r \to 0$. From this we easily deduce the a priori estimates

(3.28) u_η stays bounded in $L^2(0,T;H^1(R^n))$ and $L^\infty(0,T;L^2(R^n))$

(3.29) $\frac{1}{\eta}\int_0^T\int_{R^n-\mathcal{O}} u_\eta^2(x,t)dxdt \leq C.$

From this we deduce that $H(t)$ $(u_\eta(t))$ stays in a bounded set of $L^2(0,T;L^2(R^n))$. We now take a subsequence, again denoted u_η such that

(3.30)
$$\begin{cases} u_\eta \to w \text{ weakly in } L^2(0,T;H^1(R^n)) \text{ and } L^\infty(0,T;L^2(R^n)) \text{ weakly starred} \\ \dfrac{\partial u_\eta}{\partial t} \to \dfrac{\partial w}{\partial t} \text{ weakly in } L^2(0,T;H^{-1}(R^n)) \\ H(u_\eta) \to \xi \text{ weakly in } L^2(0,T;L^2(R^n)). \end{cases}$$

By (3.29) w is zero on $R^n-\mathcal{O}$, hence its restriction to $\mathcal{O}$ belongs to $L^2(0,T;H_0^1(\mathcal{O}))$. Since

$$\int_0^T |u_\eta(t)-w|^2_{L^2(R^n)} dt \leq C\eta + \int_0^T |u_\eta(t)-w(t)|^2_{L^2(\mathcal{O})} dt$$

and by compactness we see that

(3.31) $u_\eta \to w$ strongly in $L^2(0,T;L^2(R^n))$.

It then follows from (3.27), thanks to (3.31), that

(3.32)
$$|u_\eta(t)|^2 + \int_t^T <(A(s)-B(s))u_\eta(s),u_\eta(s)> ds +$$
$$+\frac{1}{\eta}\int_t^T\int_{R^n-\mathcal{O}} u_\eta^2(x,s)dxds \to |\bar{u}|^2 + \int_t^T (h(s),w(s))ds + \int_t^T (\xi(s),w(s))ds.$$

But, by scalar multiplication of (3.26) with w, since $w \in L^2(0,T;H_0^1)$, we have

(3.33)
$$(u_\eta(t),w(t)) + \int_t^T <(A(s)-B(s))u_\eta(s),w(s)>ds = \int_t^T (h(s),w(s))ds +$$
$$+\int_t^T (H(s)(u_\eta(s)),w(s)>ds \to \int_0^T (h(s),w(s))ds +$$

$$+\int_t^T \big(\xi(s),w(s)\big)ds = |w(t)|^2 + \int_t^T <\big(A(s)-B(s)\big)u_\eta(s),u_\eta(s)\big)>ds$$

and (3.32),(3.33) imply

$$|u_\eta(t)-w(t)|^2 + \int_t^T <\big(A(s)-B(s)\big)\big(u_\eta(s)-w(s)\big),\ u_\eta(s)-w(s)>ds$$

tends to 0 as $\eta \to 0$, for all t.

From this we deduce that

$$u_\eta \to w \text{ in } L^2\big(0,T;H^1(R^n)\big) \text{ and } C^0\big(0,T;L^2(R^n)\big)$$

since

$$\xi(t) = H(t)\big(w(t)\big)$$

and thus $w = u$. The uniqueness of the limit of every subsequence implies the result desired. □

3.6 *Continuation Operator*

We simply state the analogue of Theorem 2.3 in the parabolic case.

THEOREM 3.7: *Assume* (3.1),(1.3),(3.2),(3.3),(3.4),(3.4)' *and* (2.1),(2.28). *Then for* $h \in L^p(Q)$ *there exists one and only one solution of*

$$(3.34) \qquad \left| \begin{aligned} &-\frac{\partial u}{\partial t} + A(t)u - B(t)Pu(t) - H(t)\big(Pu(t)\big) = h \\ &\gamma\cdot\frac{\partial u}{\partial x}\Big|_\Gamma = 0,\ u \in L^p\big(0,T;W^{2,p}(\mathcal{O})\big),\ \frac{\partial u}{\partial t} \in L^p(Q)\ . \end{aligned} \right.$$

Proof: Analogous to that of Lemma 2.7. □

4. HAMILTON-JACOBI-BELLMAN VARIATIONAL INEQUALITIES

Method of Approach

Here we study the variational inequalities related to the operator $A - B$. We shall consider the case of two obstacles ψ_1,ψ_2 and the evolutionary case.

4.1 *V.I. in the Whole Space*

We shall take the conditions of Theorem 3.3. We shall also be given two functions $\psi_1(x,t)$, $\psi_2(x,t)$ satisfying

$$(4.1) \qquad \psi_i,\ \frac{\partial \psi_i}{\partial t},\ \frac{\partial \psi_i}{\partial x_j},\ \frac{\partial^2 \psi_i}{\partial x_j \partial x_k} \text{ bounded } (i = 1,2)\ ;\ \psi_1 \leqslant \psi_2$$

$$(4.2) \qquad \psi_1(x,T) \leqslant \bar{u} \leqslant \psi_2(x,T).$$

Consider the problem

$$(4.3) \qquad u \in \mathcal{W}^{2,1,p,\mu} \cap L^\infty,\ u(T) = \bar{u}$$

(4.4) $\psi_1 \leqslant u \leqslant \chi_2$

(4.5) $$\left(-\frac{\partial u}{\partial t} + \big(A(t) - B(t)\big)u - H(t;u) - h\right)\cdot\big(v-u(t)\big) \geqslant 0$$

a.e. $\forall v$ such that $\psi_1(x,t) \leqslant v \leqslant \psi_2(x,t)$ a.e.

THEOREM 4.1: *Take the assumptions of Theorem 3.1. Let* $h \in L^\infty$, $\bar{u} \in W^{2,p,\mu} \cap L^\infty$, $p \geqslant 2$, $\mu > 0$, *and let* ψ_1, ψ_2 *satisfy* (4.1),(4.2). *Then there exists a solution of* (4.3),(4.4),(4.5).

Proof: We start from the penalised scheme

(4.6) $$\left|\begin{array}{l} -\dfrac{\partial u_\varepsilon}{\partial t} + A(t)u_\varepsilon - B(t)u_\varepsilon - H(t)\big(u_\varepsilon(t)\big) - \dfrac{1}{\varepsilon}(u_\varepsilon-\psi_1)^- + \dfrac{1}{\varepsilon}(u_\varepsilon-\psi_2)^+ = h \\ u_\varepsilon(T) = \bar{u} \end{array}\right.$$

(4.7) $u_\varepsilon \in \mathcal{W}^{2,1,p,\mu} \cap L^\infty$.

There exists one and only one solution of (4.6),(4.7), by applying Theorem 3.3. In fact we notice that

(4.8) $$H(t)(\varphi) + \frac{1}{\varepsilon}(\varphi-\psi_1)^- - \frac{1}{\varepsilon}(\varphi-\psi_2)^+ = \inf_{v_1,d_2}\ \sup_{v_2,d_1}\Big\{f + \nabla\varphi\cdot g - \varphi\tilde{a}_1 +$$
$$+ \int\big[\varphi(x+z) - \varphi(x) - z\cdot\nabla\varphi\ \chi_{|z|\leqslant 1}\big]c_0c_1m(dz) -$$
$$- \frac{1}{\varepsilon}(\varphi-\psi_2)d_2 - \frac{1}{\varepsilon}(\varphi-\psi_1)d_1\Big\}$$

where d_1,d_2 satisfy the constraints

(4.9) $0 \leqslant d_1 \leqslant 1$, $0 \leqslant d_2 \leqslant 1$.

Thus we have replaced $\tilde{a}_1 + \frac{1}{\varepsilon}d_1 + \frac{1}{\varepsilon}d_2$ and f by $f + \frac{1}{\varepsilon}\psi_2 d_2 + \frac{1}{\varepsilon}\psi_1 d_1$. Next we set $v_\varepsilon = u_\varepsilon - \psi_2$. v_ε is then the solution of

(4.10) $$\left|\begin{array}{l} -\dfrac{\partial v_\varepsilon}{\partial t} + A(t)v_\varepsilon - B(t)v_\varepsilon - \dfrac{1}{\varepsilon}(v_\varepsilon+\psi_2-\psi_1)^- + \dfrac{1}{\varepsilon}v_\varepsilon^+ - H(t)(v_\varepsilon+\psi_2) = \\ \qquad = h + \dfrac{\partial\psi_2}{\partial t} - A(t)\psi_2 + B(t)\psi_2 = \tilde{h} \\ v_\varepsilon(T) = \bar{u} - \psi_2(T). \end{array}\right.$$

Now set

$$\tilde{f} = f + \nabla\psi_2\cdot g + \int\big[\psi_2(x+z) - \psi_2(x) - z\cdot\nabla\psi_2\ \chi_{|z|\leqslant 1}\big]c_0c_1m(dz) - \tilde{a}_1\psi_2\ .$$

Therefore v_ε is the solution of a problem analogous to (4.6), where ψ_1 is replaced by $\psi_1 - \psi_2$, ψ_2 by 0, h by $\tilde{h}$, f by $\tilde{f}$. It is clear that by (4.1) $\tilde{f},\tilde{h}$ are bounded.

Lastly, we can again rewrite (4.10) in the form

$$(4.11) \qquad -\frac{\partial v_\varepsilon}{\partial t} + A(t)v_\varepsilon - B(t)v_\varepsilon + \frac{1}{\varepsilon} v_\varepsilon + \frac{1}{\varepsilon}\hat{d}_1 v_\varepsilon - H(t)(v_\varepsilon+\psi_2) = \frac{1}{\varepsilon}\hat{d}_1(\psi_1-\psi_2) - \frac{1}{\varepsilon} v_\varepsilon^- + \tilde{h} \leqslant \tilde{h},$$

for a certain $\hat{d}_1(x,t)$ depending on ε. But then defining $\tilde{v}_\varepsilon$ as the solution of

$$(4.12) \qquad \left| \begin{array}{l} -\frac{\partial \tilde{v}_\varepsilon}{\partial t} + A(t)\tilde{v}_\varepsilon - B(t)\tilde{v}_\varepsilon + \frac{1}{\varepsilon}\tilde{v}_\varepsilon + \frac{1}{\varepsilon}\hat{d}_1\tilde{v}_\varepsilon - H(t)(\tilde{v}_\varepsilon+\psi_2) = \tilde{h} \\ \tilde{v}_\varepsilon(T) = 0 , \end{array} \right.$$

we have $v_\varepsilon \leqslant \tilde{v}_\varepsilon$. Moreover, $\tilde{v}_\varepsilon$ satisfies the following estimate (cf. Remark 3.2) (and noting that a_0 has been changed into $a_0 + \frac{1}{\varepsilon} + \frac{1}{\varepsilon}\hat{d}_1$ and hence we can replace, in (3.11), β_2 by $\beta_2 + \frac{1}{\varepsilon}$ and β_1 by 0 for ε small enough),

$$\|\tilde{v}_\varepsilon\|_{L^\infty} \leqslant \varepsilon\left(\|\tilde{h}\|_{L^\infty} + \sup_{v_1,v_2} \|\tilde{f}\|\right) = C\varepsilon.$$

But then $v_\varepsilon \leqslant C_\varepsilon$, hence $v_\varepsilon^+ \leqslant C\varepsilon$, so

$$(4.13) \qquad \frac{1}{\varepsilon}(u_\varepsilon - \psi_2)^+ \leqslant C.$$

Similarly we may show that

$$(4.14) \qquad \frac{1}{\varepsilon}(u_\varepsilon - \psi_1)^- \leqslant C.$$

But by equation (4.6) we then deduce the estimate

(4.15) $\quad u_\varepsilon$ remains in a bounded set of $\mathbb{W}^{2,1,p,\mu}$.

We now take a subsequence, again denoted u_ε, that is weakly convergent in $\mathbb{W}^{2,1,p,\mu}$, and we then have

$$H(t)\big(u_\varepsilon(t)\big) \to H(t)\big(u(t)\big) \text{ weakly in } L^p(0,T;L^{p,\mu}).$$

We easily deduce from this that u is the solution of (4.4),(4.5). □

Remark 4.1

We shall show by a probabilistic method that the solution of (4.3),(4.4),(4.5) is unique, at least if p is large enough. It would be interesting to give an analytic proof of uniqueness, but because of the operator B the difficulty is that energy methods are not applicable. □

Remark 4.2

We ignore what happens when the functions ψ_1, ψ_2 are only continuous, for we cannot give a weak formulation of (4.4),(4.5). □

4.2 *V.I. in a Bounded Open Set*

Here we use the conditions of Theorem 3.4; we can give Theorems that differ according to the regularity of the obstacles.

4.2.1 *Regular obstacles*

The obstacles will satisfy the properties (4.1),(4.2) and

(4.16) $$\psi_1\big|_\Sigma \leq 0 \leq \psi_2\big|_\Sigma .$$

We then have:

THEOREM 4.2: *Take the assumptions of Theorem 3.4, together with* (4.1),(4.2) *and* (4.16). *Then for* $h \in L^\infty$, *the following problem has one and only one solution:*

(4.17) $$u \in L^p(0,T;W_0^{1,p}(\mathcal{O})), \ \frac{\partial u}{\partial t} \in L^p(0,T;W^{-1,p}(\mathcal{O})),$$

(4.18) $$\psi_1 \leq u \leq \psi_2 \quad \text{a.e.}, u(x,T) = \bar{u}(x)$$

(4.19) $$\begin{cases} -\left(\frac{\partial u}{\partial t}, v-u\right) + \langle (A(t)-B(t)), u(t), v-u(t)\rangle - \\ -\left(H(t)(u(t)), v-u(t)\right) \geq (h(t), v-u(t)) \quad t. \text{ a.e.} \\ \forall v \in H \text{ such that } \psi_1(x,t) \leq v(x) \leq \psi_2(x,t) \quad \text{a.e.} \end{cases}$$

Proof: It will be noted that $\bar{u}$ is necessarily in L^∞, and $\bar{u} \in W_0^{1,p} \cap L^\infty$. It is this number p which occurs in (4.17). The method is very similar to that used to obtain the Theorem 4.1. We consider the penalised problem

(4.20) $$\begin{cases} -\frac{\partial u_\varepsilon}{\partial t} + A(t)u_\varepsilon - B(t)u_\varepsilon - H(t)(u_\varepsilon(t)) - \frac{1}{\varepsilon}(u_\varepsilon-\psi_1)^- + \frac{1}{\varepsilon}(u_\varepsilon+\psi_2)^+ = h \\ u_\varepsilon \in L^p(0,T;W_0^{1,p}), \ \frac{\partial u_\varepsilon}{\partial t} \in L^p(0,T;W^{-1,p}(\mathcal{O})), \ u_\varepsilon(T) = \bar{u}, \end{cases}$$

and look for estimates of the penalisation term.

Define $v_\varepsilon = u_\varepsilon - \psi_2$, then $\tilde{v}_\varepsilon$ analogously to (4.10),(4.12) (it is necesary to add the boundary conditions $v_\varepsilon\big|_\Sigma = -\psi_2\big|_\Sigma$ and $\tilde{v}_\varepsilon\big|_\Sigma$). Using the property (2.4) we prove that $v_\varepsilon \leq \tilde{v}_\varepsilon$ by an argument of the maximum principle type. For $\tilde{v}_\varepsilon$ we have the same estimate L^∞ as in Theorem 4.1. Thus we again have (4.13),(4.14). Beginning with (4.20), and applying Theorem 3.4, we obtain the estimates

(4.21) $$\begin{cases} \|u_\varepsilon\|_{L^\infty} \leq C, \ \|u_\varepsilon\|_{L^p(0,T;W^{1,p}(\mathcal{O}))} \leq C \\ \left\|\frac{\partial u_\varepsilon}{\partial t}\right\|_{L^p(0,T;W^{-1,p}(\mathcal{O}))} \leq C \end{cases} .$$

Let $v \in L^2(0,T;H_0^1)$ be such that $\psi_1 \leq v \leq \psi_2$ a.e.. From (4.20) we deduce that

(4.22) $$\int_t^T \left(-\frac{\partial u_\varepsilon}{\partial t}, v\right) dt + \frac{1}{2}|\bar{u}|^2 - \frac{1}{2}|u_\varepsilon(t)|^2 +$$

$$+ \int_t^T < \big(A(t)-B(t)\big)u_\varepsilon(t), v(t) - u_\varepsilon(t) > dt -$$

$$- \int_t^T \big(H(t)\big(u_\varepsilon(t)\big), v(t) - u_\varepsilon(t)\big)dt \geq \int_t^T \big(h(t), v(t) - u_\varepsilon(t)\big)dt \; .$$

From u_ε we take a subsequence weakly converging (in the sense of the spaces (4.21)) to u. We also have

$$u_\varepsilon \to u \text{ strongly in } L^2(Q),\; H(s)\big(u_\varepsilon(s)\big) \to \xi \text{ weakly in } L^2.$$

We can pass to the lower limit in (4.22), and therefore

$$\text{(4.23)} \qquad \int_t^T \Big(-\frac{\partial u}{\partial t}, v\Big)dt + \frac{1}{2}|\bar{u}|^2 - \frac{1}{2}|u(t)|^2 +$$

$$+ \int_t^T < \big(A(s)-B(s)\big)u(s), v(s) - u(s) > ds -$$

$$- \int_t^T \big(\xi(s), v(s) - u(s)\big)ds \geq \int_t^T \big(h(s), v(s) - u(s)\big)ds \; .$$

Taking $v = u$ in (4.22), and $v = u_\varepsilon$ in (4.23), we additionally deduce (using $u(T) = \bar{u}$) that

$$u_\varepsilon \to u \text{ strongly in } L^2(0,T;H_0^1).$$

From this it follows that $\xi = H(s)(u(s))$. Hence u is the solution of

$$\int_t^T -<\frac{\partial u}{\partial t}, v-u> ds + \int_t^T < \big(A(s)-B(s)\big)u(s), v(s) - u(s) > ds -$$

$$- \int_t^T \big(H(s)\big(u(s)\big), v(s) - u(s)\big)ds \geq \int_t^T \big(h(s), v(s) - u(s)\big)ds, \quad \forall t,$$

$$v \in L^2(0,T;H_0^1).$$

Taking $v(s) = u(s) + \theta(v - u(s))$ for $s \in]t, t+\delta[$, and $u(s)$ (where $v \in H_0^1$ such that $\psi_1(x,t) \leq v(x) \leq \psi_2(x,t)$ a.e. x), deviding by δ and making δ tend to 0, we obtain (4.19) a.e. t. The uniqueness follows from the coercivity of $A - B + \lambda$ for λ large enough. □

4.2.2 *Non-regular obstacles*

Here we try to reduce the regularity assumptions for obstacles. In fact, contrary to the control case (a single obstacle) which will be considered in the next section, in the games case (two obstacles) we shall not be able to reduce the regularity assumptions as much as is desirable. We shall keep some technical assumptions (because of the proof) which are probably not necessary. The case is similar to that of diffusions (cf. *B.L.*, loc. cit., Chap IV).

We shall use the assumptions of Theorem 3.4, and the following ones in addition:

$$\text{(4.24)} \qquad \frac{\partial a_{ij}}{\partial t}, \frac{\partial a_i}{\partial t}, \frac{\partial a_0}{\partial t} \in L^\infty(Q)$$

(4.25) $\frac{\partial c_0}{\partial t} \in L^\infty(Q), \ \frac{\partial^2 c_0}{\partial t \partial x_k} \in L^\infty(Q)$

(4.26) $\left|\frac{\partial c_1}{\partial t}\right| \leqslant C|z|, \ \left|\frac{\partial c_1}{\partial t}\right| \leqslant C$

(4.27) $\frac{\partial f}{\partial t}, \frac{\partial g}{\partial t}, \frac{\partial a_1}{\partial t}$ bornées

(4.28) $\left| \begin{array}{l} \bar{u} \in W_0^{1,p}(\mathcal{O}) \ (p > n+2), \ \big(A(T)-B(T)\big)\bar{u} \in L^2(\mathcal{O}) \\ h \in L^\infty(Q), \ \frac{\partial h}{\partial t} \in L^2(Q) \end{array} \right.$

(4.29) $\left| \begin{array}{l} \psi_1, \psi_2 \in C^0(\bar{Q}), \quad \psi_1 \leqslant \psi_2 \\ \psi_1|_\Sigma \leqslant 0 \leqslant \psi_2|_\Sigma \end{array} \right.$

(4.30) $\left| \begin{array}{l} \psi_i, \frac{\partial \psi_i}{\partial t} \in L^2(0,T;H_0^1), \ \frac{\partial^2 \psi_i}{\partial t^2} \in L^2(Q) \\ \left.\frac{\partial \psi_i}{\partial t}\right|_\Sigma \geqslant 0, \ \frac{\partial}{\partial t}(\psi_1 - \psi_2) \geqslant 0, \ \frac{\partial^2}{\partial t^2}(\psi_1-\psi_2) \geqslant 0 \ . \end{array} \right.$

Our objective is to prove:

THEOREM 4.3: *Take the assumptions of Theorem 3.4, and* (4.24) *to* (4.30). *Then the following problem has one and only one solution:*

(4.31) $u \in L^2(0,T;H_0^1), \ \frac{\partial u}{\partial t} \in L^2(Q), \ u \in C^0(\bar{Q})$

(4.32) $\psi_1 \leqslant u \leqslant \psi_2, \quad u(T) = \bar{u}$

(4.33) $\left| \begin{array}{l} -\left(\frac{\partial u}{\partial t}, v-u(t)\right) + \left< \big(A(t)-B(t)\big)u(t), v-u(t) \right> - \\ \quad - \Big(H(t)\big(u(t)\big), v-u(t)\Big) \geqslant \big(h(t), v-u(t)\big) \quad t. \text{ a.e.} \\ \forall v \in H_0^1, \ \psi_1(x,t) \leqslant v(x) \leqslant \psi_2(x,t) \quad x. \text{ a.e.} \end{array} \right.$ □

Remark 4.3

The assumptions (4.24),(4.25),(4.26),(4.27),(4.30) are largely technical. It is reasonable to think that they are not indispensible. □

The proof of Theorem 4.3 uses the following result:

LEMMA 4.1: *We write* B'(t) *for the operator*

(4.34) $B'(t)\varphi = \int_{R^n} \Big[\varphi(x+z) - \varphi(x) - z \cdot \nabla\varphi \ \chi_{|z| \leqslant 1}\Big] \frac{\partial c_0}{\partial t}(x,t,z)m(dz) \ ;$

then

(4.35) $\quad B'(t) \in \mathcal{L}(H_0^1;H^1)$ and $\|B'(t)\| \leq C.$

Also, let $\varphi \in L^2(0,T;H_0^1)$, $\frac{\partial\varphi}{\partial t} \in L^2(0,T;H_0^1)$, *then* $\frac{\partial}{\partial t} H(t)(\varphi(t)) \in L^2(Q)$, *and, more precisely,*

(4.36) $$\left|\frac{\partial}{\partial t}H(t)(\varphi(t))\right|_{L^2(Q)} \leq \left[1 + \|\varphi(t)\|_{H_0^1} + \|\varphi'(t)\|_{H_0^1}\right].$$

Proof: The property (4.35) is emmediate, from the assumption (4.25). The proof of (4.36) is done by the method of differential quotients. We have

$$H(t+\lambda)(\varphi(t+\lambda)) - H(t)(\varphi(t)) = \text{Inf Sup}\Big[f(x,t+\lambda) - f(x,t) +$$

$$+ (g(x,t+\lambda)-g(x,t))\cdot(\nabla\varphi(x,t+\lambda)-\nabla\varphi(x,t)) +$$

$$+ g(x,t)\cdot(\nabla\varphi(x,t+\lambda)-\nabla\varphi(x,t)) +$$

$$+ (g(x,t+\lambda)-g(x,t))\cdot\nabla\varphi(x,t) - (\tilde{a}_1(x,t+\lambda) - \tilde{a}_1(x,t))(\varphi(x,t+\lambda) - \varphi(x,t)) -$$

$$- \tilde{a}_1(x,t)(\varphi(x,t+\lambda) - \varphi(x,t)) - \varphi(x,t)(\tilde{a}_1(x,t+\lambda) - \tilde{a}_1(x,t))$$

$$+ \int_{R^n}\Big[\varphi(x+z,t+\lambda) - \varphi(x+z,t) - (\varphi(x,t+\lambda) - \varphi(x,t)) -$$

$$- z\cdot(\nabla\varphi(x,t+\lambda) - \nabla\varphi(x,t))\,\chi_{|z|\leq 1}\Big](c_0c_1(x,t+\lambda) - c_0c_1(x,t))m(dz)$$

$$+ \int_{R^n}\Big[\varphi(x+z,t) - \varphi(x,z) - z\cdot\nabla\varphi(x)\,\chi_{|z|\leq 1}\Big](c_0c_1(x,t+\lambda) - c_0c_1(x,t))m(dz)$$

$$+ \int_{R^n}\Big[\varphi(x+z,t+\lambda) - \varphi(x+z,t) - (\varphi(x,t+\lambda) - \varphi(x,t)) -$$

$$- z\cdot(\nabla\varphi(x,t+\lambda) - \nabla\varphi(x,t))\,\chi_{|z|\leq 1}\Big]c_0c_1(x,t)m(dz) +$$

$$+ f(x,t) + g(x,t)\cdot\nabla\varphi(x,t) - \tilde{a}_1(x,t)\varphi(x,t) +$$

$$+ \int_{R^n}\Big[\varphi(x+z,t) - \varphi(x,t) - z\cdot\nabla\varphi(x)\,\chi_{|z|\leq 1}\Big]c_0c_1(x,t)m(dz) -$$

$$- \text{Inf Sup}\Big[f(x,t) + g(x,t)\cdot\nabla\varphi(x,t) - \tilde{a}_1(x,t)\varphi(x,t) +$$

$$+ \int_{R^n}\Big[\varphi(x+z,t) - \varphi(x,t) - z\cdot\nabla\varphi(x)\,\chi_{|z|\leq 1}\Big]c_0c_1m(dz)\Big].$$

Therefore

$$|H(t+\lambda)(\varphi(t+\lambda))(x) - H(t)\varphi(t)(x)| \leq \lambda C\Big[1 + |\nabla\varphi(x,t+\lambda) - \nabla\varphi(x,t)| +$$

$$+ |\nabla\varphi(x,t)| + |\varphi(x,t+\lambda) - \varphi(x,t)| + |\varphi(x,t)| +$$

$$+ \int_0^1 d\theta \int_{R^n} |\nabla\varphi(x+\theta z,t+\lambda) - \nabla\varphi(x+\theta z,t)| \big(|z|^2\chi_{|z|\leq 1} + |z|\chi_{|z|\geq 1}\big) m(dz)$$

$$\int_0^1 d\theta \int_{R^n} |\nabla\varphi(x+\theta z,t)| \big(|z|^2\chi_{|z|\leq 1} + |z|\chi_{|z|>1}\big) m(dz)\Big]$$

$$+ C\Big[|\nabla\varphi(x,(t+\lambda)) - \nabla\varphi(x,t)| + |\varphi(x,t+\lambda) - \varphi(x,t)| +$$

$$+ \int_0^1 d\theta \int_{R^n} |\nabla\varphi(x+\theta z,t+\lambda) - \nabla\varphi(x+\theta z,t)| \big(|z|^2\chi_{|z|\leq 1} + |z|\chi_{|z|>1}\big) m(dz)$$

whence

$$\big|H(t+\lambda)(\varphi(t+\lambda)) - H(t)(\varphi(t))\big|_{L^2} \leq \lambda C\Big[1 + \|\varphi(t+\lambda) - \varphi(t)\|_{H_0^1} +$$

$$+ \|\varphi(t)\|_{H_0^1}\Big] + C\|\varphi(t+\lambda) - \varphi(t)\|_{H_0^1},$$

and dividing by λ, and making λ tend to 0, we deduce (4.36) for almost all t. □

Proof of Theorem 4.3: We start from the penalised scheme

$$\text{(4.37)} \quad \left| \begin{array}{l} -\dfrac{\partial u_\varepsilon}{\partial t} + A(t)u_\varepsilon - B(t)u_\varepsilon - H(t)\big(u_\varepsilon(t)\big) + \dfrac{1}{\varepsilon}(u_\varepsilon - \psi_2)^+ - \dfrac{1}{\varepsilon}(u_\varepsilon - \psi_1)^- = h \\ u_\varepsilon(T) = \bar{u},\ u_\varepsilon \in L^p(0,T;W^{2,p} \cap W_0^{1,p}),\ \dfrac{\partial u_\varepsilon}{\partial t} \in L^p(Q),\ \forall\ 2 \leq p < \infty . \end{array} \right.$$

Let $v_0 \in L^2(0,T;H_0^1)$, $\dfrac{\partial v_0}{\partial t} \in L^2(Q)$ be such that

$$\psi_1 \leq v_0 \leq \psi_2 .$$

Such a v_0 exists, thanks to (4.29). Multiply (4.37) by v_0 and integrate over x. By calculations that are now routine, we obtain

$$-\frac{1}{2}\frac{d}{dt}|u_\varepsilon(t)|^2 + \gamma\|u_\varepsilon(t)\|^2 + \frac{1}{\varepsilon}|(u_\varepsilon - \psi_2)^+|^2 + \frac{1}{\varepsilon}|(u_\varepsilon - \psi_1)^-|^2 \leq$$

$$\leq C\Big[\big(1 + \|u_\varepsilon(t)\|\big)\big(|u_\varepsilon(t)| + |v_0(t)|\big) + \Big|\Big(\frac{\partial u_\varepsilon}{\partial t}, v_0\Big)\Big| +$$

$$+ \|u_\varepsilon(t)\|\ \|v_0(t)\| + |u_\varepsilon(t)| + |v_0(t)|\Big],$$

from which we easily deduce the estimates

$$\text{(4.38)} \quad \|u_\varepsilon\|_{L^2(0,T;H_0^1)} + \|u_\varepsilon\|_{L^\infty(0,T;L^2)} \leq C$$

(Contd)

$$\left|\frac{1}{\sqrt{\varepsilon}}(u_\varepsilon-\psi_2)^+\right|_{L^2(Q)} + \left|\frac{1}{\sqrt{\varepsilon}}(u_\varepsilon-\psi_2)^-\right|_{L^2(Q)} \leq C.$$

Next, differentiate (4.37) with respect to t, and write

$$w_\varepsilon = \frac{\partial u_\varepsilon}{\partial t} ..$$

Then

$$w_\varepsilon(T) = A(T)\bar{u} - B(T)\bar{u} - h(T)$$

therefore

$$|w_\varepsilon(T)|_{L^2} \leq C.$$

Next we have

$$(4.40) \qquad -\frac{\partial w_\varepsilon}{\partial t} + A(t)w_\varepsilon - B(t)w_\varepsilon + A'(t)u_\varepsilon + B'(t)u_\varepsilon - \\ -\frac{\partial}{\partial t}H(t)\big(u_\varepsilon(t)\big) + \frac{1}{\varepsilon}\frac{\partial}{\partial t}(u_\varepsilon-\psi_2)^+ - \frac{1}{\varepsilon}\frac{\partial}{\partial t}(u_\varepsilon-\psi_1)^- = \frac{\partial h}{\partial t}.$$

Notice that

$$-\frac{1}{\varepsilon}\int_t^T\Big(\frac{\partial}{\partial t}(u_\varepsilon-\psi_1)^-, \frac{\partial}{\partial t}(u_\varepsilon-\psi_2)\Big)dt = \frac{1}{\varepsilon}\int_t^T\Big|\frac{\partial}{\partial t}(u_\varepsilon-\psi_1)^-\Big|^2dt + \\ + \frac{1}{\varepsilon}\Big((u_\varepsilon-\psi_1)^-(t), \frac{\partial}{\partial t}(\psi_1-\psi_2)(t)\Big) + \frac{1}{\varepsilon}\int_t^T\Big((u_\varepsilon-\psi_1)^-, \frac{\partial^2}{\partial t^2}(\psi_1-\psi_2)\Big)dt \\ \geq \frac{1}{\varepsilon}\int_t^T\Big|\frac{\partial}{\partial t}(u_\varepsilon-\psi_1)^-\Big|^2dt.$$

If we multiply (4.40) by $\frac{\partial}{\partial t}(u_\varepsilon - \psi_2)$ and integrate over x, from the inequality above, as well as from Lemma 4.1, we deduce that

$$(4.41) \qquad \frac{1}{2}|w_\varepsilon(t)|^2 + \gamma\int_t^T\|w_\varepsilon(t)\|^2dt + \frac{1}{\varepsilon}\int_t^T\Big|\frac{\partial}{\partial t}(u_\varepsilon-\psi_2)^+\Big|^2dt + \\ + \frac{1}{\varepsilon}\int_t^T\Big|\frac{\partial}{\partial t}(u_\varepsilon-\psi_1)^-\Big|^2dt \leq C\Big[1 + \int_t^T\|w_\varepsilon\|\,|w_\varepsilon|dt + \\ + \int_t^T|m|\,\|w_\varepsilon\|dt - \int_t^T\Big(\frac{\partial w_\varepsilon}{\partial t}, \frac{\partial\psi_2}{\partial t}\Big)dt\Big]$$

where $m \in L^2(Q)$.

Finally, using the property

$$-\int_t^T\Big(\frac{\partial u_\varepsilon}{\partial t}, \frac{\partial\psi_2}{\partial t}\Big)dt = -\Big(w_\varepsilon(T), \frac{\partial\psi_2}{\partial t}(T)\Big) + \int_t^T\Big(w_\varepsilon, \frac{\partial^2\psi_2}{\partial t^2}\Big)dt$$

we deduce from (4.41) and the assumptions that

$$(4.42)\quad \left| \begin{array}{l} \left\|\frac{\partial u_\varepsilon}{\partial t}\right\|_{L^2(0,T;H)} + \left\|\frac{\partial u_\varepsilon}{\partial t}\right\|_{L^\infty(0,T;H)} \leqslant C \\ \left|\frac{1}{\sqrt{\varepsilon}}\frac{\partial}{\partial t}(u_\varepsilon - \psi_2)^+\right|_{L^2(Q)} + \left|\frac{1}{\sqrt{\varepsilon}}\frac{\partial}{\partial t}(u_\varepsilon - \psi_2)^-\right|_{L^2(Q)} \leqslant C. \end{array}\right.$$

We can then take a subsequence, again denoted u_ε, such that

$$(4,43)\quad u_\varepsilon \to u \text{ weak star in } L^\infty(0,T;H^1_0)$$

$$\frac{\partial u_\varepsilon}{\partial t} \to \frac{u}{t} \text{ weakly in } L^2(0,T;H^1_0) \text{ and } L^\infty(0,T;L^2) \text{ weak star}$$

$$H(t)\big(u_\varepsilon(t)\big) \to \xi \text{ weakly in } L^2(Q).$$

In particular $u_\varepsilon \to u$ strongly in $L^2(Q)$. Arguing as in Theorem 4.3 we show that $\xi = H(t)(u(t))$ and $u_\varepsilon \to u$ strongly in $L^2(0,T;H^1_0)$. From this we easily deduce that u is the solution of (4.31),(4.32),(4.33), (except u is continuous); this solution is also unique (without assuming continuity).

It remains to prove that the solution u which has been found belongs to $C^0(\bar{Q})$. To do this we are going to prove an estimate that is interesting in itself, which takes the form:

LEMMA 4.2: *Let* $\psi_i, \tilde{\psi}_i$, $i = 1,2$ *be regular obstacles satisfying* (4.1),(4.2),(4.16) *respectively. Write* u *and* $\tilde{u}$ *for the solutions of* (4.17),(4.18),(4.19) *(for* $\tilde{u}$ *we replace* ψ_1, ψ_2 *by* ψ_1, ψ_2*). Assume that* $a_0 + a_1 \geqslant 0$. *Then*

$$(4.44)\quad \|u-\tilde{u}\|_{L^\infty(Q)} \leqslant \|\psi_1-\tilde{\psi}_1\|_{L^\infty(Q)} + \|\psi_2-\tilde{\psi}_2\|_{L^\infty(Q)}.$$

Proof: An argument of the maximum principle type is used; we begin from the penalised scheme (4.20). Write

$$w_\varepsilon = u_\varepsilon - \tilde{u}_\varepsilon$$

hence

$$(4.45)\quad \left| \begin{array}{l} -\frac{\partial w_\varepsilon}{\partial t} + A(t)w_\varepsilon - B(t)w_\varepsilon \quad H(t)\big(u_\varepsilon(t)\big) + H(t)\big(\tilde{u}_\varepsilon(t)\big) + \\ \quad + \frac{1}{\varepsilon}(u_\varepsilon-\psi_2)^+ - \frac{1}{\varepsilon}(\tilde{u}_\varepsilon-\tilde{\psi}_2)^+ - \frac{1}{\varepsilon}(u_\varepsilon-\psi_1)^- + \frac{1}{\varepsilon}(\tilde{u}_\varepsilon-\tilde{\psi}_2)^- = 0 \\ w_\varepsilon(T) = 0. \end{array}\right.$$

Let us set

$$k = \|\psi_1-\tilde{\psi}_1\|_{L^\infty(Q)} + \|\psi_2-\tilde{\psi}_2\|_{L^\infty(Q)}.$$

Then we have

$$\left[(u_\varepsilon-\psi_2)^+ - (\tilde{u}_\varepsilon-\tilde{\psi}_2)^+ - (u_\varepsilon-\psi_1)^- + (\tilde{u}_\varepsilon-\tilde{\psi}_2)^-\right](w_\varepsilon-k)^+ \geqslant 0.$$

Then multiply (4.45) by $(w_\varepsilon - k)^+$, and using the property (2.4) of $B(t) + H(t)$ we obtain a Gronwall inequality

$$-\frac{1}{2}\frac{d}{dt}|(w_\varepsilon-k)^+|^2 \leqslant C|(w_\varepsilon-k)^+|^2$$

and since $(w_\varepsilon(T) - k)^+ = 0$, we deduce that $w_\varepsilon \leqslant k$.

In the same way we can prove the opposite inequality, and thus (4.44). □

We can then complete the proof of Theorem 4.3. It suffices to notice that for $p > n + 2$ and regular obstacles Theorem 4.2 give us a continuous solution. In fact, for $p > n + 2$ the functions $u \in L^p(0,T;W^{1,p}(\mathcal{O}))$ such that $\frac{\partial u}{\partial t} \in L^p(0,T;W^{-1,p}(\mathcal{O}))$ belong to $C^0(\bar{Q})$. Buth then by approximating the obstacles in $C^0(Q)$ by a sequence of regular obstacles satisfying (4.1),(4.2),(4.16), and using the estimate (4.44) we obtain the desired result, for without loss of generality we can always assume $a_0 + \tilde{a}_1 \geqslant 0$.

4.2.3 Case of a single obstacle

In this case we can give a weak parabolic V.I. formulation, as in MIGNOT - PUEL [1] (cf., also, B.L. [1], loc. cit.).

We use the assumptions of Theorem 3.4, with, in addition,

(4.46) $\quad h \in L^\infty(Q),\ \bar{u} \in W_0^{1,p}(\mathcal{O}),\ p > n+2$

(4.47) $\quad \left|\begin{array}{l} \psi_1 = -\infty, \quad \psi_2 = \psi \in C^0(\bar{Q}) \\ \bar{u}(x) \leqslant \psi(x,T), \quad \psi|_\Sigma \geqslant 0. \end{array}\right.$

We write $\mathcal{K}$ for the set

(4.48) $\quad \mathcal{K} = \left\{v \mid v \in L^2(0,T;H_0^1),\ \frac{\partial v}{\partial t} \in L^2(0,T;H^{-1}),\ v \leqslant \psi\right\}.$

We then have:

THEOREM 4.4: *Take the assumptions of Theorem 3.4, and (4.46),(4.47). Then the set of functions satisfying the following conditions is non-empty and has a maximum element which is a continuous function on* $\bar{Q}$,

(4.49) $\quad u \in L^2(0,T;H_0^1), \qquad u \leqslant \psi, \qquad$ a.e.

(4.50)
$$\int_0^T\Big[-\left(\frac{\partial v}{\partial t}, v-u\right) + \langle\big(A(t)-B(t)\big)u(t), v(t) - u(t)\rangle - \big(H\big(t;u(t)\big), v(t) - u(t)\big) - \big(h(t), v(t) - u(t)\big)\Big]dt + \frac{1}{2}|v(T)-\bar{u}|^2 \geqslant 0, \quad \forall v \in \mathcal{K}.$$

Proof: This is analogous to the proof in the case of diffusions (cf. *B.L.*, loc. cit.), but here the modifications ar se from using the property (2.4). We also prove that the sequence u_ε is decreasing and converges to the maximum solution of (4.49),(4.50). As for continuity, that is proved as in Theorem 4.3, by using the analogue of (4.44) in the case of a single obstacle, so

$$\|u-\tilde{u}\|_{L^\infty} \leq \|\psi-\tilde{\psi}\|_{L^\infty} .$$

We thus obtain the desired result.□

Remark 4.4

We can treat parabolic V.I.'s in the case where B is defined by bounded measures. We can also give similar results in the case where the boundary conditions are the Neumann type (cf. Section 3.6).

5. DIFFERENTIAL CALCULUS AND STOCHASTIC INTEGRALS

Method of Approach

In this Section we give the fundamentals of stochastic integral and differential calculus for the processes having jumps which we shall consider in the remainder.

5.1 *Stochastic Processes*

An R^n-valued function defined on [0,T] (resp. [0,∞]) is called *cadlag* (1) if it is right continuous on [0,T[(resp. [0,∞[) and has a left limit (*c*ontinue *à d*roite, *l*imite *à g*auche) on]0,T[(resp. [0,∞[). We write $D([0,T];R^n)$ for the set of R^n-valued cadlag functions on [0,T]. The following properties of cadlag functions are essential. If $\xi \in D([0,T];R^n)$, then

(5.1) $|\xi(t)| \leq C \quad \forall t \in [0,T]$

(5.2) $\forall\varepsilon$, there exist at most a finite number N_ε of points

$$0 = t_0 < t_1 \dots < t_{N_\varepsilon} < T \text{ such that } |\xi(t_i)-\xi(t_{i-1})| > \varepsilon, \quad i = 1 \dots N_\varepsilon .$$

Let $(\Omega,\mathcal{Q},P)$ be a probability space and $\mathfrak{F}^t$ a filtration (an increasing family of sub-σ-algebras of $\mathcal{Q}$). An *R^n-valued stochastic process is said to be cadlag if its trajectories belong to* $D([0,\infty[;R^n)$.

Predictable processes

The σ-algebra on $\Omega \times [0,\infty[$ (resp. $\Omega \times [0,T]$) generated by the process adapted to $\mathfrak{F}^t$, whose trajectories are left-continuous is called the *predictable σ-algebra*.

A process x(t) is called *predictable* if $\omega,t \to x(t;\omega)$ is measurable with respect to the predictable σ-algebra.

Regular processes: Left quasi-continuous processes

Let T^n be an increasing sequence of stopping times converging to T a.s., where T is finite. A process x(t) is called *left quasi-continuous* if

$$x(T^n) \to x(T) \quad \text{a.s.}.$$

(1) *Translator's Note:* The French original, "cadlag", is an abbreviation, only. Rather than invent another bit of gobbledegook for the jargon, using the initials of the English translation, we have opted to let just one bit of jargon remain.

It is called *regular* when the quantities $Ex(T)$ and $\lim Ex(T)$ have a meaning (possibly equal to $+\infty$) and are equal, when T is bounded. Naturally, every positive left quasi-continuous increasing process is regular, and conversely.

The Class D: Class D.L.

A process $x(t)$ belongs to the class D if the set of R.V.'s $x(T)$, where T is a stopping time, satisfies the property

$$E|x(T)|\chi_{\{|x(T)|\geq N\}} \to 0 \text{ as } N \to \infty \quad \text{uniformly with respect to T.} \tag{5.3}$$

It is said to belong to the class D.L. if for all $a < \infty$ the property (5.3) holds for the set of stopping times such that $T \in [0,a]$.

5.2 *Properties of Super-Martingales and Martingales*

We assume that the family $\mathfrak{F}^t$ is right-continuous and $\mathfrak{F}^0$ is complete.

5.2.1 Doob-Meyer decomposition

Let $x(t)$ be a right-continuous super-martingale belonging to the class D.L.. Then it can be *uniquely* written in the form

$$x(t) = \mu(t) - \alpha(t) \tag{5.4}$$

where $\mu(t)$ is a *right-continuous martingale* and $\alpha(t)$ is a *predictable increasing* process. We say that $\alpha(t)$ is the *increasing process associated with the super-martingale* $x(t)$ in the Doob-Meyer decomposition. Furthermore, we have the following result:

> *The increasing process* $\alpha(t)$ *associated with* $x(t)$ *is continuous if and only if* $x(t)$ *is regular.*

In this case we have the following characterisation of $\alpha(t)$

$$\alpha(t) = \lim_{\delta\to 0} \sum_{\substack{t_0=0<t_1<\dots<t_{n-1}<t_n=t \\ \max(t_k-t_{k-1})\leq\delta}} E\left[x(t_{k-1})-x(t_k) \mid \mathfrak{F}^{t_{k-1}}\right] \tag{5.5}$$

with convergence holding in L^1.

5.2.2 Square-integrable martingales

Let $\mu(t)$ be a cadlag martingale. We say that it is square integrable if

$$\sup_{t>0} E|\mu(t)|^2 < \infty \tag{5.6}$$

We write $M_2 = M_2(\mathfrak{F}^t,P)$ for the set of square integrable martingales. We have the following important result: if $\mu \in M_2$ then

$$\mu(t) \to \mu(\infty) \quad \text{in } L^2 \text{ and } \quad \mu(t) = E[\mu(\infty)|\mathfrak{F}^t]. \tag{5.7}$$

We define a *scalar product* on M_2 by setting

$$(\mu,\nu) = E\mu(\infty)\nu(\infty). \tag{5.8}$$

The space M_2 is then a Hilbert space. We write M_2^c for the subspace formed by the continuous martingales.

Notice that $-\mu^2(t)$ is a super-martingale, furthermore that

$$E \sup_{t>0} |\mu(t)|^2 \leqslant 4 \sup_{t>0} E|\mu(t)|^2 < \infty$$

and hence $\mu^2(t) \in D$. If we apply the Doob-Meyer decomposition we thus have

$$(5.9) \qquad \mu^2(t) = \nu(t) + \langle\mu,\mu\rangle(t)$$

where $\nu(t)$ is a right continuous martingale and $\langle\mu,\mu\rangle(t)$ is a predictable increasing process. We say that $\langle\mu,\mu\rangle(t)$ is the *increasing process associated with the martingale* $\mu(t)$.

The notation $\langle\mu,\mu\rangle(t)$ is explained by the following considerations. Let us set

$$(5.10) \qquad \langle\mu_1,\mu_2\rangle(t) = \frac{1}{2}\langle\mu_1+\mu_2,\mu_1+\mu_2\rangle(t) - \frac{1}{2}\langle\mu_1,\mu_1\rangle(t) - \frac{1}{2}\langle\mu_2,\mu_2\rangle(t).$$

Then $\langle\mu_1,\mu_2\rangle(t)$ is the unique process satisfying the following properties

(5.11) $\langle\mu_1,\mu_2\rangle(t)$ is the difference of two predictable increasing processes.

(5.12) $\mu_1(t)\mu_2(t) - \langle\mu_1,\mu_2\rangle(t)$ is a martingale.

Further, the formula (5.10) defines a bilinear form on M_2. Also

$$(5.13) \qquad (u,\nu) = E\langle\mu,\nu\rangle(\infty).$$

We say that $\mu_1,\mu_2 \in M_2$ are *orthogonal* if $\mu_1(t)\mu_2(t)$ is a *martingale*, hence if $\langle\mu_1,\mu_2\rangle(t) = 0$. By (5.13) this implies that they are orthogonal in the Hilbert space M_2.

By the property (5.5) we can state the following:

If $\mu \in M_2$ and $\langle\mu,\mu\rangle(t)$ is *continuous*, then (in the L^1 sense)

$$(5.14) \qquad \langle\mu,\mu\rangle(t) = \lim_{\delta\to 0} \sum_{\substack{t_0=0<t_1\cdots<t_{n-1}<t_n=t \\ \max(t_k-t_{k-1})\leqslant\delta}} E\left[(\mu(t_k)-\mu(t_{k-1}))^2 \mid \mathfrak{F}^{t_{k-1}}\right]$$

Also if $\mu_1,\mu_2 \in M_2$ and $\langle\mu_1,\mu_1\rangle(t)$, $\langle\mu_2,\mu_2\rangle(t)$ are continuous, then

$$(5.15) \qquad \langle\mu_1,\mu_2\rangle(t) = \lim_{\delta\to 0} \sum_{\substack{t_0=0<t_1\cdots<t_{n-1}<t_n=t \\ \max(t_k-t_{k-1})\leq\delta}} E\left\{(\mu_1(t_k)-\mu_1(t_{k-1}))(\mu_2(t_k)-\mu_2(t_{k-1})) \mid \mathfrak{F}^{t_{k-1}}\right\}.$$

We can also prove, using subsection 5.2.1, that

(5.16) $\mu \in M_2$, then $\langle\mu,\mu\rangle(t)$ is continuous if and only if μ is left quasi-continuous.

Let us also give the following useful estimate. If $\mu \in M_2$ and $\langle\mu,\mu\rangle(t)$ are continuous, then

$$P\left[\sup_{t\in[0,T]} |\mu(t)| > \varepsilon\right] \leq \frac{N}{\varepsilon^2} + P\left[\langle\mu\ \mu\rangle(T) \geq N\right] \tag{5.17}$$

The proof of (5.17) is simple. Let

$$\tau_N = \inf\{t \in [0,T] \mid \langle\mu,\mu\rangle(t) \geq N\} .$$

Since $\langle\mu,\mu\rangle(t)$ is continuous we have $\langle\mu,\mu\rangle(\tau_N) = N$, a.s.

Now let

$$A_\varepsilon = \{\omega \mid \sup_{t\in[0,T]} |\mu(t)| > \varepsilon\}.$$

Then

$$A_\varepsilon = \left(A_\varepsilon \cap \{\tau_N < T\}\right) \cup \left(A_\varepsilon \cap \{\tau_N = T\}\right)$$

hence

$$P(A_\varepsilon) \leq P\{\tau_N < T\} + P\left(A_\varepsilon \cap \{\tau_N = T\}\right)$$

$$\leq P\{\langle\mu,\mu\rangle(T) \geq N\} + P\{\sup_{t\in[0,\tau_N]} |\mu(t)| > \varepsilon\} .$$

But

$$P\{\sup_{t\in[0,\tau_N]} |\mu(t)| > \varepsilon\} \leq \frac{1}{\varepsilon^2} E \sup_{t\in[0,\tau_N]} |\mu(t)|^2$$

$$\leq \frac{1}{\varepsilon^2} E|\mu(\tau_N)|^2 = \frac{N}{\varepsilon^2} . \ \square$$

We write M_2^r for the space of martingales $\in M_2$ that are left-quasi-continuous (thus whose associated increasing process is continuous).

5.2.3 *Locally square integrable martingales*

We shall say that $\mu(t)$ is a *locally square integrable martingale* if there exists an increasing sequence τ_n of stopping times such that

- a.s. $\tau_n \uparrow +\infty$

- $\mu(t\wedge\tau_n) \in M_2\ (\mathcal{F}^{t\wedge\tau_n})$ $\forall n$.

We denote by ℓM_2 the space of locally square integrable martingales. Similarly, we shall write ℓM_2^c for the sub-space of continuous processes.

We have the following result, which extends (5.9). Let $\mu \in \ell M_2$; we can *uniquely define a predictable increasing process* $\langle\mu,\mu\rangle(t)$ such that

$\mu^2(t\wedge\tau) - <\mu,\mu>(t\wedge\tau)$ is an $\mathfrak{F}^{t\wedge\tau}$-martingale for every stopping time τ such that $\mu(t\wedge\tau) \in M_2$.

We write ℓM_2^r for ths sub-space of left quasi-continuous martingales of M_2^r (whose increasing process is continuous). The property (5.17) is clearly valid for the elements of M_2^r. Naturally $\ell M_2^c \subset \ell M_2^r$.

5.2.4 *Representation of continuous martingales*

Let $\mu^k(t) \in \ell M_2^c$, $(k = 1\ldots m)$ also be such that $<\mu^k,\mu^k>(t)$ are *absolutely continuous* in the Lebesgue measure. Then there exists an m-dimensional Wiener process $w(t)$ with respect to a filtration $\mathfrak{F}^{\star t}$, satisfying $\mathfrak{F}^t \subset \mathfrak{F}^{\star t}$, as well as a matrix process $\psi(t)$ adapted to $\mathfrak{F}^{\star t}$ such that

$$\mu(t) = \int_0^t \psi(s)dw(s) \quad \text{a.s.} \tag{5.18}$$

If the matrix $<\mu^k,\mu^j>(t)$, $k,j = 1\ldots m$ is invertible, we can take $\mathfrak{F}^{\star t} = \mathfrak{F}^t$.

5.3 *Integral Random Measures: Martingale Measure*

Let $(\Omega,\mathcal{A},P)$ be a probability space, $\mathfrak{F}^t$ a right-continuous filtration, $\mathfrak{F}^0$-complete. Now let there be on R^p the Borel σ-algebra $\mathcal{B}$ and the class $\mathcal{B}_0$ of Borel sets whose *closure does not contain* 0 ($\mathcal{B}_0$ generates the Borel σ-algebra of $R^p - \{0\}$).

5.3.1 *Integral random measure*

We call integral random measure a random function $\nu(t,A)$, $t \in R^+$, $A \in \mathcal{B}$ such that

(5.19) $\nu(t,A)$ is cadlag, left quasi-continouos, increasing, adapted, integral valued, and has jumps of amplitude 1, for all $A \in \mathcal{B}_0$

(5.20) for t fixed, $\nu(t,A)$ is an additive function of A, and $\nu(t;\{0\}) = 0$

(5.21) if $S_\varepsilon = \{x \in R^p \mid |x| < \varepsilon\}$, then $E\nu(t;R^p - S_\varepsilon) < \infty$, for all t.

Let us give an important example of integral random measure. Let $\xi(t)$ be an R^p-valued, such that $\xi_k(t) \in M_2^r$, $k = 1\ldots p$. Set

(5.22) $\forall t > 0$, $A \in \mathcal{B}_0$

$\nu(t;A)$ = the number of jumps of the process $\xi(t)$ on $]0,t[$, whose value lies in A. Since $\xi(t)$ is cadlag, on $]0,T[$ there is at most a finite number of jumps whose value belongs to A (since the closure of A does not contain 0).

We define $\nu(t;A)$, for A belongs to the Borel σ-algebra of $R^p - \{0\}$, by continuation, since ν is a positive additive function. Now set

$$\nu(t;\{0\}) = 0, \ \nu(0,A) = 0.$$

We show that $\nu(t;A)$ is an integral random measure. The only unobvious properties are (5.21) and the fact that ν is left quasi-continuous. Now, if $\sigma \leqslant \tau$ are two stopping times we consider an excision $\sigma = \tau_0 \leqslant \tau_1 \cdots \leqslant \tau_N = \tau$, where the τ_k are stopping times, $\tau_k - \tau_{k-1} \leqslant \lambda$ (deterministic) (N is random). Let us write

$$\delta\xi(s) = \xi(s) - \xi(s^-).$$

Then we have

$$(5.23) \qquad \sum_{\substack{|\delta(s)|>\varepsilon \\ \sigma<s\leqslant\tau}} |\delta\xi(s)|^2 \leqslant \lim_{\lambda\to 0} \sum_{k=1}^{N} |\xi(\tau_k)-\xi(\tau_{k-1})|^2.$$

Since $\xi \in M_2$, for fixed

$$(5.24) \qquad E\left\{\sum_{k=1}^{N} |\xi(\tau_k)-\xi(\tau_{k-1})|^2 \,\Big|\, \mathcal{F}^\sigma\right\} \leqslant E\left\{|\xi(\tau)|^2-|\xi(\sigma)|^2 \,\Big|\, \mathcal{F}^\sigma\right\}$$

$$= \sum_{j=1}^{p} E\left\{<\xi_j,\xi_j>(\tau) - <\xi_j,\xi_j>(\sigma) \,|\, \mathcal{F}^\sigma\right\}.$$

Consequently, if $A \subset \{x \,\big|\, |x| \geqslant \varepsilon\}$ it follows from (5.23),(5.24) that

$$(5.25) \qquad E\left\{\nu(\tau,A)-\nu(\sigma;A) \,|\, \mathcal{F}^\sigma\right\} \leqslant \frac{1}{\varepsilon^2} \sum_{j=1}^{p} E\left\{<\xi_j,\xi_j>(\tau) - <\xi_j,\xi_j>(\sigma) \,|\, \mathcal{F}^\sigma\right\}.$$

Taking in particular $\sigma = 0$, $\tau = t$, $A = R^p - S$, we obtain (5.21). Further, since $\xi \in M_2^r$, and as every $A \in \mathcal{B}_0$ is contained in a set $\{x \,\big|\, |x| \geqslant \varepsilon\}$, (5.25) shows that ν is regular, hence is left quasi-continuous, since it is positive increasing.

5.3.2 *Martingale measure*

Let us consider an integral random measure; it *necessarily belongs to the class* D.L.. In fact, let $a > 0$ and τ be a stopping time such that $\tau \leqslant a$. Since ν is increasing and positive, then for $A \in \mathcal{B}_0$

$$E\,\nu(\tau;A)\, \chi_{\nu(\tau;A)\geqslant N} \leqslant E\,\nu(a,A)\, \chi_{\nu(a;A)\geqslant N} \longrightarrow 0 \quad \text{when} \quad N \longrightarrow \infty.$$

Therefore $\nu(t;A)$ is a right-continuous sub-martingale belonging to the class D.L.. Meyer's Theorem therefore provides the unique decomposition

$$(5.26) \qquad \nu(t;A) = \mu(t;A) + \pi(t;A)$$

where $\pi(t;A)$ ia a *predictable increasing process*. Since $\nu(t;A)$ is a regular sub-martingale, $\pi(t;A)$ is continuous.

Now, we have:

LEMMA 5.1:

$$(5.27) \qquad \forall A \in \mathcal{B}_0, \ \mu(t;A) \in \ell M_2^r$$

and its associated increasing process is $\pi(t;A)$.

Proof: Set

$$\tau_n = \inf\{t\ ;\ (\nu(t,A) \geqslant n) \cup (\pi(t,A) \geqslant n)\}$$

$$\nu_n(t;A) = \nu(t\wedge\tau_n;A),\ \mu_n(t;A) = \mu(t\wedge\tau_n;A),\ \pi_n(t;A) = \pi(t\wedge\tau_n;A).$$

Then $\pi_n(t;A) \leqslant n$ because π is continuous, and $\nu_n(t;A) \leqslant n$ because the jumps of $\nu(t;A)$ have amplitude 1. Thus $|\mu_n(t;A)| \leqslant n$, and thus $\mu \in \ell M_2$. Let $\alpha(t)$ be the increasing process associated with μ, then $\alpha_n(t) = \alpha(t\wedge\tau_n)$ is the increasing process associated with $\mu_n \in M_2$. We are going to show that

$$\alpha_n(t) = \pi_n(t;A) \tag{5.28}$$

which will prove (5.27). For this we first verify that α_n is continuous. By (5.16) this follows from the fact that ν is left quasi-continuous, π is continuous, hence μ is left quasi-continuous. We can then write (by(5.14))

$$\alpha_n(t) = \lim_{\delta\to 0} \sum_{k=1}^{N} E\left\{(\Delta\mu_n(t_k))^2 \middle| \mathfrak{F}^{t_{k-1}}\right\} \quad \text{in} \quad L^1 \tag{5.29}$$

where

$$\Delta\mu_n(t_k) = \mu_n(t_k)-\mu_n(t_{k-1}),\ 0 = t_0 < t_1 < t_2 \dots < t_N = t,$$

$$\delta = \max(t_k - t_{k-1}).$$

Now

$$E\left|\sum_{k=1}^{N} E\left\{(\Delta\mu_n(t_k))^2 \middle| \mathfrak{F}^{t_{k-1}}\right\} - \pi_n(t;A)\right| \leqslant$$

$$\leqslant E\left|\sum_{k=1}^{N} E\left\{(\Delta\mu_n(t_k))^2 - \Delta\nu_n(t_k) \middle| \mathfrak{F}^{t_{k-1}}\right\}\right| +$$

$$+ E\left|\sum_{k=1}^{N} E\left\{\Delta\nu_n(t_k) \middle| \mathfrak{F}^{t_{k-1}}\right\} - \pi_n(t;A)\right| = I_\delta + II_\delta.$$

By (5.5), $II_\delta \to 0$. Then

$$I_\delta \leqslant \sum_{k=1}^{N} E|(\Delta\mu_n(t_k))^2 - \Delta\nu_n(t_k)| =$$

$$= \sum_{k=1}^{N} E|(\Delta\nu_n(t_k))^2 - \Delta\nu_n(t_k) + (\Delta\pi_n(t_k))^2 - 2\Delta\pi_n(t_k)\Delta\nu_n(t_k)|)$$

$$\leqslant \sum_{k=1}^{N} E\left[(\Delta\nu_n(t_k))^2 - \Delta\nu_n(t_k)\right] + \sum_{k=1}^{N} E\ 2\Delta\pi_n(t_k)\left(\Delta\pi_n(t_k) + \Delta\nu_n(t_k)\right)$$

(Contd)

$$\leqslant \sum_{k=1}^{N} E\left[(\Delta\nu_n(t_k))^2 - \Delta\nu_n(t_k)\right] + 2 \max_k E\, \Delta\pi_n(t_k)\left(\pi_n(t) + \nu_n(t)\right).$$

Now the jumps in ν_n are equal to 1, so

$$\sum_{k=1}^{N} (\Delta\nu_n(t_k))^2 \to \nu_n(t;A) \quad \text{a.s.}$$

and

$$\sum_{k=1}^{N} (\Delta\nu_n(t_k))^2 \leqslant (\nu_n(t;A))^2.$$

Therefore

$$\sum_{k=1}^{N} E\left[(\Delta\nu_n(t_k))^2 - \Delta\nu_n(t_k)\right] \to 0 .$$

Furthermore, $\max_k \Delta\pi_n(t_k) \to 0$, therefore $I_\delta \to 0$ as $\delta \to 0$. Therefore we have (5.28). □

LEMMA 5.2: *Let* $A,B \in \mathcal{B}_0$, $A \cap B = \phi$; *then*

(5.30) $\mu(t;A\cup B) = \mu(t;A) + \mu(t;B).$

If $A \cap B = \phi$, *then*

(5.31) $\mu(t;A)$ *and* $\mu(t;B)$ *are orthogonal.*

Proof: If $A,B \in \mathcal{B}_0$, $A \cap B = \phi$, then

$$\nu(t;A\cup B) = \nu(t;A) + \nu(t;B).$$

The property (5.30) follows from the uniqueness of Doob's decomposition. Also

$$\pi(t;A\cup B) = \nu(t;A) + \nu(t;B).$$

Now set $\mu_A(t) = \mu(t;A)$. Then if $A \cap B = \phi$

$$<\mu_A,\mu_B>(t) = \frac{1}{2}<\mu_A+\mu_B,\ \mu_A+\mu_B>(t) - \frac{1}{2}<\mu_A,\mu_A>(t) - \frac{1}{2}<\mu_B,\mu_B>(t)$$

$$= \frac{1}{2}<\mu_{A\cup B},\mu_{A\cup B}>(t) - \frac{1}{2}<\mu_A,\mu_A>(t) - \frac{1}{2}<\mu_B,\mu_B>(t)$$

$$= \frac{1}{2}\pi(t;\ A\cup B) - \frac{1}{2}\pi(t;A) - \frac{1}{2}\pi(t;B) = 0$$

whence the result desired. □

For fixed t the function $\pi(t;A)$ is a positive function that is additive on $\mathcal{B}_0$, and therefore defines a (random) positive measure on $R^p - \{0\}$. We can then summarise the properties satisfied by $\mu(t;A)$.

(5.32) $\forall A \in \mathcal{B}_0$, $\mu(t;A) \in \ell M_2^r$, $\mu(0,A) = 0$

(5.33) $\forall A,B \in \mathcal{B}_0$, $A \cap B = \phi$, then

$\mu(t;A \cup B) = \mu(t;A) + \mu(t;B)$

$\mu(t;A)$ and $\mu(t;B)$ are orthogonal

(5.34) $\forall A \in \mathcal{B}_0$, if $\pi(t;A)$ denotes the continuous increasing process associated with $\mu(t;A)$, then for fixed t, $\pi(t;A)$ defines a positive measure on $R^p - \{0\}$; $E\pi(t;R^p - S_\varepsilon) < \infty$, for all t.

A random function $\mu(t;A)$, $t \in R^+$, $A \in \mathcal{B}_0$ *such that* $\mu(t;A) \in M_2^r$, *satisfying* 95.33), *and such that the continuous increasing process* $\pi(t;A)$ *associated with* $\mu(t;A)$ *defines a positive measure on* $R^p - \{0\}$ *is called a measure martingale.* We say that $\mu(t;A)$ is a *local martingale measure* if there exists a sequence $\tau_n \uparrow \infty$ a.s. of stopping times such that, $\mu_n(t;A) = \mu(t \wedge \tau_n;A)$ is a martingale measure. In the case considered above (5.32),(5.33),(5.34), we can take $\tau_n = n$.

5.4 *Stochastic Integral*

5.4.1 *Stochastic integral on* $\mathcal{L}_0$

Let $\mu(t;A)$ be a martingale measure and $\pi(t;A)$ the associated continuous increasing process. On $R^+ \times (R^p - \{0\})$ we define a measure by setting

(5.35) $$\pi(\Delta,A) = \pi(t+\Delta t,A) - \pi(t,A) \text{ si } \Delta \in]t,t+t], \text{ a Borel of } R^p - \{0\}.$$

We call $\mathcal{L}_0$ the class of process $\varphi(t,z,\omega): R^+ \times R^p \times \Omega \to R$ of the form

(5.36) $$\varphi(t,z,\omega) = \sum_{k=1}^{N} \sum_{h=1}^{M} \gamma_{kh}(\omega)\, \chi_{\Delta_k \times A_h}(t,z)$$

where $\Delta_k =]t_{k-1},t_k]$, $t_0 = 0 < t_1 \dots < t_N$, $A_h \in \mathcal{B}_0$ and γ_{kh} is $\mathcal{F}^{t_{k-1}}$-measurable and bounded, and $A_h \cap A_{h'} = \phi$, $h \neq h'$. We then set

(5.37) $$I(\varphi) = \int_0^\infty\int_{R^n} \varphi(s,z)\mu(ds,dz) = \sum_{k=1}^{N} \sum_{h=1}^{M} \gamma_{kh}\, \mu(\Delta_k,A_h)$$

where

$$\mu(\Delta_k,\Delta_h) = \mu(t_k,A_h) - \mu(t_{k-1},A_h).$$

Also we write

(5.38) $$I(\varphi)(t) = \int_0^t\int_{R^n} \varphi(s,z)\mu(ds,dz) = I(\varphi\, \chi_{]0,t]}) .$$

We shall say that $I(\varphi)$ (or $I(\varphi)(t)$) are *stochastic integrals* of φ with respect to the martingale measure $\mu(t;A)$.

THEOREM 5.1: *Let* $\mu(t;A)$ *be a martingale measure, and let* $\varphi_1, \varphi_2 \in \mathcal{L}_0$. *Then we have the following properties:*

(5.39) $$I(\gamma_1\varphi_1 + \gamma_2\varphi_2) = \gamma_1\, I(\varphi_1) + \gamma_2\, I(\varphi_2), \quad \forall \gamma_1\ \gamma_2 \text{ bounded and } \mathfrak{F}^0 \text{ measurable}$$

(5.40) $$E[I(\varphi) \mid \mathfrak{F}^0] = 0$$

(5.41) $$E[I(\varphi_1)I(\varphi_2) \mid \mathfrak{F}^0] = E\left\{ \int_0^\infty\!\!\int_{R^n} \varphi_1(s,z)\varphi_2(s,z)\pi(ds,dz) \mid \mathfrak{F}^0 \right\}$$

(5.42) $I(\varphi)(t)$ is a cadlag process belonging to M_2^r, its associated increasing process being
$$< I(\varphi), I(\varphi) > (t) = \int_0^t\!\!\int_{R^n} \varphi^2(s,z)\pi(ds,dz).$$

Proof: We remark that if $\varphi_1, \varphi_2 \in \mathcal{L}_0$ we can suitably redefine them in such a way that

$$\varphi_i(t,z) = \sum_{k=1}^{N} \sum_{h=1}^{M} \gamma^i_{k,h}\, \chi_{\Delta_k \times A_h} \qquad i = 1,2 \quad .$$

Starting from here and the properties of the martingale measure we easily deduce (5.39),(5.40),(5.41). Further,

$$I(\varphi)(t) = \sum_{j=1}^{k} \sum_{h=1}^{M} \gamma_{jh}\, \mu(\Delta_j, A_h) + \sum_{h=1}^{M} \gamma_{kh}\left[\mu(t,A_h) - \mu(t_k,A_h)\right]$$

for $t \in]t_k, t_{k+1}]$

and hence $I(\varphi)(t)$ is *cadlag*. (5.42) is now proved easily. □

5.4.2 *General case*

We now introduce the space L^2_π of *classes* of measurable functions $\varphi(t,z,\omega)$, and $\mathfrak{F}^t$ - measurable for t,z fixed, such that there exists one sequence $\varphi_n \in \mathcal{L}_0$ satisfying

(5.43) $$E\int_0^t\!\!\int_{R^n} |\varphi(s,z) - \varphi_n(s,z)|^2 \pi(ds,dz) \to 0, \quad \forall t > 0.$$

Similarly, we denote by M^2_π the space of classes of functions satisfying the same assumptions as before, where (5.43) is replaced by

(5.44) $$P\text{-lim} \int_0^t\!\!\int_{R^n} |\varphi(s,z) - \varphi_n(s,z)|^2 \pi(ds,dz) = 0, \quad \forall t > 0.$$

Remark 5.1

The processes belonging to $\mathcal{L}_0$ are *predictable*. The classes of functions belonging to L^2_π or M^2_π do not necessarily have predictable representatives. Nevertheless, if φ is cadlag and belongs to L^2_π or M^2_π, then there corresponds with it a predictable representative $\varphi(s^-,z)$, since

$$E\int_0^\infty\int_{R^n}|\varphi(s,z) - \varphi(s^-,z)|^2\pi(ds,dz) = 0$$

by the continuity of $\pi(t,A)$ at t.□

Theorem 5.1 allows us to define $I(\varphi)(t)$ for $\varphi \in L^2_\pi$ or M^2_π , and (5.42) is again satisfied with ℓM^r_2 instead of M^r_2.

Naturally, $I(\varphi)(t)$ does not depend on the representative of the class chosen. However, it is more worth considering than working with the *predictable* representative, if it exists. In fact, if, for example

$$\mu(s) = \mu(s,R^n) = \nu(s) - s$$

where ν is the Poisson process, then $\int_0^t \varphi(s)(\nu(ds) - ds)$ has a meaning as a Stieltjes' integral, for φ cadlag ($\geqslant 0$ and bounded. But then

$$\int_0^t\varphi(s)\mu(ds) = \int_0^t\varphi(s^-)\big(\nu(ds)-ds\big) \neq \int_0^t\varphi(s)\big(\nu(ds)-ds\big)$$

which shows that the good representative is the predictable representative.

Let τ be a stopping time of $\mathfrak{F}^t$, and set $\mu_\tau(t,A) = \mu(t\wedge\tau,A)$ and if $\varphi \in L^2_\pi$,

$$\varphi_\tau(t,z) = \varphi(t,z)\chi_{t<\tau}.$$

We verify on $\mathcal{L}_o$, and then by passage to the limit, that

$$\text{(5.45)} \qquad \int_0^t\int_{R^n}\varphi(s,z)\mu_\tau(ds,dz) = \int_0^t\int_{R^n}\varphi_\tau(s,z)\mu(ds,dz) = \int_0^{t\wedge\tau}\int_{R^n}\varphi(s,z)\mu(ds,dz).$$

Starting from here we verify that we can *define the stochastic integral* $I(\varphi)(t)$ for a *local martingale measure*.

Lastly, let us indicate that we can very simply characterise L^2_π and M^2_π in the case where $\pi(ds,dz) = ds\, m(dz)$, where m is a positive (deterministic) measure of R^P and $m\{0\} = 0$. We then have

$$\text{(5.46)} \qquad L^2_\pi = \Big\{\varphi \text{ measurable, } \mathfrak{F}^t\text{-measurable, } \forall \text{ fixed } t,z,\ E\int_0^t ds\int_{R^P}|\varphi(s,z)|^2 m(dz) < \infty\Big\}$$

$$\text{(5.47)} \qquad M^2_\pi = \Big\{\varphi \text{ measurable, } \mathfrak{F}^t\text{-measurable, } \forall \text{ fixed } t,z,$$
$$\int_0^t dx\int_{R^P}|\varphi(s,z)|^2 m(dz) < \infty \ \text{ p.s., } \forall t\Big\}.$$

In fact, let φ belong to the set defined on the right hand side of (5.46). We can always assume that $\varphi = 0$ for $t \geqslant T$ and for $|z| < \varepsilon$. Then let $t_j = k_j$ be a cutting up of $[0,T]$, with $k = T/N$, $N \to \infty$, $j = 0,\dots,N$. Set

$$\varphi^k(s,z) = \begin{cases} 0 \text{ if } 0 \leqslant s \leqslant t_1 \\ \frac{1}{k}\int_{t_{j-1}}^{t_j} \varphi(s,z)ds, \text{ if } t_j < s \leqslant t_{j+1},\ j \geqslant 1 \end{cases}$$

then $\varphi^k \to \varphi$ in $L^2(0,T;L^2(\Omega\times R^n,\mathcal{A}\times\mathcal{B};dP\otimes dm)$. Now φ^k, for fixed k, can be approximtated by a sequence of elements of $\mathcal{L}_0$, hence the result (5.46).

5.4.3 Properties of integral random measures

Let $\nu(t;A)$ be an integral random measure. We define a measure on $R^+\times R^p$ by setting

$$(5.48)\qquad \nu(\Delta,A) = \nu(t+\Delta t,A) - \nu(t,A), \qquad \Delta =]t,t+\Delta t], \qquad A\in\mathcal{B}.$$

We then have:

PROPOSITION 5.1: *Let* $\varphi(t,z,\omega)$ *be a measurable function for all fixed* t,z, *that is* $\mathcal{F}^t$*-measurable, and cadlag for* z,ω *fixed and positive. Then*

$$(5.49)\qquad E\int_0^\infty\int_{R^p}\varphi(t^-,z)\nu(dt,dz) = E\int_0^\infty\int_{R^p}\varphi(t,z)\pi(dt,dz).$$

Proof: The relation (5.49) is satisfied for $\varphi\in\mathcal{L}_0$. Next we take φ satisfying the assumptions and bounded as well, and zero for $t \geqslant T$ and $|z| < \varepsilon$. We then approximate φ by $\varphi_\delta(t,z) = \varphi(t_{k-1},z)$ for $t\in]t_{k-1},t_k[$. Formula (5.47) is then satisfied for such functions. In fact it is sufficient to prove the relation

$$(5.50)\qquad E\int_{R^p}\varphi(t_{k-1},z)\big(\nu(t_k,dz) - \nu(t_{k-1},dz)\big) = E\int_{R^p}\varphi(t_{k-1},z)\big(\pi(t_k,dz) - \pi(t_{k-1},dz)\big)$$

Now, in (5.50) we can replace $\nu(t;dz)$ by $\chi_{\{|z|\geqslant\varepsilon\}}\nu(t;dz)$ and $\pi(t;dz)$ by $\chi_{\{|z|\geqslant\varepsilon\}}\pi(t;dz)$, which are bounded measures on R^p, and $\varphi(t_{k-1},z)$ is integrable with respect to these measures. We then consider a cutting up of $R^p - S_\varepsilon$ and an approximation of $\varphi(t_{k-1},z)$ by step functions. We are led to the case $\mathcal{L}_0$, and thence to (5.50). Starting from (5.49) true for φ_δ, by making δ tend to 0 we obtain

$$E\int_0^\infty\int_{R^p}\varphi(t^-,z)\nu(dt,dz) = E\int_0^\infty\int_{R^p}\varphi(t^-,z)\nu(dt,dz)$$

hence (5.49) again by the continuity of π. We then obtain the general case by monotonic convergence. □

Let $\varphi\in L^2_\pi$ and let $I(\varphi)(t)$ be the stochastic integral of φ with respect to the local martingale measure associated with $\nu(t;A)$, as we saw in Subsection 5.3.2.

PROPOSITION 5.2: *We have the property*

(5.51) $$\underset{\delta\to 0}{\text{P-lim}} \sum_{\substack{t_0=0 \ldots <t_n=t \\ \max(t_k-t_{k-1})\leq\delta}} \big(I(\varphi)(t_k)-I(\varphi)(t_{k-1})\big)^2 = \int_0^t\int_{R^p} \varphi^2(s,z)\nu(ds,dz) .$$

Proof: If suffices to prove (5.51) with $\varphi \in \mathcal{L}_o$. In fact if we write

$$K_\delta(\varphi) = \sum_{t_0=0 \ldots <t_n=t} \big(I(\varphi)(t_k)-I(\varphi)(t_{k-1})\big)^2$$

then

$$EK_\delta(\varphi) = E\int_0^t\int_{R^p} \varphi^2(s,z)\pi(ds,dz) \quad \forall\delta .$$

Therefore if $\varphi_n \in \mathcal{L}_o$ converges to φ in L^2_π, then $K_\delta(\varphi - \varphi_n) \to 0$ uniformly in L^1 with respect to δ. Therefore let us assume $\varphi \in \mathcal{L}_o$.

Let us set

(5.52) $$[I(\varphi_1),I(\varphi_2)](t) = \underset{\delta\to 0}{\text{P-lim}} \sum_{t_0=0 \ldots <t_n<t} \big(I(\varphi_1)(t_k)-I(\varphi_1)(t_{k-1})\big)\big(I(\varphi_2)(t_k)-I(\varphi_2)(t_{k-1})\big)$$

when the limit of the right hand side of (5.52) exists. It is clear that if $[I(\varphi_1),K(\varphi_1)](t)$, $[I(\varphi_2),I(\varphi_2)](t)$ and $[I(\varphi_1),I(\varphi_2)](t)$ have a meaning, then $[I(\varphi_1 + \varphi_2),I(\varphi_1 + \varphi_2)](t)$ has a meaning, and

$$[I(\varphi_1+\varphi_2),I(\varphi_1+\varphi_2)](t) = [I(\varphi_1),I(\varphi_1)](t) + [I(\varphi_2),I(\varphi_2)](t) + 2[I(\varphi_1),I(\varphi_2)](t).$$

We then prove that

(5.53) $$[I(\chi_{\Delta,A}),I(\chi_{\Delta,A})](t) = \nu(\Delta,A) \text{ if } \Delta\subset [0,T]$$

$$[I(\chi_{\Delta,A}), I(\chi_{\Delta',A'})](t) = 0 \text{ if } \Delta\cap\Delta' = \phi, \quad \forall A,A'$$

$$[I(\chi_{\Delta,A}),I(\chi_{\Delta,B})](t) = 0 \text{ if } A\cap B = \phi$$

from which it easily follows that $[I(\varphi),I(\varphi)](t)$ is equal to the second member of (5.51). As for the properties (5.53), they are a consequence of the following properties ($A,B \in\mathcal{B}_o$)

(5.54) $$\lim_{\delta\to 0} \sum_k \nu^2(\Delta_k,A) = \nu(t,A) \quad \text{a.s.}$$

$$\lim_{\delta\to 0} \sum_k \pi^2(\Delta_k,A) = 0 \quad \text{a.s.}$$

$$\lim_{\delta\to 0} \sum_k \pi(\Delta_k,A)\nu(\Delta_k,A) = 0 \quad \text{a.s.}$$

$$\lim_{\delta\to 0} \sum_k \nu(\Delta_k,A)\nu(\Delta_k,B) = 0 \quad \text{a.s.} \quad \text{if } A\cap B = \phi \ ;$$

the first property of (5.54) follows from the fact that the jumps of $\nu(t,A)$ have amplitude 1, the second and the third from the fact that π and ν are increasing properties. The last relation follows from

$$\lim_{\delta\to 0} \sum_k \nu^2(\Delta_k, A\cup B) = \nu(t;A\cup B) = \nu(t;A) + \nu(t;B)$$
$$= \lim_{\delta\to 0} \sum_k \nu^2(\Delta_k,A) + \lim_{\delta\to 0} \sum_k \nu^2(\Delta_k;B)$$

but also

$$\lim_{\delta\to 0} \sum_k \nu^2(\Delta_k,A\cup B) = \left[\lim_{\delta\to 0} \sum_k \nu^2(\Delta_k,A) + \sum_k \nu^2(\Delta_k,B) + 2\sum_k \nu(\Delta_k,A)\nu(\Delta_k,B)\right]$$

whence the result desired.□

Remark 5.2

The quantity $[I(\varphi),I(\varphi)](t)$ is called the *quadratic variation* of the martingale $I(\varphi)(t)$. Note that

(5.55) $$[I(\varphi),I(\varphi)](t) - \langle I(\varphi),I(\varphi)\rangle(t) = \int_0^t\int_{R^p} \varphi^2(s,z)\mu(ds,dz) \ .$$

5.5 *Itos's Formula*

5.5.1 *Notations: Assumptions; Statement of the result*

We are given an $\mathcal{F}_0$-complete right-continuous increasing $\mathcal{F}^t$-filtration $(\Omega,\mathcal{A},P)$. and an R^n-valued normalised $\mathcal{F}^t$ Wiener process, as well as an integral random measure $\nu(t,A)$ on R^p to which corresponds a local martingale measure $\mu(t,A)$ and a continuous increasing process $\pi(t,A)$. We are also given

(5.56) $\alpha(t)$ an R^n-valued adapted process that is continuous and of bounded variation

(5.57) $\beta(t) = \int_0^t b(s).dw(s)$, $b(s)$ an adapted process with values in $(R^m;R^n)$ and a.s. $\int_0^t |b(s)|^2 ds < \infty$, for all t

(5.58) $\zeta(t) = \int_0^t\int_{R^p} \gamma(s,z)\mu(ds,dz)$ where $\gamma_k \in M^2_\pi$, for all $k = 1,\dots,n$.

Set

(5.59) $\xi(t) = \alpha(t) + \beta(t) + \zeta(t)$.

Also let $f(x,t): R^n \times R^+ \to R$ be such that

(5.60) $$f \in C^{2,1}$$

(5.61) a.s. $$\left| \begin{array}{l} \int_0^T\int_{R^p} |f(\xi(t)+\gamma(t,z),t) - f(\xi(t),t) - \nabla f(\xi(t),t)\cdot\gamma(t,z)|\pi(dt,dz) < \infty \\ \int_0^T\int_{R^p} |f(\xi(t)+\gamma(t,z),t) - f(\xi(t),t)|^2\pi(dt,dz) < \infty. \end{array}\right.$$

We shall say that $f \in E_\xi$ if (5.61) is satisfied. It is clear that if f has its first and second derivatives bounded, then $f \in E_\xi$. Ito's formula is stated in:

THEOREM 5.2: *Assume* (5.56),(5.57),(5.58),(5.60),(5.61). *Then*

(5.62) $$f(\xi(t),t) = f(\xi(0),0)+ \int_0^t \frac{\partial f}{\partial s}(\xi(s),s)ds+\int_0^t \nabla f(\xi(s),s)\cdot dx(s) +$$
$$+ \int_0^t \nabla f(\xi(s),s)\cdot b(s)dw(s) + \frac{1}{2}\int_0^t \operatorname{tr} D^2f(\xi(s),s)b(s)b^\star(s)ds +$$
$$+ \int_0^t\int_{R^p}\left[f(\xi(s)+\gamma(s,z),s) - f(\xi(s),s) - \nabla f(\xi(s),s)\cdot\gamma(s,z)\right]\pi(ds,dz) +$$
$$+ \int_0^t\int_{R^p}\left[f(\xi(s)+\gamma(s,z),s) - f(\xi(s),s)\right]\mu(ds,dz), \quad \forall t\in [0,T], \quad \text{a.s.}$$

5.5.2 Proof of Ito's formula

For $\gamma = 0$ formula (5.62) reduces to the standard formula for diffusions (cf., for example, B.L. [1], Chap. 2).

LEMMA 5.3: *Let* ζ_1,ζ_2 *be two processes such that*

(5.63) $$\zeta_i(t) = \int_0^t\int_{R^p} \gamma_i(s,z)\mu(ds,dz), \quad \gamma_i \in \mathcal{L}_0, \ i = 1,2.$$

Then

(5.64) $$\zeta_1(t)\zeta_2(t) = \int_0^t\int_{R^p} (\zeta_1(s)\gamma_2(s,z) + \zeta_2(s)\gamma_1(s,z))\mu(ds,dz) +$$
$$+ \int_0^t\int_{R^p}\gamma_1(s,z)\gamma_2(s,z)\nu(ds,dz).$$

Proof: Let δ be a subdivision of $[0,T]$, $t_0 = 0 \leqslant t_1 \cdots \leqslant t_n = t$ and $\delta = \max(t_k - t_{k-1})$ tend to 0. We set

$$\zeta_1^\delta(s) = \left| \begin{array}{ll} \zeta_1(t_{k-1}) & \text{si } s\in]t_{k-1},t_k] \quad k\leqslant N \\ 0 & \text{si } s > t. \end{array}\right.$$

Then

$$\int_0^t\int_{R^P}\left(\zeta_1^\delta(s)-\zeta_1(s)\right)^2\gamma^2(s,z)\pi(ds,dz) \to 0 \quad \text{a.s.}$$

whence it results that

$$\int_0^t\int_{R^P}\zeta_1(s)\gamma_2(s,z)\mu(ds,dz) = \text{P-}\lim_{\delta\to 0}\int_0^t\int_{R^P}\zeta_1^\delta(s)\gamma_2(s,z)\mu(ds,dz)$$

$$= \text{P}\lim_{\delta\to 0}\sum_k \zeta_1(t_{k-1})\Delta\zeta_2(t_k)$$

where

$$\Delta\zeta_2(t_k) = \zeta_2(t_k) - \zeta_2(t_{k-1}).$$

But then we easily verify that

$$\int_0^t\int_{R^P}\zeta_1(s)\gamma_2(s,z)\mu(ds,dz) + \int_0^t\int_{R^P}\zeta_2(s)\gamma_1(s,z)\mu(ds,dz) =$$

$$= \zeta_1(t)\zeta_2(t) - [\zeta_1,\zeta_2]\ (t)$$

and (5.64) is then a consequence of Proposition 5.2. □

LEMMA 5.4: *Let* ζ_1,ζ_2 *be two processes satisfying* (5.56) *and* ζ_1,ζ_2 *be as in Lemma 5.3. Set*

(5.65) $\quad \xi_i(t) = \alpha_i(t) + \zeta_i(t).$

Then

(5.66)
$$\xi_1(t)\xi_2(t) = \int_0^t\int_{R^P}\left(\xi_1(s)\gamma_2(s,z) + \xi_2(s)\gamma_1(s,z)\right)\mu(ds,dz) +$$
$$+ \int_0^t\xi_1(s)d\alpha_2(s) + \int_0^t\xi_2(s)d\alpha_1(s) + \int_0^t\int_{R^P}\gamma_1(s,z)\gamma_2(s,z)\nu(ds,dz).$$

Proof: It suffices us to prove the formula

$$\alpha(t)\zeta(t) = \int_0^t\zeta(s)d\alpha(s) + \int_0^t\int_{R^P}\alpha(s)\gamma(s,z)\mu(ds,dz).$$

Now, if we consider a subdivision of [0,T],

$$\alpha(t)\zeta(t) = \sum_{k=1}^N\left(\zeta(t_{k-1})\Delta\alpha(t_k) + \alpha(t_{k-1})\Delta\zeta(t_k)\right) + \sum_{k=1}^N\Delta\alpha(t_k)\Delta\zeta(t_k)$$

then

$$\left|\sum_k\Delta\alpha(t_k)\Delta\zeta(t_k)\right| \leq \max_k|\Delta\alpha(t_k)|\left(\int_0^t\int_{R^P}|\gamma|\nu(ds,dz) + \int_0^t\int_{R^P}|\gamma|\pi(ds,dz)\right)$$

tends to 0 a.s., hence the result desired easily follows. □

LEMMA 5.5: *Let*

$$(5.67)\qquad \zeta_i(t) = \int_0^t\int_{R^p} \gamma_i(s,z)\mu(ds,dz)$$

with $\gamma_i \in \mathcal{L}_0$. *Let* $P(x)$ *be a polynomial in* n *variables* $x = (x_1,\dots,x_n)$. *Then if* $\zeta(t)$ *denotes the vector with components* $\zeta_i(t)$,

$$(5.68)\qquad P(\zeta(t)) = P(0) + \int_0^t\int_{R^p}\left[P(\zeta(s)+\gamma(s,z)) - P(\zeta(s))\right]\mu(ds,dz) + $$
$$+ \int_0\int_{R^p}\left[P(\zeta(s)+\gamma(s,z)) - P(\zeta(s)) - \nabla P(\zeta(s))\cdot\gamma(s,z)\right]\pi(ds,dz).$$

Proof: First of all, if $n = 1$, and $P(x) = x^2$, it follows from Lemma 5.3 that

$$\zeta^2(t) = 2\int_0^t\int_{R^p}\zeta(s)\gamma(s,z)\mu(ds,dz) + \int_0^t\int_{R^p}\gamma^2(s,z)\nu(ds,dz).$$

Replacing ν by $\mu + \pi$ and rearranging the terms we immediately obtain (5.68). But then $\zeta^2(t)$ may be written in the form (5.65). We can then apply Lemma 5.4 to calculate $\zeta^3(t)$. We verify that (5.68) remains valid. We then show by induction that (5.68) holds for $P(x) = x^n$, and thus for an arbitrary polynomial in dimension 1.

Next we prove the formula for polynomials of the form

$$Q(x_1 \dots x_j) = \prod_{k-1}^{j} P_k(x_k)$$

working by induction on j, and using repetitions of (5.68) and of Lemma 5.4. □

LEMMA 5.6: *Let* $f(x) \in C^1(R^n)$ *and* $\zeta(t)$ *be as in Lemma 5.5. Then we have the formula*

$$(5.69)\qquad f(\zeta(t)) = f(0) + \int_0\int_{R^p}\left[f(\zeta(s)+\gamma(s,z)) - f(\zeta(s))\right]\mu(ds,dz) + $$
$$+ \int_0^t\int_{R^p}\left[f(\zeta(s)+\gamma(s,z)) - f(\zeta(s)) - \nabla f(\zeta(s))\cdot\gamma(s,z)\right]\pi(ds,dz).$$

Proof: There exists a sequence of polynomials $P_k(x)$ such that $P_k(x) \to f(x)$ and $\nabla P_k(x) \to \nabla f(x)$ uniformly at every point on every compact set. Then we can pass to the limit (in probability) in (5.68) written with P_k. □

Henceforward we shall write

$$(5.70)\qquad L_d(f,\zeta)(s,z) = f(\zeta(s)+\gamma(s,z),s) - f(\zeta(s),s) - \nabla f(\zeta(s),s)\cdot\gamma(s,z).$$

We then pass on to the proof of Theorem 5.2. Let us write

$$\xi(t) = \xi_c(t) + \xi(t),$$

$$\xi_c(t) = \alpha(t) + \beta(t).$$

First we assume

(5.71) $f(x,t) = f(x) \in C^2$ has bounded first and second derivatives $\gamma_i \in \mathcal{L}_0$.

$$\zeta_i(t) = \int_0^t\int_{R^P}\gamma_i(s,z)\nu(ds,dz) - \int_0^t\int_{R^P}\gamma_i(s,z)\pi(ds,dz)$$

and thus $\zeta_i(t)$ is a process with bounded variation. Consider a subdivision of $[0,T]$ into intervals of the form $\Delta_k =]t_{k-1},t_k]$, $|\Delta_k| \leq \delta$. We write

(5.72) $$f(\xi(T)) - f(\xi(0)) = \sum_{j=1}^N f(\xi(t_j)) - f(\xi(t_{j-1})) = S_1 + S_2 + S_3$$

where

$$S_1 = \sum_{j=1}^N \left[f(\xi_c(t_j)+\zeta(t_{j-1})) - f(\xi_c(t_{j-1})+\zeta(t_{j-1}))\right]$$

$$S_2 = \sum_{j=1}^N \left[f(\xi_c(t_{j-1})+\zeta(t_j)) - f(\xi_c(t_{j-1})+\zeta(t_{j-1}))\right]$$

$$S_3 = \sum_{j=1}^N \left[f(\xi_c(t_j)+\zeta(t_j)) - f(\xi_c(t_{j-1})+\zeta(t_j)) - f(\xi_c(t_j)+\zeta(t_{j-1})) + f(\xi_c(t_{j-1})+\zeta(t_{j-1}))\right].$$

First of all,

(5.73) $S_3 \to 0$ a.s. when $\delta \to 0$,

since

$$S_3 = \sum_j \int_0^1\int_0^1 d\theta d\theta' D^2 f(\xi_c(t_{j-1}) + \theta\Delta\xi_c(t_j) + \theta'\Delta\xi(t_j))\Delta\xi_c(t_j)\Delta\zeta(t_j)$$

therefore

$$|S_3| \leq C \max_j |\Delta\xi_c(t_j)| \sup_\delta \sum_j |\Delta\zeta(t_j)| \to 0 \quad \text{p.s.}$$

because ζ a.s. has bounded variation.

For S_1 we can apply Ito's formula for the case of diffusions. We obtain

$$f(\xi_c(t_j)+\zeta(t_{j-1})) - f(\xi_c(t_{j-1})+\zeta(t_{j-1})) =$$ (Contd)

$$\text{(Contd)} \quad = \int_{t_{j-1}}^{t_j} \nabla f\big(\xi_c(s)+\zeta(t_{j-1})\big)\cdot d\xi_c(s) + \frac{1}{2}\int_{t_{j-1}}^{t_j} \operatorname{tr} D^2 f\big(\xi_c(s)+\zeta(t_{j-1})\big) bb^\star ds.$$

From this it easily follows that

$$(5.74) \qquad S_1 \dashrightarrow \int_0^T \nabla f\big(\xi(s)\big)\cdot d\xi_c(s) + \frac{1}{2}\int_0^T \operatorname{tr} D^2 f\big(\xi(s)\big) bb^\star(s) ds \quad \text{a.s.}$$

Now consider S_2, for which we apply (5.69)(1). Therefore we have

$$f\big(\zeta(t_j)+\xi_c(t_{j-1})\big) - f\big(\zeta(t_{j-1})+\xi_c(t_{j-1})\big) = \int_{t_{j-1}}^{t_j}\int_{R^p} L_d\big(f,\zeta+\xi_c(t_{j-1})\big)\pi(ds,dz)$$
$$+ \int_{t_{j-1}}^{t_j}\int_{R^p}\Big[f\big(\zeta(s)+\xi_c(t_{j-1})+\gamma(s,z)\big) - f\big(\zeta(s)+\xi_c(t_{j-1})\big)\Big]\mu(ds,dz)$$

and by adding

$$S_2 = \int_0\int_{R^p}\Big[f\big(\zeta(s)+\xi_c^\delta(s)+\gamma(s,z)\big) - f\big(\zeta(s)+\xi_c^\delta(s)\big)\Big]\mu(ds,dz) +$$
$$+ \int_0^T\int_{R^p} L_d(f,\zeta+\xi_c^\delta)\ (ds,dz)$$

where

$$\xi_c^\delta(s) = \xi_c(t_{j-1}) \quad \text{for} \quad s \in \,]t_{j-1},t_j[.$$

Therefore

$$(5.75) \qquad \text{P-lim } S_2 = \int_0^T\int_{R^p} L_d(f,\zeta+\xi)\pi(dt,dz) + \int_0^T\int_{R^p}\Big[f\big(\xi(t)+\gamma(t,z)\big) - f\big(\xi(t)\big)\Big]\mu(dt,dz)$$

so that the formula (5.62) is satisfied under the assumptions (5.71). We have generalised it to the case where $\gamma = (\gamma_1,\dots,\gamma_n)$ with $\gamma_i \in M^2_\pi$. Let us assume we are in dimension 1. Take a sequence $\gamma_k \to \gamma$ in M^2_π, $\gamma_k \in \mathcal{L}_0$. Then for at least one subsequence $\xi_k \to \xi$ uniformly with respect to t, a.s.. Taking into account that

$$|L_d(f,\xi_k)| \leqslant C|\gamma_k|^2 \quad \text{and} \quad |f(\xi_k+\gamma_k) - f(\xi_k)|^2 \leqslant C|\gamma_k|^2,$$

we can pass to the limit in (5.62) written for ξ_k.

Next we generalise the formula to the case where $f \in C^2$, $f \in E_\xi$ (cf. (5.61)). Lastly we have the case where f depends on time. First, if $f \in C^{2,2}$, then by

---

(1) In fact a variant of (5.69) where $\zeta(t)$ is replaced by $\zeta(t) + \eta$, where η is an $\mathfrak{F}^0$-measurable random variable.

considering the process $\eta(t) = (\xi(t),t)$ we are reduced to applying the formula with f independent of time.

If $f \in C^{2,1}$ we consider a sequence $f_k \in C^{2,2}$ which converges uniformly to f on every compact set of $R^n \times [0,\infty[$. From here we deduce the formula in this case, then by truncation we obtain the general case. □

5.5.3 Application of Ito's formula

THEOREM 5.3: *Let*

$$\zeta(t) = \int_0^t\int_{R^p} \gamma(s,z)\mu(ds,dz) \text{ where } \gamma \in M_\pi^2.$$

where γ ⁝ M_π^2. *Assume that*

$$\pi(t;A) = \int_0^t M(s;A)ds \tag{5.76}$$

and

$$E\int_0^T \int_{R^p}|\gamma(s,z)|^2M(s,dz) \quad ds < \infty , \tag{5.77}$$
$$E\int_0^T\int_{R^p}|\gamma(s,z)|^{2m}M(s,dz)ds < \infty.$$

Then $\zeta(t)$ *has finite moments up to order* 2m, *and*

$$E|\zeta(t)|^{2m} \leqslant (A_m+B_m)\exp(1+T)K_m t \tag{5.78}$$

with

$$K_m = \sum_{k=1}^{M} (m-k)(C_m^k)^2$$

$$A_m = E\int_0 \sum_{k=1}^{m} k(C_m^k)^2 v_{2k}^{m/k}(s)ds$$

$$B_m = T\frac{(m-1)}{m} \sum_{k=2}^{m} E\int_0^T (C_m^k)^2 k v_k^{2m/k}(s)ds$$

where we have set

$$v_k(t) = \int_{R^p}|\gamma(t,z)|^k M(t,dz) .$$

Proof: First

$$v_k(t) < \infty \quad \text{a.s.}, \ 2 \leqslant k \leqslant 2m. \tag{5.79}$$

By (5.77) the property (5.79) is true for k = 2 and k = 2m. For 2 < k < 2m, we set

$$k = \alpha + \beta = \frac{2m - k}{m - 1} + \frac{(k - 2)m}{m - 1} .$$

Hölder's Inequality gives

$$v_k(t) \leqslant (v_2(t))^{\frac{2m-k}{2m-2}} (v_{2m}(t))^{\frac{k-2}{2m-2}},$$

whence (5.79).

Let $\tau = T_N = \inf\{t \in [0,T] \mid |\zeta(t)| \geqslant N\}$ and $\zeta_\tau(t) = \zeta(t\wedge\tau)$. First we verify that $f(x) = x^m \in E_{\zeta_\tau}$. Set

$$I(T) = \int_0^T\int_{R^p} |(\zeta(t\wedge\tau)+\chi_\tau(t)\gamma(t,z))^m - (\zeta(t\wedge\tau))^m|^2 \, M(t,dz)dt$$

$$J(T) = \int_0^T\int_{R^p} |(\zeta(t\wedge\tau)+\chi_\tau(t)\gamma(t,z))^m - (\zeta(t\wedge\tau))^m -$$

$$- m(\zeta(t\wedge\tau))^{m-1} \chi_\tau(t)\gamma(t,z)| \, M(t,dz)dt.$$

Then we show that

$$I(T) = \int_0^T\int_{R^p} \chi_\tau(t) \Big(\sum_{k=1}^m C_m^k \, (\zeta(t))^{m-k} (\gamma(t,z))^k \Big)^2 \, M(t,dz)dt$$

$$J(T) = \int_0^T\int_{R^p} \chi_\tau(t) \, \Big| \sum_{k=2}^m C_m^k \, \zeta^{m-k}(t) \, \gamma^k(t,z) \Big| \, M(t,dz)dt .$$

Therefore

$$I(T) \leqslant m \int_0^T\int_{R^p} \chi_\tau(t) \sum_{k=1}^m (C_m^k)^2 \, \zeta^{2m-2k}(t) \, \gamma^{2k}(t,z) \, M(t,dz)dt < \infty \quad \text{a.s.}$$

$$J(T) \leqslant \int_0^T\int_{R^p} \chi_\tau(t) \sum_{k=2}^m C_m^k |\zeta|^{m-k}(t) |\gamma|^k \, M(t,dz)dt < \infty \quad \text{a.s.}$$

by the definition of τ. We can therefore apply Ito's formula, so

$$(\zeta(t\wedge\tau))^m = \int_0^t\int_{R^p} \chi_\tau(s)\big[(\zeta(s)+\gamma(s,z))^m - \zeta^m(\varepsilon)\big]\mu(ds,dz) +$$

$$+ \int_0^t\int_{R^p} \chi_\tau(s)\big[(\zeta(s)+\gamma(s,z))^m - \zeta^m(s) - m\gamma(s,z)\zeta^{m-1}(s)\big] \, M(s,dz)ds$$

from which follows

$$E(\zeta(t\wedge\tau))^{2m} \leqslant 2E(I(t) + J^2(t))$$

and by an easy calculation,

$$EI(t) \leq K_m \int_0^t E\, \zeta^{2m}(s\wedge\tau)ds + A_m.$$

Also

$$EJ^2(t) \leq TK_m \int_0^t E\, \zeta^{2m}(s\wedge\tau)ds + B_m$$

and hence

$$E\zeta^{2m}(t\wedge\tau) \leq (A_m + B_m)\exp(1 + T)K_m t.$$

Making $N \to +\infty$, and using Fatou's Lemma, we obtain the desired result. □

We now consider a process in dimension 1 of the form

(5.80) $$\xi(t) = \int_0^t \alpha(s).dw(s) + \int_0^t\int_{R^p} (c(s,z)-1)\mu(ds,dz)$$

where

(5.81) $w(t)$ is an R^n-valued normalised Wiener process and $\alpha(t)$ is an adapted process such that

$$\int_0^t |\alpha(s)|^2 ds < \infty \quad \text{for all } t \text{ a.s.}$$

(5.82) $\mu(t;A)$ is the local martingale measure associated with $\nu(t;A)$, with increasing process $\pi(t;A)$

(5.83) $c(t,z) \geq 0$, bounded and cadlag,

$$\text{a.s.} \int_0^t\int_{R^p} |c(s,z)-1|^2\pi(ds,dz) < \infty, \int_0^t\int_{R^p} (c(s,z)-1 - \text{Log}\, c(s,z))\pi(ds,dz) < \infty$$

THEOREM 5.4: *Assume* (5.81),(5.82),(5.83). *Then the process*

(5.84) $$\eta(t) = \exp\{\xi(t) - \frac{1}{2}\int_0^t |\alpha(s)|^2 ds - \int_0^t\int_{R^p} (c(s,z)-1 - \text{Log}\, c(s,z))\nu(ds,dz)\}$$

satisfies the relation

(5.85) $$\eta(t) = 1 + \int_0^t \eta(s)\alpha(s).dw(s) + \int_0^t\int_{R^p} \eta(s)(c(s,z)-1)\mu(ds,dz).$$

Proof: Notice that $c - 1 - \text{Log}c \geq 0$, hence the expression (5.84) always has a meaning. If we replace c by $c + \varepsilon$ and make ε tend to 0, we see that it suffices to prove the result in the case $c \geq c_0 > 0$. But it is then easy to verify that $c - 1 - \text{Log}c$, $c - 1$ belong to M_π^2. Let us write $\tilde{\xi}(t)$ for the process that is the argument of exp in (5.84). Replacing $\xi(t)$ by its value (5.80) we see that

$$\tilde{\xi}(t) = \int_0^t \alpha(s).dw(s) - \frac{1}{2}\int_0 |\alpha(s)|^2 ds + \int_0^t\int_{R^p} \text{Log } c(s,z)\mu(ds,dz) -$$
$$\int_0^t\int_{R^p} \left(c(s,z) - 1 - \text{Log } c(s,z)\right)\pi(ds,dz).$$

We seek to apply Ito's formula to the process $\tilde{\xi}(t)$ and to the function exp x. It is necessary to verify that $\exp x \in E_{\tilde{\xi}}$, which easily follows from the assumptions. We easily deduce the desired result from this. □

COROLLARY 5.1: *Further assume that*

(5.86) $$\pi(t;A) = \int_0^t M(s,A)ds$$

(5.87) $$\int_{R^p} \left(c(s,z)-1\right)^2 M(s,dz) < C \qquad \textit{(non-random, independent of s)}$$

(5.88) $$|\alpha(s)| \leqslant C \qquad \textit{(non-random)}.$$

Then $\eta(t)$ *is a square integrable martingale, therefore in particular* $E\eta(t) = 1$.

Proof: Let $\tau_N = \inf\{t \geqslant 0 | \eta(t) \geqslant N\}$. We easily verify that

$$E\eta(t\wedge\tau_N)^2 \leqslant C\Big[1 + E\int_0^t \eta(s)^2 \chi_{\tau_N}(s) \; | \; \alpha(s)|^2 ds +$$
$$+ E\int_0^t\int_{R^p} \chi_{\tau_N}(s)\eta(s)^2\left(c(s,z) - 1\right)^2 M(s,dz)ds\Big]$$
$$\leqslant C'\Big[1 + E\int_0^t \eta(s\wedge\tau_N)^2 ds\Big]$$

and Gronwall's Lemma shows that $E\eta(t\wedge\tau_N)^2 < C$; making N tend to $+\infty$, we obtain the result desired. □

Remark 5.3

If $\pi(ds,dz) = ds\; m(dz)$ we can remove the assumption that $c(t,z)$ is cadlag as a result of what was seen at the end of Subsection 5.4.2.

6. STOCHASTIC DIFFERENTIAL EQUATIONS

6.1 *Poisson's Random Measures*

An integral random measure $\nu(t;A)$ is called a *Poisson measure* if the associated continuous increasing process $\pi(t;A)$ is deterministic. We can immediately prove the following properties:

THEOREM 6.1: *A Poisson measure satisfies the properties:*

(6.1) $\nu(t;A) - \nu(s;A)$ *is independent of* $\mathfrak{F}^s$, *for all* $t \geqslant s$, $A \in \mathcal{B}_0$,

(6.2) $\nu(t;A)$ *is a Poisson variable with mean* $\pi(t;A)$,

(6.3) *for all* k, $A_1 \ldots A_k \in \mathcal{B}_0$ *with* $A_i \cap A_k = \emptyset$, $i \neq k$, *the processes* $\nu(t;A_1) \ldots \nu(t;A_k)$ *are independent.*

Proof: Let

$$\mu(t;A) = \nu(t;A) - \pi(t;A).$$

Let us set

$$\zeta(t) = \sum_{j=1}^{k} \lambda_j \mu(t;A_j).$$

Now apply Ito's formula to the process $\zeta(t)$ and the function $\exp i\beta x$. Then we obtain

$$\exp i\beta\zeta(t) = \exp i\beta\zeta(s) + \int_s^t\int_{R^p}\Big[\exp i\beta\Big(\zeta(\theta) + \sum_{j=1}^{k}\lambda_j\chi_{A_j}(z)\Big) - $$
$$- \exp i\beta\zeta(\theta)\Big]\mu(d\theta,dz) + \int_s^t\int_{R^p}\Big[\exp i\beta\Big(\zeta(\theta) + \sum_{j=1}^{k}\lambda_j\chi_{A_j}(z)\Big) - \exp i\beta\zeta(\theta)$$
$$- i\beta \exp i\beta\zeta(\theta) \sum_{j=1}^{k}\lambda_j\chi_{A_j}(z)\Big]\pi(d\theta,dz)$$

hence if we set

$$j(t) = E[\exp i\beta\zeta(t)|\mathcal{F}^s]$$

we deduce

$$J(t) = \exp i\beta\zeta(s) + \int_s^t J(\theta)P(d\theta)$$

where we have set

$$P(\theta) = \int_{R^p}\Big(\exp i\beta \sum_{j=1}^{k}\lambda_j\chi_{A_j}(z) - 1 - i\beta \sum_{j=1}^{k}\lambda_j\chi_{A_j}(z)\Big)\pi(\theta,dz).$$

Consequently

$$J(t) = \exp i\beta\zeta(s)(P(t) - P(s)).$$

In particular, if $A_j \cap A_k = \emptyset$, we have

$$P(\theta) = \sum_{j=1}^{k}\big(\exp i\lambda_j\beta - 1 - i\beta\lambda_j\big)\pi(\theta,A_j)$$

from which we easily deduce the properties (6.1),(6.2),(6.3).□

When

(6.4) $\pi(t;A) = tm(A)$

where $m(A)$ is a measure on $R^p - \{0\}$, we say that the *Poisson measure is normalised*. The measure $m(A)$ is called the *associated Levy measure*.

6.2 *Assumptions: Notations: Statement of the Result*

Let us take mappings $g(x,t),\sigma(x,t),\gamma(x,t,z)$ such that

(6.5) $$\left| \begin{array}{l} g:R^n \times [0,T] \to R^n, \quad \sigma:R^n \times [0,T] \to \mathcal{L}(R^n;R^n) \\ \gamma:R^n \times [0,T] \times R^p \to R^n, \quad \text{are Borel,} \end{array} \right.$$

(6.6) $$\operatorname{tr}\sigma\sigma^*(x,t) + |g(x,t)|^2 + \int_{R^p} |\gamma(x,t,z)|^2 m(dz) \leq K(1 + |x|^2)$$

(6.7) $$\left| \begin{array}{l} \text{for all } N, \text{ there exists } C_N \text{ such that} \\ \operatorname{tr}\big(\sigma(x,t)-\sigma(y,t)\big)\big(\sigma^\star(x,t)-\sigma^\star(y,t)\big) + |g(x,t)-g(y,t)|^2 + \\ + \int_{R^p} |\gamma(x,t,z)-\gamma(y,t,z)|^2 m(dz) \leq C_N |x-y|^2 \quad \forall x,y \text{ with } |x|,\ |y| \leq N. \end{array} \right.$$

In (6.6) m is a positive measure on $R^p - \{0\}$ such that

(6.8) $$m(A) < \infty, \quad \text{for all } A \in \mathcal{B}_0.$$

Let us consider a probability space $(\Omega,\mathcal{A},P)$ and a filtration $\mathcal{F}^t$ that is right continuous and $\mathcal{F}^0$-complete. Let $w(t)$ be a normalised $\mathcal{F}^t$ Wiener process with values in R^n, and $\nu(t,A)$ a normalised Poisson measure on R^n, with Levy measure $m(dz)$. Let τ be an $\mathcal{F}^t$-stopping time, and

(6.9) η an R^n-valued R.V. that is $\mathcal{F}^\tau$-measurable.

We then set the following problem: find a process $\xi(t)$ such that

(6.10) $\xi(t)(1 - \chi_\tau(t))$ is adapted, cadlag, and R^n-valued

(6.11) $$\text{a.s.} \quad \xi(t) = \eta + \int_0^t \big(1-\chi_\tau(s)\big) g\big(\xi(s),s\big)ds + \int_0^t \big(1-\chi_\tau(s)\big)\sigma\big(\xi(s),s\big)dw(s)$$
$$+ \int_0^t\int_{R^p} \big(1-\chi_\tau(s)\big)\gamma\big(\xi(s),s,z\big)\mu(ds,dz), \quad \forall t \in [0,T].$$

In (6.11) $\mu(t,A)$ is the martingale measure

(6.12) $$\mu(t;A) = \nu(t;A) - tm(A).$$

The integrals in (6.11) are well defined, since

$$\int_0^t \big(1-\chi_\tau(s)\big) \ |\gamma\big(\xi(s),s,z\big)|^2 ds\ m(dz) < \infty \quad \text{a.s.} \quad \forall t$$

because ξ is cadlag (cf. (5.47). We shall say that (6.11) is a *stochastic differential equation.*

Our objective is to prove:

THEOREM 6.2: *Under the assumptions* (6.5),(6.6),(6.7),(6.8),(6.9), *there exists one and only one solution of* (5.10),(6.11).

6.3 *The Lipschitz Case*

We begin by considering the case where the following assumptions are satisfied

(6.13) C_N is independent of N

(6.14) $E|\eta|^2 < \infty.$

We denote the space of processes satisfying (6.10) by $\mathcal{H}$. For $\xi \in \mathcal{H}$, we write $I(t;\xi)$ for the process defined by the right hand side of (6.11), which belongs to $\mathcal{H}$ also. Next we set

(6.15) $$\mathcal{H}_2 = \{\xi \in \mathcal{H} \mid \|\xi(\cdot)\|_2 = \sup_{0\leq t\leq T} E|\xi(t)|^2 < \infty\}$$

and $\mathcal{H}_2$ is a Banach space for the norm $\|\xi(\cdot)\|_2$.

LEMMA 6.1: *For* $\xi_1, \xi_2 \in \mathcal{H}_2$

(6.16) $$E \sup_{0\leq t\leq T} |I(t;\xi_1)-I(t;\xi_2)|^2 \leq K_1 \int_0^T E|\xi_1(t)-\xi_2(t)|^2 dt.$$

Proof: Now,

$$\sup_{0\leq t\leq T} |I(t;\xi_1)-I(t;\xi_2)|^2 \leq C\Big[\int_0^T |\xi_1(t)-\xi_2(t)|^2 dt +$$
$$+ \sup_{0\leq t\leq T}\Big|\int_0^t (1-\chi_\tau(s))\big(\sigma(\xi_1(s),s) - \sigma(\xi_2(s)s)\big)dw(s)\Big|^2 +$$
$$+ \sup_{0\leq t\leq T}\Big|\int_0^t\int_{R^P} (1-\chi_\tau(s))\big(\gamma(\xi_1(s),s,z) - \gamma(\xi_2(s),s,z)\big)\mu(ds,dz)\Big|^2\Big].$$

By the properties of martingales and assumptions (6.7),(6.13), we easily obtain (6.16).□

We shall also note that thanks to (6.14) we have

(6.17) $$E \sup_{0\leq t\leq T} |I(t;\zeta)|^2 \leq C.$$

LEMMA 6.2: *There exists one and only one solution of* (6.10),(6.11) *when* (6.13), (6.14) *are satisfied.*

Proof: We define an iterative procedure $\xi^0(t) = \eta$

(6.18) $$\xi^{n+1}(t) = I(t;\xi^n).$$

Set

$$v_n(t) = E \sup_{0 \leqslant s \leqslant t} |\xi^{n+1}(s) - \xi^n(s)|^2,$$

and we then easily show that

$$v_{n+1}(t) \leqslant K_2 \int_0^t v_n(s)ds$$

hence

$$v_n(t) \leqslant V_0 K_2^n \frac{t^n}{n!} \quad \text{if} \quad V_0 = v_0(T).$$

Setting

$$\varepsilon_n = \left(\frac{V_0(K_2T)^n}{n}\right)^{1/3}$$

and using the Bienaymé Chebycheff inequality

$$P\{ \sup_{0\ t\ T} |\xi^{n+1}(t) - \xi^n(t)| > \varepsilon_n\} \leqslant \varepsilon_n$$

and since the sequence ε_n is convergent it follows that the sequence

$$\sup_{0 \leqslant t \leqslant T} |\xi^0(t)| + \sum_{k=0}^{n} \sup_{0 \leqslant t \leqslant T} |\xi^{k+1}(t) - \xi^k(t)|$$

converges almost surely. Thus $\xi^n(t)$ converges a.s. uniformly with respect to t to a process $\xi(t) \in \mathcal{H}$. By Fatou's Lemma

$$\begin{aligned} E \sup_{0 \leqslant t \leqslant T} |\xi(t) - \xi^n(t)|^2 &\leqslant E \lim_{m\to\infty} \sup_{0 \leqslant t \leqslant T} |\xi^{n+m}(t) - \xi^n(t)|^2 \\ &\leqslant E \sum_{k=n}^{\infty} \sup_{0 \leqslant t \leqslant T} |\xi^{k+1}(t) - \xi^k(t)|^2 k(k-1) \sum_{k=n}^{\infty} \frac{1}{k(k-1)} \\ &\leqslant \sum_{k=n}^{\infty} \frac{V_0(K_2T)^k}{k!} k(k-1) \sum_{k=n}^{\infty} \frac{1}{k(k-1)} \text{ tends to } 0, \end{aligned}$$

as $n \to \infty$. We then see that we can pass to the limit in (6.18). Then $\xi \in \mathcal{H}_2$ and

$$\|\xi(\cdot)\|_2 \leqslant C(1 + E|\eta|^2). \square$$

6.4 *The General Case*

We now give the proof of Theorem 6.1. Set

$$\beta_N(\lambda) = \begin{cases} \lambda & \text{if } -N \leqslant \lambda \leqslant N, \\ N & \text{if } \lambda \geqslant N, \ -N \text{ if } \lambda \leqslant -N \end{cases}$$

$$\varphi_N(x) = (\beta_N(x_1), \ldots, \beta_N(x_n))$$

$$\eta_N = \varphi_N(\eta), \quad g_N(x,t) = g(\varphi_N(x),t)$$

$$\sigma_N(x,t) = \sigma(\varphi_N(x),t), \quad \gamma_N(x,t,z) = \gamma(\varphi_N(x),t,z).$$

Notice that

$$|\varphi_N(x)| \leqslant N\sqrt{n}, \quad |\varphi_N(x) - \varphi_N(y)| \leqslant |x - y|.$$

Hence we can apply the results of Section 6.3, so that there exists one and only one solution ξ_N in $\mathcal{H}_2$ of

$$\xi_N(t) = \eta_N + \int_0^t \big(1-\chi_\tau(s)\big) g_N\big(\xi_N(s),s\big)ds + \int_0^t \big(1-\chi_\tau(s)\big)\sigma_N\big(\xi_N(s),s\big)dw(s) + \tag{6.19}$$
$$+ \int_0^t\int_{R^P} \big(1-\chi_\tau(s)\big)\gamma_N\big(\xi_N(s),s,z\big)\mu(ds,dz).$$

Set

$$\theta_N = \inf\{t \in [0,T] \,|\, |\xi_N(t)| > N\}.$$

Let us take N' > N. We are going to show that

(6.20) a.s. if $|\eta| < N$ then $\xi_N(t) = \xi_{N'}(t)$, for all $t \in [0,\theta_N[$.

For this we set

$$\chi_{\theta_N}(t) = 1 \text{ if } t < \theta_N, \quad 0 \text{ if } t \geqslant \theta_N; \quad \chi_N = 1 \text{ if } |\eta| < N,$$
$$0 \text{ otherwise.}$$

We calculate $\xi_N(t) - \xi_{N'}(t)$ using (6.19) and multiplying by $\chi_N\chi_{\theta_N}(t)$. We easily verify, using $\chi_{\theta_N}(t) = \chi_{\theta_N}(t)\chi_{\theta_N}(s)$ for $s \leqslant t$, that

$$E\, \chi_N\chi_{\theta_N}(t) \; |\xi_N(t)-\xi_{N'}(t)|^2 \leqslant 3\Big[E\,T \int_0^t \chi_N\chi_{\theta_N}(s)\,|g_N(\xi_N,s)-g_{N'}(\xi_{N'},s)|^2 ds$$
$$+ E\int_0^t \chi_N\chi_{\theta_N}(s) \; |\sigma_N(\xi_N,s)-\sigma_{N'}(\xi_{N'},s)|^2 ds +$$
$$+ E\int_0^t\int_{R^P} \chi_N\chi_{\theta_N}(s) \; |\gamma_N(\xi_N,s,z)-\gamma_{N'}(\xi_{N'},s,z)|^2 ds\; m(dz).$$

Now, in the preceding inequality we can replace $\sigma_N(\xi_N)$ by $\sigma_{N'}(\xi_N)$, similarly as for g_N, γ_N; from this it easily follows that

$$E\, \chi_N\chi_{\theta_N}(t) \; |\xi_N(t)-\xi_{N'}(t)|^2 \leqslant CC_{N'}\; E\int_0^t \chi_N\chi_{\theta_N}(s)\,|\xi_N(s)-\xi_{N'}(s)|^2 ds$$

and therefore

$$E\chi_N \chi_{\theta_N}(t)|\xi_N(t) - \xi_{N'}(t)|^2 = 0, \quad \text{for all } t$$

hence (6.20). From this we deduce

$$\{\sup_{0\leq t\leq T} |\xi_N(t) - \xi_{N+1}(t)| > 0\} \subset \{\theta_N < T\} \cup \{|\eta| \geq N\} = \Gamma_N.$$

The sequence Γ_N is decreasing, by (6.20). We are going to show that

(6.21) $\quad P(\Gamma_N) \to 0.$

If (6.21) is satisfied, then a.s. for ω there exists $N_0(\omega)$ such that $N \geq N_0$ implies $|\eta| < N$, $\theta_N = T$, therefore

$$\sup_{0\leq t\leq T} |\xi_N(t) - \xi_{N+1}(t)| = 0,$$

so

$$\xi(t) = \xi_{N_0(\omega)}(t;\omega)$$

is the solution of (6.11).

It remains to prove (6.21). Now,

$$P(\Gamma_N) \leq P(|\eta| \geq N) + P(\theta_N < T)$$

and

$$P(|\eta| \geq N) \downarrow P(|\eta| = +\infty) = 0.$$

Let us introduce the function

$$\psi(x) = \frac{1}{1 + |x|^2}.$$

We notice that

$$\psi(x)|\varphi_N(x)|^2 \leq 1.$$

Then

$$\psi(\eta) \sup_{0\leq t\leq T} |\xi_N(t)|^2 \leq C\Big[\psi(\eta)|\eta_N|^2 + \psi(\eta) \sup_{0\leq t\leq T} \Big|\int_0^t g_N(\xi_N, s)ds\Big|^2 +$$
$$+ \psi(\eta) \sup_{0\leq t\leq T} \Big|\int_0^t \sigma_N(\xi_N, s)dw(s)\Big|^2 + \psi(\eta) \sup_{0\leq t\leq T} \Big|\int_0^t\int_{R^P} \gamma_N(\xi_N, s, z)\mu(ds,dz)\Big|^2\Big].$$

From this we easily deduce

$$E\psi(\eta) \sup_{0\leq t\leq T} |\xi_N(t)|^2 \leq C\Big[1 + E\int_0^T \psi(\eta)|\xi_N(s)|^2 ds\Big]$$

and also

$$E\psi(\eta)\ |\xi_N(t)|^2 \leqslant C\left[1 + E\int_0^t \psi(\eta)\ |\xi_N(s)|^2 ds\right].$$

From this we deduce

$$P\{\sup_{0\leqslant t\leqslant T} |\xi_N(t)| \geqslant N\} \leqslant P(\psi(\eta) \leqslant \delta) + P(\psi(\eta) \sup_{0\leqslant t\leqslant T} |\xi_N(t)|^2 \geqslant N^2\delta)$$

$$\leqslant P(\psi(\eta) \leqslant \delta) + \frac{C}{N^2\delta}.$$

Since

$$\lim_{\delta\to 0} P(\psi(n) \leqslant \delta) = P(\psi(n) = 0) = 0$$

it easily follows that

$$P\{\sup_{0\leqslant t\leqslant T} |\xi_N(t)| \geqslant N\} \to 0.$$

and therefore

$$P(\theta_N < T) \to 0.$$

We have therefore proved the existence of a solution of (6.10),(6.11).

Let us prove the uniqueness. Let ξ_1,ξ_2 be two solutions. We introduce the stopping time

$$\theta_N = \inf\{t \in [0,T] | \max(\xi_1(t),\xi_2(t)) > N\}.$$

We now calculate $\chi_{\theta_N}(t)(\xi_1(t) - \xi_2(t))$. By calculations similar to those above, we show that

$$E\ \chi_{\theta_N}(t)\ |\xi_1(t)-\xi_2(t)|^2 \leqslant CC_N\ E\int_0^t |\xi_1(s)-\xi_2(s)|^2\ \chi_{\theta_N}(s)ds$$

whence

$$E\chi_{\theta_N}(t)|\xi_1(t) - \xi_2(t)|^2 = 0$$

which implies

$$\text{a.s.}\quad \xi_1(t) = \xi_2(t) \quad \text{for all } t < \theta_N.$$

Since $N \to \infty$, we have $\theta_N \to T$ a.s., therefore

$$\text{a.s.}\quad \xi_1(t) = \xi_2(t) \quad \text{for all } t < T.$$

Using the equation we see that $\xi_1(T) = \xi_2(T)$, hence the uniqueness.□

COROLLARY 6.1: *Under the assumptions of Theorem 6.1 and* $E|\eta|^2 < \infty$, *then*

(6.22) $$E \sup_{0\leq t\leq T} |\xi(t)|^2 \leq C(1 + E|\eta|^2).$$

Proof: If we consider the approximation ξ_N, we notice that (6.22) holds for ξ_N with C independent of N, thanks to (6.6). We end by applying Fatous's Lemma.□

Remark 6.1

We can construct the solution of (6.10),(6.11) on $[0,\infty[$ if the assumptions of the Theorem hold on $[0,\infty[$.

7. THE MARTINGALE PROBLEM

7.1 *Assumptions: Notations*

We set $\Omega_0 = D([0,\infty[;R^n)$ and consider the canonical process $x(t;\omega) = \omega(t)$. Let $0 \leq s \leq t$, $\mu_s^t = \sigma[x(u), s \leq u \leq t]$, $\mu_s = \mu_s^\infty$, and we are given functions $a_{ij}(x,t)$, $g_i(x,t)$ on $R^n \times [0,\infty[$ satisyfing

(7.1) $$\left|\begin{array}{l} a_{ij} = a_{ji}; \quad a \equiv a_{ij} \quad \text{is a positive matrix,} \\ a_{ij}, g_i \quad \text{are Borel and bounded.} \end{array}\right.$$

We write

(7.2) $$A(t) = -\sum_{ij} a_{ij} \frac{\partial^2}{\partial x_i \partial x_j} - \sum_i g_i \frac{\partial}{\partial x_i}.$$

We are also given a family $M(x,t;dz)$ of measures on $R^n - \{0\}$ such that

(7.3) $$\left|\begin{array}{l} M(x,t;A) \leq 0 \quad \text{for all } A \in \mathcal{B}_0,\ M(x,t;\{0\}) = 0, \\ M(x,t;A) \quad \text{is measurable in } x,t \text{ for all fixed } A. \end{array}\right.$$

(7.4) $$\sup_{x,t} \left[\int_{|z|\leq 1} |z|^2 M(x,t,dz) + \int_{|z|>1} |z| M(x,t,dz)\right] < \infty .$$

Now write

(7.5) $$B(t)\varphi(x) = \int_{R^n} \left[\varphi(x+z) - \varphi(x) - z\cdot\nabla\varphi(x)\ \chi_{|z|\leq 1}\right] M(x,t,dz).$$

A probability P^{xt} *on* Ω_0, μ_t *is the solution of the martingale problem with respect to the operator* A − B *if*

(7.6) $$P^{xt}(x(t) = x) = 1$$

(7.7) $$\text{for all } \varphi \in \mathcal{B}(R^n), \quad \varphi(x(s)) - \varphi(x(t)) + \int_t^s (A(\lambda) - B(\lambda))\varphi(x(\lambda))d\lambda$$

is a P^{xt}*-martingale with respect to the family* μ_t^s.

Remark 7.1

It is easy to verify that with the notations of Section 6.2, and if we set

$$(7.8)\qquad \left|\begin{array}{l} a = \tfrac{1}{2}\sigma\sigma^*, \quad M(x,t;A) = m\{z\,|\,\gamma(x,t,z) \in A\} \\ \tau = t, \qquad \eta = x \end{array}\right.$$

then the image measure P^{xt} of the process $\xi(s)$ which is the solution of (6.10) is the solution of the martingale problem with respect to A - B.

7.2 *Properties of the Martingale Problem*

LEMMA 7.1: *Let* $\psi(x,s) \in C_b^{2,1}(R^n \times [t,\infty[)$; *then*

$$(7.9)\qquad \psi\big(x(s),s\big) - \psi\big(x(t),t\big) + \int_t^s \Big(-\frac{\partial\psi}{\partial\lambda} + (A-B)\psi\Big)\big(x(\lambda),\lambda\big)\,d\lambda$$

is a P^{xt},μ_t^s*-martingale and*

$$(7.10)\qquad \psi\big(x(s),s\big)\exp\left[\int_t^s \frac{\Big(-\frac{\partial\psi}{\partial\lambda} + A-B)\psi\Big)\big(x(\lambda),\lambda\big)\Big)}{\psi\big(x(\lambda),\lambda\big)}\,d\lambda\right]$$

is a P^{xt},μ_t^s*-martingale if* ψ *is uniformly positive.*

Proof: For (7.9) it is clear that we can always assume $\psi \in \mathcal{D}(R^n \times]t,\infty[)$. We then have

$$(7.11)\qquad \psi\big(x(s),t\big) - \psi\big(x(t),t\big) + \int_t^s \big(A(\lambda)-B(\lambda)\big)\psi\big(x(\lambda),t\big)\,d\lambda \qquad \text{is a } P^{xt}\text{-martingale}$$

$$\frac{\partial\psi}{\partial\lambda}\big(x(s),\lambda\big) - \frac{\partial\psi}{\partial\lambda}\big(x(\lambda),\lambda\big) + \int_\lambda^s \big(A(\mu)-B(\mu)\big)\,\frac{\partial\psi}{\partial\lambda}\big(x(\mu),\lambda\big)\,d\mu \quad \text{is a } P^{xt}\text{-martingale}$$

so

$$(7.12)\qquad \frac{\partial\psi}{\partial\lambda}\big(x(s),\lambda\big) - \frac{\partial\psi}{\partial\lambda}\big(x(\lambda),\lambda\big) + \frac{\partial}{\partial\lambda}\int_\lambda^s \big(A(\mu)-B(\mu)\big)\psi\big(x(\mu),\lambda\big)\,d\mu + $$
$$+ \big(A(\lambda)-B(\lambda)\big)\psi\big(x(\lambda),\lambda\big) \quad \text{is a } P^{xt}\text{-martingale.}$$

Integrating the relation (7.12) over λ between t and s, and taking into account (7.11) and

$$\int_t^s \frac{\partial\psi}{\partial\lambda}\big(x(s),\lambda\big)\,d\lambda = \psi\big(x(s),s\big) - \psi\big(x(s),t\big)$$

we deduce (7.9)

The property (7.10) then follows from the following result (cf., for example,

B.L., Chap. 3). Let $\mu(s),\nu(s),\zeta(s)$ be processes adapted to a family $\mathcal{G}^s$ of bounded scalars such that

$$\mu(s) + \int_t^s \nu(\lambda)d\lambda \quad \text{is a } \mathcal{G}^s\text{-martingale}$$

then

$$\mu(s) \exp \int_t^s \zeta(\lambda)d\lambda + \int_t^s \big(\nu(\lambda)-\mu(\lambda)\zeta(\lambda)\big)\Big(\exp \int_t^\lambda \zeta(\mu)d\mu\Big)d\lambda$$

is a $\mathcal{G}^s$-martingale. In particular, if $\nu = \mu\zeta$, then

$$\mu(s)\exp\int_t^s \zeta(\lambda)d\lambda$$

is a $\mathcal{G}^s$-martingale. Applying this result with

$$\mu(s) = \psi(x(s),s), \qquad \nu(s) = -\frac{\partial\psi}{\partial s} + (A(s) - B(s))\phi(x(s),s),$$

we obtain (7.10).□

Let $A \in \mathcal{B}_0$, and set

(7.13) $\quad\Big|\,\eta(s,A)$ = the number of jumps on $]t,s]$ of the process $x(s)$ the value of which is in A

LEMMA 7.2: *The random function* $\eta(s,A)$ *is an integral random measure on* Ω_0,μ_t^s,P^{xt} *(cf. Subsection 5.3.1).*

Proof: Since the process $x(s)$ is cadlag, the function $\eta(s,A)$ is well defined. This is a question of verifying (5.19),(5.20),(5.21), the main thing is (5.21) and that the process is left quasi-continuous. Let $\delta > 0$, and let $\varphi_\delta(z)$ be such that

$$\varphi_\delta(z) = 0 \quad \text{for } |z| \leqslant \delta, \qquad \varphi_\delta(z) = 1 \quad \text{for } |z| \geqslant 2\delta, \qquad \varphi_\delta \text{ regular.}$$

Let $\theta \leqslant \theta'$ be two stopping times of μ_t^s. By the martingale property and Doob's Stopping Theorem,

$$E^{xt}\Big[\varphi_\delta\big(x(\theta')-x(\theta)\big) + \int_\theta^{\theta'} \big(A(\lambda)-B(\lambda)\big)\varphi_\delta\big(x(\lambda)-x(\theta)\big)d\lambda \mid \mu_t^\theta\Big] =$$
$$\varphi_\delta\big(x(\theta)-x(\theta)\big) = 0 \ .$$

Let $t \leqslant \sigma \leqslant \tau \leqslant T$ be two bounded stopping times. Consider a subdivision $\sigma = \tau_0 \leqslant \tau_1 \ldots \leqslant \tau_N = \tau$, where $\max|\tau_k - \tau_{k-1}| \leqslant \rho$, and ρ is deterministic, and tends to 0. Then

$$\int_{R^n} \varphi_\delta(z)\big(\nu(\tau;dz)-\nu(\sigma;dz)\big) = \lim_{\rho\to 0} \sum_{k=1}^N \varphi_\delta\big(x(\tau_k)-x(\tau_{k-1})\big).$$

But

$$E^{xt}\varphi_\delta\big(x(\tau_k)-x(\tau_{k-1})\big) + E^{xt}\int_{\tau_{k-1}}^{\tau_k}\big(A(\lambda)-B(\lambda)\big)\varphi_\delta\big(x(\lambda)-x(\tau_{k-1})\big)d\lambda = 0$$

and therefore

$$E^{xt}\int_{R^n}\varphi_\delta(z)\big(\nu(\tau;dz)-\nu(\sigma,dz)\big) \leqslant C_\delta\, E^{xt}(\tau-\sigma).$$

But if $A \in \mathcal{B}_0$ there exists δ such that $A \subset \{|z|\,|\,|z| \geqslant 2\delta\}$, therefore

$$E^{xt}(\nu(\tau;A) - \nu(\sigma;A)) \leqslant C_\delta E^{xt}(\tau - \sigma)$$

and hence the properties desired follow.

LEMMA 7.3: *Let* $\varphi(z)$ *be a bounded Borel function which vanishes on a neighbourhood of* 0. *Let* $\theta \in E^n$. *Then*

$$\exp\Big[i\theta\cdot\Big(x(s) - x - \int_t^s g\big(x(\lambda),\lambda\big)d\lambda\Big) + \int_{R^n}\varphi(z)\eta(s,dz) + \int_t^s \theta\cdot a\big(x(\lambda),\lambda\big)\theta d\lambda \\ - \int_t^s d\lambda\int_{R^n}\Big[\exp\big(i\theta\cdot z+\varphi(z)\big) - 1 - i\theta\cdot z\,\chi_{|z|\leqslant 1}\, M\big(x(\lambda),\lambda,dz\big)\Big] \tag{7.14}$$

is a P^{xt},μ_t^s*-martingale.*

Proof: It suffices to prove the property for $\varphi \in \mathcal{D}(R^n - \{0\})$. Further, let T be fixed, and $t, t+k,\ldots,t+Nk = T$ be a subdivision of $[t,T]$ and $\psi \in \mathcal{D}(R^n)$ such that $\psi(0) = 1$, $|\psi| \geqslant \beta > 0$. Set

$$X^j(s) = \psi\Big(x\big(s\wedge(t+jk)\big)-x\big(s\wedge\big(t+(j-1)k\big)\big)\Big)\exp\left[\int_{s\wedge(t+(j-1)k)}^{s\wedge(t+jk)}\frac{\big(A(\lambda)-B(\lambda)\big)\psi\big(x(\lambda)-x\big(t+(j-1)k\big)\big)d\lambda}{\psi\big(x(\lambda)-x\big(t+(j-1)k\big)\big)}\right]$$

$$X_N(s) = \prod_{j=1}^N X^j(s).$$

Let us assume that

$$t + k \geqslant s_2 \geqslant t + (\ell - 1)k, \quad t + mk \geqslant s_1 \geqslant t + (m - 1)k, \quad \ell \geqslant m,$$

then

$$X_N(s_2) = \prod_{j=1}^{\ell-1} X^j(t+jk)X^\ell(s_2), \quad X_N(s_1) = \prod_{j=1}^{m-1}X^j(t+jk)X^m(s_2).$$

Let us assume $\ell > m$, then

$$E^{xt}[X_N(s_2)|\mu_t^{s_1}] = \prod_{j=1}^{m-1}X^j(t+jk)E^{xt}\Big[\prod_{j=m}^{\ell-1}X^j(t+jk)X^\ell(s_2)|\mu_t^{s_1}\Big].$$

Using Lemma 7.1 (property (7.10)) we have

$$E^{xt}\left[X^{\ell}(s_2)\,|\,\mu_t^{(\ell-1)k+t}\right] = \psi(0) = 1$$

and proceeding step by step

$$E^{xt}\left[\prod_{j=m}^{\ell-1} X^{j}(t+jk)X^{\ell}(s_2)\,|\,\mu_t^{s_1}\right] = E^{xt}\left[X^{m}(t+mk)\,|\,\mu_t^{s_1}\right] = X^{m}(s_1)$$

and therefore

$$E^{xt}[X_N(s_2)\,|\,\mu_t^{s_1}] = X_N(s_1).$$

Again we have the same result with $\ell = m$. Therefore $X_N(s)$ is a P^{xt},μ_t^s-martingale (on $[t,T]$). We now apply this result with $\psi(z) = \exp(i\theta\cdot z + \varphi(z))$, from which we deduce that

$$Z_k(s) = \exp\Bigg[i\theta\cdot\big(x(s)-x\big) + \sum_{j=1}^{N}\varphi\big(x(s\wedge(t+jk)) - x\big(s\wedge\big(t+(j-1)k\big)\big)\big) -$$

$$- i\theta\cdot\int_t^s g\,d\lambda + \frac{1}{2}\int_t^s a\theta\cdot\theta d\lambda - 2i\sum_{j=1}^{N}\int_{s\wedge(t+(j-1)k)}^{s\wedge(t+jk)} a\theta\cdot D\varphi\big(x(\lambda)-x\big(t+(j-1)k\big)\big)d\lambda +$$

$$+ \sum_{j=1}^{N}\int_{s\wedge(t+(j-1)k)}^{s\wedge(t+jk)} A\varphi\big(x(\lambda)-x\big(t+(j-1)k\big)\big)d\lambda -$$

$$- \sum_{j=1}^{N}\int_{s\wedge(t+(j-1)k)}^{s\wedge(t+jk)} d\lambda\int\Big\{\exp\ i\theta\cdot z + \varphi\big(x(\lambda)-x\big(t+(j-1)k\big)+z\big) - \varphi\big(x(\lambda)-x\big(t+(j-1)k\big)\big)\big)$$

$$- 1 - z\,\chi_{|z|\le 1}\big(i\theta + D\varphi\big(x(\lambda)-x\big(t+(j-1)k\big)\big)\big)\Big\}M\big(x(\lambda),\lambda,dz\big)\Bigg]$$

is a P^{xt},μ_t^s-martingale on $[t,T]$. Now, we have

(7.15) $$\sum_{j=1}^{N}\varphi\big(x\big(s\wedge(t+jk)\big) - x\big(s\wedge\big(t+(j-1)k\big)\big)\big) \to \int_{R^n}\varphi(z)\eta(s,dz) \quad \text{a.s.}$$

(7.16) $$\sum_{j=1}^{N}\int_{s\wedge(t+(j-1)k)}^{s\wedge(t+jk)} a\theta\cdot D\varphi\big(x(\lambda)-x\big(t+(j-1)k\big)\big)d\lambda \to 0 \quad \text{a.s.},$$

since φ is zero on a neighbourhood of 0, and $x(\lambda)$ is continuous except at at most a countable number of points. Similarly

$$(7.17)\qquad \sum_{j=1}^{N}\int_{s\wedge(t+(j-1)k)}^{s\wedge(t+jk)} A\varphi\big(x(\lambda)-x\big(t+(j-1)k\big)\big)d\lambda \longrightarrow \text{a.s.}$$

Then the last term of the argument of exp in the expression for $Z_k(s)$ may be written as

$$\int_t^s\int_{R^n} Z_k(\lambda,z)M(dz)d\lambda,$$

with

$$Z_k(\lambda,z) = \exp\Big(i\theta\cdot z + \varphi\big(x(\lambda)-x\big(t+[\tfrac{\lambda-t}{k}]k\big)+z\big) - \varphi\big(x(\lambda)-x\big(t+[\tfrac{\lambda-t}{k}]k\big)\big)\Big)$$

$$- 1 - z\,\chi_{|z|\leq 1}\cdot\Big(i\theta + D\varphi\big(x(\lambda)-x\big(t+[\tfrac{\lambda-t}{k}]k\big)\big)\Big)$$

$$\longrightarrow \exp\big(i\theta\cdot z+\varphi(z)\big) - 1 - i\,\chi_{|z|\leq 1}\;z\cdot\theta$$

for all z, except for a set of points λ, independent of z, that is at most countable. Furthermore,

$$|Z_k(\lambda,z)| \leq \chi_{|z|>1}M_1|z| + |z|^2 M_2\,\chi_{|z|\leq 1}$$

where M_1,M_2 are constants. From this we deduce

$$\int_t^s\int_{R^n} Z_k(\lambda,z)M(dz)d\lambda \longrightarrow \int_t^s\int_{R^n}\Big[\exp\big(i\theta\cdot z+\varphi(z)\big) - 1 - i\chi_{|z|\leq 1}z\cdot\theta\Big]M(dz)d\lambda$$

a.s.. Lastly, since $Z_k(s)$ is uniformly bounded in all its variables, we deduce the result (7.14).□

Next, for $A \in \mathcal{B}_0$ we set

$$(7.18)\qquad \zeta(s,A) = \eta(s,A) - \int_t^s\int_A M\big(x(\lambda),\lambda,dz\big)d\lambda\ .$$

LEMMA 7.4: $\zeta(s,A)$ *is the local martingale measure associated with the integral random measure* $\eta(s,A)$. *If* $\varphi(z)$ *is Borel on* $R^n - \{0\}$ *and satisfies* $|\varphi(z)| \leq C|z|$, φ *bounded, then* $\varphi \in M_\pi^2$ *(cf.* (5.44)*), and then*

$$(7.19)\qquad \exp\Big[i\theta\cdot\big(x(s) - x - \int_t^s g d\lambda\big) + \int_{R^n}\varphi(z)\zeta(s,dz) + \int_t^s a\theta\cdot\theta d\lambda -$$

$$- \int_t^s d\lambda\int_{R^n}\big(\exp\big(i\theta\cdot z+\varphi(z)\big) - 1 - i\theta\cdot z\;\chi_{|z|\leq 1} - \varphi(z)\big\}M(dz)\Big]$$

is a P^{xt},μ_t^s*-martingale.*

Proof: Let $\varphi_\varepsilon(z) = \varphi(z)\chi_{|z|>\varepsilon}$. By Lemma 7.3, the process

$$X_\varepsilon(s) = \exp\Big[i\theta\cdot\Big(x(s) - x - \int_t^s g\,d\lambda\Big) + \int_{R^n}\varphi_\varepsilon(z)\eta(s,dz) + \int_t^s a\theta\cdot\theta\,d\lambda -$$
$$- \int_t^s d\lambda\int_{R^n}\{\exp\big(i\theta\cdot z+\varphi_\varepsilon(z)\big) - 1 - i\theta\cdot z\,\chi_{|z|\leqslant 1}\}M(dz)$$

is a P^{xt},μ_t^s-martingale. Let us set

$$M_\varepsilon(s) = \int_{R^n}\varphi_\varepsilon(z)\eta(s,dz) - \int_t^s\int_{R^n}\varphi_\varepsilon(z)M(dz)d\lambda\ .$$

Then

$$X_\varepsilon(s) = \exp\Big[i\theta\Big(x(s) - x - \int_t^s g\,d\lambda\Big) + M_\varepsilon(s) + \int_t^s a\theta\cdot\theta\,d\lambda -$$
$$- \int_t^s d\lambda\int_{R^n}\{\exp\big(i\theta\cdot z+\varphi_\varepsilon(z)\big) - 1 - i\theta\cdot z\,\chi_{|z|\leqslant 1} - \varphi_\varepsilon(z)\}M(dz)\Big]\ .$$

Taking $\theta = 0$ and replacing φ by $\beta\varphi$, we deduce that

$$Y_\varepsilon^\beta(s) = \exp[\beta M_\varepsilon(s) - \int_t^s\int_{R^n}\{\exp\ \beta\varphi_\varepsilon(z) - 1 - \beta\varphi_\varepsilon(z)\}M(dz)\,]$$

is a martingale. By differentiating with respect to β, and setting $\beta = 0$, we find that $M_\varepsilon(s)$ is a martingale. Differentiating a second time and setting $\beta = 0$, we obtain that

$$M_\varepsilon^2(s) - \int_t^s\int_{R^n}\varphi_\varepsilon^2(z)M(dz)d\lambda$$

is a martingale. If $A \in \mathcal{B}_0$, there exists ε such that $A \subset \{z\,|\,|z| \geqslant \varepsilon\}$. Taking $\varphi(z) = \chi_A(z)$, we have $\varphi_\varepsilon(z) = \varphi(z)$, therefore

$$\eta(s,A) - \int_t^s\int_A M\big(x(\lambda),\lambda,dz\big)d\lambda$$

is a martingale. This shows that $\zeta(s,A)$ is the martingale measure associated with the integral random measure $\eta(s,A)$ and the associated increasing process is

$$\int_t^s\int_{R^n}M(x(\lambda),\lambda,dz)d\lambda.$$

Therefore we have

$$M_\varepsilon(s) = \int_t^s\int_{R^n}\varphi_\varepsilon(z)\zeta(d\lambda,dz).$$

It is clear that $\varphi \in M^2_\pi$. To prove (7.19) it suffices to prove that

$$\int_{R^n} \varphi_\varepsilon(z)\zeta(s,dz) \longrightarrow \int_{R^n} \varphi(z)\zeta(s,dz) \quad \text{in} \quad L^2 .$$

Now,

$$\int_t^s\int_{R^n}\big(\varphi_\varepsilon(z)-\varphi(z)\big)^2 M(dz)d\lambda = \int_t^s\int_{R^n}\big(\varphi(z)\big)^2 \chi_{|z|\leq\varepsilon} M(dz)d\lambda \longrightarrow 0$$

whence the result desired.□

LEMMA 7.5: *The process*

$$(7.20) \qquad \exp\Big[i\theta\cdot\Big(x(s) - x - \int_t^s g d\lambda - \int_{R^n} z\, \chi_{|z|\leq 1}\, \zeta(s,dz) - \int_{R^n} z\, \chi_{|z|>1}\, \eta(s,dz)\Big) + \int_t^s a\theta\cdot\theta d\lambda\Big]$$

is a P^{xt}, μ_t^s*-martingale*

Proof: Let R > 1, and let it tend to +∞. Take

$$\varphi(z) = \varphi_R(z) = - i\theta\cdot z\chi_{z\leq R}$$

in (7.19). Therefore

$$\exp\Big[i\theta\cdot\Big(x(s) - x - \int_t^s g d\lambda - \int_{R^n} z\, \chi_{|z|\leq 1}\, \zeta(s,dz) - \int_{R^n} z\, \chi_{1<|z|\leq R}\, \eta(s,dz)\Big) + \int_t^s a\theta\cdot\theta d\lambda - \int_t^s\int_{R^n}\big(\exp\, i\theta\cdot z\, \chi_{|z|>R} - 1\big)M(dz)d\lambda\Big]$$

is a P^{xt}, μ_t^s-martingale. But

$$\int_{R^n} z\big(\chi_{1<|z|\leq R} - \chi_{1<|z|}\big)\eta(s,dz) = \int_{R^n} z\, \chi_{|z|>R}\, \eta(s,dz) \longrightarrow 0 \quad \text{in} \quad L^1$$

and

$$\Big|\int_t^s\int_{R^n}\big(\exp\, i\theta\cdot z\, \chi_{|z|>R} - 1\big)M(dz)d\lambda\Big| \leq \int_t^s\int_{R^n}|\theta||z|\, \chi_{|z|>R} M(dz)d\lambda \longrightarrow 0 \quad \text{a.s.}$$

from whence we easily deduce the property desired.□

LEMMA 7.6: *The process*

$$(7.21) \qquad \theta\cdot\Big(x(s) - x - \int_t^s g d\lambda - \int_{R^n} z\, \chi_{|z|\leq 1}\, \zeta(s,dz) - \int_{R^n} z\, \chi_{|z|>1}\, \eta(s,dz)\Big)$$

is a continuous P^{xt}, μ_t^s*-martingale with associated increasing process* $2\int_t^s a\theta\cdot\theta d\lambda$.

Proof: In (7.20) we replace θ by $\beta\theta$, differentiate with respect to β, and make $\beta = 0$, and obtain the first part of the result. Similarly, differentiating twice by β and making $\beta = 0$, we obtain the second part, with the exception of the continuity property. To obtain this we set

$$z(s) = x(s) - \int_{R^n} z\, \chi_{|z|\leq 1}\, \zeta(s,dz) - \int_{R^n} z\, \chi_{|z|>1}\, \eta(s,dz)$$

$$z_\varepsilon(s) = x(s) - \int_{R^n} z\, \chi_{\varepsilon<|z|\leq 1}\, \zeta(s,dz) - \int_{R^n} z\, \chi_{|z|>1}\, \eta(s,dz)\ .$$

Now,

$$E \sup_{t\leq s\leq T} |z_\varepsilon(s)-z(s)|^2 = E \sup_{t\leq s\leq T} \left|\int_{R^n} z\, \chi_{|z|\leq\varepsilon}\, \zeta(s,dz)\right|^2$$
$$\leq 4E \int_t^T\int_{R^n} |z|^2\, \chi_{|z|\leq\varepsilon}\, M\big(x(\lambda),\lambda,dz\big)d\lambda$$

which tends to 0 when $\varepsilon \to 0$. Therefore for fixed T we can pick a subsequence $z_{\varepsilon_k}(s)$ converging a.s. uniformly with respect to s to z(s), $s \in [t,T]$. But

$$z_\varepsilon(s) = x(s) - \int_{R^n} z\, \chi_{|z|>\varepsilon}\, \eta(s,dz) + \int_t^s\int_{R^n} z\, \chi_{\varepsilon<|z|\leq 1}\, M(dz)d\lambda.$$

By construction, on [t,T] $z_\varepsilon(s)$ does not have jumps larger than ε. But then

$$\sup_{t\leq s\leq T} |z(s)-z(s^-)| \leq \sup_{t\leq s\leq T} \Big[|z(s)-z_\varepsilon(s)| + |z_\varepsilon(s)-z_\varepsilon(s^-)| +$$
$$+ |z_\varepsilon(s^-)-z(s^-)|\Big] \leq \varepsilon + 2 \sup_{t\leq s\leq T} |z(s)-z_\varepsilon(s)|.$$

Taking $\varepsilon = \varepsilon_k$, it follows that

$$\sup_{t\leq s\leq T} |z(s) - z(s^-)| = 0 \quad \text{a.s.}.$$

Since T is arbitrary, the process z(s) is certainly a.s. continuous, hence the result desired.□

We define $\sigma(x,t)$ so that

$$\text{(7.22)} \qquad \tfrac{1}{2}\sigma^2(x,t) = a(x,t), \qquad \sigma = \sigma^*.$$

LEMMA 7.7: *Assume* a(x,t) *is invertible for all* x,t. *Then there exists a normalised* μ_t^s*-Wiener process with values in* R^n *such that*

$$\text{(7.23)} \qquad x(s) = x + \int_t^s g\big(x(\lambda),\lambda\big)d\lambda + \int_t^s \sigma\big(x(\lambda),\lambda\big)dw(\lambda) + \qquad \text{(Contd)}$$

(Contd) $$+ \int_{R^n} z\, \chi_{|z|\leqslant 1}\, \zeta(s,dz) + \int_{R^n} z\, \chi_{|z|>1}\, \eta(s,dz) .$$

Proof: This follows from the result on the representation of continuous martingales whose associated increasing process is absolutely continuous with respect to the Lebesgue measure (cf. Subsection 5.2.4).□

7.3 *Uniqueness*

We are going to give a uniqueness result for the solution of the martingale problem with respect to A - B, with the assumptions

(7.24) $$\sum a_{ij}\xi_i\xi_j \geqslant \alpha|\xi|^2, \qquad \text{for all } \xi \in R^n, \quad \alpha > 0$$

(7.25) $$\frac{\partial a_{ij}}{\partial x_k} \in L^\infty(R^n \times]0,T[)$$

(7.26) $$\left| \begin{array}{l} M(x,t,dz) = c_0(x,t,z)m(dz), \\ c_0 \text{ measurable}, \quad 0 \leqslant c_0 \leqslant 1 \end{array} \right.$$

(7.27) m a positive measure on $R^n - \{0\}$, $m\{0\} = 0$,

$$\int_{|z|\leqslant 1} |z|^2 m(dz) + \int_{|z|>1} |z| m(dz) < \infty.$$

We then have:

THEOREM 7.1: *Under the assumptions* (7.1),(7.24) to (7.27) *there exists a single probability* P^{xt} *on* Ω_0, μ_t *which is the solution of the martingale problem with respect to the operator* A - B.

We begin with some preliminary results. Let P^{xt} be a solution of the martingale problem with respect to A - B. By what we have seen in Section 7.1 there corresponds with to an integral random measure $\eta(s,A)$ and a local martingale measure $\zeta(s,A)$ with associated increasing process

$$\int_t^s \int_{R^n} M(x(\lambda),\lambda,dz)d\lambda.$$

Set

(7.28) $$\zeta_\varepsilon(s,A) = \int_A \chi_{|z|>\varepsilon}\, \zeta(s,dz), \quad \eta_\varepsilon(s,A) = \int_A \chi_{|z|>\varepsilon}\, \eta(s,dz) .$$

Then $\eta_\varepsilon(s,A)$ is an integral random measure, $\zeta_\varepsilon(s,A)$ the associated local martingale measure with increasing process

$$\int_t^s \int_{R^n} \chi_{|z|>\varepsilon} M(dz)d\lambda.$$

We define $\xi_\varepsilon(s)$ to be the solution of

$$(7.29)\quad \left|\begin{array}{l} \xi_\varepsilon(s) = x + \int_t^s \sigma\big(\xi_\varepsilon(\lambda),\lambda\big)dw(\lambda) + \int_t^s g\big(x(\lambda),\lambda\big)d\lambda + \\ \qquad + \int_{R^n} z\,\chi_{|z|\leq 1}\,\zeta_\varepsilon(s,dz) + \int_{R^n} z\,\chi_{|z|>1}\,\eta_\varepsilon(s,dz) \\ \xi_\varepsilon \text{ adapted, cadlag, } E \sup_{t\leq s\leq T} |\xi_\varepsilon(s)|^2 \leq C_\varepsilon . \end{array}\right.$$

The process ξ_ε exists and is uniquely defined by the theory of Section 6 to within a variant(1).

Further, we have

$$x(s) - \xi_\varepsilon(s) = \int_t^s \big(\sigma\big(x(\lambda),\lambda\big) - \sigma\big(\zeta_\varepsilon(\lambda),\lambda\big)\big)dw(\lambda) + \int_{R^n} z\,\chi_{|z|\leq\varepsilon}\,\zeta(s,dz)$$

from which we deduce

$$(7.30)\quad E^{xt} \sup_{t\leq s\leq T} |x(s)-\xi_\varepsilon(s)|^2 \leq C_T \int_{R^n} |z|^2\,\chi_{|z|\leq\varepsilon}\,m(dz) \longrightarrow 0$$

when $\varepsilon \to 0$.

LEMMA 7.8: *Let* $f \in L^p(R^n \times]0,T[)$, $p > n+1$, *then*

$$(7.31)\quad \left|E^{xt}\int_t^T f\big(\xi_\varepsilon(s),s\big)ds\right| \leq C_\varepsilon |f|_{L^p} .$$

Proof: It suffices to prove (7.31) with $f \in \mathcal{D}(R^n \times]0,T[)$ and $f \geq 0$. Define the sequence of stopping times

$$\tau_0 = t,\ \dots,\ \tau_{j+1} = \inf\{T \geq s \geq \tau_j \mid \eta_\varepsilon(s,R^n) > \eta_\varepsilon(\tau_j,R^n)\}.$$

Since

$$E^{xt}\eta_\varepsilon(T;R^n) \leq Tm_\varepsilon(R^n) \quad \text{where} \quad m_\varepsilon(A) = \int_A \chi_{|z|>\varepsilon}m(dz),$$

then $\tau_j = T$ a.o. for j large enough. Consequently,

$$E^{xt}\int_t^T f\big(\xi_\varepsilon(s),s\big)ds = \sum_{j=1}^\infty E^{xt}\int_{\tau_j}^{\tau_{j+1}} f\big(\xi_\varepsilon(s)\big)ds.$$

---

(1) Equation (7.29) does not appear at all in stochastic differential equations because of the last integral, which we can write in the form

$$\int z\chi_{|z|>1}\zeta_2(s,dz) + \iint z\chi_{|z|>1}c_0 d\lambda m_\varepsilon(dz),$$

as none of these integrals has a meaning (the sum has a meaning). In fact, we have a simple additive term that does not depend upon ξ^ε. The iterative method of Theorem 6.1 is easily altered.

We define the function $\varphi(x,t)$ as the solution of

$$-\frac{\partial\varphi}{\partial t}-\sum_{ij}a_{ij}(x,t)\frac{\partial^2\varphi}{\partial x_i\partial x_j}=f(x,t),\ \varphi(x,T)=0$$
$$\varphi\in W^{2,1,p}(R^n\times]0,T[).$$

For $p > n + 1$ we have

$$\|\varphi\|_{C^0},\ \left\|\frac{\partial\varphi}{\partial x_i}\right\|_{C^0}\leqslant C|f|_{L^p}\ .$$

Now, on the interval $[\tau_j,\tau_{j+1}[$, the process $\xi_\varepsilon(s)$ is the solution of

$$\xi_\varepsilon(s) = x+\int_t^s\sigma dw+\int_t^s g d\lambda-\int_t^s\int_{R^n}z\ c_0\ \chi_{|z|\leqslant 1}\ m_\varepsilon(dz)d\lambda$$

and therefore Ito's formula for diffusions, which is applicable to φ and to the process $\xi_\varepsilon(s)$, is

$$\varphi\big(\xi^\varepsilon(\tau_{j+1}-0),\tau_{j+1}\big)-\varphi\big(\xi^\varepsilon(\tau_j),\tau_j\big) = -\int_{\tau_j}^{\tau_{j+1}}f\big(\xi^\varepsilon(s),s\big)ds +$$
$$+\int_{\tau_j}^{\tau_{j+1}}\nabla\varphi\big(\xi^\varepsilon(s),s\big)\cdot g\big(x(s),s\big)ds-\int_{\tau_j}^{\tau_{j+1}}\nabla\varphi\big(\xi^\varepsilon(s),s\big)\cdot$$
$$\cdot\int_{R^n}z\ \chi_{|z|\leqslant 1}\ c_0\big(x(s),s,z\big)m_\varepsilon(dz)ds+\int_{\tau_j}^{\tau_{j+1}}\nabla\varphi\big(\xi^\varepsilon(s),s\big)\cdot\sigma\big(\xi_\varepsilon(s),s\big)dw(s).$$

From this we deduce, in particular, that

$$E^{xt}\Big[\int_{\tau_j}^{\tau_{j+1}}f\big(\xi^\varepsilon(s),s\big)ds\ \big|\ \mu_t^{\tau_j}\Big]\leqslant C|f|_{L^p}\ \chi_{\tau_j<T}\quad \text{a.s.}$$

Therefore

$$E^{xt}\int_t^T f\big(\xi_\varepsilon(s),s\big)ds\leqslant C|f|_{L^p}\sum_{j=1}^\infty E^{xt}\ \chi_{\tau_j<T}\ .$$

Since

$$\sum_{j=1}^\infty E^{xt}\ \chi_{\tau_j<T}\leqslant 1+E^{xt}\sum_{j=1}^\infty\big(\eta_\varepsilon(\tau_j;R^n)-\eta_\varepsilon(\tau_{j-1};R^n)\big)$$
$$= 1+E^{xt}\ \nu_\varepsilon(T;R^n)\leqslant 1+Tm_\varepsilon(R^n),$$

we obtain (7.31).□

We are going to prove that the estimate (7.31) can be made independent of ε.

LEMMA 7.9: *In* (7.31) *the constant can be made independent of* ε.

Proof: We now consider φ_λ the solution of

$$-\frac{\partial\varphi_\lambda}{\partial t} - \sum_{ij} a_{ij}(x,t)\frac{\partial^2\varphi_\lambda}{\partial x_i \partial x_j} + \lambda\varphi_\lambda = f,\ \varphi_\lambda(x,T) = 0,$$
$$\varphi_\lambda \in W^{2,1,p}(R^n \times]0,T[).$$

We know that for $\lambda > 0$,

$$|\varphi_\lambda|_{L^p} \leqslant \frac{C|f|_{L^p}}{\lambda},\quad \|\varphi_\lambda\|_{W^{2,1,p}} \leqslant C|f|_{L^p},\quad \|\varphi_\lambda\|_{W^{1,0,p}} \leqslant \frac{C|f|_{L^p}}{\sqrt{\lambda}}.$$

On the other hand, by regularisation, thanks to Lemma 7.8, we easily show that Ito's formula may be applied to φ_λ and the process $\xi_\varepsilon(s)$, for example in the integrated form. Therefore we have

$$E^{xt}\,\varphi_\lambda(\xi_\varepsilon(T),T)\,e^{-\lambda(T-t)} = \varphi_\lambda(x,t) +$$
$$+ E^{xt}\int_t^T\Big(\frac{\partial\varphi_\lambda}{\partial s} + \sum a_{ij}\frac{\partial^2\varphi_\lambda}{\partial x_i\partial x_j} - \lambda\varphi_\lambda\Big)(\xi_\varepsilon(s))\,e^{-\lambda(s-t)}ds$$
$$+ E^{xt}\int_t^T\int_{R^n}\Big(\varphi_\lambda(\xi_\varepsilon(s)+z,s) - \varphi_\lambda(\xi_\varepsilon(s),s) -$$
$$- \chi_{|z|\leqslant 1}\, z\cdot\nabla\varphi_\lambda(\xi_\varepsilon(s),s)\Big)\,c_0\, e^{-\lambda(s-t)}m_\varepsilon(dz)$$

and hence

$$E^{xt}\int_t^T f(\xi_\varepsilon(s),s)\,e^{-\lambda(s-t)}dt \leqslant \varphi_\lambda(x,t) +$$
$$+ E^{xt}\int_t^T\int_{R^n}|\varphi_\lambda(\xi_\varepsilon(s)+z) - \varphi_\lambda(\xi_\varepsilon(s)) - z\cdot\nabla\varphi_\lambda\,\chi_{|z|\leqslant 1}| \times e^{-\lambda(s-t)}m_\varepsilon(dz).$$

Let us set

$$L_\lambda^\varepsilon(f) = E^{xt}\int_t^T f(\xi_\varepsilon(s),s)\,e^{-\lambda(s-t)}d\lambda.$$

From the preceding inequality it follows that

$$L_\lambda^\varepsilon(f) \leqslant \varphi_\lambda(x,t) + L_\lambda^\varepsilon(h_\lambda),$$

where we have set

$$h_\lambda(x,s) = \int_{R^n}|\varphi_\lambda(x+z,s) - \varphi_\lambda(x,s) - \chi_{|z|\leqslant 1}\, z.\nabla\varphi_\lambda(x,s)|\,m(dz).$$

But as in Lemma 1.1, we verify that

$$|h_\lambda|_{L^p} \leqslant \sigma(r)\,\|\varphi_\lambda\|_{W^{2,1,p}} + \tau(r)\,\|\varphi_\lambda\|_{W^{1,0,p}}$$

with $\sigma(r) \to 0$ when $r \to 0$. Consequently

$$|h_\lambda|_{L^p} \leq C\left(\sigma(r)|f|_{L^p} + \frac{\tau(r)}{\sqrt{\lambda}}|f|_{L^p}\right).$$

But Lemma 7.8 shows that L_λ^ε is a linear operator contained on L^p. Therefore (for $p > n + 1$) we can write

$$L_\lambda^\varepsilon(f) \leq C|f|_{L^p}\left[1 + \|L_\lambda^\varepsilon\|\left(\sigma(r) + \frac{\tau(r)}{\sqrt{\lambda}}\right)\right]$$

Hence

$$\|L_\lambda^\varepsilon\| \leq C\left[1 + \|L_\lambda^\varepsilon\|\left(\sigma(r) + \frac{\tau(r)}{\sqrt{\lambda}}\right)\right].$$

Thus, choosing r small enough and λ large enough, $\lambda \geq \lambda_1$, we have

$$\|L_\lambda^\varepsilon\| \leq C, \quad \text{independent of } \varepsilon,\lambda .$$

Now,

$$L_0^\varepsilon(f) = E^{xt}\int_t^T f(\xi_\varepsilon(s),s)e^{-\lambda_1(s-t)}e^{\lambda_1(s-t)}ds$$

$$\leq e^{\lambda_1 T} L_{\lambda_1}^\varepsilon(f) \leq C\, e^{\lambda_1 T}\|f\| \quad ;$$

and so we obtain that the constant in (7.31) can be made independent of ε.□

Proof of Theorem 7.1: From (7.30) and Lemma 7.9 it easily follows that

$$(7.32) \qquad E^{xt}\int_t^T f(x(s),s)ds| \leq C_{T,p}|f|_{L^p} \qquad \forall f \in L^p,\ p > n+1 \ .$$

This property allows us to apply Ito's formula (5.62), at least in the integrated form, to the process x(s), satisfying (7.23) and with functions belong to $\mathcal{W}^{2,1,p}(R^n \times]0,T[)$, $p > n + 1$.

In particular, let $\varphi \in \mathcal{D}(R^n)$, and let us consider

$$(7.33) \qquad \left|\begin{array}{l} -\dfrac{\partial\psi}{\partial t} + A(t)\psi - B(t)\psi = 0 \\ \psi(x,T) = \varphi, \quad \psi \in \mathcal{W}^{2,1,p} \qquad \forall p \geq 2,\ p < \infty \end{array}\right.$$

which by Theorem 3.2 has one and only one solution. From this we deduce the formula

$$(7.34) \qquad \psi(x,t) = E^{xt}\,\varphi(x(T)) = \int_{R^n}\varphi(y)P(x,t;T,dy)$$

if $P(x,t;T,dy)$ denotes the probability measure on R^n, the image of P^{xt}, for the variable $x(T)$. The formula (7.34) shows that $P(x,t;T,dy)$ is uniquely defined, when P^{xt} varies, in the set of solutions of martingale problems. Now, from (7.34) and equation (7.33) we can easily show that $P(x,t;T,dy)$ is a Markov transition probability. Thus, to it there corresponds a single probability on the canonical space.□

Remark 7.2

The Markov process defined by P^{xt} is Fellerian, as is easily shown from equation (7.33).□

7.4 *Existence of a Solution of the Martingale Problem*

Our objective is to prove:

THEOREM 7.2: *Under the assumptions of Theorem 7.1 there exists a solution of the martingale problem.*

We shall begin by constructing a solution of the martingale problem in the case where m(dz) is replaced by $m_\varepsilon(dz)$. To do this we shall use the method of stochastic differential equations and a Girsanov type of method. Next we shall show that the family P_ε^{xt} so contructed is relatively compact, and we shall establish a priori estimates that allow us to pass to the limit.

7.4.1 Approximate solution

Let there be given arbitrary $\Omega, \mathcal{Q}, P$, an $\mathfrak{F}^0$-complete right continuous filtration $\mathfrak{F}^t$, a normalised R^n-valued $\mathfrak{F}^t$-Wiener process, an integral random *Poisson* measure $\nu_\varepsilon(dt,dz)$, with Levy measure m (dz). We denote by $\mu_\varepsilon(dt,dz)$ the associated martingale measure. We begin by solving the equation

$$(7.35)\quad \left|\begin{array}{l} \xi_\varepsilon(s) = x + \int_t^s \sigma\left(\xi_\varepsilon(\lambda),\lambda\right)dw(\lambda) + \int_t^s\int_{R^n} z\,\chi_{|z|\leqslant 1}\,\mu_\varepsilon(d\lambda,dz) + \\ \quad + \int_t^s\int_{R^n} z\,\chi_{|z|>1}\,\nu_\varepsilon(d\lambda,dz) \\ \xi_\varepsilon(s) \text{ is adapted, cadlag, and } E\sup_{t\leqslant s\leqslant T}|\xi_\varepsilon(s)| \quad C(1+|x|) \end{array}\right.$$

which has one and only one solution.

We now consider a process $\alpha(s)$ (which will be specified later) such that

$$(7.36)\quad \alpha(s) \text{ is adapted and } R^n\text{-valued, } |\alpha(s)| \leqslant C.$$

Next we consider the process $\eta_\varepsilon(s)$ defined by

$$(7.37)\quad \eta_\varepsilon(s) = 1 + \int_t^s \eta_\varepsilon(\lambda)\alpha(\lambda).dw(\lambda) + \\ + \int_t^s\int_{R^n} \eta_\varepsilon(\lambda)\left(c_0\left(\xi_\varepsilon(\lambda),\lambda,z\right)-1\right)\mu_\varepsilon(d\lambda,dz)$$

which has a solution, by Theorem 5.4, given explicitly by

$$(7.38)\quad \eta_\varepsilon(s) = \exp\left\{\int_t^s \alpha(\lambda)\cdot dw(\lambda) + \int_t^s\int_{R^n}\left(c_0\left(\xi_\varepsilon(\lambda),\lambda,z\right)-1\right)\mu_\varepsilon(d\lambda,dz) - \\ - \frac{1}{2}\int_t^s|\alpha(\lambda)|^2 d\lambda - \int_t^s\int_{R^n}\left(c_0\left(\xi_\varepsilon(\lambda),\lambda,z\right) - 1 - \text{Log } c_0\right)\nu_\varepsilon(d\lambda,dz)\right\} .$$

Furthermore, we have

(7.39) $\eta_\varepsilon(s) \geqslant 0, \quad E\eta_\varepsilon(s) = 1 \quad$ for all s.

We can therefore define a probability on $\Omega,\mathcal{A}$, by setting

(7.40) $$\left.\frac{d\tilde{P}^\varepsilon}{dP}\right|_{\mathcal{F}^s} = \eta_\varepsilon(s).$$

Next we write P^{xt} for the probability measure on Ω_0,μ_t, which is the image of $\tilde{P}^\varepsilon$ for the R.V. $\xi_\varepsilon(\cdot)$. We recall that we wrote $B_\varepsilon(t)$ for the operator $B(t)$ where $M(x,t,dz)$ was replaced by

$$M_\varepsilon(x,t,dz) = c_0(x,t,z)m_\varepsilon(dz).$$

We then have:

LEMMA 7.10: *The measure* P_ε^{xt} *is the solution of the martingale problem with respect to the operator* $A(t) - B_\varepsilon(t)$.

Proof: Let $\varphi \in \mathcal{D}(R^n)$, and set

$$f(x,z) = \varphi(x)z.$$

Apply Ito's formula to the pair $\xi_\varepsilon(s),\eta_\varepsilon(s)$ and to the function $f(x,z)$. We must take some precautions with the process $\xi_\varepsilon(s)$, which does not have a stochastic differential in the standard form (cf. the footnote on p. 290). In fact we consider the process

$$\xi_\varepsilon^R(s) = x + \int_t^s \sigma\big(\xi_\varepsilon(\lambda),\lambda\big)dw(\lambda) + \int_t^s\int_{R^n} z\,\chi_{|z|\leqslant R}\,\mu_\varepsilon(ds,dz) + $$
$$+ \int_t^s\int_{R^n} z\,\chi_{|z|>1}\,d\lambda m_\varepsilon(dz)$$

and notice that

$$\xi_\varepsilon^R(s) - \xi_\varepsilon(s) = -\int_t^s\int_{R^n} z\,\chi_{|z|>R}\,\nu_\varepsilon(d\lambda,dz) - \int_t^s\int_{R^n} z\,\chi_{|z|>R}\,d\lambda m_\varepsilon(dz)$$

therefore

(7.41) $$E \sup_{t\leqslant s\leqslant T} |\xi^R(s) - \xi(s)| \to 0 \quad \text{when} \quad R \to +\infty.$$

We can, however, apply Ito's formula to the pair $\xi_\varepsilon^R,\eta_\varepsilon$. We then make R tend to $+\infty$. Thanks to (7.41) we obtain

(7.42) $$\varphi\big(\xi_\varepsilon(s)\big)\eta_\varepsilon(s) = \varphi(x) + \int_t^s \operatorname{tr} D^2\varphi\big(\xi_\varepsilon(\lambda)\big)a\big(\xi_\varepsilon(\lambda),\lambda\big)\eta_\varepsilon(\lambda)d\lambda + \quad \text{(Contd)}$$

(Contd)
$$+ \int_t^s \eta_\varepsilon(\lambda) D\varphi\big(\xi_\varepsilon(\lambda)\big)\cdot\sigma\big(\xi_\varepsilon(\lambda)\big)\alpha(\lambda)d\lambda + \int_t^s \eta_\varepsilon(\lambda) D\varphi\big(\xi_\varepsilon(\lambda)\big)\cdot\sigma\big(\xi_\varepsilon(\lambda)\big)dw(\lambda)$$
$$+ \int_t^s \eta_\varepsilon(\lambda)\varphi\big(\xi_\varepsilon(\lambda)\big)\alpha(\lambda)\cdot dw(\lambda) - \int_t^s\int_{R^n} \eta_\varepsilon(\lambda) D\varphi\big(\xi_\varepsilon(\lambda)\big)\cdot z\, \chi_{|z|\leq 1}\, d\lambda m_\varepsilon(dz) +$$
$$+ \int_t^s\int_{R^n}\Big[\varphi\big(\xi_\varepsilon(\lambda)+z\big) - \varphi\big(\xi_\varepsilon(\lambda)\big)\Big]\eta_\varepsilon(\lambda)c_0\big(\xi_\varepsilon(\lambda),\lambda,z\big)d\lambda m_\varepsilon(dz) +$$
$$+ \int_t^s\int_{R^n}\Big[\varphi\big(\xi_\varepsilon(\lambda)+z\big)c_0 - \varphi\big(\xi_\varepsilon(\lambda)\big)\Big]\eta_\varepsilon(\lambda)\mu_\varepsilon(d\lambda,dz).$$

We now precisely specify the choice of $\alpha(s)$. We take

$$\alpha(\lambda) = \sigma^{-1}\big(\xi_\varepsilon(\lambda),\lambda\big)\Big[g\big(\xi_\varepsilon(\lambda),\lambda\big) + \int_{R^n}\big(1 - c_0\big(\xi_\varepsilon(\lambda),\lambda,z\big)\big)z\,\chi_{|z|\leq 1}\, m_\varepsilon(dz)\Big]. \tag{7.43}$$

We can then rewrite (7.42) in the form

$$\varphi\big(\xi_\varepsilon(s)\big)\eta_\varepsilon(s) - \varphi(x) + \int_t^s \eta_\varepsilon(\lambda)\Big[\big(A(\lambda)-B_\varepsilon(\lambda)\big)\varphi\big(\xi_\varepsilon(\lambda)\big)\Big]d\lambda = \tag{7.44}$$
$$= \int_t^s \eta_\varepsilon(\lambda)\Big(D\varphi\big(\xi_\varepsilon(\lambda)\big)\cdot\sigma\big(\xi_\varepsilon(\lambda)\big)dw(\lambda) + \varphi\alpha(\lambda)\cdot dw(\lambda)\Big) +$$
$$+ \int_t^s \eta_\varepsilon(\lambda)\int_{R^n}\Big[\varphi\big(\xi_\varepsilon(\lambda)+z\big)c_0 - \varphi\big(\xi_\varepsilon(\lambda)\big)\Big]\mu_\varepsilon(d\lambda,dz).$$

With the definition (7.40) of $\tilde{P}^\varepsilon$, from (7.44) we deduce that for H a bounded $\mathcal{F}^{s_1}$-measurable R.V. we have

$$\tilde{E}^\varepsilon\Big\{\Big[\varphi\big(\xi_\varepsilon(s_2)\big) + \int_{s_1}^{s_2}\big(A(\lambda)-B_\varepsilon(\lambda)\big)\varphi\big(\xi_\varepsilon(\lambda)\big)d\lambda\Big]H\Big\} = \tilde{E}^\varepsilon\, \varphi\big(\xi_\varepsilon(s_1)\big)H. \tag{7.45}$$

which can be done by passing to the image measure P^{xt} (with H $\mu_t^{s_1}$-measurable). The desired result follows from this.□

Remark 7.2

The technique for constructing P_ε^{xt} uses a probability measure transformation of the Girsanov type (cf. also B.L. [1]).

7.4.2 Compactness

We are going to study how the family P_ε^{xt} varies as $\varepsilon \to 0$. We shall show in this Subsection that it is relatively compact. We begin with:

LEMMA 7.11: *We have the estimate*

$$\tilde{E}^\varepsilon\Big(\sup_{t\leq s\leq t+h}|\xi_\varepsilon(s)-x| > \eta\Big) \leq \tag{7.46}$$
$$2n\exp\Big\{-\frac{\lambda}{n}(\eta-k_0h-A) + \frac{\lambda^2}{2}k_1h\big(1+e^{|\lambda|}\big)\Big\} + \frac{k_1h}{A}$$

where k_0 *is a constant such that* $|g| \leqslant k_0$, k_1 *is a constant depending only on the bounds of the* a_{ij} *and the integrals* (7.27), λ *is an arbitrary real number, and* A *is an arbitrary positive number.*

Proof: Let us set

$$M(s) = M_\varepsilon(s) = \xi_\varepsilon(s) - x - \int_t^s g\big(\xi_\varepsilon(u),u\big)\,du - \int_t^s\int_{R^n} z\,\chi_{|z|>1}\,\nu_\varepsilon(du,dz). \tag{7.47}$$

Let θ be real. We show that $\theta\cdot M(s)$ is a locally square integrable $\tilde{P}^\varepsilon, \mathfrak{F}^s$-martingale whose associated increasing process is

$$\langle \theta M, \theta M\rangle(s) = \int_t^s 2\theta\cdot a\big(\xi_\varepsilon(u),u\big)\theta\,du + \int_t^s du\int_{R^n}(\theta.z)^2\,\chi_{|z|\leqslant 1}\;c_0\big(\xi_\varepsilon(u),u,z\big)m_\varepsilon(dz)\;. \tag{7.48}$$

For this we first remark that

$$\theta\cdot M(s) = \int_t^s \theta\cdot\sigma\big(\xi_\varepsilon(u),u\big)\,dw(u) - \int_t^s \theta\cdot g\big(\xi_\varepsilon(u),u\big)\,du + \int_t^s\int_{R^n}\theta\cdot z\;\chi_{|z|\leqslant 1}\;\mu_\varepsilon(du,dz). \tag{7.49}$$

Next we apply Ito's formula to the function $f(x,z) = xz$ on R^n and to the pair of processes $\theta\cdot M(s), \eta_\varepsilon(s)$, then to the function $f(x,z) = x^2z$, and to the pair $\theta\cdot M(s)$, $\eta_\varepsilon(s)$. By simple reductions, using the value of $\alpha(s)$ we obtain

$$\theta\cdot M(s)\eta_\varepsilon(s) = \int_t^s \theta\cdot M(u)\eta_\varepsilon(u)\alpha(u)\cdot dw(u) + \int_t^s\int_{R^n} \eta_\varepsilon(u)\Big[\theta\cdot M(u)\big(c_0\big(\xi_\varepsilon(u),u,z\big) - 1\big) + \theta.z\;\;\chi_{|z|\leqslant 1}\;c_0\Big]\mu_\varepsilon(du,dz)\;. \tag{7.50}$$

From this we easily deduce that if H is $\mathfrak{F}^{s_1}$-measurable and bounded, then

$$\tilde{E}^\varepsilon[\theta\cdot(M(s_2) - M(s_1))H] = 0.$$

Hence $\theta\cdot M(s)$ is a $\tilde{P}^\varepsilon, \mathfrak{F}^s$-martingale. Then we have

$$\big(\theta.M(s)\big)^2\eta_\varepsilon(s) = \int_t^s 2\eta_\varepsilon(u)\theta\cdot a\big(\xi_\varepsilon(u),u\big)\theta\;du + \int_t^s\int_{R^n}(\theta-z)^2\;\chi_{|z|\leqslant 1}\;\eta_\varepsilon(u)c_0\big(\xi_\varepsilon(u),u,z\big)\,du\,dm_\varepsilon(dz) + \text{stochastic integrals} \tag{7.51}$$

thus $\theta\cdot M(s)$ is locally square integrable, and (7.48) holds.

We next write

$$X(s) = \lambda\, \theta\cdot M(s) - \lambda^2\int_t^s \theta\cdot\theta\, du - \tag{7.52}$$
$$- \int_t^s\int_{R^n}\left(\exp \lambda\cdot\theta\, \chi_{|z|\leqslant 1} - 1 - \lambda\theta\cdot z\, \chi_{|z|\leqslant 1}\right) c_0\left(\xi_\varepsilon(u),u,z\right) du m_\varepsilon(dz)$$

and apply Ito's formula to the pair $X(s),\eta_\varepsilon(s)$ and to the function $f(x,z) = z \exp x$. The formula may be applied because

$$\int_t^T\int_{R^n}\left|\exp\left(X(u)+\lambda\theta\cdot z\, \chi_{|z|\leqslant 1}\right)\eta_\varepsilon(u)c_0(u) - \exp X(u)\eta_\varepsilon(u)\right|^2 du dm_\varepsilon(z) < \infty \quad \text{a.s.}$$

and

$$\int_t^T\int_{R^n}\Big[\exp\left(X(u)+\lambda\theta\cdot z\, \chi_{|z|\leqslant 1}\right)\eta_\varepsilon(u)c_0(u) - \exp X(u)\eta_\varepsilon(u) -$$
$$- \exp X(u)\eta_\varepsilon(u)\lambda\theta\cdot z\, \chi_{|z|\leqslant 1} - \exp X(u)\eta_\varepsilon(u)\left(c_0(u) - 1\right)\Big] du\, m_\varepsilon(dz) < \infty \quad \text{a.s.}$$

By simple reductions we then verify that

$$\exp X(s)\eta_\varepsilon(s) = 1 + \int_t^s \eta_\varepsilon(u)\exp X(u)\left[\lambda\theta - \sigma\left(\xi_\varepsilon(u),u\right) + \alpha(u)\right] dw(u) + \tag{7.53}$$
$$+ \int_t^s\int_{R^n} \exp X(u)\eta_\varepsilon(u)\left[\exp\left(\theta\cdot z\, \chi_{|z|\leqslant 1}\right) c_0\left(\xi_\varepsilon(u),u,z\right) - 1\right]\mu_\varepsilon(du,dz)$$

and therefore

(7.54) $\exp X(s)$ is a $\tilde{P}^\varepsilon,\mathcal{F}^s$-martingale and $\tilde{E}^\varepsilon \exp X(s) = 1$.

Now, by (7.52), and if $t \leqslant s \leqslant t+h$ we have

$$\lambda\theta\cdot M(s) \leqslant X(s) + \frac{\lambda^2}{2} k_1|\theta|^2 h + \frac{1}{2}k_1\lambda^2\, |\theta|^2 \left(\exp|\lambda|\,|\theta|\right)h \tag{7.55}$$

where k_1 is a constant such that

$$k_1 \geqslant 2\|a\|, \quad k_1 \geqslant \int_{R^n}|z|^2\, \chi_{|z|\leqslant 1}\, m(dz).$$

Now let η be a constant. Then

$$\tilde{P}^\varepsilon\left[\sup_{t\leqslant s\leqslant t+h} \theta\cdot M(s) \geqslant \eta\right] = \tilde{P}^\varepsilon\left[\sup_{t\leqslant s\leqslant t+h} \lambda\theta\cdot M(s) \geqslant \lambda\eta\right] \leqslant$$
$$\leqslant \tilde{P}^\varepsilon\left[\sup_{t\leqslant s\leqslant t+h} X(s) \geqslant \lambda\eta - \lambda^2 k_1 h|\theta|^2 - \frac{1}{2}k_1\lambda^2 h\, |\theta|^2 \exp|\lambda|\,|\theta|\right]$$

(Contd)

(Contd) $$= \tilde{P}^{\varepsilon}\left[\sup_{t\leqslant s\leqslant t+h} \exp X(s) \geqslant \exp\left[\lambda\eta - \lambda^2 k_1 h|\theta|^2 - \frac{1}{2}k_1\lambda^2 h|\theta|^2 \exp|\lambda||\theta|\right]\right]$$

and, since $\exp X(s)$ is a $\tilde{P}^{\varepsilon}$-martingale, it follows that

$$\tilde{P}^{\varepsilon}\left[\sup_{t\leqslant s\leqslant t+h} \theta\cdot M(s) \geqslant \eta\right] \leqslant \exp\left\{-\lambda\eta + \frac{\lambda^2}{2} k_1 h|\theta|^2\left(1+\exp|\lambda\theta|\right)\right\} \times \tilde{E}^{\varepsilon}\exp X(t+h$$

$$= \exp\left\{-\lambda\eta + \frac{\lambda^2}{2} k_1 h|\theta|^2\left(1+\exp|\lambda||\theta|\right)\right\} .$$

Changing θ into $-\theta$, we deduce that

$$\tilde{P}^{\varepsilon}\left[\sup_{t\leqslant s\leqslant t+h} |\theta\cdot M(s)| \geqslant \eta\right] \leqslant 2\exp\left\{-\lambda\eta + \frac{\lambda^2}{2} k_1 h|\theta|^2\left(1+\exp|\lambda||\theta|\right)\right\}.$$

Next we take θ in vector coordinates. We easily deduce the estimate

$$\tilde{P}\left[\sup_{t\leqslant s\leqslant t+h} |M(s)| \geqslant \eta\right] \leqslant 2n \exp\left\{-\frac{\lambda\eta}{n} + \frac{\lambda^2}{2} k_1 h\left(1+\exp|\lambda|\right)\right\}. \tag{7.56}$$

With the estimate (7.56) and the relation (7.47) we are going to be able to obtain the desired estimate (7.46). However, before that we shall have to obtain an estimate for the last integral on the right hand side of (7.47). In fact, we have

$$\tilde{E}^{\varepsilon}\int_t^s\int_{R^n}|z|\, \chi_{|z|>1}\, \nu_{\varepsilon}(du,dz) = \tilde{E}^{\varepsilon}\int_t^s\int_{R^n}|z|\, \chi_{|z|>1}\, c_0\left(\xi_{\varepsilon}(u),u,z\right)du\, dm_{\varepsilon}(z), \tag{7.57}$$

This expression is obtained by applying Ito's formula to the pair $K_R(s),\eta_{\varepsilon}(s)$ where

$$K_R(s) = \int_t^s\int_{R^n}|z|\, \chi_{R\geqslant|z|>1}\, \nu_{\varepsilon}(du,dz).$$

From this we deduce that equation (7.57) is valid with $\chi_{|z|>1}$ replaced by $\chi_{R\geqslant|z|>1}$. Next, it suffices to make R tend to $+\infty$, whence (7.57). From (7.57) it follows, in particular, that

$$\tilde{E}^{\varepsilon}\int_t^{t+h}\int_{R^n}|z|\, \chi_{|z|>1}\, \nu_{\varepsilon}(du,dz) \leqslant k_1 h . \tag{7.58}$$

Now, by (7.47) we have

$$\sup_{t\leqslant s\leqslant t+h} |\xi_{\varepsilon}(s)-x| \leqslant \sup_{t\leqslant s\leqslant t+h} |M(s)| + k_0 h + \int_t^{t+h}\int_{R^n}|z|\, \chi_{|z|>1}\, \nu_{\varepsilon}(du,dz)$$

therefore

$$\tilde{P}^{\varepsilon}\left[\sup_{t\leqslant s\leqslant t+h}|\xi_2(s)-x|>\eta\right]\leqslant \tilde{P}^{\varepsilon}\left[\sup_{t\leqslant s\leqslant t+h}|M(s)|>\eta-k_0h-1\right] +$$
$$+\tilde{P}^{\varepsilon}\left[\int_t^{t+h}\int_{R^n}|z|\ \chi_{|z|>1}\ \nu_{\varepsilon}(du,dz)>A\right].$$

From (7.58) and (7.56) we easily deduce the estimate (7.46).□

Remark 7.3

Let τ be an $\mathfrak{F}^s$-stopping time, $s \geqslant t$ and $\tau \leqslant T$, T deterministic a.e.. A proof identical with that of Lemma 7.11 allows us to state that

(7.59) $$\tilde{P}^{\varepsilon}\left[\sup_{0\leqslant s\leqslant h}|\xi_{\varepsilon}(s+\tau)-\xi_{\varepsilon}(\tau)|>\eta\right]\leqslant$$
$$\leqslant 2n\exp\left\{-\frac{\lambda}{n}(\eta-k_0h-A)+\frac{\lambda^2}{2}k_1h\left(1+e^{|\lambda|}\right)\right\}+\frac{k_1h}{A}.$$

Remark 7.4

Let h,β,η be given, and take $A=\frac{2}{\beta}k_1h$, and considering β and η as fixed let us set

$$Z_h(\lambda) = \exp\left\{-\frac{\lambda}{n}(\eta-k_0h-\frac{2}{\beta}k_1h)+\frac{\lambda^2}{2}k_1h(1+e^{|\lambda|})\right\}.$$

For fixed h with $\eta - k_0h - \frac{2}{\beta}k_1h > 0$, the function $Z_h(\lambda)$ reaches its minimum in λ at the point $\lambda(h)$, the unique solution of the equation

$$\lambda(1+e^{\lambda})+\frac{\lambda^2e^{\lambda}}{2}=\frac{1}{n}\ \frac{\left(\eta-k_0h-\frac{2}{\beta}k_1h\right)}{k_1h}.$$

As $h\to 0$, it is clear that $\lambda(h)\to+\infty$ and $\lambda(h)^2e^{\lambda(h)}h$ stays bounded, therefore $Z_h(\lambda(h))\to 0$. Consequently we can find $h(\beta,\eta)$ small enough so that

$$Z_{h(\beta,\eta)}(\lambda(h))\leqslant\frac{\beta}{4n}.$$

We then deduce from (7.59) that for all fixed η,β there exists h_{η}^{β} such that

(7.60) $$h\leqslant h_{\eta}^{\beta}\ \text{ implies }\ \tilde{P}^{\varepsilon}[\sup_{0\leqslant s\leqslant h}|\xi_{\varepsilon}(s+\tau)-\xi_{\varepsilon}(\tau)|>\eta]\leqslant\beta.$$ □

Beginning from (7.60) we are going to prove that the family P_{ε}^{xt} is relatively compact.

LEMMA 7.12: *The family* P_{ε}^{xt} *on* Ω_0,μ_t *is relatively compact.*

Proof: We use a compactness criterion given in BILLINGSLEY [1], (cf. also LEPELTIER - MARCHAL [1]). This criterion is the following: For fixed $\beta,\eta>0$ and for every integer p, there exists

(7.61) (i) m_p^β such that $\tilde{P}^\varepsilon\{\sup_{t\leq s\leq p}|\xi_\varepsilon(s)|>m_p^\beta\}\leq\beta$

(ii) δ_η^β such that $\tilde{P}^\varepsilon\{\sup_{t\leq s_1,s_2\leq t+\delta_\eta^\beta}|\xi_\varepsilon(s_2)-\xi_\varepsilon(s_1)|>\eta\}\leq\beta$

(iii) $\delta_{\eta p}^\beta$ such that $\tilde{P}^\varepsilon\{\sup_{\sigma_{\eta p}^\beta\leq s_1<s_2<p}|\xi_\varepsilon(s_2)-\xi_\varepsilon(s_1)|>\eta\}\leq\beta$

(iv) $\delta_{\eta p}^\beta$ such that $\tilde{P}^\varepsilon\{\sup_{\substack{t\leq s_1<s_2\geq p\\ s_2-s_1\leq h_{\eta p}^\beta}}\sup_{s_1\leq s\leq s_2}\min(|\xi_\varepsilon(s)-\xi_\varepsilon(s_1)|,$

$|\xi_\varepsilon(s)-\xi_\varepsilon(s_2)|>\eta\}\leq\beta$.

The property (i) does not follows from (7.60). But re-using the end of the proof of Lemma 7.11, we have

$$\sup_{t\leq s\leq t+h}|\xi_\varepsilon(s)|\leq\sup_{t\leq s\leq t+h}|M(s)|+|x|+k_0h+\int_t^{t+h}\int_{R^n}|z|\;\chi_{|z|>1}\;\nu_\varepsilon(du,dz)$$

therefore

$$\tilde{P}^\varepsilon\{\sup_{t\leq s\leq t+h}|\xi_\varepsilon(s)|>\eta\}\leq 2n\;\exp\{-\frac{\lambda}{n}(\eta-k_0h-|x|-A)+$$
$$+\frac{\lambda^2}{2}k_1h(1+e^{|\lambda|})\}+\frac{k_1h}{A}\;.$$

Let us choose $h=p-t$, $A=\dfrac{2k_1h}{\beta}$ and $\lambda>0$ fixed, arbitrarily; clearly there exists $\eta=m_p^\beta$ large enough so that

$$2n\;\exp\{-\frac{\lambda}{n}(\eta-k_0h-|x|-A)+\frac{\lambda^2}{2}k_1h(1+e^{|\lambda|})\}<\frac{\beta}{2}$$

which proves (i). Let us prove (ii). Now,

$$\sup_{t\leq s_1,s_2<t+\delta}|\xi_\varepsilon(s_2)-\xi_\varepsilon(s_1)|\leq 2\sup_{t\leq s<t+\delta}|\xi_\varepsilon(s)-x|$$

and, by (7.60), we can find δ_η^β such that

$$\tilde{P}^\varepsilon\{\sup_{t\leq s<t+\delta}|\xi_\varepsilon(s)-x|>\frac{\eta}{2}\}\leq\beta$$

whence (ii). Let us now prove (iii). Now, we have

$$\sup_{\sigma\leq s_1<s_2<p}|\xi_\varepsilon(s_2)-\xi_\varepsilon(s_1)|\leq 2\sup_{\sigma\leq s<p}|\xi_\varepsilon(s)-\xi_\varepsilon(\sigma)|.$$

By taking $\sigma=p-h_{\eta/2}^\beta$ and applying (7.60) we obtain (iii). It remains to verify property (iv). Define a sequence of stopping times

$$\tau_0 = t,\ \tau_n = \inf\{s > \tau_{n-1} \mid |\xi_\varepsilon(s)-\xi_\varepsilon(\tau_{n-1})| > \tfrac{\eta}{2}\} \quad (\tau_n = +\infty \text{ if }$$
$$|\xi_\varepsilon(s)-\xi_\varepsilon(\tau_{n-1})| \leqslant \tfrac{\eta}{2}\ \forall\ s > \tau_{n-1}).$$

Let us fix p, then if

$$\sup_{\substack{t\leqslant s_1<s_2\leqslant p\\ s_2-s_1\leqslant h}}\ \sup_{s_1\leqslant s\leqslant s_2} \min\big(|\xi_\varepsilon(s)-\xi_\varepsilon(s_1)|,\ |\xi_\varepsilon(s)-\xi_\varepsilon(s_2)|\big) > \eta.$$

there then exist s_1, s, s_2 with $s_1 \leqslant s \leqslant s_2 \leqslant p$ such that

$$|\xi_\varepsilon(s)-\xi_\varepsilon(s_1)| > \eta,\ |\xi_\varepsilon(s)-\xi_\varepsilon(s_2)| > \eta.$$

Let us assume that $\tau_{n-1} \leqslant s_1 < \tau_n$, then necessarily $s \geqslant \tau_n$ and $s_2 \geqslant \tau_{n+1}$. In fact,

$$\eta < |\xi_\varepsilon(s)-\xi_\varepsilon(s_1)| \leqslant |\xi_\varepsilon(s)-\xi_\varepsilon(\tau_{n-1})| + |\xi_\varepsilon(\tau_{n-1})-\xi_\varepsilon(s_1)|$$
$$\leqslant \tfrac{\eta}{2} + |\xi_\varepsilon(s)-\xi_\varepsilon(\tau_{n-1})|, \text{ then } s \geqslant \tau_n.$$

If $s \geqslant \tau_{n+1}$, then $s_2 \geqslant \tau_{n+1}$, otherwise $\tau_n \leqslant s < \tau_{n+1}$, therefore

$$\eta < |\xi_\varepsilon(s)-\xi_\varepsilon(s_2)| \leqslant |\xi_\varepsilon(s)-\xi_\varepsilon(\tau_n)| + |\xi_\varepsilon(\tau_n)-\xi_\varepsilon(s_2)|$$
$$\leqslant \tfrac{\eta}{2} + |\xi_\varepsilon(\tau_n)-\xi_\varepsilon(s_2)|$$

and thus

$$|\xi_\varepsilon(s_2)-\xi_\varepsilon(\tau_n)| > \tfrac{\eta}{2},$$

hence $s_2 \geqslant \tau_{n+1}$. Consequently $\tau_{n-1} \leqslant s_1 < \tau_n$ implies $\tau_{n+1} \leqslant s_2$, therefore

$$\tau_{n+1} - \tau_n \leqslant s_2 - s_1 \leqslant h.$$

Therefore, ω a.s., there will also exist an integer n' such that $\tau_{n'+1} - \tau_{n'} \leqslant h$.

Let us set

$$\Lambda_p^\eta = \inf_n\{\tau_n - \tau_{n-1} | \tau_{n-1} < p\}.$$

By the preceding, we can state that the probability of the left hand side of (iv) is less than or equal to $\tilde{P}^\varepsilon[\Lambda_p^\eta < h]$; thus to prove (iv) it suffices us to show that

$$\tilde{P}^\varepsilon[\Lambda_p^\eta < h_{\eta p}^\beta] < \beta.$$

But

$$\{\inf_n(\tau_n-\tau_{n-1} \mid \tau_{n-1}<p)<h\} = \bigcup_{n\geq 0}\{\tau_1-\tau_0\geq h,\ \tau_2-\tau_1\geq h,\dots,\tau_n-\tau_{n-1}\geq h,$$
$$\tau_{n+1}-\tau_n<h,\ \tau_n<p\}$$
$$\subset \bigcup_{j=1,\dots,K}\{\tau_1-\tau_0\geq h,\ \tau_2-\tau_1\geq h,\dots,$$
$$\tau_{j-1}-\tau_{j-2}\geq h,\ \tau_j-\tau_{j-1}<h,\ \tau_{j-1}<p\}\cup\{\tau_K<p\}$$

for every integer $K \geqslant 1$. Therefore

$$\tilde{P}^\varepsilon[\Lambda_p^\eta<h] \leqslant \sum_{j=1}^K \tilde{P}^\varepsilon[\tau_j-\tau_{j-1}<h,\ \tau_{j-1}<p]+\tilde{P}^\varepsilon[\tau_K<p]$$
$$\leqslant K\tilde{P}^\varepsilon\Big[\sup_{0\leqslant s\leqslant h}|\xi_\varepsilon(s+\tau)-\xi_\varepsilon(\tau)|>\frac{\eta}{2}\Big]$$
$$+\tilde{P}^\varepsilon\Big[\sup_{t\leqslant s\leqslant p}|\xi_\varepsilon(s)-\xi_\varepsilon(t)|>K\frac{\eta}{2}\Big].$$

Now,

$$\tilde{P}^\varepsilon\Big[\sup_{t\leqslant s\leqslant p}|\xi_\varepsilon(s)-\xi_\varepsilon(t)|>K\frac{\eta}{2}\Big]\leqslant 2n\ \exp\{-\frac{\lambda}{n}(K\frac{\eta}{2}-k_0(p-t)-A)\ +$$
$$+\frac{\lambda^2}{2}k_1(p-t)(1+e^{|\lambda|})\}+k_1\frac{(p-t)}{A}.$$

Choosing $A = K\frac{\eta}{4}$, $\lambda > 0$ fixed, we deduce that

$$\tilde{P}^\varepsilon\Big[\sup_{t\leqslant s\leqslant p}|\xi_\varepsilon(s)-\xi_\varepsilon(t)|>K\frac{\eta}{2}\Big]\longrightarrow 0, \quad \text{when } K\to\infty,$$

for p,η fixed, uniformly with respect to ε. This and the property (7.60) easily imply (iv).□

7.4.3 A priori estimates

LEMMA 7.13: *Let* $f \in L^p(R^n\times]0,T[)$, $p > \frac{n}{2}+1$. *Then*

$$(7.62)\qquad |E_\varepsilon^{xt}\int_t^T f(x(s),s)\,ds| \leqslant C_{p,T}|f|_{L^p}.$$

Proof: In fact, we can consider that (7.62) is a consequence of P_ε^{xt} being a solution of the martingale problem with respect to $A - B_\varepsilon$ (Lemma (7.10), and of the properties proved in Section 7.2 (cf. (7.32)). First let us take $f \in \mathcal{D}(R^n\times]0,T[)$ $\geqslant 0$ and consider the problem

$$(7.63)\qquad -\frac{\partial\psi_\varepsilon}{\partial t}+A(t)\psi_\varepsilon - B_\varepsilon(t)\psi_\varepsilon = f,\ \psi_\varepsilon(T) = 0,\ \psi_\varepsilon\in \mathbb{W}^{2,1,p}.$$

As was done in the proof of Theorem 7.1 (cf. (7.33)) we show that

$$(7.64)\qquad \psi_\varepsilon(x,t) = E_\varepsilon^{xt}\int_t^T f(x(s),s)\,ds.$$

But then, for $p > \frac{n}{2} + 1$, we have

$$|\psi_\varepsilon(x,t)| \leq C \|\psi_\varepsilon\|_{W^{2,1,p}} \leq C|f|_{L^p}$$

where the constants are independent of ε, hence we have (7.62).

If, however, we wish to avoid using the a priori properties of solutions of the martingale problem with respect to $A - B_\varepsilon$, we can proceed directly for ending up with (7.64). It is a matter of showing that

$$\psi_\varepsilon(x,t) = \tilde{E}^\varepsilon \int_t^T f(\xi_\varepsilon(s),s)ds.$$

To do this we have to be able to apply Ito's formula to the function $\psi_\varepsilon(x,t)z$ and the pair $\xi_\varepsilon(s),\eta_\varepsilon(s)$. Since ψ_ε is not C^2, we must proceed by approximation. Therefore it suffices to have a priori an estimate of the type (7.62), with a constant possibly depending on ε, and a number p that is finite, but arbitrary. First we note that

$$\tilde{E}^\varepsilon \int_t^T f(\xi_\varepsilon(s),s)ds = E\,\eta_\varepsilon(T) \int_t^T f(\xi_\varepsilon(s),s)ds$$

$$\leq (E\,\eta_\varepsilon(T)^2)^{\frac{1}{2}} \sqrt{T}(E \int_t^T f^2(\xi_\varepsilon(s),s)ds)^{\frac{1}{2}}.$$

Since, by Ito's formula,

$$E\,\eta_\varepsilon(s)^2 = 1 + E\int_t^s |\eta_\varepsilon(\lambda)|^2\ |\alpha(\lambda)|^2 d\lambda + E\int_t^s\int_{R^n} |\eta_\varepsilon(\lambda)|^2 (c_0-1)^2 d\lambda dm_\varepsilon(z)$$

and (cf. (7.43))

$$|\alpha(\lambda)| \leq C\left[1 + \int_{R^n} |z| m_\varepsilon(dz)\right]$$

we easily show, using Gronwall, that $E\eta_\varepsilon(T)^2 \leq C_{\varepsilon,T}$.

Thus we are led to the study of $E\int_t^T f^2(\xi_\varepsilon(s),s)ds$. The proof is very similar to that of Lemma 7.8. We define the sequence of stopping times

$$\tau_0 = t, \ldots, \tau_{j+1} = \inf\{s,\ \tau_j \leq s \leq T,\ \nu_\varepsilon(s,R^n) > \nu_\varepsilon(\tau_j,R^n)\}$$

and we use the fact that between two instants τ_j,τ_{j+1}, $\xi_\varepsilon(s)$ is a diffusion, and thus Ito's formula is applicable with functions belonging to $W^{2,1,p}$ for p large enough.

7.4.4 Proof of the existence

We are now in a position to prove Theorem 7.2. We are going to show that every limit point of P_ε^{xt} is a solution. Therefore let P^{xt} be such a point. Let $\varphi \in \mathcal{D}(R^n)$, and let H be a $\mu_t^{s_1}$-measurable bounded continuous(1) R.V.. For $t \leqslant s_1 \leqslant s_2$ we have

$$E_\varepsilon^{xt}\{[\varphi(x(s_2)) - \varphi(x(s_1)) + \int_{s_1}^{s_2}(A(\lambda)-B_\varepsilon(\lambda))\varphi(x(\lambda))d\lambda]H\} = 0 .$$

Therefore

$$(7.65)\qquad E_\varepsilon^{xt}\{[\varphi(x(s_2)) - \varphi(x(s_1)) + \int_{s_1}^{s_2}(A(\lambda)-B(\lambda))\varphi(x(\lambda))d\lambda]H\} =$$
$$= E_\varepsilon^{xt}\{\Big[\int_{s_1}^{s_2}(B_\varepsilon(\lambda)-B(\lambda))\varphi(x(\lambda))d\lambda\Big]H\}.$$

It is easy to see that since $\varphi \in \mathcal{D}(R^n)$ we have

$$\left|\int_{s_1}^{s_2}(B_\varepsilon(\lambda)-B(\lambda))\varphi(x(\lambda))d\lambda\right| \leqslant C(\varphi)\int_{|z|\leqslant\varepsilon}|z|^2 m(dz)$$

where the constant $C(\varphi)$ is not random. Hence the right hand side of (7.65) tends to 0. To pass to the limit in the left hand side we meet a technical difficulty, namely that the function $(A(s) - B(s))\varphi(x)$ is not continuous on $R^n \times [0,T]$, although $\varphi \in \mathcal{D}(R^n)$. This is owed to certain coefficients not being continuous. But $(A(s) - B(s))\varphi(x) \in L^p(R^n \times]0,T[)$ for all $p \geqslant 2$, $p < \infty$. We can find a sequence of bounded continuous functions ψ_k such that

$$\psi_k \to (A(s) - B(s))\varphi \quad \text{in} \quad L^p.$$

By Lemma 7.13 it follows that

$$E_\varepsilon^{xt}\int_t^T |(A(s)-B(s))\varphi(x(s)) - \psi_k(x(s),s)|ds \longrightarrow 0,$$

uniformly in ε as $k \to \infty$.

Furthermore, outside a countable set of instants $s \geqslant t$ the function $x(\cdot) \to x(s)$ is continuous on Ω_0 a.s. P^{xt} (property of the topology of Ω_0, cf. LEPELTIER - MARCHAL [1]). Hence for fixed s_1, s_2 outside a countable set, the functional

$$x(\cdot) \longrightarrow [\varphi(x(s_2)) - \varphi(x(s_1)) + \int_{s_1}^{s_2}\psi_k(x(s),s)ds]H$$

is a.s. P^{xt}, continuous, and bounded, therefore for fixed k

$$E_\varepsilon^{xt}\{[\varphi(x(s_2)) - \varphi(x(s_1)) + \int_{s_1}^{s_2}\psi_k(x(s),s)ds]H\} \longrightarrow \qquad \text{(Contd)}$$

(1) $H(\omega)$ is continuous from $\Omega_0 \to R$.

(Contd) $\to E^{xt}\{[\varphi(x(s_2)) - \varphi(x(s_1)) + \int_{s_1}^{s_2} \psi_k(x(s),s)ds]H\}.$

From the preceding considerations we deduce that the left hand side of (7.65) converges to

(7.66) $$E^{xt}\{[\varphi(x(s_2)) - \varphi(x(s_1)) + \int_{s_1}^{s_2} (A(s)-B(s))\varphi(x(s))ds]H\} = 0$$

for all $t \leqslant s_1 \leqslant s_2$ outside a countable set of instants.

Since x(s) is right continuous, it is clear that (7.66) holds for all s_1, s_2. Further, from

$$P^{xt}[x(t) = x] = 1$$

we deduce

$$P^{xt}[x(t) = x] = 1.$$

which completes the proof of the result desired.□

7.5 *The Case of Bounded Measures*

Here we consider the martingale problem without the density assumption (7.26). However, we shall assume that the *measures are bounded*. Moreover, because of the control problems that we shall consider in the remainder it is useful to assume that the derivative terms g(x,t), as well as the measure M(x,t;A) may be random. Thus we assume (7.1),(7.24),(7.25) and

(7.67) $g(x,s) = g(x,s,\omega)$ is measurable, bounded, and adapted to μ_t^s, for fixed x,

(7.68)
$M(x,s,\omega;A)$ is measurable for all fixed A, is μ_t^s-measurable for fixed x,s,A

$M(x,s,\omega;A) \geqslant 0$, $\int_{R^n-\{0\}} M(x,t,\omega;dz) \leqslant C$ (independent of x,t,ω)

$M(x,t,\omega;\{0\}) = 0.$

Our objective is to prove:

THEOREM 7.3: *Assume* (7.1),(7.24),(7.25),(7.67),(7.68). *Then there exists a solution* P^{xt} *of the martingale problem* (7.6),(7.7).□

Remark 7.5

We do not state an uniqueness theorem, because since g and M are random the measure P^{xt} will not correspond to a Markov process. Nevertheless, all the properties stated in Section 7.2, as well as the L^p estimates, stay valid. Therefore we shall have uniqueness if the data do not depend upon ω.

Proof of Theorem 7.3: We use the proof of STROOCK [1]. Set

$$\Omega = \Omega_0 \times [0,\infty[^N \quad \text{where} \quad N = \{0,1,\dots,n,\dots\}, \qquad \tilde{\omega} = (\omega;\alpha_0,\dots,\alpha_n,\dots).$$

Next, define

$$x(t;\tilde{\omega}) = x(t;\omega), \qquad \tau_n(\tilde{\omega}) = \alpha_n$$

and then

$$\tilde{\mu}^s_{t,n} = \sigma\big(x(\lambda),\tau_k \ ;\ t \leqslant \lambda \leqslant s,\ 0 \leqslant k \leqslant n\big) \ ;\ \tilde{\mu}_{t,n} = \tilde{\mu}^{\infty}_{t,n} \ .$$

For x,t fixed we write Q_{xt} for the probability on Ω_0,μ_t such that x(s) is a diffusion with derivative

$$g(x,s) - \int_{R^n} z\chi_{|z|\leqslant 1} M(x,s,dz)$$

and with diffusion matrix $\sigma(x,s)$. The construction of Q_{xt} is standard, using an argument of the Girsanov type to treat the derivative term. Thus, if $\varphi \in \mathcal{D}(R^n)$, then

$$\text{(7.69)} \quad \left|\begin{array}{l} \varphi\big(x(s)\big) + \displaystyle\int_t^s A(\lambda)\varphi\big(x(\lambda)\big)d\lambda + \int_t^s\int_{R^n} D\varphi\big(x(\lambda)\big)\cdot z\ \chi_{|z|\leqslant 1} M\big(x(\lambda),\lambda,dz\big)d\lambda \\ \text{is a } Q_{xt},\mu_t^s\text{-martingale} \end{array}\right.$$

Next, set

$$\text{(7.70)} \qquad M'(x,s,dz) = \frac{M(x,s,dz)}{M(x,s,R^n - 0)} = \frac{M(x,s,dz)}{M(x,s)} \ .$$

Let ω,s be fixed as parameters, and define a probability on Ω_0,μ_t by the formula

$$\text{(7.71)} \qquad Q_{\omega,s} = \delta_\omega \otimes \int Q_{x(s-o,\omega)+z,s}\ M'\big(x(s-0),s,\omega,dz\big).$$

This measure is defined exactly by

$$\text{(7.72)} \qquad Q_{\omega,s}(\Gamma \cap \Delta) = \chi_\Gamma(\omega) \int Q_{x(s-o,\omega)+z,s}(\Delta) M'\big(x(s-o),s,\omega,dz\big)$$

for

$$\Gamma \in \mu_t^{s-} = \sigma\Big(\bigcup_{t\leqslant\lambda<s} \mu_t^\lambda\Big) \quad \text{and} \quad \Delta \in \mu_s.$$

For $\tilde{\omega}$ and n fixed we now define on $\tilde{\Omega},\tilde{\mu}_{t,n}$ the probability measure

$$\text{(7.73)} \qquad \tilde{Q}^n_{\tilde{\omega}} = Q_{\omega,\tau_n(\tilde{\omega})} \otimes \delta_{\tau_0(\tilde{\omega})} \otimes \dots \otimes \delta_{\tau_n(\tilde{\omega})}.$$

Also, for $\tilde{\omega}$,n fixed, we define a probabilty on $\tilde{\Omega},\sigma(\tau_n)$ by setting

$$(7.74)\qquad \mu_{n,\tilde{\omega}}([s,\omega[) = \exp - \int_{s\wedge\tau_{n-1}(\tilde{\omega})}^{s} M(x(\lambda),\lambda,\omega)d\lambda, \quad n\geq 1.$$

We are now going to define a sequence of probabilities $\tilde{P}^n_{xt}$ on $\tilde{\Omega},\tilde{\mu}_{t,n}$ by recurrence. We begin by setting

$$(7.75)\qquad \tilde{P}^0_{xt} = Q_{xt}\otimes\delta_t$$

Knowing P^n_{xt}, we start by defining $\tilde{P}^n_{xt}$ on $\tilde{\Omega},\tilde{\mu}_{t,n+1}$ by the formula

$$(7.76)\qquad P^n_{xt} = P^n_{xt}\otimes\mu_{n+1},$$

thus $\tilde{P}^n_{xt}$ is the measure such that for $\Gamma\in\tilde{\mu}_{t,n}$ and $\Delta\in\sigma(\tau_{n+1})$ we have

$$(7.77)\qquad \widetilde{P}^n_{xt}(\Gamma\cap\Delta) = E^{P^n_{xt}}[\chi_\Gamma(\widetilde{\omega})\mu_{n+1,\widetilde{\omega}}(\Delta)].$$

For $\tilde{\Gamma}\in\tilde{\mu}_{t,n+1}$ we then set

$$(7.78)\qquad P^{n+1}_{xt}(\widetilde{\Gamma}) = E^{\widetilde{P}^n_{xt}}[\widetilde{Q}^{n+1}_{\widetilde{\omega}}(\widetilde{\Gamma})].$$

We begin by proving some properties of the family P^n_{xt}.

LEMMA 7.14: *Let* $\varphi\in\mathcal{B}(R^n)$, $\Gamma\in\mu_t^{s_1}$, $t\leq s_1\leq s_2$, *then*

$$(7.79)\qquad E^{P^n}\left[\varphi(x(s_2))\chi_\Gamma\chi_{\tau_n>s_2}\right] = E^{P^n}\left[\varphi(x(s_1))\chi_\Gamma\chi_{\tau_n>s_1}\right] -$$

$$- E^{P^n}\left[\chi_\Gamma\int_{s_1}^{s_2}\left\{\chi_{\tau_n>\lambda}\left(A\varphi(x(\lambda)) + D\varphi(x(\lambda))\cdot\int_{R^n} z\,\chi_{|z|\leq 1}M(dz)\right) -\right.\right.$$

$$\left.\left. - \chi_{\tau_{n-1}>\lambda}\int_{R^n}\left(\varphi(x(\lambda)+z) - \varphi(x(\lambda))\right)M(dz)\right\}d\lambda\right] -$$

$$- E^{P^n}\left[\chi_\Gamma\int_{s_1}^{s_2}\chi_{\tau_{n-1}\leq\lambda<\tau_n}M(x(\lambda),\lambda)\varphi(x(\lambda))d\lambda\right], \quad n\geq 1.$$

Proof: By induction: first, by definition

$$E^{P^1}\left[\varphi(x(s_2))\chi_\Gamma\chi_{\tau_1>s_2}\right] = E^{\widetilde{P}^0}E^{\widetilde{Q}^1_{\widetilde{\omega}}}\left[\varphi(x(s_2))\chi_\Gamma\chi_{\tau_1>s_2}\right]$$

and

$$E^{\widetilde{Q}^1_{\widetilde{\omega}}}\left[\varphi(x(s_2))\chi_\Gamma\chi_{\tau_1>s_2}\right] = \chi_{\tau_1>s_2}E^{Q_{\omega,\tau_1}}\left[\chi_\Gamma\varphi(x(s_2))\right] = \chi_{\tau_1>s_2}\chi_\Gamma\varphi(x(s_2))$$

and then

$$E^{\tilde{P}^0} \chi_{\tau_1 > s_2} \chi_\Gamma \varphi(x(s_2)) = E^{P^0} \chi_\Gamma \varphi(x(s_2)) \exp - \int_{s_2 \wedge \tau_0}^{s_2} M(x(\lambda),\lambda)\, d\lambda$$

$$= E^{Q_{xt}} \chi_\Gamma \varphi(x(s_2)) \exp - \int_t^{s_2} M(x(\lambda),\lambda)\, d\lambda .$$

But from (7.69) we deduce that

$$\varphi(x(s)) \exp - \int_t^s M(x(\lambda),\lambda)\, d\lambda + \int_t^s \varphi(x(\lambda)) M(x(\lambda),\lambda) \left(\exp - \int_t^\lambda M(x(\mu),\mu)\, d\mu\right) d\lambda +$$

$$+ \int_t^s \left\{A\varphi(x(\lambda)) + D\varphi(x(\lambda)) \cdot \int_{R^n} z\, \chi_{|z|\le 1}\, M(dz)\right\} \left(\exp - \int_t^\lambda M(x(\mu),\mu)\, d\mu\right) d\lambda$$

is also a Q_{xt}, μ_t^s-martingale. Therefore we finally obtain that

$$E^{P^1}\left[\varphi(x(s_2)) \chi_\Gamma \chi_{\tau_1 > s_2}\right] = E^{Q_{xt}} \chi_\Gamma \varphi(x(s_1)) \exp - \int_t^{s_1} M(x(\lambda),\lambda)\, d\lambda \, -$$

$$- E^{Q_{xt}}\Big[\chi_\Gamma \int_{s_1}^{s_2} \left\{A\varphi(x(\lambda)) + D\varphi(x(\lambda)) \int_{R^n} z\, \chi_{|z|\le 1}\, M(dz) + \varphi(x(\lambda)) M(x(\lambda),\lambda)\right\} \times$$

$$\times \left(\exp - \int_t^\lambda M(x(\mu),\mu)\, d\mu\right) d\lambda\Big]$$

and by similar calculations, from it we deduce

$$= E^{P^1}\left[\varphi(x(s_1)) \chi_\Gamma \chi_{\tau_1 > s_2}\right] - E^{P^1}\left[\chi_\Gamma \int_{s_1}^{s_2} \chi_{\tau_1 > \lambda}\, M(x(\lambda),\lambda) \varphi(x(\lambda))\, d\lambda\right]$$

$$- E^{P^1}\left[\chi_\Gamma \int_{s_1}^{s_2} \chi_{\tau_1 > \lambda} \left\{A\varphi(x(\lambda)) + D\varphi(x(\lambda)) \cdot \int_{R^n} z\, \chi_{|z|\le 1}\, M(dz)\right\} d\lambda\right]$$

and noticing that also

$$E^{P^1}[\tau_0 = t] = E^{\tilde{P}^0}[\tau_0 = t] = E^{P^0}[\tau_0 = t] = 1$$

We now assume that (7.79) is satisfied at the stage n and we prove it for the case n + 1. Thus we have

=

$$E^{P^{n+1}}\left[\varphi(x(s_2)) \chi_\Gamma \chi_{\tau_{n+1} > s_2}\right] = E^{\tilde{P}^n} E^{\tilde{Q}^{n+1}_{\tilde{\omega}}} \left[\quad\right] =$$

$$= E^{\tilde{P}^n}\left[\quad\right] = E^{P^n}\left[\varphi(x(s_2)) \chi_\Gamma \exp - \int_{\tau_n \wedge s_2}^{s_2} M(x(\lambda),\lambda)\, d\lambda\right]$$

$$= E^{P^n}\left[\varphi(x(s_2)) \chi_\Gamma \chi_{\tau_n > s_2}\right] +$$

$$+ E^{P^n}\left[\varphi(x(s_2)) \chi_\Gamma \chi_{\tau_n \le s_2} \exp - \int_{\tau_n}^{s_2} M(x(\lambda),\lambda)\, d\lambda\right] = I + II .$$

But

$$II = E^{\tilde{P}^{n-1}} E^{\tilde{Q}^n_{\tilde{\omega}}}\left[\quad\right] = E^{\tilde{P}^{n-1}} \chi_{\tau_n \leqslant s_2} E^{Q_{\omega,\tau_n}}\left[\varphi(x(s_2))\chi_\Gamma \exp - \int_{\tau_n}^{s_2} M(x(\lambda),\lambda)\,d\lambda\right]$$

$$= E^{\tilde{P}^{n-1}} \chi_{\tau_n \leqslant s_1} E^{Q_{\omega,\tau_n}}\left[\varphi(x(s_2))\chi_\Gamma \exp - \int_{\tau_n}^{s_2} M(x(\lambda))\,d\lambda\right] +$$

$$+ E^{\tilde{P}^{n-1}} \chi_{s_1 < \tau_n \leqslant s_2} \chi_\Gamma E^{Q_{\omega,\tau_n}}\left[\varphi(x(s_2)) \exp - \int_{\tau_n}^{s_2} M(x(\lambda))\,d\lambda\right] .$$

We then use the definition (7.71) as well as (7.69), and we obtain

$$II = E^{\tilde{P}^{n-1}} \chi_{\tau_n \leqslant s_1} E^{Q_{\omega,\tau_n}}\Big[\chi_\Gamma \varphi(x(s_1)) \exp - \int_{\tau_n}^{s_1} M(x(\lambda))\,d\lambda -$$

$$- \chi_\Gamma \int_{s_1}^{s_2}\Big\{A\varphi(x(\lambda)) + D\varphi(x(\lambda))\cdot\int_{R^n} z\,\chi_{|z|\leqslant 1}\, M(dz) +$$

$$+\varphi\,(x(\lambda))M(x(\lambda),\lambda)\Big\}\Big(\exp - \int_{\tau_n}^{\lambda} M(x(\mu),\mu)\,d\mu\Big)\,d\lambda\Big] +$$

$$+ E^{\tilde{P}^{n-1}} \chi_{s_1 < \tau_n \leqslant s_2} \chi_\Gamma\Big[\int \varphi(x(\tau_n - 0)+z)M'(x(\tau_n-0),\tau_n,dz) -$$

$$- E^{Q_{\omega,\tau_n}} \int_{\tau_n}^{s_2}\Big\{A\varphi(x(\lambda)) + D\varphi(x(\lambda))\cdot\int_{R^n} z\,\chi_{|z|\leqslant 1} M(x(\lambda),\lambda,dz) +$$

$$+ \varphi(x(\lambda))M(x(\lambda),\lambda)\Big\}\Big(\exp - \int_{\tau_n}^{\lambda} M(x(\mu),\mu)\,d\mu\Big)\,d\lambda\Big] =$$

$$= E^{P^{n+1}} \chi_\Gamma \chi_{\tau_n \leqslant s_1 < \tau_{n+1}} \varphi(x(s_1)) -$$

$$- E^{P^{n+1}} \int_{s_1}^{s_2} \chi_\Gamma \chi_{\tau_n \leqslant \lambda < \tau_{n+1}}\Big\{A\varphi(x(\lambda)) + D\varphi(x(\lambda))\cdot\int_{R^n} z\,\chi_{|z|\leqslant 1} M(x(\lambda),\lambda,dz) +$$

$$+ \varphi(x(\lambda))M(x(\lambda),\lambda)\Big\}\,d\lambda +$$

$$+ E^{\tilde{P}^{n-1}} \chi_{s_1 < \tau_n \leqslant s_2} \chi_\Gamma \int \varphi(x(\tau_n-0)+z)M'(x(\tau_n-0),\tau_n,dz) .$$

Now

$$E^{\tilde{P}^{n-1}} \chi_{s_1 < \tau_n \leqslant s_2} \chi_\Gamma \int \varphi(x(\tau_n-0)+z)M'(x(\tau_n-0),\tau_n,dz) =$$

$$= E^{P^{n-1}} \int_{\tau_{n-1}}^{+\infty} \chi_{s_1 < \lambda \leqslant s_2} \chi_\Gamma\Big(\int \varphi(x(\lambda)+z)M'(x(\lambda),\lambda,dz)\Big)M(x(\lambda),\lambda)\Big(\exp - \int_{\tau_{n-1}}^{\lambda} M\,d\mu\Big)\,d\lambda$$

$$= E^{P^n} \int_{s_1}^{s_2} \chi_\Gamma \chi_{\tau_{n-1} \leqslant \lambda < \tau_n}\Big(\int \varphi(x(\lambda)+z)M(x(\lambda),\lambda,dz)\Big)\,d\lambda = \qquad \text{(Contd)}$$

(Contd) $$= E^{P^{n+1}} \int_{s_1}^{s_2} \chi_\Gamma \chi_{\tau_{n-1} \leq \lambda < \tau_n} \left(\int \varphi(x(\lambda)+z) M(x(\lambda),\lambda,dz) \right) d\lambda .$$

We may also evaluate I by using the induction hypothesis, and by noticing that we can replace E^{P^n} by $E^{P^{n+1}}$. Collecting the results for I + II, we obtain the relation (7.79) for n + 1.□

LEMMA 7.15: *We have the relations*

(7.80) $$E^{P^n}\left[\varphi(x(s_2))\chi_\Gamma \chi_{s_1 < \tau_n \leq s_2}\right] =$$
$$= E^{P^n}\left[\chi_\Gamma \int_{s_1}^{s_2} \chi_{\tau_{n-1} \leq \lambda < \tau_n} \left(\int_{R^n} \varphi(x(\lambda)+z) M(x(\lambda),\lambda,dz) \right) d\lambda \right] -$$
$$- E^{P^n}\left[\chi_\Gamma \chi_{s_1 < \tau_n} \int_{s_1}^{s_2} \chi_{\tau_n \leq \lambda} \left\{ A\varphi(x(\lambda)) + D\varphi(x(\lambda)) \cdot \int_{R^n} z\, \chi_{|z| \leq 1} M(x(\lambda),\lambda,dz) \right\} d\lambda \right],$$

(7.81) $$E^{P^n}\left[\varphi(x(s_2))\chi_\Gamma \chi_{\tau_n \leq s_1}\right] = E^{P^n}\left[\varphi(x(s_1))\chi_\Gamma \chi_{\tau_n \leq s_1}\right] -$$
$$- E^{P^n}\left[\chi_\Gamma \chi_{\tau_n \leq s_1} \int_{s_1}^{s_2} \left\{ A\varphi(x(\lambda)) + D\varphi(x(\lambda)) \cdot \int_{R^n} z\, \chi_{|z| \leq 1} M(x(\lambda),\lambda,dz) \right\} d\lambda \right].$$

Proof: This is analogous to that of the preceding lemma, and is actually simpler.□

We can then conclude the proof of Theorem 7.3. First, from the relations (7.79),(7.80),(7.81) we deduce

(7.82) $$E^{P^n}\left[\chi_\Gamma \left\{ \varphi(x(s_2)) + \int_{s_1}^{s_2} \left(A\varphi(x(\lambda)) + D\varphi(x(\lambda)) \cdot \int_{R^n} z\, \chi_{|z|\leq 1} M(x(\lambda),\lambda,dz)\right) d\lambda \right.\right.$$
$$\left.\left. - \int_{s_1}^{s_2} \chi_{\tau_n > \lambda} \left(\int_{R^n} \left(\varphi(x(\lambda)+z) - \varphi(x(\lambda))\right) M(x(\lambda),\lambda,dz) \right) d\lambda \right\}\right] =$$
$$= E^{P^n} \chi_\Gamma \varphi(x(s_1)).$$

Starting from P^n, we define P^∞ on $\tilde{\Omega}, \tilde{\mu}_{t,\infty}$ in the following way: Let

$$\tilde{\Gamma} \in \sigma\left(x(\lambda_1), x(\lambda_2), \ldots, x(\lambda_h)\ ;\ \tau_{k_0}, \tau_{k_1}, \ldots, \tau_{k_p}\right).$$

(Assuming that $t \leq \lambda_1 \ldots \leq \lambda_h$; $0 \leq k_0 \leq k_1 \ldots \leq k_p$)

(7.83) $$P_{\lambda_1,\ldots,\lambda_h\ ;\ k_0 \ldots k_p}(\tilde{\Gamma}) = \lim_{n\to\infty} P^n\left(\tilde{\Gamma} \cap \{\tau_n > \lambda_h\}\right).$$

This definition is permissible. In fact, for $n \geq k_p$, the event $\tilde{\Gamma} \cap \{\tau_n > \lambda_h\} \subset \tilde{\mu}_{t,n}$. Further, the limit of the right hand side of (7.83) exists. In fact

$$P^n(\tilde{\Gamma}\cap\{\tau_n>\lambda_h\}) = \tilde{P}^{n-1}(\tilde{\Gamma}\cap\{\tau_n>\lambda_h\}) =$$

$$= E^{P^{n-1}} \chi_{\tilde{\Gamma}} \exp - \int_{\tau_{n-1}\wedge\lambda_h}^{\lambda_h} M(x(\lambda),\lambda)\,d\lambda \geqslant$$

$$\geqslant E^{P^{n-1}} \chi_{\tilde{\Gamma}}\chi_{\{\tau_{n-1}>\lambda_h\}} = E^{P^{n-1}}(\tilde{\Gamma}\cap\{\tau_{n-1}>\lambda_h\}),$$

which proves that the sequence $P^n(\tilde{\Gamma}\cap\{\tau_n > \lambda_h\})$ is increasing.

Next we show that the family of marginal probabilities defined by (7.83) is a compatible family. Therefore let

$$\tilde{\Gamma}\in\sigma(x(\lambda_1),\dots,\ x(\lambda_{h'})\ ;\ \tau_{k_0},\dots,\ \tau_{k_{p'}}),\ \text{avec}\ \lambda_{h'}\leqslant\lambda_h,\ k_{p'}\leqslant k_p,$$

and we have to prove that

$$\lim_{n\to\infty} P^n(\tilde{\Gamma}\cap\{\tau_n>\lambda_h\}) = \lim_{n\to\infty} P^n(\tilde{\Gamma}\cap\{\tau_n>\lambda_{h'}\}).$$

Now,

$$P^n(\tilde{\Gamma}\cap\{\tau_n>\lambda_{h'}\}) - P^n(\Gamma\cap\{\tau_n>\lambda_h\}) = P^n(\Gamma\cap\{\lambda_{h'}<\tau_n\leqslant\lambda_h\})$$

$$\leqslant P^n(\tau_n\leqslant\lambda_h).$$

Now, for fixed n we show by induction that

$$P^n(\tau_n\leqslant s)\leqslant E^{P^{n-k}}\left[\chi_{\tau_{n-k}\leqslant s}\left(1-\sum_{j=0}^{k-1}\frac{\bar{M}^j(s-\tau_{n-k})^j}{j!}\exp -\bar{M}(s-\tau_{n-k})\right)\right]$$

where $\bar{M}$ is a constant such that $M(x,\lambda,\omega) \leqslant \bar{M}$. In particular, for $k = n$, we obtain

$$P^n(\tau_n\leqslant s)\leqslant 1 - \sum_{j=0}^{n-1}\frac{\bar{M}^j(s-t)^j}{j!}\exp -\bar{M}(s-t) \to 0$$

as $n \to \infty$. Thus the compatibility condition is proved.

By a standard extension result (cf., for example, NEVEU [1]) the family (7.83) is the projection of an unique probability P^∞ on $(R^n)^{[0,\infty]}\times R^N$ provided with the σ-algebra that is the product of the Borel σ-algebras. It is clear that $P^\infty = P^n$ on $\sigma(x(\lambda),t \leqslant \lambda < \tau_n;\tau_0,\dots,\tau_n)$, and since $\tau_n \uparrow \infty$, a.s. P^∞, from this it follows that P^∞ is concentrated on $\tilde{\Omega},\tilde{\mu}_{t,\infty}$.

Now let P^{xt} be the restriction of P^∞ to Ω_0,μ_t. We notice that in (7.82) we can replace P^n by P^∞; we make n tend to ∞ and noticing that in the result we can replace P^∞ by P^{xt} we obtain the P^{xt} is a solution of the martingale problem. The proof of Theorem 7.3 is therefore complete.□

8. INTERPRETATION OF SOLUTIONS OF EQUATIONS

Method of Approach

In this Section we are going to give the probabilistic interpretation of the equations treated in Section 1,2, and 3. To shorten what we could write we restrict ourselves to parabolic equations. The adaptations to the elliptic case are easy.

8.1 *Linear Equations*

We shall be interested in Equation (3.14) in the linear case ($H(t) = 0$). However, we shall modify our notation slightly to make it compatible with that of the martingale problem. We shall write (cf. Section 1.2)

$$(8.1) \qquad A(t) = -\sum \frac{\partial}{\partial x_i} a_{ij}(x,t) \frac{\partial}{\partial x_j} + \sum a_i \frac{\partial}{\partial x_i}$$

so that we now denote by $A(t) + a_0$ the operator called $A(t)$ in Section 3.1. Furthermore, we write

$$(8.2) \qquad g_i(x,t) = -a_i(x,t) + \sum_j \frac{\partial a_{ij}}{\partial x_j}$$

and in this case $A(t)$, given by (8.1), coincides with the operator $A(t)$ occurring in the formulation of the martingale problem (cf. (7.2))[(1)]. Here we make the assumptions (3.1),(3.2), and (1.3). Let $h \in L^\infty$, $\bar{u} \in W^{2,p,\mu} \cap L^\infty$, $2 \leqslant p < \infty$, $\mu > 0$, then by Theorem 3.3 there exists one and only one solution u of the problem

$$(8.3) \qquad \left| \begin{array}{l} -\dfrac{\partial u}{\partial t} + A(t)u - B(t)u + a_0 u = h, \quad u(x,T) = \bar{u}(x) \\ u \in W^{2,1,p,\mu} \cap L^\infty . \end{array} \right.$$

Also, by Theorems 7.1 and 7.2 there exists one and only one solution P^{xt} of the martingale problem with respect to the operator $A(t) - B(t)$ (cf. (7.6),(7.7)). We then have:

THEOREM 8.1: *Under the assumptions* (1.3),(3.1),(3.2), *the solution* u *of* (8.3) *is given explicitly by*

$$(8.4) \qquad u(x,t) = E^{xt}\Big[\int_t^T h(x(s),s)\Big(\exp - \int_t^s a_0(x(\lambda),\lambda)\,d\lambda\Big)ds + \\ + \bar{u}(x(T))\exp - \int_t^T a_0(x(s),s)\,ds\Big].$$

Proof: Let us note that the principle of the regularisation of parabolic operators implies that $u(x,t)$ is regular for $t < T$, in particular it is continuous, and therefore $u(x,t)$ is defined unambiguously. Further, at least if $h \in L^q(R^n \times]0,T[)$, where q is large enough, the formula (8.4) results from the consideration developed in Sections 7.2,7.3, in particular (7.34) and (7.64). Its extension to the case $a_0 \neq 0$ is a consequence of the fact

(1) We can always reduce to the case where $a_{ij} = a_{ji}$, by possibly modifying a_i, and this is possible because the a_{ij} are Lipschitz continuous.

$$\varphi(x(s))\exp - \int_t^s a_0(x(\lambda),\lambda)\,d\lambda \; - \varphi(x(t)) + \int_t^s (A-B+a_0)\,\varphi(x(\lambda))\left(\exp - \int^\lambda a_0 d\mu\right) d\lambda$$

is a P^{xt},μ_t^s-martingale. The formula (8.4) is thus valid at least if h satisfies $h \in L^\infty \cap L^q$, for q large enough. If $h \in L^\infty$ it suffices to replace h by $h\varphi_R$, where φ_R has compact support, $0 \leq \varphi_R \leq 1$, is regular, and $\varphi_R = 1$ on the ball of radius R. It is clear that we can pass to the limit when $R \to \infty$, whence the result.□

We now consider the case of equations on a bounded open set. Therefore we assume (2.1),(3.1),(3.2),(3.16). Let $h \in L^p(Q)$, $\bar{u} \in W_0^{1,p}(\mathcal{O})$, $2 \leq p < \infty$, then by Theorem 3.4 there exists one and only one solution of the problem

$$(8.5) \qquad \left| \begin{array}{l} -\dfrac{\partial u}{\partial t} + A(t)u - B(t)u = h, \; u(x,T) = \bar{u}(x) \\ u \in L^p(0,T;W_0^{1,p}(\mathcal{O})), \; \dfrac{\partial u}{\partial t} \in L^p(0,T;W^{-1,p}(\mathcal{O})). \end{array} \right.$$

We shall assume that $p > n + 2$. In this case

$$\{z \mid z \in L^p(0,T;W^{1,p}(\mathcal{O})), \; \frac{\partial z}{\partial t} \in L^p(0,T;W^{-1,p}(\mathcal{O}))\} \subset C^0(\bar{Q}).$$

Our objective is to given an explicit form for u(x,t). To simplify the calculation we shall take $a_0 = 0$.

We again consider the probability measure P^{xt} on Ω_0,μ_t the solution of the martingale problem with respect to the operator A - B. However, it will be useful to modify μ_t^s into

$$\bar{\mu}_t^s = \mu_t^{s+0} \text{ the completion of } \mu_t \text{ by the sets of measure zero (for } P^{xt}),$$

and P^{xt} is again the solution of the martingale problem by replacing μ_t^s with $\bar{\mu}_t^s$ The usefulness of this is the following: Let us set

$$(8.6) \qquad \tau_t = \inf\{s \geq t \mid x(s) \notin \mathcal{O}\}$$

$$(8.7) \qquad \hat{\tau}_t = \inf\{s \geq t \mid x(s) \notin \bar{\mathcal{O}}\}.$$

Then $\hat{\tau}_t$ is a $\bar{\mu}_t^s$-stopping time, since x(s) is right continuous. However, τ_t is not necessarily a stopping time. Later we shall see that

$$(8.8) \qquad \hat{\tau}_t = \tau_t \quad \text{a.s. } P^{xt},$$

so that τ_t will also be a $\bar{\mu}_t^s$-stopping time.

Our objective is to prove:

THEOREM 8.2: *Assume* (2.1),(3.1),(3.2),(3.16), *and also* $a_0 = 0$. *Then for* $h \in L^p(Q)$, $\bar{u} \in W_0^{1,p}(\mathcal{O})$, *the solution* u(x,t) *of* (8.5) *is given explicitly by the formula*

$$(8.9)\qquad u(x,t) = E^{xt}\left[\int_t^{T\wedge\tau_t} h(x(s),s)\,ds + \bar{u}(x(T))\chi_{T<\tau_t}\right].$$

Here we meet a technical difficulty which is the following. Since $B(t) \notin \mathcal{L}(W^{2,p}(\mathcal{O});L^p(\mathcal{O}))$, if we consider a sequence of regular functions u_n approximating u, $B(t)u_n$ converges only in $L^p(0,T;W^{-1,p}(\mathcal{O}))$. Now, since we have no estimate for the martingale problem of the type (7.32) for functions in $L^p(0,T;W^{-1,p}(\mathcal{O}))$, it is not possible to generalise Ito's formula to the functions of $L^p(0,T;W_0^{1,p}(\mathcal{O}))$. This is why we shall proceed differently, using the method of penalisation of the domain described in Section 3.5. We define

$$(8.10)\qquad \tilde{u}(x,t) = E^{xt}\left[\int_t^{T\wedge\hat{\tau}_t} h(x(s),s)\,ds + \bar{u}(x(T))\chi_{T<\hat{\tau}_t}\right].$$

We consider the penalisation problem on the whole space

$$(8.11)\qquad \left|\begin{array}{l} -\dfrac{\partial u_\eta}{\partial t} + A(t)u_\eta - B(t)u_\eta + \dfrac{1}{\eta}\, q\, u_\eta = h \\ u_\eta(T) = \bar{u},\quad u_\eta \in W^{2,1,p}(R^n\times]0,T[),\quad u_\eta \in C^0(R^n\times[0,T]). \end{array}\right.$$

By Theorem 8.1, we have the formula

$$(8.12)\qquad u_\eta(x,t) = E^{xt}\left[\int_t^T h(x(s),s)\left(\exp -\frac{1}{\eta}\int_t^s q(x(\lambda))\,d\lambda\right)ds + \right.$$

$$\left. + \bar{u}(x(T))\exp -\frac{1}{\eta}\int_t^T q(x(s))\,ds\right].$$

LEMMA 8.1: $u_\eta(x,t) \to \tilde{u}(x,t)$ *for all* x,t *when* $\eta \to 0$.

Proof: In fact,

$$(8.13)\qquad \exp -\frac{1}{\eta}\int_t^s q(x(\lambda))\,d\lambda \to \left|\begin{array}{ll} 0 & \text{if } s > \tau_t, \\ 1 & \text{if } s \leqslant \hat{\tau}_t. \end{array}\right.$$

Let us prove (8.13). Since the process $x(s)$ is right continuous, for $s > \hat{\tau}_t$, there exists $s_0 < s$ such that $x(s_0) \notin \bar{\mathcal{O}}$, hence there exists $\delta > 0$ (depending on ω) such that $x(\lambda) \notin \bar{\mathcal{O}}$, for $s_0 \leqslant \lambda \leqslant s_0 + \delta \leqslant s$. But then

$$\exp -\frac{1}{\eta}\int_t^s q\,d\lambda \leqslant \exp -\frac{\delta}{\eta} \to 0.$$

However, if $s \leqslant \hat{\tau}_t$, then

$$\exp -\frac{1}{\eta}\int_t^s q\big(x(\lambda)\big)\,d\lambda = 1.$$

But (8.13) again means

$$\exp -\frac{1}{\eta}\int_t^s q\big(x(\lambda)\big)\,d\lambda \longrightarrow \chi_{s\leqslant\hat{\tau}_t} \qquad \forall s.$$

From this it follows by (8.12) that $u_\eta \to \tilde{u}$, at least if h is bounded. Now, we can always consider a sequence $h_k \in \mathcal{D}(R^n \times]0,T[)$, such that $h_k \to h$ in $L^p(R^n \times]0,T[)$. Noticing that

$$E^{xt}\left|\int_t^T (h-h_k)\big(x(s),s\big)\Big(\exp -\frac{1}{\eta}\int_t^s q\,d\lambda\Big)ds\right| \leqslant$$

$$\leqslant E^{xt}\int_t^T |h-h_k|\big(x(s),s\big)\,ds \leqslant |h-k_k|_{L^p} \longrightarrow 0,$$

uniformly with respect to η, we deduce the same result in the general case.□

LEMMA 8.2: *We have*

(8.14) $\tilde{u}(x,t) = u(x,t)$ *for all* x,t.

Proof: Since by Theorem 3.6, $u_\eta \to u$ in $L^2(0,T;H^1)$ and $C^0(0,T;L^2)$ we can affirm that for all t $\tilde{u}(x,t) = u(x,t)$, x a.e.. Let s be fixed, $t < s \leqslant T$. Analogously to Theorem 8.1 we show that

$$(8.15)\qquad u_\eta(x,t) = E^{xt}\Big[\int_t^s h\big(x(\lambda),\lambda\big)\Big(\exp -\frac{1}{\eta}\int_t^\lambda q\big(x(\mu)\big)\,d\mu\Big)d\lambda + u_\eta\big(x(s),s\big)\exp -\frac{1}{\eta}\int_t^s q\big(x(\mu)\big)\,d\mu\Big].$$

Since

$$u_\eta\big(x(s),s\big) \longrightarrow \tilde{u}\big(x(s),s\big) \qquad \forall s,\omega$$

$$\exp -\frac{1}{\eta}\int_t^\lambda q\big(x(\mu)\big)\,d\mu \longrightarrow \chi_{\lambda\leqslant\hat{\tau}_t} \qquad \forall t\leqslant\lambda\leqslant 0,\ \omega$$

we can pass to the limit in (8.15), whence it follows that

$$(8.16)\qquad \tilde{u}(x,t) = E^{xt}\Big[\int_t^{s\wedge\hat{\tau}_t} h\big(x(\lambda),\lambda\big)\,d\lambda + \tilde{u}\big(x(s),s\big)\chi_{s\leqslant\hat{\tau}_t}\Big].$$

But

$$\Big|E^{xt}\Big(\tilde{u}\big(x(s),s\big) - u\big(x(s),s\big)\Big)\chi_{s\leqslant\hat{\tau}_t} \leqslant E^{xt}|\tilde{u}-u|\big(x(s),s\big) = 0$$

since $(\tilde{u} - u)(x,s) = 0$, for x a.e. (s fixed and > t) (the principle of the regularisation of parabolic operators). Therefore

$$\tilde{u}(x,t) = E^{xt}\left[\int_t^{s\wedge\hat{\tau}_t} h(x(\lambda),\lambda)\,d\lambda + u(x(s),s)\chi_{s\leq\hat{\tau}_t}\right], \quad \forall s>t,$$

and $s \leq T$. But then by subtraction we have

$$(8.17)\qquad E^{xt}\, u(x(s),s)\chi_{s\leq\hat{\tau}_t} = E^{xt}\left[\int_{s\wedge\hat{\tau}_t}^{T\wedge\hat{\tau}_t} h(x(\lambda),\lambda)\,d\lambda + \bar{u}(x(T))\chi_{T<\hat{\tau}_t}\right] \quad t<s\leq T.$$

Let us then consider a sequence $s_k \downarrow t$; since x(s) is right continuous and u is a continuous function, we have

$$u(x(s_k),s_k) \to u(x(t),t) \quad \text{a.s.}.$$

Moreover, since $s_k > t$,

$$\chi_{s_k \leq \hat{\tau}_t} = \chi_{s_k\leq\hat{\tau}_t}\,\chi_{t<\hat{\tau}_t} \to 1 \quad \text{a.s.}.$$

Therefore we can pass to the limit in (8.17), whence we deduce (8.14), using the definition of $\tilde{u}$.□

Proof of Theorem 8.2: Everything reduces now to proving (8.8). First we note that if θ is a stopping time (of $\bar{\mu}_t^s$), then Ito's formula for u_η gives

$$u_\eta(x,t) = E^{xt}\left[\int_t^{T\wedge\theta} h(x(s),s)\left(\exp -\frac{1}{\eta}\int_t^s q(x(\lambda))\,d\lambda\right)ds + \right.$$
$$\left. + u_\eta(x(T\wedge\theta),T\wedge\theta)\exp -\frac{1}{\eta}\int_t^{T\wedge\theta} q(x(s))\,ds\right]$$

a formula in which it is permissible to pass to the limit, thanks to Lemmas 8.1 and 8.2. From this we deduce

$$(8.18)\qquad u(x,t) = E^{xt}\left[\int_t^{T\wedge\hat{\tau}_t\wedge\theta} h(x(s),s)\,ds + u(x(T\wedge\hat{\tau}_t\wedge\theta),\ T\wedge\hat{\tau}_t\wedge\theta)\chi_{T\wedge\theta\leq\hat{\tau}_t}\right].$$

We then consider a sequence of open sets $\mathcal{O}_k$ such that

$$\mathcal{O}_k \subset \mathcal{O}_{k+1},\ \bigcup_k \mathcal{O}_k = \mathcal{O},\quad \mathcal{O}_k \subset \mathcal{O}.$$

We write τ_t^k for the exit time from $\mathcal{O}_k$ of the process x(s). Naturally, we have

$$\tau_t^k \leq \tau_t^{k+1} \leq \cdots \leq \tau_t \leq \hat{\tau}_t.$$

In (8.18) we take h = 1, u = 0 and apply the formula with $\mathcal{O} = \tau_t^k$, and obtain

$$u(x,t) = E^{xt}\left[(T\wedge\tau_t^k - t) + u\left(x(\tau_t^k),\tau_t^k\right)\chi_{\tau_t^k<T}\right].$$

But $x(\tau_t^k) \notin \mathcal{O}_k$. Since u is uniformly continuous on $R^n \times [0,T]$ and $u = 0$ for $x \notin \mathcal{O}$ it is clear that

$$\sigma_k = \sup_{\substack{x\notin\mathcal{O}'_k \\ t\in[0,T]}} |u(x,t)| \to 0$$

and therefore

$$\left|E^{xt}[T\wedge\tau_t^k - t] - u(x,t)\right| \leq \sigma_k .$$

Now,

$$u(x,t) = E^{xt}[T\wedge\hat{\tau}_t - t]$$

whence

$$0 \leq E^{xt}(T\wedge\hat{\tau}_t - T\wedge\tau_t^k) \leq \sigma_k$$

so

$$E^{xt}(T\wedge\hat{\tau}_t - T\wedge\tau_t) \leq \sigma_k$$

and therefore

$$T\wedge\hat{\tau}_t = T\wedge\tau_t \quad \text{a.s. } P^{xt}.$$

Since T is arbitrary, we deduce (8.8), and therefore (8.9).□

8.2 *Non-Linear Equations*

We work under the conditions of Theorem 3.3, and therefore assume (3.1), (3.2), (3.3), (3.4), (3.4)', and furthermore that

(8.19) $\mathcal{V}_1, \mathcal{V}_2$ are metric compact sets

(8.20) $f(x,t,v_1,v_2)$, $g(x,t,v_1,v_2)$, $\tilde{a}_1(x,t,v_1,v_2)$, $c_1(x,t,v_1,v_2)$

are continuous in v_1,v_2 and Borel; $c_0(x,t)$ is Borel.

We define H(t) and B(t) as in Section 3.1, and then A(t) by (8.1). We shall take

(8.21) $$a_i = \sum_j \frac{\partial a_{ij}}{\partial x_j}, \quad a_0 = 0$$

which produces no loss of generality, even though we add to $g(v_1,v_2)$ the vector defined in (8.2) and change a_1 into $a_1 + a_0$. Therefore

$$(8.22)\qquad A(t) = -\sum_{ij} a_{ij}(x,t)\,\frac{\partial^2}{\partial x_i \partial x_j}\ .$$

Let $h \in L^\infty$, $\bar{u} \in L^\infty \cap W^{w,p,\mu}$ where $p \geqslant n$, $\mu > 0$ are fixed. By Theorem 3.3, there exists one and only solution of

$$(8.23)\qquad \left|\begin{array}{l} -\dfrac{\partial u}{\partial t} + A(t)u - B(t)u - H(t)\big(u(t)\big) = h \\[2mm] u(T) = \bar{u},\ u \in \mathcal{W}^{2,1,p,\mu} \cap L^\infty \end{array}\right.$$

Our objective is to interpret the solution $u(x,t)$. We shall write

$$(8.24)\qquad L_\varphi(x,t,v_1,v_2) = f + D\varphi(x)\cdot g - \varphi(x)\tilde{a}_1 + \\ + \int_{R^n}\Big[\varphi(x+z) - \phi(x) - z\cdot\nabla\varphi\,\chi_{|z|\leqslant 1}\Big]\,c_0 c_1\ m(dz)$$

which has a meaning for $\varphi \in W^{1,p,\mu} \cap L^\infty$. The function u, the solution of (8.23), is continuous, as well as its first derivatives. The function $L_u(x,t,v_1,v_2)$ is a Borel function, continuous in v_1, v_2. Thus we can find Borel mappings $\hat{v}_1(x,t)$, $\hat{v}_2(x,t,v_1)$ such that

$$(8.25)\qquad H(t)\big(u(t)\big)(x) = L_u\big(x,t\ ;\ \hat{v}_1(x,t)\ ;\ \hat{v}_2(x,t,\hat{v}_1)\big)\big)\quad \forall x,t$$

$$= \sup_{v_2} L_u\big(x,t\ ;\ \hat{v}_1(x,t),v_2\big)$$

and

$$(8.26)\qquad L_u\big(x,t,v_1,\hat{v}_2(x,t,v_1)\big) = \sup_{v_2} L_u(x,t,v_1,v_2)\quad \forall x,t,v_1\ .$$

We now define a *problem in differential games*. The control of player number 1 (the *minimizer*) is a stochastic process $v_1(s)$ *adapted* to μ_t^s, and $\mathcal{V}_1$-valued. The control of player number 2 (the *maximizer*) is a stochastic process $v_2(s,v_1)$ depending in measure on all the arguments adapted to μ_t^s and $\mathcal{V}_2$-valued. Player number 2 can observe the control applied by player number 1. The control which it *effectively* applies is the process $v_2(s,v_1(s))$, denoted $v_2(s)$ by abuse of language. It will be convenient to set

$$(8.27)\qquad \left|\begin{array}{l} f_{v_1,v_2}(s) = f\big(x(s),s,v_1(s),v_2(s)\big) + h\big(x(s),s\big) \\ a_{1,v_1,v_2}(s) = \tilde{a}_1\big(x(s),s,v_1(s),v_2(s)\big) \\ g_{v_1,v_2}(s) = g\big(x(s),s,v_1(s),v_2(s)\big) \\ c_{1,v_1,v_2}(s,z) = c_1\big(x(s),s,v_1(s),v_2(s),z\big). \end{array}\right.$$

We also note that

$$(8.28)\qquad \alpha_{v_1,v_2}(s) = \sigma^{-1}(x(s),s)\Big[g_{v_1,v_2}(s) - \int_{R^n} z\,\chi_{|z|\leq 1}\, c_0(x(s),s,z)\,C_{1,v_1,v_2}(s,z)m(dz)\Big] .$$

Let P^{xt} be the solution of the martingale problem on Ω_0,μ_t with respect to the operator A - B. By Lemma 7.7, taking (8.21) into account, the canonical process (for the measure P^{xt}) is given by

$$(8.29)\qquad x(s) = x + \int_t^s \sigma(x(\lambda),\lambda)dw(\lambda) + \int_{R^n} z\,\chi_{|z|\leq 1}\,\zeta(s,dz) + \int_{R^n} z\,\chi_{|z|>1}\,\eta(s,dz)$$

where $\eta(s,A)$ is the integral random measure corresponding to the jumps of the process $x(s)$, and $\zeta(s,A)$ is the associated local martingale measure, with increasing process

$$\int_t^s\int_A c_0(x(\lambda),\lambda,z)m(dz)d\lambda .$$

We then define the process

$$(8.30)\qquad \chi_{v_1,v_2}(s) = \exp\Big\{\int \alpha_{v_1,v_2}(\lambda)\cdot dw(\lambda) - \frac{1}{2}\int_t^s |\alpha_{v_1,v_2}(\lambda)|^2 d\lambda + \int_t^s\int_{R^n} c_{1,v_1,v_2}(\lambda,z)\zeta(d\lambda,dz) - \int_t^s\int_{R^n} [c_1 - \mathrm{Log}(1+c_1)]\,\eta(d\lambda,dz)\Big\}.$$

By Theorem 5.4 and Corollary 5.1, $E^{xt}\chi_{v_1,v_2}(s) = 1$, therefore we can set

$$(8.31)\qquad \frac{dQ^{xt}_{v_1,v_2}}{dP^{xt}}\bigg|_{\mu_t^s} = \chi_{v_1,v_2}(s).$$

Lastly, we define

$$(8.32)\qquad J^{xt}(v_1(\cdot),v_2(\cdot,v_1)) = E^{Q^{xt}_{v_1,v_2}}\Big[\int_t^T f_{v_1,v_2}(s)\Big(\exp - \int_t^s a_{1,v_1,v_2}(\lambda)d\lambda\Big)ds + \bar{u}(x(T))\exp - \int_t^T a_{1,v_1,v_2}(s)ds\Big].$$

To the mappings $\hat{v}_1(x,t),\hat{v}_2(x,t,v_2)$ defined in (8.25),(8.26) we make correspond controls $\hat{v}_1(s),\hat{v}_2(s,v_1)$ given by

$$(8.33)\qquad \begin{cases} \hat{v}_1(s) = \hat{v}_1(x(s),s), \\ \hat{v}_2(s,v_1) = \hat{v}_2(x(s),s,v_1). \end{cases}$$

We can then state:

THEOREM 8.3: *Assume* (3.1),(3.2),(3.4),(3.4)',(8.18),(8.19),(8.20) *and* (8.21)[1]. *Then the solution* u *of* (8.22) *is given explicitly by*

$$(8.34)\qquad u(x,t) = J^{xt}(\hat{v}_1(\cdot),\ \hat{v}_2(\cdot,v_1)) =$$

$$= \underset{v_1(\cdot)}{\mathrm{Min}}\ \underset{v_2(\cdot,v_1)}{\sup}\ J^{xt}(v_1(\cdot),v_2(\cdot,v_1)) =$$

$$= \underset{v_2(\cdot,v_1)}{\mathrm{Max}}\ \underset{v_1(\cdot)}{\mathrm{Inf}}\ J^{xt}(v_1(\cdot),\ v_2(\cdot,v_1)) =$$

$$= \underset{v_1(\cdot)}{\mathrm{Min}}\ J^{xt}(v_1(\cdot),\ \hat{v}_2(\cdot,v_1)) =$$

$$= \underset{v_2(\cdot,v_1)}{\mathrm{Max}}\ J^{xt}(\hat{v}_1(\cdot),\ v_2(\cdot,v_1)).$$

We begin with:

LEMMA 8.3: *Let* $\psi \in L^p(R^n \times]0,T[)$, $p > 2n+2$. *Then*

$$(8.35)\qquad \left| E^{Q^{xt}_{v_1,v_2}} \int_t^T \psi(x(s),s)\,ds \right| \leqslant C|\psi|$$

where C *is a constant depending neither on* ψ, *nor on the controls* v_1, v_2.

Proof: For $\psi \in \mathcal{D}(R^n \times]0,T[) \geqslant 0$,

$$(8.36)\qquad E^{Q^{xt}_{v_1,v_2}} \int_t^T \psi(x(s),s)\,ds = E^{xt}\, \chi_{v_1,v_2}(T) \int_t^T \psi(x(s),s)\,ds\ ,$$

$$\leqslant E^{xt}\left[(\chi_{v_1,v_2}(T))^2\right]^{\frac{1}{2}} \sqrt{T}\left[E^{xt}\int \psi^2(x(s),s)\,ds\right]^{\frac{1}{2}}.$$

By (7.32), for $q > n+1$ we have

$$(8.37)\qquad E^{xt}\int_t^T \psi^2(x(s),s)\,ds \leqslant C\left|\psi^2\right|_{L^q} \leqslant C|\psi|^2_{L^{2q}}\ .$$

Further, by Ito's formula, we have

$$(8.38)\qquad E^{xt}\,\chi_{v_1,v_2}(T)^2 = 1 + E^{xt}\int_t^T \chi^2(s)|\alpha_{v_1,v_2}(s)|^2 ds + E^{xt}\int_t^T\int_{R^n} \chi^2(s)c_1^2c_0\ m(dz)\,ds$$

$$\leqslant 1 + C\,E^{xt}\int_t^T \chi(s)^2 ds$$

(1) This last assumption is solely for simplification.

where C does not depend on v_1, v_2. By Gronwall's Lemma it follows that

$$E^{xt}(\chi_{v_1,v_2}(T))^2 \leq C_T. \tag{8.39}$$

This and (8.37), applied in (8.36) imply

$$E^{Q^{xt}_{v_1,v_2}} \int \psi(x(s),s)\,ds \leq C|\psi|_{L^{2q}}, \quad \forall q > n+1, \tag{8.40}$$

whence the result.□

Proof of Theorem 8.3: First we take the case where we use the additional assumptions

$$|f| \leq \bar{f} \in L^2(R^n) \cap L^\infty(R^n), \tag{8.41}$$

$$h \in L^2 \cap L^\infty, \quad \bar{u} \in \mathcal{D}(R^n). \tag{8.42}$$

Then in (8.23) we can take $\mu = 0$, and then

$$u \in W^{2,1,q}(R^n \times]0,T[), \quad \infty > q \geq 2 \text{ arbitrary},$$

therefore u, $\frac{\partial u}{\partial x_i}$ are continuous functions on $R^n \times [0,T]$ and are bounded. As in Theorem 8.1, we see that Ito's formula can be applied to the function u(x,t)z and to the pair of processes $x(s), \chi_{v_1,v_2}(s)$. We obtain

$$\begin{aligned}
&E^{Q^{xt}_{v_1,v_2}} u(x(T),T)\exp - \int_t^T a_{1,v_1,v_2}(s)\,ds = u(x,t) + \\
&+ E^{Q^{xt}_{v_1,v_2}} \int_t^T \left(\frac{\partial u}{\partial s} - A(s)u + B(s)u\right)(x(s),s)\left(\exp - \int_t^s a_{1,v_1,v_2}(\lambda)\,d\lambda\right) ds + \\
&+ E^{Q^{xt}_{v_1,v_2}} \int_t^T ds \Big\{- a_{1,v_1,v_2}(s)u(x(s),s) + Du(x(s),s)\cdot g_{v_1,v_2}(s) + \\
&+ \int_{R^n} \left[u(x(s)+z) - u(x(s),s) - z\,\chi_{|z|\leq 1} Du(x(s))\right] \\
&\qquad c_0 c_{1,v_1,v_2}(s,z)m(dz)\Big\} \exp - \int_t^s a_{1,v_1,v_2}\,d\lambda
\end{aligned}$$

so

$$\begin{aligned}
u(x,t) &= E^{Q^{xt}_{v_1,v_2}} \bar{u}(x(T))\exp - \int_t^T a_{1,v_1,v_2}(s)\,ds + \\
&+ E^{Q^{xt}_{v_1,v_2}} \int_t^T \left(-\frac{\partial u}{\partial s} + (A-B)u\right)(x(s),s)\left(\exp - \int_t^s a_{1,v_1,v_2}(\lambda)\,d\lambda\right) ds + \\
&+ E^{Q^{xt}_{v_1,v_2}} \int_t^T \left[- L_u(x(s),s,v_1(s),v_2(s)) + f_{v_1,v_2}(s)\right]\left(\exp - \int_t^s a_{1,v_1,v_2}(\lambda)\,d\lambda\right) ds
\end{aligned}$$

(Contd)

(Contd) $$= J^{xt}\big(v_1(\cdot),v_2(\cdot,v_1)\big) +$$

$$+ E^{Q^{xt}_{v_1,v_2}} \int_t^T ds\Big[H(s)\big(u(s)\big)\big(x(s)\big) - L_u\big(x(s),s,v_1(s),v_2(s)\big)\Big] \exp - \int_t^s a_{1,v_1,v_2}($$

and thanks to (8.25),(8.33), it follows that

(8.43) $$u(x,t) = J^{xt}(\hat{v}_1(\cdot),\hat{v}_2(\cdot,\hat{v}_1)).$$

Further, by (8.29),(8.30) we also have

$$u(x,t) \geqslant J^{xt}\big(\hat{v}_1(\cdot),\ v_2(\cdot,\ v_1)\big)$$

$$u(x,t) \leqslant J^{xt}\big(v_1(\cdot),\ \hat{v}_2(\cdot,\ v_1)\big)$$

so that the desired result follows, at least if (8.41),(8.42) are satisfied.

We shall no longer make these assumptions. We approximate f and h by $f\varphi_R$ and $h\varphi_R$, where φ_R converges to 1, and φ_R is regular with compact support, and then u by $\bar{u}_R$ with compact support, and $\bar{u}_R \to \bar{u}$ in $C^0 \cap W^{2,p,\mu}$. We write u_R for the corresponding solution of (8.23). Naturally $u_R \to u$ in $\mathcal{W}^{2,1,p,\mu}(R^n \times]0,T[)$. From this we easily deduce that for all $v_1(\cdot),v_2(\cdot)$

(8.44) $$\left|\begin{array}{l} L_{u_R}\big(x(s),s,v_1(s),v_2(s)\big) \to L_u\big(x(s),s,v_1(s),v_2(s)\big) \\ H(s)\big(u_R(s)\big)\big(x(s)\big) \to H(s)\big(u(s)\big)\big(x(s)\big) \quad , \ \forall s, \end{array}\right.$$

and u_R, Du_R are bounded on every compact set (in fact u_R is bounded).

Also, let

$$\tau_N = \inf\{s \geqslant t \,|\, |x(s)| > N\};$$

applying Ito's formula to u_R gives

(8.45) $$u_R(x,t) = E^{Q^{xt}_{v_1,v_2}} u_R\big(x(T\wedge\tau_N),T\wedge\tau_N\big) \exp - \int_t^{T\wedge\tau_N} a_{1,v_1,v_2}(s)ds +$$
$$+ E^{Q^{xt}_{v_1,v_2}} \int_t^{T\wedge\tau_N} f_{v_1v_2}(s)\,\varphi_R\big(x(s)\big)\Big(\exp - \int_t^s a_{1,v_1,v_2}(\lambda)d\lambda\Big)ds +$$
$$+ E^{Q^{xt}_{v_1,v_2}} \int_t^{T\wedge\tau_N}\Big[H(s)(u_R)\big(x(s)\big) - Lu_R\big(x(s),s,v_1(s),v_2(s)\big)\Big]$$
$$\Big(\exp - \int_t^s a_{1,v_1,v_2}(\lambda)d\lambda\Big)ds.$$

Since $|x(s)| \leqslant N$ for $s < T\wedge\tau_N$, we can make R tend to $+\infty$ in (8.45), using (8.44). From this we deduce

$$(8.46)\quad u(x,t) = E^{Q^{xt}_{v_1,v_2}} u\big(x(T\wedge\tau_N), T\wedge\tau_N\big) \exp - \int_t^{T\wedge\tau_N} a_{1,v_1,v_2}(s)ds +$$
$$+ E^{Q^{xt}_{v_1,v_2}} \int_t^{T\wedge\tau_N} f_{v_1,v_2}(s)\Big(\exp - \int_t^s a_{1,v_1,v_2}(\lambda)d\lambda\Big)ds +$$
$$+ E^{Q^{xt}_{v_1,v_2}} \int_t^{T\wedge\tau_N} \Big[H(s)(u)\big(x(s)\big) - L_u\big(x(s),s,v_1(s),v_2(s)\big)\Big]$$
$$\Big(\exp - \int_t^s a_{1,v_1,v_2}(\lambda)d\lambda\Big)ds$$

therefore in particular

$$(8.47)\quad u(x,t) = E^{Q^{xt}_{\hat v_1,\hat v_2}} u\big(x(T\wedge\tau_N), T\wedge\tau_N\big) \exp - \int_t^{T\wedge\tau_N} a_{1,v_1,v_2}(s)ds +$$
$$+ E^{Q^{xt}_{\hat v_1,\hat v_2}} \int_t^{T\wedge\tau_N} f_{\hat v_1,\hat v_2}(s)\Big(\exp - \int_t^s a_{1,\hat v_1,\hat v_2}(\lambda)d\lambda\Big)ds$$

and

$$(8.48)\quad u(x,t) \geq E^{Q^{xt}_{\hat v_1,v_2}} u\big(x(T\wedge\tau_N), T\wedge\tau_N\big) \exp - \int_t^{T\wedge\tau_N} a_{1,v_1,v_2}(s)ds +$$
$$+ E^{Q^{xt}_{\hat v_1,v_2}} \int_t^{T\wedge\tau_N} f_{\hat v_1,v_2}(s)\Big(\exp - \int_t^s a_{1,\hat v_1,v_2}(\lambda)d\lambda\Big)ds$$

$$(8.49)\quad u(x,t) \leq E^{Q^{xt}_{v_1,\hat v_2}} u\big(x(T\wedge\tau_N), T\wedge\tau_N\big) \exp - \int_t^{T\wedge\tau_N} a_{1,v_1,\hat v_2}(s)ds +$$
$$+ E^{Q^{xt}_{v_1,\hat v_2}} \int_t^{T\wedge\tau_N} f_{v_1,\hat v_2}(s)\Big(\exp - \int_t^s a_{1,v_1,\hat v_2}(\lambda)d\lambda\Big)ds.$$

Now, when $N \to \infty$, we have $T\wedge\tau_N = T$ for $N \geqslant N(\omega)$ large enough, thus $T\wedge\tau_N \to T$ and $x(T\wedge\tau_N) \to x(T)$. Since u and f are bounded, we can pass to the limit at N in (8.47),(8.48),(8.49). We easily deduce the desired result from this.□

We now consider the case of a bounded open set, and take the assumptions (2.1), (3.1),(3.2),(3.3),(3.4),(3.4)',(3.16).

For $h \in L^p(Q)$, $u \in W_0^{1,p}(\mathcal{O})$, consider the equation (cf. Theorem 3.4)

$$(8.50)\quad \left|\begin{array}{l} -\dfrac{\partial u}{\partial t} + A(t)u - B(t)u - H(t)\big(u(t)\big) = h \\ u(x,T) = \bar u(x), \\ u \in L^p\big(0,T;W_0^{1,p}(\mathcal{O})\big),\ \dfrac{\partial u}{\partial t} \in L^p\big(0,T;W^{-1,p}(\mathcal{O})\big) \end{array}\right.$$

which has one and only one solution. We shall also assume (8.19),(8.20), and (8.21) to simplify matters. Moreover, we assume $p > n + 2$, which implies $u \in C^0(\bar Q)$; also (which reinforces (8.19))

(8.51) $\mathcal{V}_1, \mathcal{V}_2$ are compact sets contained in R^d.

The reason for the assumption (8.51) is the following: let us consider $L_u(x,t,v_1,v_2)$ (cf. (8.24) with $\varphi = u$). Since Du is not continuous, this function is not a priori Borel. Nevertheless, it is a Caratheodory function (for almost all x,t it is continuous with respect to v_1,v_2, and for all fixed v_1,v_2 it is Lebesgue measurable at x,t). It follows from this (cf., for example, EKELAND - TEMAM [1], Chap. VIII) that there exists a (Lebesgue) measurable mapping $\hat{v}_2(x,t,v_1)$ such that (8.26) is satisfied for all x,t,v_1. Similarly,

$$\sup_{v_2} L_u(x,t,v_1,v_2) = \operatorname{Max}_{v_2} L_u(x,t,v_1,v_2)$$

is also a Caratheodory function, since for almost all x,t it is continuous in v_1, and for all v_1 it is measurable at x,t. Hence we have the existence of a measurable $\hat{v}_1(x,t)$ such that (8.25) still holds.

In the remainder we shall use the Borel representatives of $\hat{v}_1$ and $\hat{v}_2(x,t,v_1)$. The results will not depend upon the choice of representative (as will be shown).

We define controls $v_1(s)$ and $v_2(s,v_1)$ as above, where v_1 is a process adapted to $\bar{\mu}_t^s$, and $v_2(s,v_1)$ is a process adapted to $\bar{\mu}_t^s$, depending through the measure on all the arguments. We also define $\hat{v}_1(s), \hat{v}_2(s,v_1)$ as above, and introduce f_{v_1,v_2}, $\ldots, c_{1,v_2,v_2}$ through (8.27). Note that f is no longer bounded, since h is not bounded. Lastly, we define χ_{v_1,v_2}, and then the measure $Q^{xt}_{v_1,v_2}$ by (8.30),(8.31), and set

$$(8.52) \qquad J^{xt}\big(v_1(\cdot),\, v_2(\cdot,v_1)\big) = E^{Q^{xt}_{v_1,v_2}}\Big[\int_t^{T\wedge\tau_t} f_{v_1,v_2}(s)\Big(\exp - \int_t^s a_{1,v_1,v_2}(\lambda)d\lambda\Big)ds + \bar{u}\big(x(T)\big)\Big(\exp - \int_t^T a_{1,v_1,v_2}(s)ds\Big)\chi_{T<\tau_t}\Big] .$$

Our objective is to prove:

THEOREM 8.4: *Assume* (2.1),(3.1),(3.2),(3.3),(3.4),(3.4)',(3.16) *together with* (8.20),(8.51) *(and* (8.21) *to simplify). Consider* (8.50) *with* $h \in L^p(Q)$, $u \in W_0^{1,p}(\mathcal{O})$, $p > n + 2$. *Then the conclusions of Theorem 8.3 hold for* u(x,t) *and for the cost function* $J^{xt}(v_1(\cdot),v_2(\cdot,v_1))$ *defined by the relation* (8.52).

Proof: First we notice that the relation

$$(8.53) \qquad u(x,t) = J^{xt}(v_1(\cdot),v_2(\cdot,v_1)) \quad \text{for all } x,t$$

follows from the linear theory. In fact,

$$H(t)\big(u(t)\big)(x) = L_u\big(x,t\ ;\ \hat{v}_1(x,t)\ ;\ \hat{v}_2(x,t,\hat{v}_1)\big) \quad \forall x,t$$

and thus u is the solution of a linear equation similar to (8.5) with the following

$$h \longrightarrow h + f(x,t,\hat{v}_1,\hat{v}_2)$$

$$a_i \longrightarrow a_i - g_i(x,t,\hat{v}_1,\hat{v}_2)$$

$$a_0 \longrightarrow a_0 + \tilde{a}_1(x,t,\hat{v}_1,\hat{v}_2)$$

$$c_0 \longrightarrow c_0\big(1 + c_1(x,t,\hat{v}_1,\hat{v}_2)\big)$$

and applying Theorem 8.2 then leads to (8.53).

Next we introduce the operators

(8.54) $$L^1(t)(\varphi)(x) = \sup_{v_2}\Big\{f\big(x,t,\hat{v}_1(x,t),v_2\big) + D\varphi(x)\cdot g\big(x,t,\hat{v}_1(x,t),v_2\big) - \varphi(x)\tilde{a}_1\big(x,t,\hat{v}_1(x,t),v_2\big) + \int_{R^n}\Big[\varphi(x+z)-\varphi(x)-z\, D\varphi(x)\ \chi_{|z|\leq 1}\Big]c_0c_1\big(x,t,\hat{v}_1(x,t),v_2,z\big)m(dz)\Big\}.$$

(8.55) $$L^2(t)(\varphi)(x) = \inf_{v_1}\Big\{f\big(x,t,v_1,\hat{v}_2(x,t),v_1\big) + D\varphi(x)\cdot g\big(x,t,v_1,\hat{v}_2(x,t,v_1)\big) - \varphi(x)\tilde{a}_1\big(x,t,v_1,\hat{v}_2(x,t,v_1)\big) + \int_{R^n}\Big[\varphi(x+z) - \varphi(x) - z\cdot D\varphi(x)\chi_{|z|\leq 1}\Big]c_0c_1\big(x,t,v_1,\hat{v}_2(x,t,v_1)\big)m(dz)\Big\}.$$

Naturally,

(8.56) $$\left|\begin{aligned} L^1(t)(u)(x) &= H(t)\big(u(t)\big)(x) \\ L^2(t)(u)(x) &= H(t)\big(u(t)\big)(x)\,. \end{aligned}\right.$$

The operators $L^1(t), L^2(t)$ have properties identical to those of H(t). Also, we can always consider that $f(x,t,v_1,v_2) = 0$ for $x \notin \mathcal{O}$. This modifies neither the solution u of (8.50) nor the expression for $J^{xt}(v_1(\cdot),v_2(\cdot,v_1))$. Similarly we can take $h = 0$, $\bar{u} = 0$ outside $\mathcal{O}$. We then introduce the penalised equation (penalisation of the domain)

(8.57) $$\left|\begin{aligned} & -\frac{\partial u_\eta}{\partial t} + A(t)u_\eta - B(t)u_\eta - L^1(t)\big(u_\eta(t)\big) + \frac{1}{\eta}q\,u_\eta = h \\ & u_\eta(T) = \bar{u}, \quad u_\eta \in \mathbb{W}^{2,1,p,0}\,. \end{aligned}\right.$$

By Theorem 3.6, (where h(t) is replaced by $L^1(t)$), we have

$$u_\eta \to z \quad \text{in} \quad L^2(0,T;H^1(R^n)) \quad \text{and} \quad C^0(0,T;L^2(R^n)),$$

where z is the solution of

$$\text{(8.58)}\quad \left|\begin{array}{l} -\dfrac{\partial z}{\partial t} + A(t)z - B(t)z - L^1(t)(z) = h \\ z(x,T) = \bar{u},\ z \in L^p(0,T;W_0^{1,p}(\mathcal{O})),\ \dfrac{\partial z}{\partial t} \in L^p(0,T;W^{-1,p}(\mathcal{O})). \end{array}\right.$$

Now, by (8.56) u is the solution of (8.58), and therefore

$$\text{(8.59)}\quad u_\eta \to u \quad \text{in} \quad L^2(0,t:H^1(R^n)) \quad \text{and} \quad C^0(0,T;L^2(R^n)).$$

By Theorem 8.3 we can write

$$\text{(8.60)}\quad u_\eta(x,t) = \sup_{v_2(\cdot,v_1)} J_\eta^{xt}(v_1(\cdot),v_2(\cdot,v_1))$$

where we have set

$$\text{(8.61)}\quad J_\eta^{xt}(v_1(\cdot),v_2(\cdot,v_1)) = E^{Q^{xt}_{v_1,v_2}} \int_t^T f_{v_1,v_2}(s)\left(\exp - \int_t^s a_{1,v_1,v_2}(\lambda)d\lambda\right) \times$$

$$\times \left(\exp - \int_t^s \frac{1}{\eta}\, q(x(\lambda))d\lambda\right) ds +$$

$$+ \bar{u}(x(T))\left(\exp - \int_t^T a_{1,v_1,v_2}(s)ds\right) \exp - \frac{1}{\eta}\int_t^T q(x(\lambda))d\lambda\Big].$$

As the measur $Q^{xt}_{v_1,v_2}$ is absolutely continuous with respect to P^{xt}, we have $\hat{\tau}_t = \tau_t$, $Q^{xt}_{v_1,v_2}$ a.s.. We can therefore write

$$J_\eta^{xt}(v_1(\cdot),v_2(\cdot,v_1)) - J^{xt}(v_1(\cdot),v_2(\cdot,v_1)) =$$

$$= E^{Q^{xt}_{v_1,v_2}}\left[\int_{T\wedge\hat{\tau}_t}^T f_{v_1,v_2}(s)\left(\exp - \int_t^s a_{1,v_1,v_2}(\lambda)d\lambda\right)\left(\exp - \frac{1}{\eta}\int_{T\wedge\tau_t}^s q(x(\lambda))d\lambda\right) ds + \right.$$

$$\left. + \bar{u}(x(T))\left(\exp - \int_t^T a_{1,v_1,v_2}(\lambda)d\lambda\right)\left(\exp - \frac{1}{\eta}\int_t^T q(x(\lambda)d\lambda\right)\chi_{T>\hat{\tau}_t}\right]$$

$$= E^{xt}\ \chi_{v_1,v_2}(T)\Big[\qquad\Big].$$

By what we saw in Lemma 8.3, $E^{xt}(\chi_{v_1,v_2}(T))^2 \leqslant C_T$. From this it follows that

$$\text{(8.62)}\quad |J_\eta^{xt}(v_1(\cdot),v_2(\cdot,v_1)) - J^{xt}(v_1(\cdot),v_2(\cdot,v_1))| \leqslant C_T' \times$$

$$\times \left(E^{xt}\left[\int_{T\wedge\hat{\tau}_t}^T \left(\exp - \frac{1}{\eta}\int_{T\wedge\hat{\tau}_t}^s q(x(\lambda))d\lambda\right) ds + \right.\right.$$

$$\left.\left. + \chi_{T>\hat{\tau}_t} \exp - \frac{1}{\eta}\int_t^T q(x(\lambda))d\lambda\right]^2\right)^{\frac{1}{2}} \longrightarrow 0$$

as $\eta \to 0$ *uniformly* with respect to v_1,v_2. From this it follows that

$$(8.63)\qquad u_\eta(x,t) \longrightarrow \sup_{v_2(\cdot,v_1)} J^{xt}(\hat{v}_1(\cdot),\, v_2(\cdot,\hat{v}_1)) = \tilde{u}(x,t) \quad \forall x,t\ .$$

Using (8.59) we deduce

$$(8.64)\qquad \tilde{u}(x,t) = u(x,t) \quad \text{for all } t,\ x \text{ a.e..}$$

Let $\theta > t$, $\theta \leqslant T$. Consider the measure $Q^{xt}_{\hat{v}_1,v_2}$ and apply Ito's formula to the function u_η and to the pair $x(s)$, $\chi_{\hat{v}_1,v_2}(s)$. We obtain

$$E^{Q^{xt}_{\hat{v}_1,v_2}} u_\eta(x(s),s)\left(\exp - \int_t^s a_{1,\hat{v}_1,v_2}\, d\lambda\right)\left(\exp - \int_t^s \frac{q(x(\lambda))d\lambda}{\eta}\right) =$$

$$E^{Q^{xt}_{\hat{v}_1,v_2}} \overline{u}(x(T))\left(\exp - \int_t^T a_{1,\hat{v}_1,v_2}\, d\lambda\right)\left(\exp - \int_t^T \frac{q(x(\lambda))d\lambda}{\eta}\right) +$$

$$+ E^{Q^{xt}_{\hat{v}_1,v_2}} \int_s^T f_{\hat{v}_1,v_2}(\lambda)\left(\exp - \int_t^\lambda a_{1,\hat{v}_1,v_2}\, d\mu\right)\left(\exp - \int_t^\lambda \frac{q(x(\mu))d\mu}{\eta}\right) d\lambda$$

$$+ E^{Q^{xt}_{\hat{v}_1,v_2}} \int_s^T \left(L^1(\lambda)(u_\eta(\lambda)) - L_{u_\eta}(x(\lambda),\lambda,\hat{v}_1(\lambda),v_2(\lambda))\right)\left(\exp - \int_t^\lambda a_{1,\hat{v}_1,v_2} d\mu\right) \times$$

$$\times\left(\exp - \int_t^\lambda \frac{q(x(\mu))d\mu}{\eta}\right) d\lambda$$

and thus by the definition (8.54) of L^1, it follows that

$$(8.65)\qquad E^{Q^{xt}_{\hat{v}_1,v_2}} u_\eta(x(s),s)\left(\exp - \int_t^s a_{1,\hat{v}_1,v_2}\, d\lambda\right) \exp - \int_t^s \frac{q(x(\lambda))d\lambda}{\eta} \geqslant$$

$$E^{Q^{xt}_{\hat{v}_1,v_2}} \overline{u}(x(T))\left(\exp - \int_t^T a_{1,\hat{v}_1,v_2}\, d\lambda\right)\left(\exp - \int_t^T \frac{q(x(\lambda))d\lambda}{\eta}\right) +$$

$$+ E^{Q^{xt}_{\hat{v}_1,v_2}} \int_s^T f_{\hat{v}_1,v_2}(\lambda)\left(\exp - \int_t^\lambda a_{1,\hat{v}_1,v_2}\, d\mu\right)\left(\exp - \int_t^\lambda \frac{q(x(\mu))d\mu}{\eta}\right) d\lambda .$$

We can pass to the limit in η in (8.65), and we obtain

$$(8.66)\qquad E^{Q^{xt}_{\hat{v}_1,v_2}} \tilde{u}(x(s),s)\left(\exp - \int_t^s a_{1,\hat{v}_1,v_2}\, d\lambda\right) \chi_{s\leqslant\hat{\tau}_t} \geqslant$$

$$E^{Q^{xt}_{\hat{v}_1,v_2}} \overline{u}(x(T))\left(\exp - \int_t^T a_{1,\hat{v}_1,v_2}\, d\lambda\right) \chi_{T<\hat{\tau}_t} +$$

$$+ E^{Q^{xt}_{\hat{v}_1,v_2}} \int_s^T f_{\hat{v}_1,v_2}(\lambda)\, \chi_{\lambda\leqslant\hat{\tau}_t} \left(\exp - \int_t^\lambda a_{1,\hat{v}_1,v_2}\, d\mu\right) d\lambda .$$

Since $\tilde{u}(x,s) = u(x,s)$ a.e. for fixed s, because $s > t$ we have

(8.67) $$E^{Q^{xt}_{\hat{v}_1,v_2}} \tilde{u}(x(s),s)\left(\exp - \int_t^s a_{1,\hat{v}_1,v_2}\, d\lambda\right) \chi_{s\leq\hat{\tau}_t} =$$

$$= E^{Q^{xt}_{\hat{v}_1,v_2}} u(x(s),s)\left(\exp - \int_t^s a_{1,\hat{v}_1,v_2}\, d\lambda\right) \chi_{s\leq\hat{\tau}_t}.$$

Next consider a sequence $s_k \downarrow t$. Working as in Lemma 8.2, we deduce from (8.66), (8.67) that

$$u(x,t) \geq J^{xt}(v_1(\cdot),v_2(\cdot,v_1)).$$

Taking (8.53) into account, it follows that

$$u(x,t) = \underset{v_2(\cdot,v_1)}{\text{Max}} \; J^{xt}(\hat{v}_1(\cdot),v_2(\cdot,v_1)).$$

Similarly, working with the operator L^2 instead of L^1, we obtain

$$u(x,t) = \underset{v_1(\cdot)}{\text{Min}} \; J^{xt}\left(v_1(\cdot), \hat{v}_2(\cdot,v_1)\right)$$

from which we easily deduce the result desired.□

To finish with, we treat the case of bounded measures, i.e., the interpretation of the equation (3.25). Hence we use the framework of Theorem 3.5, with the assumptions (2.1),(3.1),(3.23),(3.24). Let $h \in L^p(Q)$, $\bar{u} \in W^{2,p}(\mathcal{O}) \cap W_0^{1,p}(\mathcal{O})$, $p > n+1$. Let u be the solution of the problem

(8.68) $$\left| \begin{array}{l} -\dfrac{\partial u}{\partial t} + A(t)u - H(t)\big(u(t)\big) = h \\ u(T) = \bar{u},\; u \in L^p\big(0,T;W^{2,p}(\mathcal{O}) \cap W_0^{1,p}(\mathcal{O})\big),\; \dfrac{\partial u}{\partial t} \in L^p(Q). \end{array}\right.$$

It will be noted that $u, \dfrac{\partial u}{\partial x_k} \in C^0(\bar{Q})$ because $p > n+1$.

We now make two additional assumptions, (8.19) and

(8.69) f,g,a_1 are continuous in v_1,v_2, and are Borel,

(8.70) $\int_{R^n} \varphi(z)M(x,t,v_1,v_2,dz)$ is continuous with respect to v_1,v_2, for φ Borel and bounded, x,t fixed.

We shall write

$$L_\varphi(x,t,v_1,v_2) = f + D\varphi\cdot g - \varphi\tilde{a}_1 +$$

$$+ \int_{R^n}\left[\varphi(x+z) - \varphi(x) - z\cdot D\varphi(x)\; \chi_{|z|\leq 1}\right]M(x,t,v_1,v_2,dz).$$

When u is the solution of (8.68), L_u is Borel and continuous in v_1,v_2. There will exist Borel mappings $\hat{v}_1(x,t),\hat{v}_2(x,t,v_1)$ such that (8.25),(8.26) are satisfied.

The differential game is defined as in Section 8.2. Next we set

(8.72) $$\left| \begin{array}{l} f_{v_1,v_2}(s) = f\big(x(s),s,v_1(s),v_2(s)\big) + h\big(x(s),s\big) \\ a_{1,v_1,v_2}(s) = \tilde{a}_1\big(x(s),s,v_1(s),v_2(s)\big) \\ g_{v_1,v_2}(s) = g\big(x(s),s,v_1(s),v_2(s)\big) \end{array} \right.$$

(8.73) $$M_{v_1,v_2}(x,s,dz) = M\big(x,s,v_1(s),v_2(s),dz\big)$$

which is thus a random measure satisfying (7.68). By Theorem 7.3, there exists a solution $Q^{xt}_{v_1,v_2}$ of the martingale problem (7.6),(7.7) with the data above.

We then define the functional $J^{xt}(v_1(\cdot),v_2(\cdot,v_1))$ by (8.52)

We then have:

THEOREM 8.5: *Assume* (2.1),(3.1),(3.23),(3.24),(8.19),(8.69),(8.70). *Let* $h \in L^p(Q)$, $\bar{u} \in W^{2,p}(\mathcal{O}) \cap W^{1,p}_0(\mathcal{O})$, $p > n + 1$. *Then the conclusions of Theorem 8.3 hold for the function* u *that is the solution of* (8.68), *and for the functional* $J^{xt}(v_1,(\cdot),v_2(\cdot,v_1))$ *defined by* (8.52).

Proof: Note that for $p > n + 1$, $\varphi \in L^p$, we have

(8.74) $$\left| E^{Q^{xt}_{v_1,v_2}} \int_t^T \varphi\big(x(s),s\big)ds \right| \leq C_{T,p}|\varphi|_{L^p}$$

where the constant depends neither upon v_1 nor v_2. We can then apply Ito's formula to functions satisfying the regularity of u (8.68)). Then we can proceed as in Theorem 8.3, and obtain the desired result.□

9. INTERPRETATION OF VARIATIONAL INEQUALITIES

Method of Approach

We now interpret the problems treated in Section 4. However, in order not to weigh down our exposition too much we shall content ourselves with considering the case of linear operators. The non-linear case will be interpreted by drawing upon what was done in Section 8.2.

9.1 *The Case of the V.I. in the Whole Space*

We assume (3.1),(3.2) and (1.3), and define g_i by (8.2), and (to simplify matters) we take

(9.1) $$a_0 = 0.$$

Next we take

(9.2) $$h \in L^\infty, \quad \bar{u} \in W^{2,p,\mu} \cap L^\infty, \quad 2 \leq p < \infty, \quad \mu > 0.$$

Consider two obtacles ψ_1,ψ_2 satisfying (4.1),(4.2). Now define the problem

(9.3)
$$\left|\begin{array}{l} u \in \mathcal{W}^{2,1,p,\mu} \cap L^{\infty}, \quad u(T) = \bar{u} \\ \psi_1 \leq u \leq \psi_2 \\ \left(-\frac{\partial u}{\partial t} + \big(A(t)-B(t)\big)u - h\right). \ \big(v-u(t)\big) \qquad \text{a.e.} \quad \forall v \text{ such that} \\ \qquad \psi_1(x,t) \leq v \leq \psi_2(x,t) \end{array}\right.$$

We write P^{xt} for *the* solution of the martingale problem with respect to the operator $A(t) - B(t)$.

Let θ_1, θ_2 be two $\bar{\mu}_t^s$-stopping times. We define the functional

(9.4)
$$J^{xt}(\theta_1,\theta_2) = E^{xt}\Big[\int_t^{T\wedge\theta_1\wedge\theta_2} f\big(x(s),s\big)\,ds + $$
$$+ \psi_1\big(x(\theta_1),\theta_1\big)\ \chi_{\theta_1\leq\theta_2,\theta_1<T} +$$
$$+ \psi_2\big(x(\theta_2),\theta_2\big)\ \chi_{\theta_2<\theta_1\wedge T} + \bar{u}\big(x(T)\big)\ \chi_{T\leq\theta_1\wedge\theta_2}\Big] .$$

We are going to give the interpretation of u in the form of a differential game with stopping times. For $p > n + 1$, u is continuous on $R^n \times [0,T]$. We then define the open sets of $R^n \times [0,T]$:

(9.5)
$$\left|\begin{array}{l} C_1 = \{x,t \mid u(x,t) > \psi_1(x,t)\} \\ C_2 = \{x,t \mid u(x,t) < \psi_2(x,t)\} \end{array}\right.$$

and we set

(9.6)
$$\hat{\theta}_1 = \inf\{s \geq t \mid u\big(x(s),s\big) = \psi_1\big(x(s),s\big)\}$$
$$\hat{\theta}_2 = \inf|s \geq t \mid u\big(x(s),s\big) = \psi_2\big(x(s),s\big)\} .$$

Since $x(s)$ is cadlag and left quasi-continuous, $\hat{\theta}_1, \hat{\theta}_2$ are $\bar{\mu}_t^s$-stopping times. We then have:

THEOREM 9.1: *Under the assumptions* (3.1),(3.2),(1.3),(9.2),(4.1),(4.2), *and* (9.1) *(to simplify) and* $p > n + 1$, *the solution* u *of* (9.3) *is unique and satisfies*

(9.7)
$$J^{xt}(\theta_1,\hat{\theta}_2) \leq u(x,t) = J^{xt}(\hat{\theta}_1,\hat{\theta}_2) \leq J^{xt}(\hat{\theta}_1,\theta_2), \ \forall\ \theta_1,\ \theta_2$$

(9.8)
$$u(x,t) = \min_{\theta_2}\ \sup_{\theta_1}\ J^{xt}(\theta_1,\theta_2) = \max_{\theta_1}\ \inf_{\theta_2}\ J^{xt}(\theta_1,\theta_2) .$$

Proof: It is useful to note that if $\varphi \in L^{p,\mu}$, $p > n + 1$, $\mu > 0$, then it follows from the estimate (7.32) that

(9.9)
$$\left|E^{xt}\int_t^{T\wedge\tau_R} \varphi\big(x(s),s\big)\,ds\right| \leq C_R|\varphi|_{L^{p,\mu}}$$

$R > 0$, and τ_R denoting the exit time of the process $x(s)$ from the ball B_R with centre 0 and radius R. From this it follows that if θ is a stopping time of $\bar{\mu}_t^s$ then

$$(9.10) \qquad u(x,t) = E^{xt}\left[\int_t^{T\wedge\theta\wedge\tau_R} \left(-\frac{\partial u}{\partial s} + (A-B)u\right)\left(x(s)\right)ds + \right. \\ \left. + u\left(x(T\wedge\theta\wedge\tau_R), T\wedge\theta\wedge\tau_R\right)\right],$$

since $u \in W^{2,1,p,\mu}$ with $p > n + 1$, $\mu > 0$, and from applying Ito's formula. In (9.10) we take $\theta = \hat{\theta}_1\wedge\hat{\theta}_2$. For $s < \hat{\theta}_1\wedge\hat{\theta}_2$, by (9.3) we have

$$-\frac{\partial u}{\partial s} + A(s)u - B(s)u = h.$$

Consequently we deduce from (9.10) that

$$u(x,t) = E^{xt}\left[\int_t^{T\wedge\hat{\theta}_1\wedge\hat{\theta}_2\wedge\tau_R} h\left(x(s)\right)ds + u\left(x(T\wedge\hat{\theta}_1\wedge\hat{\theta}_2\wedge\tau_R), T\wedge\hat{\theta}_1\wedge\hat{\theta}_2\wedge\tau_R\right)\right].$$

Now $\tau_R \uparrow +\infty$ as $R \uparrow +\infty$, since that process $x(s)$ stays bounded in every bounded time interval Thus, making R tend to $+\infty$,

$$(9.11) \qquad u(x,t) = E^{xt}\left[\int_t^{T\wedge\hat{\theta}_1\wedge\hat{\theta}_2} h\left(x(s)\right)ds + u\left(x(T\wedge\hat{\theta}_1\wedge\hat{\theta}_2), T\wedge\hat{\theta}_1\wedge\hat{\theta}_2\right)\right].$$

But

$$u\left(x(T\wedge\hat{\theta}_1\wedge\hat{\theta}_2), T\wedge\hat{\theta}_1\wedge\hat{\theta}_2\right) = \bar{u}\left(x(T)\right) \chi_{T\leqslant\hat{\theta}_1\wedge\hat{\theta}_2} + \\ + u\left(x(\hat{\theta}_1),\hat{\theta}_1\right) \chi_{\hat{\theta}_1\leqslant\hat{\theta}_2,\hat{\theta}_1<T} + u\left(x(\hat{\theta}_2),\hat{\theta}_2\right) \chi_{\hat{\theta}_2<\hat{\theta}_1\wedge T}.$$

However, $x(\hat{\theta}_2) \notin C_2$, therefore

$$u(x(\hat{\theta}_2),\hat{\theta}_2) = \psi_2(x(\hat{\theta}_2),\hat{\theta}_2).$$

Similarly, $x(\hat{\theta}_1) \notin C_1$, therefore

$$u(x(\hat{\theta}_1),\hat{\theta}_1) = \psi_1(x(\hat{\theta}_1),\hat{\theta}_1).$$

Consequently,

$$u(x,t) = J^{xt}(\hat{\theta}_1,\hat{\theta}_2).$$

Now let θ_2 be an arbitrary stopping time of $\bar{\mu}_t^s$. Apply (9.10) with $\theta = T\wedge\hat{\theta}_1\wedge\hat{\theta}_2\wedge\tau_R$. But for $s < T\wedge\hat{\theta}_1\wedge\hat{\theta}_2\wedge\tau_R$, by (9.3) we have

$$\left(-\frac{\partial u}{\partial s} + (A(s) - B(s))u\right)(x(s)) \leqslant h(x(s),s).$$

Consequently we obtain

$$u(x,t) \leqslant E^{xt}\left[\int_t^{T\wedge\hat{\theta}_1\wedge\theta_2\wedge\tau_R} h\big(x(s)\big)ds + u\big(x(T\wedge\hat{\theta}_1\wedge\theta_2\wedge\tau_R),T\wedge\hat{\theta}_1\wedge\theta_2\wedge\tau_R\big)\right]$$

and making R tend to $+\infty$ this yields

$$u(x,t) \leqslant E^{xt}\left[\int_t^{T\wedge\hat{\theta}_1\wedge\theta_2} h\big(x(s)\big)ds + u\big(x(T\wedge\theta_1\wedge\theta_2),\ T\wedge\theta_1\wedge\theta_2\big)\right] .$$

Further, we have

$$\begin{aligned} u\big(x(T\wedge\hat{\theta}_1\wedge\theta_2),T\wedge\hat{\theta}_1\wedge\theta_2\big) &= \bar{u}\big(x(T)\big)\ \chi_{T\leqslant\hat{\theta}_1\wedge\theta_2} + \\ &+ u\big(x(\hat{\theta}_1),\hat{\theta}_1\big)\ \chi_{\hat{\theta}_1\leqslant\theta_2,\hat{\theta}_1<T} + u\big(x(\theta_2),\theta_2\big)\ \chi_{\theta_2<\hat{\theta}_1\wedge T} \\ &\leqslant u\big(x(T)\big)\ \chi_{T\leqslant\hat{\theta}_1\wedge\theta_2} + \psi_1\big(x(\hat{\theta}_1),\hat{\theta}_1\big)\ \chi_{\hat{\theta}_1\leqslant\theta_2\wedge\hat{\theta}_1<T} + \\ &+ \psi_2\big(x(\theta_2),\theta_2\big)\ \chi_{\theta_2<\hat{\theta}_1\wedge T} . \end{aligned}$$

We deduce the second part of (9.7) from this. The left part is deduced similarly.

The property (9.8) is an immediate consequence of (9.7). From this we deduce the uniqueness of the solution of the problem (9.3).□

9.2 *The Case of the V.I. in a Bounded Open Set*

We shall distinguish the case of regular obstacles and that of obstacles that are only continuous.

9.2.1 Regular obstacles

Here we assume (1.3),(2.1),(3.1),(3.2), and as in Section 9.1, g_i is defined by (8.2), and (to simplify) we assume (9.1). We also assume (3.16) and that the obstacles satisfy (4.1),(4.2),(4.16). We are given $h \in L^\infty(Q)$, $\bar{u} \in W_0^{1,p}(\mathcal{O}) \cap L^\infty$, $p > n + 2$ (satisfying (4.2)). We consider the V.I.

$$u \in L^p\big(0,T;W_0^{1,p}(\mathcal{O})\big),\ \frac{\partial u}{\partial t} \in L^p\big(0,T:W^{-1,p}(\mathcal{O})\big) \tag{9.12}$$

$$\psi_1 \leqslant u \leqslant \psi_2\ ,\quad u(x,T) = \bar{u}(x) \tag{9.13}$$

$$\left|\begin{array}{l} -\left(\frac{\partial u}{\partial t},\ v-u\right) + \big<\big(A(t)-B(t)\big)u(t),\ v-u(t)\big> \geqslant \big(h,v-u(t)\big) \\ t \text{ a.e.},\quad v \in H_0^1 \quad \text{such that } \psi_1(x,t) \leqslant v(x) \leqslant \psi_2(x,t) \quad \text{a.e..} \end{array}\right. \tag{9.14}$$

Let P^{xt} be the solution of the martingale problem with respect to the operator $A(t) - B(t)$, and τ_t the exit time of the process $x(t)$ from the open set. For θ_1,θ_2 we define the functional

(Contd)

$$J^{xt}(\theta_1,\theta_2) = E^{xt}\left[\int_t^{T\wedge\theta_1\wedge\theta_2\wedge\tau_t} h\big(x(s),s\big)ds + \bar{u}\big(x(T)\big)\ \chi_{T\leqslant\theta_1\wedge\theta_2\leqslant\tau_t} + \right. \tag{9.15}$$

(Contd) $$+ \psi_1(x(\theta_1),\theta_1)\ \chi_{\theta_1\leq\theta_2,\theta_1<T\wedge\tau_t} + \psi_2(x(\theta_2),\theta_2)\ \chi_{\theta_2<\theta_1\wedge T\wedge\tau_t}\Big].$$

Lastly we define C_1, C_2 by (9.5) and $\hat\theta_1, \hat\theta_2$ by (9.6). We shall also, in order to use the penalised equation associated with (9.14), let

(9.16) $$\left| \begin{array}{l} -\frac{\partial u_\varepsilon}{\partial t} + A(t)u_\varepsilon - B(t)u_\varepsilon + \frac{1}{\varepsilon}(u_\varepsilon - \psi_2)^+ - \frac{1}{\varepsilon}(u_\varepsilon - \psi_1)^- = h \\ u_\varepsilon(x,T) = \bar{u}(x) \\ u_\varepsilon \in L^p(0,T;W_0^{1,p}(\mathcal{O})),\ \frac{\partial u_\varepsilon}{\partial t} \in L^p(0,T;W^{-1,p}(\mathcal{O})). \end{array} \right.$$

In the course of proving Theorem 4.2 we saw that

(9.17) $$\left| \begin{array}{l} u_\varepsilon \to u \text{ weakly in } L^p(0,T;W_0^{1,p}(\mathcal{O})) \\ \frac{\partial u_\varepsilon}{\partial t} \to \frac{\partial u}{\partial t} \text{ weakly in } L^p(0,T;W^{-1,p}(\mathcal{O})) \end{array} \right.$$

Since the injection from

$$Z = \{\varphi \mid \varphi \in L^p(0,T;W_0^{1,p}(\mathcal{O})),\ \frac{\partial\varphi}{\partial t} \in L^p(0,T;W^{-1,p}(\mathcal{O}))\}$$

into $C^0(\bar{Q})$ is *compact* for $p > n + 2$ it follows from (9.17) that we also have

(9.18) $$u_\varepsilon \to u \text{ strongly in } C^0(\bar{Q}).$$

We can also state:

THEOREM 9.2: *Assume* (1.3),(2.1),(3.1),(3.2),(3.16),(4.1),(4.2),(4.16). *For* $h \in L^\infty(Q)$, $\bar{u} \in W_0^{1,p}(\mathcal{O})$, $p > n + 2$, *the solution* u *of* (9.12),(9.13),(9.14) *is given explicitly by* (9.7),(9.8) *where the functional* $J^{xt}(\theta_1,\theta_2)$ *is defined by* (9.15).

Proof: Let θ be a stopping time of $\bar{\mathcal{F}}_t^s$. Theorem 8.2 (the interpretation of the equations) allows us to write, using (9.16),

(9.19) $$u_\varepsilon(x,t) = E^{xt}\Big[\int_t^{T\wedge\theta\wedge\tau_t} \big(h(x(\lambda),\lambda) - \frac{1}{\varepsilon}(u_\varepsilon - \psi_2)^+(x(\lambda),\lambda) + \frac{1}{\varepsilon}(u_\varepsilon - \psi_1)^-(x(\lambda),\lambda)\big)d\lambda + u_\varepsilon(x(T\wedge\theta\wedge\tau_t), T\wedge\theta\wedge\tau_t)\Big].$$

If $\theta = T$, (9.19) follows immediately from Theorem 8.2. In the general case we proceed as in the proof of Theorem 8.2. Let (x,t) be fixed such that $u(x,t) > \psi_1(x,t)$, and δ such that $u(x,t) > \psi_1(x,t) + \delta$. We define

$$\theta_1^\delta = \inf\{s \geq t \mid u(x(s),s) < \psi_1(x(s),s) + \frac{\delta}{2}\}$$

which is a stopping time of $\mathcal{F}_t^s$. Take $\theta = \theta_2 \wedge \theta_1^\delta$ in (9.12), and let

$$(9.20) \qquad u_\varepsilon(x,t) = E^{xt}\Big[\int_t^{T\wedge\theta_2\wedge\theta_1^\delta\wedge\tau_t} \Big(h\big(x(\lambda),\lambda\big) - \frac{1}{\varepsilon}(u_\varepsilon-\psi_2)^+\big(x(\lambda),\lambda\big) + \frac{1}{\varepsilon}(u_\varepsilon-\psi_1)^-\big(x(\lambda),\lambda\big)\Big)d\lambda$$

$$+ u_\varepsilon\big(x(T\wedge\theta_2\wedge\theta_1^\delta\wedge\tau_t),\ T\wedge\theta_2\wedge\theta_1^\delta\wedge\tau_t\big)\Big].$$

By definition

$$s \in [t, \theta_1^\delta \wedge \tau_t \wedge T[\text{ implies } x(s) \in \mathcal{O} \text{ and } u\big(x(s),s\big) \geq \psi_1\big(x(s),s\big) + \frac{\delta}{2}.$$

Since $u_\varepsilon \to u$ in $C^0(\bar{Q})$, we can find ε_δ such that for $\varepsilon \leq \varepsilon_\delta$

$$\|u_\varepsilon - u\|_{C^0(\bar{Q})} \leq \frac{}{2}.$$

From this it follows that

$$\varepsilon \leq \varepsilon_\delta,\ s \in [t, \theta_1^\delta\wedge\tau_t\wedge T[\text{ implies } u_\varepsilon\big(x(s),s\big) \geq \psi_1\big(x(s),s\big) + \frac{\delta}{4}$$

and therefore

$$(u_\varepsilon - \psi_1)^-(x(s),s) = 0.$$

Consequently we deduce from (9.20) that

$$(9.21) \qquad u_\varepsilon(x,t) \leq E^{xt}\Big[\int_t^{T\wedge\theta_2\wedge\theta_1^\delta\wedge\tau_t} h\big(x(\lambda),\lambda\big)d\lambda +$$

$$+ u_\varepsilon\big(x(T\wedge\theta_2\wedge\theta_1^\delta\wedge\tau_t),\ T\wedge\theta_2\wedge\theta_1^\delta\wedge\tau_t\big)\Big].$$

Next, make ε tend ot 0. From (9.21) it follows that

$$(9.22) \qquad u(x,t) \leq E^{xt} \int_t^{T\wedge\theta_2\wedge\theta_1^\delta\wedge\tau_t} h\big(x(\lambda),\lambda\big)d\lambda + u\big(x(T\wedge\theta_2\wedge\theta_1^\delta\wedge\tau_t), T\wedge\theta_2\wedge\theta_1^\delta\wedge\tau_t\big).$$

Next, make δ tend to 0. Then $\theta_1^\delta \uparrow \hat{\theta}_1$. In fact the sequence θ_1^δ increases as $\delta \uparrow 0$, and $\theta_1^\delta \leq \hat{\theta}_1$, thus $\theta_1^\delta \uparrow \Lambda$. By the definition of θ_1^δ we have

$$u\big(x(\theta_1^\delta),\theta_1^\delta\big) \leq \psi_1\big(x(\theta_1^\delta),\theta_1^\delta\big) + \frac{\delta}{2}.$$

As the process x(s) is left quasi-continuous, since u is continuous we have

$$u(x(\Lambda),\Lambda) \leq \psi_1(x(\Lambda),\Lambda)$$

and since $u \geq \psi_1$,

$$u(x(\Lambda),\Lambda) = \psi_1(x(\Lambda),\Lambda)$$

therefore $\Lambda \geqslant \hat{\theta}_1$, hence $\Lambda = \hat{\theta}_1$.

Using the left quasi-continuity of x(s) again, it then follows from (9.22) that

$$u(x,t) \leqslant E^{xt}\left[\int_t^{T\wedge\theta_2\wedge\hat{\theta}_1\wedge\tau_t} h(x(\lambda),\lambda)\,d\lambda + u(x(T\wedge\theta_2\wedge\hat{\theta}_1\wedge\tau_t),\ T\wedge\theta_2\wedge\hat{\theta}_1\wedge\tau_t)\right]$$

therefore

$$u(x,t) \leqslant J^{xt}(\hat{\theta}_1,\theta_2) \quad \text{for all } \theta_2. \tag{9.23}$$

Therefore (9.23) is true, at least if x,t satisfy $u(x,t) > \psi_1(x,t)$. If $u(x,t) = \psi_1(x,t)$, then $\hat{\theta}_1 = t < T$, and thus

$$J^{xt}(\hat{\theta}_1,\tilde{\theta}_2) = \psi_1(x,t)$$

so that (9.23) is again satisfied, at least if $t < T$. If $t = T$, then

$$J^{xt}(\theta_1,\theta_2) = \bar{u}(x) \quad \text{and} \quad u(x,t) = \bar{u}(x),$$

therefore (9.23) is always satisfied. We identically show that

$$J^{xt}(\theta_1,\hat{\theta}_2) \leqslant u(x,t) \quad \text{for all } \theta_1,$$

whence the result desired.□

9.2.2 Non-regular obstacles

Here we shall take the assumptions (1.3),(2.1),(3.1),(3.2),(3.16). We add (4.24),(4.25). The data $\bar{u},h,\psi_1,\psi_2$ will satisfy the assumptions (4.28),(4.29), (4.50). We then consider the problem

$$(9.24)\qquad \begin{aligned} &u\in L^2(0,T;H_0^1),\ \frac{\partial u}{\partial t}\in L^2(Q),\ u\in C^0(\bar{Q}),\\ &\psi_1\leqslant u\leqslant \psi_2,\ u(x,T)=\bar{u}(x)\ , \end{aligned}$$

$$(9.25)\qquad \begin{aligned} &-\left(\frac{\partial u}{\partial t},v-u(t)\right) + \langle (A(t)-B(t))u(t),v-u(t)\rangle \geqslant\\ &\qquad \geqslant (h(t),v-u(t)) \quad t \text{ a.e.}\\ &\forall v\in H_0^1,\ \psi_1(x,t)\leqslant v(x)\leqslant \psi_2(x,t) \quad x \text{ a.e.}. \end{aligned}$$

Again we consider $J^{xt}(\theta_1,\theta_2)$ defined by (9.15), and C_1,C_2 by (9.5), and $\hat{\theta}_1,\hat{\theta}_2$ by (9.6). We can state:

THEOREM 9.8: *Assume* (1.3),(2.1),(3.1),(3.2),(3.16),(4.24),(4.25),(4.28),(4.29), (4.30). *Then the solution* u *of* (9.22),(9.23),(9.24) *is again given explicitly by* (9.7),(9.8), *where* J^{xt} *is given by* (9.15).

Proof: We work by regularising the data. Let ψ_1^N, ψ_2^N be such that

$$\left|\begin{array}{l} \psi_i^N \in W^{2,1,\infty}(Q) \\ \psi_1^N \leqslant \psi_2^N \text{ on } \bar{Q},\ \psi_1^N\big|_\Sigma \leqslant 0 \leqslant \psi_2^N\big|_\Sigma \\ \psi_1^N(x,T) \leqslant \bar{u}(x) \leqslant \psi_2^N(x,T) \\ \psi_i^N \to \psi_i \text{ in } C^0(\bar{Q}). \end{array}\right.$$

We denote by u^N the solution of (9.12),(9.13),(9.14) corresponding to the ostacles ψ_1^N, ψ_2^N.

Next, let u^ε be the solution of the penalised problem

$$\text{(9.26)} \quad \left|\begin{array}{l} -\dfrac{\partial u_\varepsilon}{\partial t} + A(t)u_\varepsilon - B(t)u_\varepsilon + \dfrac{1}{\varepsilon}(u_\varepsilon - \psi_2)^+ - \dfrac{1}{\varepsilon}(u_\varepsilon - \psi_1)^- = h \\ u_\varepsilon\big|_{\bar{\Sigma}} = 0, \quad u_\varepsilon(x,T) = \bar{u}(x) \\ u_\varepsilon \in L^p(0,T;W_0^{1,p}(\mathcal{O})),\ \dfrac{\partial u_\varepsilon}{\partial t} \in L^p(0,T;W^{-1,p}(\mathcal{O})) \end{array}\right.$$

and let u_ε^N be the solution of the corresponding problem with the obstacles ψ_i changed into ψ_i^N. Let $J^{x,t,N}$ be the functional defined by (9.15), with ψ_i^N replacing ψ_i. By formula (9.15) we easily see that

$$\text{(9.27)} \quad |J^{x,t,N}(\theta_1,\theta_2) - J^{x,t}(\theta_1,\theta_2)| \leqslant \max\left(\|\psi_1^N - \psi_1\|_{C^0(\bar{Q})}, \|\psi_1^N - \psi_1\|_{C^0(\bar{Q})}\right) = \mu_N$$

thus necessarily we have

$$\text{(9.28)} \quad u(x,t) = \inf_{\theta_2} \sup_{\theta_1} J^{xt}(\theta_1,\theta_2) = \sup_{\theta_1} \inf_{\theta_2} J^{xt}(\theta_1,\theta_2).$$

On the other hand, we can give the interpretation of u_ε by applying Theorem 8.4. If $v_1(s), v_2(s)$ are scalar processes adapted to $\bar{\mathcal{F}}_t^s$ such that $0 \leqslant v_i(s) \leqslant 1$, then we define

$$\text{(9.29)} \quad J^{x,t}(v_1,v_2) = E^{xt}\Big[\int_t^T \Big(h(x(s),s) + \frac{1}{\varepsilon} v_1(s)\psi_1(x(s),s) + \frac{1}{\varepsilon} v_2(s)\psi_2(x(s),s)\Big)\Big(\exp -\frac{1}{\varepsilon}\int_t^s (v_1(\lambda) + v_\varepsilon(\lambda))d\lambda\Big) ds + \bar{u}(x(T))\exp -\frac{1}{\varepsilon}\int_t^T (v_1(s) + v_2(s))ds\Big].$$

By applying Theorem 8.4 we then have

$$(9.30)\qquad u_\varepsilon(x,t) = \min_{v_2}\sup_{v_1} J_\varepsilon^{x,t}(v_1,v_2) =$$

$$= \max_{v_1}\inf_{v_2} J_\varepsilon^{xt}(v_1,v_2) = J_\varepsilon^{xt}(\hat{v}_1^\varepsilon,\hat{v}_2^\varepsilon)$$

where

$$(9.31)\qquad \begin{cases} \hat{v}_1^\varepsilon(s) = \begin{vmatrix} 1 \text{ if } u_\varepsilon\big(x(s),s\big) \leqslant \psi_1\big(x(s),s\big) \\ 0 \text{ otherwise.}\end{vmatrix} \\ \hat{v}_2^\varepsilon(s) = \begin{vmatrix} 1 \text{ if } u_\varepsilon\big(x(s),s\big) \geqslant \psi_2\big(x(s),s\big) \\ 0 \text{ otherwise.}\end{vmatrix} \end{cases}$$ (1)

But then using (9.29) we immediately see, as for (9.27), that

$$(9.32)\qquad |J_\varepsilon^{xt}(v_1,v_2) - J_\varepsilon^{xt,N}(v_1,v_2)| \leqslant \mu_N \ .$$

From (9.27),(9.28) and (9.30),(9.32) we deduce

$$(9.33)\qquad |u^N(x,t)-u(x,t)| \leqslant \mu_N$$

$$|u_\varepsilon^N(x,t)-u_\varepsilon(x,t)| \leqslant \mu_N \ .$$

Now, for fixed N we know by (9.18) that

$$u_\varepsilon^N \to u^N \quad \text{in} \quad C^0(\bar{Q})$$

The estimates (9.33) show that

$$(9.34)\qquad u_\varepsilon \to u \quad \text{in } C^0(\bar{Q}).$$

From here on the proof is identical to that of Theorem 9.2.□

Remark 9.1

In the case of a single obstacle we can intepret the solution of the weak V.I. studied in Theorem 4.4 as the lower bound of a control problem. The proof is very similar to that of Theorem 9.3, the essential point being to show that $u_\varepsilon \to u$ in $C^0(\bar{Q})$.□

10. COMMENTS

Under the assumption (1.3) the set $\{|z| \leqslant 1\}$ can be replaced by an arbitrary ball and its complement. This remark also applies to the operators B and $\tilde{B}$.

The introduction of the two operators B and $\tilde{B}$, which do not have quite the same properties, is owed to operating in R^n with weighted spaces. It is the non-local character of the operators which implies some difficulties when we use weights.

---

(1) It is no use to introduce here processes of the type $v_2(\cdot,v_1)$.

Lemma 1.1 shows that B or $\tilde{B}$ are comparable to second order differential operators, with, however, the great advantage that we can make the norm small. The presence of a non-degenerate operator then allows us to solve problems such as (1.22), since $\tilde{B},\tilde{H}$ (or B,H) appear as perturbations of A.

It would be interesting to consider the case of a degenerate operator A and to generalise the results of J.L. MENALDI [1],[2] for diffusions in this framework.

Lemma 1.6 is a Maximum Principle for the operator A - B - H.

It would be interesting to show that in Theorems 1.2 and 1.4 we can take $\mu = 0$.

The case of a bounded open set is very different from the case R^n, because of the non-local character of the operator B.

The Dirichlet problem for A - B - H is not solved when none of the assumptions (2.5),(2.10) is satisfied, whilst the problem in R^n remains well set. It would be interesting to extend the cases of solving the Dirichlet problem. Naturally, we can use the continuation operator method, expounded in Section 2.4, with the Dirichlet boundary conditions, but the solution is not zero outside $\partial\mathcal{O}$.

Condition (2.30) means that jumps are only possible towards the interior of $\mathcal{O}$. We then understand that we can impose a reflection condition at the boundary, as in the jump-free case. Without this condition equation (2.29) cannot have a probabilistic interpretation. The technique of penalisation of the domain described in Section 3.5 is owed to A. BENSOUSSAN, J.L. MENALDI [1].

We could consider integral random measures not satisfying (5.19), for which the increasing process $\pi(t;A)$ defined in (5.16) would not longer be continuous.

Theorem 5.4, and especially its Corollary 5.1, allow us to make changes in probability similar to those of Girsanov's Theorem for diffusions.

We can solve stochastic differential equations corresponding to monotonic or multivariable operators, as is done for diffusions, cf. B.L. [1], loc. cit.. The proof of the uniqueness of the solution of the martingale problem is different from that of D. STROOCK [1] or J.P. LEPELTIER - B. MARCHAL [1]. It approaches that of T. KOMATSU [1]. It consists of identifying a Markovian probability transition function which stays the same for every solution of the martingale problem. The estimate of the type (7.32) plays a fundamental role in this approach.

The approximation process for the existence also is different from what was done in the works previously cited. But the compactness Lemma 7.11 is greatly inspired by J.P. LEPELTIER - B. MARCHAL [1], loc. cit..

The case of bounded measures treated in Section 7.4 is rather special, since the measures can even be random, provided that they are uniformly bounded. Under the assumptions of Theorem 9.1 the case of non-regular obstacles is open.

Chapter 4
Quasi-variational inequalities of elliptic type

INTRODUCTION

The object of this Chapter is to present the principal analytical results about elliptic quasi-variational inequalities. The problem is presented in essentially the form

$$a(u,v - u) \geqslant (f,v - u), \quad u \in V,$$

$$\text{for all } v \in V \text{ such that } v \in K(u),\ u \in K(u),$$

where V is a Hilbert space, a(u,v) a continuous bilinear form, and $K(\varphi)$ a family of closed convex sets in V. In short, this is a matter of a variational inequality (V.I.) where the convex set depends upon the solution. This explains the terminology.

An important particular case of a Q.V.I. is that of the implicit obstacle problem, i.e.,

$$K(\varphi) = \{v | v \leqslant M(\varphi)\},$$

where M is a certain non-linear operator.

In general, if we call $T(\varphi) = \psi$ the solution of the Q.V.I. (depending on φ)

$$a(\psi,v - \psi) \geqslant (f,v - \psi)$$

$$\text{for all } v \in V,\ v \in K(\varphi),\ \psi \in K(\varphi),$$

the problem of the Q.V.I. is reduced to that of finding a fixed point of the mapping T. This is why two types of method can be considered, the continuity methods (founded upon the theorems of Schauder or Kakutani) and the monotonicity methods (Birkhoff). For applications to control (impulse control, cf. Chap. VI) the most useful are those of monotonicity. Let us consider the implicit obstacle problem; the operator M is increasing in the following sense

$$M\varphi_1 \geqslant M\varphi_2 \quad \text{if} \quad \varphi_1 \geqslant \varphi_2 \text{ (pointwise)}.$$

However, it does not have continuity properties in suitable spaces. The increasing property of the solution of the obstacle problem with respect to the obstacle implies that the mapping T is increasing. Section 1 is devoted to the implicit obstacle problem, the corresponding Q.V.I. is called the impulse control Q.V.I. (since the operator M which occurs is that for impulse control and will be motivated in Chap. VI (cf. Chap. 1, already)). We study the problem of the existence of a solution with the help of recurrence methods, or by using an abstract theorem of G. BIRKHOFF [1] generalised by L. TARTAR [1]. We thus show the existence of a maximum solution and a minimum solution (in a certain interval of functions). We establish a dual symmetry when recurrence methods are used. A decreasing procedure easily leads to the maximum solution. However, the increasing procedure which must lead to the minimum solution poses difficulties. Of course, under reasonable assumptions the solution of the Q.V.I. becomes unique. This is the case if we look for bounded solutions.

Section 2 is devoted to the reularity problem once the uniqueness of a bounded and continuous solution in V is established. If $V = H_0^1(\mathcal{O})$ (or $H^1(\mathcal{O})$), we seek to learn whether the solution u is in $W^{2,p}(\mathcal{O})$ (always in the case of the impulse control Q.V.I.). The problem is completely solved in the Dirichlet case. In the Neumann case the regularity results at our disposal are (on the whole) local in nature.

Section 3 is devoted to a problem where the (linear) operator A, defined by the form

$$a(u,v) = \langle Au,v\rangle$$

(the Dirichlet problem), is a second order differential operator, and is replaced by a non-linear operator

$$A - H,$$

where H(x,u,Du) is an operator bringing in derivatives of order 1 at most, but non-linear (H is called a Hamiltonian). In fact, when $H(x,\lambda,p)$ increases less than quadratically in p, $A - H + \lambda$ becomes monotonic for λ large enough. In this case the Q.V.I. is solved identically to the case where $H = 0$. The case where H increases quadratically poses very much more complex problems. We study such a case motivated by a stochastic control problem where there appear at the same time a continuous control and an impulse control.

Section 5 is devoted to a weakened formulation of the implicit obstacle problem called the semi-group formulation. Here we consider a set of functions satisfying

u bounded and continuous,

$$u \leqslant \int_0^t e^{-\alpha t}\Phi(s)f\,dt + e^{-\alpha t}\Phi(t)u, \quad \text{for all } t \geqslant 0,$$

$$u \leqslant M(u),$$

where $\Phi(t)$ is a Markov semi-group (for example, a diffusion semi-group whose infinitesimal generator in a suitable sense is $-A$). The "good" function is the "maximum" solution of the set above. We study this problem with the help of various approximation procedures (discretisation or penalisation).

Section 6 is devoted to systems of Q.V.I.'s. Again, this is a situation in which the monotonicity method is applied, although the problem is different from the implicit obstacle problem. Essentially we look for N functions $(u^1,\ldots,u^N)$ satisfying

$$a^m(u^m, v - u^m) \geqslant (f^m, v - u^m),$$

$$\text{for all } v \leqslant k + \operatorname*{Inf}_{\mu\neq m} u^\mu, \qquad u^m \leqslant k + \operatorname*{Inf}_{\mu\neq m} u^\mu,$$

where the $a^m(u,v)$ are bilinear forms of the same type as $a(u,v)$. We study the problems of existence, uniqueness, and regularity, and we consider a semi-group formulation of the problem.

Section 7 is devoted to Q.V.I.'s where monotonicity methods are not applied. Essentially we use continuity methods. We can give general existence theorems by making assumptions about the multi-mapping $\varphi \to K(\varphi)$. In general, however, it is better to apply the "idea" of the continuity method to the example, rather than the general theorem. In fact we establish that there may be various ways of defining a mapping whose fixed points are solutions of the Q.V.I.. We give several examples of such a situation.

Section 8 is devoted to Q.V.I.'s arising not in the optimal control problem, but in games situations. For example, we establish the double obstacle V.I.

$$\left| \begin{array}{l} a(u,v-u) \geqslant (f,v-u), \\ \text{for all } v|\psi_1 \leqslant v \leqslant \psi_2, \quad \psi_1 \leqslant u \leqslant \psi_2, \end{array} \right.$$

is related to a Q.V.I. (in fact to a system of Q.V.I.'s).

The preceding double Q.V.I. is related to a games problems in which the decision variables are stopping times. It is a game with zero outcome. We can consider games with zero outcome which lead to very much more complex Q.V.I.'s. We also show that the search for Nash points under constraints leads naturally to Q.V.I.'s and even often to "decreasing" Q.V.I.'s. If, for example, we take the implicit obstacle problem, by that we shall understand a situation in which the operator M is decreasing instead of increasing. We study such a case. We finish with a Q.V.I. arising from impulse games, a situation in which there are two implicit obstacles $M_1(u), M_2(u)$.

1. THE Q.V.I. FOR IMPULSE CONTROL

1.1 *Notations: Assumptions.*

We are given functions $a_{ij}(x), a_i(x), a_o(x)$, $i,j = 1,\dots,n$ satisfying the relations

$$(1.1) \qquad \left| \begin{array}{l} a_{ij} = a_{ji} \in W^{1,\infty}(R^n) \quad , \quad a_i, a_o \in L^\infty(R^n) \\ \sum a_{ij}\, \xi_i\, \xi_j \geqq \alpha |\xi|^2,\ \forall\ \xi \in R^n,\ \alpha > 0;\ a_o(x) \geqq \beta > 0. \end{array} \right.$$

We define the second order differential operator

$$(1.2) \qquad A\,\varphi = -\sum_{i,j} \frac{\partial}{\partial x_i}\, a_{ij}\, \frac{\partial\varphi}{\partial x_j} + \sum_i a_i\, \frac{\partial\varphi}{\partial x_i} + a_o\, \varphi .$$

Let $\mathcal{O}$ be a bounded open set of R^n. For $u,v \in H^1(\mathcal{O})$ we set

$$(1.3) \qquad a(u,v) = \sum_{i,j} \int_{\mathcal{O}} a_{ij}(x) \frac{\partial u}{\partial x_j} \frac{\partial v}{\partial x_i} dx + \sum_i \int_{\mathcal{O}} a_i \frac{\partial u}{\partial x_i} v\, dx + \int_{\mathcal{O}} a_o u v\, dx .$$

Next we are given

$$(1.4) \qquad f \in L^2(\mathcal{O}), \quad f \geqslant 0,$$

$$(1.5)\quad \left|\begin{array}{l} k > 0, \quad C_o : (R^n)^+ \to R^+, \text{ continuous}, \quad C_o(0) = 0, \\ C_o \text{ sublinear and non-decreasing, } C_o(\xi) \leqslant d|\xi|^\gamma \\ \text{where } d,\gamma \text{ are positive constants.} \end{array}\right.$$

Define the non-linear operator

$$(1.6)\qquad M\varphi(x) = k + \inf_{\substack{\xi \geqq 0 \\ x+\xi \in \bar{\mathcal{O}}}} [\varphi(x+\xi) + C_o(\xi)]$$

and the expression (1.6) has a meaning for φ bounded from below.

We shall call the following problem a "*Quasi-Variational Inequality* (Q.V.I.) *of impulse control* :

$$(1.7)\quad \left|\begin{array}{l} a(u,v-u) \geqq (f,v-u) , \quad u \in V , \quad u \geqq 0 \\ \forall\, v \in V , \text{ such that } v \leqq Mu \text{ a.e. } ; u \leqq Mu \text{ a.e.} \end{array}\right.$$

In (1.7) V *is the Hilbert space* $H_o^1(\mathcal{O})$ or $H^1(\mathcal{O})$ (and boundary conditions are of the Dirichlet or Neumann type).

Remark 1.1

The terminology "impulse control" will be justified later on.□

Remark 1.2

The problem (1.7) can be considered as an obstacle problem with obstacle dependent upon the solution, which explains the terminology quasi-variational inequality.□

Remark 1.3

The operator M does not preserve the regularity H^m for $m \geqslant 1$. Nevertheless, it has the following increasing property:

$$M\varphi_1(x) \leqslant M\varphi_2(x) \quad \text{for all } \varphi_1,\varphi_2 \text{ such that } \varphi_1 \leqslant \varphi_2 \text{ a.e.}$$

This property will play a fundamental role in the study of (1.7).□

1.2 *Existence*

Here we assume that

(1.8) $\mathcal{O}$ is a regular bounded open set.

Under these conditions the equation

$$(1.9)\qquad a(u^o,v) = (f,v) \quad \text{for all } v \in V,\ u^o \in V$$

has one and only one solution $u^o \geqslant 0$.

In fact, by Fredholm's alternative the number of solutions of (1.9) is the same as that of the equation corresponding to $f = 0$. Furthermore, when $f \in L^\infty$ the solution $u^o \in V$ is in fact regular, $u^o \in W^{2,p}$, $2 \leqslant p < \infty$, thus in particular $u^o \in L^\infty$.

Now, the solution in $V \cap L^\infty$ is unique (cf., for example, B.L. [1] Chap. 2).

We are interested in the solutions of (1.7) that satisfy the following property:

(1.10) $u(x) \leqslant u^o(x)$ a.e..

We shall prove:

THEOREM 1.1: *Under the assumptions* (1.1),(1.4),(1.5),(1.8) *the set of solutions* (1.7),(1.10) *is non-empty and has a minimum element and a maximum element.* □

Before proving Theorem 1.1, we are going to give some preliminary results. First of all, there exists $\lambda \geqslant 0$ such that

(1.11) $$a(v,v) + \lambda |v|^2 \geq \gamma \|v\|^2 \qquad \forall v \in V, \quad \gamma > 0.$$

Let

$$z \in H = L^2(\mathcal{O}), \quad z \geqslant 0.$$

Consider the problem

(1.12) $$\left| \begin{array}{l} a(\zeta_\lambda, v-\zeta_\lambda) + \lambda(\zeta_\lambda, v-\zeta_\lambda) \geq (f+\lambda z, v-\zeta_\lambda) \\ \forall v \in V, \quad v \leq Mz, \quad \zeta_\lambda \leq Mz, \quad \zeta_\lambda \in V. \end{array} \right.$$

The problem (1.2) is a variational inequality (V.I.). Thanks to (1.11) it has one and only one solution ζ_λ, furthermore, $\zeta_\lambda \geqslant 0$. Set

(1.13) $$T_\lambda z = \zeta_\lambda,$$

which defines a mapping of H^+ into itself (H^+ = the set of elements $h \in H$, $h \geqslant 0$).

We begin by giving some properties of T_λ.

LEMMA 1.1: T_λ *is an increasing operator, i.e.,*

(1.14) $$T_\lambda z_1 \leq T_\lambda z_2 \text{ p.p. } \textit{if } z_1 \leq z_2 \text{ a.e.}$$

(1.15) $$T_\lambda z \geq 0, \quad \forall z \in H^+,$$

(1.16) $$T_\lambda z \leq u^o, \forall z \in H^+ \textit{ such that } z \leq u^o.$$

Proof: The property (1.14) results from the fact that M is an increasing operator, as well as from comparison theorems about V.I.'s. The property (1.15) corresponds to $\zeta_\lambda \geqslant 0$. Let us prove (1.16). For that let us take

$$v = \zeta_\lambda - (\zeta_\lambda - u^o)^+ \quad \text{in (1.12)},$$

therefore

$$-a(\zeta_\lambda, (\zeta_\lambda - u^o)^+) - \lambda(\zeta_\lambda, (\zeta_\lambda - u^o)^+) \geq -(f+\lambda z, (\zeta_\lambda - u^o)^+)$$

and

$$a(u^o, (\zeta_\lambda - u^o)^+) = (f, (\zeta_\lambda - u^o)^+).$$

By addition this yields

$$-a(\zeta_\lambda - u^o, (\zeta_\lambda - u^o)^+) - \lambda(\zeta_\lambda, (\zeta_\lambda - u^o)^+) \geqq -\lambda(z, (\zeta_\lambda - u^o)^+)$$

$$-a(\zeta_\lambda - u^o, (\zeta_\lambda - u^o)^+) - \lambda(\zeta_\lambda - u^o, (\zeta_\lambda - u^o)^+) \geqq$$
$$\geqq \lambda(u^o - z, (\zeta_\lambda - u^o)^+) \geqq 0$$

and consequently

$$\zeta_\lambda - u^o \leqslant 0,$$

i.e., (1.16).□

If we set

$$K = \{ z \in H \mid 0 \leqq z \leqq u^o \}$$

then (1.15),(1.16) express that T_λ maps K into itself.

We notice that the solutions of the Q.V.I. (1.7) correspond to the *fixed points* of T_λ. In this view it is natural to consider the following iterations

(1.17) $\quad u^{n+1} = T_\lambda u^n, \quad u^o$ a solution of (1.9),

(1.18) $\quad u_{n+1} = T_\lambda u_n, \quad u^o = 0.$

We shall say that (1.17) is the *decreasing* approximation procedure, and that (1.18) is the *increasing* approximation procedure. In fact,

(1.19) $$\begin{cases} 0 = u_o \leqq u_1 \ldots \leqq u_n \leqq \ldots \leqq u^o \\ u^o \geqq u^1 \geqq \ldots \geqq u^n \geqq \ldots \geqq 0 . \end{cases}$$

The properties (1.19) are an easy consequence of Lemma 1.1. We then have:

LEMMA 1.2: *The decreasing procedure* u^n *converges to* $\bar{u}$ *the maximum solution of the Q.V.I.* (1.7),(1.10).

Proof: It is clear that by (1.19)

(1.20) $\quad u^n \downarrow \bar{u}, \quad \bar{u} \in K.$

Also, in (1.12), for $z \geqslant 0$, $Mz \geqslant 0$, and thus we may take $v = 0$ as test function. From this it follows that

$$a(\zeta_\lambda, \zeta_\lambda) + \lambda|\zeta_\lambda|^2 \leqq (f + \lambda z, \zeta_\lambda) .$$

whence, by (1.11),

(1.21) $\quad \gamma\|\zeta_\lambda\| \leqq |f + \lambda z| .$

We then apply (1.21) with $z = u^n$, and obtain

$$\gamma \| u^{n+1} \| \leq |f + \lambda u^n| \leq |f| + \lambda \; |u^n|$$
$$\leq |f| + \lambda \; |u^o|$$

by (1.19). Hence u^n stays bounded in V. Consequently we can complete (1.20) by

(1.22) $\quad u^n \to \bar{u}$ weakly in V.

Let us show that $\bar{u}$ is the solution of the Q.V.I. (1.7). First we show that

(1.23) $\quad \bar{u} \leqslant M\bar{u}$.

In fact, from

$$u^{n+1} \leqslant Mu^n$$

it follows that

$$u^{n+1}(x) \leq k + c_o(\xi) + u^n(x+\xi) \quad \forall\ \xi \geq 0 \text{ with } x + \xi \in \bar{\mathcal{O}}$$

and thus at the limit

$$\bar{u}(x) \leq k + c_o(\xi) + \bar{u}(x+\xi) \quad \forall\ \xi \geq 0 \text{ with } x+\xi \in \bar{\mathcal{O}}$$

whence (1.23). Now let $v \leqslant M\bar{u}$, therefore $v \leqslant Mu^u$ for all n. We can therefore take v as test function for

$$a(u^{n+1}, v-u^{n+1}) + \lambda\ (u^{n+1}, v-u^{n+1}) \geq (f + \lambda\ u^n, v-u^{n+1}).$$

Using (1.20),(1.22) and the weak lower semi-continuity of a(v,v), at the limit we easily obtain

$$a(\bar{u}, v-\bar{u}) \geq (f, v-\bar{u}) \quad \forall\ v \leq M\,\bar{u}$$

which, with (1.23) shows that $\bar{u}$ is the solution of the Q.V.I..

Let us show that $\bar{u}$ is the maximum solution. In fact, let u be a solution of (1.7) satisfying (1.10). Then

$$u = T_\lambda u \leqslant T_\lambda u^o = u^1$$

and by induction

$$u \leqslant u^n \quad \text{for all } n,$$

whence $u \leqslant \bar{u}$.□

Remark 1.4

The sequence u_n converges to $\underline{u}$ a.e., and weakly in V. It is very probable that $\underline{u}$ is the minimum solution of the Q.V.I.. Nevertheless, under the assumptions of Theorem 1.1 we do not know how to prove this property.

However, it is easy to prove that $\underline{u} \leqslant M\underline{u}$, and

$$a(\underline{u}, v - \underline{u}) \geqslant (f, v - \underline{u}) \quad \text{for all } v \leqslant Mu_n \quad (n \text{ fixed})$$

but this is not sufficient. The difficulty is that if $v \leq M\underline{u}$, then v is not necessarily an admissible test function for the problem u_{n+1}.

We shall see later on that if f satisfies additional assumptions, then we shall be able to prove that $\underline{u}$ is the minimum solution.

Proof of Theorem 1.1: By Lemma 1.2 we have that the set of solutions of (1.7), (1.10) is non-empty and possesses a maximum element. It is therefore necessary for us to prove that this set has a minimum element u*. Now, let us write

$$S = \text{the set of solutions of (1.7),(1.10)}$$

which is non-empty. Also, let

$$\Sigma = \{ z \in H,\ 0 \leqq z \leqq u^o,\ z \leqq T_\lambda z,\ z \leqq u \quad \forall u \in S \} .$$

Then Σ is non-empty, since $0 \in \Sigma$. Also, T_λ maps Σ into itself. In fact, if $z \in \Sigma$, then

$$0 \leqq T_\lambda z \leqq u^o ,\quad T_\lambda z \leqq T_\lambda (T_\lambda z) \text{ et } T_\lambda z \leqq T_\lambda u = u ,$$

for all $u \in S$. We are going to show that

(1.24) Σ has a maximal element u*.

If (1.24) is true, then

$$T_\lambda u^* \in \Sigma \quad \text{and} \quad T_\lambda u^* \geqslant u^*.$$

Since u* is maximal, then $u^* = T_\lambda u^*$ necessarily holds, therefore u* is a solution, and since $u^* \leqslant u$ for all $u \in S$, u* is the minimal solution. It remains to prove (1.24). This is a consequence of Zorn's Lemma, if we prove that every totally ordered family z_α of Σ has an upper bound. Now, if z_α is such a sequence, since $0 \leqslant z_\alpha \leqslant u^o$ and $H = L^2$, z_α converges in H to z (1), and $z_\alpha \leqslant z$.

Moreover, $z \in \Sigma$, since $0 \leqslant z_\alpha \leqslant u^o$ implies $0 \leqslant z \leqslant u^o$, $z_\alpha \leqslant u$ implies $z \leqslant u$. Lastly, $z_\alpha \leqslant T_\lambda z_\alpha \leqslant T_\lambda z$, thus $z \leqslant T_\lambda z$, and so z is certainly the upper bound of the family z_α, for if $\zeta \in \Sigma$, $z_\alpha \leqslant \zeta$, then $z \leqslant \zeta$ is necessarily true. The proof of Theorem 1.1 is therefore complete.□

Remark 1.5

In the definition of Σ we can replace S by the set of upper-solutions in the interval o,u^o, i.e.

$$S = \{z \in H, o \leqslant z \leqslant u^o, z \geqslant T_\lambda z\}$$

which is non-empty, because u^o belongs to it. This allows us to prove the exist-

---

(1) $(z_\alpha, u^o - z_{\alpha_o})$ is an increasing family of real numbers, bounded from above, and therefore they converge. Furthermore, $|z_\alpha - z_\beta|^2 \leqslant (z_\beta - z_\alpha, u^o - z_{\alpha_o})$ for $\alpha_o \leqslant \alpha \leqslant \beta$, and thus z_α is continuous.

ence of a minimum solution without knowing a priori that the set of solutions is non-empty. We can also prove, similarly, the existence of a maximum solution.□

The technique used in the proof of Theorem 1.1 is owed to L. TARTAR [2], where more general results will be found about the fixed points of increasing mappings on a Hilbert space (cf., also, C. BAIOCCHI, A. CAPELO [1]). The proof of Lemma 1.2 is owed to A. BENSOUSSAN, M. GOURSAT, J.L. LIONS [1].

1.3 *Uniqueness*

We begin with a property of T_λ.

LEMMA 1.3: *The operator* T_λ *is concave on* H^+.

Proof: Let

$$z_1, z_2 \in H^+ \quad \text{et} \quad \zeta_1 = T_\lambda z_1, \ \zeta_2 = T_\lambda z_2, \ \zeta = T_\lambda (\theta z_1 + (1-\theta))$$

Write

$$\zeta_\theta = \theta \zeta_1 + (1-\theta)\zeta_2 \quad , \quad z_\theta = \theta z_1 + (1-\theta) z_2 \quad , \quad \theta \in [0,1]$$

Notice that the operator M is itself concave, since

$$M(\theta z_1 + (1-\theta) z_2) \geqq \theta M(z_1) + (1-\theta) M(z_2) . \tag{1.25}$$

Because

$$\zeta_1 \leqq M(z_1) \ , \quad \zeta_2 \leqq M(z_2)$$

and with (1.25) we have

$$\zeta_\theta \leqq M(\theta z_1 + (1-\theta) z_2)$$

and thus ζ_θ is admissible for the problem which has ζ as solution. From this it follows that $\zeta + (\zeta - \zeta_\theta)^-$ is admissible for this problem, therefore

$$a(\zeta,(\zeta-\zeta_\theta)^-) + \lambda(\zeta,(\zeta-\zeta_\theta)^-) \geqq (f + \lambda z_\theta , (\zeta-\zeta_\theta)^-) . \tag{1.26}$$

Also, we can take $\zeta_1 - (\zeta - \zeta_\theta)^-$ as test function in the problem, where ζ_1 is the solution, and $\zeta_2 - (\zeta - \zeta_\theta)^-$ can be taken as test function in the problem whose solution is ζ_2.

From this we deduce that

$$-a(\zeta_1,(\zeta-\zeta_\theta)^-) - \lambda(\zeta_1, (\zeta-\zeta_\theta)^-) \geqq -(f+\lambda z_1 , (\zeta-\zeta_\theta)^-) \tag{1.27}$$

$$-a(\zeta_2,(\zeta-\zeta_\theta)^-) - \lambda(\zeta_2,(\zeta-\zeta_\theta)^-) \geqq -(f+\lambda z_2 , (\zeta-\zeta_\theta)^-) . \tag{1.28}$$

Now multiply (1.27) by θ, and (1.28) by $(1 - \theta)$. Then addition yields

$$- a(\zeta_\theta,(\zeta-\zeta_\theta)^-) - \lambda(\zeta_\theta,(\zeta-\zeta_\theta)^-) \geqq - (f + \lambda z_\theta ,(\zeta-\zeta_\theta)^-) ,$$

which when added to (1.26) leads to

$$a(\zeta-\zeta_\theta,\ (\zeta-\zeta_\theta)^-) + \lambda(\zeta-\zeta_\theta\ ,\ (\zeta-\zeta_\theta)^-) \geqq 0$$

whence it follows that

$$(\zeta - \zeta_\theta)^- = 0,$$

therefore

$$T_\lambda(\theta z_1 + (1-\theta) z_2) \geqq \theta\ T_\lambda(z_1) + (1-\theta)\ T_\lambda(z_2)$$

which proves the concaveness of T_λ.□

In the remainder we assume that

$$f \in L^p(\mathcal{O})\ ,\quad p > \frac{n}{2}\ .$$

From this it follows that $u^o \in W^{2,p}(\mathcal{O})$, $p > n/2$, hence $u^o \in C^o(\bar{\mathcal{O}})$.

LEMMA 1.4: *Assume* (1.29) *and one or the other of the following assumptions*

(1.30) β *large enough,*

(1.31) $f \geqslant f_o$ *where* f_o *is a positive constant.*

Let

$$z,\zeta \in K \quad \textit{and} \quad \gamma \in [0,1]$$

be such that

(1.32) $z(x) - \zeta(x) \leqq \gamma\ z(x)$ *a.e.*

Then if (1.31) *is satisfied, we have*

(1.33) $T_\lambda\ z(x) - T_\lambda\ \zeta\ (x) \leqq \gamma(1-\mu)\ T_\lambda\ z(x)$ *a.e.*

where

(1.34) $$0 \leqq u \leqq \min\left(\frac{h}{\|u^o\|}\ ,\ \frac{fo}{\lambda\|u^o\| + fo}\right).$$

If (1.30) *is satisfied, then* (1.33) *is satisfied for* $\lambda = 0$ *and* μ *such that*

(1.35) $$0 \leqq \mu \leqq \min\left(\frac{k}{\|u^o\|}\ ,\ 1\right).$$

Proof: By (1.32),

$$(1 - \gamma)z \leqslant \zeta.$$

As the mapping T_λ is increasing, it follows that

$$T_\lambda(\zeta) \geqq T_\lambda((1-\gamma)\ z).$$

Since T_λ is concave, we deduce

(1.36) $T_\lambda(\zeta) \geqq T_\lambda((1-\gamma)z + \gamma o) \geqq (1-\gamma)T_\lambda(z) + \gamma T_\lambda(0)$.

We are going to show that in both cases of the Lemma we have

(1.37) $T_\lambda(0) \geqslant \mu u^o$.

But from (1.36),(1.37) we then deduce

$$T_\lambda(\zeta) \geqq (1-\gamma)T_\lambda(z) + \gamma\mu u^o.$$

But since $z \in K$, $z \leqslant u^o$, whence

$$T_\lambda(z) \leqslant u^o,$$

therefore

$$T_\lambda(\zeta) \geqq (1-\gamma)T_\lambda(z) + \gamma\mu T_\lambda(z)$$

from which we deduce (1.33). It remains to verify (1.37).

First we consider the case where (1.31) is satisfied. Recall that

$$T_\lambda(0) = u_1.$$

Since $M(0) = k$, it follows from the choice of μ that

$$\mu u^o \leqslant M(0).$$

Therefore μu^o is admissible for the problem of which u_1 is the solution. From this it follows that $u_1 + (u_1 - \mu u^o)^-$ is also admissible for this problem, therefore

$$a(u_1,(u_1-\mu u^o)^-) + \lambda(u_1,(u_1-\mu u^o)^-) \geqq (f,(\mu_1 - \mu u^o)^-)$$

now,

$$a(\mu u^o, (u_1 - \mu u^o)^-) = (\mu f, (u_1 - \mu u^o)^-)$$

hence

(1.38) $$a(u_1 - \mu u^o, (u_1 - \mu u^o)^-) + \lambda(u_1 - \mu u^o,(u_1 - \mu u^o)^-) \geqq (f(1-\mu) - \lambda\mu u^o, (u_1 - \mu u^o)^-).$$

But

$$f(1-\mu) - \lambda\mu u^o \geqq f_o(1-\mu) - \lambda\mu \|u^o\| \geqq 0$$

by the choice of μ. But from (1.38) we then immediately deduce that

$$(u_1 - \mu u^o)^- = 0,$$

and thus (1.37).

If (1.30) is now satisfied, with β large enough for the form $a(u,v)$ to be coercive, then by taking $\lambda = 0$ and $u_1 = T_o(0)$, we see as above that

$$a(u_1-\mu u^o\ ,\ (u_1-\mu u^o)^-) \geqq (f(1-\mu)\ ,\ (u_1-\mu u^o)^-) \geqq 0$$

by the choice of μ. Since a is coercive, we again deduce (1.37).

LEMMA 1.5: *Take the conditions of Lemma 1.4. Then the solution of the Q.V.I. (1.7),(1.10) is unique. Moreover, the limit of the increasing procedure $\underline{u}$ coincides with the unique solution* u *of the Q.V.I..*

Proof: In the case (1.30) we consider T_λ only for $\lambda = 0$. In all cases there exists a constant $0 \leqslant \mu < 1$ such that (1.33) is satisfied. Let $u,\tilde{u}$ be two solutions of the Q.V.I. (1.7),(1.10). These are fixed points of T_λ. Therefore by taking $z = u$, $\zeta = \tilde{u}$ in (1.33) we have

$$u - \tilde{u} \leqslant (1 - \mu)u$$

since $u - \tilde{u} \leqslant u$. Doing this again with $\gamma = 1 - \mu$, we obtain

$$u-\tilde{u} \leqq (1-\mu)^2\ u$$

and quite generally

$$u-\tilde{u} \leqq (1-\mu)^n\ u \leqq (1-\mu)^n\ u^o \leqq (1-\mu)^n\ \|u^o\|$$

and making n tend to ∞ yields $u \leqslant \tilde{u}$.

Interchanging the roles of u and $\tilde{u}$, we obtain $u = \tilde{u}$. Consider the increasing procedures u_n and decreasing procedures u^n. Apply (1.33) with

$$z = u^o, \quad \zeta = u_o, \quad \gamma = 1,$$

therefore

$$T_\lambda(u^o) - T_\lambda\ (u_o) \leqq (1-\mu)\ T_\lambda\ (u^o)$$

so

$$u^1 - u_1 \leqq (1-\mu)\ u^1$$

and applying (1.33) again, this yields

$$u^2 - u_2 \leqq (1-\mu)^2\ u^2$$

and quite generally we have

$$u^m - u_m \leqq (1-\mu)^m\ u^m \leqq (1-\mu)^m\ u^o \leqq (1-\mu)^m\ \|u^o\|$$

and thus

$$u^m - u_m \to 0 \quad \text{in } L^\infty,$$

from which it follows that $u_m \to L^\infty$ to $\underline{u} = u$ the unique solution of the Q.V.I..□

Remark 1.6

The method of proving Lemma 1.4 is owed to B. HANOUZET - J.L. JOLY [1][4].□

We are now going to give a continuity result for the solution u. First we have:

LEMMA 1.6: *Let* $C^o_{+o}(\bar{\mathcal{O}})$ *be the space of continuous functions on* $\bar{\mathcal{O}}$*, that are positive and zero on* Γ*. Then* M *maps* C^o_{+o} *into* C^o_+*. Moreover, if we assume that*

$$\text{(1.39)} \qquad \forall\, x \in \bar{\mathcal{O}}\,, \quad \xi \in (R^n)^+\,,\ \exists\, P(x,\xi) \in (R^n)^+ \text{ such that}$$
$$x + P(x,\xi) \in \bar{\mathcal{O}}\,, \quad P(x,\xi) = \xi \text{ if } x + \xi \in \bar{\mathcal{O}}$$
$$\text{and } x \to P(x,\xi) \text{ continuous for all } \xi,$$

then M *sends* $C^o(\bar{\mathcal{O}})$ *into itself.*

Proof: $\varphi \in C^o_{+o}$, then

$$M\varphi(x) = k + \inf_{\xi \geqq 0} (c_o(\xi) + \varphi(x+\xi)) \qquad \forall\, x \in \bar{\mathcal{O}}$$

and φ is continued by 0 outside $\bar{\mathcal{O}}$. In fact,

$$k + \inf_{\substack{\xi \geqq 0 \\ x+\xi \notin \bar{\mathcal{O}}}} \{c_o(\xi) + \varphi(x+\xi)\} = k + \inf_{\substack{\xi \geqq 0 \\ x+\xi \notin \bar{\mathcal{O}}}} c_o(\xi) \geqq k + \inf_{\substack{\xi \geqq 0 \\ x+\xi \in \Gamma}} c_o(\xi) \equiv \gamma(x)$$

since for all $\xi \geqslant 0$ such that $z + \xi \notin \bar{\mathcal{O}}$ there exists $0 \leqslant \theta \leqslant 1$ such that

$$x + \theta\xi \in \Gamma \text{ et } c_o(\xi) \geqq c_o(\theta\xi) \geqq \inf_{\substack{\xi \geqq 0 \\ x+\xi \in \Gamma}} c_o(\xi) = \gamma(x).$$

Now,

$$\gamma(x) = k + \inf_{\substack{\xi \geqq 0 \\ x+\xi \in \Gamma}} (c_o(\xi) + \varphi(x+\xi)) \geqq M_\varphi(x)$$

whence the result.

We now take $\varphi \in C^o(\bar{\mathcal{O}})$ and make the assumption (1.39). Let $x_n \to x$, $x_n \in \bar{\mathcal{O}}$, $x \in \bar{\mathcal{O}}$. Since φ is continuous there exists $\xi_n \in (R^n)^+$ such that

$$M\varphi(x_n) = k + c_o(\xi_n) + \varphi(x_n + \xi_n)\ , \quad x_n + \xi_n \in \bar{\mathcal{O}}.$$

If we pick a convergent subsequence (which is possible because x_n, ξ_n stay bounded) then

$$x_n \to x \ , \ \xi_n \to \xi^* \ , \quad \xi^* \geqq 0 \quad , \quad x + \xi^* \in \bar{\mathcal{O}}$$

and by continuity

$$M\varphi(x_n) \to k + c_o(\xi^*) + \varphi(x + \xi^*) \geqq M\varphi(x)$$

therefore

$$\underline{\lim}\, M\varphi(x_n) \geqq M\varphi(x) \,.$$

But, also, let ξ, fixed, $\xi \geqslant 0$, be such that $x + \xi \in \bar{\mathcal{O}}$, then

$$M\varphi(x_n) \leqq k + c_o\,(P(x_n,\xi)) + \varphi\,(x_n + P(x_n\,,\,\xi))$$

$$\to k + c_o(P(x,\xi)) + \varphi\,(x + P(x,\xi))$$

$$= k + c_o\,(\xi) + \varphi\,(x + \xi)$$

by assumption (1.39), therefore

$$\overline{\lim}\, M\varphi(x_n) \leqq k + c_o\,(\xi) + \varphi\,(x + \xi) \quad \forall\ \xi \geqq 0$$

for all $\xi \geqslant 0$ such that $x + \xi \in \bar{\mathcal{O}}$

which implies

$$\overline{\lim}\, M\varphi(x_n) \leqq M\varphi(x)$$

which implies the desired result.□

Remark 1.7

The assumption (1.37) is satisfied if $\mathcal{O}$ is a regular convex set. It suffices to take the point or the segment x, $x + P(x,\xi)$, intersecting Γ as $P(x,\xi) + x$.□

LEMMA 1.7: *Assume the conditions of Lemma 1.4. In the Dirichlet case* ($V = H_0^1$) *the solution of the Q.V.I.* (1.7),(1.10) *belongs to* $C^o(\bar{\mathcal{O}})$.

Furthermore, if we assume (1.39), *then the same conclusion is also true in the case of the Neuman problem* ($V = H^1$).

Proof: Consider the decreasing process u_n (to fix our ideas). If $u_n \in C^o(\bar{\mathcal{O}})$, then it follows from Lemma 1.6 that $Mu_n \in C^o(\bar{\mathcal{O}})$. By a result on V.I.'s (the solution is continuous when the obstacle is continuous, cf., for example, B.L., loc. cit.), from which we deduce that $u_{n+1} \in C^o(\bar{\mathcal{O}})$. Now, $u_n \to u$ in L^∞, whence $u \in C^o(\bar{\mathcal{O}})$.□

THEOREM 1.2: *Under the assumptions of Theorem 1.1, together with* (1.29) *and* (1.39), *if* $V = H^1$, *then there exists one and only one solution of the problem* (1.7) *which belongs to* $L^\infty(\mathcal{O})$. *Furthermore, the solution* $u \in C^o(\bar{\mathcal{O}})$ *and* $u \leqslant u^o$.

Proof: Let $z \in L^\infty(\mathcal{O})$, $z \geqslant 0$. Consider the Q.V.I.

(1.40) $$\left|\, a\,(u_\lambda\,,\, v - u_\lambda) + \lambda\,(u_\lambda\,,\, v - u_\lambda) \geqq (f + \lambda z\,,\, v - u_\lambda) \right.$$

(Contd) $|\forall\ v \leqq Mu_\lambda\ ,\quad u_\lambda \leqq Mu_\lambda \qquad\qquad ,\ u_\lambda \geqq 0\ ,\ u_\lambda \in V$

where λ is as in (1.11). We write u^o_λ for the solution of

(1.41) $\quad a(u^o_\lambda, v) + \lambda\,(u^o_\lambda, v) = (f + \lambda z, v)\ .$

Let us notice that every solution of (1.40) satisfies

(1.41)' $\quad u_\lambda \leqslant u^1_\lambda.$

In fact, we take $v = u_\lambda - (u_\lambda - u^o_\lambda)^+$ in (1.40) and $v = (u_\lambda - u^o_\lambda)^+$ in (1.41), then by addition,

$$- a(u_\lambda - u^o_\lambda, (u_\lambda - u^o_\lambda)^+) - \lambda\,(u_\lambda - u^o_\lambda\ ,\ (u_\lambda - u^o_\lambda)^+) \geqq 0$$

whence we deduce $u_\lambda \leqslant u^o_\lambda$. By Lemmas 1.5 and 1.7, (1.40) has one and only one solution such that $u_\lambda \leqslant u^o_\lambda$ and $u_\lambda \in C^o(\bar{\mathcal{O}})$. Since every solution u satisfies (1.41)', we see that (1.39) has one and only one solution u_λ, which also satisfies (1.41) and belongs to $C^o(\bar{\mathcal{O}})$. We have thus defined a mapping Σ_λ of $L^\infty_+(\mathcal{O})$ into itself (L^∞_+ = the subspace of L^∞ of positive functions). We are going to prove the estimate

(1.42) $\quad \| \Sigma_\lambda(z_1) - \Sigma_\lambda(z_2) \| \leqq \dfrac{\lambda}{\lambda + \beta}\ \| z_1 - z_2 \|\ .$

If (1.42) is true, then $\Sigma_\lambda(z)$ is a contraction in L^∞_+. Consequently it has one and only one fixed point in L^∞_+. Since the set of solutions of (1.7) which belong to L^∞ are fixed points of Σ_λ, it follows that there exists one and only one solution of the Q.V.I. (1.7) which belongs to L^∞.

Let u be this solution; $u = \Sigma_\lambda(u)$ is also continuous on $\bar{\mathcal{O}}$. We are going to prove that:

(1.43) $\quad u \leqslant u^o$

which will complete the proof of the Theorem, at least if (1.42) is satisfied. We write $S_\lambda(z)$ for the mapping of L^∞_+ into itself defined by

$$u_\lambda = S_\lambda(z).$$

It is clear that S_λ satisfies the same estimate as Σ_λ, and hence S_λ is a contraction in L^∞_+. It is evident that u^o is the unique fixed point of S_λ. Now (1.41)' means that

$$\Sigma_\lambda(z) \leqq S_\lambda(z) \quad \forall\ z \in L^\infty_+.$$

Since Σ_λ is increasing, which is easily verified by approximation procedures for the Q.V.I., we deduce that

$$\Sigma_\lambda^2(z) \leq \Sigma_\lambda(S_\lambda(z)) \leq S_\lambda^2(z)$$

and quite generally,

$$\Sigma_\lambda^n(z) \leq S_\lambda^n(z)$$

then passing to the limit in n we deduce (1.43).

To show (1.42) we consider the increasing approximation procedure u_n of u_λ; more precisely, $u_{n+1} = T_o u_n$, where in the definition of T_o, the form a(u,v) is replaced by a(u,v) + λ(u,v). Therefore

$$a(u_{n+1}, v-u_{n+1}) + \lambda(u_{n+1}, v-u_{n+1}) \geq (f+\lambda z_1, v-u_{n+1})$$

$$\forall\, v \leq Mu_n, \quad u_{n+1} \leq Mu_n, \quad u_o = 0$$

$$a(\tilde{u}_{n+1}, v-\tilde{u}_{n+1}) + \lambda(\tilde{u}_{n+1}, v-\tilde{u}_{n+1}) \geq (f+\lambda z_2, v-\tilde{u}_{n+1})$$

$$\forall\, v \leq M\tilde{u}_n, \quad \tilde{u}_{n+1} \leq M\tilde{u}_n, \quad \tilde{u}_o = 0.$$

then

$$(1.44) \qquad \|u_{n+1} - \tilde{u}_{n+1}\|_{C^o} \leq \max\left(\frac{\lambda\|z_1 - z_2\|_{C^o}}{\lambda+\beta}, \|Mu_n - M\tilde{u}_n\|_{C^o}\right).$$

The estimate (1.44) results from properties of V.I.'s (cf. B.L. [1]). Let us briefly recall the method for obtaining (1.44). Let K be the constant in the right hand side of (1.44). We can take

$$v = u_{n+1} - (u_{n+1} - \tilde{u}_{n+1} - K)^+$$

in the problem u_{n+1}, and

$$v = \tilde{u}_{n+1} + (u_{n+1} - \tilde{u}_{n+1} - K)^+$$

in the problem $\tilde{u}_{n+1}$ (since $K \geq Mu_n - M\tilde{u}_n$). By addition, and thanks to the choice of K, we easily obtain

$$(u_{n+1} - \tilde{u}_{n+1} - K)^+ = 0,$$

hence (1.44).

Since M is Lipschitz from L^∞ into itself with a constant equal to 1, we deduce from (1.44) that

$$\|u_{n+1} - \tilde{u}_{n+1}\| \leq \max\left(\frac{\lambda\|z_1 - z_2\|}{\lambda+\beta}, \|u_n - \tilde{u}_n\|\right).$$

Therefore if

$$\|u_n - \tilde{u}_n\| \leq \frac{\lambda \| z_1 - z_2 \|}{\lambda + \beta} ,$$

the same inequality is satisfied at step n + 1. Now, $u_0 - \tilde{u}_0 = 0$ clearly satisfies this inequality, therefore

$$\|u_n - \tilde{u}_n\| \leq \frac{\lambda \| z_1 - z_2 \|}{\lambda + \beta} .$$

Now, in this case Lemma 1.5 applies, and therefore

$$u_n \to \Sigma_\lambda (z_1), \quad \tilde{u}_n \to \Sigma_\lambda (z_2) \quad \text{in} \quad L^\infty ,$$

hence the result (1.42), which completes the proof of the Theorem.□

Let us give another proof of the uniqueness of an L^∞ solution of the Q.V.I. (1.7), owed to T. LAETSCH [1], already mentioned for V.I.'s in B.L. [1], loc. cit.. Let us first of all assume that (1.31) is satisfied. Let u and ũ be two solutions of (1.7) which belong to $L^\infty(\mathcal{O})$. Let γ be the largest number lying between 0 and 1 such that

(1.45) $\quad \gamma u(x) \leqslant \tilde{u}(x) \quad$ a.e..

We are going to prove that $\gamma = 1$. Let us assume that $0 \leqslant \gamma < 1$; we are going to construct a number γ' such that

(1.46) $\quad \gamma' u(x) \leq \tilde{u}(x) \quad$ a.e. and $\gamma < \gamma' \leq 1$

which will imply a contradiction. Now, $\gamma' u$ satisfies

$$a(\gamma' u, v - \gamma' u) + \lambda(\gamma' u, v - \gamma' u) \geq (\gamma' f + \lambda \gamma' u, v - \gamma u)$$

$$\forall v \leq \gamma' Mu, \qquad \gamma' u \leq \gamma' Mu$$

and u satisfies

$$a(\tilde{u}, w - \tilde{u}) + \lambda(\tilde{u}, w - \tilde{u}) \geq (f + \lambda \tilde{u}, w - \tilde{u})$$

$$\forall w \leq M\tilde{u} \quad , \quad \tilde{u} \leq M\tilde{u} .$$

By the comparison theorems for V.I.'s we will have (1.46), if

(1.47) $\quad \gamma' f + \lambda \gamma' u \leq f + \lambda \tilde{u}$

(1.48) $\quad \gamma' Mu \leq M\tilde{u} .$

Taking account of (1.45) it suffices that

$$\gamma' f + \lambda \gamma' u \leq f + \gamma u$$

is satisfied, which will be so if

$$\frac{1 - \gamma'}{\gamma' - \gamma} \geq \frac{\|u\|}{f_0} .$$

To prove (1.48), by (1.45) it suffices to have

$$\gamma' M u \leqslant M \gamma u$$

which will be satisfied if

$$\frac{1-\gamma'}{\gamma'-\gamma} \geqslant \frac{\|u\|}{k}.$$

We have introduced the parameter λ in a way that reduces us to the case of a coercive form $a(u,v) + \lambda(u,v)$, which allows us to obtain more easily a comparison result for V.I.'s. In fact, the comparison result is valid if only (1.47),(1.48) are satisfied with $\lambda = 0$, in which case (1.47) is automatically satisfied, without which $f \geqslant f_o$ would have to be assumed.□

1.4 *Characterisation of the Maximum Solution of the Q.V.I. as the Envelope of Sub-Solutions*

We shall say that z is a sub-solution of (1.7), (1.10) if

(1.49)
$$\left| \begin{array}{l} z \in V \quad 0 \leqq z \leqq u^o \\ a(z,v) \leqq (f,v) \ \forall v \in V,\ v \geqq 0 \\ z \leqq Mz \quad \text{a.e.} \end{array} \right.$$

Let us notice that every solution of the Q.V.I. is a sub-solution, because if $\varphi \in V, \varphi \geqslant 0$, we can take $v = u - \varphi$ as test function in (1.7), from which it follows that;

$$a(u,\varphi) \geqslant (f,\varphi).$$

We are going to prove:

THEOREM 1.3: *Take the assumptions of Theorem 1.1. Then the maximum solution* $\bar{u}$ *is also the maximum element of the set of sub-solutions of* (1.7),(1.10).

Proof: It suffices to prove by induction that

(1.50) $$u^n \geqslant z$$

where u^n is the increasing process $u^{n+1} = T\lambda(u^n)$. The property (1.50) is clearly satisfied for $n = 0$. Let us assume that it is satisfied at step n. Then

(1.51)
$$\left| \begin{array}{l} a(u^{n+1}, v-u^{n+1}) + \lambda\,(u^{n+1}, v-u^{n+1}) \geqq (f + \lambda u^n, v-u^{n+1}) \\ \forall v \leqq Mu^n, \end{array} \right.$$

(1.52) $$a(z,\varphi) + \lambda\,(z,\varphi) \leqq (f,\varphi) + \lambda\,(z,\varphi) \ \forall \varphi \geqq 0$$

Now, we can take $v = u^{n+1} + (z - u^{n+1})^+$ as test function in (1.51), since $z \leqslant Mz \leqslant Mu^n$, because $z \leqslant u^n$. We take $\varphi = (z - u^{n+1})^+$ in (1.52), then by subtraction we deduce that

$$a(u^{n+1} - z, (z-u^{n+1})^+) + \lambda(u^{n+1} - z, (z-u^{n+1})^+) \geqq$$

$$\geqq \lambda(u^n - z, (z-u^{n+1})^+) \geqq 0$$

from which we easily deduce that $(z - u^{n+1})^+ = 0$, whence the result desired.□

1.5 *Penalised Problem*

We can associate with the Q.V.I. (1.7) a penalised problem defined as follows:

$$\left| \begin{array}{l} a(u_\varepsilon, v) + \frac{1}{\varepsilon}((u_\varepsilon - Mu_\varepsilon)^+, v) = (f,v) \\ \forall\, v \in V,\ u_\varepsilon \in V,\ u_\varepsilon \geq 0. \end{array} \right. \tag{1.53}$$

The problem (1.53) is actually studied in a way parallel to that used for the Q.V.I.. For $z \in H^+$ we define

$\zeta_{\lambda\varepsilon}$ is the solution of

$$a(\zeta_{\lambda\varepsilon}, v) + \lambda(\zeta_{\lambda\varepsilon}, v) + \frac{1}{\varepsilon}((\zeta_{\lambda\varepsilon} - Mz)^+, v) = (f + \lambda z, v) \tag{1.54}$$

which is none other than the penalised problem associated with the V.I. (1.12). We set

$$T_\lambda^\varepsilon(z) = \zeta_{\lambda\varepsilon}. \tag{1.55}$$

By the properties of V.I.'s we have

$$\left| \begin{array}{l} T_\lambda^\varepsilon(z) \geq T_\lambda(z)\ \forall\, z \\ T_\lambda^\varepsilon(z) \downarrow T_\lambda(z) \quad \text{when} \quad \varepsilon \to 0. \end{array} \right. \tag{1.56}$$

The operator T_λ^ε has the same properties as T_λ. It is a concave increasing operator which maps K into K. These properties are proved similarly to those of T_λ. We define a decreasing procedure u_ε^n and an increasing procedure $u_{\varepsilon n}$ similarly to u^n and u_n. We are going to prove:

THEOREM 1.4: *Take the assumptions of Theorem 1.1. Then the set of solutions of* (1.53) *which belong to* K *is non-empty, and it has a minimum element* $\underline{u}_\varepsilon$ *and a maximum element* $\bar{u}_\varepsilon$. *Moreover,* $\bar{u}_\varepsilon \downarrow \bar{u}$ *a.e., and this holds in* V.

Proof: The existence of a minimum solution and a maximum solution is shown as in Theorem 1.1. We have the properties

$$u_\varepsilon^n \geqslant \bar{u}_\varepsilon, \quad u_\varepsilon^n \downarrow \bar{u}_\varepsilon \quad \text{a.e. when } n \to \infty, \tag{1.57}$$

which are the analogues of

$$u^n \geqslant \bar{u}, \quad u^n \downarrow \bar{u} \quad \text{a.c..} \tag{1.58}$$

We also have

$$u_\varepsilon^n \geqslant u^n, \quad u^n \downarrow u^n \quad \text{a.e. when } \varepsilon \downarrow 0. \tag{1.59}$$

In fact,

$$u_\varepsilon^0 = u^0,$$

and let us assume

$$u_\varepsilon^n \geqslant u^n,$$

then

$$u_\varepsilon^{n+1} = T_\lambda^\varepsilon (u_\varepsilon^n) \geqq T_\lambda^\varepsilon (u^n) \geqq T_\lambda (u^n) = u^{n+1} .$$

Let us also assume $u_\varepsilon^n \downarrow u^n$ at step n. Let $\varepsilon_0 \leqslant \varepsilon_1$, then $u_{\varepsilon_0}^n \leqslant u_{\varepsilon_1}^n$, therefore

$$u_{\varepsilon_0}^{n+1} = T_\lambda^{\varepsilon_0} (u_{\varepsilon_0}^n) \leqq T_\lambda^{\varepsilon_0} (u_{\varepsilon_1}^n) \leqq T_\lambda^{\varepsilon_1} (u_{\varepsilon_1}^n) = u_{\varepsilon_1}^{n+1} .$$

To prove the second part of (1.59) we notice the following property of T_λ^ε:

(1.60) if $z_\varepsilon \downarrow z$, then $T_\lambda^\varepsilon(z_\varepsilon) \downarrow T_\lambda(z)$ when $\varepsilon \to \downarrow 0$.

In fact, let us set

$$\zeta_\varepsilon = T_\lambda^\varepsilon(z_\varepsilon),$$

therefore

(1.61) $$a(\zeta_\varepsilon, v) + \lambda(\zeta_\varepsilon, v) + \frac{1}{\varepsilon}((\zeta_\varepsilon - Mz_\varepsilon)^+, v) = (f + \lambda z_\varepsilon, v) .$$

We already know that ζ_ε decreases, therefore $\zeta_\varepsilon \downarrow \zeta$ pointwise, and weakly in V. By the properties of M we have $Mz_\varepsilon \downarrow Mz$, hence $(\zeta_\varepsilon - Mz_\varepsilon)^+ \to (\zeta - Mz)^+$ pointwise.

Now, multiplying (1.61) by ε and taking into account that ζ_ε is bounded in V, by making ε tend to 0 we obtain

$$((\zeta - Mz)^+, v) = 0 \quad \text{for all } v,$$

whence

$$\zeta \leqslant Mz.$$

Also, let $v \leqslant Mz$, hence $v \leqslant Mz_\varepsilon$, then it follows from (1.61) that

$$a(\zeta_\varepsilon, v - \zeta_\varepsilon) + \lambda(\zeta_\varepsilon, v - \zeta_\varepsilon) = (f + \lambda z_\varepsilon, v - \zeta_\varepsilon) -$$

$$- \frac{1}{\varepsilon}((\zeta_\varepsilon - Mz_\varepsilon)^+, v - Mz_\varepsilon + Mz_\varepsilon - \zeta_\varepsilon)$$

$$\geqq (f + \lambda z_\varepsilon, v - \zeta_\varepsilon)$$

and therefore as $\varepsilon \downarrow 0$ we obtain

$$a(\zeta,v-\zeta)+\lambda(\zeta,v-\zeta) \geqq (f+\lambda z, v-\zeta)$$

which proves that

$$\zeta = T_\lambda(z).$$

Let us now show that

(1.62) $\bar{u}_\varepsilon \downarrow \bar{u}.$

First, let $\varepsilon_0 \leqslant \varepsilon_1$, then

$$u^n_{\varepsilon_0} \leqslant u^n_{\varepsilon_1},$$

and on making n tend to ∞ it follows that $\bar{u}_{\varepsilon_0} \leqslant \bar{u}_{\varepsilon_1}$, thus u_ε decreases, and therefore

$$u_\varepsilon \downarrow w.$$

Now,

(1.63) $w \geqslant \bar{u},$

since from (1.59) it follows that

$$\bar{u}_\varepsilon \geqslant \bar{u},$$

whence (1.63).

Let us show that $w = u$. Assume that at a point x, on a set of measure zero we have

$$w(x) > \bar{u}(x),$$

then for n large enough, since $u^n(x) \downarrow \bar{u}(x)$, we have

$$w(x) > u^n(x) > \bar{u}(x),$$

and since $u^n_\varepsilon \downarrow u^n$, for ε small enough we have

$$w(x) > u^n_\varepsilon(x),$$

which is impossible, because

$$u^n_\varepsilon \geqslant \bar{u}_\varepsilon \geqslant w.$$

Therefore $w = \bar{u}$, whence (1.62).

We can also directly prove that w is the maximum solution, or, what is easier, that w is the maximum element in the set of sub-solutions lying between 0 and u^0. Let us prove this latter point, for example. Since $\bar{u}_\varepsilon \downarrow w$, we have $M(\bar{u}_\varepsilon) \downarrow M(w)$, and since $\bar{u}_\varepsilon$ stays bounded in V, and $\bar{u}_\varepsilon - M(\bar{u}_\varepsilon) \to w - Mw$, it follows from (1.53), after multiplying by ε and passing to the limit, that

$$(w - Mw)^+ \quad \text{or} \quad w \leqslant Mw.$$

Further, if $\varphi \in V, \varphi \geqslant 0$, then from (1.53) it immediately follows that

$$a(\bar{u}_\varepsilon, \varphi) \leqslant (f, \varphi)$$

and therefore in the limit

$$a(w, \varphi) \leqslant (f, \varphi),$$

thus w is a sub-solution. But if z is a sub-solution, we know (cf. (1.50)) that $z \leqslant u^n$, and thus by (1.59) $z \leqslant u^n_\varepsilon$, whence $z \leqslant \bar{u}_\varepsilon$, and therefore we necessarily have $z \leqslant w$.□

In the case where $f \in L^p(\mathcal{O})$, $p > n/2$, we verify, as in Theorem 1.2, that the problem (1.53) has one and only one solution in L^∞. Furthermore, this solution is $u^\varepsilon \in C^0(\mathcal{O})$, and $u^\varepsilon \leqslant u^0$. Moreover, $u^\varepsilon \in W^{2,p}(\mathcal{O})$. Since $u^\varepsilon \downarrow u$, and since $u \in C^0(\bar{\mathcal{O}})$, it follows from Dini's Theorem that $u^\varepsilon \to u$ in $C^0(\bar{\mathcal{O}})$.

Lastly, let us remark that (at least for the Dirichlet problem) the problem (1.53) fits into the framework of the problems studied in Chapter 3, Section 2.3. In fact it may be written in the form

$$\text{(1.64)} \qquad Au_\varepsilon - H(u_\varepsilon) = f$$

with

$$H(u_\varepsilon) = -\frac{1}{\varepsilon}(u_\varepsilon - Mu_\varepsilon)^+ = \inf_{\substack{0 \leqq d \leqq 1 \\ \xi \geqq 0 \\ : x+\xi \in \bar{\mathcal{O}}}} \frac{1}{\varepsilon} d\,[k + c_0(\xi) + u_\varepsilon(x+\xi) - u_\varepsilon(x)]$$

so

$$\text{(1.65)} \qquad H(\varphi)(x) = \inf_v \Big[f(x,v) + \nabla\varphi(x) \,.\, g(x,v) - \varphi(x)\, a_1(x,v) + \int_{R^n} (\varphi(x+z) - \varphi(x) - z.\nabla\varphi\, \chi_{|z| \leqq 1})\, M(x,v,dz) \Big]$$

where

$$v = (d, \xi), \qquad 0 \leqq d \leqq 1, \qquad \xi \in R^n \quad \xi \geqq 0$$

$$\text{(1.66)} \qquad f(x,v) = \frac{1}{\varepsilon} d\,(k + c_0(\xi)), \qquad a_1 = 0$$

$$\text{(1.67)} \qquad M(x,v,dz) = \begin{cases} \frac{1}{\varepsilon} d\,\delta_\xi & \text{if} \quad x + \xi \in \bar{\mathcal{O}} \\ 0 & \text{if} \quad x + \xi \notin \bar{\mathcal{O}} \end{cases}$$

where δ_ξ denotes the Dirac mass at the point ξ,

$$(1.68)\qquad g(x,v) = \begin{cases} 0 & \text{if } x+\xi \notin \bar{\mathcal{O}} \text{ where } |\xi| > 1 \\ \frac{1}{\epsilon}\, d\xi & \text{if } x+\xi \in \bar{\mathcal{O}} \text{ and } |\xi| \leq 1 \end{cases}$$

1.6 *Some Variants*

To fix our ideas, here we take $V = H_0^1$ (Dirichlet's problem), then

$$a(u,v) = \langle Au,v\rangle \quad \text{for all } u,v \in V$$

since

$$A \in \mathcal{L}(H_0^1, H^{-1})$$

The set of sub-solutions (cf. (1.49)) is then the set of functions z such that

$$(1.69)\qquad \begin{cases} z \in H_0^1, & 0 \leq z \leq u^o \\ Az \leq f & \text{in } H^{-1} \\ z \leq Mz & \text{a.e.} \end{cases}$$

and we shall write u for the maximum element of the set of sub-solutions (this is the maximum solution of the Q.V.I. written as $\bar{u}$ above). We assume that $f \in L^p(\mathcal{O})$, $p > n/2$. In this case, $u \in C^o(\bar{\mathcal{O}})$, and $Mu \in C^o(\bar{\mathcal{O}})$, therefore

$$(1.70)\qquad C = \{ x \in \bar{\mathcal{O}} \mid u(x) < Mu(x)\} \quad \text{is open in } \bar{\mathcal{O}}.$$

Remark 1.8

If $c_o = 0$, then $Mu = k$, since $Mu \geq k$ and $Mu(x) \leq \underset{x+\xi\in\Gamma}{k} + u(x+\xi) = k$. In this case the Q.V.I. reduces to a V.I. whose obstacle is the constant k.□

THEOREM 1.5: *Under the assumptions of Theorem 1.2 the unique solution* u *of the Q.V.I.* (1.7) *is the solution of the problem*

$$(1.71)\qquad \begin{cases} 0 \leq u \leq Mu, & u \in C^o(\bar{\mathcal{O}}) \cap H_0^1 \\ Au \leq f & \textit{in } H^{-1} \\ Au = f & \textit{in the sense of distributions in } C\cap\mathcal{O}, \end{cases}$$

where C *is defined by* (1.70).

Proof: The only property to prove is the last one stated in (1.71). Now, let $\varphi \in \mathcal{D}(C\cap\mathcal{O})$, $\varphi \geq 0$. We can find ε small enough > 0 such that $u + \varepsilon\varphi \leq Mu$. In fact it suffices to take ε such that $\varepsilon\|\varphi\| \leq \sup_{x\in\operatorname{supp}\varphi} (Mu - u)$, a quantity which is strictly positive, because $Mu - u$ is a positive continuous function on the compact support $\operatorname{supp}\varphi$, therefore

$$(1.72)\qquad a(u,\varphi) \geq (f,\varphi),$$

and since

$$a(u,\varphi) \geqslant (f,\varphi)$$

because $\varphi \geqslant 0$, we have

(1.73) $$a(u,\varphi) = (f,\varphi),$$

i.e.,

$$Au = f \text{ in the sense of distirbutions in } C \cap \mathcal{O}.$$

A first important variant for applications consists of replacing A by

$$A - B - H,$$

where B,H are operators satisfying the properties of Section 2.1 of Chapter 3. We leave it to the reader to prove that the results of the preceding Theorems for A carry over when A is replaced by A - B - H. Essentially speaking, we have seen in Chapter 3 that the results for V.I.'s corresponding to the operator A carry over to the case A - B - H. Starting from that point, the arguments allowing us to study the Q.V.I. remain valid.□

Let us give another variant consisting of putting the obstacle on the boundary (a Signorini type of condition). Consider the problem

(1.74) $$\left| \begin{array}{l} a(u,v-u) \geqq (f,v-u) \qquad u \geqq 0, \qquad u \in C^{0}(\bar{\mathcal{O}}) \cap H^{1}(\mathcal{O}) \\ u - Mu \leqq 0 \quad \text{on } \Gamma, \quad \forall v \in H^{1}(\mathcal{O}) \text{ such that} \\ \quad v \leqq Mu \quad \text{on } \Gamma. \end{array} \right.$$

We are going to prove:

THEOREM 1.6: *Take the assumptions of Theorem 1.2. Then there exists one and only one solution of* (1.74).

Proof: Consider first the case where a is *coercive*. Every solution of (1.74) satisfies $u \leqslant u^{0}$, where u^{0} is the solution of

(1.75) $$a(u^{0},v) = (f,v) \qquad \forall\ v \in H^{1}.$$

Therefore u^{0} is the solution of the Neumann problem and $u^{0} \in C^{0}(\bar{\mathcal{O}})$. Next we define a mapping T of C^{0}_{+} into itself by setting

(1.76) $$\zeta = Tz$$

where ζ is the solution of

(1.77) $$\left| \begin{array}{l} a(\zeta, v-\zeta) \geqq (f, v-\zeta), \quad \zeta \geqq 0, \quad \zeta \in C^{0}(\bar{\mathcal{O}}) \cap H^{1}(\mathcal{O}) \\ \forall\ v \in H^{1}(\mathcal{O}) \text{ such that } v \leqq Mz \quad \text{on } \Gamma, \ \zeta \leqq Mz \quad \text{on } \Gamma. \end{array} \right.$$

If $z \in C^{0}_{+}$, then $Mz \in C^{0}_{+}$. Under these conditions ζ exists and is defined uniquely. The mapping T is increasing, concave, and satisfies

$$T(0) \geqslant \mu u^{0}, \quad 0 \leqslant \mu < 1.$$

In fact, let us set $u_1 = T(0)$, then

(1.78) $$\left| \begin{array}{l} a(u_1, v-u_1) \geqq (f, v-u_1) \\ \forall v \leqq k|_\Gamma, \ u_1 \leqq k|_\Gamma . \end{array} \right.$$

Take

$$\mu < \min\left(\frac{k}{\|u^o\|}, 1\right),$$

then μu^o is admissible for (1.78), thus $u_1 + (\mu u^o - u_1)^+$ also, hence

$$a(u_1, (\mu u^o - u_1)^+) \geqq (f, (\mu u^o - u_1)^+)$$

and

$$a(\mu u^o, (\mu u^o - u_1)^+) = (\mu f, (\mu u^o - u_1)^+)$$

which, using the coercivity of a, proves that

$$u_1 \geqslant \mu u^o.$$

The mapping T therefore satisfies (cf. Lemma 1.4):

(1.79) If $z, \zeta \in K$ and $\gamma \in [0,1]$ are such that $z - \zeta \leqslant \gamma z$, then

$$Tz - T\zeta \leqslant \gamma(1 - \mu)Tz.$$

Starting from this, and arguing as in Lemma 1.5, we obtain the existence and uniqueness of a fixed point of T in C_+^o. Next we pass from the coercive case to the general case, as in the proof of Theorem 1.2.□

Remark 1.9

$$\Gamma \subset C \quad \text{since } u|_\Gamma = 0, \quad \text{and}$$
$$Mu|_\Gamma = k.$$

Remark 1.10

Here is a variant with a different character. In the defintion of $u(x)$ given in Chapter 1, take the lower bound of the cost function with respect to *all* possible jumps that are compatible with the constraints, which are of two types:

(a) if $x \in \mathcal{O}$, we must have $x + \xi \in \mathcal{O}$;

(b) for example, $\xi_i \geqslant 0$ for all i.

We can *add* another constraint, namely:

(1.80) $$\left| \begin{array}{l} \sum_\alpha |\xi_\alpha| \leqslant q, \ \alpha \text{ running over the set of jumps,} \\ \text{where } q \text{ is given and } \geqslant 0. \end{array} \right.$$

We can then introduce:

(1.81) $u(x,q)$, $(x \in \mathcal{O},\ q \geqslant 0)$, denotes the optimal cost function under the constraint (1.80).

We then obtain for $u(x,q)$ the family of Q.V.I.'s

$$(1.82)\quad \left|\begin{array}{l} A_x u(x,q) - f(x) \leq 0, \quad u(x,q) \geq 0, \\ u(x,q) - M(q, u(x,q)) \leq 0, \\ [A_x u(x,q) - f(x)][u(x,q) - M(q, u(x,q))] = 0 \quad \text{in} \quad \mathcal{O} \end{array}\right.$$

where A_x means that we take the derivatives with respect to x, with q playing the role of parameter in the first condition (1.82), and where

$$(1.83)\quad M(q, u(x,q)) = \inf_{\xi_i \geq 0,\ x+\xi \in \mathcal{O},\ |\xi| \leq q} [k + c_0(\xi) + u(x+\xi, q-|\xi|)].$$

If $q = 0$, we have

$$u(x,0) < M(0; u(x,0))$$

so that (1.82) reduces to

$$(1.84)\quad Au(x,0) = f(x).$$

The conditions at the boundaries for $x \in \partial\mathcal{O}$ are of Dirichlet or Neumann type. So we have:

(1.85) $u(x,q)$ *decreases* if q *increases*,

(1.86) $u(x,q) \to u$ in $H^1(\mathcal{O})$ when $q \to +\infty$, where u is the solution of the Q.V.I. studied in the Sections above.

2. REGULARITY

The regularity problem consists of knowing whether the solution of the Q.V.I. (1.7) defined in Theorem 1.2 is more regular than $C^o(\bar{\mathcal{O}}) \cap H^1(\mathcal{O})$. Numerous authors have worked on this problem. The first results were obtain by J.L. JOLY, U. MOSCO, G.M. TROIANIELLO [1], then J.L. JOLY, U. MOSCO [2], L.A. CAFFARELLI, A. FRIEDMAN [1], followed by works by B. HANOUZET, J.L. JOLY [4], U. MOSCO [1],[2] (cf., also, M.G. GARRONI, G.M. TROIANIELLO [1]).

First we shall take the case $V = H_0^1(\mathcal{O})$; the case $V = H^1(\mathcal{O})$ will be studied in Section 2.3 below.

2.1 *The Lipschitz Property*

We assume

$$(2.1)\quad f \in L^p(\mathcal{O}), \quad p > n,$$

and prove:

LEMMA 2.1: *Under the assumptions of Theorem 1.1, together with* (2.1), *if* u *denotes the solution of the Q.V.I.* (1.7) *in* L^∞, *then*

(2.2) $\quad Mu \in W^{1,\infty}(\bar{\mathcal{O}})$

Proof: We have seen in Theorem 1.5 that u satisfies

$$Au = f \text{ in the sense of distributions in } C \cap \mathcal{O},\ u|_\Gamma = 0.$$

By AGMON-DOUGLIS-NIRENBERG [1], then by using $\Gamma \subset C$, it follows that

$$u \in W^{2,p}(D) \text{ for all open subsets } D \text{ of } \mathcal{O} \text{ such that } \bar{D} \subset C.$$

Since $p > n$, it follows that $u \in W^{1,\infty}(\bar{D})$. Since u is continued by 0 and $Mu|_\Gamma = k$, if we write

$$\mathcal{C} = \{x \in R^n \mid u(x) < Mu(x)\}$$

then we also have

(2.3) $\quad u \in W^{1,\infty}(\bar{F})$, for all open sets F of R^n such that $\bar{F} \subset \mathcal{C}$.

Further, there exists a Borel mapping $x \to \xi(x)$ from $R^n \to (R^n)^+$ such that

(2.4) $$\left| \begin{array}{l} Mu(x) = k + c_o(\xi(x)) + u(x + \xi(x)) \\ \xi(x) = 0 \quad \forall x \notin \mathcal{O},\ x + \xi(x) \in \bar{\mathcal{O}} \quad \text{if } x \in \bar{\mathcal{O}}. \end{array} \right.$$

But by the sublinearity of c_o we have

$$\inf_{\xi \geq 0} [c_o(\xi) + u(x+\xi)] \leq \inf_{\eta \geq 0} [c_o(\xi(x)+\eta) + u(x+\xi(x)+\eta)]$$

$$\leq c_o(\xi(x)) + \inf_{\eta \geq 0} [c_o(\eta) + u(x+\xi(x)+\eta)]$$

therefore

(2.5) $\quad Mu(x) \leq c_o(\xi(x)) + Mu(x + \xi(x))$

and thus by (2.4) it follows from (2.5) that

(2.6) $\quad u(x+\xi(x)) \leq Mu(x+\xi(x)) - k$.

We then take

$$F = \{ y \in R^n \mid u(y) < Mu(y) - \frac{k}{2} \},$$

and $\bar{F} \subset \mathcal{C}$. By (2.6) we see that

$$x + \xi(x) \in F, \text{ for all } x.$$

Since the function u - Mu is uniformly continuous on R^n, there exists ρ_o such that

(2.7) $$|y-y'|\leq\rho_o \text{ implies } |Mu(y')-u(y')-(Mu(y)-u(y))|\leq\frac{k}{4}$$

Let χ be a unit vector of R^n, θ a real number, and x_o be fixed in R^n. For $|\theta|\leqslant\rho_o$, if we set

$$y = x_o + \xi(x_o), \quad y' = y + \theta\chi,$$

it follows from (2.7) that

$$u(y')-Mu(y')\leq u(y)-Mu(y)+\frac{k}{4}\leq -k+\frac{k}{4} < -\frac{k}{2}$$

and thus $y,y' \in F$. By (2.3) we can then affirm that

(2.8) $$|u(x_o+\xi(x_o)+\theta\chi)-u(x_o+\xi(x_o))|\leq C_F|\theta|$$

for all θ such that $|\theta|\leqslant\rho_o$,

where

$$C_F = \sup_{\substack{y,y'\in\bar{F}\\ y\neq y'}} \frac{|u(y)-u(y')|}{|y-y'|}.$$

But

$$Mu(x_o) = k + c_o(\xi(x_o)) + u(x_o+\xi(x_o))$$

$$Mu(x_o+\theta\chi)\leq k + c_o(\xi(x_o)) + u(x_o+\theta\chi+\xi(x_o))$$

and therefore by (2.8) we deduce

(2.9) $$\frac{Mu(x_o+\theta\chi)-Mu(x_o)}{|\theta|}\leq C_F \quad \forall x_o, \ \forall\chi$$

for all unit vectors χ of R^n, and for all θ such that $|\theta|\leqslant\rho_o$. But when $\theta\to 0$, the left hand side of (2.9) converges in the sense of distributions in R^n to $\frac{\partial Mu}{\partial\chi}$, and therefore $\frac{\partial}{\partial\chi}Mu\leqslant C_F$ in the sense of distributions in R^n, for all χ. This implies $Mu\in W^{1,\infty}(R^n)$, and hence the result desired.

Now make the assumption

(2.10) $$a_{ij}\in W^{2,\infty}, \quad a_i\in W^{1,\infty}, \quad a_o\in W^{1,\infty}.$$

From this we deduce:

THEOREM 2.1: *Take the assumptions of Lemma 2.1 and* (2.10). *Then the solution of the Q.V.I.* (1.7) *in* L^∞ *satisfies the property*

(2.11) $u \in W^{1,\infty}(\bar{\mathcal{O}})$.

Proof: This follows from Lemma 2.1, and from a result on V.I.'s (cf. Lemma 2.2).□

LEMMA 2.2: *Assume* (1.1) *and* (2.1),(2.10). *Let* $\psi \in W^{1,\infty}(\bar{\mathcal{O}})$, $\psi|_\Gamma \geqslant 0$. *Consider the V.I.*

$$(2.12) \qquad \begin{vmatrix} a(u,v-u) \geqq (f,v-u) \quad \forall\, v \in H_0^1 \;,\; v \leqq \psi \\ u \in H_0^1 \;,\; u \leqq \psi \end{vmatrix}$$

then

(2.13) $u \in W^{1,\infty}(\bar{\mathcal{O}})$.

Proof: Let z be the solution of

$$a(z,v) = (f,v) \quad \forall\, v \in H_0^1 \;,\quad z \in H_0^1$$

and θ the solution of

$$a(\theta,v) = (1,v) \quad \forall\, v \in H_0^1 \;,\quad \theta \in H_0^1$$

Then $\tilde{u} = u - z + k\theta$ is the solution of the V.I. (2.12), where f is replaced by

$$\tilde{f} = k,$$

and ψ is replaced by

(2.14) $\tilde{\psi} = \psi - z + k\theta$.

Now $z \in W^{2,p}(\mathcal{O})$, thus $z \in W^{1,\beta}(\bar{\mathcal{O}})$, $\theta \in C^2(\bar{\mathcal{O}})$, so $\tilde{\psi}$ is Lipschitz continuous. Furthermore, if ν is the normal to Γ oriented to the exterior, then $\left.\frac{\partial\theta}{\partial\nu}\right|_\Gamma < 0$, therefore

$$\theta(x) \geq \text{Min}(\rho_1|x-y_x|\;,\rho_2),\; x \in \bar{\mathcal{O}}$$

where y_x = the point of Γ with minimal distance to x, and ρ_1,ρ_2 are constants > 0. Moreover, since $\psi - z$ is Lipschitz and positive on Γ, we have

$$\psi(x) - z(x) \geq -\sigma|x-y_x|$$

and therefore we can find k so that $\tilde{\psi} \leqslant 0$. Thus, without loss of generality, we are reduced to the case where

(2.15) $f \in W^{1,\infty} \quad f \geq 0, \quad \psi \in W^{1,\infty} \quad \psi \geq 0$.

Consider a sequence β_ε of regular functions from R → R such that

$$\beta'_\varepsilon \geqslant 0, \quad \beta_\varepsilon = 0 \text{ for } x \leqslant 0, \quad \beta_\varepsilon \to \infty \text{ when } \varepsilon \downarrow 0 \text{ for } x > 0.$$

First we assume that $f,\psi,a_{ij},a_i,a_0 \in C^4(\bar{\mathcal{O}})$, and write A in the non-divergent form

$$A\phi = -a_{ij}\frac{\partial^2\phi}{\partial x_i\,\partial x_j} + b_i\frac{\partial\phi}{\partial x_i} + a_0\,\phi$$

where $b_i = a_i - a_{ij,x_j}$. For $\varepsilon > 0$, we consider $u^\varepsilon \in C^3(\bar{\mathcal{O}})$ the unique solution of

$$(2.16)\qquad Au^\varepsilon + \beta_\varepsilon(u^\varepsilon - \psi) = f \ \text{ in } \ \mathcal{O}, \quad u^\varepsilon|_\Gamma = 0\,.$$

By the Maximum Principle

$$(2.17)\qquad 0 \le u^\varepsilon \le \frac{1}{\beta}\,\|f\|_{L^\infty}\,.$$

Furthermore, let z be the solution already introduced above, of

$$Az = f, \quad z|_\Gamma = 0,$$

then

$$A(u^\varepsilon - z) \le 0 \qquad u^\varepsilon - z|_\Gamma = 0$$

whence

$$u^\varepsilon(x) \leqslant z(x).$$

Therefore if x_0 is a point of Γ,

$$0 \le u^\varepsilon(x) - u^\varepsilon(x_0) \le z(x) - z(x_0) \le C|x - x_0|\ \ \|f\|_{L^\infty}$$

since

$$\|z\|_{W^{2,p}} \le C\ \|f\|_{L^\infty}$$

therefore

$$(2.18)\qquad \|Du^\varepsilon\|_{L^\infty(\Gamma)} \le C\ \|f\|_{L^\infty}\,.$$

Next we set $\mu > 0$,

$$w(x) = |Du^\varepsilon(x)|^2 + \mu|u^\varepsilon(x)|^2$$

and let $x_0 \in \bar{\mathcal{O}}$ be such that w is a maximum at the point x_0,

$$w(x_0) = \|w\|_{L^\infty}.$$

We are going to show that if $x_0 \in \mathcal{O}$ and $|Du^\varepsilon(x_0)|^2 > |D\psi(x_0)|^2$, then

(2.19) $$w(x_0) \le C \; \|f\|^2_{W^{1,\infty}} \; .$$

Now,

(2.20) $$\begin{aligned} 2\beta \; w(x_0) &\le - a_{ij} \, w_{x_i x_j} + b_i \, w_{x_i} + 2\, a_0 w \\ &= - 2a_{ij} \, u_{x_k x_i} \, u_{x_k x_j} + \\ &+ 2u_{x_k} \, [-a_{ij} \, u_{x_i x_j x_k} + b_i \, u_{x_k x_i} + a_0 \, u_{x_k} \,] \\ &- 2\,\mu\, a_{ij} \, u_{x_i} \, u_{x_j} + 2\,\mu\, u \,[- a_{ij} \, u_{x_i x_j} + b_i \, u_{x_i} + a_0 u] \\ &= - 2\, a_{ij} \, u_{x_k x_i} \, u_{x_k x_j} + 2\, u_{x_k} \, [a_{ij,x_k} \, u_{x_i x_j} - \\ &\quad - b_{i,x_k} \, u_{x_i} - a_{0,x_k} \, u + f_{x_k} - \beta'(u-\psi)(u_{x_k} - \psi_{x_k})] \\ &\quad - 2\,\mu\, a_{ij} \, u_{x_i} \, u_{x_j} + 2\,\mu\, u \; [f - \beta(u-\psi)] \; . \end{aligned}$$

But

$$\begin{aligned} - \beta'(u-\psi)(u_{x_k} - \psi_{x_k}) \, u_{x_k} &\le - \frac{1}{2} \beta'(u-\psi) \; [|Du|^2 - |D\psi|^2] \\ &\le 0 \end{aligned}$$

by the assumption. Also taking into account that $u \geqslant 0$, $\beta^\varepsilon \geqslant 0$, from (2.20) we deduce that

$$\begin{aligned} 2\, \beta w(x_0) &\le - 2\, a_{ij} \, u_{x_k x_i} \, u_{x_k x_j} + 2u_{x_k} \, [a_{ij,x_k} \, u_{x_i x_j} - \\ &\quad - b_{i,x_k} \, u_{x_i} - a_{0,x_k} \, u + f_{x_k} \,] - 2\,\mu\, a_{ij} \, u_{x_i} u_{x_j} \\ &\quad + 2\,\mu\, u\, f \\ &\le - 2\, a_{ij} \, u^\varepsilon_{x_k x_i} \, u^\varepsilon_{x_k x_j} + \eta \sum_{k=1}^{n} |a_{ij,x_k} \, u^\varepsilon_{x_i x_j}|^2 \\ &\quad + \frac{1}{\eta} \; |Du^\varepsilon|^2 - 2\,\mu\,\alpha\, |Du^\varepsilon|^2 + C \; \|f\|^2_{W^{1,\infty}} \end{aligned}$$

Next we use the estimate (owed to OLEINIK - RADKEVIC, cf. D. STROOCK - S.R.S. VARADHAN [2])

$$\sum_{k=1}^{n} |a_{ij,x_k} \, u^\varepsilon_{x_i x_j}|^2 \le C \, a_{ij} \, u^\varepsilon_{x_i x_k} \, u^\varepsilon_{x_j x_k}$$

where C depends on upon $\|a_{ij}\|_{W^{2,\infty}}$, and choosing η small enough, μ large enough, we obtain (2.19). In all the cases we have

$$\|Du^\varepsilon\|_{L^\infty(\bar{\mathcal{O}})} \le C \max \left(\|\psi\|_{W^{1,\infty}} \; , \; \|f\|_{W^{1,\infty}} \right) .$$

By approximation we deduce from this the same estimate under the Lemma's assumptions, and passing to the limit in ε, we deduce the desired result in the case (2.15).□

Remark 2.1

The proof of Lemma 2.2 is inspired by L.C. EVANS and J.L. MENALDI [1].□

2.2 $W^{2,p}$-*Regularity*

Set

$$\gamma(x) = k + \inf_{\substack{\xi \geq 0 \\ x+\xi\in\Gamma}} c_0(\xi) \quad , \quad x \in \bar{\mathcal{O}}$$

and assume

(2.21) $\quad \gamma \in W^{2,\infty}(\mathcal{O})$, $f \in C^{0,\infty}(\bar{\mathcal{O}})$

(2.22) $\quad a_{ij} \in C^{1,\alpha}(\bar{\mathcal{O}})$, $a_i \in C^{0,\alpha}(\bar{\mathcal{O}})$, $a_0 \in C^{0,\alpha}(\bar{\mathcal{O}})$, $\alpha > 0$.

We have:

LEMMA 2.3: *Take the assumptions of Theorem 1.1 and* (2.21),(2.22). *Then*

(2.23) $\quad A(Mu) \geqslant - c$

in the sense of distributions in $\mathcal{O}$, *where* c *is a constant.*

Proof: By Lemma 2.1, A(Mu) has a meaning in $H^{-1}(\mathcal{O})$. On the other hand, by (2.21), (2.22) we can affirm (cf. O.A. LADYZHENSKAYA - N.N. URAL'TSEVA [1]) that

(2.24) $\quad u \in W^{2,\infty}(D)$, for every open subset D of $\mathcal{O}$ such that $\bar{D} \subset C$.

Notice that

(2.25) $\quad Mu(x) \leq \gamma(x) \quad \forall x \in \bar{\mathcal{O}}$, $Mu(x) = \gamma(x) = k$ on Γ .

If $Mu(x) = \gamma(x)$ for all $x \in \mathcal{O}$, (2.23) follows immediately from the first assumption (2.21). Otherwise there exists a point x_0 of $\mathcal{O}$ such that

$$Mu(x_0) < \gamma(x_0).$$

We then set

$$\mathcal{O}^1_\varepsilon = \{x \in \mathcal{O} \mid Mu(x) < \gamma(x) - \frac{\varepsilon}{2}\}$$

which is non-empty for ε small enough. Also let

$$\mathcal{O}^2_\varepsilon = \{x \in \mathcal{O} \mid Mu(x) > \gamma(x) - \varepsilon\}$$

then

$$\mathcal{O} = \mathcal{O}^1_\varepsilon \cup \mathcal{O}^2_\varepsilon .$$

Let us set

$$\psi_\varepsilon(x) = Mu(x) \wedge (\gamma(x)-\varepsilon) \ . \tag{2.26}$$

We are going to prove that

(2.27) $A\psi_\varepsilon \geqslant - c$ in the sense of distributions in $\mathcal{O}$ (c independent of ε).

It suffices to prove (2.27) in the sense of distributions in $\mathcal{O}_\varepsilon^1$ and $\mathcal{O}_\varepsilon^2$ respectively. But on $\mathcal{O}_\varepsilon^2$ we have

$$\psi_\varepsilon(x) = \gamma(x)-\varepsilon$$

and therefore the result is true by the assumption (2.22). Therefore we have to show that

(2.28) $A\psi_\varepsilon \geqslant - c$ in the sense of distributions in $\mathcal{O}_\varepsilon^1$.

We define

$$\mathcal{E}(x) = \{\xi \geq 0 \mid c_0(\xi) + u(x+\xi) = \inf_{\eta\geq 0} (c_0(\eta) + u(x+\eta))\}$$

which is closed for all x and $x + \mathcal{E}(x) \in \bar{\mathcal{O}}$, if $x \in \bar{\mathcal{O}}$. Therefore $x + \mathcal{E}(x)$ is compact for all $x \in \bar{\mathcal{O}}$. Let us set

$$d^\varepsilon = \inf_{\{x \in \bar{\mathcal{O}} \mid Mu(x) \leq \gamma(x) - \frac{\varepsilon}{2}\}} d(x+ \mathcal{E}(x),\Gamma) \tag{2.29}$$

and show that

$$d^\varepsilon > 0. \tag{2.30}$$

If this were not so there would exist a sequence $x_n \in \bar{\mathcal{O}}$, $Mu(x_n) \leqslant \gamma(x_n) - \frac{\varepsilon}{2}$ such that

$$d(x_n + \mathcal{E}(x_n),\Gamma) \rightarrow 0 \ .$$

Now, there exists $\xi_n \in \mathcal{E}(x_n)$, $y_n \in \Gamma$ such that $x_n + \xi_n \in \bar{\mathcal{O}}$ and

$$d(x_n + \mathcal{E}(x_n),\Gamma) = |x_n + \xi_n - y_n| \ .$$

If we consider an extracted subsequence $x_n \to x$, $\xi_n \to \xi$, $y_n \to y$ we obtain (since u is uniformly continuous)

$$x \in \bar{\mathcal{O}} \ , \xi \geq 0, \ x + \xi \in \bar{\mathcal{O}} \ , \ y \in \Gamma \ , \ \xi \in \mathcal{E}(x) \ ,$$

$$Mu(x) \leq \gamma(x) - \frac{\varepsilon}{2} \ ,$$

and

$$|x + \xi - y| = 0,$$

which is impossible, for then

$$Mu(x) = k + c_0(\xi) + u(x+\xi) = k + c_0(\xi) \geq \gamma(x) .$$

Therefore we have (2.30). As in Lemma 2.1 we again take

$$F = \{ y \in R^n \ |u(y) < Mu(y) - \frac{k}{2}\}$$

where ρ_0 is such that

$$|y-y'| \leq \rho_0 \quad \text{implies} \quad |Mu(y)-u(y)-(Mu(y')-u(y'))| \leq \frac{k}{4} .$$

Let $D = F \cap \mathcal{O}$. Then D satisfies (2.24).

Let χ be a unit vector of R^n, and θ a real number such that

$$(2.31) \qquad |\theta| \leq \frac{d^\varepsilon}{2} \wedge \rho_0 .$$

Let $x \in \mathcal{O}_\varepsilon^1$, then by (2.29) $d(x + \xi(x),\Gamma) \geqslant d^\varepsilon$, and by what we saw in Lemma 2.1, $x + \xi(x) \in F$.

Furthermore, for $0 \leqslant |\lambda| \leqslant 1$, $x + \lambda\theta\chi + \xi(x)$ satisfies

$$d(x+\lambda\theta\chi+\xi(x),\Gamma) \geq \frac{d^\varepsilon}{2} ,$$

by (2.31), and thus belongs to $\mathcal{O}$ and as was seen in Lemma 2.1, $x + \lambda\theta\chi + \xi(x) \in F$. Thus for all $x \in \mathcal{O}_\varepsilon^1$ and θ satisfying (2.31), and χ a unit vector of R^n, if $|\lambda| \leqslant 1$ then

$$x+\xi(x) \in D, \quad x+\xi(x) + \lambda\theta\chi \in D .$$

Now,

$$(2.32) \qquad \begin{aligned} &\frac{1}{\theta^2} [Mu(x+\theta\chi)+ Mu(x-\theta\chi)-2Mu(x)] \leq \\ &\frac{1}{\theta^2} [u(x+\xi(x)+\theta\chi)+ u(x+\xi(x)-\theta\chi) - 2u(x+\xi(x))] = \\ &= \frac{1}{|\theta|} \int_0^1 (Du(x+\xi(x)+\lambda\theta\chi)- Du(x+\xi(x)-\lambda\theta\chi)).\chi d\lambda \\ &\leq C_D \end{aligned}$$

where we have set

$$C_D = \sup_{\substack{y,y' \in \bar{D} \\ y \neq y'}} \frac{|Du(y)-Du(y')|}{|y-y'|} .$$

From (2.32) and from the regularity of the function γ, it follows that

(2.33) $$\frac{1}{\theta^2}\,[\psi_\varepsilon(x+\theta\chi)+\psi_\varepsilon(x-\theta\chi) - 2\psi_\varepsilon(x)] \le c$$

where

$$c = \max\,(C_D,\ \|D^2\gamma\|).$$

Now let $\beta_\varepsilon \in \mathcal{D}(\mathcal{O}^1)$, $\beta_\varepsilon \geqslant 0$. For $x \in R^n$ we have

(2.34) $$\begin{aligned}&\frac{1}{\theta^2}[\beta_\varepsilon\psi_\varepsilon(x+\theta\chi)+\beta_\varepsilon\,\psi_\varepsilon(x-\theta\chi)- 2\beta_\varepsilon\,\psi_\varepsilon(x)] = \\ &\frac{1}{\theta^2}\,\beta_\varepsilon(x)[\psi_\varepsilon(x+\theta\chi)+\psi_\varepsilon(x-\theta\chi) - 2\psi_\varepsilon(x)] + \\ &+\frac{1}{\theta^2}(\beta_\varepsilon(x+\theta\chi)-\beta_\varepsilon(x))(\psi_\varepsilon(x+\theta\chi)-\psi_\varepsilon(x)) + \\ &+\frac{1}{\theta^2}(\beta_\varepsilon(x-\theta\chi)-\beta_\varepsilon(x))(\psi_\varepsilon(x-\theta\chi)-\psi_\varepsilon(x)) + \\ &+\frac{1}{\theta^2}\,\psi_\varepsilon(x)[\beta_\varepsilon(x+\theta\chi)+\beta_\varepsilon(x-\theta\chi) - 2\,\beta_\varepsilon(x)] \\ &\le C\ \beta_\varepsilon(x) + 3 \text{ other terms of the right hand side of (2.34).}\end{aligned}$$

Also, let $\rho_\delta(x)$ be a regularizing function

$$\rho_\delta(x) = \frac{1}{\delta^n}\,\rho\left(\frac{x}{\delta}\right)$$

with $\rho \in \mathcal{D}(R^n)$ with support in the unit ball, and

$$\int_{R^n}\rho(x)dx = 1,\ \rho > 0 \text{ in } \{x \mid |x| < 1\},\quad \rho = 0 \text{ outside.}$$

Now write

$$\zeta_\delta(x) = \rho_\delta * (\beta_\varepsilon\,\psi_\varepsilon).$$

From (2.34) we easily deduce that

(2.35) $$\begin{aligned}\frac{\partial^2}{\partial\chi^2}\,\zeta_\delta(x) &\le c\,\rho_\delta * \beta_\varepsilon + 2\int \rho_\delta(x-y)\,\frac{\partial\beta_\varepsilon}{\partial\chi}(y)\,\frac{\partial\psi_\varepsilon}{\partial\chi}(y)\,dy + \\ &+\int \rho_\delta(x-y)\psi_\varepsilon(y)\,\frac{\partial^2\beta_\varepsilon}{\partial\chi^2}(y)dy, \quad \forall\, x,\ \forall\, \chi.\end{aligned}$$

We fix x_0 and denote by $\chi_1(x_0),\ldots,\chi_n(x_0)$ the eigenvectors of the matrix $a_{ij}(x_0)$ corresponding to the eigenvalues $\lambda_1(x_0),\ldots,\lambda_n(x_0)$. Then

(2.36) $$A\zeta_\delta(x_0) = a_0\,\zeta_\delta - a_{ij}(x_0)\frac{\partial^2\zeta_\delta}{\partial x_i\,\partial x_j} - \frac{\partial a_{ij}}{\partial x_i}\frac{\partial\zeta_\delta}{\partial x_j}$$
$$= a_0\,\zeta_\delta - \sum_{ij}\frac{\partial a_{ij}}{\partial x_i}\frac{\partial\zeta_\delta}{\partial x_j} - \sum_i \lambda_i(x_0)\frac{\partial^2}{\partial \chi_i^2}\zeta_\delta(x_0)$$

and by (2.35)

$$\geq a_0\,\zeta_\delta - \sum_{i,j}\frac{\partial a_{ij}}{\partial x_i}\frac{\partial\zeta_\delta}{\partial x_j} - c\sum_i \lambda_i\,\rho_\delta * \beta_\varepsilon -$$
$$- 2\sum_i \lambda_i(x_0)\int \rho_\delta(x_0-y)\frac{\partial\beta_\varepsilon}{\partial\chi_i}(y)\frac{\partial\psi_\varepsilon}{\partial\chi_i}(y)\,dy -$$
$$- \sum_i \lambda_i(x_0)\int\rho_\delta(x_0-y)\,\psi_\varepsilon(y)\frac{\partial^2\beta_\varepsilon}{\partial\chi_i^2}(y)\,dy$$
$$= a_0\,\zeta_\delta(x_0) - \sum_{i,j}\frac{\partial a_{ij}}{\partial x_i}(x_0)\frac{\partial\zeta_\delta}{\partial x_j}(x_0) - c\sum_{i,j} a_{ij}(x_0)\rho_\delta * \beta_\varepsilon\;(x_0)$$
$$- 2\sum_{i,j}\int\rho_\delta(x_0-y)a_{ij}(x_0)\frac{\partial\beta_\varepsilon}{\partial x_j}(y)\frac{\partial\psi_\varepsilon}{\partial x_i}(y)\,dy -$$
$$- \sum_{i,j}\int\rho_\delta(x_0-y)\,\psi_\varepsilon(y)\frac{\partial^2\beta_\varepsilon}{\partial x_i\partial x_j}(y)\,dy\;.$$

Now, when $\delta \to 0$, $A(\zeta_\delta) \to A(\beta_\varepsilon\psi_\varepsilon)$ in H^{-1}, and the right hand side of (2.36) converges in L^∞ weakly to

$$a_0\,\beta_\varepsilon\psi_\varepsilon - \sum_{i,j}\frac{\partial a_{ij}}{\partial x_i}\frac{\partial}{\partial x_j}(\beta_\varepsilon\psi_\varepsilon) - c\,\beta_\varepsilon\sum_{i,j} a_{ij} -$$
$$- 2\sum_{i,j} a_{ij}\frac{\partial\beta_\varepsilon}{\partial x_j}\frac{\partial\psi_\varepsilon}{\partial x_i} - \sum_{i,j} a_{ij}\,\psi_\varepsilon\frac{\partial^2\beta_\varepsilon}{\partial x_i\partial x_j}\;.$$

But

$$\beta_\varepsilon\,A\psi_\varepsilon = A(\beta_\varepsilon\psi_\varepsilon) + 2\sum a_{ij}\frac{\partial\beta_\varepsilon}{\partial x_i}\frac{\partial\psi_\varepsilon}{\partial x_j} + \sum\frac{\partial a_{ij}}{\partial x_i}\frac{\partial\beta_\varepsilon}{\partial x_j}\psi_\varepsilon +$$
$$+ \sum a_{ij}\frac{\partial\beta_\varepsilon}{\partial x_i\partial x_j}\psi_\varepsilon$$

and therefore we have obtained

$$\beta_\varepsilon\,A\,\psi_\varepsilon \geq a_0\,\beta_\varepsilon\,\psi_\varepsilon - c_1\,\beta_\varepsilon - \sum_{i,j}\frac{\partial a_{ij}}{\partial x_i}\frac{\partial\psi_\varepsilon}{\partial x_j}\beta_\varepsilon$$

which means

$$A\psi_\varepsilon \geq a_0 \psi_\varepsilon - c_1 - \sum_{i,j} \frac{\partial a_{ij}}{\partial x_i} \frac{\partial \psi_\varepsilon}{\partial x_j}$$

as a distribution in $\mathcal{O}^1_\varepsilon$, from which (2.27) follows. In (2.27) we make $\psi_\varepsilon \to 0$, and notice that $\psi_\varepsilon \to Mu$ in H^1, whence the result desired.□

We shall now deduce the regularity of the solution u of the Q.V.I. as a consequence of the Lewy-Stampacchia inequality, which we shall give again for the convenience of the reader (cf., for example, U. MOSCO [3] or B.L., loc. cit. [1]). (We could also use the method of homographic penalisation, as in Chapter 2).

LEMMA 2.4: *Assume* (1.1). *Let* $f \in L^2(\mathcal{O})$, $\psi \in H^1(\mathcal{O})$, $\psi|_\Gamma \geqslant 0$, $A\psi \geqslant g$ *(in the sense of* H^{-1}*) where* $g \in L^2(\mathcal{O})$. *Let* u *be the solution of the V.I.*

(2.37) $$a(u,v-u) \geq (f,v-u)$$
$$\forall\, v \in H_0^1 \quad v \leq \psi \quad , \quad u \in H_0^1 \,, \quad u \leq \psi \quad .$$

Then

(2.38) $$f \geq Au \geq f \wedge g \quad .$$

Proof: We notice that it suffices to prove (2.38) with β large enough for a(u,v) to be coercive. In fact, let us assume the result to be true in the coercive case. Then we can write (2.37) in the form

$$f + \lambda u \geq Au + \lambda u \geq (f + \lambda u) \wedge (g + \lambda\psi)$$

and therefore we can write

$$f \geq Au \geq f \wedge (g + \lambda\psi - \lambda u) \geq f \wedge g \;.$$

Therefore we assume a to be coercive. Set $f \wedge g = h$ and let z be the solution of

(2.39) $$a(z,v-z) \geq (h,v-z) \qquad \forall\, v \in H_0^1 \,, \; v \geq u$$
$$z \in H_0^1, \quad z \geq u \quad .$$

We shall show that z = u, from which the desired result will immediately follow. To show that z = u, it suffices to show that

(2.40) $$z \leqslant \psi.$$

In fact, if (2.40) holds, then

$$a(u,z-u) \geqslant (f,z-u) \geqslant (h,z-u)$$

and

$$a(z,u-z) \geqslant (h,u-z),$$

whence z = u.

To prove (2.40) we take $v = z - (z - \psi)^+$ in (2.39), which is permissible, hence

$$a(z,(z-\psi)^+) \le (h,(z-\psi)^+) \le \langle A\psi,(z-\psi)^+\rangle = a(\psi,(z-\psi)^+) \; .$$

whence (2.40).□

We can therefore state:

THEOREM 2.2: *Take the assumptions of Theorem 1.1 together with* (2.21), (2.22). *Then the solution* u *of the Q.V.I.* (1.7) *in* L^∞ *satisfies the property*

(2.41) $$u \in W^{2,p}(\mathcal{O}) \;, \quad \forall\, p \ge 2 \quad p < \infty \quad , \; Au \in L^\infty \; .$$

Proof: By Lemmas 2.1 and 2.3, the Levy-Stampacchia inequality can be applied, and

$$f \geqslant Au \geqslant f \wedge (-c),$$

whence $Au \in L^\infty$, hence (2.41) holds by the result of S. AGMON - A. DOUGLIS - L. NIRENBERG [1].□

2.3 *Regularity for the Neumann Problem*

Here we assume that the open set $\mathcal{O}$ is defined by

(2.42) $$\mathcal{O} = \{x \mid \Phi(x) < 0\}$$

where Φ is a C^3 function that is bounded as well as its derivatives, is convex, and $\Phi(x) = 0$ on Γ.

We begin by giving some additional properties of $P(x,\xi)$ (cf. (1.39)). Now,

(2.43) $$P(x,\xi) = (\rho \wedge 1)\; \xi$$

where ρ is the *unique* solution of

(2.43') $$\Phi(x+\rho\xi) = 0 \qquad 0 \le \rho \; .$$

LEMMA 2.5: *The function* $x \to \rho(x,\xi)$ *is continuous on* $\bar{\mathcal{O}}$, *uniformly with respect to* ξ *on* $m \leqslant |\xi| \leqslant M$, $m > 0$. *Moreover,* ρ *is twice continuously differentiable at* x *on* $\mathcal{O}$, *for* $\xi \neq 0$ *and* ρ *is concave.*

Proof: Let $x_n \to x$ in $\bar{\mathcal{O}}$, $m \leqslant |\xi_n| \leqslant M$, $\xi_n \to \xi$. Then ρ_n is bounded, since $x_n + \rho_n\xi_n \in \bar{\mathcal{O}}$, which is therefore bounded, hence $\rho_n\xi_n$ is bounded by $|\rho_n\xi_n| \leqslant \Delta$, where

$$\Delta = \sup_{y,y' \in \bar{\mathcal{O}}} |y-y'| .$$

Therefore

$$\rho_n \leqslant \frac{\Delta}{m} \; .$$

If we pick a subsequence $\rho_n \to \rho$, it is clear that $\Phi(x+\rho\xi) = 0$. Therefore $\rho(x,\xi)$ is continuous with respect to the pair x,ξ on $\bar{\mathcal{O}} \times \{\xi \mid m \leqslant |\xi| \leqslant M\}$, whence the first part of the Lemma. Next, by differentiating (2.43') formally, and writing $y = x + \rho\xi$, we have

$$(2.44)\qquad D\Phi(y) + D\Phi(y).\xi \;\; D_x\rho \;=\; 0\ .$$

But

$$\Phi(x) - \Phi(y) \geq D\Phi(y).\ (x-y)$$

whence

$$\Phi(x) \geq -\rho \;\; D\Phi(y).\xi$$

therefore

$$(2.45)\qquad D\,\Phi(y).\xi \;\geq\; \frac{-\Phi(x)}{\rho} > 0 \quad \text{if } x \in \mathcal{O}.$$

Since

$$\rho\,|\xi| \leq \Delta,$$

we deduce that

$$(2.46)\qquad |D_x\rho| \;\leq\; \frac{|D\Phi(y)|\Delta}{|\Phi(x)|\;|\xi|} \;\leq\; \frac{\Delta\,\|D\Phi\|_{L^\infty(\Gamma)}}{|\xi|\;|\Phi(x)|}\ .$$

Differentiating (2.44) again, we obtain

$$(2.47)\qquad D^2\Phi(y) + D^2\Phi(y)\xi \;\; D_x\rho^* + D_x\rho\;(D^2\Phi(y)\xi)^* + $$
$$+ D^2\Phi(y)\xi.\xi \;\; D_x\rho \;\; D_x\rho^* + D\Phi(y).\xi \;\; D_x^2\rho = 0$$

from which we deduce that if $v \in R^n$, then

$$(2.48)\qquad D_x^2\,\rho v.v = \frac{-D^2\Phi(y)(\xi D\Phi(y).v+v\,D\Phi(y).\xi)(\xi D\Phi(y).v+v\,D\Phi(y).\xi)}{(D\Phi(y).\xi)^2}$$
$$\leq 0$$

and

$$(2.49)\qquad \|D_x^2\,\rho\| \;\leq\; \frac{4\,\Delta^2\;\|D^2\Phi\|_{L^\infty(\Gamma)}\;\|D\Phi\|^2_{L^\infty(\Gamma)}}{\Phi^2(x)}$$

whence the result.□

CORROLARY 2.1: *The function* $P(x,\xi)$ *satisfies*

$$(2.50)\qquad |P(x,\xi) - P(x',\xi)| \leq \Delta|x-x'|\;\|D\Phi\|_{L^\omega(\Gamma)}\int_0^1 \frac{d\theta}{|\Phi(x+\theta(x'-x)|}$$
$$\forall\, x,\, x' \in \bar{\mathcal{O}}\ ,\ \forall\,\xi\ .$$

Proof: This results immediately from (2.46) and (2.43).□

We shall consider the set C defined in (1.70), and shall show that the solution of the Q.V.I. (1.7) satisfies

(2.51) $Au = f$ in the sense of distributions on $C \cap \mathcal{O}$

$$\frac{\partial u}{\partial \nu_A} = 0 \quad \text{on } C \cap \partial\mathcal{O}.$$

In fact, let $\phi \in \mathcal{D}(C)$, $\phi \geqslant 0$, then, as in Theorem 1.5, we can find ε small enough that $u + \varepsilon\phi \leqslant Mu$, therefore

$$a(u,\phi) = (f,\phi) \quad \forall\, \phi \in \mathcal{D}(C)$$

whence the first relation of (2.51) by taking $\phi \in \mathcal{D}(C \cap \mathcal{O})$.

Let us set $\partial_1 C = C \cap \partial\mathcal{O}$; the boundary of C is made up of two parts, $\partial C = \partial_1 C \cup \partial_2 C$, where $\partial_2 C$ is the part of ∂C inside $\mathcal{O}$ — and about which we have no information (in particular, about its regularity). However, $\partial_1 C$ is included in Γ, which is regular, and therefore each connected component of $\partial_1 C$ is regular; then since $u \in H^1(\mathcal{O})$ (hence $u \in H^1(C)$) and $Au = f \in L^2(C)$ on $C \cap \mathcal{O}$, and $\frac{\partial u}{\partial \nu_A}$ has a meaning on each connected component $\partial_1 C^{(k)}$ of $\partial_1 C$ and belongs to $H^{-1/2}(\partial_1 C^{(k)})$.

Now take $\phi \in \mathcal{D}(\bar{C})$, $\phi \equiv 0$ in the neighbourhood of ∂C *except* in the neighbourhood of $\partial_1 C^{(k)}$. We again have $a(u,\phi) = (f,\phi)$ and by Green's formula we have:

$$\int_{\partial_1 C^{(k)}} \frac{\partial u}{\partial \nu_A} \phi \, d\Gamma = 0 \quad \forall\, \phi \quad \text{as above,}$$

from which the second property of (2.51) results.

If we take the assumptions (2.22) as well as

(2.52) $$f \in C^{0,\alpha}(\mathcal{O}) \quad , \quad c_0 \in C_b^2 \, .$$

then the result of O.A. LADYZHENSKAYA, N. URAL'TSEVA [1] for the Neumann boundary problem allows us to affirm that the property (2.24) is again satisfied.

LEMMA 2.6: *Let* $\mathcal{O}_\varepsilon = \{x \mid \Phi(x) < -\varepsilon\}$. *Then if* χ *is a unit vector of* R^n *and* θ *is a real number, with* $|\theta| \leqslant \beta^\varepsilon$, *then*

(2.53) $$\frac{1}{|\theta|}(Mu(x+\theta\chi) - Mu(x)) \leq \frac{C}{\varepsilon} \quad \forall\, x \in \bar{\mathcal{O}}_\varepsilon$$

where C *is a constant independent of* $x \in \bar{\mathcal{O}}_\varepsilon, \theta, \chi$.

Proof: Define

$$F = \{y \in \bar{\mathcal{O}} \mid u(y) < Mu(y) - \frac{k}{2}\} \quad , \text{ thus } \bar{F} \subset C.$$

Noticing that we again have

$$Mu(x) \leq c_0(\xi(x)) + Mu(x+\xi(x))$$

therefore

$$u(x+\xi(x)) \leq Mu(x+\xi(x))-k$$

and we have

$$x+\xi(x) \in F \quad \forall x .$$

Set

$$y = x+\xi(x) \quad \text{and} \quad y' = x+\theta\chi + P(x+\theta\chi,\xi(x)) .$$

Then for $x \in \mathcal{O}_\varepsilon$, since $P(x,\xi(x)) = \xi(x)$, by (2.50) we have

$$|y-y'| \leq |\theta| + \frac{\Delta\ |\theta|\ \|D\Phi\|_{L^\infty(\Gamma)}}{\varepsilon - |\theta|\ \|D\Phi\|_{L^\infty}} .$$

We then take

$$|\theta| < \beta_\varepsilon = \text{Min}\left(\frac{\varepsilon}{2\|D\Phi\|}, \frac{\varepsilon\rho_0}{\varepsilon+2\Delta\|}\right)$$

where ρ_0 is such that

(2.54) $$|z-z'| \leq \rho_0,\ z,z' \in \bar{\mathcal{O}} \text{ implies } |Mu(z')-u(z')-(Mu(z)-u(z))| \leq \frac{k}{4} .$$

From this it follows that

$$\Phi(x+\theta\chi) \leq \Phi(x) + |\theta|\ \|D\Phi\|_{L^\infty}$$

hence

$$\leq -\varepsilon + \frac{\varepsilon}{2} \leq 0, \quad \text{thus } x+\theta\chi \in \bar{\mathcal{O}}$$

$$y' \in \mathcal{O} \text{ and } |y-y'| \leq \rho_0,$$

therefore $y,y' \in F$.

Furthermore, using (2.24) with $D = F$, we have

$$|u(y') - u(y)| \leq C_F \frac{|\theta|}{\varepsilon}(\varepsilon + 2\Delta\ \|D\Phi\|_{L^\infty})$$

and

$$|c_0(P(x+\theta\chi,\xi(x)))- c_0(\xi(x))| \leq 2\ C_0\ |\theta|\ \|D\Phi\| \frac{\Delta}{\varepsilon} .$$

But

$$Mu(x+\theta\chi)- Mu(x) \leq u(x+\theta\chi+ P(x+\theta\chi,\xi(x))-u(x+\xi(x))$$
$$+ c_0(P(x+\theta\chi,\xi(x)) - c_0(\xi(x)) \leq \frac{C}{\varepsilon}$$

whence the result desired.□

LEMMA 2.7: *With the notations of Lemma 2.6 we have*

(2.55) $$\frac{1}{\theta^2}(Mu(x+\theta\chi)+ Mu(x-\theta\chi)-2Mu(x)) \le \frac{C}{\varepsilon^2}$$

for

$$|\theta| \le \gamma^{\varepsilon} \quad , \quad \forall x \in \bar{\mathcal{O}} \quad , \varepsilon \le \varepsilon_0 .$$

Proof: As the function $P(x,\xi)$ is not C^2, we can no longer proceed as in Lemma 2.6. We are going to consider two cases. First of all

(2.56) $$\Phi(x+\xi(x)) < - \varepsilon^3 .$$

We take

$$|\theta| \le \gamma_1^{\varepsilon} = \min(\rho_0 , \frac{\varepsilon}{\|D\Phi\|}) .$$

Then

$$\Phi(x+\theta\chi+\xi(x)) < - \varepsilon^3 + |\theta| \, \|D\Phi\| \le 0$$

and therefore

$$x + \xi(x) + \theta\chi \in \mathcal{O} .$$

Moreover,

$$x+\xi(x), \ x +\theta\chi +\xi(x) \in F.$$

Therefore

$$\frac{1}{\theta^2}(Mu(x+\theta\chi) + Mu(x-\theta\chi)-2Mu(x)) \le \frac{1}{\theta^2} [u(x+\theta\chi+\xi(x)) +$$
$$+ u(x-\theta\chi+\xi(x))- 2u(x+\xi(x))]$$
$$= \frac{1}{\theta} \int_0^1 (Du(x+\xi(x)+\lambda\theta\chi)- Du(x+\xi(x)-\lambda\theta\chi)).\chi d\lambda$$
$$\le C_F'$$

where

$$C_F' = \sup_{\substack{y,y'\in\bar{F} \\ y\neq y'}} \frac{|Du(y)- Du(y')|}{|y-y'|} .$$

We now assume

(2.57) $$0 \ge \Phi(x+\xi(x)) \ge - \varepsilon^3 .$$

We denote by $\rho_\varepsilon(x,\xi)$ the unique solution of

(2.58) $$\Phi(x+\rho_\varepsilon \xi) = - \varepsilon^3 \quad , \quad \rho_\varepsilon \ge 0 .$$

Then

$$\Phi(x+\rho_\varepsilon\ \xi)-\Phi(x+\xi) \geq D\Phi(x+\xi).\xi\ (\rho_\varepsilon-1)$$

so

$$0 \geq D\Phi(x+\xi).\xi\ \ (\rho_\varepsilon-1)$$

and

$$\Phi(x) - \Phi(x+\xi) \geq - D\Phi(x+\xi).\xi$$

so

$$-\varepsilon+\ \varepsilon^3 \geq - D\Phi(x+\xi).\xi$$

hence if $\varepsilon > 1$ we have

(2.59) $\quad 0 \leqslant \rho_\varepsilon \leqslant 1.$

Furthermore,

$$\Phi(x+\rho_\varepsilon\ \xi)-\Phi(x+\rho\xi) \geq D\Phi(x+\rho\xi).\xi\ (\rho_\varepsilon-\rho)$$

$$-\varepsilon^3 \geq D\Phi(x+\rho\xi).\xi\ (\rho_\varepsilon-\rho)$$

and since by (2.45)

$$D\Phi(x+\rho\xi).\xi \geq 0,$$

we obtain

(2.60) $\quad \rho_\varepsilon \leqslant \rho.$

However,

$$\Phi(x+\rho\xi)-\Phi(x+\rho_\varepsilon\xi) \geq D\Phi(x+\rho_\varepsilon\xi).\xi\ (\rho-\rho_\varepsilon)$$

so

$$\varepsilon^3 \geq D\Phi(x+\rho_\varepsilon\xi).\xi\ (\rho-\rho_\varepsilon)\ .$$

But

$$\Phi(x) - \Phi(x+\rho_\varepsilon\xi) \geq - D\Phi(x+\rho_\varepsilon\xi).\xi\ \ \rho_\varepsilon$$

$$-\varepsilon + \varepsilon^3 \geq - D\Phi(x+\rho_\varepsilon\xi).\xi\ \ \rho_\varepsilon\ .$$

Therefore

$$\varepsilon^3 \geq (\rho-\rho_\varepsilon)\ \frac{(\varepsilon-\varepsilon^3)}{\rho_\varepsilon}$$

whence by using (2.60) and (2.59),

(2.61) $$\rho - \rho_\varepsilon \leq \frac{\varepsilon^2}{1-\varepsilon^2} ,$$

therefore (if (2.57) is satisfied) we necessarily have

(2.62) $$1 \leq \rho \leq 1 + \frac{\varepsilon^2}{1-\varepsilon^2} .$$

Let $z \in \bar{\mathcal{O}}_{\varepsilon/2}$, and define

(2.63) $$\sigma(z,x) = \rho(z-(1-\rho(x,\xi(x)))\xi(x),\xi(x))\xi(x)$$

This expression is well defined if $\varepsilon \leqslant \varepsilon_0$. In fact

(2.64) $$\Phi(z+(\rho-1)\xi) \leq \Phi(z) + ||D\Phi|| \, ||\xi|| \, (\rho-1)$$
$$\leq -\frac{\varepsilon}{2} + ||D\Phi|| \, \Delta \frac{\varepsilon^2}{1-\varepsilon^2} \leq -\frac{\varepsilon}{4}$$

and therefore $z + (\rho - 1)\xi \in \bar{\mathcal{O}}$. Furthermore,

(2.65) $$z + \sigma(z,x) \in \bar{\mathcal{O}}$$

since $z \in \bar{\mathcal{O}}$, and by the definition of σ we have $z + \sigma + (\rho - 1)\xi \in \bar{\mathcal{O}}$. By (2.47), (2.49), $z \to \sigma(z,x) \in C^2(\bar{\mathcal{O}}_{\varepsilon/2})$, and by using (2.64) we verify the estimates

(2.66) $$|D_z \sigma| \leq \frac{4 \, \Delta \, ||D\Phi||}{\varepsilon}$$
$$||D_z^2 \sigma|| \leq \frac{4^3 \, \Delta^4 \, ||D^2\Phi|| \, ||D \Phi||^2}{\varepsilon^2} .$$

As in Lemma 2.6, we show that if $|z - x| \leqslant \beta^\varepsilon$, then $z + \sigma(z,x) \in \bar{F}$, $z \in \bar{\mathcal{O}}_{\varepsilon/2}$. But then the function $z \to u(z + \chi(z,x)) \in W^{2,\infty}(B(x,\beta^\varepsilon))$, with a norm bounded by C/ε^2, where C depends only on C_F' and constants appearing in (2.66). Then taking $|\theta| \leqslant \beta^\varepsilon$ and using

$$Mu(x+\theta\chi) \leq k+u(x+\theta\chi+\sigma(x+\theta\chi,x))+ c_0(\sigma(x+\theta\chi),x)$$

we can finish as in the first part.

Collecting the results together, we see that for $|\theta| \leqslant \gamma^\varepsilon = \min(\gamma_1^\varepsilon,\beta^\varepsilon)$ we have (2.55).

We can then state:

THEOREM 2.3: *Take the assumptions of Theorem 1.1, together with* (2,2),(2.42), (2.52). *Then the solution* u *of the Q.V.I.* (1.7) *in* L^∞ *corresponding to* $V = H^1$ *(Neumann's problem) satisfies the local regularity property*

(2.67) $$u \in W^{2,p}_{loc}(\mathcal{O}), \quad 2 \leqslant p < \infty.$$

Proof: By Lemma 2.6, $Mu \in W^{1,\infty}_{loc}(\mathcal{O})$. Thus we can consider AMu as an element of $H^{-1}_{loc}(\mathcal{O})$.

Using Lemma 2.7, a proof similar to that for Lemma 2.3 shows that

$$AMu \leqslant -\frac{C}{\varepsilon^2} \text{ in the sense of distributions on } \mathcal{O}_\varepsilon .$$

By the Levy-Stampacchia inequality corresponding to the Neumann problem (cf. Chap. 2, Theorem 5.8),

$$f \geq Au \geq \text{Min}\,(f, -\frac{C}{\varepsilon^2})$$

in the sense of distributions in $\mathcal{O}_\varepsilon$. Hence $Au \in L^\infty_{loc}(\mathcal{O})$. This concludes the proof of (2.67).

We can also give some global results if we strengthen the assumptions about the open set. We now assume that the function $\Phi(x)$ satisfies

$$(2.68) \qquad D^2\Phi(x) \geq \delta I \;, \quad \delta > 0, \qquad \forall\, x \in \bar{\mathcal{O}} .$$

LEMMA 2.8: *For all* $x \in \mathcal{O}$, $\xi \neq 0$, *we have the estimates*

$$(2.69) \qquad \rho |D_x\rho(x,\xi)| \leq \frac{2\,\|D\Phi\|_{L^\infty(\Gamma)}}{\delta|\xi|^2}$$

$$(2.70) \qquad \rho^2 \|D_x^2\rho(x,\xi)\| \leq \frac{4\,\|D^2\Phi\|_{L^\infty(\Gamma)}\,\|D\Phi\|^2_{L^\infty(\Gamma)}}{\delta^2\;|\xi|^2} .$$

Proof: In fact, notice that (setting $y = x + \rho\xi$)

$$\Phi(x) = \Phi(x+\rho\xi) - \rho\, D\Phi(y).\xi + \frac{1}{2}\rho^2\, D^2\Phi(x+\lambda\rho\xi)\xi$$

and therefore

$$\rho\, D\Phi(y).\xi \geq \frac{1}{2}\rho^2\delta\, |\xi|^2$$

and $\rho \neq 0$, since $x \in \mathcal{O}$, $x + \rho\xi \in \Gamma$, therefore

$$D\Phi(y).\xi \geq \frac{1}{2}\rho\,\delta\, |\xi|^2$$

which, with the formulae (2.44),(2.48) immediately gives the estimates (2.69), (2.70).

Again consider the set C defined in (1.70), and let

$$C = \{x \in \bar{\mathcal{O}} \mid u(x) < Mu(x)\}$$

and

$$F = \{x \in \bar{\mathcal{O}} \mid u(x) < Mu(x) - \frac{k}{2}\} \quad , \quad \bar{F} \subset C .$$

We can find an open set G of $\bar{\mathcal{O}}$ such that

$$G \supset \bar{\mathcal{O}} - C \quad \text{and} \quad \bar{G} \cap \bar{F} = \emptyset .$$

We shall take a second open set G_1 of $\bar{\mathcal{O}}$ such that

$$\bar{G} \subset G_1 \quad \text{and} \quad \bar{G}_1 \cap \bar{F} = \emptyset .$$

and set

(2.71) $\quad d = \text{dist}(\bar{G}_1, \bar{F}).$

Also let d_0 be such that

(2.72) $\quad x, x' \in \bar{\mathcal{O}}, \; x \in \bar{G}, \quad |x-x'| < d_0$

implies

$$x' \in G_1, \quad d_0 = \text{dist}\,(\bar{G}, \bar{\mathcal{O}} - G_1).$$

We can then prove:

LEMMA 2.9:

(2.73) $\quad \|Mu\|_{W^{1,\infty}(G)} \le C$

(2.74) $\quad AMu \geq -c$ *in the sense of distributions in* G

Proof: Since

$$x + \xi(x) \in \bar{F}, \quad \forall x,$$

it follows from this that for $x \in \bar{G}_1$

$$|\xi(x)| \geq d.$$

Thus by (2.69)

(2.75) $\quad |D_x \rho(x,\xi(x))| \le \dfrac{2\,\|D\Phi\|_{L^\infty(\Gamma)}}{\delta\, d^2} = c_0 \quad \forall x \in \bar{G}_1 .$

From this we deduce

(2.76) $\quad |P(x',\xi(x)) - \xi(x)| \le c_0 |x'-x| , \quad \forall x, x' \in \bar{G}_1 .$

Then let $x \in G \cap \mathcal{O}_\varepsilon$, θ be a real number, and χ a unit vector of R^n. For $|\theta| < d_0$, we have $x + \theta\chi \in G_1$. Further, by Lemma 2.6, for $|\theta| \leq \frac{\varepsilon}{2\|D\Phi\|}$, we have $x + \theta\chi \in \bar{\mathcal{O}}$, hence by (2.76)

$$|P(x+\theta\chi, \xi(x)) - \xi(x)| \le c_0\,|\theta|$$

and taking

$$|\theta| \leqslant \frac{\rho_0}{1 + c_0}$$

we obtain

$$y = x+\xi(x), \; y' = x+\theta\chi + P(x+\theta\chi, \xi(x)) \in F \; .$$

We then obtain

$$\frac{1}{|\theta|}(Mu(x+\theta\chi) - Mu(x)) \leq C_F + \|c_0'\| \; c_0 \; , \tag{2.77}$$

for

$$|\theta| \leq \text{Min}\left(\frac{\rho_0}{1+c_0}, d_0, \frac{\varepsilon}{2\,\|D\Phi\|}\right) \text{ and } x \in G \cap \mathcal{O}_\varepsilon \; .$$

From this we deduce (2.73), since the constant on the right of (2.77) does not depend on ε.

We now prove the analogue of Lemma 2.7; let

$$\frac{1}{\theta^2}(Mu(x+\theta\chi) + Mu(x-\theta\chi) - 2Mu(x)) \leq C \tag{2.78}$$

for

$$|\theta| \leq \delta^\varepsilon \; , \forall x \in G \cap \mathcal{O}_\varepsilon$$

where the constant C does not depend on ε.

If (2.56) is satisfied, then $C = C_F'$ for $|\theta| \leqslant \gamma_1^\varepsilon$ (cf. Lemma 2.7). Therefore we assume (2.57). Then for $z \in \bar{\mathcal{O}}_{\varepsilon/2}$ we consider the function $\sigma(z,x)$ defined in (2.63), which is well defined, as was seen in Lemma 2.7) (for $\varepsilon \leqslant \varepsilon_0$). Let us write $\rho_x = \rho(x,\xi(x))$. Then by (2.62)

$$|z+(\rho_x-1)\xi(x)-x| \leq |z-x| + \frac{\varepsilon^2}{1-\varepsilon^2}\Delta$$

and hence if

$$|z-x| < d_0 - \frac{\varepsilon^2\Delta}{1-\varepsilon^2}$$

then

$$z+(\rho_x-1)\xi(x) \in G_1 .$$

On the other hand, by Lemma 2.5 the function

$\rho(z + (\rho_x - 1)\xi(x),\xi(x))$ converges uniformly to ρ_x with respect to x when $z \to x$

(since $|\xi(x)| \leqslant d$); since $\rho_x \geqslant 1$, we deduce that for $|z - x| \leqslant \delta$ fixed and small enough (independent of x) we then have

$$\rho(z + (\rho_x - 1)\xi(x), \xi(x)) \geq \frac{1}{2} .$$

But then by (2.70)

$$\|D_z^2 \sigma(z,x)\| \leq \frac{16 \|D^2\Phi\|_{L^\infty(\Gamma)} \|D\Phi\|^2_{L^\infty(\Gamma)}}{\delta^2 d^2} = C_1$$

for

$$|z-x| \leq \text{Min}\ (\delta, d_0, \frac{\varepsilon^2 \Delta}{1-\varepsilon^2}, \frac{\varepsilon}{2\|D\Phi\|}) , \quad x \in G \cap \mathcal{O}_\varepsilon .$$

Also

$$|D_z \sigma(z,x)| \leq \frac{4 \|D\Phi\|_{L^\infty(\Gamma)}}{\delta d^2}$$

and thus for

$$|z-x| (1 + \frac{4 \|D\Phi\|_{L^\infty(\Gamma)}}{\delta d^2}) \leq \rho_0$$

we have

$$x + \xi(x) \in F , \quad z + \sigma(z,x) \in F.$$

We finish as for the proof of Lemma 2.7, in order to obtain (2.78), and with the constant C being independent of ε. From this we deduce (2.74), as indicated in Theorem 2.3.

We can then state:

THEOREM 2.4: *Take the assumptions of Theorem 2.3 together with* (2.68). *Then*

(2.79) $\quad Au \in L^\infty.$

Proof: It suffices to consider a partition of unity ϕ_G, ϕ_C subordinate to the covering G,C of $\mathcal{O}$, then to notice that u is the solution of the V.I. for the obstacle

$$\psi = \phi_G Mu + \phi_C(u+k)$$

and

$$\psi \in W^{1,\infty}(\mathcal{O}), \quad A\psi \geqslant - c \quad \text{in the sense of distributions on } \mathcal{O}.$$

From this we deduce (2.79), thanks to the Levy-Stampacchia inequality.□

Remark 2.2

We refer to B. HANOUZET - J.L. JOLY [1] for global regularity results for the solution u of the Q.V.I. for the Neumann problem in these particular cases (i.e., particular form of the open set of dimension 2).□

3. Q.V.I. WITH QUADRATIC HAMILTONIAN

In Section 1.6 we said that the theorems on existence and uniquenss extend to the case where A is replace by A - H, where H is a Hamiltonian satisfying the properties of Chapter 3.

Essentially we require that the function $H(x,\lambda,p): R^n \times R \times R^n \times R$ is globally Lipschitz continuous with respect to p. Under these conditions $A - H(\cdot,u,\nabla u) + \lambda u$ becomes a monotonic operator for λ large enough. The case where $H(x,\lambda,p)$ increases quadratically in p poses problems which we are now going to study. The problem has been studied by J. FREHSE - U. MOSCO [1] and A. BENSOUSSAN - J. FREHSE - U. MOSCO [1].

3.1 *Notations: Assumptions*

We take the assumptions of Theorem 1.1 together with (2.1). We also take the case $V = H^1_0$. Further, we are given a mapping $H(x,\lambda,p)$ satisfying the following properties

(3.1) $$|H(x,\lambda,p)-H(x,\mu,q)| \leq C(|\lambda-\mu|+|p-q|(1+|p|+|q|+|\lambda|^{\frac{1}{2}}+|\mu|^{\frac{1}{2}})$$

(3.2) $$H(x,\lambda,p) \leq h^*(x) + \lambda c^*(x) + p.\, g^*(x)$$

with

$$h^* \geqslant 0, \quad h^*,c^*,g^* \text{ bounded},$$

$$c^*(x) + \beta \leqslant a_0(x),$$

(3.3) $$H(x,0,0) \geq 0$$

(3.4) $$\mu H(x,\lambda,p) \leq H(x,\mu\lambda,\mu p) \qquad \forall \quad 0 \leq \mu \leq 1$$

$$H(x,\lambda+\mu,p) \leq H(x,\lambda,p) + \mu(a_0(x)-\beta) \qquad \forall \mu \geq 0 .$$

In addition assume that there exists a family $H_m(x,\lambda,p)$ satisfying the same properties as H, i.e. (3.1) to (3.4) with constants independent of m, and

(3.5) $$|H_m(x,\lambda,p) - H_m(x,\mu,q)| \leq C_m(|\lambda-\mu| + |p-q|)$$

(3.6) $$H_m(x,\lambda,p) \downarrow H(x,\lambda,p) \quad \forall\, x,\lambda,p \quad \text{when} \quad m \uparrow +\infty,\ C_m \uparrow .$$

A Hamiltonian satisfying the preceding properties arises naturally in optimal control when quadratic functionals occur. Thus we give

(3.7) $$Q(x) \in L^\infty(R^n; \mathcal{L}(R^m;R^m))$$
Q symmetric and positive definitive
$$Q^{-1}(x) \in L^\infty(R^n; \mathcal{L}(R^m;R^m))$$

(3.8) $$B(x) \in L^\infty(R^n; \mathcal{L}(R^m;R^n))$$

(3.9) $$h(x,v) \in L^\infty(R^n \times R^m), \quad h \geq 0$$
$$g(x,v) \in L^\infty(R^n \times R^m; R^n)$$
$$c(x,v) \in L^\infty(R^n \times R^m)$$

(3.10) $$a_0(x) - c(x,v) \geq \beta > 0 .$$

We then set

(3.11) $$H(x,\lambda,p) = \inf_v [h(x,v) + \lambda c(x,v) + Q(x)v.v + p.g(x,v) + 2p.\, Bv].$$

LEMMA 3.1: *The function* H *defined by* (3.11) *satisfies all the properties* (3.1) *to* (3.6).

Proof: It is clear that without loss of generality we can restrict the control in (3.11) to the set

$$\{v \mid Q(x)v.v + p.g(x,v) + 2p.Bv + \lambda c(x,v) \leq h(x,0) + \lambda c(x,0) + p.g(x,0)\}$$

therefore

$$Qv.v + 2p.Bv \leq C(1 + |p| + |\lambda|)$$

whence it follows that

$$|v| \leq C(1 + |p| + |\lambda|^{\frac{1}{2}}) .$$

Next consider (λ,p) and (μ,q). We can consider just as well for $H(x,\lambda,p)$ as for $H(x,\mu,q)$ that the control is subjected to the constraint

$$|v| \leq C(1 + |p| + |q| + |\lambda|^{\frac{1}{2}} + |\mu|^{\frac{1}{2}}) .$$

Using assumptions (3.8),(3.9), we easily deduce from the property (3.1) from (3.11). Also

$$H(x,\lambda,p) \leq h(x,0) + \lambda c(x,0) + p.g(x,0)$$

and thus by setting

$$h^*(x) = h(x,0),\ c^*(x) = c(x,0),\ g^*(x) = g(x,0),$$

we obtain (3.2), thanks to the assumptions (3.9),(3.10).

Since

$$H(x,0,0) = \inf [h(x,v) + Q(x)v.v]$$

we have (3.3), by (3.9) and (3.7). The properties (3.4) result immediately from (3.9),(3.10). We next define the sequence $H_m(x,\lambda,p)$ by

(3.12) $$H_m(x,\lambda,p) = \inf_{|v| \leq m} [h(x,v)+ \lambda c(x,v) + Q(x)v.v + p.g(x,v) + 2p.Bv]$$

and verify that H_m has the properties (3.1) to (3.4) with the same constants as H. Moreover, we have (3.5). It is clear that H_m decreases for fixed x,λ,p.

By (3.12) we have

$$H_m(x,\lambda,p) = H(x,\lambda,p)$$

for

$$m > C(1 + |p| + |\lambda|^{\frac{1}{2}})$$

whence (3.6).□

3.2 *Study of a Variational Inequality*

We are given an obstacle ψ satisfying

(3.13) ψ bounded and measurable on $\mathcal{O}$, $\psi \geqslant 0$,

$$\psi(z) \leqslant \psi(x) = d|x - z|^{\gamma}, \quad \forall\, z \leqslant x \in \mathcal{O}.$$

We consider the problem

(3.14)
$$a(u,v-u) - (H(.,u,\nabla u),v-u) \geq (f,v-u)$$
$$\forall\, v \in H_0^1 \cap L^{\infty} \text{ such that } v \leq \psi \quad \text{a.e.}$$
$$u \in H_0^1 \cap L^{\infty}, \quad u \leq \psi \quad \text{a.e.}$$

with

(3.15) $f \in L^p(\mathcal{O})$, $p > \frac{n}{2}$, $f \geq 0$

and $a(u,v)$ is the form (1.3).

Remark 3.1

Contrary to what is normally done for V.I.'s, we insist that the test functions are in L^{∞}. This is owed to the fact that if $u \in H^1$, then $H(\cdot,u,\nabla u) \in L^1$, and therefore we can give a meaning to (3.14).□

THEOREM 3.1: *Assume* (1.1),(1.8),(3.1) *to* (3.6),(3.13),(3.15). *Then the set of solutions of* (3.14) *is non-empty and has a maximum element denoted* u. *Furthermore,* $u \geqslant 0$ *and* $u \in C^{0,\delta}(\bar{\mathcal{O}})$, $0 < \delta < 1$, $\|u\|_{C^{0,\delta}} \leqslant M$, *and the constants* δ *and* M *are independent of the obstacle* ψ *inside the class of obstacles satisfying* (3.13).

Remark 3.2

It would be interesting to specify the structure of the set of solutions, if there were no uniqueness. For example, would there exist a minimum element of the positive solutions.□

We are going to approximate (3.14) by a sequence of V.I.'s related to the Hamiltonian H_m. More precisely, we consider

(3.16)
$$a(u_m,v-u_m)-(H_m(.,u_m,\nabla u_m),v-u_m) \geq (f,v-u_m)$$
$$\forall\, v \in H_0^1, \quad v \leq \psi, \quad u_n \in H_0^1 \quad u_m \leq \psi, \quad u_m \subset L^{\infty}.$$

We have:

LEMMA 3.1: *There exists one and only one solution* u_m *of* (3.16). *Furthermore,*

(3.17) $\quad u_m \geq 0 \quad , \|u_m\|_{L^\infty} \leq C \quad , \quad \|u_m\|_{H_0^1} \leq C$

where the constants do not depend on m.

Proof: If we fix $\delta > 0$, we can choose β_m in such a way that

(3.18)
$$\alpha|p-q|^2 - c_m|\lambda-\mu|^2 - c_m|\lambda-\mu|\ |p-q| + \beta_m\ |\lambda-\mu|^2 \geq$$
$$\geq \delta(|\lambda-\mu|^2 + |p-q|^2)\ , \quad \forall \lambda,\mu \in R,\ p,\ q \in R^n .$$

Under these conditions the operator $A - H_m + \beta_m$ is strongly monotonic from H_0^1 into H^{-1}, i.e.,

(3.19)
$$< Au_1 - H_m(.,u_1,\nabla u_1) + \beta_m u_1 - (Au_2 - H_m(.,u_2,\nabla u_2) + \beta_m u_2), u_1 - u_2 >$$
$$\geq \delta\ \|u_1-u_2\|^2 \qquad \forall\ u_1,u_2 \in H_0^1 .$$

Then let $z \in L^\infty$, and we solve uniquely

(3.20)
$$a(\zeta_m, v-\zeta_m) - (H_m(.,\zeta_m,\nabla\zeta_m), v-\zeta_m) + \beta_m(\zeta_m, v-\zeta_m) \geq$$
$$\geq (f + \beta_m z, v-\zeta_m) \qquad \forall\ v \in H_0^1\ , \quad v \leq \psi$$
$$\zeta_m \in H_0^1\ , \quad \zeta_m \in H_0^1 .$$

Consider also the solution w_m of

(3.21)
$$Aw_m + \beta_m w_m - c^* w_m - g^*. \nabla w_m = f + \beta_m z + h^* \quad ;$$
$$w_m \in H_0^1$$

then

(3.22 $\quad \zeta_m \leqslant w_m$.

In fact, if we take $v = \zeta_m - (\zeta_m - w_m)^+$ in (3.20), we obtain

$$-a(\zeta_m,(\zeta_m-w_m)^+) + (H_m(.,\zeta_m,\nabla\zeta_m),(\zeta_m-w_m)^+) -$$
$$- \beta_m(\zeta_m,(\zeta_m-w_m)^+) \geq -(f + \beta_m z,(\zeta_m-w_m)^+)$$

and using (3.21) we have

$$a(w_m,(\zeta_m-w_m)^+) + \beta_m(w_m,(\zeta_m-w_m)^+) -$$
$$- (c^* w_m,(\zeta_m-w_m)^+) - (g^* .\nabla w_m,(\zeta_m-w_m)^+) = (f+ \beta_m z+h^* ,(\zeta_m-w_m)^+) ;$$

then by adding and using

$$H_m(x,\lambda,p) - c^* (x)\lambda - g^*(x).p \leq h^*(x)$$

we obtain

$$-a((\zeta_m-w_m)^+,\ \zeta_m-w_m) - \beta_m|\ (\zeta_m-w_m)^+|^2 \geq 0$$

whence (3.22).

Similarly, if $z \geqslant 0$, taking $v = \zeta_m^+$ in (3.20), which is permissible since $\psi \geqslant 0$, we obtain

(3.23) $\qquad \zeta_m \geqslant 0.$

On the other hand, since $f \in L^p(\mathcal{O})$, $p > n/2$, the solution $w_m \in W^{2,p}(\mathcal{O})$, and thus belongs to $C^0(\bar{\mathcal{O}})$. We have also defined two mappings T_m, S_m of $L^\infty(\mathcal{O})$ into $H_0^1 \cap L^\infty$ by setting

$$T_m z = \zeta_m, \qquad S_m z = w_m.$$

These are contractions in L^∞. In fact, if $z^1, z^2 \in L^\infty$ and $\zeta_m^i = T_m z^i$, then

(3.24) $$||\zeta_m^1 - \zeta_m^2||_{L^\infty} \leq ||z^1-z^2||_{L^\infty} \frac{\beta_m}{\beta_m + \beta}$$

is an identical estimate for $S_m z^i$. This estimate is proved by a technique similar to that carried out earlier (cf., also, Sections 1 and 2, Chap. 3). If c_m is the constant occurring in (3.24) we take $v = \zeta_m^1 - (\zeta_m^1 - \zeta_m^2 - c_m)^+$ in the V.I. (3.20) with respect to ζ_m^1, and $v = \zeta_m^2 + (\zeta_m^1 - \zeta_m^2 - c_m)^+$ in that with respect to ζ_m^2. Adding and using the second assumption (3.4) which implies

$$H_m(x,\zeta_m^2 + c_m,\nabla\zeta_m^2) \leq H_m(x,\zeta_m^2,\nabla\zeta_m^2) + c_m(a_0-\beta)$$

together with

$$|(H_m(x,\zeta_m^1,\nabla\zeta_m^1) - H_m(x,c_m+ \zeta_m^2,\nabla\zeta_m^2))(\zeta_m^1 -\zeta_m^2- c_m)^+| \leq$$
$$\leq c_m(\ |(\zeta_m^1-\zeta_m^2- c_m)^+|^2 + |(\zeta_m^1-\zeta_m^2- c_m)^+||\ \nabla(\zeta_m^1-\zeta_m^2- c_m)^+|$$

and by the choice of β_m this yields

$$\delta\ ||(\zeta_m^1- \zeta_m^2- c_m)^+||^2 \leq 0$$

therefore

$$\zeta_m^1 - \zeta_m^2 \leqslant c_m.$$

Similarly,

$$\zeta_m^1 - \zeta_m^2 + c_m \geqslant 0,$$

whence (3.24).□

The mappings T_m, S_m are increasing, as is easily verified by taking $v = \zeta_m^1 - (\zeta_m^1 - \zeta_m^2)^+$ in the V.I. with respect to ζ_m^1, and $v = \zeta_m^2 + (\zeta_m^1 - \zeta_m^2)^+$ in the V.I. with respect to ζ_m^2.

The fixed points of T_m are the solutions of (3.16). Similarly S_m has a fixed point which does not depend on m, and is the solution w of the problem

$$(3.25)\qquad Aw - c^* w - g^* . \nabla w = f + h^* \quad , \quad w \in H_0^1 .$$

Since $a_0 - c^* \geqslant \beta$, (3.25) is an elliptic problem, and by regularity $w \in W^{2,p}(\mathcal{O})$, thus $w \in C^0(\bar{\mathcal{O}})$.

By (3.22)

$$(3.26)\qquad T_m z \leqslant S_m z.$$

Iterating and using the increasing property of T_m, we have

$$T_m^k z \leqslant S_m^k z$$

and therefore

$$(3.27)\qquad u_m \leqslant w.$$

This proves the first part of (3.17). Next, taking v = 0 in (3.16) we obtain

$$a(u_m,u_m) \leq (H_m(.,u_m,\nabla u_m),u_m) + (f,u_m) .$$

But $u_m \geqslant 0$, which with (3.2) implies

$$a(u_m,u_m) \leq C\ (1+|u_m|^2 + |u_m||\nabla u_m|)$$

and since $\|u_m\|_{L^\infty}$ is bounded, we deduce the second part of (3.17).□

3.3 *Proof of Theorem 3.1*

The essential step is going to be to obtain an estimate in a Hölder space. We introduce some notations and take $n \geqslant 3$ (cf. Remark 3.4 for n = 2).

Let Q be a ball such that $Q \supset\supset \mathcal{O}$. Let $x_0 \in \mathcal{O}$. Define the Green's function(1) with respect to x_0, $G = G^{x_0}$, as the solution of

$$(3.28)\qquad \sum_{i,j}\int_Q a_{ij}\frac{\partial\phi}{\partial x_j}\frac{\partial G}{\partial x_i}\,dx = \phi(x_0) \quad \forall\,\phi \in \mathcal{D}(Q), \quad G \in W_0^{1,s}(Q), \quad 1 \leqslant s \leqslant \frac{n}{n-1} .$$

(1) Recall that $a_{ij} = a_{ji}$.

We shall use the fact that G satisfies

(3.29) $$c_0|x-x_0|^{2-n} \le G(x) \le c_1|x-x_0|^{2-n}$$
for all x belonging to a neighbourhood of x_0, where c_0, c_1 are constants.

We write $B_\rho(x_0)$ for the ball with centre x_0 and radius ρ, and

(3.30) $$\phi_\rho = \frac{1}{|B_\rho(x_0)|}\int_{B_\rho(x_0)} \phi(x)dx .$$

We also define $G_\rho = G_\rho^{x_0}$ by

(3.31) $$\sum_{i,j}\int_Q a_{ij}\frac{\partial\phi}{\partial x_j}\frac{\partial G_\rho}{\partial x_i}dx = \phi_\rho \qquad \forall\, \phi \in \mathcal{D}(Q),\ G_\rho \in H_0^1(Q) .$$

Then G regularises G and

(3.32) $$G_\rho \to G \text{ in } L^r(Q), \quad 1 \leqslant r < \frac{n}{n-2}$$
$$G_\rho \to G \text{ weakly in } W_0^{1,s}(Q), \quad 1 \leqslant s < \frac{n}{n-1}$$
$$G_\rho \to G \text{ pointwise for all } x \neq x_0.$$

LEMMA 3.2: *Let* $R \leqslant 1$ *and* $x_0 \in \mathcal{O}$. *Then we have the estimate*

(3.33) $$\int_{B_R(x_0)} \chi_{\mathcal{O}} |\nabla u_m|^2\, |x-x_0|^{2-n}dx \le$$
$$\le K\Big[\int_{B_{4R}(x_0)-B_R(x_0)} \chi_{\mathcal{O}} |\nabla u_m|^2\, |x-x_0|^{2-n}\,dx + R^k\Big]$$

where K,k *are constants independent of* x_0,m,R, $k \in]0,2[$.

Proof: By (3.29) everything reduces to showing that

(3.34) $$\int_{B_R}\chi_{\mathcal{O}}|\nabla u_m|^2\, G\,dx \le K\Big[\int_{B_{4R}-B_R}\chi_{\mathcal{O}}|\nabla u_m|^2\, G\,dx + R^k\Big] .$$

Let $\tau \equiv \tau_R$ be such that $\tau_R = 1$ on B_R, $\tau_R = 0$ on $R^n - B_{2R}$, $0 \leqslant \tau_R \leqslant 1$, τ_R regular and

(3.35) $$|\nabla\tau_R(x)| \le \frac{C}{R}, \qquad \|D^2\tau_R(x)\| \le \frac{C}{R^2} .$$

Write

(3.36) $$T_R(x_0) = \{x \mid x_i \le x_{0i} - 2R\} \cap B_{4R} \cap \mathcal{O}\}u_R^m = \frac{1}{|T_R(x_0)|}\int_{T_R} u_m(x)\,dx .$$

Set

$$(3.37)\quad \begin{aligned} &\xi_R^m = 0 \quad \text{if} \quad B_{4R} \cap (R^n - \mathcal{O}) \neq \emptyset \\ &\xi_R^m = u_R^m - d(6R)^\gamma \quad \text{if} \quad B_{4R} \subset \mathcal{O} \quad . \end{aligned}$$

The constant ξ_R^m satisfies the property

$$(3.38)\qquad \xi_R^m \leq \psi(y) \qquad \forall\, y \in B_{2R} \, .$$

This is obvious in the first case (3.37). In the second case we notice that if $x \in T_R$ and $y \in B_{2R}$, then $x \leqslant y$, and thus by (3.13)

$$\psi(x) \leq \psi(y) + d|x-y|^\gamma \leq \psi(y) + d(6R)^\gamma$$

but since $u_m \leqslant \psi$, we deduce

$$u_R^m \leq \psi(y) + d(6R)^\gamma$$

hence (3.38).

Let $q \geqslant 1$ and consider

$$v_m = u_m - \varepsilon \, |u_m - \xi_R^m|^{q-1} (u_m - \xi_R^m) \, G_\rho \, \tau^2 \quad .$$

Then v_m is an admissible test function for (3.16), at least for ε small enough (depending on all the parameters). In fact, writing $\xi_m = \xi_R^m$, we have

$$v_m = u_m [1 - \varepsilon |u_m - \xi_m|^{q-1} G_\rho \tau^2] + \varepsilon \, |u_m - \xi_m|^{q-1} G_\rho \tau^2 \xi_m$$

and for

$$\varepsilon \, |u_m - \xi_m|^{q-1} G_\rho \tau^2 \leq 1,$$

we obtain

$$\begin{aligned} &v_m = u_m \quad \text{on} \quad \mathcal{O} - B_{2R}, \\ &v_m \leqslant \psi \quad \text{on} \quad B_{2R}, \end{aligned}$$

thanks to the choice of ξ_m. Moreover, $v_m \in H_0^1$. Therefore

$$(3.39)\quad \begin{aligned} a(u_m, |u_m-\xi_m|^{q-1}(u_m-\xi_m)G_\rho\tau^2) &\leq (H_m(.,u_m,\nabla u_m), |u_m-\xi_m|^{q-1}(u_m-\xi_m)G_\rho\tau^2) + \\ &+ (f, |u_m-\xi_m|^{q-1}(u_m-\xi_m)G_\rho\tau^2). \end{aligned}$$

First consider the term

$$X = \Sigma \int a_{ij} \frac{\partial u_m}{\partial x_j} \frac{\partial}{\partial x_i}(|u_m-\xi_m|^{q-1}(u_m-\xi_m)G_\rho\tau^2)\,dx = \qquad \text{(Contd)}$$

(Contd) $$= q \Sigma \int a_{ij}|u_m-\xi_m|^{q-1} \frac{\partial u_m}{\partial x_j}\frac{\partial u_m}{\partial x_i} G_\rho \tau^2 \, dx +$$

$$+ \frac{1}{q+1}\Sigma \int a_{ij} \frac{\partial G_\rho}{\partial x_i}\frac{\partial}{\partial x_j}(\tau^2|u_m-\xi_m|^{q+1})dx -$$

$$- \frac{1}{q+1}\Sigma \int a_{ij}|u_m-\xi_m|^{q-1} \frac{\partial G_\rho}{\partial x_i}\frac{\partial \tau^2}{\partial x_j} dx +$$

$$+ \Sigma \int a_{ij} \frac{\partial u_m}{\partial x_j}|u_m-\xi_m|^{q-1}(u_m-\xi_m) G_\rho \frac{\partial \tau^2}{\partial x_i} dx .$$

By (3.28) the second term in the expression for X is positive. Recalling that $|u_m - \xi_m|^{q+1}\tau$ vanishes on Γ we deduce from this that

$$X \geq q\,\alpha \int |u_m-\xi_m|^{q-1}|\nabla u_m|^2 G_\rho \tau^2 dx +$$

$$+ \frac{2}{q+1}\Sigma \int G_\rho [\frac{\partial a_{ij}}{\partial x_i}|u_m-\xi_m|^{q-1}\tau \frac{\partial \tau}{\partial x_j} + a_{ij}(q+1)|u_m-\xi_m|^{q-1}(u_m-\xi_m)\frac{\partial u_m}{\partial x_i}\frac{\partial \tau}{\partial x_j}$$

$$+ |u_m-\xi_m|^{q+1} a_{ij} (\frac{\partial \tau}{\partial x_i}\frac{\partial \tau}{\partial x_j} + \tau \frac{\partial^2 \tau}{\partial x_i \partial x_j})] dx +$$

$$+ 2\Sigma \int a_{ij} \frac{\partial u_m}{\partial x_j}|u_m-\xi_m|^{q-1}(u_m-\xi_m) G_\rho \tau \frac{\partial \tau}{\partial x_i} dx$$

$$= (1) + (2) + (3) + (4) + (5) .$$

Taking into account that u_m is bounded, together with (3.35), we have

$$|(2)| \leq \frac{C_q}{R^2}\int_{B_{2R}-B_R} \chi_{\mathcal{O}} G_\rho |u_m-\xi_m|^2 dx$$

$$|(3)| \leq C_q \int_{B_{2R}-B_R} \chi_{\mathcal{O}} G_\rho |\nabla u_m|^2 + \frac{C_q}{R^2}\int_{B_{2R}-B_R} \chi_{\mathcal{O}} G_\rho |u_m-\xi_m|^2 dx$$

$$|(4)| \leq \frac{C_q}{R^2}\int_{B_{2R}-B_R} \chi_{\mathcal{O}} G_\rho |u_m-\xi_m|^2 dx$$

$$|(5)| \leq C_q \int_{B_{2R}-B_R} \chi_{\mathcal{O}} G_\rho |\nabla u_m|^2 dx + \frac{C_q}{R^2}\int_{B_{2R}-B_R} \chi_{\mathcal{O}} G_\rho |u_m-\xi_m|^2 dx .$$

Next we have

$$|(H_m(.,u_m,\nabla u_m), |u_m-\xi_m|^{q-1}(u_m-\xi_m)G_\rho \tau^2)| \leq$$
$$C [\int_{\mathcal{O}} |\nabla u_m|^2 |u_m-\xi_m|^q G_\rho \tau^2 dx] + C_q \int_{\mathcal{O}} G_\rho \tau^2 dx$$

and we can absorb the first and zero-th order terms of the form a(u,v) into the preceding expression. Notice that the constant does not depend on q here.

Lastly, we have

$$|(f,|u_m-\xi_m|^{q-1}(u_m-\xi_m)G_\rho\tau^2)| \le C_q(\int_{\mathcal{O}} (G_\rho\tau^2)^{p'}dx)^{1/p'}$$

where p' is the conjugate of p.

Collecting together the preceding results we obtain the estimate

$$(3.40)\qquad q\,\alpha \int_{\mathcal{O}} |\nabla u_m|^2|u_m-\xi_m|^{q-1}\, G_\rho\tau^2\, dx \le C \int_{\mathcal{O}}|\nabla u_m|^2|u_m-\xi_m|^q\, G_\rho\tau^2\, dx + \frac{C_q}{R^2}\int_{B_{2R}-B_R} \chi\, G_\rho\, |u_m-\xi_m|^2\, dx + C_q \int_{B_{2R}-B_R} \chi\, G_\rho|\nabla u_m|^2\, dx + C_q|G_\rho\tau^2|^2_{L^p(\mathcal{O})}$$

In the estimate (3.40) the constants depend neither on ρ nor R, nor on m, and only the constants written C_q depend on q.

We begin by applying (3.40) with q = 1. We obtain

$$(3.41)\qquad \int_{\mathcal{O}} |\nabla u_m|^2\, G_\rho\tau^2 dx \le C \int_{\mathcal{O}} |\nabla u_m|^2\, |u_m-\xi_m|\, G_\rho\tau^2\, dx + \frac{C}{R^2}\int_{B_{2R}-B_R} \chi_{\mathcal{O}}\, G_\rho|u_m-\xi_m|^2 dx + C\int_{B_{2R}-B_R} \chi_{\mathcal{O}}\, G_\rho|\nabla u_m|^2\, dx + C\,|G_\rho\tau^2|_{L^{p'}(\mathcal{O})}.$$

Next we write

$$\int_{\mathcal{O}} |\nabla u_m|^2\, |u_m-\xi_m|\, G_\rho\tau^2 dx \le \ell\int_{\mathcal{O}}|\nabla u_m|^2\, G_\rho\tau^2\, dx + \int_{|u_m-\xi_m|>\ell} |\nabla u_m|^2\, |u_m-\xi_m|\, G_\rho\tau^2\, dx$$

$$\le \ell \int_{\mathcal{O}}|\nabla u_m|^2\, G_\rho\tau^2 dx + \ell^{2-q}\int_{\mathcal{O}}|\nabla u_m|^2|u_m-\xi_m|^{q-1}\, G_\rho\tau^2\, dx$$

for all $q \geqslant 2$. We then take $\ell = 1/2$ and $q = \bar{q}$ such that

$$\bar{q}\,\alpha - C\,\|u_m-\xi_m\| \ge \delta>0.$$

Then we deduce from (3.40) and (3.41) the estimate

$$(3.42)\qquad \int_{\mathcal{O}} |\nabla u_m|^2\cdot G_\rho\tau^2 dx \le \frac{C}{R^2}\int_{B_{2R}-B_R} \chi_{\mathcal{O}} G_\rho\, |u_m-\xi_m|^2 dx + C\int_{B_{2R}-B_R} \chi_{\mathcal{O}}\, G_\rho|\nabla u_m|^2\, dx + C\,|G_\rho\tau^2|_{L^{p'}(\mathcal{O})}.$$

Since $p > n/2$, we have $p' < \frac{n}{n-2}$. Let s be such that $\frac{n}{n-2} > s > p'$, then

$$(3.43)\qquad |G_\rho \tau^2|_{L^{p'}(\mathcal{O})} \le |G_\rho|_{L^s(\mathcal{O})} |B_{2R}|^{\frac{s-p'}{sp'}} \le C_s R^{n(\frac{1}{p'}-\frac{1}{s})} .$$

We can then make ρ tend to 0 in (3.42). We use the fact that the integrals on the right hand side of (3.2) converge, because the singularity at x_0 is avoided. Fatou's Lemma allows us to obtain

$$(3.44)\qquad \int_{B_R} \chi_{\mathcal{O}} |\nabla u_m|^2 G\,dx \le \frac{C}{R^2}\int_{B_{2R}-B_R} \chi_{\mathcal{O}} G |u_m-\xi_m|^2 dx + C\int_{B_{2R}-B_R} \chi_{\mathcal{O}} G |\nabla u_m|^2 dx + C_s R^{n(\frac{1}{p'}-\frac{1}{s})} .$$

Now, by (3.37) and (3.39)

$$\int_{B_{2R}-B_R} \chi_{\mathcal{O}} G |u_m-\xi_m|^2 dx \le C[\int_{B_{2R}-B_R} \chi_{\mathcal{O}} G|u_m-u_R^m|^2 dx + R^{2\gamma+2}] .$$

But

$$\int_{B_{2R}-B_R} \chi_{\mathcal{O}} G|u_m-u_R^m|^2 dx \le \int_{B_{4R}-B_R} \chi_{\mathcal{O}} G|u_m-u_R^m|^2 dx$$

and since

$$\int_{T_R} (u_m-u_R^m)\,dx = 0$$

it follows from Poincaré's inequality that

$$\int_{B_{4R}-B_R} \chi_{\mathcal{O}} G|u_m-u_R^m|^2 dx \le CR^{2-n} \int_{B_{4R}-B_R} \chi_{\mathcal{O}}|u_m-u_R^m|^2 dx$$
$$\le CR^{4-n} \int_{B_{4R}-B_R} \chi_{\mathcal{O}} |\nabla u_m|^2 dx$$
$$\le CR^2 \int_{B_{4R}-B_R} \chi_{\mathcal{O}} |\nabla u_m|^2 G\,dx$$

hence we deduce from (3.44) that

$$\int_{B_R} \chi_{\mathcal{O}} |\nabla u_m|^2 G\,dx \le C_s[\int_{B_{4R}-B_R} \chi_{\mathcal{O}} |\nabla u_m|^2 G\,dx + R^{\min(2\gamma,n(\frac{1}{p'}-\frac{1}{s}))}] .$$

This completes the proof of (3.33), and we have

$$\text{(3.45)} \qquad k = \min(2\gamma\ ,\ n(\tfrac{1}{p'} - \tfrac{1}{s})) \ \in\ ]0,2[\quad .$$

Remark 3.3

The constants K and k only depend on ψ through the parameters α,γ, since the norm L^∞ of u_m does not depend on ψ (cf. (3.27)).□

LEMMA 3.3: *We have the estimate*

$$\text{(3.46)} \qquad \int |\nabla u_m|^2\ |x-x_0|^{2-n}\ dx \le M \qquad \forall\ x \in \mathcal{O} \quad .$$

where M *depends neither on* x_0, *nor* m, *nor upon the obstacle* ψ *inside the class defined by* (3.13).

Proof: This is similar to that of Lemma 3.2, and, indeed, simpler. Take as test function

$$v_m = u_m - G\ u_m^q, \qquad q \geq 1,$$

which is permissible, since $u_m \geq 0$. We obtain

$$a(u_m,\ u_m^q\ G_\rho\ ) \le (H_m(.,u_m,\nabla u_m),\ u_m^q\ G_\rho\ ) + (f,u_m^q\ G_\rho) \quad .$$

Now,

$$\int_{\mathcal{O}} a_{ij}\ \frac{\partial u_m}{\partial x_j}\ \frac{\partial}{\partial x_i}(u_m^q\ G_\rho\ )dx \ge q \int_{\mathcal{O}} a_{ij}\ \frac{\partial u_m}{\partial x_j}\ \frac{\partial u_m}{\partial x_i}\ u_m^{q-1} G_\rho\ dx$$

and

$$|(H_m(.,u_m,\nabla u_m),u_m^q\ G_\rho)| \le C\ [\int_{\mathcal{O}} |\nabla u_m|^2\ u_m^q\ G_\rho\ \ dx] + C_q \ .$$

Therefore

$$q\ \alpha \int_{\mathcal{O}} |\nabla u_m|^2\ u_m^{q-1}\ G_\rho\ dx \le C \int_{\mathcal{O}} |\nabla u_m|^2\ u_m^q\ G_\rho\ dx + C_q \ .$$

Working as in Lemma 3.2, we deduce

$$\int_{\mathcal{O}} |\nabla u_m|^2\ G_\rho\ dx \le C$$

hence with Fatou's Lemma

$$\int_{\mathcal{O}} |\nabla u_m|^2\ G\ \ dx \le C,$$

and therefore (3.46) holds.□

LEMMA 3.4: *We have the estimate*

$$(3.47)\qquad \int_{B_R(x_0)} \chi_{\mathcal{O}}\ |\nabla u_m|^2\ |x-x_0|^{2-n}\,dx \ \le\ C_\mu\ R^\mu\ ,\qquad R \le 1$$

$$\forall\qquad \mu \le k \quad \textit{and} \quad \theta 4^\mu < 1 \ \textit{where}\ \theta = \frac{K}{K+1}\ ,$$

$$C_\mu = \frac{\theta}{1-\theta 4^\mu} + \frac{M}{\theta}\ .$$

Proof: Define

$$H_R^m = \int_{B_R} \chi_{\mathcal{O}}\ |\nabla u_m|^2 |x-x_0|^{2-n}\,dx\ .$$

By Lemma 3.3, H_R^m is bounded by a constant independent of m,R. Furthermore, by Lemma 3.2 we have

$$H_R^m \ \le\ \theta(H_{4R}^m + R^k)$$

so, since $R \leqslant 1$, $\mu \leqslant k$,

$$(3.48)\qquad H_R^m \ \le\ \theta(H_{4R}^m + R^\mu) \qquad \forall\quad R \le 1\ ,$$

therefore

$$(3.49)\qquad H_R^m \ \le\ \frac{M}{\theta}\,R^\mu \qquad \text{for}\quad \frac{1}{4} < R \le 1\ .$$

When $\frac{1}{4^2} < R \leqslant \frac{1}{4}$, then $\frac{1}{4} < 4R \leqslant 1$, and therefore by (3.49)

$$H_{4R}^m \ \le\ \frac{M}{\theta}(4R)^\mu$$

which with (3.48) shows that

$$H_R^m \ \le \theta R^\mu + M(4R)^\mu \qquad \text{for}\quad \frac{1}{4^2} < R \le \frac{1}{4}\ .$$

Iterating, we obtain

$$H_R^m \ \le\ \theta R^\mu + \theta^2(4R)^\mu + \ldots + \theta^j(4^{j-1}R)^\mu + M\,\theta^{j-1}(4R)^{j\mu}$$
$$\text{for}\qquad \frac{1}{4^{j+1}} < R \le \frac{1}{4^j}$$

whence the desired result, as is easily seen.□

We can also give the proof of Theorem 3.1. Take $\mu < 1$ satisfying the conditions of Lemma 3.4. Then by the Lemma in C.B. MORREY [1], we deduce from Lemma 3.4 that

(3.50) $\qquad \| u_m \|_{C^{0,\delta}} \leq C \quad , \qquad \delta = \frac{\mu}{2} \ .$

Moreover, the sequence u_m is decreasing in m. In fact, if we again take the mapping T_m defined by (3.20), then

(3.51) $\qquad u_{m'} \leq T_m u_{m'} \qquad$ for $\quad m' \geq m$.

In fact, if $\zeta_m = T_m u_{m'}$, then by taking $v = u_{m'} - (u_{m'} - \zeta_m)^+$ in $(3.16)_{m'}$ and $v = \zeta_m + (u_{m'} - \zeta_m)^+$ in (3.20), we have

$$-a((u_{m'}-\zeta_m)^+,(u_{m'}-\zeta_m)^+) - \beta_m \,|(u_{m'}-\zeta_m)^+|^2 +$$
$$+ (H_{m'}(.,u_{m'},\nabla u_{m'}) - H_m(.,\zeta_m,\nabla\zeta_m),(u_{m'}-\zeta_m)^+) \geq 0 \ .$$

But

$$H_{m'}(.,u_{m'},\nabla u_{m'}) \leq H_m(.,u_{m'},\nabla u_{m'})$$

and by the choice of β_m, we deduce (3.51). By the increasing property of T_m it follows that

$$u_{m'} \leqslant T_m^k u_{m'},$$

hence at the limit $u_{m'} \leqslant u_m$. Consequently the sequence u_m satisfies

(3.52) $\qquad u_m \leqslant u \;$ in C^o and weakly in H_0^1.

In (3.52) we used the compactness of $C^{0,\delta}$ in C^o, and the estimates (3.50) and (3.17).

Next consider $(3.16)_m$. Taking $v = u \leqslant \psi$ we have

$$a(u_m,u-u_m)-(H_m(.,u_m,\nabla u_m),u-u_m) \geq (f,u-u_m) \quad .$$

But $H_m(\cdot,u_m,\nabla u_m)$ stays bounded in L^1, which with (3.52) implies

$$\overline{\lim}\; a(u_m,u_m) \leq a(u,u)$$

and since

$$a(u,u) \leq \underline{\lim}\; a(u_m,u_m)$$

we see that

(3.53) $\qquad u_m \to u \;$ strongly in H_0^1.

By the property (3.1) we have

$$|H_m(x,u_m,\nabla u_m) - H_m(x,u,\nabla u)| \leq C(|u_m-u| + $$
$$+ |\nabla u_m-\nabla u| \quad (1+|\nabla u_m| + |\nabla u| + |u_m|^{1/2} + |u|^{1/2}))$$

and since $u_m \to u$ in H_0^1 and C^0, we deduce

$$H_m(x,u_m,\nabla u_m) - H_m(x,u,\nabla u) \to 0 \quad \text{in} \quad L^1 .$$

Since also

$$H_m(x,u,\nabla u) \downarrow H(x,u,\nabla u) \quad \text{and in} \quad L^1$$

we deduce that

$$H_m(x,u_m,\nabla u_m) \to H(x,u,\nabla u) \text{ in } L^1.$$

We then easily pass to the limit in (3.16) and therefore u is a solution of (3.14). Let us show that the solution obtained thus is the maximum solution.

Let $\tilde{u}$ be a solution of (3.14). We are going to show that

$$(3.54) \qquad T_m\tilde{u} \geqslant \tilde{u}.$$

In fact, let $\zeta_m = T_m\tilde{u}$, and take $v = \tilde{u} - (\tilde{u} - \zeta_m)^+$ in (3.14) written for $\bar{u}$, and and $v = \zeta_m + (\tilde{u} - \zeta_m)^+$ in (3.20) (with $z = \tilde{u}$). From this we deduce

$$-a((\tilde{u}-\zeta_m)^+,(\tilde{u},\zeta_m)^+) - \beta_m \; |(\tilde{u}-\zeta_m)^+|^2 \; +$$
$$+ \; (H(.,\tilde{u},\nabla\tilde{u}) - H_m(.,\zeta_m,\nabla\zeta_m), \; (\tilde{u},\zeta_m)^+) \geq 0$$

but

$$H(x,\tilde{u},\nabla\tilde{u}) \leq H_m(x,\tilde{u},\nabla\tilde{u})$$

and by the choice of β_m it follows that

$$(\tilde{u} - \zeta_m)^+ = 0.$$

We then deduce from (3.54) that

$$u_m \geqslant \tilde{u}$$

hence

$$u \geqslant \tilde{u}.$$

The proof of Theorem 3.1 is therefore complete.□

Remark 3.4

In Lemma 3.2 and the following ones we have assumed that $n \geqslant 3$. For $n = 2$ we can prove a result analogous to that of Lemma 3.4 (cf. (3.47)) by replacing $|x - x_0|^{2-n}$ by $\text{Log}|x - x_0|$. The proofs are identical, for we may note that in the estimate of the Green's function (3.29) we can replace $|x - x_0|^{2-n}$ by $\text{Log}|x - x_0|$. Using this we again have (3.50), whence the result.□

3.4 *Study of the Q.V.I.*

We now introduce the (Q.V.I.) problem

$$(3.55)\qquad \begin{aligned} & a(u,v-u) - (H(.,u,\nabla u),v-u) \ge (f,v-u) \\ & \forall\, v \in H_0^1(\mathcal{O}) \cap L^\infty(\mathcal{O}) \;,\; v \le Mu \quad \text{a.e.} \\ & u \in H_0^1(\mathcal{O}) \cap L^\infty(\mathcal{O}) \;,\; u \le Mu \quad \text{a.e.} \end{aligned}$$

We are going to prove:

THEOREM 3.2: *Assume* (1.1),(1.5),(1.8),(3.1) *to* (3.6),(3.15). *Then the set of solutions of* (3.55) *is non-empty and has a maximum element denoted* u. *Moreover,* $u \geqslant 0$ *and* u *is Hölderian on* $\bar{\mathcal{O}}$.

Proof: Consider the approximated Q.V.I.

$$(3.56)\qquad \begin{aligned} & a(u_m,v-u_m) - (H_m(.,u_m,\nabla u_m),v-u_m) \ge (f,v-u_m) \\ & \forall\, v \in H_0^1 \cap L^\infty \;,\; v \le Mu_m \;,\quad u_m \in H_0^1 \cap L^\infty \;,\; u_m \le Mu_m \;. \end{aligned}$$

This Q.V.I. comes within the framework of the Q.V.I.'s studied in Section 1, since $A - H_m + \beta_m$ is monotonic from $H_0^1 \to H^{-1}$. We can therefore affirm the existence and uniqueness of a solution of (3.56) $u_m \geqslant 0$. We then have

$$(3.57)\qquad u_{m'} \leqslant u_m \quad \text{if} \quad m' \geqslant m.$$

Let us write $\sigma_m(\psi)$ for the solution of the V.I. (3.16). In the course of the proof of Theorem 3.1 we have seen that

$$(3.58)\qquad \sigma_{m'}(\psi) \leqslant \sigma_m(\psi) \quad \text{if} \quad m' \geqslant m.$$

If we consider a decreasing process

$$(3.59)\qquad u_m^{n+1} = \sigma_m(M\,u_m^n) \;,\; u_m^0 = w$$

then

$$u_m^n \downarrow u_m \quad \text{when } n \to \infty.$$

By induction, from (3.59),(3.58), and the increasing property of the operator M,

$$u_{m'}^n \leqslant u_m^u,$$

hence the property (3.57) when $n \to \infty$.

Let us show that the obstacle Mu_m belongs to the class (3.13) with constants d and γ independent of m. If we let $z \leqslant x \in \mathcal{O}$, then

$$\inf_{\substack{\xi \ge 0 \\ x+\xi \in \bar{\mathcal{O}}}} (u_m(x+\xi) + c_0(\xi)) = \inf_{\substack{\xi \ge 0 \\ x+\xi \in \bar{\mathcal{O}}}} (u_m(z+x-z+\xi) + c_0(\xi))$$

and by the sublinearity of c_0

$$\geq \inf_{\substack{\xi \geq 0 \\ x+\xi \in \bar{\mathcal{O}}}} u_m(z+x-z+\xi) + c_0(x-z+\xi) - c_0(x-z)$$

$$\geq \inf_{\substack{\eta \geq 0 \\ z+\eta \in \bar{\mathcal{O}}}} u_m(z+\eta) + c_0(\eta) - d|x-z|^\gamma$$

which is the result desired.

By Lemmas 3.1 and 3.4 we can then state that

$$(3.60) \qquad \|u_m\|_{L^\infty} \leq C \qquad \|u_m\|_{H_0^1} \leq C \quad , \quad \|u_m\|_{C^{0,\delta}} \leq C .$$

Therefore

$$u_m \downarrow u \text{ strongly in } C^0 \text{ and weakly in } H_0^1.$$

Since $u \leqslant u_m$, $Mu \leqslant Mu_m$, and therefore if $v \leqslant Mu$, it satisfies (3.56), therefore by taking $v = u$ first,

$$a(u_m,u_m) \leq a(u_m,u) - (H_m(.,u_m,\nabla u_m),u-u_m) - (f,u-u_m)$$

therefore

$$\overline{\lim}\, a(u_m,u_m) \leq a(u,u)$$

hence

$$u_m \to u \text{ strongly in } H_0^1.$$

Next, we easily deduce from this that u is the solution of the Q.V.I. (3.55). It is the maximum solution, for if $\tilde{u}$ is the solution, then

$$\tilde{u} \leqslant w$$

by what we saw in Theorem 3.1 ($u \leqslant w$ for every obstacle ψ the solution of (3.14)). Therefore

$$\tilde{u} \leqslant u_m^0.$$

If we assume $\tilde{u} \leqslant u_m^n$, then

$$(3.61) \qquad \tilde{u} \leqslant \sigma_m(M\tilde{u});$$

this results from the fact that u is the unique solution of the V.I. (3.14) corresponding to the obstacle Mu, and therefore by the approximation procedure (3.16) for the V.I. (3.14) we certainly have (3.61). Hence

$$\tilde{u} \leq \sigma_m(M\,\tilde{u}) \leq \sigma_m(M\,u_m^n) = u_m^{n+1}$$

and therefore $\tilde{u} \leqslant u_m$, whence we finally have $\tilde{u} \leqslant u$.

Remark 3.5

The solution u_m will in general be more regular than $H_0^1 \cap C^{0,\delta}$, at least if we take the assumptions of Section 2. But naturally the estimates depend on m.□

Remark 3.6

To show the existence of a solution of the V.I. (3.14) or of the Q.V.I. (3.55) we needed $C^{0,\delta}$ estimates and we proved the existence of a $C^{0,\delta}$ solution. In fact the methods employed also show that *every positive solution* of (3.14) or (3.15) belong to the same space $C^{0,\delta}$. In fact every positive solution of (3.14) or (3.15) stays between 0 and w and behaves as in Lemmas 3.2,3.3,3.4; and we again obtain the estimate

$$\int_{B_R(x_0)} \chi_{\mathcal{O}} \, |\nabla u|^2 \, |x-x_0|^{2-n} \, dx \le C_\mu \, R^\mu$$

with the same constants as in (3.47), Hence $u \in C^{0,\delta}$, $\delta = \mu/2$.

We restrict ourselves to the positive solutions, so as to have an L^∞ estimate independent of the obstacle ψ.□

3.5 *Regularity of* Mu

In this Section we are going to prove that if $u \geqslant 0$, $u \in H_0^1(\mathcal{O}) \cap C^{0,\delta_0}(\bar{\mathcal{O}})$ is the solution of the Q.V.I. (3.55), then $Mu \in W^{1,\infty}(\bar{\mathcal{O}})$.

We begin by recalling an important result (F.W. GEHRING [1], W.G. MEYERS and A. ELCRAT [1], M. GIAQUINTA and G. MODICA [1]). First of all, if h is summable, we set

$$(3.62) \qquad \fint_{B_R(x_0)} h \, dx = \frac{1}{|B_R|} \int_{B_R(x_0)} h \, dx$$

Let $h \geqslant 0$, and h be defined on a cube Q of R^n, with $h = 0$ outside Q. Assume that $h \in L^\lambda(Q)$, $\lambda > 1$. Also let $\ell \in L^\mu(Q)$, $\ell \geqslant 0$, $\ell = 0$ outside Q, $\mu > \lambda$. Assume the following estimate is satisfied:

$$(3.63) \qquad \fint_{B_R(x_0)} h^\lambda(x) dx \le b \; [(\fint_{B_{2R}(x_0)} h(x) dx)^\lambda + \fint_{B_{2R}(x_0)} \ell^\lambda(x) \, dx]$$

where $b > 1$, and depends neither on x_0, nor on R.

Then *there exists* $\varepsilon > 0$ *depending only on* λ, b, n, μ such that

$$h \in L^{\lambda+\varepsilon}(Q)$$

and, furthermore, we have the estimate

$$(3.64) \qquad (\fint_{B_R(x_0)} h^{\lambda+\varepsilon}(x) dx)^{\frac{1}{\lambda+\varepsilon}} \le C \; [(\fint_{B_{2R}(x_0)} h^\lambda(x) dx)^{\frac{1}{\lambda}} + (\fint_{B_{2R}(x_0)} \ell^{\lambda+\varepsilon}(x) dx)^{\frac{1}{\lambda+\varepsilon}}]$$

where c *depends only on* λ, b, n, μ.

We now return to the Q.V.I. (3.55). Let u be a positive solution, therefore

$u \in C^{0,\delta_0}$. Write

(3.65) $$C = \{x \in \bar{\mathcal{O}} \mid u(x) < Mu(x)\} \ .$$

Then we have:

LEMMA 3.5: *Let* $\phi \in \mathcal{D}(C)$. *Then* $u\phi \in W^{1,r}(R^n)$, *where* $r > 2$ *depends neither on* u *nor on* ϕ *(*u *and* ϕ *are continued by* 0 *outside their support).*

Proof: Since $u\phi$ is bounded on $\bar{\mathcal{O}}$ and zero outside of $\bar{\mathcal{O}}$, this is a question of proving an estimate for the gradient.

Let x_0 be fixed and $B_R(x_0)$ be the ball with centre x_0 and radius R. Let τ_R be regular, $\tau_R = 1$ on B_R, $\tau_R = 0$ on $R^n - B_{2R}$, $|\nabla \tau_R| \leqslant \frac{\tau}{R}$, $0 \leqslant \tau_R \leqslant 1$. Also write $\tau \equiv \tau_R$, and set

$$\xi \equiv \xi_R = \begin{cases} 0 & \text{if} \quad B_{2R} \cap (R^n - \mathcal{O}) \neq \emptyset \\ u_R & \text{if} \quad B_{2R} \subset \mathcal{O} \end{cases} .$$

where

$$u_R = \frac{1}{|B_R|} \int_{B_R} u(x)dx = \fint_{B_R} u(x)dx \ .$$

Let $q \geqslant 1$. Consider

$$v = u - \varepsilon \, |u - \xi|^{q-1} (u-\xi) \, \tau^2 \phi^2 \ .$$

as test function in the Q.V.I. (3.55). At least for ε small enough v is admissible, since $v \in H_0^1$ and

$$v = u \qquad \text{outside the support of } \phi,$$

$$v \leqslant u + \varepsilon C_\phi \qquad \text{on the support of } \phi,$$

and since $Mu - u \geqslant d_0 > 0$ on the support of ϕ, we have $v \leqslant Mu$ for ε small enough. Therefore we obtain

(3.66) $$a(u, |u-\xi|^{q-1}(u-\xi)\tau^2\phi^2) \leq (H(.,u,\nabla u), |u-\xi|^{q-1}(u-\xi)\tau^2\phi^2) + (f, |u-\xi|^{q-1} (u-\xi)\tau^2\phi^2).$$

We first consider the term

$$X = \Sigma \int a_{ij} \frac{\partial u}{\partial x_j} \frac{\partial}{\partial x_i} (|u-\xi|^{q-1}(u-\xi)\tau^2\phi^2) \, dx =$$

$$= q \, \Sigma \int a_{ij} \frac{\partial u}{\partial x_j} \frac{\partial u}{\partial x_i} |u-\xi|^{q-1} \tau^2\phi^2 \, dx + 2 \, \Sigma \int a_{ij} \frac{\partial u}{\partial x_j} |u-\xi|^{q-1} (u-\xi) \Big(\frac{\tau \partial \tau \phi^2}{\partial x_i} + \tau^2 \phi \frac{\partial \phi}{\partial x_i}\Big) \, dx \ .$$

From this we easily deduce

$$X \geq (q-\tfrac{1}{2})\,\alpha \int_{R^n} |\nabla u|^2\, |u-\xi|^{q-1}\, \tau^2\phi^2\, dx - \tag{3.67}$$
$$- \frac{C_q}{R^2} \int_{B_{2R}} |u-\xi|^2\, \phi^2 dx - C_q \int_{R^n} \tau^2 |\nabla\phi|^2 \chi_{\mathcal{O}}\, dx$$

where the C_q denote constants depending only on q.

Next we have

$$|(H(.,u,\nabla u),\ |u-\xi|^{q-1}(u-\xi)\ \tau^2\xi^2)| \leq$$
$$\leq C \int_{R^n} |\nabla u|^2\, |u-\xi|^q\, \tau^2\phi^2 dx + C_q \int_{\mathbb{R}^n} \chi_{\mathcal{O}}\, \tau^2\phi^2 dx$$

and we can absorb the first and zero-th order terms of the form a(u,v) in the preceding expression. The constant denoted C does not depend on q. Finally,

$$|(f, |u-\xi|^{q-1}(u-\xi)\ \tau^2\phi^2)| \leq C_q \int_{R^n} |f|\ \tau^2\phi^2\ \chi_{\mathcal{O}}\ dx .$$

Therefore we have the estimate

$$(q-\tfrac{1}{2})\alpha \int_{R^n} |\nabla u|^2\, |u-\xi|^{q-1}\, \tau^2\phi^2\, dx \leq C \int |\nabla u|^2\, |u-\xi|^q\, \tau^2\phi^2\, dx + \tag{3.68}$$
$$+ \frac{C_q}{R^2} \int_{B_{2R}} |u-\xi|^2\, \phi^2\, dx + C_q \int \chi_{\mathcal{O}}\, \tau^2(|f|\phi^2 + \phi^2 + |\nabla\phi|^2)\, dx.$$

Working as in Lemma 3.2 for the choices of q, we finally deduce the estimate

$$\int_{R^n} |\nabla u|^2\, \tau^2\phi^2\, dx \leq \frac{C}{R^2} \int_{R^n} |u-\xi|^2\, \phi^2 dx + C \int_{R^n} \chi_{\mathcal{O}}\, \tau^2(|f|\phi^2 + \phi^2 + |\nabla\phi|^2) dx. \tag{3.69}$$

Further, let us write

$$\tilde{\xi} = \begin{cases} 0 & \text{if } B_{2R} \cap (R^n - \mathcal{O}) \neq \emptyset \\ (u\phi)_R = \displaystyle⨍_{B_R(x_0)} u\phi\, dx, & \text{if } B_{2R} \subset \mathcal{O}. \end{cases}$$

We easily see that

$$|\tilde{\xi} - \xi\phi| \leq C\ R\|\nabla\phi\| \quad \text{on} \quad B_{2R}$$

and thus we deduce from (3.69) that

$$\int_{R^n} |\nabla(u\phi)|^2 \tau^2 dx \le \frac{C}{R^2} \int_{B_{2R}} |u\phi - \tilde{\xi}|^2 dx +$$

$$+ C \int_{B_{2R}} \chi_{\mathcal{O}} (|f|\phi^2 + \phi^2 + \|\nabla\phi\|^2) dx$$

so again by setting

$$m = \chi_{\mathcal{O}}(|f|\phi^2 + \phi^2 + \|\nabla\phi\|^2)$$

we have

$$\int_{B_R} |\nabla(u\phi)|^2 dx \le \frac{C}{R^2} \int_{B_{2R}} |u\phi - \tilde{\xi}|^2 dx + C \int_{B_{2R}} m\, dx .$$

By the Sobolev-Poincaré inequality

$$\int_{B_{2R}} |u\phi - \tilde{\xi}|^2 dx \le C\left(\int_{B_{2R}} |\nabla(u\phi)|^\theta dx\right)^{\frac{2}{\theta}} , \qquad \theta = \frac{2n}{n+2}$$

which, on dividing by R^n, implies

$$⨍_{B_R} |\nabla(u\phi)|^2 dx \le C\left(⨍_{B_{2R}} |\nabla(u\phi)|^\theta dx\right)^{\frac{2}{\theta}} + C ⨍_{B_{2R}} m\, dx .$$

Let us set

$$h = |\nabla(u\phi)|^\theta , \quad \ell = m^{\theta/2} , \quad \lambda = \frac{2}{\theta} > 1,$$

then this yields

$$⨍_{B_R} h^\lambda dx \le C\left[\left(⨍_{B_{2R}} h\, dx\right)^\lambda + ⨍_{B_{2R}} \ell^\lambda dx\right] .$$

and for $\mu = \lambda p > \lambda$ the assumption (3.63) is satisfied for arbitrary Q containing $\bar{\mathcal{O}}$. From this it follows that $h \in L^{\lambda+\varepsilon}(R^n)$, hence $u\phi \in W^{1,r}$ with $r > 2$, and r does not depend on ϕ, since ϕ arises only in the definition of ℓ.

Let us now recall that $u \in C^{0,\delta_0}$.

LEMMA 3.6: *Let* G *be an arbitrary open set such that* $\bar{G} \subset C$. *Then if* $p \leqslant \frac{r}{2}$ *or if* $p \leqslant \frac{n}{\delta_0}$ *we have* $u \in W^{2,p}(G)$. *If* $p > \frac{n}{\delta_0}$ *and* $p > \frac{r}{2}$, *let* h *be an integer* $\geqslant 0$ *such that*

$$\frac{2}{r} - \frac{h\,\delta_0}{n} > \frac{1}{p} \geq \frac{2}{r} - \frac{(h+1)\delta_0}{n}$$

if

$$\frac{2}{r} - \frac{(h+1)}{n}\,\delta_0 > 0$$

then

$$u \in W^{2,p}(G)$$

and if

$$\frac{2}{r} - \frac{(h+1)}{n}\,\delta_0 \leq 0$$

then

$$u \in W^{2,\frac{nr}{2n-hr\delta_0}}(G) \,.$$

Proof: Since we can always choose $\phi = 1$ on G, we see that $u \in W^{1,r}(G)$ and $u \in C^{0,\delta_0}$. Furthermore,

(3.70) $\quad Au - H(x,u,\nabla u) = f \quad \text{on} \quad C\,.$

We then use an interpolation inequality between spaces of Holderian functions and Sobolev spaces (cf. E. GAGLIARDO [1], C. MIRANDA [1] and L. NIRENBERG [1]).

Let

$$\psi \in C^{0,\beta} \quad , \quad \psi \in W^{2,j} \qquad (j > 1),$$

then

$$\psi \in W^{1,\bar{j}} \,,$$

where

$$\frac{1}{\bar{j}} = \frac{1}{2j} - \frac{1}{2}\,\frac{\beta}{n} \,.$$

We apply this result to (3.70) in the following way: Let us set $g = f + H(x,u,\nabla u)$. Then $g \in L^{p\wedge\frac{r}{2}}$. Therefore by S. AGMON, A. DOUGLIS and L. NIRENBERG [1], $u \in W^{2,p\wedge\frac{r}{2}}$. We therefore set up a recurrence by setting $j_0 = p\wedge\frac{r}{2}$, $\beta = \delta_0$. If $p \leqslant \frac{r}{2}$, $j_0 = p$ and $u \in W^{2,p}$ necessarily. If $p > \frac{r}{2}$, then $j_0 = \frac{r}{2}$. Let h be as in the statement; if $p \leqslant \frac{n}{\delta_0}$, then

$$\frac{2}{r} - \frac{(h+1)\delta_0}{n} > 0.$$

Therefore

$$\bar{j}_1 = \frac{2j_0 n}{n-j_0\,\delta_0} > 0 \quad \bar{j}_1 = \frac{2r\ n}{2n - r\ \delta_0} \text{ and } u \in W^{1,\bar{j}_1},$$

hence

$$g \in L^{p \wedge \frac{\bar{j}_1}{2}} = L^{j_1}.$$

If $h = 0$, then $j_1 = p$.

If $h = 1$, then

$$j_1 = \frac{r\ n}{2n-r\ \delta_0} \text{ and } \frac{1}{j_1} = \frac{2}{r} - \frac{\delta_0}{n} > 0,$$

therefore

$$\bar{j}_2 = \frac{2r\ n}{2n-2r\ \delta_0} \text{ and } u \in W^{1,\bar{j}_2},$$

hence

$$g \in L^{p\wedge \frac{\bar{j}_2}{2}} = L^{j_2}, \text{ and } j_2 = p.$$

And so we establish that $u \in W^{2,p}$ for all h.

If $p > \frac{r}{2}$ and $p > \frac{n}{\delta_0}$, we also easily verify the assertions.

COROLLARY 3.1: *Let* $p > n$, *then there exists* $\delta_1 > 0$ *such that* $u \in C^{1,\delta_1}(\bar{G})$ *for every open set* G *with* $\bar{G} \subset C$.

Proof: Either $u \in W^{2,p}(G)$, $p > n$, or

$$u \in W^{2,\frac{nr}{2n-hr\delta_0}} \text{ and } \frac{2}{r} - \frac{h\delta_0}{n} \leq \frac{\delta_0}{n},$$

hence

$$\frac{nr}{2n-hr\delta_0} \geq \frac{n}{\delta_0} > n.$$

In all cases, by Sobolev's inclusion theorems we have $u \in C^{1,\delta_1}(\bar{G})$. □

Remark 3.5

The methods of proving Lemmas 3.5 and 3.6 are owed to J. FREHSE [3] (cf., also, A. BENSOUSSAN, J. FREHSE, U.MOSCO [1]). The classical interpretation of Sobolev would give us $u \in W^{1,\bar{j}_1}$, with $\frac{1}{\bar{j}_1} = \frac{1}{j_0} - \frac{1}{n}$. Since we know that $u \in W^{1,r}$, the method would only have progressed if $\bar{j}_1 > r$, so $\frac{1}{j_0} - \frac{1}{n} < \frac{1}{r}$. Since $\frac{1}{j_0} > \frac{2}{r}$, then we must necessarily have $r > n$, which has no reason to be satisfied.□

COROLLARY 3.2: *The obstacle* Mu *belongs to* $W^{1,\infty}(\bar{\mathcal{O}})$.

Proof: This is very similar to Lemma 2.1. The function u is continued by 0 outside $\bar{\mathcal{O}}$. We consider

$$F = \{y \in R^n \mid u(y) < Mu(y) - \frac{k}{2}\}$$

and the function $\xi(x)$ defined in (2.4). Then $x + \xi(x) \in F$, for all x. By Corollary 3.1, u is differentiable at the point $x + \xi(x)$ and

$$|Du(x + \xi(x))| \leq C_F.$$

Now, arguing as in Lemma 2.1, we easily show that for every unit vector χ of R^n

$$\frac{\partial}{\partial \chi} Mu(x) = Du(x+\xi(x)).\chi$$

whence the result.□

4. UNBOUNDED OPEN SETS

Method of Approach

We are going to consider the Q.V.I. analogous to (1.7) in the case of an unbounded open set. The important point is that we can no longer guarantee uniqueness. We shall obtain a maximum solution and a minimum solution. But the techniques of Section 1 will not be able to be applied completely, since we are no longer working with bounded functions. In particular, the Hanouzet-Joly technique (cf. Lemma 1.4) is very closely connected with the L^∞ framework. However, we shall use certain techniques of Section 3 to obtain estimates in spaces of Hölderian functions.

4.1 *Assumptions: Notations*

We consider the coefficients a_{ij}, a_i, a_o satisfying (1.1), and the operator A defined in (1.2). We also introduce weighted Sobolev spaces. For $\mu > 0$ set

$$\beta_\mu(x) = \exp - \mu(|x|^2 + 1)^{1/2} .$$

We are given

(4.1) $\mathcal{O}$ an unbounded, regular, open set of R^n.

Set

$$L^{p,\mu}(\mathcal{O}) = |\phi \mid \int \beta_\mu^p |\phi|^p dx < \infty \}$$

and

$$W^{1,p,\mu}(\mathcal{O}) = \{ \phi \in L^{p,\mu} \mid \frac{\partial \phi}{\partial x_i} \in L^{p,\mu}\}$$
$$W^{2,p,\mu}(\mathcal{O}) = \{ \phi \in L^{p,\mu} \mid \frac{\partial \phi}{\partial x_i} , \frac{\partial^2 \Phi}{\partial x_i \partial x_j} \in L^{p,\mu}\} .$$

We shall write

$$H^{\mu} = L^{2,\mu} , \quad V^{\mu} = W_0^{1,2,\mu}(\mathcal{O}) .$$

By $(\, , \,)_\mu$, $((\, , \,))_\mu$ we denote the scalar products in H^μ and V^μ respectively. Next, for $u,v \in V_\mu$ we define

$$a_\mu(u,v) = \int_{\mathcal{O}} a_{ij} \frac{\partial u}{\partial x_j} \frac{\partial v}{\partial x_i} \beta_\mu^2 \, dx + \int_{\mathcal{O}} (a_i - \frac{\sum_j a_{ji} x_j \mu}{(1+|x|^2)^{1/2}}) \frac{\partial u}{\partial x_i} v \beta_\mu^2 \, dx$$
$$+ \int_{\mathcal{O}} a_0 \, u \, v \, \beta_\mu^2 \, dx$$

so that if $u \in W^{2,2,\mu}$ and $v \in V^\mu$ we have

$$(4.3) \qquad a_\mu(u,v) = \int_{\mathcal{O}} Au \, v \, \beta_\mu^2 \, dx .$$

We are given f(x) satisfying

$$(4.4) \qquad f \text{ measurable}, \quad 0 \leqslant f(x) \leqslant f_0(1 + |x|^s),$$

and c_0,k as in (1.5). We define $M\phi$ by (1.6).

The problem that interests us is the following: find the solution u of

$$(4.5) \qquad a_\mu(u,v-u) \geq (f,v-u)_\mu \quad , \quad u \in V^\mu \quad , \quad u \geq 0$$
$$\forall v \in V^\mu \text{ such that } v \leq Mu \text{ a.e.}, \; u \leq Mu \text{ a.e.}$$

We shall notice that for all $f \in H^\mu$, $v > 0$. Moreover, there exists a positive such that

$$(4.6) \qquad a_\mu(v,v) + \lambda |v|_\mu^2 \geq \gamma \, \| v \|_\mu^2 \quad \forall v \in V^\mu .$$

4.2 *Existence*

Set

$$(4.7) \qquad u^0(x) = F^0 \beta_{-\mu_0}(x) = F^0 \exp \mu_0(|x|^2 + 1)^{1/2} .$$

LEMMA 4.1: $F^0 > 0$ *and* $\mu_0 > 0$ *can be chosen such that for* $\mu > \mu_0$ *there holds* .

$$(4.8) \qquad a_\mu(u^0,v) \geqslant (f,v)_\mu \quad \forall v \in V^\mu, \; v \geqslant 0.$$

Proof: By (4.3), everything reduces to showing that

$$Au^0 \geqslant f.$$

But

$$Au^0 = F^0\beta_{-\mu_0}\left[-\mu_0^2 \Sigma \frac{a_{ij}\, x_i\, x_j}{1+|x|^2} - \mu_0\Sigma \frac{a_{ij}\,\delta_{ij}}{(1+|x|^2)^{1/2}} + \mu_0\, \Sigma \frac{a_{ij}\, x_i\, x_k}{(1+|x|^2)^{3/2}} \right.$$
$$\left. + (a_i - \sum_j \frac{\partial}{\partial x_j} a_{ji})\, \mu_0 \frac{x_i}{(1+|x|^2)^{1/2}} + a_0 \right]$$

therefore

$$Au^0 \geq F^0\beta_{-\mu_0}\left[- \mu_0^2 \frac{M|x|^2}{1+|x|^2} - \mu_0 \frac{n\, M}{(1+|x|^2)^{1/2}} + \mu_0\alpha \frac{|x|^2}{(1+|x|^2)^{3/2}} + \right.$$
$$\left. + (a_i - \sum_j \frac{\partial}{\partial x_j} a_{ji})\, \mu_0 \frac{x_i}{(1+|x|^2)^{1/2}} + \beta \right]$$
$$\geq F^0\, \beta_{-\mu_0}[- \mu_0^2 M - \mu_0 M' + \beta]$$

and if we choose $\mu_0 > 0$ such that

$$- \mu_0^2 M - \mu_0 M' + \beta > 0 ,$$

and then F^0 such that

$$F^0 \geq \frac{f_0}{\beta - \mu_0^2 M - M'} \sup_\gamma [(1+|x|^s)\exp - \mu_0(|x|^2 + 1)^{1/2}]$$

we obtain the desired result.□

In the remainder μ will always be assumed to satisfy the condition of Lemma 4.1. Let $z \in H^\mu$ be such that $0 \leqslant z \leqslant u^0$.

Consider the problem

(4.9) $$a_\mu(\zeta_\lambda, v-\zeta_\lambda) + \lambda(\zeta_\lambda, v-\zeta_\lambda)_\mu \geq (f+\lambda z, v-\zeta_\lambda)_\mu$$
$$\forall\, v \in V^\mu , \quad v \leq Mz , \quad \zeta_\lambda \leq Mz , \quad \zeta_\lambda \in V^\mu .$$

Thanks to (4.6) the problem (4.9) defines one and only one solution ζ_λ. We set

$$K = \{z \in H^\mu \mid 0 \leq z \leq u^0 \} .$$

Then on setting $T_\lambda z = \zeta_\lambda$ we have:

LEMMA 4.2: T_λ *is an increasing operator of* K *into* K.

Proof: This is analogous to Lemma 1.1. Let us simply verify that $\zeta_\lambda \leqslant u^0$.

In (4.9) take

$$v = \zeta_\lambda - (\zeta_\lambda - u^0)^+,$$

therefore

$$-a_\mu(\zeta_\lambda, (\zeta_\lambda - u^0)^+) - \lambda(\zeta_\lambda, (\zeta_\lambda - u^0)^+)_\mu \geq -(f + \lambda z, (\zeta_\lambda - u^0)^+)_\mu$$

and by (4.8)

$$a_\mu(u^0, (\zeta_\lambda - u^0)^+) \geq (f, (\zeta_\lambda - u^0)^+)_\mu .$$

By addition

$$a_\mu(u^0 - \zeta_\lambda, (\zeta_\lambda - u^0)^+) - \lambda(\zeta_\lambda - u^0, (\zeta_\lambda - u^0)^+)_\mu \geq \lambda(u^0 - z, (\zeta_\lambda - u^0)^+)_\mu$$

from which it easily follows that

$$(\zeta_\lambda - u^0)^+ = 0. \square$$

THEOREM 4.1: *Under the assumptions* (1.1),(1.5),(4.1),(4.4) *the set of solutions of* (4.5) *which belong to* K *is non-empty and has a minimum element and a maximum element.*

Proof: It suffices to apply Tartar's Theorem, which has already been used in the proof of Theorem 1.1. The details are left to the reader. $\square$

In the remainder we consider increasing and decreasing approximation procedures, denoted respectively by

(4.10) $\quad u^{n+1} = T u^n , \quad u^0$ defined by (4.7)

(4.11) $\quad u_{n+1} = T u_n , \quad u_0 = 0 .$

LEMMA 4.3: *The decreasing procedure* u^n *converges to* $\bar{u}$, *the maximum solution of* (4.5) *(in* K*).*

Proof: Identical to that of Lemma 1.2.

4.3 *Convergence of the Increasing Procedure*

In order to obtain the convergence of the increasing procedure, we first prove the following result:

PROPOSITION 4.1: *Under the assumptions of Theorem 4.1, for all* $z \in K$, $T_\lambda z$ *satisfies the Hölder property* (4.23), *regardless of the function* z *in* K.

We are motivated by Section 3. Set

$$\mathcal{O}_H = \{x \in \mathcal{O} \mid |x| < H\}$$

Let $Q_H \supset\supset \mathcal{O}_H$; where Q_H is a ball. For $x_0 \in \mathcal{O}_H$ we consider the Green's function

$G \equiv G_H^{x_0}$ defined as the solution of

$$(4.12) \qquad \sum_{i,j} \int_{Q_H} a_{ij} \frac{\partial \phi}{\partial x_j} \frac{\partial G}{\partial x_i} dx = \phi(x_0) \quad \forall \phi \in \mathcal{D}(Q_H),\ G_H \in W^{1,s}(Q_H) \quad 1 \le s < \frac{n}{n-1}$$

and (for $n \geqslant 3$) G_H satisfies

$$(4.13) \qquad c_{0H}|x-x_0|^{2-n} \le G_H(x) \le c_{1H}|x-x_0|^{2-n} .$$

Next we define $G_\rho \equiv G_{\rho,H}^{x_0}$ by

$$(4.14) \qquad \sum_{i,j} \int_{Q_H} a_{ij} \frac{\partial \phi}{\partial x_j} \frac{\partial G\rho}{\partial x_i} dx = \phi_\rho = \fint_{B_\rho(x_0)} \phi(x)dx = \frac{1}{|B_\rho|} \int_{B_\rho} \phi(x)dx$$

$$\forall \phi \in \mathcal{D}(Q_H),\quad G_\rho \in H_0^1(Q_H).$$

We shall write

$$\psi = Mz$$

and show (cf. the proof of Theorem 3.2) that ψ satisfies

$$(4.15) \qquad \psi(x) \le \psi(z) + d|x-z|^\gamma \qquad \forall\ x \le z \in \bar{\mathcal{O}} .$$

We shall write $\zeta = \zeta_\lambda$.

Proof of Proposition 4.1: We are given $\Theta_H \in \mathcal{D}(Q_H)$ and $\bar{\Theta}_H = 1$ on $\bar{\mathcal{O}}_H$, $0 \leqslant \Theta_H \leqslant 1$. We write $B_R(x_0)$ for the ball with centre x_0 and radius $R \leqslant 1$, and $\tau_R \equiv \tau$ is defined as in Lemma 3.2. We define $T_R(x_0)$ as in (3.36), and

$$\xi \equiv \xi_R = \begin{cases} 0 & \text{if } B_{4R} \cap (R^n - \mathcal{O}) \neq \emptyset \\ \zeta_R - d(GR)^\gamma & \text{if } B_{4R} \subset \mathcal{O},\ \zeta_R = \dfrac{1}{|T_R(x_0)|} \displaystyle\int_{T_R} \zeta(x)dx . \end{cases}$$

As in (3.38), we have

$$\xi_R \le \psi(y) \quad \forall\ y \in B_{2R} .$$

We take as test function in (4.9),

$$v = \zeta - \varepsilon\ (\zeta - \xi_R) G_\rho \tau^2\ \theta_H^2 .$$

For ε small enough v is admissible. Therefore

$$a_\mu(\zeta, (\zeta-\xi)G_\rho \tau^2 \theta_H^2) \le (f + \lambda z - \lambda\zeta, (\zeta-\xi)G_\rho\tau^2\theta_H^2)_\mu ,$$

so

$$(4.16)\qquad \int a_{ij}\frac{\partial\xi}{\partial x_j}\frac{\partial}{\partial x_i}[(\zeta-\xi)G_\rho\tau^2\phi_H^2]\,\beta_\mu^2 dx + \int \tilde{a}_i\frac{\partial\xi}{\partial x_i}(\zeta-\xi)G_\rho\tau^2\phi_H^2\,\beta_\mu^2\,dx$$
$$\le \int (f+\lambda z-\lambda\zeta-a_0\zeta)(\zeta-\xi)G_\rho\tau^2\theta_H^2\,\beta_\mu^2\,dx$$

where we have set

$$\tilde{a}_i = a_i - \sum_j \frac{\partial\, a_{ji}\, x_j\, \mu}{(1+|x|^2)^{1/2}} .$$

Next, we have

$$X = \int a_{ij}\frac{\partial\zeta}{\partial x_j}\frac{\partial}{\partial x_i}[(\zeta-\xi)G_\rho\tau^2\theta_H^2]\,\beta_\mu^2\,dx =$$
$$= \int a_{ij}\frac{\partial\zeta}{\partial x_j}\frac{\partial\zeta}{\partial x_i}G_\rho\tau^2\varepsilon_H^2\,\beta_\mu^2 dx + \int a_{ij}\frac{\partial\zeta}{\partial x_j}(\zeta-\xi)\frac{\partial}{\partial x_i}(G_\rho\tau^2\theta_H^2)\,\beta_\mu^2\,dx$$
$$= \int a_{ij}\frac{\partial\zeta}{\partial x_j}\frac{\partial\zeta}{\partial x_i}G_\rho\tau^2\theta_H^2\,\beta_\mu^2 dx + \frac{1}{2}\int a_{ij}\frac{\partial G_\rho}{\partial x_i}\frac{\partial}{\partial x_j}[(\zeta-\xi)^2\,\tau^2\theta_H^2\,\beta_\mu^2]\,dx$$
$$-\frac{1}{2}\int a_{ij}\frac{\partial G_\rho}{\partial x_i}(\zeta-\xi)^2\frac{\partial}{\partial x_j}(\tau^2\theta_H^2\,\beta_\mu^2)dx + \int a_{ij}\frac{\partial\zeta}{\partial x_j}(\zeta-\xi)G_\rho\frac{\partial}{\partial x_i}(\tau^2\theta_H^2)\beta_\mu^2$$

Since $(\zeta-\xi)^2\tau^2\theta_H^2\beta_\mu^2 \in H_0^1(Q_H)$, and is positive, it follows from (4.14) that the second integral is positive, therefore

$$X \ge \alpha\int |\nabla\zeta|^2\,G_\rho\tau^2\theta_H^2\,\beta_\mu^2\,dx +$$
$$+\frac{1}{2}\int G_\rho\Big[\frac{\partial a_{ij}}{\partial x_i}(\zeta-\xi)^2\frac{\partial}{\partial x_j}(\tau^2\theta_H^2\,\beta_\mu^2) + 2a_{ij}(\zeta-\xi)\frac{\partial\zeta}{\partial x_i}\frac{\partial}{\partial x_j}(\tau^2\theta_H^2\beta_\mu^2) +$$
$$+ a_{ij}(\zeta-\xi)^2\frac{\partial^2}{\partial x_i\partial x_j}(\tau^2\theta_H^2\beta_\mu^2)\Big]\,dx +$$
$$+\int a_{ij}\frac{\partial\zeta}{\partial x_j}(\zeta-\xi)G_\rho\frac{\partial}{\partial x_i}(\tau^2\theta_H^2)\,\beta_\mu^2\,dx =$$
$$= (1) + (2) + (3) + (4) + (5) .$$

Now,

$$|(2)| \le \frac{C}{R}\int_{B_{2R}-B_R}(\zeta-\xi)^2 G_\rho\,\theta_H^2\,\beta_\mu^2\,dx + C\int(\zeta-\xi)^2\tau^2 G_\rho|\nabla(\theta_H^2\,\beta_\mu^2)|dx$$

$$|(3)| \le C\int_{B_{2R}-B_R}|\nabla\zeta|^2 G_\rho\theta_H^2\beta_\mu^2 dx + \frac{C}{R^2}\int_{B_{2R}-B_R}(\zeta-\xi)^2 G_\rho\,\theta_H^2\,\beta_\mu^2\,dx +$$
$$+ C\,\varepsilon\int|\nabla\zeta|^2 G_\rho\,\tau^2\,\beta_\mu^2\,\theta_H^2 dx + \frac{C}{\varepsilon}\int(\zeta-\xi)^2 G_\rho\tau^2|\nabla(\theta_H\beta_\mu)|^2 dx$$

where ε is arbitrary, and

$$|④| \le \frac{C}{R^2}\int_{B_{2R}-B_R} (\zeta-\xi)^2 G_\rho \theta^2 \beta_\mu^2 \, dx + C\int (\zeta-\xi)^2 G_\rho \tau^2 \, |\nabla(\theta_H\beta_\mu)|^2 \, dx$$

$$+ C\int (\zeta-\xi)^2 \, G_\rho \tau^2 \, \| D^2(\theta_H^2\beta_\mu^2) \| \, dx$$

$$|⑤| \le C\int_{B_{2R}-B_R} |\nabla\zeta|^2 \, G_\rho \tau^2 \theta_H^2 \beta_\mu^2 \, dx + \frac{C}{R^2}\int_{B_{2R}-B_R} (\zeta-\xi)^2 G_\rho \theta_H^2 \, \beta_\mu^2 \, dx +$$

$$+ C\,\varepsilon\int |\nabla\zeta|^2 G_\rho \tau^2 \beta_\mu^2 \theta^2 \, dx + \frac{C}{\varepsilon}\int (\zeta-\xi)^2 G_\rho \tau^2 \beta_\mu^2 |\nabla\,(\theta_H)|^2 dx \,,$$

and the constants denoted C do not depend on R, nor H, nor ρ.

Also

$$\left|\int \tilde{a}_i \, \frac{\partial\xi}{\partial x_i}(\zeta-\xi) G_\rho \tau^2 \theta_H^2 \beta_\mu^2 \, dx\right| \le C\varepsilon\int |\nabla\zeta|^2 G_\rho \tau^2 \theta_H^2 \beta_\mu^2 \, dx +$$

$$+ \frac{C}{\varepsilon}\int (\zeta-\xi)^2 G_\rho \tau^2 \theta_H^2 \beta_\mu^2 \, dx \,.$$

On the other hand, it is easy to establish that

$$0 \le \xi \le C\,(1+|x|\,)^s \quad \forall \; x \in B_{2R}$$

(recalling that $R \leqslant 1$). From this we deduce

$$|(f+\lambda z-\lambda\zeta-a_0\zeta)(\zeta-\xi)|\beta_\mu^2 \le C \,.$$

Collecting together the preceding results, we obtain (by choosing ε suitably)

$$\text{(4.17)} \quad \int_{B_{2R}} |\nabla\zeta|^2 G_\rho \tau^2 \theta_H^2 \beta_\mu^2 \, dx \le \frac{C}{R^2}\int_{B_{2R}-B_R} (\zeta-\xi)^2 G_\rho \theta_H^2 \beta_\mu^2 \, dx +$$

$$+ C\int_{B_{2R}-B_R} |\nabla\zeta|^2 \, G_\rho \tau^2 \theta_H^2 \beta_\mu^2 \, dx + C_H\int_{B_{2R}} G_\rho \tau^2 \, dx$$

where C_H depends on H, but not on R. Also

$$\left|\int_{B_{2R}} G\rho\tau^2 \, dx\right| \le C_H \, R^{2-\nu}$$

for all $\nu > 0$, $\nu < 2$. We can then make ρ tend to 0 in (4.17), and we obtain

$$\text{(4.18)} \quad \int_{B_{2R}} |\nabla\zeta|^2 G \, \tau^2 \theta_H^2 \beta_\mu^2 \, dx \le \frac{C}{R^2}\int_{B_{2R}-B_R} (\zeta-\xi)^2 G\theta_H^2 \, \beta_\mu^2 dx + \qquad \text{(Contd)}$$

(Contd) $$+ C \int_{B_{2R}-B_R} |\nabla\zeta|^2 G_\rho \tau^2 \theta_H^2\, \beta_\mu^2\, dx + C_H\, R^{2-\nu} .$$

Let us set

$$\xi_H = \frac{1}{T_R} \int_{T_R} \zeta\, \theta_H \beta_\mu\, dx .$$

Then

$$\frac{1}{R^2} \int_{B_{2R}-B_R} (\zeta-\xi)^2 G\, \theta_H^2\, \beta_\mu^2\, dx \le C_H\, R^{-n} \int_{B_{2R}-B_R} (\zeta-\xi)^2\, \theta_H^2\, \beta_\mu^2\, dx$$

$$\le C_H R^{-n}\, [R^{n+2\gamma} + R^{n+2} + \int_{B_{2R}-B_R} (\zeta\theta_H\beta_\mu - \xi_H)^2\, dx]$$

and by Poincaré's inequality

$$\le C_H\, [R^{2\gamma \wedge 2} + R^{2-n} \int_{B_{4R}-B_R} |\nabla(\zeta\theta_H\beta_\mu)|^2 dx] .$$

Collecting the results together, we obtain

(4.19) $$\int_{B_R} G|\nabla(\zeta\theta_H)|^2 dx \le C_H\, [\int_{B_{4R}-B_R} G|\nabla(\zeta\theta_H)|^2 dx + R^{(2-\nu)\wedge\, 2\gamma}] .$$

Next, we take

$$v = \zeta - G_\rho \zeta \theta_H^2,$$

whence

$$a_\mu(\zeta, \zeta G_\rho\, \theta_H^2) \le (f+\lambda z-\lambda\xi,\ \zeta\, G_\rho\, \theta_H^2)_\mu .$$

From this we deduce

(4.20) $$\int_{\mathcal{O}} G|\nabla(\zeta\theta_H)|^2\, dx \le M_H .$$

Recalling that ζ is zero outside $\mathcal{O}$, we easily deduce from (4.19) and (4.20), as in Lemma 3.4, that

(4.21) $$\int_{B_{R(x_0)}} |\nabla(\zeta\theta_H)|^2 dx \le C_H\, R^{n-2+2\delta_H}$$

where $0 < \delta_H < 1$, and C_H depends neither on x_0 nor R. From this we deduce

(4.22) $$\int_{B_R} \chi_{\mathcal{O}_H} |\nabla\zeta|^2 dx \leq C_H R^{n-2+2\delta_H}$$

whence, after C.B. MORREY [1],

(4.23) $$\| \xi \|_{C^{0,\delta_H}(\bar{\mathcal{O}}_H)} \leq C_H . \square$$

THEOREM 4.2: *Under the assumptions of Theorem 4.1, the increasing procedure* u_n *converges to* u. *Moreover,* u *is the minimum solution of* (4.5) *in* K, *and* u *is continuous on* $\bar{\mathcal{O}}$.

Proof: Since $u_n \leqslant u^0$ and u_n is increasing, we have $u_n(x) \uparrow \underline{u}(x)$. By Proposition 4.1

$$\|u_n\|_{C^{0,\delta_H}(\bar{\mathcal{O}}_H)} \leq C_H \quad \forall\ H .$$

Therefore

(4.24) $$\|u_n - \underline{u}\|_{C^0(\bar{\mathcal{O}}_H)} \to 0 , \quad \forall\ H \text{ fixed when } \quad n \to \infty .$$

Consequently $\underline{u}$ is continuous on $\bar{\mathcal{O}}$. Next we have

$$\begin{aligned} Mu_n(x) &\leq k + u_n(x) \leq k + u^0(x) \\ &= k + F^0 \beta_{-\mu_0}(x) . \end{aligned}$$

Thus for fixed x we can add the following constraint on the values of ξ:

$$c_0(\xi) \leq F^0 \beta_{-\mu_0}(x) .$$

Therefore there exists a Borel map $\xi_n(x)$ such that

$$Mu_n(x) = k + c_0(\xi_n(x)) + u_n(x+\xi_n(x)) \qquad \forall\ x ,$$

$$\xi_n(x) \geq 0, \quad \text{and}$$

$$c_0(\xi_n(x)) \leq F^0 \beta_{-\mu_0}(x) \qquad \forall\ x .$$

Therefore $|x| \leqslant H$ implies $|\xi_n(x)| \leqslant \lambda(H)$. By (4.24), therefore,

$$|u_n(x) + \xi_n(x)) - \underline{u}\ (x+ \xi_n(x))| \leq \rho_n(H + \lambda(H))$$

where ρ_n is a function which tends to 0 when n tends to $+\infty$ for every fixed value of the argument. Consequently we obtain

$$Mu_n(x) \geq k + c_0(\xi_n(x)) + \underline{u}(x+\xi_n(x)) - \rho_n(H + \lambda(H))$$
$$\forall x \in \mathcal{O}_H$$

whence, again,

$$Mu_n(x) \geq M\underline{u}(x) - \rho_n(H + \lambda(H))$$

and since

$$Mu_n(x) \leqslant M\underline{u}(x),$$

we obtain

$$(4.25) \qquad \|Mu_n - M\underline{u}\|_{C^0(\bar{\mathcal{O}}_H)} \to 0 \qquad \forall \text{ fixed } H \text{ when } n \to \infty.$$

We now consider the following functions

$$0 \leqslant \phi_H \leqslant 1, \quad \phi_H = 1 \text{ on } B_H(0),$$

ϕ_H compactly supported and regular, for example $\phi_H(x) = \tau(\frac{x}{H})$,

where $\tau = 1$ on B_1, $\tau = 0$ on $R^n - B_2$,

$$m_\varepsilon(x) = \begin{cases} 1 & \text{if } d(x,\Gamma) > 2\varepsilon, \quad x \in \mathcal{O} \\ \frac{1}{\varepsilon}(d(x,\Gamma)-\varepsilon) & \text{if } \varepsilon < d(x,\Gamma) \leq 2\varepsilon \\ 0 & \text{if } d(x,\Gamma) \leq \varepsilon . \end{cases}$$

If $v \in V^\mu = H_0^{1,\mu}$ then $vm_\varepsilon\phi_H \to v\phi_H$ in V^μ when $\varepsilon \to 0$ and $v\phi_H \to v$ in V^μ when $H \to \infty$.

We note also that

$$(v-\varepsilon)m_\varepsilon \phi_H \to v\phi_H \quad \text{in} \quad V^\mu \quad \text{when } \varepsilon \to 0.$$

We are now given $v \in V^\mu$ such that $v \leqslant M\underline{u}$. Then

$$\begin{aligned}(v-\varepsilon)m_\varepsilon\phi_H &\leq (M\underline{u} - \varepsilon)m_\varepsilon \phi_H \\ &= Mu_n m_\varepsilon \phi_H + (M\underline{u} - Mu_n - \varepsilon)m_\varepsilon \phi_H\end{aligned}$$

and

$$(M\underline{u}(x) - Mu_n(x))\,\phi_H(x) \leq \rho_n(2H + \lambda(2H))\phi_H(x) .$$

Having fixed ε and H, we can find $N(\varepsilon,H)$ such that for $n \geqslant N$ we have $\rho_n \leqslant \varepsilon$. Thus for $n \geqslant N(\varepsilon,H)$,

$$(v-\varepsilon)m_\varepsilon \phi_H \leq Mu_n m_\varepsilon \phi_H \leq Mu_n .$$

Consequently,

$$a_\mu(u_{n+1},(v-\varepsilon)m_\varepsilon\,\phi_H-u_{n+1}) + \lambda(u_{n+1},(v-\varepsilon)m_\varepsilon\,\phi_H-u_{n+1})_\mu \geq \geq (f+\lambda u_n,(v-\varepsilon)m_\varepsilon\,\phi_H- u_{n+1})_\mu \quad . \tag{4.26}$$

Now, $u_n \to \underline{u}$ in V^μ weakly, and strongly in H^μ. Making n tend to ∞ in (4.26), we obtain

$$a_\mu(\underline{u},(v-\varepsilon)m_\varepsilon\,\phi_H - \underline{u}) \geq (f,(v-\varepsilon)m_\varepsilon\,\phi_H - \underline{u})_\mu \quad .$$

We then make ε tend successively to 0, and then H to ∞, in order to deduce

$$a_\mu(\underline{u}, v-\underline{u}) \geq (f,v-\underline{u})_\mu$$

and since $\underline{u} \leqslant M\underline{u}$, we have shown that $\underline{u}$ is the solution of (4.5).

Let us show that $\underline{u}$ is the minimum solution. If u is the solution in K, then

$$u = T_\lambda u, \quad u \geqslant u_0 = 0.$$

If $u \geqslant u_n$, then

$$u = T_\lambda u \geqslant T_\lambda u_n = u_{n+1}.$$

Thus passing to the limit this yields $u \geqslant \underline{u}$.□

Remark 4.1

The maximum solution $\bar{u}$ is also a continuous function.

Remark 4.2

We can prove local regularity properties for $\underline{u}$ and $\bar{u}$ analogous to those proved in the bounded case.□

5. SEMI-GROUPS APPROACH

Method of Approach

In this Section we are going to consider V.I.'s and Q.V.I.'s corresponding to general semi-groups. This generalises what was done in the preceding Sections, in the measure that, at least implicitly, we worked with the semi-group whose infinitesimal generator (in a suitable sense) is the differential operator A.

5.1 *Variational Inequalities*

5.1.1 Notations: Assumptions

Let E be a topological space and ξ the Borel σ-algebra. We write

$B \equiv B(E)$ = space of bounded Borel functions on E,

$C \equiv C(E)$ = space of uniformly continuous and bounded functions on E.

We provide B with the Banach norm

$$||f|| = \sup_{x \in E} |f(x)|$$

and C is a Banach subspace of B.

We are now given a linear semi-group $\Phi(t)$ on B such that

$$\Phi(t+s) = \Phi(t)\,\Phi(s) \tag{5.1}$$

$$\Phi(0) = I \tag{5.2}$$

$$\|\Phi(t)1\| \leqq 1 \tag{5.3}$$

$$\Phi(t)f \geqq 0 \quad \forall\, f \in B,\ f \geqq 0\,. \tag{5.4}$$

From (5.3),(5.4) we immediately deduce that

$$\|\Phi(t)f\| \leqq \|f\| \quad \forall\, f \in B\,. \tag{5.5}$$

We also assume

$$\Phi(t): C \to C \tag{5.6}$$

$$\Phi(t)f \to f \quad \text{in } C \text{ when } t \downarrow 0,\ \forall\, f \in C. \tag{5.7}$$

It follows from (5.7) that $t \to \Phi(t)f$ is continuous from $[0,\infty[$ into C for all fixed f in C. We are also given

$$\alpha > 0, \qquad \psi \in C, \tag{5.8}$$

$$L \in B, \qquad t \to \Phi(t)L \quad \text{is measurable from } [0,\infty[\text{ into } C. \tag{5.9}$$

We call V.I. (variational inequality) by abuse of language, the system

$$\left|\begin{array}{l} u \in C\,, \quad u \leqq \psi \\ u \leqq \displaystyle\int_0^t e^{-\alpha s}\,\Phi(s)L\,ds + e^{-\alpha t}\,\Phi(t)u\,, \quad \forall\, t \geqq 0\,. \end{array}\right. \tag{5.10}$$

Our objective is to prove that the set of functions u satisfying (5.11) is non-empty and has a *maximum element*.

Remark 5.1

The terminology V.I. is improper, but it is kept by analogy with what was done in Section 1. We shall make this analogy explicit in the examples. Really, (5.11) is rather a set of *sub-solutions*.□

5.1.2 Preliminary results

LEMMA 5.1: *Let* $g \in B$ *be such that* $t \to \Phi(t)g$ *is measurable with values in* C, *and let* $w \in C$ *be such that*

$$w \leq \int_0^t e^{-\alpha s}\,\Phi(s)g\,ds + e^{-\alpha t}\,\Phi(t)w \quad \forall\, t \geqq 0\,. \tag{5.11}$$

Then for all $\beta > 0$ *we have*

$$w \leqq \int_0^t e^{-\beta s}\,\Phi(s)(g + (\beta-\alpha)w)\,ds + e^{-\beta t}\,\Phi(t)w \quad \forall\, t \geqq 0\,. \tag{5.12}$$

Finally, the function

$$w = \int_0^\infty e^{-\alpha t}\,\Phi(t)g\,dt \tag{5.13}$$

is the unique solution of the problem

(5.14)
$$\left| \begin{array}{l} w \in C \\ w = \int_0^t e^{-\alpha s}\Phi(s)g\,ds + e^{-\alpha t}\Phi(t)w \quad \forall\, t \geq 0 . \end{array} \right.$$

Proof: Let us prove (5.12). Multiply (5.11) by $e^{-(\beta-\alpha)t}$ and integrate over (0,T). We obtain

$$[1 - e^{-(\beta-\alpha)T}]w \leq \int_0^T e^{-\beta t}\Phi(t)w(\beta-\alpha)dt + $$
$$+ \int_0^T (\beta-\alpha)\, e^{-(\beta-\alpha)t}\left(\int_0^t e^{-\alpha s}\Phi(s)g\,ds\right)dt$$

and integrating the second integral by parts we obtain

$$w \leq \int_0^T e^{-\beta t}\,\Phi(t)(g + (\beta-\alpha)w)dt + e^{-(\beta-\alpha)T}\left(w - \int_0^T e^{-\alpha t}\Phi(t)g\,dt\right).$$

Now, by (5.11)

$$w - \int_0^T e^{-\alpha t}\,\Phi(t)g\,dt \leq e^{-\alpha T}\Phi(T)w$$

therefore

$$w \leq \int_0^T e^{-\beta t}\,\Phi(t)(g + (\beta-\alpha)w)dt + e^{-\beta T}\Phi(T)w ,$$

from which we deduce (5.12).

Furthermore, the integral (5.13) is well defined and $w \in C$. We further have

$$w = \int_0^t e^{-\alpha s}\Phi(s)g\,ds + \int_t^{+\infty} e^{-\alpha s}\Phi(s)g\,ds \quad =$$
$$= \int_0^t e^{-\alpha s}\Phi(s)g\,ds + \int_0^{\infty} e^{-\alpha(t+s)}\,\Phi(t+s)g\,ds$$

and by the semi-group property we deduce that w is the solution of (5.14). This solution is unique, for if $t \to +\infty$ in (5.14) we obtain the formula (5.13).□

Let $g \in C$. We shall say that g is a *regular function* if there exists G in C such that

(5.15)
$$g = \int_0^\infty e^{-\alpha t}\Phi(t)\;G\,dt .$$

By (5.14) we then have

$$g = \int_0^t e^{-\alpha s}\Phi(s)G\,ds + e^{-\alpha t}\Phi(t)g$$

so

$$\frac{g - \Phi(t)g}{t} = \frac{1}{t}\int_0^t e^{-\alpha s}\Phi(s)G\,ds + \frac{(e^{-\alpha t} - 1)\Phi(t)g}{t}$$

and by (5.7) it follows that

$$\frac{g - \Phi(t)g}{t} \to G - \alpha g \quad \text{in} \quad C .$$

Therefore g being regular means that g belongs to the *domain of the infinitesimal generator* of $\Phi(t)$ in C. If we write $-\mathcal{A}$ for this infinitesimal generator, we thus have

(5.16) $\mathcal{A}g + \alpha g = G \quad , \quad g \in D(\mathcal{A}) .$

We have:

LEMMA 5.2: *Let* $g \in C$. *Then there exists a sequence* g_n *of regular functions which converge to* g *in* C *(therefore* $D(\mathcal{A})$ *is dense in* C*).*

Proof: We set

$$g_n = n\int_0^\infty e^{-nt}\Phi(t)g\,dt = \int_0^\infty e^{-s}\Phi(\tfrac{s}{n})g\,ds$$

and

$$\|g_n - g\| \leq \int_0^\infty e^{-s}\|\Phi(\tfrac{s}{n})g - g\|ds .$$

But by (5.7),

$$\|\Phi(\tfrac{s}{n})g - g\| \to 0 \ , \quad \forall s \ ,$$

and remains bounded by $2\|g\|$. Lebesgue's Theorem then shows us that $\|g_n - g\| \to 0$. By Lemma 5.1 we can again write

$$g_n = \int_0^\infty e^{-\alpha t}\Phi(t)(ng + (\alpha - n)g_n)dt$$

and setting

$$G_n = ng + (\alpha - n)g_n \ ,$$

we see that g_n is a regular function.

THEOREM 5.1: *Assume* (5.1),(5.2),(5.3),(5.4),(5.6),(5.7),(5.8),(5.9). *Then the set of solutions of* (5.10) *is non-empty and has a maximum element.*□

The proof of Theorem 5.1 will be carried out in several steps.

5.1.3 Penalised Scheme

We now introduce the *penalised scheme* defined by

$$(5.17) \qquad u_\varepsilon = \int_0^\infty e^{-\alpha t}\,\Phi(t)\,\left(L - \frac{1}{\varepsilon}(u_\varepsilon - \psi)^+\right) dt\;, \quad u_\varepsilon \in C\;.$$

The integral in the second member of (5.17) has a meaning, because if $u_\varepsilon \in C$ then by (5.8) $(u_\varepsilon - \psi)^+ \in C$, whence by (5.7) $\Phi(t)(u_\varepsilon - \psi)^+$ is continuous from $[0,\infty[$ to values in C, and is bounded.

LEMMA 5.3: *There exists one and only one solution of* (5.17).

Proof: By Lemma 5.1, the equation (5.17) is equivalent to

$$(5.18) \qquad u_\varepsilon = \int_0^\infty e^{-(\alpha+\frac{1}{\varepsilon})t}\,\Phi(t)\,\left(L + \frac{1}{\varepsilon}u_\varepsilon - \frac{1}{\varepsilon}(u_\varepsilon - \psi)^+\right) dt$$
$$= \int_0^\infty e^{-(\alpha+\frac{1}{\varepsilon})t}\,\Phi(t)\,\left(L + \frac{1}{\varepsilon}u_\varepsilon \wedge \psi\right)dt\;.$$

We then define a mapping $T_\varepsilon : C \to C$ by

$$(5.19) \qquad T_\varepsilon\, z = \int_0^\infty e^{-(\alpha+\frac{1}{\varepsilon})t}\,\Phi(t)\,\left(L + \frac{1}{\varepsilon}z \wedge \psi\right)dt\;.$$

It is clear that the fixed points of T_ε coincide with the solutions of (5.18). It suffices to show that T_ε is a contraction. In fact,

$$T_\varepsilon\, z_1 - T_\varepsilon\, z_2 = \int_0^\infty e^{-(\alpha+\frac{1}{\varepsilon})t}\,\Phi(t)\,\left(\frac{1}{\varepsilon}z_1 \wedge \psi - \frac{1}{\varepsilon}z_2 \wedge \psi\right)dt$$

whence

$$\|T_\varepsilon\, z_1 - T_\varepsilon\, z_2\| \leq \frac{1}{\varepsilon}\,\|z_1 - z_2\| \int_0^\infty e^{-(\alpha+\frac{1}{\varepsilon})t}dt = \frac{\|z_1 - z_2\|}{1+\alpha\varepsilon}$$

whence the desired result.

LEMMA 5.4: *Lemma*

$$(5.20) \qquad u_\varepsilon \leqslant u_{\varepsilon'}\,, \quad \textit{if} \quad \varepsilon \leqslant \varepsilon'$$

and

$$(5.21) \qquad \|u_\varepsilon\| \leqslant K.$$

Proof: Now,

$$T_\varepsilon\, u_{\varepsilon'} = \int_0^\infty e^{-(\alpha+\frac{1}{\varepsilon})t}\,\Phi(t)\,\left(L + \frac{1}{\varepsilon}u_{\varepsilon'} \wedge \psi\right)dt$$

and by Lemma 5.1

$$u_{\varepsilon'} = \int_0^\infty e^{-(\alpha+\frac{1}{\varepsilon})t}\,\Phi(t)\,\left(L + \frac{1}{\varepsilon}u_{\varepsilon'} - \frac{1}{\varepsilon'}(u_{\varepsilon'}-\psi)^+\right)dt$$
$$= \int_0^\infty e^{-(\alpha+\frac{1}{\varepsilon})t}\,\Phi(t)\,\left(L + \frac{1}{\varepsilon}u_{\varepsilon'} \wedge \psi + (\frac{1}{\varepsilon}-\frac{1}{\varepsilon'})(u_{\varepsilon'}-\psi)^+\right)dt$$

and therefore

$$T_\varepsilon u_{\varepsilon'} - u_{\varepsilon'} = \int_0^\infty e^{-(\alpha+\frac{1}{\varepsilon})t}\,\Phi(t)\,(\frac{1}{\varepsilon'}-\frac{1}{\varepsilon})\,(u_{\varepsilon'}-\psi)^+dt \leqq 0 \quad \text{if} \quad \varepsilon \leqq \varepsilon' .$$

Since T_ε is monotonically increasing, from this we deduce that

$$T_\varepsilon^n u_{\varepsilon'} \leqslant u_{\varepsilon'},$$

and by making n tend to ∞ we obtain (5.20). Furthermore,

$$u_\varepsilon \leqq \int_0^\infty e^{-\alpha t}\Phi(t)\,L\,dt \leqq \frac{\|L\|}{\alpha}$$

and therefore in order to prove (5.21) it suffices us to prove an inequality in the opposite direction. Now,

$$T_\varepsilon 0 = \int_0^\infty e^{-(\alpha+\frac{1}{\varepsilon})t}\,\Phi(t)\,(L - \frac{1}{\varepsilon}\psi^-)dt$$
$$\geqq - \int_0^\infty e^{-(\alpha+\frac{1}{\varepsilon})t}\,\Phi(t)\,(L^- + \frac{1}{\varepsilon}\psi^-)dt$$
$$\geqq - \frac{\varepsilon\|L^-\|+\|\psi^-\|}{1+\varepsilon\alpha} \geqq - (\|\psi^-\| + \frac{\|L^-\|}{\alpha}) = -k .$$

Let us now assume that $z \geqslant -k$. Since $\psi \geqslant -k$ we deduce that

$$T_\varepsilon z \geqq \int_0^\infty e^{-(\alpha+\frac{1}{\varepsilon})t}\,\Phi(t)\,(-\|L^-\| - \frac{1}{\varepsilon}k)dt \geqq - \frac{\|L^-\|}{\alpha+\frac{1}{\varepsilon}} - \frac{k}{\varepsilon(\alpha+\frac{1}{\varepsilon})} =$$
$$= - \frac{\varepsilon\|L^-\|+k}{\alpha\varepsilon+1}$$
$$= -\left(\frac{\|L^-\|}{\alpha} + \frac{\|\psi^-\|}{\alpha\varepsilon+1}\right) \geqq -k .$$

Therefore

$$T_\varepsilon^n\, 0 \geqslant -k,$$

and in the limit

$$u_\varepsilon \geqslant -k. \square$$

LEMMA 5.5: *Let* u *be a solution of* (5.10). *Then*

(5.22) $$u \leqslant u_\varepsilon .$$

Proof: Applying Lemma 5.1 (property (5.12)), we have

$$u \leqq \int_0^\infty e^{-(\alpha+\frac{1}{\varepsilon})t}\, \Phi(t)\, (L + \frac{1}{\varepsilon} u)dt .$$

Also, by (5.19),

$$T_\varepsilon u = \int_0^\infty e^{-(\alpha+\frac{1}{\varepsilon})t}\, \Phi(t)\, (L + \frac{1}{\varepsilon} u \wedge \psi)dt$$

$$T_\varepsilon u - u \geqq \int_0^\infty e^{-(\alpha+\frac{1}{\varepsilon})t}\, \Phi(t)\, \frac{1}{\varepsilon}(u \wedge \psi - u)dt .$$

But $u \leqslant \psi$, therefore $u \wedge \psi = u$. Consequently,

$$T_\varepsilon u - u \geqslant 0,$$

and

$$T_\varepsilon^n u \geqslant u,$$

and making n tend to ∞, we obtain (5.22).□

LEMMA 5.6: *Let* $\tilde{\psi}$ *satisfy the same assumptions as* ψ, *and* $\tilde{u}_\varepsilon$ *be the solution of* (5.17) *corresponding to* $\tilde{\psi}$ *instead of* ψ. *Then*

(5.23) $$\|u_\varepsilon - \tilde{u}_\varepsilon\| \leqq \|\psi - \tilde{\psi}\| .$$

Proof: Let $z, \tilde{z} \in C$ satisfy

(5.24) $$\|z - \tilde{z}\| \leqq \|\psi - \tilde{\psi}\| .$$

If $\tilde{T}_\varepsilon$ is the mapping (5.19), with $\tilde{\psi}$ replacing ψ, then

(5.25) $$\|T_\varepsilon z - \tilde{T}_\varepsilon \tilde{z}\| \leqq \|\psi - \tilde{\psi}\| .$$

In fact,

$$T_\varepsilon z - \tilde{T}_\varepsilon \tilde{z} = \int_0^\infty e^{-(\alpha+\frac{1}{\varepsilon})t}\, \Phi(t)\, \frac{1}{\varepsilon}(\psi \wedge z - \tilde{\psi} \wedge \tilde{z})dt .$$

But from (5.24) it follows that

$$\|\psi \wedge z - \tilde{\psi} \wedge \tilde{z}\| \leqq \|\psi - \tilde{\psi}\|$$

and hence (5.25). Now,

$$T_\varepsilon\, 0 - \tilde{T}_\varepsilon\, 0 = \int_0^\infty e^{-(\alpha+\frac{1}{\varepsilon})t}\, \Phi(t)\, \frac{1}{\varepsilon}(\tilde{\psi}^- - \psi^-)dt$$

and therefore

$$\|T_\varepsilon 0 - \tilde{T}_\varepsilon 0\| \leq \|\psi - \tilde{\psi}\| .$$

On applying (5.25) with $z = T_\varepsilon 0$, $\tilde{z} = \tilde{T}_\varepsilon 0$, we obtain

$$\|T_\varepsilon^2 0 - \tilde{T}_\varepsilon^2 0\| \leq \|\psi - \tilde{\psi}\|$$

and quite generally we have

$$\|T_\varepsilon^n 0 - \tilde{T}_\varepsilon^n 0\| \leq \|\psi - \tilde{\psi}\|$$

then by making n tend to ∞ we deduce (5.23).□

LEMMA 5.7:

(5.26) $u_\varepsilon \downarrow u$ *and* $\|u_\varepsilon - u\| \to 0.$

Proof: By Lemma 5.4 we see that $u_\varepsilon \downarrow u$ pointwise, and $u \in B$. To prove the uniform convergence it suffices to consider the case where ψ is regular. In fact, let us assume the result is proved for $\psi \in D(\mathcal{A})$. Let $\psi \in C$. By Lemma 5.2 there exists a sequence of regular functions ψ_n such that

$$\|\psi_n - \psi\| \to 0 .$$

Let $u_{\varepsilon n}$ be the solution of the penalised problem corresponding to ψ_n. Then

$$u_{\varepsilon n} \downarrow u_n, \quad u_n \in B.$$

Furthermore, by Lemma 5.6 we have

$$\|u_{\varepsilon n} - u_\varepsilon\| \leq \|\psi_n - \psi\|$$

and therefore, making ε tend to 0,

$$\|u_n - u\| \leq \|\psi_n - \psi\| .$$

Now,

(5.27) $$\|u_\varepsilon - u\| \leq \|u_\varepsilon - u_{\varepsilon n}\| + \|u_{\varepsilon n} - u_n\| + \|u_n - u\| \leq 2\|\psi_n - \psi\| + \|u_{\varepsilon n} - u_n\| .$$

Because ψ_n is regular,

$$\|u_{\varepsilon n} - u_n\| \to 0 \quad \text{when } \varepsilon \to 0, \text{ with n fixed.}$$

From this and (5.27) we easily deduce the second part of (5.26). Now we assume that $\psi \in D(\mathcal{A})$, so

$$\psi = \int_0^\infty e^{-\alpha t} \, \Phi(t) \wedge dt \qquad = \int_0^t e^{-ds} \, \Phi(s) \wedge ds + e^{-\alpha t} \Phi(t)\psi .$$

Consequently,

$$u_\varepsilon - \psi = \int_0^\infty e^{-\alpha t}\Phi(t)\,(L - \wedge - \frac{1}{\varepsilon}(u_\varepsilon - \psi)^+)dt \,.$$

Let us set

$$\tilde{L} = L - \Lambda, \quad \tilde{u}_\varepsilon = u_\varepsilon - \psi,$$

then $\tilde{u}_\varepsilon$ is the solution of the equation

$$(5.28) \qquad \begin{aligned} \tilde{u}_\varepsilon &= \int_0^\infty e^{-\alpha t}\Phi(t)\,(\tilde{L} - \frac{1}{\varepsilon}\tilde{u}_\varepsilon^{\,+})dt \\ &= \int_0^\infty e^{-(\alpha+\frac{1}{\varepsilon})t}\Phi(t)\,(\tilde{L} + \frac{1}{\varepsilon}\tilde{u}_\varepsilon - \frac{1}{\varepsilon}\tilde{u}_\varepsilon^{\,+})dt \\ &\leqq \int_0^\infty e^{-(\alpha+\frac{1}{\varepsilon})t}\Phi(t)\,\tilde{L}^+dt \leqq \frac{\varepsilon\|\tilde{L}^+\|}{1+\alpha\varepsilon} \end{aligned}$$

therefore

$$\frac{\tilde{u}_\varepsilon}{\varepsilon} \leqslant \|\tilde{L}^+\|,$$

and consequently, also,

$$(5.29) \qquad \frac{(\tilde{u}^\varepsilon)^+}{\varepsilon} \leqslant \|\tilde{L}^+\|.$$

But (cf. Lemma 5.4)

$$T_\varepsilon u_{\varepsilon'} - u_{\varepsilon'} = \int_0^\infty e^{-(\alpha+\frac{1}{\varepsilon})t}\,\Phi(t)\,(\frac{1}{\varepsilon'} - \frac{1}{\varepsilon})\,\tilde{u}_{\varepsilon'}^{\,+}\,dt$$

and by (5.29)

$$(5.30) \qquad T_\varepsilon u_{\varepsilon'} - u_{\varepsilon'} \geqq -\varepsilon'\|\tilde{L}^+\| \,.$$

Next we have

$$(5.31) \qquad \begin{aligned} T_\varepsilon^2 u_{\varepsilon'} - \psi &= \int_0^\infty e^{-(\alpha+\frac{1}{\varepsilon})t}\,\Phi(t)\,(\tilde{L} - \frac{1}{\varepsilon}\psi + \frac{1}{\varepsilon}T_\varepsilon u_{\varepsilon'} \wedge \psi)\,dt \\ &= \int_0^\infty e^{-(\alpha+\frac{1}{\varepsilon})t}\,\Phi(t)\,(\tilde{L} - \frac{1}{\varepsilon}(T_\varepsilon u_{\varepsilon'} - \psi)^-)dt \,. \end{aligned}$$

Now, by (5.30)

$$T_\varepsilon u_{\varepsilon'} - \psi \geqq \tilde{u}_{\varepsilon'} - \varepsilon'\|\tilde{L}^+\|$$

which is negative, by (5.29). Consequently

$$-(T_\varepsilon u_{\varepsilon'} - \psi)^- \geqq \tilde{u}_{\varepsilon'} - \varepsilon'\|\tilde{L}^+\|$$

whence

$$(5.32)\qquad T_\varepsilon^2 u_{\varepsilon'} - \psi \geqq \int_0^\infty e^{-(\alpha+\frac{1}{\varepsilon})t}\,\Phi(t)\,(\tilde{L} + \frac{1}{\varepsilon}\tilde{u}_{\varepsilon'} - \frac{\varepsilon'}{\varepsilon}\|\tilde{L}^+\|)dt\ .$$

But by (5.28)

$$\tilde{u}_{\varepsilon'} = \int_0^\infty e^{-(\alpha+\frac{1}{\varepsilon})t}\,\Phi(t)\,(\tilde{L} + \frac{1}{\varepsilon}\tilde{u}_{\varepsilon'} - \frac{1}{\varepsilon'}\tilde{u}_{\varepsilon'}{}^+)dt$$

therefore

$$\int_0^\infty e^{-(\alpha+\frac{1}{\varepsilon})t}\,\Phi(t)\,(\tilde{L} + \frac{1}{\varepsilon}\tilde{u}_{\varepsilon'})dt \geqq \tilde{u}_{\varepsilon'}\ .$$

From (5.32) it then follows that

$$\begin{aligned} T_\varepsilon^2 u_{\varepsilon'} - \psi &\geqq \tilde{u}_{\varepsilon'} - \frac{\varepsilon'}{\varepsilon}\int_0^\infty e^{-(\alpha+\frac{1}{\varepsilon})t}\,\Phi(t)\,\|\tilde{L}^+\|dt \\ &\geqq \tilde{u}_{\varepsilon'} - \varepsilon'\|\tilde{L}^+\| \end{aligned}$$

so again

$$T_\varepsilon^2 u_{\varepsilon'} - u_{\varepsilon'} \geqq -\varepsilon'\|\tilde{L}^+\|\ .$$

We can then iterate and show that

$$T_\varepsilon^k u_{\varepsilon'} - u_{\varepsilon'} \geqq -\varepsilon'\|\tilde{L}^+\|\ .$$

Making k tend to $+\infty$, we obtain

$$(5.33)\qquad u_\varepsilon - u_{\varepsilon'} \geqq -\varepsilon'\|\tilde{L}^+\| \quad \text{if} \quad \varepsilon < \varepsilon'\ .$$

We now make ε tend to 0, leaving ε' fixed, and this yields

$$u \geqq u_{\varepsilon'} - \varepsilon'\|\tilde{L}^+\|\ .$$

Hence we have proved that in the case where $\psi \in D(A)$ we have the following estimate

$$(5.34)\qquad \|u_\varepsilon - u\| \leqq \varepsilon\|\tilde{L}^+\|$$

which completes the proof of (5.26).□

Proof of Theorem 5.1: Because of the uniform convergence, $u \in C$. Now make ε tend to 0 in (5.17) after multiplying by ε. We then obtain

$$\int_0^\infty e^{-\alpha t}\Phi(t)\ (u-\psi)^+ dt = 0$$

therefore

$$\frac{1}{t}\int_0^t e^{-\alpha s}\Phi(s)\ (u-\psi)^+ dt = 0$$

and making t tend to 0, and using the continuity of $\Phi(s)(u - \psi)^+$ with respect to s, we obtain

$$(u - \psi)^+ = 0.$$

Furthermore, from (5.17) it follows that

$$u_\varepsilon = \int_0^t e^{-\alpha s}\Phi(s)\ (L - \frac{1}{\varepsilon}(u_\varepsilon - \psi)^+)ds + e^{-\alpha t}\Phi(t)u_\varepsilon$$

$$\leq \int_0^t e^{-\alpha s}\Phi(s)\ L\,ds + e^{-\alpha t}\Phi(t)u_\varepsilon$$

and making ε tend to 0 we obtain

$$u \leq \int_0^t e^{-\alpha s}\Phi(s)\ L\,ds + e^{-\alpha t}\Phi(t)u$$

and thus u is an element of the set (5.10). By Lemma 5.5 it is the maximum element.□

Remark 5.2

Theorem 5.1 completes some results of BENSOUSSAN-LIONS [1], M. ROBIN [3], J.-L. MENALDI [2] in which probabilistic methods are used. It is given in A. BENSOUSSAN [1].□

Remark 5.3

In the applications, it is useful to make the following remark. Let C_o be a closed subspace of C. Assume that

(5.35) $\Phi(t) : C_o \to C_o$

(5.36) $t \to \Phi(t)L$ is measurable, with values in C_o

(5.37) $\forall\ \varphi \in C_o\ ,\ \ (\varphi - \psi)^+ \in C_o$.

Under these conditions the maximum solution of (5.10) belongs to C_o. In fact it suffices to show that $u_\varepsilon \in C_o$. Now, the mapping T_ε maps C_o into C_o, thanks to the assumptions, therefore its unique fixed point belongs to C_o, since it is a contraction.

5.1.4 Examples

Let $a_{ij}(x), a_i(x)$ be functions defined on R^n such that

(5.38)
$$\begin{cases} a_{ij}\,,\ a_i \text{ are Borel and bounded} \\ a_{ij} \in W^{1,\infty}(R^n)\,, \quad a_{ij} = a_{ji} \\ \sum a_{ij}(x)\,\xi_i\,\xi_j \geq \gamma|\xi|^2 \quad \forall\, \xi \in R^n\,, \quad \gamma > 0\,. \end{cases}$$

Set

(5.39)
$$A = -\sum \frac{\partial}{\partial x_i}\, a_{ij}\, \frac{\partial}{\partial x_j} + \sum a_i\, \frac{\partial}{\partial x_i}\,.$$

Let $\mathcal{O}$ be a regular, bounded, open set of R^n.

Dirichlet's Problem

Take $E = \bar{\mathcal{O}}$. Let $f \in B$. We solve the non-homogeneous Dirichlet problem

(5.40)
$$\begin{cases} \dfrac{\partial z}{\partial t} + Az = 0 \\ z|_\Sigma = f|_\Sigma \qquad\qquad \Sigma = \partial\mathcal{O} \times\ ]0,T[\\ z(x,0) = f(x)\,. \end{cases}$$

In fact we know how to solve (5.40) for $f \in \mathcal{B}(\bar{\mathcal{O}})$, and we set

(5.41)
$$z(x,t) = \Phi(t)f(x);$$

$z(\cdot,t) \in C$ in particular. Next we define (5.41) for $f \in C$, by showing that if $f_n \in \mathcal{D}(\bar{\mathcal{O}})$ converges to f in C, then $\Phi(t)f_n \to \Phi(t)f$ in C. We have defined thus a semi-group on C, which satisfies all the properties (5.1) to (5.7). Since

(5.42)
$$\Phi(t)f(x) = \int_E P(x,t;dy)f(y)$$

where $P(x,t;dy)$ is a probability measure, the formula (5.42) extends to $f \in B$. Let us set $C_o = \{f \in C \mid f|_{\partial\mathcal{O}} = 0\}$, which is a Banach subspace of C. It is clear that by (5.40) $\Phi(t): C_o \to C_o$.

Let $f \in B$, and let us set

(5.43)
$$L = f\chi_{\mathcal{O}}\,.$$

By (5.40),

$$z(x,t) = \Phi(t)\,f\chi_{\mathcal{O}}(x) = \zeta(x,t) \quad \forall\, t > 0$$

where ζ is the solution of the homogeneous Dirichlet problem

(5.44)
$$\begin{cases} \dfrac{\partial \zeta}{\partial t} + A\zeta = 0 \\ \zeta|_\Sigma = 0\,, \quad \zeta(x,0) = f(x) \end{cases}$$

and for $t > 0$, $\zeta(\cdot,t) \in C_o$.

Next we are given

$$\psi \in C, \quad \psi|_\Gamma \geq 0,$$

so that (5.37) is satisfied.

For $t > 0$ equation (5.44) defines a mapping

$$\tilde{\Phi}(t) : B \to C_o , \quad \tilde{\Phi}(t)f(x) = \zeta(x,t) . \tag{5.45}$$

Now,

$$\tilde{\Phi}(t)f = \Phi(t)f\chi_{\mathcal{O}} \tag{5.46}$$

so that $\tilde{\Phi}$ and Φ coincide on C_o. Thus we have solved the problem

$$\left| \begin{aligned} & u \in C_o , \quad u \leq \psi \\ & u \leq \int_0^t e^{-\alpha s} \tilde{\Phi}(s) f\, ds + e^{-\alpha t} \tilde{\Phi}(t) u \quad \forall\, t > 0 . \end{aligned} \right. \tag{5.47}$$

The set of u's satisfying (5.47) is non-empty and has a maximum element.

Neumann's Problem

We take $E = \bar{\mathcal{O}}$. Let $f \in B$. We solve the Neumann problem

$$\left| \begin{aligned} & \frac{\partial z}{\partial t} + Az = 0 \qquad \left.\frac{\partial z}{\partial \nu_A}\right|_\Sigma = 0 , \\ & z(x,0) = f(x) . \end{aligned} \right. \tag{5.48}$$

For $f \in B$, $z(x,t) \in C$ for all $t > 0$, we set

$$z(x,t) = \Phi(t)f(x).$$

We easily prove all the desired properties for $\Phi(t)$.

Problem on the Whole Space

We take $E = R^n$, and for $f \in B$ we solve

$$\frac{\partial z}{\partial t} + Az = 0 , \quad z(x,0) = f(x) . \tag{5.49}$$

If $f \in C$ (i.e., f is uniformly continuous and bounded), then $z(\cdot,t) \in C$ for all $t \geq 0$. We prove this, at least with the additional assumption

$$a_{ij} \in W^{2,\infty} , \quad a_i \in W^{1,\infty} \tag{5.50}$$

and using the probabilistic interpretation. In fact, if (5.50) is satisfied, then we can consider the Ito equation

$$dy = g(y)dt + \sigma(y)dw , y(0) = x \tag{5.51}$$

where

$$\frac{\sigma\sigma^*}{2} = a \equiv a_{ij}, \qquad g_j = \sum_i \frac{\partial a_{ij}}{\partial x_i} - a_j .$$

We then have

$$z(x,t) = Ef(y_x(t)),$$

and we easily verify that

$$E|y_x(t) - y_{x'}(t)|^2 \leq C_0|x - x'|^2 e^{C_1 t} .$$

Let

$$\rho(\delta) = \sup_{|x-x'| \leq \delta} |f(x) - f(x')| \to 0 \quad \text{when} \quad \delta \to 0 .$$

Then

$$|z(x,t) - z(x',t)| \leq \rho(\delta) + \frac{2\|f\|}{\delta^2} E|y_x(t) - y_{x'}(t)|^2$$
$$\leq \rho(\delta) + \frac{2\|f\|}{\delta^2} C_0|x - x'|^2 e^{C_1 t}$$

whence the result.

Relations with the Usual V.I.'s

We prove this in the case of the Dirichlet problem (5.47). First let $g \in D(\mathcal{A})$, therefore there exists $G \in C$ such that

$$g = \int_0^\infty e^{-\alpha t} \Phi(t) G\, dt = \int_0^\infty e^{-\alpha t} z(x,t) dt$$

where

$$\frac{\partial z}{\partial t} + Az = 0, \quad z|_\Sigma = G|_\Sigma, \quad z(x,0) = G .$$

From this it easily follows that g is the solution of the problem

$$\text{(5.52)} \quad \left| \begin{array}{lll} Ag + \alpha g = G & \text{in} & \mathcal{O} \\ \alpha g = G & \text{on} & \partial\mathcal{O} . \end{array} \right.$$

But the penalised problem is then the following:

$$\text{(5.53)} \quad \left| \begin{array}{lll} Au_\epsilon + \alpha u_\epsilon + \frac{1}{\epsilon}(u_\epsilon - \psi)^+ - f & \text{in} & \mathcal{O} \\ u_\epsilon = 0 & \text{on} & \partial\mathcal{O} . \end{array} \right.$$

We rediscover, thus, the penalised problem which approximates the V.I.

$$\text{(5.54)} \quad \left| \begin{array}{l} a(u,v-u) + \alpha(u,v-u) \geq (f,v-u) \\ \forall\ v \in H_0^1 , \quad v \leq \psi , \quad u \in H_0^1 , \quad u \leq \psi \end{array} \right.$$

where

$$a(u,v) = \langle Au,v \rangle.$$

Therefore the maximum solution of (5.47) coincides with the solution of (5.54).□

5.2 *Quasi-Variational Inequality (Q.V.I.)*

We now study the case where the obstacle ψ in (5.10) is replaced by the variable obstacle Mu. We therefore have the Q.V.I.

$$(5.55)\quad \begin{cases} u \in C , \quad u \leq Mu \\ u \leq \int_0^t e^{-\alpha} \Phi(s) L\, ds + e^{-\alpha t} \Phi(t) u . \end{cases}$$

where we have assumed (5.1),...,(5.4),(5.6),(5.7),(5.9). We further assume

$$(5.57)\quad \begin{cases} M \text{ is an increasing operator (in the sense } M\varphi_1 \leqslant M\varphi_2 \text{ if } \varphi_1 \leqslant \varphi_2), \\ \text{and is concave, } M:C \to C \text{ is Lipschitz continuous,} \end{cases}$$

$$(5.57)\quad \begin{cases} L \geq 0 , \quad \alpha > 0 \\ M(0) \geq k > 0 . \end{cases}$$

We have:

THEOREM 5.2: *Take the assumptions* (5.1),...,(5.4),(5.6),(5.7),(5.9) *and* (5.56), (5.57). *Then the set of solutions of* (5.55) *is non-empty and has a maximum element, which is a positive function.*

Proof: We adopt the methods used in Section 1 (in particular, Section 1.3).

Let $z \in C$, and define $Tz = \zeta$ as the *maximum* solution of

$$(5.58)\quad \begin{cases} \zeta \in C , \quad \zeta \leq Mz \\ \zeta \leq \int_0^t e^{-\alpha s} \Phi(s) L\, ds + e^{-\alpha t} \Phi(t) \zeta . \end{cases}$$

The mapping T is well defined, by Theorem 5.1.

First of all we show that

(5.59) T is increasing and concave.

It will be convenient to set

(5.60) $T = \sigma o M$

where $\sigma(\psi):C \to C$ is the maximum solution of (5.10). Then (5.59) follows from the assumptions on M and from

(5.61) σ is increasing and concave.

Let

$$\psi_1, \psi_2 \in C, \quad \text{and} \quad u_1 = \sigma(\psi_1), \quad u_2 = \sigma(\psi_2).$$

Let us assume that $\psi_1 \leqslant \psi_2$, then

$$u_1 \leqq \psi_2 \ , \quad u_1 \leqq \int_0^t e^{-\alpha s}\Phi(s)L\,ds + e^{-\alpha t}\Phi(t)u_1$$

therefore

$$u_1 \leqslant \sigma(\psi_2) = u_2.$$

Moreover,

$$\theta u_1 + (1-\theta)u_2 \leqq \theta\psi_1 + (1-\theta)\psi_2$$

$$\theta u_1 + (1-\theta)u_2 \leqq \int_0^t e^{-\alpha s}\Phi(s)L\,ds + e^{-\alpha t}\Phi(t)(\theta u_1 + (1-\theta)u_2)$$

and therefore

$$\theta u_1 + (1-\theta)u_2 \leqq \sigma(\theta\psi_1 + (1-\theta)\psi_2)$$

which proves (5.61).

Next we define

$$u^o = \int_0^\infty e^{-\alpha t}\Phi(t)L\,dt \tag{5.62}$$

and $u^o \in C$, $u^o \geqslant 0$. Let $1 > \mu > 0$ be such that

$$\mu\|u^o\| \leqslant k. \tag{5.63}$$

Then

$$T(o) \geqslant \mu u^o. \tag{5.64}$$

In fact,

$$\mu u^o \leqslant M(o),$$

and by (5.62)

$$\mu u^o = \int_0^t e^{-\alpha s}\Phi(s)\,\mu L\,ds + e^{-\alpha t}\Phi(t)\mu u^o$$
$$\leqq \int_0^t e^{-\alpha s}\Phi(s)L\,ds + e^{-\alpha t}\Phi(t)\,\mu u^o$$

since $L \geqslant 0$, whence we deduce (5.64).

From this we then deduce, as in Lemma 1.4, the following property

(5.65) Let $z, \tilde{z} \in C$ et $\gamma \in [0,1]$ such that $z(x) - \tilde{z}(x) \leqq \gamma z(x)\ \forall\, x$.
Then $Tz(x) - T\tilde{z}(x) \leqq \gamma(1-\mu)Tz(x) \quad \forall\, x$.

After having noted that

(5.66) $\quad Tz \leqslant u^o$ for all z,

we deduce from this, as in Lemma 1.5, that T has one and only one fixed point. This fixed point is obviously an element of (5.55). It is the maximum element, for if $\tilde{u}$ satisfies (5.55), then

$$\tilde{u} \leqslant T\tilde{u}, \quad \text{and} \quad \tilde{u} \leqslant u^o,$$

therefore

$$\tilde{u} \leqslant Tu^o,$$

$$\tilde{u} \leqslant T\tilde{u} \leqslant T^2u^o,$$

whence

$$\tilde{u} \leqslant T^n u^o \to u \quad \text{when } n \to \infty, \text{ and } u \geqslant 0.$$

Remark 5.4

If (5.35),(5.36) are satisfied, together with

(5.67) $\quad (\varphi - M\tilde{\varphi})^+ \in C_o \quad \forall\, \varphi \in C_o \;,\; \tilde{\varphi} \in C \;,\; \varphi, \tilde{\varphi} \geqq 0$

then $u \in C_o$. In particular this is the case in the example of the Dirichlet problem.□

5.3 *Discretisation*

We are going to study the analogue of the problems (5.10) and (5.55) in discrete time, and prove some results on convergence, due to A. BENSOUSSAN and M. ROBIN [1].

5.3.1 V.I. in discrete time

Here we take the assumptions of Theorem 5.1, replacing (5.9) by

(5.68) $\quad L \in C.$ (1)

Let h be a parameter that tends to 0. We give the name of discretised V.I. to the problem

(5.69) $\quad u_h = \text{Min}\,[\psi,\, hL + e^{-\alpha h}\Phi(h)u_h],\; u_h \in C\,.$

We can immediately verify that the equation (5.69) has one and only one solution, because the mapping

$$z \to \text{Min}\,[\psi,\, hL + e^{-\alpha h}\Phi(h)z]$$

(1) If this assumption is not satisfied, in what follows it is necessary to replace L by $L_h = \frac{1}{h}\int_0^h e^{-\alpha t}\Phi(t)L\,dt$. We obtain the same results.

is a contraction on C.

Our objective is to prove:

THEOREM 5.3: *Assume* (5.1),(5.2),(5.3),(5.4),(5.6),(5.7),(5.8),(5.68). *Then the solution* u_h *of* (5.69) *converges in* C *to* u *the maximum solution of* (5.10).□

We begin with some preliminary results. We write as u_h^o the solution of the linear equation

(5.70) $$u_h^o = hL + e^{-\alpha h}\Phi(h)u_h^o\ ,\quad u_h^o \in C\ .$$

LEMMA 5.8:

(5.71) $$u_h^o \to u^o \quad \textit{in } C.$$

Proof: It is clear that

$$u_h^o = \sum_{n=0}^{\infty} e^{-n\Phi h}\,\Phi(nh)hL$$

and therefore

$$u^o - u_h^o = \sum_{n=0}^{\infty} he^{-n\alpha h}\,\Phi(nh)\ \frac{\left[\int_0^h \left(e^{-\alpha s}\Phi(s)L - L\right)ds\right]}{h}$$

so

$$\|u^o - u_h^o\| \le \frac{\left\|\int_0^h \left(e^{-\alpha s}\Phi(s)L - L\right)ds\right\|}{h}\ \frac{h}{1-e^{-\alpha h}}$$

$$\to 0 \quad \text{when } h \to 0,$$

since $s \to e^{-\alpha s}\Phi(s)L$ is continuous from $[0,\infty[$ with values in C. □

We now define a penalised problem related to the discretised problem (5.69), namely

(5.72) $$u_h^\varepsilon = hL - \frac{h}{\varepsilon}(u_h^\varepsilon - \psi)^+ + e^{-\alpha h}\Phi(h)u_h^\varepsilon\ ,\quad u_h^\varepsilon \in C\ .$$

First we study the problem (5.72) and its properties (analogues of those obtained in the continuous case). We introduce the mapping

(5.73) $$T_h^\varepsilon z = \sum_{n=0}^{\infty} \left(\frac{\varepsilon}{h+\varepsilon}\right)^{n+1} e^{-n\alpha h}\,\Phi(nh)\left[hL + h\,\frac{\psi \wedge z}{\varepsilon}\right]$$

on C. We show that T^ε is a contraction, and thus has one and only one fixed point solution of (5.72). In fact,

$$\|T_h^\varepsilon z_1 - T_h^\varepsilon z_2\| \leq \sum_{n=0}^{\infty} \left(\frac{\varepsilon}{h+\varepsilon}\right)^{n+1} e^{-n\alpha h} \frac{h}{\varepsilon} \|z_1 - z_2\|$$

$$\leq \frac{h\|z_1 - z_2\|}{h+\varepsilon-\varepsilon e^{-\alpha h}} = \frac{\|z_1 - z_2\|}{1+\varepsilon \dfrac{1-e^{-\alpha h}}{h}}$$

and since

$$\frac{1-e^{-\alpha h}}{h} \geqq \alpha e^{-\alpha h} \geqq \alpha e^{-\alpha} \quad \text{for} \quad h < 1 ,$$

we obtain

$$\|T_h^\varepsilon z_1 - T_h^\varepsilon z_2\| \leq \|z_1 - z_2\| \frac{1}{1+\varepsilon \dfrac{e^{-\alpha}}{\alpha}} .$$

By (5.70) the fixed point u_h^ε is the solution of

$$u_h^\varepsilon = \left(hL + h \frac{\psi \wedge u_h^\varepsilon}{\varepsilon}\right) \frac{\varepsilon}{h+\varepsilon} + e^{-\alpha h} \frac{\varepsilon \Phi(h)}{h+\varepsilon} u_h^\varepsilon \tag{5.74}$$

$$= \frac{\varepsilon hL}{h+\varepsilon} + \frac{h}{h+\varepsilon} u_h^\varepsilon - \frac{h}{h+\varepsilon} (u_h^\varepsilon - \psi)^+ + \frac{\varepsilon}{h+\varepsilon} e^{-\alpha h} \Phi(h) u_h^\varepsilon .$$

whence we easily deduce (5.72).

LEMMA 5.9:

$$u_h^\varepsilon \leqslant u_h^{\varepsilon'} \quad \textit{if } \varepsilon \leqslant \varepsilon' \tag{5.75}$$

and

$$\|u_h^\varepsilon\| \leqslant K. \tag{5.76}$$

Proof: Now,

$$T_h^\varepsilon u_h^{\varepsilon'} = \sum_{n=0}^{\infty} \left(\frac{\varepsilon}{h+\varepsilon}\right)^{n+1} e^{-n\alpha h} \Phi(nh) \left[hL + h \frac{\psi \wedge u_h^{\varepsilon'}}{\varepsilon}\right] .$$

Also, by (5.72)

$$u_h^{\varepsilon'} \frac{\varepsilon}{h+\varepsilon} = \frac{\varepsilon hL}{h+\varepsilon} - \frac{h}{\varepsilon'} \frac{\varepsilon}{h+\varepsilon} (u_h^{\varepsilon'} - \psi)^+ + \frac{\varepsilon}{h+\varepsilon} e^{-\alpha h} \Phi(h) u_h^{\varepsilon'}$$

so

$$u_h^{\varepsilon'} = \frac{\varepsilon hL}{h+\varepsilon} + \frac{h}{h+\varepsilon} u_h^{\varepsilon'} - \frac{h}{h+\varepsilon} \frac{\varepsilon}{\varepsilon'} (u_h^{\varepsilon'} - \psi)^+ + \frac{\varepsilon}{h+\varepsilon} e^{-\alpha h} \Phi(h) u_h^{\varepsilon'} =$$

$$= \frac{\varepsilon hL}{h+\varepsilon} + \frac{h}{h+\varepsilon} \psi \wedge u_h^{\varepsilon'} + \frac{h}{h+\varepsilon} \left(1 - \frac{\varepsilon}{\varepsilon'}\right) (u_h^{\varepsilon'} - \psi)^+ +$$

$$+ \frac{\varepsilon}{h+\varepsilon} e^{-\alpha h} \Phi(h) u_h^{\varepsilon'}$$

and therefore

$$u_h^{\varepsilon'} = \sum_{n=0}^{\infty} \left(\frac{\varepsilon}{h+\varepsilon}\right)^{n+1} e^{-n\alpha h} \Phi(nh)\left[hL + \frac{h}{\varepsilon}\psi \wedge u_h^{\varepsilon'} + h\left(\frac{1}{\varepsilon} - \frac{1}{\varepsilon'}\right)(u_h^{\varepsilon'} - \psi)^+\right] \tag{5.77}$$

hence

$$T_h^{\varepsilon} u_h^{\varepsilon'} \leqq u_h^{\varepsilon'} \quad \text{for} \quad \varepsilon \leqq \varepsilon'$$

and iterating we deduce

$$(T_h^{\varepsilon})^n u_h^{\varepsilon'} \leqq u_h^{\varepsilon'}$$

hence

$$u_h^{\varepsilon} \leqq u_h^{\varepsilon'} .$$

By (5.72) and (5.70) we have

$$u_h^0 - u_h^{\varepsilon} \geqq e^{-\alpha h}\Phi(h)\,(u_h^0 - u_h^{\varepsilon})$$

and therefore by iterating,

$$u_h^0 - u_h^{\varepsilon} \geqq e^{-n\alpha h}\,\Phi(nh)(u_h^0 - u_h^{\varepsilon}) \to 0$$

so

$$u_h^0 \geqslant u_h^{\varepsilon}.$$

Now,

$$u_h^0 \leqq \|L^+\| \frac{h}{1 - e^{-\alpha h}} \leqq \frac{e^{\alpha}}{\alpha}\|L^+\| . \tag{5.78}$$

Furthermore,

$$\begin{aligned} T_h^{\varepsilon} 0 &= \sum_{n=0}^{\infty} \left(\frac{\varepsilon}{h+\varepsilon}\right)^{n+1} e^{-n\alpha h}\,\Phi(nh)\,(hL - \frac{h}{\varepsilon}\psi^-) \\ &\geqq - \sum_{n=0}^{\infty} \left(\frac{\varepsilon}{h+\varepsilon}\right)^{n+1} e^{-n\alpha h}\,(h\|L^-\| + \frac{h}{\varepsilon}\|\psi^-\|) \\ &= - \frac{h}{h+\varepsilon(1-e^{-\alpha h})}\,(\varepsilon\|L^-\| + \|\psi^-\|) \\ &\geqq - \|\psi^-\| - \frac{h}{1-e^{-\alpha h}}\,\|L^-\| = -k . \end{aligned}$$

Let us assume that $z \geqslant -k$, then

$$T_h^\epsilon z \geq \sum_{n=0}^{\infty} \left(\frac{\epsilon}{h+\epsilon}\right)^{n+1} e^{-n\alpha h} \Phi(nh) \left[- h\|L^-\| - \frac{h}{\epsilon} k\right]$$

$$\geq - \frac{\epsilon}{h+\epsilon(1-e^{-\alpha h})} \left[h \|L^-\| + \frac{h}{\epsilon} k\right]$$

$$\geq - k .$$

From this we deduce

$$u_h^\epsilon \geq - \|\psi^-\| - \frac{h}{1-e^{-\alpha h}} \|L^-\| \geq - \|\psi^-\| - \frac{e^\alpha}{\alpha} \|L^-\| \tag{5.79}$$

whence (5.75).□

LEMMA 5.10:

$$u_h^\varepsilon \geqslant u_h. \tag{5.80}$$

Proof: Now,

$$T_h^\epsilon u_h = \sum_{n=0}^{\infty} \left(\frac{\epsilon}{h+\epsilon}\right)^{n+1} e^{-n\alpha h} \Phi(nh) \left[hL + h \frac{\psi \wedge u_h}{\epsilon}\right]$$

and

$$u_h \leq hL + e^{-\alpha h} \Phi(h) u_h$$

$$\frac{\epsilon}{h+\epsilon} u_h \leq \frac{\epsilon hL}{h+\epsilon} + \frac{\epsilon}{h+\epsilon} e^{-\alpha h} \Phi(h) u_h$$

so

$$u_h \leq \frac{\epsilon hL}{h+\epsilon} + \frac{h}{h+\epsilon} u_h + \frac{\epsilon}{h+\epsilon} e^{-\alpha h} \Phi(h) u_h$$

hence

$$u_h \leq \sum_{n=0}^{\infty} \left(\frac{\epsilon}{h+\epsilon}\right)^{n+1} e^{-n\alpha h} \Phi(nh) \left[hL + \frac{h}{\epsilon} u_h\right]$$

and consequently

$$T_h^\epsilon u_h - u_h \geq \sum_{n=0}^{\infty} \left(\frac{\epsilon}{h+\epsilon}\right)^{n+1} e^{-n\alpha h} \Phi(nh) \frac{h}{\epsilon} \left[\psi \wedge u_h - u_h\right] = 0 \text{ since } u_h \leq \psi$$

since $u_h \leqslant \psi$, whence (5.80).□

LEMMA 5.11: *Let* $\tilde{\psi} \in C$ *and let* $\tilde{u}_h^\varepsilon$ *be the solution of* (5.72) *with* $\tilde{\psi}$ *instead of* ψ. *Then*

$$\|u_h - \tilde{u}_h\| \leqslant \|\psi - \tilde{\psi}\|. \tag{5.81}$$

Proof: Let $z, \tilde{z} \in C$ be such that

$$\|z - \tilde{z}\| \leq \|\psi - \tilde{\psi}\|$$

then by (5.73),

$$\|T_h^\epsilon z - T_h^\epsilon \tilde{z}\| \leq \sum_{n=0}^{\infty} \left(\frac{\epsilon}{h+\epsilon}\right)^{n+1} e^{-n\alpha h} \frac{h}{\epsilon} \|\psi - \tilde{\psi}\| \leq \|\psi - \tilde{\psi}\|$$

whence (5.81).□

We are now given $\psi \in D(\mathcal{A})$, therefore

$$(5.82) \qquad \psi = \int_0^\infty e^{-\alpha t} \Phi(t) \wedge dt \ , \quad \wedge \in C$$
$$= \int_0^h e^{-\alpha t} \Phi(t) \wedge dt + e^{-\alpha h} \Phi(h) \psi .$$
$$= h \wedge_h + e^{-\alpha h} \Phi(h) \psi .$$

We are going to prove:

LEMMA 5.12: *If* (5.82) *is satisfied, then*

$$(5.83) \qquad \|u_h^\epsilon - u_h\| \leq \epsilon \ (\|L\| + \|\wedge\|) \ .$$

Proof: Let us set

$$\tilde{L}_h = L - \wedge_h \ , \quad \tilde{u}_h^\epsilon = u_h^\epsilon - \psi$$

therefore by taking the difference between (5.72) and (5.82) we obtain

$$(5.84) \qquad \tilde{u}_h^\epsilon = h L_h - \frac{h}{\epsilon} (\tilde{u}_h^\epsilon)^+ + e^{-\alpha h} \Phi(h) \tilde{u}_h^\epsilon ;$$

therefore, as already seen,

$$\tilde{u}_h^\epsilon = \sum_{n=0}^{\infty} \left(\frac{\epsilon}{h+\epsilon}\right)^{n+1} e^{-n\alpha h} \Phi(nh) [h \tilde{L}_h - \frac{h}{\epsilon} (\tilde{u}_h^\epsilon)^-]$$
$$\leq \sum_{n=0}^{\infty} \left(\frac{\epsilon}{h+\epsilon}\right)^{n+1} e^{-n\alpha h} \Phi(nh) h \tilde{L}_h$$
$$\leq \frac{h\epsilon \|\tilde{L}_h^+\|}{h+\epsilon - \epsilon e^{-\alpha h}} \leq \epsilon \|\tilde{L}_h^+\|$$

therefore

$$(5.85) \qquad \frac{(\tilde{u}_h^\epsilon)^+}{\epsilon} \leq \|\tilde{L}_h^+\| \ .$$

By (5.76) and (5.77), for $\varepsilon < \varepsilon'$ we therefore have

(5.86) $$T_h^\epsilon u_h^{\epsilon'} - u_h^{\epsilon'} = -\sum_{n=0}^{\infty} \left(\frac{\epsilon}{h+\epsilon}\right)^{n+1} e^{-n\alpha h} \Phi(nh)h\left(\frac{1}{\epsilon} - \frac{1}{\epsilon'}\right)(u_h^{\epsilon'} - \psi)^+$$

and by (5.85), after a calculation already performed above,

$$\geqq - \epsilon' \|\tilde{L}_h^+\| .$$

Notice that

(5.87) $$\psi = \sum_{n=0}^{\infty} \left(\frac{\epsilon}{h+\epsilon}\right)^{n+1} e^{-n\alpha h}[h\Lambda_h + \frac{\psi h}{\epsilon}]$$

and thus

$$(T_h^\epsilon)^2 u_h^{\epsilon'} - \psi = \sum_{n=0}^{\infty} \left(\frac{\epsilon}{h+\epsilon}\right)^{n+1} e^{-n\alpha h}[h\, \tilde{L}_h - \frac{h}{\epsilon}(T_h^\epsilon u_h^{\epsilon'} - \psi)^-]$$

and by (5.86)

$$-(T_h^\epsilon u_h^{\epsilon'} - \psi)^- \geqq \tilde{u}_h^{\epsilon'} - \epsilon' \|\tilde{L}_h^+\|$$

hence

(5.88) $$(T_h^\epsilon)^2 u_h^{\epsilon'} - \psi \geqq \sum_{n=0}^{\infty} \left(\frac{\epsilon}{h+\epsilon}\right)^{n+1} e^{-n\alpha h}[h\, \tilde{L}_h + h\frac{u_h^{\epsilon'}}{\epsilon} - \frac{h\epsilon'}{\epsilon}\|\tilde{L}_h^+\|] .$$

But from (5.77) and (5.87) it follows that

(5.89) $$\tilde{u}_h^{\epsilon'} = \sum_{n=0}^{\infty} \left(\frac{\epsilon}{h+\epsilon}\right)^{n+1} e^{-n\alpha h}[h\, \tilde{L}_h + \frac{h}{\epsilon}\tilde{u}_h^{\epsilon'} - \frac{h}{\epsilon'}(\tilde{u}_h^{\epsilon})^+]$$

which with (5.88) implies

$$(T_h^\epsilon)^2 u_h^{\epsilon'} - \psi \geqq \tilde{u}_h^{\epsilon'} - \frac{h\epsilon'}{\epsilon} \sum_{n=0}^{\infty} \left(\frac{\epsilon}{h+\epsilon}\right)^{n+1} e^{-n\alpha h} \|\tilde{L}_h^+\| \geqq \tilde{u}_h^{\epsilon'} - \epsilon'\|\tilde{L}_h^+\|$$

whence

$$(T_h^\epsilon)^2 u_h^{\epsilon'} - u_h^{\epsilon'} \geqq -\epsilon'\|\tilde{L}_h^+\|$$

and by iteration we deduce

(5.90) $$u_h^\epsilon - u_h^{\epsilon'} \geqq -\epsilon' \|\tilde{L}_h^+\| \quad \text{for} \quad \epsilon < \epsilon' .$$

Now,

(5.91) $$u_h^\epsilon \downarrow u_h .$$

In fact, if u_h^* denotes the decreasing limit of u_h^ϵ, then by (5.90) we have

$$0 \geqslant u_h^* - u_h^{\varepsilon'} \geqslant -\varepsilon' \|L_h^+\|,$$

and therefore $u_h \to u_h^*$ in C. Starting from (5.72) we easily show that $u_h^* = u_h$, therefore, finally,

$$\|u_h^{\varepsilon} - u_h\| \leq \varepsilon \|\tilde{L}_h^+\| \leq \varepsilon \ (\|L\| + \|\Lambda\|)$$

i.e., (5.83).□

LEMMA 5.13:

(5.92) $u_h^{\varepsilon} \to u^{\varepsilon}$ *in* C *when* $h \to 0$.

Proof: Now,

$$u^{\varepsilon} - u_h^{\varepsilon} = \sum_{n=0}^{\infty} e^{-n\alpha h} \Phi(nh)\left[\int_0^h e^{-\alpha s}\Phi(s)(L + \frac{1}{\varepsilon}\psi \wedge u^{\varepsilon}) e^{-\frac{nh}{\varepsilon}} ds - \right.$$
$$\left. - \left(\frac{\varepsilon}{\varepsilon + h}\right)^{n+1} h(L + \frac{1}{\varepsilon}\psi \wedge u_h^{\varepsilon})\right]$$
$$= \sum_{n=0}^{\infty} e^{-n\alpha h} \Phi(nh)\left[\int_0^h e^{-\alpha s}\Phi(s)(L + \frac{1}{\varepsilon}\psi \wedge u^{\varepsilon}) e^{-\frac{nh}{\varepsilon}} ds - \right.$$
$$\left. - \left(\frac{\varepsilon}{\varepsilon + h}\right)^{n+1} h(L + \frac{1}{\varepsilon}\psi \wedge u^{\varepsilon})\right] +$$
$$+ \sum_{n=0}^{\infty} e^{-n\alpha h} \Phi(nh) \left(\frac{\varepsilon}{\varepsilon + h}\right)^{n+1} \frac{h}{\varepsilon} (\psi \wedge u^{\varepsilon} - \psi \wedge u_h^{\varepsilon})$$
$$= I + II .$$

And

$$|II| \leq \frac{\|u^{\varepsilon} - u_h^{\varepsilon}\|}{1 + \varepsilon\left(\frac{1 - e^{-\alpha h}}{h}\right)}$$

$$|I| = \sum_{n=0}^{\infty} e^{-n\alpha h}\|L + \frac{1}{\varepsilon}\psi \wedge u^{\varepsilon}\| \left| e^{-\frac{nh}{\varepsilon}} - \left(\frac{1}{1 + \frac{h}{\varepsilon}}\right)^{n+1} \right| +$$
$$+ \sum_{n=0}^{\infty} e^{-n\alpha h} \left\| \frac{\int_0^h (e^{-\alpha s}\Phi(s) - I)(L + \frac{1}{\varepsilon}\psi \wedge u^{\varepsilon}) ds}{h} \right\|$$

$$\leq \frac{h}{\varepsilon} \|L + \frac{1}{\varepsilon}\psi \wedge u^{\varepsilon}\| \sum e^{-n\alpha h}(n+2) +$$
$$+ \sum_{n=0}^{\infty} h e^{-n\alpha h} \left\| \frac{\int_0^h (e^{-\alpha s}\Phi(s) - I)(L + \frac{1}{\varepsilon}\psi \wedge u^{\varepsilon}) ds}{h} \right\|$$
$$= \mathcal{O}_{\varepsilon}(h) \to 0 \quad \text{when} \quad h \to 0 \quad \text{and } \varepsilon \text{ fixed.}$$

But then we obtain

$$\|u^\varepsilon - u_h^\varepsilon\| \leq \left(1 + \frac{h}{\varepsilon(1-e^{-\alpha h})}\right) 0_\varepsilon(h) \leq (1 + \frac{e^\alpha}{\varepsilon\alpha}) 0_\varepsilon(h)$$

whence the desired result.□

We can now give the proof of Theorem 5.3.

We approximate ψ by $\psi^N \in D(A)$, add the index N in the quantities $u, u_h, u^\varepsilon, u_h^\varepsilon$ when these quantities are related to ψ^N instead of ψ, and write

$$u - u_h = u - u^N + u^N - u^{N,\varepsilon} + u^{N,\varepsilon} - u_h^{N,\varepsilon} + u_h^{N,\varepsilon} - u_h^N + u_h^N - u_h;$$

then we have

$$\|u - u^N\| \leq \rho_N$$
$$\|u^N - u^{N,\varepsilon}\| \leq C_N \varepsilon$$
$$\|u^{N,\varepsilon} - u_h^{N,\varepsilon}\| \leq O_{N,\varepsilon}(h) \quad \text{(Lemma 5.13)}$$
$$\|u_h^{N,\varepsilon} - u_h^N\| \leq C_N \varepsilon \quad \text{(Lemma 5.12)}$$
$$\|u_h^N - u_h\| \leq \rho_N$$

this latter inequality is deduced from (5.81) by passing to the limit in ε. Making successively h tend to 0, then ε to 0, then $N \to \infty$, from this we deduce that $\|u - u_h\| \to 0$ with h.

5.3.2 Q.V.I. in discrete time

We are now going to approximate the maximum solution u of (5.55). We define the discretised Q.V.I. as

$$u_h = \text{Min}[Mu_h \,,\, hL + e^{-\alpha h}\Phi(h)u_h] \,, \quad u_h \in C \,. \tag{5.93}$$

THEOREM 5.4: *Assume* (5.1),...,(5.4),(5.6),(5.7),(5.68),(5.56),(5.57). *Then the problem* (5.93) *has one and only one solution, and* $u_h \to u$ *the maximum solution of* (5.55) *when* $h \to 0$, *in the sense of* C.

We shall carry out the proof in several steps. We set

$\sigma_h(\psi)$ = the solution of (5.69), $\sigma_h : C \to C$.

LEMMA 5.14: *The operator* σ_h *is increasing and concave.*

Proof: First we show that if $\zeta \in C$ satisfies

$$\zeta \leq \psi \,, \quad \zeta \leq hL + e^{-\alpha h}\Phi(h)\zeta \tag{5.94}$$

then $\zeta \leqslant \sigma_h(\psi)$. Let us set

$$S_h(z) = \text{Min}[\psi, hL + e^{-\alpha h}\Phi(h)z]$$

which is a contraction on C. Let $\zeta_h = \sigma_h(\psi)$, then ζ_h is the fixed point of S_h. Now (5.94) implies

$$\zeta \leqslant S_h(\zeta),$$

and therefore by iterating (since S_h is increasing)

$$\zeta_h = S_h^n(\zeta_h) \quad , \quad \zeta \leq S_h^n(\zeta)$$

whence

$$\zeta_h - \zeta \geq S_h^n(\zeta_h) - S_h^n(\zeta) \to 0 \quad \text{when} \quad n \to \infty ,$$

and thus $\zeta_h \geqslant \zeta$. From this we deduce that

$$\theta\sigma_h(\psi_1) + (1-\theta)\sigma_h(\psi_2) \leq \theta\psi_1 + (1-\theta)\psi_2$$
$$\theta\sigma_h(\psi_1) + (1-\theta)\sigma_h(\psi_2) \leq hL + e^{-\alpha h}\Phi(h)[\theta\sigma_h(\psi_1) + (1-\theta)\sigma_h(\psi_2)]$$

whence

$$\theta\sigma_h(\psi_1) + (1-\theta)\sigma_h(\psi_2) \leq \sigma_h(\theta\psi_1 + (1-\theta)\psi_2)$$

thus σ_h is concave. Similarly we show that σ_h is increasing.□

Proof of Theorem 5.4: Define

$$T_h = \sigma_h o M$$

therefore

(5.95) T_h is increasing and concave.

The solutions of (5.93) are the fixed points of T_h. Furthermore, we have (cf. (5.78))

$$\|u_h^o\| \leq \frac{e^\alpha}{\alpha} \|L\| .$$

We can therefore find $\mu \in]0,1[$ such that $\mu\|u_h^o\| \leqslant k$. We then have

$$T_h(o) \geqslant \mu u_h^o.$$

By considering increasing and decreasing schemes we easily deduce that

$$u_h^{n+1} = T_h(u_h^n)$$
$$u_{h,n+1} = T_h(u_{h,n}) \qquad u_{h,o} = 0$$
$$\|u_h^n - u_{h,n}\| \leq (1-\mu)^n \frac{e^\alpha}{\alpha} \|L\| .$$

From this it follows that u_h^n and $u_{h,n}$ converge to u_h the unique fixed point of T_h, and then

(5.96) $$\|u_h^n - u_h\| \leq (1-\mu)^n \frac{e^\alpha}{\alpha} \|L\| .$$

Further, by recurrence on n, by Theorem 5.4 we show that

$$\|u_h^n - u^n\| \to 0 \quad \text{when } h \to 0, \text{ for fixed } n.$$

Also, since

$$\|u^n - u\| \to 0,$$

we easily deduce from it, by the uniform estimate (5.96), that

$$u_h \to u \quad \text{in } C. \square$$

5.4 *The Case of Non-Continuous Data*

It will be useful to consider the V.I. (5.10) in the case of non-continuous data. Take $\Phi(t)$ satisfying (5.1),...,(5.4) and

(5.97) $x,t \to \Phi(t)f(x)$ is measurable $\forall\, f \in B$,

(5.98) $t \to \Phi(t)f(x)$ is continuous from $]0,\infty[\, \to R$, $\forall$ fixed x, $\forall\, f \in B$.

We are also given

(5.99) $L \in B, \quad \psi \in B.$

Consider the problem:

(5.100) $$\left| \begin{array}{l} u \in B\,, \quad u \leq \psi \\ u \leq \displaystyle\int_0^t e^{-\alpha s}\Phi(s)\,L\,ds + e^{-\alpha t}\Phi(t)u \quad \forall\, t \geq 0 . \end{array} \right.$$

We are going to prove:

THEOREM 5.5: *Assume* (5.1),...,(5.4) *together with* (5.97),(5.98),(5.99). *Then the set of solutions of* (5.100) *is non-empty and has a maximum element. This element coincides with that of* (5.10) *if we take the assumptions of Theorem 5.1.*

We begin by recalling the following property of $\Phi(t)$ (cf., for example, E. DYNKIN [1]):

(5.101) $$\left| \begin{array}{l} \text{if } f_n \in B,\ f_n(x) \to f(x) \quad \forall\, x \quad \text{and} \quad \|f_n\| \leq \text{constant}, \\ \text{then} \quad \Phi(t)f_n(x) \to \Phi(t)f(x) \quad \forall \quad x\,, \quad \forall\, t \geq 0 . \end{array} \right.$$

Let us now give the essential elements that lead to (5.101). Let $\mathcal{M}$ be the space of bounded measures on (E,ε), which is a Banach space. We write

$$(m,\varphi) = \int_E \varphi(x)\,dm(x) \quad \text{if} \quad \varphi \in B\,, \quad m \in \mathcal{M} .$$

Then we have

(5.102) $\left| \begin{array}{l} f_n \in B, \quad f_n(x) \to f(x) \quad \forall\, x \quad \text{and} \quad \|f_n\| \leqslant \text{constant is equivalent to} \\ \qquad (m,f_n) \to (m,f) \quad \forall\, m \in \mathcal{M}. \end{array} \right.$

Furthermore we define an operator on $\mathcal{M}$ by the formula

$$U(t)m(\Gamma) = (m, \Phi(t)\chi_\Gamma) \quad \forall\, \Gamma \in \varepsilon .$$

We verify that $U(t)$ is a contraction semi-group on $\mathcal{M}$ and that we have the relation

(5.103) $(U(t)m,f) = (m, \Phi(t)f) \quad \forall\, f \in B, \quad \forall\, m \in \mathcal{M} .$

From (5.102) and (5.103) the property (5.101) immediately follows.

Proof of Theorem 5.5: We consider an approximation by discretisation:

(5.104) $u_h = \text{Min}[\psi, \int_0^h e^{-\alpha t}\Phi(t)\,L\,dt + e^{-\alpha h}\Phi(h)u_h], \quad u_h \in B .$

For $z \in B$ we set

$$T_h z = \text{Min}[\psi, \int_0^h e^{-\alpha t}\Phi(t)\,L\,dt + e^{-\alpha h}\Phi(h)\,z]$$

and so T_h is a contraction, thus (5.104) has one and only one fixed point. Let us notice that

(5.105) $z \leqslant T_h z \quad \text{and} \quad z \in B \quad \text{implies} \quad z \leqslant u_h,$

since

$$z \leqslant T_h^n z \to u_h .$$

Also, from

$$u_h \leqq \int_0^h e^{-\alpha t}\Phi(t)\,L\,dt + e^{-\alpha h}\Phi(h)u_h$$

we deduce

$$\begin{aligned} e^{-\alpha h}\Phi(h)u_h &\leqq \int_0^h e^{-\alpha(t+h)}\Phi(t+h)\,L\,dt + e^{-2\alpha h}\Phi(2h)u_h \\ &= \int_h^{2h} e^{-\alpha t}\Phi(t)\,L\,dt + e^{-2\alpha h}\Phi(2h)u_h \end{aligned}$$

therefore

$$u_h \leqq \int_0^{2h} e^{-\alpha t}\Phi(t)\,L\,dt + e^{-2\alpha h}\Phi(2h)u_h$$

which proves that

$$u_h \leqslant T_{2h} u_h,$$

and therefore by (5.105)

(5.106) $\quad u_h \leqslant u_{2h}.$

We then take $h = \dfrac{1}{2^q}$ and set

$$u_q = u_h \quad \text{with} \quad h = \frac{1}{2^q} .$$

It follows from (5.106) that

$$u_q \leqslant u_{q-1}.$$

Further, let

$$K = \mathrm{Max}[\|\psi^-\|, \ \|\frac{L^-}{\alpha}\|] .$$

Then we have

(5.107) $\quad u_h \geqslant - K.$

In fact,

$$- K \leqslant \psi$$

and

$$- K \leqq \int_0^h e^{-\alpha t}\Phi(t)\, L\, dt - e^{-\alpha h}\Phi(h)\, K$$

whence

$$- K \leqslant u_h$$

by (5.105). Therefore

$$u_q \downarrow u \quad \text{et} \quad u \in B \ , \ u \geqq - K \ .$$

It is clear that $u \leqslant \psi$. Also we verify, as for (5.106), that

$$u_h \leqq \int_0^{ph} e^{-\alpha t}\Phi(t)\, L\, dt + e^{-p\alpha h}\, \Phi(ph) u_h \quad \forall \text{ integers}$$

therefore for $q \geqslant \ell$, if we take $h = \dfrac{1}{2^q}$, $p = m2^{q-\ell}$, then

$$u_q \leq \int_0^{\frac{m}{2^\ell}} e^{-\alpha t}\Phi(t)\, L\, dt + e^{-\frac{\alpha m}{2\ell}}\, \Phi(\frac{m}{2^\ell}) u_q$$

and making q tend to +∞, then using (5.101), we obtain

(5.108) $$u \leq \int_0^{\frac{m}{2^\ell}} e^{-\alpha s} \Phi(s) L\, ds + e^{-\frac{\alpha m}{2^\ell}} \Phi\left(\frac{m}{2^\ell}\right) u .$$

Let t > 0, apply (5.108) with

$$m = [t2^\ell] + 1$$

and make ℓ tend to +∞. Using (5.98) we obtain

$$u \leq \int_0^t e^{-\alpha s} \Phi(s) L\, ds + e^{-\alpha t} \Phi(t) u .$$

Therefore u satisfies (5.100). It is the maximum element. In fact, if $\tilde{u}$ is the solution of (5.100), then

$$\tilde{u} \leq \int_0^h e^{-\alpha s} \Phi(s) L\, ds + e^{-\alpha h} \Phi(h) \tilde{u}$$

therefore

$$\tilde{u} \leqslant T_h \tilde{u},$$

which, with (5.105), hows that $\tilde{u} \leqslant u_h$, and therefore $\tilde{u} \leqslant u$.□

Remark 5.5

If

$$\psi \in C \text{ and } \int_0^\infty e^{-\alpha s} \Phi(s) L\, ds \in C ,$$

then in the statement of Theorem 5.5 we can replace the assumption (5.98) by the weaker assumption

(5.109) $t \to \Phi(t) f(x)$ is continuous from $]0,\infty[\to R$, ∀ fixed x, ∀ f ∈ C.

In fact we always have (5.108) which is applied with $m = [t2^\ell] + 1$. We can no longer make ℓ tend to +∞, but we notice that

$$u \leq \int_0^{\frac{m}{2^\ell}} e^{-\alpha s} \Phi(s) L\, ds + e^{-\frac{\alpha m}{2^\ell}} \Phi\left(\frac{m}{2^\ell}\right) u_q \quad \forall q \text{ fixed.}$$

By the assumptions about ψ and L, u_q is continuous. Then applying (5.109) we obtain

$$u \leq \int_0^t e^{-\alpha s} \Phi(s) L\, ds + e^{-\alpha t} \Phi(t) u_q \quad \forall q .$$

We then make q tend to +∞ using (5.101), whence the result.

If we take the assumptions of Theorem 5.1, we have $u_h \to u$ in C, which is the maximum solution of (5.10), and this completes the proof.□

6. SYSTEMS OF Q.V.I.'s

6.1 *Assumptions: Notations*

We shall be given operators A^m satisfying the same properties as a (cf. (1.2)), $m = 1,\dots,N$. The bilinear form $a^m(u,v)$ on $H^1(\mathcal{O})$ (cf. (1.3)) is associated with the operator A^m. There exists $\lambda \geqslant 0$ such that

(6.1) $$a^m(v,v) + \lambda|v|^2 \geq \gamma \|v\|^2 \quad \forall\, v \in V\,, \quad \gamma > 0\,.$$

Next we are given functions

(6.2) $$f^m \in L^2(\mathcal{O})\,, \quad f^m \geq 0\,.$$

Consider the following problem. Find $u = (u^1,\dots,u^N)$ such that

(6.3) $$\left|\begin{array}{l} u \in (H^1_o(\mathcal{O}))^N\,, \quad u^m \geq 0 \\ a^m(u^m, v-u^m) \geq (f^m, v-u^m) \quad \forall\, v \in H^1_o \ \text{ such that} \\ \qquad v \leq k + \operatorname*{Inf}_{\mu \neq m} u^\mu \\ \quad u^m \leq k + \operatorname*{Inf}_{\mu \neq m} u^\mu\,. \end{array}\right.$$

6.2 *Existence*

If we take the explicit form

$$A^m = -\sum_{i,j} \frac{\partial}{\partial x_i}\, a^m_{ij} \frac{\partial}{\partial x_j} + a^m_i \frac{\partial}{\partial x_i} + a^m_o\,,$$

the assumptions for the coefficients are

(6.4) $$\left|\begin{array}{l} a^m_{ij} = a^m_{ji} \in W^{1,\infty}(R^n)\,, \quad a^m_i\,,\ a^m_o \in L^\infty(R^n) \\ \sum a^m_{ij}\, \xi_i\, \xi_j \geq \alpha\, |\xi|^2, \quad \forall\, \xi \in R^n,\ \alpha > 0\,, a^m_o(x) \geq \beta > 0\,. \end{array}\right.$$

We also assume

(6.5) $\mathcal{O}$ is a regular bounded open set.

We also define the functions $u^{m,o}$ by

(6.6) $$a^m(u^{m,o},v) = (f^m,v) \quad \forall\, v \in H^1_o\,, \quad u^{m,o} \in H^1_o\,.$$

We can then state:

THEOREM 6.1: *Assume* (6.2),(6.4),(6.5). *Then the set of solutions* $u = (u^1,\dots,u^N)$

of (6.3) *such that* $0 \leqslant u^m \leqslant u^{m,o}$ *for all* m *is non-empty and has a minimum element and a maximum element.*

Proof: Very similar to that in Section (cf. Theorem 1.1).□

If we add the assumption

$$f^m \in L^p(\mathcal{O}) \ , \quad p > \frac{n}{2} \ , \tag{6.7}$$

we then have:

THEOREM 6.2: *Take the assumptions of Theorem 6.1 together with* (6.7). *Then there exists one and only one solution of* (6.3) *belonging to* $(L^\infty(\mathcal{O}))^N$. *Moreover,* $u \in (C^o(\bar{\mathcal{O}}))^N$ *and* $u^m \leqslant u^{m,o}$.

Proof: Analogous to that of Theorem 1.2.□

6.3 *Regularity*

We give the following regularity result due to M.G. GARRONI - B. HANOUZET - J.L. JOLY [1]. Assume

$$f^m \in C^{o,\alpha}(\bar{\mathcal{O}}) \tag{6.8}$$

$$a^m_{ij} \in C^{1,\alpha}(\bar{\mathcal{O}}) \ , \quad a^m_i \in C^{o,\alpha}(\bar{\mathcal{O}}) \ , \quad a^m_o \in C^{o,\alpha}(\bar{\mathcal{O}}) \ , \quad \alpha > 0 \ . \tag{6.9}$$

THEOREM 6.3: *Take the assumptions of Theorem 6.1 together with* (6.8),(6.9). *Then,*

$$u \in [W^{2,p}(\mathcal{O})]^N \ , \quad \forall\, p, \ 2 \leq p < \infty \ . \tag{6.10}$$

Proof: To simplify our notation we take N = 3. Introduce the open sets

$$C^m = \{x \in \bar{\mathcal{O}} \mid u^m(x) < k + \inf_{\mu \neq m} u^\mu(x)\}$$

and the compact sets

$$S^{mp} = \{x \in \bar{\mathcal{O}} \mid u^m(x) = k + u^p(x)\} \ , \quad \begin{matrix} m,p = 1 \ldots 3 \ . \\ m \neq p \end{matrix}$$

Then

$$S^{mp} \subset C^p \quad \text{if } m \neq p.$$

In fact, if

$$u^m(x) = k + u^p(x),$$

then

$$u^p(x) \leq u^m(x) \leq k + \inf_{\mu \neq m} u^\mu(x)$$

hence

$$u^p(x) \leq k + \inf_{\mu \neq p} u^\mu(x) .$$

Also,

$$A^m u^m = f^m \text{ on } C^m \text{ in the sense of distributions,}$$

therefore

(6.11) $$u^m \in W^{2,\infty}(D) \quad \forall\, D \text{ an open set of } \mathcal{O} \text{ such that } \bar{D} \subset C^m,$$

recalling that $\Gamma \subset C^m$.

From what we saw above, C^1, C^2, C^3 is an open covering of $\bar{\mathcal{O}}$. Let C'^2, C'^3 be two open sets such that

$$S^{1,2} \subset C'^2 \subset C^2, \qquad S^{1,3} \subset C'^3 \subset C^3.$$

Notice that

$$C^1 , C^2 \cap C^3 , C^2 - \bar{C}'^3 , C^3 - \bar{C}'^2$$

is an open covering of φ, to which we associate a partition of unity $\varphi^1, \varphi^{23}, \varphi^2, \varphi^3$.

Now we introduce the new obstacle

$$\psi^1 = \varphi^1(k+u^1) + \varphi^2(k+u^2) + \varphi^3(k+u^3) + \varphi^{23}(k+u^2 \wedge u^3) .$$

We can then show that u^1 is the solution of the V.I.

(6.12) $$\left| \begin{array}{l} u^1 \in H^1_0 , \quad u^1 \leq \psi^1 \\ \forall\, v \in H^1_0 , \quad v \leq \psi^1 , \text{ then } \quad a^1(u^1, v-u^1) \geq (f^1, v-u^1) . \end{array} \right.$$

In fact

$$u^1 \leqslant \psi^1,$$

since

$$u^1 \leqslant k + u^2 \wedge u^3.$$

Furthermore,

$$\psi^1 \in H^1,$$

and if we consider the function

$$m^\epsilon(x) = \left| \begin{array}{ll} 1 & \text{if } d(x,\Gamma) \geq 2\epsilon \\ \dfrac{d(x,\Gamma) - \epsilon}{\epsilon} & \text{if } \epsilon \leq d(x,\Gamma) \leq 2\epsilon \\ 0 & \text{if } d(x,\Gamma) \leq \epsilon \end{array} \right.$$

then

$$m^{\varepsilon} v \to v \quad \text{in} \quad H_o^1 \,, \quad \forall\, v \quad \text{in} \quad H_o^1 \,.$$

Since $u^1 - u^2$ is continuous on $\bar{\mathcal{O}}$ and zero on Γ, we have

$$|u^1(x) - u^2(x)| \leqq \rho(d(x,\Gamma)) \quad \text{où} \quad \rho(\delta) \to 0 \quad \text{when} \quad \delta \to 0 \,.$$

Thus at least for ε small enough, $S^{1,2}$ is contained in $\{x | d(x,\Gamma) \geqslant 2\varepsilon\}$, therefore $m^{\varepsilon}(x)$ on $S^{1,2}$. Similarly we can assume that $m^{\varepsilon} = 1$ on $S^{1,3}$. Let $v \leqslant \psi^1$, then

$$a^1(u^1 \,, (v - \psi^1)m^{\varepsilon}) \geqq (f^1 \,, (v - \psi^1)m^{\varepsilon}) \tag{6.13}$$

since

$$A^1 u^1 - f^1 \leqq 0 \quad \text{in} \quad H^{-1}$$

and

$$(v - \psi^1)\, m^{\varepsilon} \leqq 0 \quad \text{and belongs to} \quad H_o^1 \,.$$

Next we show that

$$a^1(u^1 \,, (\psi^1 - u^1)m^{\varepsilon}) = (f^1 \,, (\psi^1 - u^1)m^{\varepsilon}) \,. \tag{6.14}$$

Since $S^{1,2} \cup S^{1,3} = \bar{\mathcal{O}} - C^1$ and $A^1 u^1 = f^1$ on C^1, it suffices to show that

$$(\psi^1 - u^1)m^{\varepsilon} = 0 \quad \text{on} \quad S^{1,2} \cup S^{1,3}$$

or again by the choice of ε,

$$\psi^1 - u^1 = 0 \quad \text{on} \quad S^{1,2} \cup S^{1,3} \,. \tag{6.15}$$

Now,

$$\psi^1 - u^1 = \varphi^1(k + u^1 - u^1) + \varphi^2(k + u^2 - u^1) + \varphi^3(k + u^3 - u^1) + $$
$$+ \varphi^{23}(k + u^2 \wedge u^3 - u^1)$$

and

$$x \in S^{1,2} - S^{1,2,3} \quad \text{implies} \quad u^1 = k + u^2 \,,\ u^2 < u^3 \,,\ \varphi^1 = 0 \,,\ \varphi^3 = 0$$
$$x \in S^{1,3} - S^{1,2,3} \quad \text{implies} \quad u^1 = k + u^3 \,,\ u^3 < u^2 \,,\ \varphi^1 = 0 \,,\ \varphi^2 = 0$$
$$x \in S^{1,2,3} \quad \text{implies} \quad u^1 = k + u^2 = k + u^3 \,,\ \varphi^1 = 0$$

from which we immediately deduce (6.15). From (6.13) and (6.14) it follows that

$$a^1(u^1, (v - u^1)m^{\varepsilon}) \geqq (f^1, (v - u^1)m^{\varepsilon})$$

and making ε tend to 0, we deduce

$$a^1(u^1, v - u^1) \geqslant (f^1, v - u^1)$$

whence (6.12).

The obstacle $\psi^1 \in H^1$, and

$$A^1\psi^1 = \sum_{i=1}^{3} A^1(\varphi^i(k+u^i)) + A^1\varphi^{23}(k+u^2 \wedge u^3).$$

But

$$\varphi^i(k+u^i) \in C^2(\mathcal{O})$$

and

$$A^1\varphi^{23}(k+u^2 \wedge u^3) = A^i \operatorname{Min}[\varphi^{23}(k+u^2) \;, \varphi^{23}(k+u^3)] \;,$$

and

$$\varphi^{23}(k+u^2) \;, \varphi^{23}(k+u^3) \in C^2(\mathcal{O}) \;.$$

As in Lemma 2 we deduce from this that

$A^1\varphi^1 \geqslant - C$, where C is a constant (in the sense of distributions in $\mathcal{O}$).

By the Levy-Stampacchia inequality (cf. Lemma 2.4) we deduce that

$$A^1u^1 \in L^\infty,$$

from which we get the result

$$u^1 \in W^{2,p}(\mathcal{O}) \quad \forall\, p, \quad 2 \leqslant p < \infty.$$

Therefore (6.10) holds.□

6.4 *Semi-Group Formulation*

We shall use certain notations of Section 5.1 again. We take a family $\Phi^m(t)$ of semi-groups on B satisfying (5.1),...,(5.4),(5.6),(5.7). We then set the following problem

$$(6.16) \quad \left| \begin{array}{l} u^m \in C \;, \quad u^m \leqq k + \operatorname*{Inf}_{u \neq m} u^\mu \\ u^m \leqq \displaystyle\int_0^t e^{-\alpha s}\Phi^m(s)\,L^m\,ds + e^{-\alpha t}\Phi^m(t)u^m \end{array} \right.$$

where

$$(6.17) \quad \left| \begin{array}{l} L^m \geqq 0 \;, \quad \alpha > 0 \\ L^m \in B \;, \quad t \to \Phi^m(s)\,L^m \text{ is measurable from }]0,\infty[\quad \text{in} \quad C \;. \end{array} \right.$$

We then have:

THEOREM 6.4: *Assume that* $\Phi^m(t)$, $m = 1,\ldots,N$ *satisfies the properties* (5.1),..., (5.4),(5.6),(5.7) *and that we also assume* (6.17). *Then the set of* $u = (u^1,\ldots,u^N)$ *satisfying* (6.16) *is non-empty and has a maximum element consisting of positive functions.*

Proof: Analogous to that of Theorem 5.2.□

We now direct our attention to what happens when $k \to 0$. We introduce the following problem

$$(6.18)\qquad \left|\begin{array}{l} u \in B\ ,\quad u \ \text{u.s.c.} \\ u \leq \displaystyle\int_0^t e^{-\alpha s}\Phi^m(s)\,L^m\,ds + e^{-\alpha t}\Phi^m(t)u\ ,\quad \forall\, m = 1 \ldots N\ ,\quad \forall\, t\ . \end{array}\right.$$

We shall write u_k^m for the maximum solution of (6.16). We have:

THEOREM 6.5: *Take the assumptions of Theorem 6.4. Then* $u_k^m \downarrow u$ *when* $k \downarrow 0$, *where* u *is the maximum solution of* (6.18). *Furthermore,* $u \geq 0$.

Proof: Here we introduce a notation: let $u \in B^N$, then write

$$M_k^m(u) = k + \inf_{\mu \neq m} u^\mu$$

and

$$M_k(u) = (M_k^1(u)\ ,\ldots,\ M_k^N(u)) \in B^N\ .$$

Let $\psi \in C$, then we write $\sigma^m(\psi)$ for the maximum solution of

$$\left|\begin{array}{l} v \in C\ ,\quad v \leq \psi \\ v \leq \displaystyle\int_0^t e^{-\alpha s}\Phi^m(s)\,L^m\,ds + e^{-\alpha t}\Phi^m(t)v\ . \end{array}\right.$$

Lastly, if $u \in C^N$ we define

$$T_k^m(u) = \sigma^m \circ M_k^m(u)$$
$$T_k(u) = (T_k^1(u)\ ,\ldots,\ T_k^N(u))\ .$$

Then u, the maximum solution of (6.16), is the fixed point of the mapping T_k. This point is the limit of the iteration

$$(6.19)\qquad T_k(u_k^{(n)}) = u_k^{(n+1)}\ ,\qquad u_k^{(0)} = u^{(0)} = (u^{01}\ ,\ldots,\ u^{0N})$$

where

$$u^{0m} = \int_0^\infty e^{-\alpha t}\Phi^m(t)\,L^m\,dt\ .$$

Let us write u_k for the fixed point of T_k, then we verify that

$$(6.20)\qquad u_k \leq u_{k'}\quad \text{if}\ \ k \leq k'\quad (\text{i.e.}\ \ u_k^m \leq u_{k'}^m,\ \ \forall\, m)\ .$$

In fact T_k is increasing with respect to k; this follows from the fact that M_k is increasing with respect to k, and σ^m is increasing with respect to the argument. Therefore

$$T_k(u^{(0)}) \leqq T_{k'}(u^{(0)})$$

so

$$u_k^{(1)} \leqq u_{k'}^{(1)} .$$

Next, if

$$u_k^{(n)} \leqq u_{k'}^{(n)} ,$$

we have

$$T_k(u_k^{(n)}) \leqq T_{k'}(u_k^{(n)}) \leqq T_{k'}(u_{k'}^{n)}) = u_{k'}^{(n+1)} .$$

So by making n tend to $+\infty$, (6.23) follows.

Since $u_k^m \geqslant 0$, we have

$$u_k^m \downarrow u^m \quad \text{when} \quad k \downarrow 0 , \quad u^m \in B \quad \text{et} \quad u^m \quad \text{u.s.c.}$$

as the decreasing limit of continuous functions. But,

$$u_k^m \leqq k + u_k^\mu \quad \forall\, \mu \neq m .$$

Making $k \downarrow 0$, we obtain

$$u^m \leqq u^\mu \quad \forall\, \mu \neq m$$

and thus $u^m = u \in B$ necessarily. Furthermore, by (6.16) we have

$$u_k^m \leqq \int_0^t e^{-\alpha s} \, \Phi^m(s) L^m \, ds + e^{-\alpha t} \Phi^m(t) u_k^m \quad \forall\, t \geqq 0 .$$

Making k tend to 0, we can pass to the limit, from which it follows that

$$u \leqq \int_0^t e^{-\alpha s} \Phi^m(s) L^m \, ds + e^{-\alpha t} \Phi^m(t) u .$$

Moreover, u is the maximum solution of (6.18), for if $\tilde{u}$ is a solution, then by setting

$$\tilde{u}^m = \tilde{u},$$

we have

$$\tilde{u}^m \leqslant k + \operatorname*{Inf}_{u \neq m} \tilde{u}^\mu,$$

and thus $\tilde{u}^m$ satisfies (6.16). Now, $\tilde{u}^m \leqslant u^{om}$, thus $\tilde{u}^m \leqslant M_k^m(u^{(0)})$. By Theorem 5.5, completed by Remark 5.5 (in particular $M_k^m(u^{(0)})$ is continuous), we can state that $\tilde{u} \leqslant u_k^{(1)}$. By induction we show that $\tilde{u} \leqslant u_k^{(n)}$, and therefore $\tilde{u} \leqslant u_k$. Consequently $\tilde{u} \leqslant u$.

We can give another proof of the existence of a maximum solution of (6.18) using an approximation by discretisation, namely,

$$(6.21) \qquad u_h = \underset{m}{\mathrm{Min}} \left[\int_0^h e^{-\alpha s} \Phi^m(s)\, L^m\, ds + e^{-\alpha h} \Phi^m(h) u_h \right], \quad u_h \in C .$$

THEOREM 6.6: *Assume that* Φ^m *satisfies the properties* (5.1),...,(5.4),(5.6) (5.109). *Assume* (6.17) *also. Then* $u_{1/2^q} \downarrow u$ *is the maximum solution of* (6.18) *when* $q \to +\infty$.

Proof: For $z \in B$ we define

$$T_h z = \underset{m}{\mathrm{Min}} \left[\int_0^h e^{-\alpha s} \Phi^m(s)\, L^m\, ds + e^{-\alpha h} \Phi^m(h) z \right] .$$

Then T_h is a contraction, and so has one and only one fixed point. For $z \in C$, $T_h z \in C$, hence the fixed point $u_h \in C$. We again have

$$(6.22) \qquad z \leqslant T_h z \quad \text{and} \quad z \in B \quad \text{implies } z \leqslant u_h.$$

As in the proof of Theorem 5.5, we have

$$u_h \leqslant T_{2h} u_h,$$

whence

$$(6.23) \qquad u_h \leqslant u_{2h}.$$

Furthermore,

$$(6.24) \qquad u_h \geqq -K , \quad K = \underset{m}{\mathrm{Max}} \frac{\|L^{m-}\|}{\alpha} .$$

Therefore we have

$$(6.25) \qquad u_q = u_{1/2^q} \downarrow u \in \text{ u.s.c.}$$

then

$$u_q \leqq \int_0^{j/2^\ell} e^{-\alpha s} \Phi^m(s)\, L^m\, ds + e^{-\alpha \frac{j}{2^\ell}} \Phi^m\!\left(\frac{j}{2^\ell}\right) u_q \quad \text{for } q \geqq \ell ,$$

and in the limit

$$u \leq \int_0^{j/2^\ell} e^{-\alpha s}\Phi^m(s)\,L^m\,ds + e^{-\alpha \frac{j}{2^\ell}}\Phi^m(\tfrac{j}{2^\ell})u$$

$$\leq \int_0^{j/2^\ell} e^{-\alpha s}\Phi^m(s)\,L^m\,ds + e^{-\alpha \frac{j}{2^\ell}}\Phi^m(\tfrac{j}{2^\ell})u_q \ .$$

Taking $j = [t2^\ell] + 1$ and making ℓ tend to $+\infty$, we obtain

$$u \leq \int_0^t e^{-\alpha s}\Phi^m(s)\,L^m\,ds + e^{-\alpha t}\Phi^m(t)u_q$$

then making q tend to $+\infty$, we obtain that u is the solution of (6.18). This is the maximum solution, for if $\tilde{u}$ satisfies (6.18), then

$$\tilde{u} \leqslant T_h\tilde{u},$$

therefore

$$\tilde{u} \leqslant u_h,$$

whence the result.□

Remark 6.1

If, in addition to the assumptions of Theorem 6.6, Φ^m satisfies (5.35) and L^m satisfies (5.36), and if

(6.26) $\quad f \wedge g \in C_o \quad$ if $\quad f, g \in C_o$

then

$$u_h \in C_o \ \text{ and } \ u \in B_o \ . \qquad \square$$

6.5 *Regularity of the Maximum Solution*

We now give some regularity results for u, the maximum solution of (6.18). Assume that

(6.27) $\quad |L^m(x) - L^m(y)| \leq K\,|x-y|^\delta \quad 0 < \delta \leq 1\ , \quad \forall\ m$

and

(6.28) $\quad \forall\ g \in C^{o,\delta}(E)\ , \quad$ i.e $\quad |g(x) - g(y)| \leq \ \|g\|_\delta\ |x-y|^\delta$

then

$$|\Phi^m(t)g(x) - \Phi^m(t)g(y)| \leq e^{\lambda t}\,\|g\|_\delta\,|x-y|^\delta\ , \quad \lambda \geq 0\ .$$

THEOREM 6.7: *Take the assumptions of Theorem 6.6 together with* (6.27),(6.28). *Then if* $\alpha > \lambda$, $u \in C^{o,\delta}(E)$.

Proof: Let $z \in C^{o,\delta}(E)$. Let us fix x_o in E, then there exists m_o (dependent on x_o)

such that

$$T_h z(x_0) = \int_0^h e^{-\alpha s} \Phi^{m_0}(s) L^{m_0}(x_0)ds + e^{-\alpha h} \Phi^{m_0}(h) z(x_0) .$$

Let x be arbitrary in E, then

$$T_h z(x) \le \int_0^h e^{-\alpha s} \Phi^{m_0}(x) L^{m_0}(x)ds + e^{-\alpha h} \Phi^{m_0}(h) z(x)$$

and so by subtraction

$$T_h z(x) - T_h z(x_0) \le \int_0^h e^{-\alpha s}(\Phi^{m_0}(s) L^{m_0}(x) - \Phi^{m_0}(s) L^{m_0}(x_0))ds +$$
$$+ e^{-\alpha h}(\Phi^{m_0}(h) z(x) - \Phi^{m_0}(h) z(x_0))$$

and by the assumptions (6.27),(6.28) we deduce

$$\le \int_0^h e^{-\alpha s} e^{\lambda s} K|x-x_0|^\delta ds + e^{-\alpha h} e^{\lambda h} ||z||_\delta |x-x_0|^\delta$$

and hence

$$\frac{T_h z(x) - T_h z(x_0)}{|x-x_0|^\delta} \le K \frac{1 - e^{-(\alpha-\lambda)h}}{\alpha - \lambda} + e^{-(\alpha-\lambda)h} ||z||_\delta$$

and since x, x_0 are arbitrary, we obtain

$$(6.29) \qquad ||T_h z|| \le K \frac{1-e^{-(\alpha-\lambda)h}}{\alpha - \lambda} + e^{-(\alpha-\lambda)h} ||z||_\delta ,$$

and by iterating we obtain

$$||T_h^k z||_\delta \le K \frac{1-e^{-(\alpha-\lambda)h}}{\alpha - \lambda} (1+e^{-(\alpha-\lambda)h} + \dots + e^{-(\alpha-\lambda)(k-1)h}) +$$
$$+ e^{-k(\alpha-\lambda)h} ||z||_\delta ;$$

making k tend to $+\infty$, we deduce from this that

$$||u_h||_\delta \le \frac{K}{\alpha-\lambda}$$

but then taking $h = \frac{1}{2^q}$ and making q tend to $+\infty$, we deduce that

$$(6.30) \qquad ||u||_\delta \le \frac{K}{\alpha-\lambda} ,$$

whence the desired result.□

We are going to give another regularity result. Assume that $\Phi^m(t)$ satisfies (5.1),...,(5.4),(5.97),(5.109) and

(6.31) $\Phi^m(t):C \to C.$

LEMMA 6.1: *The maximum solution* u *of* (6.18) *is also the maximum solution of*

(6.32)
$$\begin{cases} u \in B, \quad u \text{ u.s.c.} \\ u \le \int_0^t e^{-\beta s}\, \Phi^m(s)(L^m+(\beta-\alpha)u)ds + e^{-\beta t}\, \Phi^m(t)\, u, \\ \qquad \forall m, \quad \forall t \end{cases}$$

for all $\beta > \alpha$.

Proof: First let us provisionally show that (6.32) has a maximum solution, denoted $\tilde{u}$. In fact, we proceed analogously to the proof of Theorem 6.6. For $z \in B$ we define the mapping

$$\Theta_h z = \underset{m}{\text{Min}}\, [\int_0^h e^{-\beta s}\, \Phi^m(s)(L^m+(\beta-\alpha)z)ds + e^{-\beta h}\, \Phi^m(h)z] \ .$$

This is a contraction, since

$$\|\Theta_h z_1 - \Theta_h z_2\| \le \|z_1 - z_2\| (e^{-\beta h} + \frac{\beta-\alpha}{\beta}(1-e^{-\beta h}))$$
$$= \|z_1-z_2\| (\frac{\beta-\alpha(1-e^{-\beta h})}{\beta}) \ .$$

For $z \in C$, $\Theta_h z \in C$. We write $\tilde{u}_h$ for the fixed point, and $\tilde{u}_h \in C$ and $\tilde{u}_h \geqslant 0$, since $\Theta_h z \geqslant 0$ if $z \geqslant 0$. As in Theorem 6.6 we establish that $\tilde{u}_h \leqslant \tilde{u}_{2h}$. Set

$$\tilde{u}_q = \tilde{u}_{1/2^q} \quad \text{and} \quad \tilde{u}_q \downarrow \tilde{u} \quad \text{u.s.c.}$$

Then

$$\tilde{u}_q \le \int_{0}^{j/2^\ell} e^{-\beta s}\, \Phi^m(s)(L^m+(\beta-\alpha)\tilde{u}_q)ds + e^{-\alpha j/2^\ell}\, \Phi^m(1/2^\ell)\, \tilde{u}_q$$

and as in Theorem 6.6 we show that $\tilde{u}$ is a solution of (6.32), and that it is the maximum solution, since every other solution v satisfies

$$v \leqslant \Theta_k v,$$

and therefore

$$v \leqslant \tilde{u}_h.$$

Let us show that

$$u = \tilde{u},$$

where u denotes the maximum solution of (6.18). In fact, by Lemma 5.1 we also have

$$\tilde{u} \le \int_0^t e^{-\alpha s}\, \Phi^m(s)L^m ds + e^{-\alpha t}\, \Phi^m(t)\tilde{u}$$

and therefore $\tilde{u} \leqslant u$. But also by Lemma 5.1 we always have

$$u \le \int_0^t e^{-\beta s}\, \Phi^m(s)(L^m+(\beta-\alpha)u)ds + e^{-\beta t}\, \Phi^m(t)u$$

and therefore $u \leqslant \tilde{u}$.

THEOREM 6.8: *Assume* (5.1),...,(5.4),(5.97),(5.109),(6.31) *and* (6.27),(6.28). *Then the maximum solution* u *of* (6.18) *belongs to* C.

Proof: Let $z \in C$, $z \geqslant 0$, and $\zeta \in B$ be the u.s.c. maximum solution of

(6.33)
$$\left|\begin{array}{l} \zeta \in B \;, \zeta \text{ u.s.c.} \\ \zeta \le \int_0^t e^{-\beta s}\, \Phi^m(s)(L^m+(\beta-\alpha)z)ds + e^{-\beta t}\, \Phi^m(t)\zeta \\ \qquad \forall m, \quad \forall t \;. \end{array}\right.$$

The set (6.33) certainly has a maximum element, by Theorem 6.6. We have thus defined a mapping $S:C^+ \to B^+$, $\zeta = S(z)$. By Theorem 6.6 we have

(6.34) $$S(z) = \lim \downarrow S_{\frac{1}{2^q}}(z)$$

where $\zeta_h = S_h(z)$ is defined by

(6.35)
$$\left|\begin{array}{l} \zeta_h = \underset{m}{\text{Min}} \left[\int_0^h e^{-\beta s}\, \Phi^m(s)(L^m+(\beta-\alpha)z)ds + e^{-\beta h}\, \Phi^m(h)\zeta_h\right] . \\ \zeta_h \in C . \end{array}\right.$$

We easily prove the estimate

(6.36) $$\|S_h(z_1) - S_h(z_2)\| \le \frac{\beta-\alpha}{\beta}\|z_1-z_2\|$$

from which we deduce the well known inequality

(6.37) $$\|S(z_1) - S(z_2)\| \le \frac{\beta-\alpha}{\beta}\|z_1-z_2\| \;.$$

Also, as in (6.22), we have

(6.38) $$u \leqslant S_h(u)$$

on noticing that $S_h:B \to B$.

We now set

$$u^n = S^n(0), \quad u_h^n = S_h^n(0);$$

also we assume $\beta > \lambda$.

Let us note that if $z \in C^{0,\delta}$, then by Theorem 6.7 we have $S(z) \in C^{0,\delta}$, therefore $u^n \in C^{0,\delta}$. By (6.37) we have

$$\|u^{n+1} - u^n\| \le \left(\frac{\beta-\alpha}{\beta}\right)^n \|u^1\|$$

and therefore

$$u^n \to w \quad \text{in } C.$$

Further, $S_h^n(0) \to w_h$ the fixed point of S_h, and

$$\|u^n - w\| \le \frac{\left(\frac{\beta-\alpha}{\beta}\right)^n \|u^1\| \;\beta}{\alpha} \tag{6.39}$$

$$\|u_h^n - w_h\| \le \left(\frac{\beta-\alpha}{\beta}\right)^n \frac{\beta}{\alpha} \max_m \frac{\|L^m\|}{\beta}\,. \tag{6.40}$$

By (6.38) we have

$$u \leqslant w_h. \tag{6.41}$$

By induction on n we show that

$$u_h^n \leqslant u_{2h}^n,$$

and therefore

$$u_q^n = u_{1/2^q}^n \downarrow \tilde{u}^n \quad \text{when } q \to +\infty.$$

By induction on n we next show that $\tilde{u}^n = u^n$. Thus

$$u_{1/2^q}^n \downarrow u^n \quad \text{when } q \to \infty \quad \forall\, n. \tag{6.42}$$

From (6.39), (6.40) and (6.42) it follows that

$$w_q(x) \downarrow w(x) \quad \forall\, x,$$

which with (6.41) implies

$$u \leqslant w. \tag{6.43}$$

Let us notice that

$$w_h \le \int_0^h e^{-\beta s}\, \Phi^m(s)\,(L^m + (\beta-\alpha)w_h)\,ds + e^{-\beta h}\, \Phi^m(h)\, w_h$$

from which we deduce that

$$w_h \leq \int_0^{ph} e^{-\beta s}\, \Phi^m(s)(L^m+(\beta-\alpha)w_h)ds + e^{-p\beta h}\, \Phi^m(ph)w_h$$

then

$$w_q \leq \int_0^{j/2^\ell} e^{-\beta s}\, \Phi^m(s)(L^m+(\beta-\alpha)w_q)ds + e^{-\beta j/2^\ell}\, \Phi^m(j/2^\ell)w_q ,$$

for $q \geqslant \ell$. From this we easily deduce that w satisfies

$$w \leq \int_0^t e^{-\beta s}\, \Phi^m(s)(L^m+(\beta-\alpha)w)ds + e^{-\beta t}\, \Phi^m(t)w ,$$

$\forall\, t, \ \forall\, m$.

Therefore by Lemma 6.1 $w \leqslant u$, which with (6.43) shows that $w = u$, and therefore $u \in C$.

Remark 6.2

The results of this Section are due to A. BENSOUSSAN and M. ROBIN [1]. For similar considerations, cf. M. NISIO [1].

Example: Verification of (6.28)

Consider the case of the whole space (cf. (5.49)) with the assumption (5.50). Thus we have

$$(6.44) \qquad \frac{\partial z}{\partial t} + Az = 0 \quad , \quad z(x,0) = g(x) .$$

We write

$$(6.45) \qquad A = - a_{ij} \frac{\partial^2}{\partial x_i \partial x_j} + b_i \frac{\partial}{\partial x_i}$$

and by (5.50)

$$a_{ij} \in W^{2,\infty} \quad , \quad b_i \in W^{1,\infty} .$$

We are going to prove the property

$$(6.46) \qquad |z(x,t) - z(y,t)| \leq e^{\lambda t}\, ||Dg||\, |x-y|$$

for λ large enough and fixed, which allows (6.28) to be satisfied. To show this let us set

$$w(x,t) = |Dz(x,t)|^2\, e^{-2\lambda t} .$$

We are going to show that we can choose λ so that

$$(6.47) \qquad 0 \leq w(x,t) \leq ||Dg||^2 .$$

First of all we assume that the coefficients are very regular. We are going to calculate

$$\frac{\partial}{\partial t} w + Aw$$

and show that this quantity is negative, and since

$$w(x,0) = \|Dg\|^2$$

it will follows from (6.47). Therefore,

$$\frac{\partial}{\partial t} w + Aw = w_t + b_i w_{x_i} - a_{ij} w_{x_i x_j}$$
$$= e^{-2\lambda t} [-2\lambda |Dz|^2 + 2z_{x_k} (z_{x_{kt}} + b_i z_{x_k x_i} -$$
$$- a_{ij} z_{x_k x_i x_j}) - 2 a_{ij} z_{x_k x_j} z_{x_k x_i}] .$$

But by equation (6.44) we have

$$z_{x_k t} - a_{ij,k} z_{x_i x_j} - a_{ij} z_{x_k x_i x_j} + b_{ix_k} z_{x_i} + b_i z_{x_i x_k} = 0$$

whence

$$\frac{\partial w}{\partial t} + Aw = e^{-2\lambda t}[-2\lambda |Dz|^2 + 2z_{x_k} (a_{ijx_k} z_{x_i x_j} - b_{ix_k} z_{x_i}) -$$
$$- 2 a_{ij} z_{x_k x_j} z_{x_k x_i}]$$
$$\leq e^{-2\lambda t} [-2\lambda |Dz|^2 + \eta \sum_k (a_{ij,x_k} z_{x_i x_j})^2 +$$
$$+ \frac{C}{\eta} |Dz|^2 - 2 a_{ij} z_{x_k x_j} z_{x_k x_i}]$$

and since

$$\sum_k (a_{ij,x_k} z_{x_i x_j})^2 \leq C a_{ij} z_{x_k x_j} z_{x_k x_i}$$

we easily deduce that λ can be chosen fixed and large enough so as to obtain

$$\frac{\partial w}{\partial t} + Aw \leqslant 0,$$

whence the result.

From this we also easily deduce that $z(\cdot,t) \in C$ if $g \in C$.□

We refer to N. KRYLOV [1], P.L. LIONS - J.L. MENALDI [1], P.L. LIONS [1],[2] for the complete study of the problem (6.18) when the semi-groups $\Phi^m(t)$ correspond to diffusions in the whole space or on a bounded open set.

We can also study the V.I.'s and Q.V.I.'s corresponding to (6.18).□

7. CONTINUITY METHODS AND OTHER METHODS

Method of Approach

In the preceding Sections we have studied Q.V.I.'s with the help of monotonic methods. In this Section we are going to see other examples of Q.V.I.'s for which monotonic methods will not work. We shall use other techniques (essentially continuity methods).

7.1 *An Implicit Signorini Problem*

We here present an example studied in U. MOSCO [3].

On $H^1(\mathcal{O})$, where $\mathcal{O}$ is a regular bounded set of R^n, and $\Gamma = \partial\mathcal{O}$, we are given

$$a(u,v) = \Sigma \int_{\mathcal{O}} a_{ij} \frac{\partial u}{\partial x_j} \frac{\partial v}{\partial x_i} dx + \Sigma \int_{\mathcal{O}} a_i \frac{\partial u}{\partial x_i} v \, dx + \int_{\mathcal{O}} a_0 u \, v \, dx \tag{7.1}$$

where

$$a_{ij} \in L^\infty(R^n) \quad , \quad \Sigma\, a_{ij}\, \xi_i\, \xi_j \geq \alpha|\xi|^2, \ \forall \xi \in R^n, \ \alpha > 0 \tag{7.2}$$
$$a_i \in L^\infty(R^n) \ , \quad a_0 \in L^\infty \ , \quad a_0 \geq 0$$

and we assume

$$a(v,v) \geq \alpha \| v \|^2 \qquad \forall\, v \in H^1(\mathcal{O}) \ . \tag{7.3}$$

We are given

$$f \in L^2(\mathcal{O}) \tag{7.4}$$
$$h \in H^1(\mathcal{O}) \ , \tag{7.5}$$
$$\phi \in H^{1/2}(\Gamma) \ , \quad \phi \geq 0 \ . \tag{7.6}$$

Let

$$A = - \sum_{i,j} \frac{\partial}{\partial x_i} a_{ij} \frac{\partial}{\partial x_j} + \sum_i a_i \frac{\partial}{\partial x_i} + a_0$$

and

$$\frac{\partial}{\partial \nu_A} = \Sigma\, a_{ij} \frac{\partial}{\partial x_j} \nu_i$$

where ν denotes the exterior normal of Γ.

Let us recall (cf. J.L. LIONS - E. MAGENES [1]) that if $u \in H^1(\mathcal{O})$ and $Au \in L^2$, then

$$\frac{\partial u}{\partial \nu_A} \in H^{-1/2}(\Gamma)$$

and Green's formula

$$a(u,v) = \left< v, \frac{\partial u}{\partial \nu_A} \right> + \int Au \, v \, dx$$

holds for all $v \in H^1$, where $<\ ,\ >$ denotes the scalar product between $H^{1/2}(\Gamma)$ and $H^{-1/2}(\Gamma)$. We then consider the Q.V.I.

(7.7)
$$\left|\begin{array}{l} a(u,v-u) \geq (f,v-u) \quad , \quad u \in H^1, \quad Au \in L^2 \\ \forall\, v \in H^1 \text{ such that } v \geq h - \langle \phi , \dfrac{\partial u}{\partial \nu_A} \rangle \quad \text{on } \Gamma \\ \quad u \geq h - \langle \phi , \dfrac{\partial u}{\partial \nu_A} \rangle \quad \text{on } \Gamma \ . \end{array}\right.$$

We are going to prove:

THEOREM 7.1: *Assume* (7.2),(7.3),(7.4),(7.5),(7.6). *Then the problem* (7.7) *has a solution.*□

Remark 7.1

The number of solutions is an open problem.□

Remark 7.2

If we define

$$M(u) = h - \langle \phi , \frac{\partial u}{\partial \nu_A} \rangle,$$

we therefore have an implicit obstacle $u \geqslant M(u)$. But the operator $M(u)$ is not monotonic.□

Proof of Theorem 7.1: Let $\lambda \geqslant 0$, and consider the problem

(7.8)
$$\left|\begin{array}{l} a(u_\lambda,v-u_\lambda) \geq (f,v-u_\lambda) \quad , \quad u_\lambda \in H^1 \\ \forall\, v \in H^1 \text{ such that } v \geq h - \lambda \quad \text{on } \Gamma \\ \quad u_\lambda \geq h - \lambda \quad \text{on } \Gamma \ . \end{array}\right.$$

This is the classical Signorini problem. The solution u_λ exists and is unique. Moreover, since we can take

$$v = u_\lambda \pm \phi \text{ where } \phi \in \mathcal{D}(\mathcal{O})$$

we can deduce from (7.8) that

$$Au_\lambda = f \quad \text{in the sense of distributions in } \mathcal{O},$$

and therefore $\partial u_\lambda/\partial \nu_A$ has a meaning in $H^{-1/2}(\Gamma)$. Furthermore,

$$\frac{\partial u_\lambda}{\partial \nu_A} \geq 0 \tag{7.9}$$

for if we take $v = u_\lambda + \psi$, $\psi \in H^1$, $\psi \geqslant 0$, which is permissible, from (7.8) we deduce that

$$a(u_\lambda,\psi) \geqslant (f,\psi),$$

so

$$\langle \psi, \frac{\partial u_\lambda}{\partial \nu_A} \rangle + \int Au_\lambda \, \psi \, dx \geq (f,\psi)$$

and therefore

$$<\psi, \frac{\partial u_\lambda}{\partial \nu_A}> \geq 0,$$

from whence we deduce (7.9).

We can therefore define

$$(7.10) \qquad F(\lambda) = <\phi , \frac{\partial u_\lambda}{\partial \nu_A}> \quad .$$

and F is a mapping from $R^+ \to R^+$. We are going to show that F has a fixed point. Let us show that F is continuous. To do that we notice that h is admissible for (7.8), and therefore

$$a(u_\lambda, h-u_\lambda) \geq (f, h-u_\lambda)$$

from whence it follows that

$$(7.11) \qquad \|u_\lambda\|_{H^1} \leq C .$$

Also, Au_λ stays equal to f, therefore

$$(7.12) \qquad \left\|\frac{\partial u_\lambda}{\partial \nu_A}\right\|_{H^{-1/2}(\Gamma)} \leq C$$

and therefore

(7.13) $F(\lambda)$ stays bounded when $\lambda \to +\infty$.

If $\lambda_n \to \lambda$, then u_{λ_n} stays bounded in H^1 and $\partial u_{\lambda_n}/\partial \nu_A$ stays bounded in $H^{-1/2}$. Let a subsequence, again denoted u_{λ_n}, be picked out that converges weakly in H^1 to u^*. Then

$$(7.14) \qquad F(\lambda_n) \to <\phi, \frac{\partial u^*}{\partial \nu_A}> .$$

Let v be such that $v \in H^1$, $v \geq h - \lambda$ on Γ. Then $v + \lambda - \lambda_n$ is admissible for the problem with index n, whence

$$a(u_{\lambda_n}, v+ \lambda- \lambda_n - u_{\lambda_n}) \geq (f, v-u_{\lambda_n})$$

and on passing to the limit,

$$a(u^*, v-u^*) \geq (f, v-u^*) \ ,$$

and therefore $u^* = u_\lambda$, therefore with (7.14) we have $F(\lambda_n) \to F(\lambda)$.

Since $F(0) \geqslant 0$ it follows from (7.13) and from the continuity of F that there exists a fixed point of F. In fact,

(7.15) $\quad F(\lambda) \to 0 \quad \text{when } \lambda \to +\infty.$

So, let $\bar{u}$ be a solution of

(7.16) $\quad a(\bar{u},v) = (f,v) \quad \forall\, v \in H^1, \quad \bar{u} \in H^1 .$

Let us set $\tilde{u}_\lambda = u_\lambda - h$. Then $\tilde{u}_\lambda$ is a solution of the V.I.

(7.17)
$$\left| \begin{array}{l} a(\tilde{u}_\lambda, v-\tilde{u}_\lambda) \geq (f, v-\tilde{u}_\lambda) - a(h, v-\tilde{u}_\lambda) \\ \forall\, v \in H^1, \quad v \geq -\lambda \text{ on } \Gamma, \quad \tilde{u}_\lambda \geq -\lambda \text{ on } \Gamma . \end{array} \right.$$

Let $v \in \mathcal{D}(\mathcal{O})$, then if λ is large enough $v \geqslant -\lambda$ on Γ. Passing to the limit in (7.17) we deduce

$$a(\tilde{u}, v - \tilde{u}) \geqslant (f, v - \tilde{u}) - a(h, v - \tilde{u}) \quad \forall\, v \in \mathcal{D}(\bar{\mathcal{O}}),$$

hence for all $v \in H^1$ also. But then $\tilde{u} + h = \bar{u}$, therefore

$$u_\lambda \to \bar{u} \text{ weakly in } H^1.$$

Now,

$$\left.\frac{\partial \bar{u}}{\partial \nu_A}\right|_\Gamma = 0$$

whence (7.15).□

We now to find a way of suppressing the assumption $\phi \geqslant 0$. We notice that F is a mapping from $R \to R$ and (7.13), and even (7.15), stay true without the assumption $\phi \geqslant 0$. $F(\lambda)$ is still continuous, but we no longer necessarily have $F(0) \geqslant 0$, and therefore $F(\lambda)$ may remain less than λ for all λ. We are going to give a sufficient condition for a fixed point to exist.

We begin by studying the behaviour of u_λ when $\lambda \to -\infty$. Since $u_\lambda \geqslant h - \lambda$, it is clear that u_λ is not bounded. Recalling that λ is negative, we set

$$z_\lambda = \frac{u_\lambda}{\lambda} .$$

We now multiply (7.8) by $\frac{1}{\lambda^2}$. This yields

$$a(z_\lambda, \frac{v}{\lambda} - z_\lambda) \geq (\frac{f}{\lambda}, \frac{v}{\lambda} - z_\lambda)$$
$$\forall\, v \geq h - \lambda \ , \quad z_\lambda \leq \frac{h}{\lambda} - 1$$

therefore z_λ is a solution of

(7.18)
$$\left| \begin{array}{l} a(z_\lambda, v-z_\lambda) \geq (\frac{f}{\lambda}, v-z_\lambda) \\ \forall\, v \in H^1, \quad v \leq \frac{h}{\lambda} - 1 , \quad z_\lambda \leq \frac{h}{\lambda} - 1 . \end{array} \right.$$

Now take $v = \frac{h}{\lambda} - 1$ in (7.18), whence

$$a(z_\lambda, z_\lambda) \le a(z_\lambda, \frac{h}{\lambda} - 1) - (\frac{f}{\lambda}, \frac{h}{\lambda} - 1) + (\frac{f}{\lambda}, z_\lambda)$$

from which we deduce

$$\|z_\lambda\|_{H^1} \leqslant \text{constant.}$$

If we pick a convergent subsequence out of z_λ we have

$$z_\lambda \to z \quad \text{weakly in } H^1$$

and

(7.19) $\quad z \leqslant -1 \quad \text{on } \Gamma.$

Let $v \in H^1$, $v|_\Gamma \leqslant -1$, and take $v + \frac{h}{\lambda}$ in (7.18), therefore

(7.20) $$a(z_\lambda, v + \frac{h}{\lambda} - z_\lambda) \ge (\frac{f}{\lambda}, v + \frac{h}{\lambda} - z_\lambda)$$

and passing to the limit, we obtain

$$a(z, v-z) \ge 0 \qquad \forall v \in H^1, \quad v|_\Gamma \le -1,$$

which, with (7.19), shows that $z_\lambda \to z$ weakly in $H^1(\mathcal{O})$, where z is a solution of

(7.21) $$\left| \begin{array}{lll} Az = 0 & z|_\Gamma \le -1 & \frac{\partial z}{\partial \nu_A}\Big|_\Gamma \le 0 \\ \langle \frac{\partial z}{\partial \nu_A}, z+1 \rangle = 0 & , \quad z \in H^1(\mathcal{O}) . & \end{array} \right.$$

Since

$$\frac{F(\lambda)}{\lambda} = \langle \phi, \frac{\partial z_\lambda}{\partial \nu_A} \rangle$$

we have thus proved:

LEMMA 7.1: $\frac{u_\lambda}{\lambda} \to z$ *weakly in* H^1, *where* z *is a solution of* (7.21), *and*

$$\frac{F(\lambda)}{\lambda} \to \langle \phi, \frac{\partial z}{\partial \nu_A} \rangle \quad \text{when} \quad \lambda \to -\infty . \qquad \Box$$

From this we deduce:

COROLLARY 7.1: *A sufficient condition for* (7.7) *to have a solution, without using the assumption* $\phi \geqslant 0$, *is that*

$$\langle \phi, \frac{\partial z}{\partial \nu_A} \rangle < 1 .$$

Proof: We therefore have

$$\frac{F(\lambda)}{\lambda} < 1 \quad \text{for } \lambda \text{ negative and small enough.}$$

Therefore

$$F(\lambda) > \lambda \quad \text{for } \lambda \text{ negative and small enough.}$$

Since we necessarily have $F(\lambda) < \lambda$ for λ positive and large enough, the curve $F(\lambda)$ intersects the first bisectrix, hence the result.□

7.2 *Example of a Q.V.I. with a Finite Number of Solutions*

In the standard examples of Q.V.I.'s already studied the condition $u \geqslant 0$ (cf. (1.7)) was indispensible for limiting the size of the set of solutions, and, in particular, to ensure uniqueness. We can therefore ask if there exist negative solutions. We are going to give an example of this case (cf. B.L. [1]).

Consider the problem that has the notations of Section 7.1, and is of the form

$$\text{(7.22)} \quad \left| \begin{array}{lll} a(u,v-u) \geq (f,v-u) & & \\ \forall\, v \in H^1 , & v \leq M(u) & \text{on } \Gamma \\ \quad u \in H^1 , & u \leq M(u) & \text{on } \Gamma \end{array} \right.$$

where

$$\text{(7.23)} \quad M(v) = k + \inf_{i=1\ldots m} \int_\Gamma g_i v \, d\Gamma , \quad g_i \in L^2(\Gamma) , \; g_i \geq 0 .$$

Let us assume

$$\text{(7.24)} \quad f \geqslant 0, \quad f \in L^\infty.$$

Then the general theory, (cf. Lemma 1.4) allows us to state:

THEOREM 7.2: *Assume* (7.2),(7.3),(7.23),(7.24). *Then there exists one and only one positive bounded solution of* (7.22). *This solution is continuous on* $\bar{\mathcal{O}}$.

We are now going to investigate whether there exist other (non-positive) solutions. We reduce matters, as in Section 7.1, to a fixed point problem for a mapping of $R \to R$. Let $\lambda \in R$, and define u_λ by

$$\text{(7.25)} \quad \left| \begin{array}{lll} a(u_\lambda,v-u_\lambda) \geq (f,v-u_\lambda) & & \\ \forall\, v \in H^1 , & v \leq \lambda & \text{on } \Gamma \\ \quad u \in H^2 , & u_\lambda \leq \lambda & \text{on } \Gamma. \end{array} \right.$$

The problem (7.25) has one and only one solution. Next we set

$$\text{(7.26)} \quad F(\lambda) = M(u_\lambda) = k + \inf_{i=1\ldots m} \int_\Gamma g_i \, u_\lambda \, d\Gamma$$

and thus we are reduced to studying the fixed points of the mapping $F(\lambda)$. First we have the following properties of F:

LEMMA 7.2: *The mapping* F *is continuous, increasing, concave, and*

$$F(\lambda) = M(w) \quad , \quad \textit{if} \quad \lambda \geq \sup_{\Gamma} w = \lambda_0 \tag{7.27}$$

where w *is a solution of*

$$a(w,v) = (f,v) \quad \forall\, v \in H^1, \quad w \in H^1$$

and

$$F(\lambda) = k + \mu\lambda \quad \textit{if} \quad \lambda \leq 0 ,$$

where

$$\mu = \sup_i \int_\Gamma g_i \, d\Gamma .$$

Proof: The continuity is easy. The increasing property and concaveness are proved by recalling that u_λ can be characterised as the maximum element of all the sub-solutions, i.e., if

$$\left| \begin{array}{l} w \in H^1 \text{ satisfies } a(w,\phi) \leq (f,\phi) \quad \forall\, \phi \in H^1,\ \phi \geq 0 \\ \qquad\qquad\qquad\quad w \leq \lambda \\ \text{then } w \leq u \quad . \end{array} \right. \tag{7.29}$$

Therefore if $\lambda_1 < \lambda_2$, then u_{λ_1} is a sub-solution of the problem for λ_2, therefore $u_{\lambda_1} < u_{\lambda_2}$. Similarly, $\theta u_{\lambda_1} + (1 - \theta)u_{\lambda_2}$ is a sub-solution of the problem for $\theta\lambda_1 + (1 - \theta)\lambda_2$, therefore

$$\theta u_{\lambda_1} + (1-\theta)u_{\lambda_2} \leq u_{\theta\lambda_1 + (1-\theta)\lambda_2} .$$

These properties and equation (7.26) immediately imply the corresponding properties of F. Further, let w be the solution of the Neumann problem (7.27). Then since f is bounded, w is bounded. Therefore $w \leqslant \lambda$ for λ large enough. It is clear that for $\lambda \geqslant \lambda_0$ we have $u_\lambda = w$, whence (7.27).

Lastly, let ψ_0 and ψ_1 be solutions respectively of

$$\begin{array}{llll} A\,\psi_0 = f & \text{in } \mathcal{O} , & \psi_0|_\Gamma = 0 \\ A\,\psi_1 = 0 & \text{in } \mathcal{O} , & \psi_1|_\Gamma = 1 . \end{array}$$

Then

$$A(\psi_0 + \lambda\psi_1) = f \quad , \quad \psi_0 + \psi_1 = \lambda \quad \text{on} \quad \Gamma$$

and since

$$\frac{\partial\psi_1}{\partial\nu_A}\Big|_\Gamma \geq 0, \quad \frac{\partial\psi_0}{\partial\nu_A}\Big|_\Gamma \leq 0,$$

we have

$$\frac{\partial}{\partial \nu_A}(\psi_0+\lambda\psi_1)\Big|_\Gamma \leq 0 \quad \text{for} \quad \lambda \leq 0 .$$

It is clear that

$$u_\lambda = \psi_0 + \psi_1 \quad \text{for} \quad \lambda \leq 0$$

and therefore

$$M(u_\lambda) = k + \mu\lambda$$

whence (7.28).□

THEOREM 7.3: *Take the assumptions of Theorem 7.2. Then if* $\mu \leqslant 1$, *the problem* (7.25) *has one and only one solution in* H^1. *If* $\mu > 1$, *the problem* (7.25) *has two solutions.*

Proof: By Lemma 7.2, and taking $F(0) = k > 0$ into account, the curve $F(\lambda)$ intersects the bisectrix in an unique point for $\lambda > 0$. For $\lambda \leqslant 0$ there can only be an intersection if

$$\lambda = k + \mu\lambda$$

which implies

$$\mu > 1, \quad \text{and} \quad \lambda = \frac{k}{1 - \mu} .$$

Therefore if $\mu > 1$, there are two solutions, the non-positive solution being

$$\psi_0 + \frac{k}{1 - \mu}\psi_1.$$

Remark 7.3

If we no longer assume that the $g_i \geqslant 0$, then we lose the increasing property and the concaveness, but not the properties of continuity, nor (7.27),(7.28). Moreover, $F(0) = k > 0$.

Therefore in these cases we can state that the Q.V.I. (7.22) has at least one solution u such that $M(u) \geqslant 0$.

The solutions u such that $M(u) \leqslant 0$ are necessarily reduced to

$$\psi_0 + \frac{k}{1-\mu}\psi_1 \text{ if } \mu > 1 .$$

They do not exist if $\mu \leqslant 1$.□

Remark 7.4

R.T. VESCAN [1] has considered certain non-standard examples of Q.V.I.'s for which he has given existence theorems. In particular, he has considered the problem (7.7) in the case where

$$(7.30) \qquad a(u,v) = \int_{\mathcal{O}} \frac{\partial u}{\partial x_i}\frac{\partial v}{\partial x_i}\, dx + \int_{\mathcal{O}} u\, v\, dx$$

and ϕ is not necessarily $\geqslant 0$. If we introduce the constant K_0 such that

$$\|\frac{\partial v}{\partial n}\|_{H^{-1/2}(\Gamma)} \leq K_0(\|v\|^2_{H^1} + \|-\Delta v + v\|^2_{L^2})^{1/2}$$

then Vescan has given (as a consequence of an abstract general theory) the following sufficiency condition for the existence of a solution of (7.7)

(7.30') $$\|\phi\|_{H^{1/2}(\Gamma)} < \frac{1}{K_0 m} \quad \text{où} \quad m = (\int_{\mathcal{O}} dx)^{1/2} .$$

Let us show that this condition implies (7.22). In fact

$$\begin{aligned} <\phi, \frac{\partial z}{\partial n}> &\leq \|\phi\|_{H^{1/2}} \|\frac{\partial z}{\partial n}\|_{H^{-1/2}} \\ &\leq \|\phi\|_{H^{1/2}} K_0 \|z\|_{H^1} \end{aligned}$$

and taking v = -1 in (7.20), which is admissible, we obtain

$$\|z\|^2 \leq -\int_{\mathcal{O}} z\, dx \leq \|z\| \, m$$

whence

$$\|z\| \leqslant m$$

and therefore (7.30') holds.□

Let us consider another of Vescan's examples. For $v \in H^1(\mathcal{O})$ we define

(7.31) $$M(v) = h + \int_{\mathcal{O}} g\, v\, dx$$

where

(7.32) $$\left| \begin{array}{l} h \in H^1(\mathcal{O}) , \quad h \geq k > 0, \quad k \text{ constant} \\ g \in L^2(\mathcal{O}) , \quad f \in L^2(\mathcal{O}) . \end{array} \right.$$

Consider the problem

(7.33) $$\left| \begin{array}{l} a(u,v-u) \geq (f,v-u) , \quad u \in H^1 \\ \forall\, v \in H^1 , \quad v \leq M(u) , \quad u \leq M(u) , \end{array} \right.$$

and a(u,v) is again given by (7.30).

We define $F(\lambda)$ as

(7.34) $$F(\lambda) = \int_{\mathcal{O}} g\, u_\lambda\, dx$$

where u_λ is a solution of

(7.35)
$$\left| \begin{array}{l} a(u_\lambda, v-u_\lambda) \geq (f, v-u_\lambda) \quad , \quad u_\lambda \in H^1 \\ \forall\, v \in H^1 \, , \quad v \leq h + \lambda \, , \quad u_\lambda \leq h + \lambda \quad \text{a.e.} \end{array} \right.$$

One shows that $F(\lambda)$ is continuous, and that

(7.36) $$F(\lambda) \to \int_{\mathcal{O}} g\, w\, dx \qquad \text{when} \quad \lambda \to +\infty$$

where w is a solution of

$$a(w,v) = (f,v) \quad \forall v \in H^1 \, , \quad w \in H^1 \, .$$

We also show that if we set $z_\lambda = \dfrac{u_\lambda}{\lambda}$, then for $\lambda \leqslant 0$, z_λ is a solution of the problem

(7.37)
$$\left| \begin{array}{l} a(z_\lambda, v-z_\lambda) \geq (\frac{f}{\lambda}, v-z_\lambda) \\ \forall\, v \in H^1, \quad v \geq \frac{h}{\lambda} + 1, \quad z_\lambda \geq \frac{h}{\lambda} + 1 \end{array} \right.$$

and $z_\lambda \to z$ weakly in H^1 when $\lambda \to -\infty$, where z is a solution of

(7.38) $$a(z, v-z) \geq 0 \quad \forall\, v \geq 1, \quad z \geq 1, \quad \text{so } z = 1.$$

By (7.36) and the continuity of F, a *sufficient* condition for there to exist a fixed point of F is that

(7.39) $$\exists\, \lambda < 0 \text{ such that } \frac{F(\lambda)}{\lambda} < 1 \, .$$

Now,

$$\frac{F(\lambda)}{\lambda} \to \int_{\mathcal{O}} g\, z\, dx \quad \text{when} \quad \lambda \to -\infty \, .$$

Therefore a sufficient condition for (7.39) to hold is that

(7.40) $$\int_{\mathcal{O}} g\, z\, dx < 1 \, .$$

For this it suffices (since $z = 1$) that

(7.41) $$|g| m < 1,$$

which is a first sufficient condition provided by Vescan. More generally, we can exploit (7.39) differently. We have to find

(7.42) $$\lambda < 0 \quad \text{such that} \quad \int_{\mathcal{O}} g\, z_\lambda\, dx < 1.$$

Let us use the assumption $h \geqslant k > 0$, which has not been used up to now. For $\lambda < 0$,

$$\frac{k}{\lambda} + 1 \geqslant \frac{h}{\lambda} + 1,$$

therefore we can take

$$v = \frac{k}{\lambda} + 1$$

in (7.37). We obtain

$$a(z_\lambda, \frac{k}{\lambda} + 1 - z_\lambda) \geq (\frac{f}{\lambda}, \frac{k}{\lambda} + 1 - z_\lambda)$$

so

$$(7.43) \qquad \|z_\lambda\|^2 \leq \int z_\lambda \left(\frac{k}{\lambda} + 1 + \frac{f}{\lambda}\right)dx - \left(\frac{k}{\lambda} + 1\right) \int \frac{f}{\lambda}\, dx .$$

A simple calculation allows us to deduce from (7.43) that

$$\begin{aligned} \frac{1}{2} |z_\lambda|^2 &\leq \frac{1}{2} \left|\frac{k}{\lambda} + 1 + \frac{f}{\lambda}\right|^2 - \left(\frac{k}{\lambda} + 1\right) \int \frac{f}{\lambda}\, dx \\ &= \frac{1}{2} \left(\frac{k}{\lambda} + 1\right)^2 m^2 + \frac{1}{2} \left|\frac{f}{\lambda}\right|^2 \end{aligned}$$

so finally,

$$|z_\lambda|^2 \leq \left(\frac{k}{\lambda} + 1\right)^2 m^2 + \left|\frac{f}{\lambda}\right|^2 .$$

To make sure that (7.42) holds it suffices to find, therefore,

$$(7.44) \qquad \lambda < 0 \text{ such that } |g| \sqrt{m^2\left(\frac{k}{\lambda} + 1\right)^2 + \left|\frac{f}{\lambda}\right|^2} < 1 .$$

Let us assume, for example,

$$|g| \leq \frac{1}{m} \qquad |f| \leq km$$

then it suffices us to find $\lambda < 0$ such that

$$\sqrt{2 \frac{k^2}{\lambda^2} + \frac{2k}{\lambda} + 1} < 1$$

and we can choose $\lambda < -k$ arbitrarily (cf. R. VESCAN [1], A. BENSOUSSAN - J.L. LIONS [13]).□

7.3 *A General Existence Theorem for Solutions of a Q.V.I.*

Let V be a Hilbert space, and let us be given a form

$$a(u,v) \text{ is bilinear and continuous on } V, \text{ such that}$$

$$(7.45) \qquad a(v,v) \geqslant \alpha\|v\|^2 \quad \forall\, v \in V.$$

We are again given

(7.46) K_0 a non-empty closed convex subset of V,

(7.47) K(u) a correspondence that associates a non-empty closed convex set of V with $u \in K_0$.

We make the following assumptions about the mapping K(u)

(7.48) K is closed in the weak topology of V, i.e., if $u_k \to u$ weakly in V and if $v_k \in K(u_k)$ and $v_k \to v$ weakly in V, then $v \in K(u)$,

(7.49) K is l.s.c. in the weak topology of V, i.e. if $u_k \to u$ weakly in V and if $v \in K(u)$, then there exists $v_k \in K(u_k)$, and $v_k \to v$ *strongly* in V.

We also assume

(7.50) $\exists\ \bar{u}$ such that $\bar{u} \in K(u) \quad \forall\, u \in K_0$.

We then consider the Q.V.I. problem

(7.51)
$$\begin{aligned} & a(u,v-u) \geq L(v-u) \\ & \forall\, v \in K(u) \\ & \text{et } u \in K_0,\quad u \in K(u) \end{aligned}$$

where

(7.52) $L \in V'$.

We assume that

(7.53)
$$\left| \begin{aligned} & \forall\, z \in K_0, \text{ the solution } w \text{ of the V.I.} \\ & a(w,v-w) \geqslant L(v-w) \quad \forall\, v \in K(z),\ w \in K(z), \text{ belongs to } K_0. \end{aligned} \right.$$

We can state:

THEOREM 7.4: *Assume* (7.45),(7.46),(7.47),(7.48),(7.49),(7.50),(7.52) *and* (7.53). *Then there exists a solution* u *of* (7.51).

Proof: For all $z \in K_0$, K(z) being a non-empty closed convex set of V, we can solve the V.I. (assumption (7.53))

(7.54)
$$\left| \begin{aligned} & a(w,v-w) \geq L(v-w) \\ & \forall\, v \in K(z), \quad w \in K_0 \cap K(z). \end{aligned} \right.$$

We have thus defined a mapping $S: K_0 \to K_0$. By (7.50) we can take $v = \bar{u}$ in (7.54), therefore

$$a(w,\bar{u} - w) \geqslant L(\bar{u} - w)$$

from which it easily follows that

(7.55) $\|S(z)\| \leq M \quad \forall z \in K_0$.

Let us show that S is continuous in the weak topology of V. In fact, let $z_k \to z$ weakly in V, and

$$w_k = S(z_k).$$

By (7.55) w_k stays in a bounded set of V, therefore we can pick out a subsequence

$$w_{k_p} \to w \text{ weakly in } V.$$

Clearly $w \in K_0$. Since $w_{k_p} \in K(z_{k_p})$, it follows from (7.48) that $w \in K(z)$. Now let $v \in K(z)$, by (7.49) there exists $v_{k_p} \in K(z_{k_p})$ such that

$$v_{k_p} \to v \text{ strongly in } V.$$

We then have

$$a(w_{k_p}, v_{k_p} - w_{k_p}) \geq L(v_{k_p} - w_{k_p})$$

and we can pass to the limit, from which it follows that

$$a(w, v-w) \geq L(v-w)$$

and therefore w is a solution of (7.54), so $w = S(z)$. We then restrict S to the set

$$\tilde{K} = \{z \in K_0 , \quad \|z\| \leq M\}$$

which is a weak compact convex set of V. By (7.55) $S: \tilde{K} \to \tilde{K}$, and S is continuous, therefore S has a fixed point by Schauder's Theorem.□

Remark 7.5

We can generalise Theorem 7.4 a great deal, in particular to consider V.I.'s corresponding to monotonic operators. We shall refer the reader to U. MOSCO [3], C. BAIOCCHI - CAPELO [1], J.P. AUBIN [1]). However, the assumptions (7.48),(7.49), which are well known to be very restrictive, can be weakened but with difficulty.□

Remark 7.6

We also see that we have two types of methods of studying Q.V.I.'s, namely the continuity methods and the monotonic methods. They each correspond to two different types of fixed point theorems. Theorem 7.4 must be considered as a prototype.

Let us give an example of an application of Theorem 7.4. We are going to rediscover Theorem 7.1. Therefore we take

$$V = H^1(\mathcal{O}) \quad , \quad L(v) = (f,v)$$
$$K_0 = \{v \in H^1(\mathcal{O}), \quad Av = f \quad , \quad \frac{\partial v}{\partial \nu_A}\Big|_\Gamma \geq 0\} \quad ,$$
$$K(z) = \{v \in H^1, \quad v \geq h - \langle \phi, \frac{\partial z}{\partial \nu_A} \rangle \quad \text{on } \Gamma\} \quad .$$

Thanks to the assumption $\phi \geqslant 0$ we easily show that all the assumptions for applying Theorem 7.4 are satisfied, whence the existence of a solution of (7.51) which is equivalent to (7.7).□

8. Q.V.I.'s ARISING IN GAMES

Method of Approach

In this Section we study other Q.V.I.'s motivated by games problems, although the Q.V.I.'s corresponding to implicit obstacle problems arise in control problems.

8.1 *Double V.I. and Systems of Q.V.I.'s*

In this Section we present an interesting equivalence due to O. NAKOULIMA [1]. We assume (1.1) and we define $a(u,v)$ by (1.3).

We are given

(8.1) $$f \in L^2(\mathcal{O}),$$

and two functions ψ_1,ψ_2 such that

(8.2) $$\left| \begin{array}{l} \psi_1, \psi_2 \ : \ \bar{\mathcal{O}} \to \bar{R}, \quad \psi_1 \le \psi_2 \quad \text{a.e.} \\ \text{and} \\ K_{\psi_1}^{\psi_2} = \{v \in V, \quad \psi_1 \le v \le \psi_2 \ \text{ a.e.}\} \\ \text{is non-empty.} \end{array} \right.$$

In (8.2) V denotes H_0^1 or H^1. To simplify matters we assume

(8.3) $$a(v,v) \ge \gamma \|v\|^2 \qquad \forall\, v \in V, \quad \gamma > 0 .$$

Consider the double obstacle

(8.4) $$a(u,v-u) \ge (f,v-u), \quad \forall\, v \in K_{\psi_1}^{\psi_2}, \quad u \in K_{\psi_1}^{\psi_2} .$$

By the general theory of V.I.'s (cf. J.L. LIONS - G. STAMPACCHIA 1) it has one and only one solution. We shall see that it is possible to associate with it a system of Q.V.I.'s of a type very similar to that studied in Section 6.

We assume that we have the decomposition

(8.5) $$f = f^1 - f^2 , \quad f^1, f^2 \in L^2(\mathcal{O}) .$$

We then consider the Q.V.I.: find u^1,u^2 such that

(8.6) $$\left| \begin{array}{ll} a(u^1,v-u^1) \ge (f^1,v-u^1) & \forall\, v \in V, \quad v \le u^2 + \psi^2 \\ \quad u^1 \in V, \quad u^1 \le u^2 + \psi^2 & \\ a(u^2,v-u^2) \ge (f^2,v-u^2) & \forall\, v \in V, \quad v \le u^1 - \psi^1 \\ \quad u^2 \in V, \quad u^2 \le u^1 - \psi^1 . & \end{array} \right.$$

We then have:

THEOREM 8.1: *Assume* (1.1),(8.1),(8.2),(8.3). *Then for every decomposition* (8.5), *if* u^1,u^2 *is a solution of* (8.6),

(8.7) $u^1 - u^2 = u,$

where u *is the solution of the double Q.V.I.* (8.2).

Proof: Let u^1, u^2 be a solution of (8.6) and u be defined by (8.7). First of all we have

$$u^1 - u^2 \le \psi^2 \quad , \quad u^1 - u^2 \ge \psi^1 \quad .$$

Now let $v_0 \in K_{\psi_1}^{\psi_2}$. Take $v = u^2 + v_0$ in the first V.I. (8.6) and $v = u^1 - v_0$ in the second, which is permissible, since

$$v \le u^2 + \psi^2 \quad , \quad v \le u^1 - \psi^1 \quad .$$

From this we deduce

$$a(u^1, v_0 - u) \ge (f^1, v_0 - u)$$

$$a(u^2, u - v_0) \ge (f^2, u - v_0)$$

and by addition we see that u satisfies

$$a(u, v_0 - u) \geqslant (f, v_0 - u)$$

so that u certainly is a solution of (8.4), and therefore it is the solution.□

We can study the Q.V.I. (8.6) by monotonic methods. Let us make this more precise. Take the assumption

(8.8) $\exists\ u_0^1, u_0^2 \in V$ such that $a(u_0^i, v) \le (f^i, v), \ \forall v \in V, v \ge 0, \ i=1,2$

and $\psi^1 \le u_0^1 - u_0^2 \le \psi^2$.

Also define

(8.9) $u^{1,0}, u^{2,0} \in V$ such that $a(u^{i,0}, v) = (f^i, v) \quad \forall v \in V$.

Since

$$a(u_0^i - u^{i,0}, v) \le 0 \quad \forall v \ge 0$$

and taking

$$v = (u_0^i - u^{i,0})^+,$$

we establish that

$$u_0^i \le u^{i,0} \quad .$$

We then have:

THEOREM 8.2: *Take the assumptions of Theorem 8.1 together with* (8.8). *Then the set of solutions* u^i, $i = 1,2$, *of* (8.6) *such that*

$$u_0^i \leqslant u^i \leqslant u^{i,0} \tag{8.10}$$

is non-empty, and has a maximum element and a minimum element.

Proof: Write $\sigma(\psi,f)$ for the solution of the obstacle problem

$$\left| \begin{array}{l} a(u,v-u) \geq (f,v-u) \\ \forall\, v \in V, \quad v \leq \psi, \text{ et } u \in V, \quad u \leq \psi . \end{array} \right. \tag{8.11}$$

Let us notice that the assumption (8.8) implies, in particular, that the convex set $K_{\psi_1}^{\psi_2}$ is non-empty. Therefore for $v^1,v^2 \in V\times V$, the convex sets $\{v | v \in v^1 + \psi^2\}$ and $\{v | v \in v^1 - \psi^1\}$ are non-empty.

We can therefore uniquely define the solution of (8.11) with

$$f = f^1, \quad \psi = v^2 + \psi^2 \quad \text{and} \quad f = f^2, \quad \psi = v^1 - \psi^1$$

and thus $\sigma(v^2 + \psi^2,f^1)$, $\sigma(v^1 - \psi^1,f^2)$. We thus define a mapping

$$S : (v^1,v^2) \rightarrow \sigma(v^2 + \psi^2,f^1), \quad \sigma(v^1 - \psi^1,f^2) .$$

Naturally, the fixed point of S are the solutions of (8.6). Consider the set

$$K = \{(v_1,v_2), \quad v^i \in L^2 \quad ; \quad u_0^i \leq v^i \leq u^{i,0}\} .$$

Then S maps K into itself and is increasing. For this it suffices to show that

$$\left| \begin{array}{l} S(u^{1,0},u^{2,0}) \leq (u^{1,0},u^{2,0}) \\ S(u_0^1,u_0^2) \geq (u_0^1,u_0^2) . \end{array} \right. \tag{8.12}$$

Now,

$$S(u^{1,0},u^{2,0}) = \sigma(u^{2,0} + \psi^2,f^1), \quad \sigma(u^{1,0} - \psi^1,f^2)$$

and

$$\begin{array}{l} \sigma(u^{2,0} + \psi^2,f^1) \leq u^{1,0} \\ \sigma(u^{1,0} - \psi^1,f^2) \leq u^{2,0} \end{array}$$

since

$$\begin{array}{l} \sigma(\psi,f_1) \leq u^{1,0} \quad \forall\ \psi \\ \sigma(\psi,f_2) \leq u^{2,0} \quad \forall\ \psi . \end{array}$$

Moreover,

$$S(u_0^1,u_0^2) = \sigma(u_0^2+\psi^2,f^1), \quad \sigma(u_0^1-\psi^1,f^2)$$

and

$$u_0^1 \le u_0^2+\psi^2, \quad a(u_0^1,v) \le (f^1,v) \quad \forall v \in V, \quad v \ge 0$$

implies

$$u_0^1 \le \sigma(u_0^2+\psi^2,f^1).$$

Similarly

$$u_0^2 \le u_0^1-\psi^1, \quad a(u_0^2,v) \le (f^2,v) \quad \forall v \in V, \quad v \ge 0$$

implies

$$u_0^2 \le \sigma(u_0^1-\psi^1,f^2),$$

whence (8.12). It suffices to invoke Tartar's general result (cf. the proof of Theorem 1.1) to complete the proof.□

We can give an uniqueness theorem for the Q.V.I. (8.6), and at the same time give a sufficient condition for (8.8) to be satisfied. We assume

(8.13) $\quad f \in L^\infty(\mathcal{O})$,

(8.14) $\quad \left| \begin{array}{l} \psi_1, \psi_2 \in C^0(\bar{\mathcal{O}}) \text{ and } \psi_1 < \psi_2 \text{ on } \bar{\mathcal{O}}. \\ \text{If } V = H_0^1 \text{ then } \psi_1|_\Gamma < 0 < \psi_2|_\Gamma . \end{array} \right.$

THEOREM 8.3: *Assume* (1.1),(8.3) *and* (8.13),(8.14). *There exist* u_0^1,u_0^2 *such that the Q.V.I.* (8.6) *has one and only one solution in the interval*

$$K = \{v^i \in L^2,\ u_0^i \le v^i \le u^{i,0}\}$$

and $u^1,u^2 \in C^0(\bar{\mathcal{O}})$.

Proof: We begin by verifying the hypothesis (8.8). Let us take the case $V = H_0^1$, and let us set

$$m = \min_{x \in \bar{\mathcal{O}}} (\psi_2(x)-\psi_1(x))$$

$$m_1 = -\sup_{x\in\Gamma} \psi_1(x), \quad m_2 = \inf_{x\in\Gamma} \psi_2(x)$$

and m,m_1,m_2 are > 0. We write

$$0 < \eta \le \min(m,2m_1,2m_2).$$

Therefore

$$\psi_1 < \frac{\eta}{4} + \psi_1 < \psi_2 - \frac{\eta}{4} < \psi_2$$

and

$$\psi_1 < -\frac{\eta}{4} \qquad \psi_2 > \frac{\eta}{4} \quad \text{on} \quad \Gamma .$$

Since ψ_1, ψ_2 are continuous on $\bar{\mathcal{O}}$, we can find $\hat{\psi}_1, \hat{\psi}_2 \in C^2(\bar{\mathcal{O}})$ such that

$$\| \hat{\psi}_1 - \psi_1 - \frac{\eta}{8} \| \le \frac{\eta}{16} \qquad \|\hat{\psi}_2 - \psi_2 + \frac{\eta}{8}\| \le \frac{\eta}{16} ,$$

therefore

(8.15) $$\psi_1 < \psi_1 + \frac{\eta}{16} \le \hat{\psi}_1 \le \psi_1 + \frac{3\eta}{16} < \psi_1 + \frac{\eta}{4} < \psi_2 - \frac{\eta}{4} < \psi_2 - \frac{3\eta}{16} <$$
$$\le \hat{\psi}_2 \le \psi_2 - \frac{\eta}{16} < \psi_2 \quad \text{on} \quad \bar{\mathcal{O}} ,$$

and

(8.16) $$\hat{\psi}_1 < -\frac{\eta}{16} , \qquad \hat{\psi}_2 > \frac{\eta}{16} \quad \text{on} \quad \Gamma .$$

We write $\hat{u}$ for the solution of (8.4) corresponding to the obstacles $\hat{\psi}_1, \hat{\psi}_2$ instead of ψ_1, ψ_2. Since the obstacles are regular, we know (cf. B.L. [1]) that

$$\hat{u} \in W^{2,p}(\mathcal{O}) \quad \forall\, 2 \le p < \infty \quad , \text{and } A\,\hat{u} \in L^\infty .$$

We then have

(8.17) $$A\, u_0^1 = -(A\,\hat{u})^- + f_1 \wedge f_2 \quad , \quad u_0^1|_\Gamma = 0 ,$$

(8.18) $$A\, u_0^2 = -(A\,\hat{u})^+ + f_1 \wedge f_2 \quad , \quad u_0^2|_\Gamma = 0 .$$

Therefore

$$u_0^1 - u_0^2 = +\,\hat{u}$$

whence

(8.19) $$\psi_1 + \frac{\eta}{16} \le \hat{\psi}_1 \le u_0^1 - u_0^2 \le \hat{\psi}_2 \le \psi_2 - \frac{\eta}{16}$$

and by (8.18) we also have

$$a(u_0^1, v) \le (f_1, v) \quad , \quad a(u_0^2, v) \le (f_2, v) \qquad \forall\, v \in H_0^1, \quad v \ge 0 .$$

Let us now assume that $V = H^1$, then we take $\eta = m$ and construct $\hat{\psi}_1, \hat{\psi}_2 \in C^2(\bar{\mathcal{O}})$ as above. Thus we again have (8.15). We then define u_0^1, u_0^2 by

$$(8.20) \qquad a(u_0^1,v) = -\int_{\mathcal{O}} (A\,\hat{\psi}_1)^- v\,dx - \int_\Gamma \left(\frac{\partial\hat{\psi}_1}{\partial\nu_A}\right)^- v\,d\Gamma \qquad \forall\, v \in H^1$$

$$(8.21) \qquad a(u_0^2,v) = -\int_{\mathcal{O}} (A\,\hat{\psi}_1)^+ v\,dx - \int_\Gamma \left(\frac{\partial\hat{\psi}_1}{\partial\nu_A}\right)^+ v\,d\Gamma \qquad \forall\, v \in H^1$$

therefore

$$u_0^1 - u_0^2 = \hat{\psi}_1$$

and (8.19) is again satisfied. Therefore the assumption (8.8) is certainly satisfied. Moreover, u_0^1, u_0^2 and $u^{1,0}, u^{2,0}$ are bounded functions. By (8.19) there exists a number $\lambda \in]0,1[$ such that

$$(8.22) \qquad \left|\begin{array}{l} \lambda(u^{1,0} - u_0^1) + u_0^1 \le u_0^2 + \psi_2 \\ \lambda(u^{2,0} - u_0^2) + u_0^2 \le u_0^1 - \psi_1 \,. \end{array}\right.$$

Using this we can make use of the argument of B. HANOUZET - J.L. JOLY [3] (cf. Section 1.3). In fact, S is increasing and concave (since σ is concave). We then show that if w^i and $\tilde{w}^i$ satisfy

$$(8.23) \qquad w^i - \tilde{w}^i \le \beta\,(w^i - u_0^i) \qquad \beta \in [0,1] \quad , i=1,2$$

then by writing the vector (w^1,w^2) as w we have

$$(8.24) \qquad S(w) - S(\tilde{w}) \le (1-\lambda)\;\beta(S(w)-u_0)$$

where $u_0 = (u_0^1,u_0^2)$. Let us prove (8.24). From the concaveness and increasing nature of S, from (8.23) we deduce that

$$(8.25) \qquad S\,\tilde{w} \ge (1-\beta)\,S\,w + \beta\;S\,u_0$$

but

$$(8.26) \qquad S\,u_0 \ge \lambda\,u^0 + (1-\lambda)u_0$$

since by (8.22)

$$u_0^2 + \psi^2 \ge \lambda\,u^{0,1} + (1-\lambda)u_{01}$$

and

$$\begin{aligned} a(\lambda u^{0,1} + (1-\lambda)u_{01}, v) &= \lambda(f^1,v) + (1-\lambda)a(u_{01},v) \\ &\le (f^1,v) \qquad \forall\, v \ge 0 \end{aligned}$$

therefore

$$\lambda u^{0,1} + (1-\lambda)u_{01} \leq \sigma(u_0^2 + \psi^2, f^1) .$$

Similarly,

$$\lambda\, u^{0,2} + (1-\lambda)u_{02} \leq \sigma(u_0^1 - \psi^1, f^2)$$

whence (8.25). Lastly, since

$$u^0 \geqslant Sw,$$

it follows from (8.25),(8.26) that (8.24) certainly holds. Therefore S is a contraction on K, and as a consequence has one and only one fixed point in K.□

8.2 *Generalisation: The Case of Several Obstacles*

We know (cf. A. FRIEDMAN [3], B.L. [1]) that the V.I. (8.4) arises in games problems with two players.

Here we are going to consider a case of several players which lead directly to a Q.V.I.. The latter reduces to the V.I. (8.4) as a particular case. Here we follow A. BENSOUSSAN and A. FRIEDMAN [1].

We are given a(u,v) defined by (1.3), and we assume (8.3). We are also given some functions

$$f^i \in L^\infty(\mathcal{O}), \quad i=1,\dots,N \tag{8.27}$$

$$\left|\begin{array}{l} \phi^i,\ \psi^i \in C^0(\bar{\mathcal{O}}),\quad i=1,\dots,N \\ \phi^i|_\Gamma \geq 0,\quad \psi^i|_\Gamma \leq 0,\quad \psi^i \leq \phi^i . \end{array}\right. \tag{8.28}$$

Let $u = (u^1,\dots,u^N) \in V^N$, where $V = H_0^1(\mathcal{O})$, and we define the sets

$$K^i(\hat{u}^k) = \{v \in V | v \leqslant \phi^i \text{ a.e., and for x a.e. if } u^j(x) \geqslant \phi^j(x) \text{ for a certain } j \neq i \text{ then } v(x) = \phi^i(x)\}. \tag{8.29}$$

We then pose the problem: Find $u \in V^N$ such that

$$\left|\begin{array}{l} u^i \in K^i(\hat{u}^i) \\ a(u^i, v-u^l) \geq (f^i, v-u^i) \quad \forall\, v \in K^i(\hat{u}^i),\quad i=1,\dots,N . \end{array}\right. \tag{8.30}$$

Let us first of all show how this problem generalises the double V.I.. Consider the case N = 2, and

$$\begin{array}{ll} f^1 = f, & f^2 = -f \\ \phi^1 = \phi, & \phi^2 = -\phi \\ \psi^1 = \psi, & \psi^2 = -\psi . \end{array}$$

Therefore

$$u^1 \leq \sigma, \quad u^2 \leq -\psi \tag{8.31}$$

and

$$u^2 = -\psi \quad \text{implies} \quad u^1 = \psi.$$

Then

$$(8.32)\quad \left|\begin{array}{l} a(u^1,v-u^1) \geq (f,v-u^1) \quad \forall v \leq \phi \text{ such that} \\ \qquad v = \psi \text{ if } u^2 = -\psi \\ a(u^2,v-u^2) \geq -(f,v-u^2) \quad \forall v \leq -\psi \text{ such that} \\ \qquad v = -\phi \text{ if } u^1 = \phi . \end{array}\right.$$

But then if u is a solution of

$$(8.33)\quad \left|\begin{array}{l} a(u,v-u) \geq (f,v-u) \quad \forall \psi \leq v \leq \phi \\ \text{and } \psi \leq u \leq \phi , \end{array}\right.$$

then $u^1 = u$, $u^2 = -u$ is a solution of (8.21),(8.32). In fact, (8.33) may be interpreted as

$$\begin{array}{lll} Au = f & \text{if} & \psi < u < \phi \\ Au \leq f & \text{if} & u = \phi \\ Au \geq f & \text{if} & u = \psi \end{array}$$

and therefore let $v \leqslant \phi$ be such that $v = \psi$ if $u = \psi$, then

$$a(u,v-u) = \int Au(v-u)dx = \int_{\{u>\psi\}} Au(v-u)dx$$

$$= \int_{\{\psi<u<\phi\}} f(v-u)dx + \int_{\{u=\phi\}\cap\{u>\psi\}} Au(v-u)dx \geq \int f(v-u)dx$$

and similarly we prove the second part of (8.32).

On V^N we define the set

$$(8.34)\quad K(u) = \{v \in V^N \mid v^i \in K^i(\hat{u}^i)\}$$

which is a closed convex set of V^N. Rewrite (8.30) in the form

$$(8.35)\quad \left|\begin{array}{l} a(u,v-u) \geq (f,v-u) \\ \quad \forall v \in K(u),\ u \in K(u) \end{array}\right.$$

where, naturally, we have set

$$(8.36)\quad \left|\begin{array}{l} a(u,v) = \sum_i a(u^i,v^i) \\ f = (f^1,\ldots,f^N). \end{array}\right.$$

We certainly have a Q.V.I., but one that does not fit any of the frameworks met up to now. Nevertheless, let us notice the following property of K(v)

(8.37) if $v \geqslant v'$ (in the sense $v^i(x) \geqslant v'^i(x)\ \forall i$ a.e.), then $K(v) \subset K(v')$.

We are going to use this property in a particular case. Let us now assume $N = 2$ and the restrictive hypothesis

$$(8.38) \qquad \psi^i \in V,\ a(\psi^i,\chi) \le (f^i,\chi) \qquad \forall\, \chi \in V,\ \chi \ge 0,\ i=1,2\ .$$

We then have:

THEOREM 8.4: *Assume* (1.1),(8.3),(8.27),(8.28),(8.38) *together with* $N = 2$. *If we define the interval* I *of* $H \times H$ *as*

$$I = \{v^i \in H\ ,\quad \psi^i \le v^i \le \phi^i\ ,\quad i=1,2\}$$

then the set of solutions of the Q.V.I. (8.35) *belonging to* K *is non-empty, and has a maximum element and a minimum element.*

Proof: Let $w^1,w^2 \in V$, and set

$$(8.39) \qquad \begin{aligned} K^2(w^1) &= \{v \in V \mid v \le \phi^2 \text{ a.e. and if } w^1 \ge \phi^1 \text{ then } v = \psi^2\} \\ K^2(w^2) &= \{v \in V \mid v \le \phi^1 \text{ a.e. and if } w^2 \ge \phi^2 \text{ then } v = \psi^1\}\ . \end{aligned}$$

Next we define two mappings T^1,T^2 on V in the following way. If $w^1 \in V$ is given, then $u^2 = T^2w^1$ is a solution of the V.I.

$$(8.40) \qquad a(u^2,v-u^2) \ge (f^2,v-u^2) \qquad \forall\, v \in K^2(w^1),\ u^2 \in K^2(w^1)\ .$$

Similarly we define $u^1 = T^1w^2$ as a solution of the V.I.

$$(8.41) \qquad a(u^1,v-u^1) \ge (f^1,v-u^1) \qquad \forall\, v \in K^1(w^2),\ u^1 \in K^1(w^2)\ .$$

We shall set

$$(8.42) \qquad u^1 = \Theta^1 w^1 = T^1\, T^2\, w^1$$

and we similarly define

$$\Theta^2 = T^2\, T^1.$$

If u^1 is a fixed point of Θ^1, then (u^1,T^2,u^1) is a solution of the Q.V.I. (8.35). Moreover, T^2u^1 is a fixed point of Θ^2. Similarly, if u^2 is a fixed point of Θ^2, the T^2u^2 is a fixed point of Θ^1, and (T^1u^2,u^2) is a solution of the Q.V.I.. The fundamental point will therefore consist of proving that

$$(8.43) \qquad \Theta^1 \text{ (resp. } \Theta^2) \text{ is increasing.}$$

First we show that

$$(8.44) \qquad \begin{aligned} u^2 &= T^2\, w^1 \ge \psi^2 \quad \forall\, w^1 \\ u^1 &= T^1\, w^2 \ge \psi^1 \quad \forall\, w^2\ . \end{aligned}$$

Let us notice that $\psi^1 \in K^1(w^2)$ for all w^2, therefore $\psi^1 \vee u^1 \in K^1(w^2)$, and we can therefore take

$$v = u^1 + (u^1-\psi^1)^- = \mathrm{Max}(\psi^1,u^1)$$

in (8.41), whence

$$a(u^1,(u^1-\psi^1)^-) \geq (f^1,(u^1-\psi^1)^-)$$

and by (8.38)

$$a(\psi^1,(u^1-\psi^1)^-) \leq (f^1,(u^1-\psi^1)^-)$$

from which it follows that

$$a((u^1-\psi^1)^-,(u^1-\psi^1)^-) = 0,$$

therefore $u^1 \geqslant \psi^1$. A similar argument shows that $u^2 \geqslant \psi^2$.

Now let

$$w^1 \leq w'^1 \text{ and } u^2 = T^2w^1, \quad u^1 = T^1T^2w^1 = \Theta^1w^1 .$$

We define u'^2 and u'^1 for w'^1 similarly.

First of all we show that

$$u^2 \geqslant u'^2. \tag{8.45}$$

Since $w^1 \leqslant w'^1$, we have

$$K^2(w^1) \supset K^2(w'^1).$$

Therefore $u^2,u'^2 \in K^2(w^1)$, therefore $\mathrm{Max}(u^2,u'^2) \in K^2(w^1)$ also.

Further, from $u'^2 \in K^2(w'^1)$ it follows that if $w'^1 \geqslant \phi^1$, then $u'^2 = \psi^2$, therefore $\mathrm{Min}(u^2,u'^2) \leqslant \psi^2$. But by (8.44) we also have $\mathrm{Min}(u^2,u'^2) \geqslant \psi^2$, whence $\mathrm{Min}\ (u^2,u'^2) = \psi^2$ if $w'^1 \geqslant \phi^1$. Since $\mathrm{Min}(u^2,u'^2) \leqslant u'^2 \leqslant \phi^2$, we see that $\mathrm{Min}(u^2,u'^2) \in K^2(w'^1)$.

Consequently we can take $\mathrm{Max}(u^2,u'^2)$ as test function in (8.40) and $\mathrm{Min}(u^2,u'^2)$ as test function of (8.40') (i.e. (8.40) written with w'^1 instead of w^1).

Now,

$$\mathrm{Max}(u^2,u'^2) = u^2 + (u'^2- u^2)^+$$

and

$$\mathrm{Min}(u^2,u'^2) = u'^2 - (u'^2- u^2)^+$$

and consequently

$$a(u^2,(u'^2- u^2)^+) \geq (f^2,(u'^2-u^2)^+)$$

$$a(u'^2,(u'^2-u^2)^+) \geq -(f^2,(u'^2-u^2)^+)$$

and by addition

$$a((u'^2-u^2)^+,(u'^2-u^2)^+) \le 0$$

therefore

$$(u'^2 - u^2)^+ = 0,$$

whence (8.45).

From (8.45) it next follows that $K^1(u^2) \subset K^2(u'^2)$. Therefore $u^1,u'^1 \in K^1(u'^2)$, hence $\mathrm{Max}(u^1,u'^1) \in K^1(u'^2)$ also.

Using (8.44) again we show that $\mathrm{Min}\,(u^1,u'^1) \in K^1(u^2)$. We use $\mathrm{Min}\,(u^1,u'^1)$ as test function in the V.I. with respect to u'^1 and $\mathrm{Min}(u^1,u'^1)$ as test function in the V.I. relative to u^1. From this we easily deduce that $u^1 \leqslant u'^1$, which proves (8.43).

Also, by (8.44) we have

$$\Theta^1\psi^1 \geqslant \psi^1 \tag{8.46}$$

therefore ψ^1 is a super-solution. By the construction of Θ^1 and T^1 we also have

$$\Theta^1\phi^1 \leqslant \phi^1 \tag{8.47}$$

therefore ϕ^1 is a sub-solution. Since $\psi^1 \leqslant \phi^1$, Tartar's abstract result allows us to state that Θ^1 has fixed points in the interval $[\psi^1,\phi^1]$, and also a minimum fixed point and a maximum fixed point. Let us write $\underline{u}^1$ and $\bar{u}^1$ for the minimum fixed point and the maximum fixed point. We define

$$\bar{u}^2 = T^2\,\underline{u}^1\ ,\quad \underline{u}^2 = T^2\,\bar{u}^1,$$

therefore $(\underline{u}^1,\bar{u}^2)$ and $(\bar{u}^1,\underline{u}^2)$ are two solutions of the Q.V.I..

If u^1,u^2 is now a solution, then by (8.44) we have

$$\psi^1 \le u^1 \le \phi^1 \quad , \quad \psi^2 \le u^2 \le \phi^2$$

and

$$u^1 = \Theta^1 u^1,$$

therefore

$$\underline{u}^1 \le u^1 \le \bar{u}^1$$

and therefore

$$T^2\,\bar{u}^1 \le u^2 \le T^2\,\underline{u}^1$$

also. Consequently $(\underline{u}^1,\bar{u}^2)$ and $(\bar{u}^1,\underline{u}^2)$ are minimum and maximum solutions of the Q.V.I..□

Remark 8.1

The matter of regularity is completely unresolved. It would be very interesting to prove continuity of the solution (u^1,u^2).□

8.3 *Nash Points*

In this Section we show that the problem of finding Nash points leads to the formulation of Q.V.I.'s. We begin with a general theorem (cf., for example, U. MOSCO [3]).

Let X be a locally convex T.V.S.. We are given

(8.48) $\quad C$ a compact convex subset of X,

(8.49) $\quad Q(v)$ a correspondence which associates with $v \in C$ a non-empty closed convex subset $Q(v) \subset C$, and $Q(v)$ has a fixed graph,

(8.50) $\quad J(v,w): C \times C \to R$ is l.s.c., $w \to J(v,w)$ is convex for all $v \in C$,

(8.51) $\quad \beta(v) = \inf_{w \in Q(v)} J(v,w)$ is u.s.c. on C.

THEOREM 8.5: *Under the assumptions* (8.48),(8.49),(8.50),(8.51) *the set of* $u \in C$ *such that*

$$u \in Q(u), \quad J(u,u) \le \inf_{w \in Q(u)} J(u,w) \tag{8.52}$$

is non-empty.

Proof: Assume first of all that

$$\forall\, v \in C, \quad w \to J(v,w) \text{ is strictly convex.} \tag{8.53}$$

Then for all $v \in C$ there exists one and only one point, denoted $S(v)$, which reaches the minimum of $J(v,w)$ for $w \in Q(v)$. Now, $S(v) \in Q(v) \subset C$. Thus we have defined a mapping S of C into itself. Let us prove that

(8.54) $\quad S$ is continuous.

If (8.54) is true, since C is convex and compact, from Schauder's Theorem it follows that there exists u such that $u = S(u)$, so (8.52) holds. Therefore let us show that (8.54) holds. Consider $v_n \to v$. Let us set

$$w_n = S(v_n).$$

Since $w_n \in C$, we can, even if a subsequence is picked, assume that $w_n \to w \in C$. Since Q has closed graph, and $w_n \in Q(v_n)$, we have

$$w \in Q(v). \tag{8.55}$$

Further, by the first part of (8.50) we have

$$J(v,w) \le \underline{\lim}\ J(v_n,w_n) = \underline{\lim}\ \beta(v_n) \le \overline{\lim}\ \beta(v_n)$$

and by (8.51) we also deduce from this that

$$J(v,w) \leqslant \beta(v),$$

which together with (8.55) shows that $w = S(v)$, whence (8.54).

Thus we have proved that the set of u's satisfying (8.52) is non-empty, at least if we assume (8.53). Let us prove that this result is still true even if the assumption (8.53) is no longer satisfied.

In this case, for all fixed v in C the set of points which realised the minimum of $J(v,w)$ for $w \in Q(v)$ is non-empty, but no longer is reduced to a point as before. Let us again write this set as $S(v)$. To prove that the set of u's satisfying (8.52) is non-empty comes down to proving the existence of a fixed point for the correspondence S, so

(8.56) $\quad u \in S(u).$

We can invoke Kakutani's theorem (cf., for example, J.P. AUBIN [1]). For this it suffices to show

(8.57) $\quad$ S has closed graph,

(8.58) $\quad S(v)$ is a non-empty closed convex set of C.

Let us establish (8.57). Let $v_n \to v$ and $w_n \in S(v_n)$, $w_n \to w$. Since $w_n \in Q(V_n)$ and Q has closed graph, we have $w \in Q(v)$. Proceeding as above we also see that

$$J(v,w) \leqslant \beta(v),$$

and therefore $w \in S(v)$, whence (8.57). Moreover, $S(v)$ is non-empty and closed, since $w_n \in S(v)$, and $w_n \to w$, then $w_n \in Q(v)$ implies $w \in Q(v)$, then

$$J(v,w_n) \leq J(v,w^0) \qquad \forall\, w^0 \in Q(v)$$

implies by lower semi-continuity that

$$J(v,w) \leqslant J(v,w^0),$$

and therefore $w \in S(v)$.

Lastly, $S(v)$ is convex, since $w \to J(v,w)$ is convex. In fact if $w_1,w_2 \in S(v)$, then

$$J(v,w_1) \leqslant J(v,w)$$

$$J(v,w_2) \leq J(v,w) \qquad \forall\, w \in Q(v) \text{ and } w_1,w_2 \in Q(v)$$

therefore

$$\theta w_1 + (1-\theta)w_2 \in Q(v)$$

and

$$J(v, \theta w_1 + (1-\theta)w_2) \leq J(v,w) \qquad \forall\, w \in Q(v). \qquad \square$$

Assumption (8.51) is clearly the most restrictive in practice. Let us give a sufficient condition for (8.51) to be satisfied.

LEMMA 8.1: *Assume that*

(8.59) J *is u.s.c.,*

(8.60) *if* $v_n \to v$, *then* $\forall\, z \in Q(v)$ *there exists* $z_n \in Q(v_n)$ *such that* $z_n \to z$.

Then (8.51) *is satisfied.*

Proof: Let $v_n \to v$ and $z \in Q(v)$. Then let z_n be as in (8.60). Then

$$\beta(v_n) \leqslant J(v_n, z_n),$$

and therefore

$$\overline{\lim}\ \beta(v_n) \leq \overline{\lim}\ J(v_n, z_n) \leq J(v,z)$$

and since z is arbitrary in Q(v) we see that

$$\overline{\lim}\ \beta(v_n) \leq \beta(v) . \qquad \square$$

Remark 8.2

If we take the assumptions (8.48),(8.49),(8.50) together with (8.59),(8.60), and if X is a Banach space, then in the proof of Theorem 8.5 we can avoid using Kakutani's Theorem. In fact, if we replace $J(v,w)$ by

$$J_\varepsilon(v,w) = J(v,w) + \varepsilon|w|^2$$

then J_ε is continuous and strictly convex in w for fixed v. Thus the assumption (8.53) is satisfied for J_ε, so that the result is true for J_ε, so there exists $u_\varepsilon \in C$ such that

$$u_\varepsilon \in Q(u_\varepsilon) , \qquad J_\varepsilon(u_\varepsilon, u_\varepsilon) \leq J_\varepsilon(u_\varepsilon, w_\varepsilon) \quad \forall\, w_\varepsilon \in Q(u_\varepsilon) .$$

Let us make ε tend to 0, then u_ε stays in a compact set, therefore even if we pick out a subsequence we can assume that $u_\varepsilon \to u$. But then

$$u \in Q(u),$$

but if $z \in Q(u)$ there exist

$$z_\varepsilon \to z \quad \text{and} \quad z_\varepsilon \in Q(u_\varepsilon),$$

whence

$$J(u_\varepsilon, u_\varepsilon) + \varepsilon\,|u_\varepsilon|^2 \leq J(u_\varepsilon, z_\varepsilon) + \varepsilon|z_\varepsilon|^2$$

therefore in the limit

$$J(u,u) \leq J(u,z) \quad \forall\, z \in Q(u) .$$

and hence the result again. $\square$

Remark 8.3

The set of u's satisfying (8.52) is closed. In fact, let u_n satisfy (8.52) and $u_n \to u$, then

$$u_n \in Q(u_n) \quad \text{and} \quad J(u_n,u_n) \le \beta(u_n)$$

whence

$$u \in Q(u)$$

and

$$J(u,u) \le \underline{\lim}\, J(u_n,u_n) \le \underline{\lim}\ \beta(u_n) \le \overline{\lim}\, \beta(u_n) \le \beta(u) .$$

Therefore u satisfies (8.52).□

Let us now show how Theorem 8.5 and Lemma 8.1 can be applied to the problem of Nash points of functionals subject to global constraints. Let us be given locally convex T.V.S.'s $X_1,\dots,X_N$ and

(8.61) C_i a non-empty compact subset of X_i,

(8.62) K a non-empty closed concave subset of $X_1 \times \cdots \times X_N = X$.

For $v = (v_1,\dots,v_N) \in C = C_1 \times \cdots \times C_N$

(8.63) $Q_i(v) = \{ z_i \in C_i \mid v_i,\dots,v_{i-1},z_i,v_{i+1}\cdots v_N \in K \}$

and we assume

(8.64) $Q_i(v)$ is non-empty for all $v \in C$.

Let us note that (8.61),(8.62) imply that $Q_i(v)$ is a closed convex subset in C_i. We set

(8.65) $Q(v) = Q_1(v) \times \dots \times Q_N(v)$.

It is easy to show that (8.49) is satisfied. We also assume

(8.66) $\forall\, v \in C$, if $v^n \to v$, then for $i = 1,\dots,N$; $\forall\, J_i \in C_i$ such that $(v_1,\dots,v_{i-1},z_i,v_{i+1},\dots,v_N) \in K$, there exist $z_i^n \to C_i$ such that $(v_1^n,\dots,v_{i-1}^n,z_i^n,v_{i+1}^n,\dots,v_N^n) \in K$, and $z_j^n \to z_i$.

Therefore (8.60) is satisfied. We are now given functionals

$$J_i(v_1,\dots,v_N) = J_i(v)$$

such that

(8.67) J_i is continuous on C,

$v_i \to J_i(v_i,\dots,v_{i-1},v_i,v_{i+1},\dots,v_N)$ is convex.

Nash's equilibrium problem is as follows: Find $u = (u_1,\ldots,u_N)$ such that

$$(8.68)\qquad \left| \begin{array}{l} u \in K \quad , \quad u_i \in C_i \\ J_i(u) \leq J_i(u_1,\ldots,u_{i-1},v_i,u_{i+1},\ldots,u_N) \\ \forall\, v_i \in C_i \ \text{ such that} \\ \qquad (u_1,\ldots,u_{i-1},v_i,u_{i+1},\ldots,u_N) \in K, \quad i=1\ldots N\,. \end{array} \right.$$

If we define

$$J(v,w) = \sum_{i=1}^{N} J_i(v_1,\ldots,v_{i-1},w_i,v_{i+1},\ldots,v_N)$$

then the problem (8.68) is equivalent to the problem (8.52).

We say that u is a *Nash equilibrium for the functionals* J_i, *under the global constraints* K. Theorem 8.5 and Lemma 8.1 allow us to state:

THEOREM 8.6: *Take assumptions* (8.61),(8.62),(8.64),(8.66) *and* (8.67). *Then there exists a Nash equilibrium for the functionals* J_i *under the global constraints* K.

Remark 8.4

Let us give another example where we can establish (8.51) without assuming J is continuous. We assume (8.48),(8.49) and

(8.69) $\quad J(w):C \to R$ is u.s.c. and convex,

(8.70) $\quad$ if $v_n \to v$, then $\forall\, z \in Q(v)$ there exist $z_n \in Q(V_n)$ such that $J(z_n) \to J(z)$.

Then (8.51) is satisfied. In fact, let $v_n \to v$, then

$$\beta(v_n) \leqslant J(z_n),$$

therefore

$$\overline{\lim}\ \beta(v_n) \leq J(z) \qquad \forall\, z \in Q(v)$$

whence

$$\overline{\lim}\ \beta(v_n) \leq \beta(v)\ .$$

As an application consider a bounded convex subset C_i of H (Hilbert space equipped with a positive cone). For $v = (v_1,\ldots,v_N)$, v_i C_i, we set

$$Q_i(v) = \{z_i \mid z_i \in C_i,\ L_i(z_i) + \sum_{j\neq i} c_j v_j \leq b\}$$

where the c_j are constants, the L_i *are convex homeomorphisms on* H *and* L_i^{-1} *is completely continuous.*

$$J_i(w_i) = a_i(w_i,w_i) - 2(f_i,w_i)$$

where a_i is a symmetric continuous bilinear form on E.

We consider the following problem: find $u = (u_1,\dots,u_N)$ such that

(8.71)
$$\left| \begin{array}{l} J_i(u_i) \le J_i(w_i) \\ \forall\ w_i \text{ such that } L_i(w_i) + \sum_{j\neq i} c_j u_j \le b \ ,\quad w_i \in C_i \\ L_i(u_i) + \sum_{j\neq i} c_j u_j \le b \ ,\quad u_i \in C_i \ . \end{array} \right.$$

We assume that $Q_i(v)$ is a closed convex set that is non-empty for all fixed v.

Next we define

$$J(w) = \sum_i J_i(w_i)$$

and (8.69) is satisfied. We easily show that $Q(v) = (Q_1(v),\dots,Q_N(v))$ is a closed graph correspondence in $C = C_1 \times \cdots \times C_N$.

It remains to verify assumption (8.70). So let

$$v^n = (v_1^n,\dots,v_N^n) \to v = (v_1,\dots,v_N).$$

We define ζ_i^n by

$$L_i(\zeta_i^n) = L_i(z_i) + \sum_{j\neq i} c_j v_j - \sum_{j\neq i} c_j v_j^n$$

and

$$z_i^n = P\ \zeta_i^n = \text{projection of } \zeta_i^n \text{ on the ball } C_i \ .$$

Since L_i is a homeomorphism and L_i^{-1} is completely continuous, $\zeta_i^n \to z_i$ strongly in H, therefore

$$z_i^n \to z_i \text{ strongly in H,}$$

whence

$$J_i(z_i^n) \to J_i(z_i)$$

and therefore (8.70) is satisfied.

Therefore we have the existence of a Nash point for (8.71).□

Remark 8.5

In the case where $L_i(w_i) = c_i w_i$, we cannot apply the preceding work directly. But in this case we notice that if u minimizes

$$J(w) = \sum_i J_i(w_i)$$

under the constraints

$$w_i \in C_i, \qquad \sum_i c_i w_i \leq b$$

then u is a Nash point. Thus we have directly the existence of a Nash point, and non-uniqueness is clear, because we can replace J(w) by

$$\sum \lambda_i J_i(w_i)^* \qquad (\lambda_i > 0);$$

u is a Loreto point.

8.4 *Decreasing Q.V.I.'s*

In (8.35) we have seen an example of a Q.V.I. where the correspondence $v \to K(v)$ is *decreasing*. Here we are going to consider a general case where this eventuality is produced, and study what happens to the recurrence procedures. However, we shall need a *continuity hypothesis*, as in Section 7. We follow A. BENSOUSSAN - J.L. LIONS [10].

We are given two Hilbert spaces V and H, and

(8.72) $V \subset H$, V is dense in H, and the injection of V into H is compact.

We assume that there exist an ordering relation $\geqslant 0$ on H, induced by a closed cone P. We assume that H is reticulated. More precisely, for $u_1,u_2 \in H$, there exists an element of H, denoted $\sup(u_1,u_2)$, such that

(8.73) $\sup(u_1,u_2) \geq u_1, \quad \sup(u_1,u_2) \geq u_2$

(8.74) $\sup(u_1,u_2) = u_2 + \sup(0,u_1-u_2)$.

We set

$$u^+ = \sup(u,0) \quad , \quad u^- = \sup(0,-u)$$
$$\operatorname{Inf}(u_1,u_2) = -\sup(-u_1,-u_2), \qquad \bar{u}^- = -\operatorname{Inf}(0,u)$$

therefore

$$\sup(u_1,u_2) = u_2 + (u_2-u_1)^-$$
$$\operatorname{Inf}(u_1,u_2) = u_1 - (u_2-u_1)^-$$
$$u = u^+ - u^-.$$

Furthermore, we assume that

(8.75) $u_1,u_2 \in V \Rightarrow \sup(u_1,u_2) \in V$.

We are also given a coercive continuous bilinear form $a(v_1,v_2)$ on V,

(8.76) $a(v,v) \geq \alpha \|v\|^2 \qquad \forall\, v \in V,$

and which also satisfies

(8.77) $a(v^+,v^-) = 0 \qquad \forall\, v \in V.$

Lastly we are given a mapping $v \to M(v)$ such that

(8.78) $v \to M(v)$ is continuous from $H \to V$,

(8.79) M is decreasing, i.e. $u \geqslant v \Rightarrow M(u) \leqslant M(v)$.

Let $f \in H$. If $f \geqslant 0$ it will be possible to have in place of (8.79) the assumption (of a different nature)

(8.80) $u \geq 0 \Rightarrow M(u) > 0$.

Let us give an example. Take $V = H^1(\mathcal{O})$, $H = L^2(\mathcal{O})$, and $a(u,v)$ defined by (7.1). Also take

(8.81) $$M(v) = k - \int_{\mathcal{O}} v\,dx\ , \quad k \text{ constant.}$$

This is therefore the example (7.31) with $h = k$, $g = -1$. Assumption (8.79) is satisfied.

We then have:

THEOREM 8.7: *Assume* (8.72),(8.73),(8.74),(8.75),(8.76),(8.77),(8.78),(8.79). *Then there exists a solution of the Q.V.I.*

(8.82) $$\left|\begin{array}{l} a(u,v-u) \geqslant (f,v-u)\ \forall\, v \in V \text{ such that } v \leqslant M(u), \\ u \leqslant M(u). \end{array}\right.$$

If R *is identified with a subspace of* V, *and if* $M(v) \in R$, *then the solution of* (8.82) *is unique.*

Proof: Define u^0 to be a solution of the variational problem

(8.83) $$a(u^0,v) = (f,v) \quad \forall\, v \in C,\quad u^0 \in V\ .$$

Next define u^1 as the solution of the V.I.

(8.84) $$\left|\begin{array}{l} a(u^1,v-u^1) \geq (f,v-u^1) \quad \forall\, v \in V,\ v \leq M(u^0) \\ u^1 \leq M(u^0)\ . \end{array}\right.$$

Naturally, since u^1 is the solution of an obstacle problem,

(8.85) $u^1 \leqslant u^0$.

Next define

(8.86) $$K = \{v \mid v \in H\ ,\quad u^1 \leq v \leq u^0\}\ ,$$

then the mapping $S = \sigma \circ M$, where $\sigma(\psi) = z$ is the solution of the obstacle problem ($\psi \in H$)

$$a(z,v-z) \geq (f,v-z) \quad \forall\, v \leq \psi,\quad z \leq \psi\ .$$

The essential point (where the decreasing nature of M occurs) is the following

(8.87) $S:K \to K$.

Let $w \in K$ and $z = S(w)$. Then it is clear that $z \leqslant u^0$.

Let us show that $z \geqslant u^1$. But $w \leqslant u^0$ implies $M(w) \geqslant M(u^0)$, therefore since σ is increasing, $z \geqslant u^1$. Further, if $w \in K$ and $z = S(w)$, then

(8.88) $$a(z,v-z) \geq (f,v-z) \quad \forall\, v \leq M(w), \quad z \leq M(w) .$$

We can then take $v = u^1$ as test function, since

$$u^1 \leq M(u^0) \leq M(w),$$

whence

$$a(z,u^1-z) \geq (f,u^1-z)$$

and therefore

(8.89) $$\|S(w)\| \leqslant \text{constant}, \ \forall\, w \in K.$$

Lastly, the mapping S is continuous from $H \to H$. In fact, let $w_n \to w$ in H, therefore

$$M(w_n) \to M(w) \quad \text{in } V.$$

By (8.89), writing $z_n = S(w_n)$, we have $\|z_n\| \leqslant$ constant, therefore we can pick out a subsequence, again denoted z_n, such that

$$z_n \to z \quad \text{weakly in } V.$$

In (8.88), for fixed v_0 such that $v_0 \leqslant M(w)$ we take

$$v = v_0 + M(w_n) - M(w) \leq M(w_n)$$

therefore

$$a(z_n,v_0+ M(w_n)-M(w)-z_n) \geq (f,v_0+ M(w_n)-M(w)-z_n)$$

and we can pass to the limit, from which we deduce

$$a(z,v_0-z) \geq (f,v_0-z) .$$

Since $z_n \leqslant M(w_n)$, we also have $z \leqslant M(w)$, and since v_0 is arbitrary, $v_0 \leqslant M(w)$, we have $z = S(w)$.

By Schauder's Theorem S has a fixed point, and therefore (8.82) has a solution.

Let us establish uniqueness in the case where $M(u) \in R$. In fact, let u_1,u_2 be two possible solutions. If $M(u_1) = M(u_2)$, then $u_1 = u_2$. Therefore let us assume $M(u_1) > M(u_2)$; then since σ is an increasing mapping, $u_1 \leqslant u_2$, therefore $M(u_1) \geqslant M(u_2)$, which is a contradiction.□

Remark 8.5

If, instead of (8.79), we assume (8.80) together with $f \geqslant 0$, then we have the same result as in Theorem 8.7. In fact, if $w \geqslant 0$, then $M(w) \geqslant 0$, which together with $f \geqslant 0$ implies $S(w) \geqslant 0$. But then 0 is admissible for (8.88), therefore we again have

$$\|S(w)\| \leqslant \text{constant.}$$

We can then finish as in Theorem 8.7.□

Remark 8.6

In the case of example (8.81) the sufficient condition (7.40), say $\int_{\mathcal{O}} g dx < 1$, is satisfied, and therefore the result of Theorem 8.7 already has been established.□

It is interesting to study what is given by the recurrence method

$$u^{n+1} = S(u^n) \tag{8.90}$$

in the case of the decreasing Q.V.I.. Starting from u^0 defined by (8.83), the arguments that follow do not use the fact that the injection of $V \to H$ is compact, nor assumption (8.78), but the weaker assumption

$$M \text{ is continuous from } V \to V. \tag{8.91}$$

We easily verify the following properties.

$$\left| \begin{array}{l} u^0 \geq u^2 \geq \dots \geq u^{2n-2} \geq u^{2n} \geq \dots \\ u^1 \leq u^3 \leq \dots \leq u^{2n-1} \leq u^{2n+1} \leq \dots \\ \quad u^{2n+1} \leq u^{2n} . \end{array} \right. \tag{8.92}$$

By (8.92),

$$\|u^{2n}\| \leqslant \text{constant}, \qquad \|u^{2n+1}\| \leqslant \text{constant}. \tag{8.93}$$

THEOREM 8.8: *Take the assumptions of Theorem 8.7, with* (8.91) *instead of* (8.78), *and the continuous (instead of compact) injection* $V \to E$. *Then*

$$u^{2n} \downarrow \phi, \quad u^{2n+1} \uparrow \psi \quad \textit{weakly in } V$$

where (ϕ,ψ) *is a solution of*

$$\left| \begin{array}{l} \psi \leq \phi, \ a(\phi,v-\phi) \geq (f,v-\phi) \quad \forall\, v \leq M(\psi), \phi \leq M(\psi) \\ a(\psi,v-\psi) \geq (f,v-\psi) \quad \forall\, v \leq M(\phi), \ \psi \leq M(\phi). \end{array} \right. \tag{8.94}$$

Proof: The first relation of (8.94) is obvious. Consider a convex combination of $u^1,u^3,\dots,u^{2j+1}$, denoted s^{2j+1}, such that

$$s^{2j+1} \to \psi \text{ strongly in } V.$$

Then

$$s^{2j+1} \leqslant u^{2j+1},$$

therefore

$$M(s^{2j+1}) \geq M(u^{2j+1}) \geq u^{2j+2} .$$

When $j \to \infty$, $u^{2j+2} \to \phi$ weakly in V and $M(s^{2j+1}) \to M(\psi)$ strongly in V, hence $\phi \leqslant M(\psi)$. Now let $v \leqslant M(\psi)$. Since $u^{2n-1} \leqslant \psi$, $M(u^{2n-1} \geqslant M(\psi)$ holds, therefore $v \leqslant M(u^{2n-1})$, therefore

$$a(u^{2n},v-u^{2n}) \geq (f,v-u^{2n})$$

and therefore in the limit

$$a(\phi,v-\phi) \geq (f,v-\phi)$$

which proves the second inequality of (8.94).

Let us now prove the last inequality of (8.94). Since

$$u^{2n+1} \leq M(u^{2n}) \leq M(\phi),$$

we have $\psi \leqslant M(\phi)$. Now let $v \leqslant M(\phi)$.

Consider a convex combination of $u^0,\ldots,u^{2j}$, denoted s^{2j}, such that

$$s^{2j} \to \phi \quad \text{strongly in V.}$$

Then

$$s^{2j} \geqslant u^{2j},$$

therefore

$$M(s^{2j}) \leqslant M(u^{2j}).$$

Let us set

$$v^{2j} = v + M(s^{2j}) - M(\phi) \leq M(u^{2j}) \cdot$$

Consequently

$$a(u^{2j+1},v^{2j}- u^{2j+1}) \geq (f,v^{2j}-u^{2j+1})$$

and since $v^{2j} \to v$ strongly in V, we deduce from this that

$$a(\psi,v-\psi) \geq (f,v-\psi)$$

which completes the proof.□

Remark 8.7

Other examples have been studied by L. BOCCARDO - I. CAPUZZO DOLCETTA [1].□

8.5 *Double Q.V.I.*

Here we consider the analogue of the Q.V.I. (1.7), but with bilateral constraints.

We are given the form a(u,v) on $H^1(\mathcal{O})$ defined by (1.3), with the assumptions (1.1). We shall further assume

(8.95) $a(v,v) \geq \gamma \|v\|^2 \quad \forall v \in H^1(\mathcal{O}), \quad \gamma > 0$.

Next we are given

(8.96) $f \in L^\infty(\mathcal{O})$

(8.97) $k_1, k_2 > 0$ such that $k_1 \geqslant \frac{\|f^-\|}{\beta}, \quad k_2 \geqslant \frac{\|f^+\|}{\beta}$

where $\beta > 0$ is such that $a_0(x) \geqslant \beta$ (cf. (1.1)).

Write

(8.98) $$M_1\phi(x) = k_1 + \inf_{\substack{\xi \geq 0 \\ x+\xi \in \bar{\mathcal{O}}}} \phi(x+\xi)$$

(8.99) $$M_2\phi(x) = -k_2 + \sup_{\substack{\xi \geq 0 \\ x+\xi \in \bar{\mathcal{O}}}} \phi(x+\xi) \quad {}^{(1)} \; .$$

We are interested in the Q.V.I.

(8.100) $$\left| \begin{array}{lll} a(u,v-u) \geq (f,v-u) & & \\ \forall v \in H_0^1 , & M_2 u \leq v \leq M_1 u , \\ \quad u \in H_0^1 , & M_2 u \leq u \leq M_1 u . \end{array} \right.$$

We are also given constants L_1 and L_2 such that

(8.101) $$\frac{\|f^-\|}{\beta} \leq L_1 \leq k_1 \quad , \quad \frac{\|f^+\|}{\beta} \leq L_2 \leq k_2 .$$

We then have:

THEOREM 8.9: *Assume* (1.1),(8.95),(8.96),(8.97). *Then the set of solutions* u *of* (8.100) *such that* $u \geqslant -L_1$, $u \leqslant L_2$, *is non-empty, and has a maximum element and a minimum element.*

Proof: Let $z \in L^\infty$ be such that

$$-L_1 \leqslant z \leqslant L_2.$$

Let us show that

(8.102) $M_2 z \leqslant 0 \leqslant M_1 z.$

In fact,

$$M_1 z(x) = k_1 + \inf_{\substack{\xi > 0 \\ x+\xi \in \bar{\mathcal{O}}}} z(x+\ ) \geq k_1 - L_1 \geq 0$$

(1) We could add cost variables $c_1(\xi), c_2(\xi)$ analogous to $c_0(\xi)$ (cf. (1.5)).

$$M_2 z(x) = -k_2 + \sup_{\substack{\xi \geq 0 \\ x+\xi \in \mathcal{O}}} z(x+\ ) \leq -k_2 + L_2 \leq 0\ .$$

We can therefore consider the V.I.

$$(8.103)\quad \left|\begin{array}{l} a(w,v-w) \geq (f,v-w) \\ \forall\ M_2 z \leq v \leq M_1 z \quad , \quad M_2 z \leq w \leq M_1 z \\ \quad w \in H_0^1 \ , \quad v \in H_0^1 \end{array}\right.$$

which has one and only one solution. Let us show that

$$- L_1 \leqslant w \leqslant L_2.$$

In fact $v = w - (w - L_2)^+ = \text{Min}(w,L_2)$ is admissible as test function in (8.103), since $v \in H_0^1$ and

$$\text{Min}(w,L_2) \leqslant w \leqslant M_1 z$$

and

$$\text{Min}(w,L_2) \geqslant M_2 z$$

since $w \geqslant M_2 z$ and $L_2 \geqslant M_2 L_2 \geqslant M_2 z$, because $z \leqslant L_2$. Therefore we deduce from (8.103) that

$$- a(w,(w - L_2)^+) \geqslant - (f,(w - L_2)^+) \geqslant - (f^+,(w - L_2)^+)$$

and consequently

$$0 \geqslant a((w - L_2),(w - L_2)^+) + \int_{\mathcal{O}} (L_2 a_0 - f^+)(w - L_2)^+ dx \geqslant 0$$

by the choice of L_2. From this it follows that $(w - L_2)^+ = 0$, so $w \leqslant L_2$. A similar argument shows that $w \geqslant - L_1$. We have therefore defined a mapping S of the set

$$K = \{z \in L^2(\mathcal{O}) | - L_1 \leqslant z \leqslant L_2\}$$

into itself.

Let us show that S is increasing. Let $z \leqslant \tilde{z}$ and $\tilde{w} = S(\tilde{z})$, $w = S(z)$. We can take $\text{Min}(w,\tilde{w})$ as test function in (8.103) since

$$\tilde{w} \geqslant M_2 \tilde{z} \geqslant M_2 z,$$

and $\text{Max}(w,\tilde{w})$ as test function in the problem of which w is a solution, since

$$w \leqslant M_1 z \leqslant M_1 \tilde{z};$$

by addition it follows from this that $(w - w)^+ = 0$, hence $S(z) \leqslant S(z)$.

By Tartar's abstract result the mapping S has fixed points on K, and it has a maximum fixed point and a minimum fixed point, moreover.□

Remark 8.8

If $k_2 \geqslant k_1 + \frac{\|f^+\|}{\beta}$, then the solution of the unilateral Q.V.I.

$$a(u,v - u) \geqslant (f,v - u)$$

$$\forall\, v \in H_0^1, \quad v \leqslant M_1 u, \quad u \in H_0^1, \quad u \leqslant M_1 u,$$

in the interval $-L_1, \frac{\|f^+\|}{\beta}$, where L_1 satisfies (8.101), is a solution of (8.100), since necessarily

$$u \geqslant - L_1 \geqslant - k_1 \geqslant - k_2 + \frac{\|f^+\|}{\beta}$$

$$\geqslant - k_2 + \sup_{\substack{\xi \geqslant 0 \\ x+\xi \in \mathcal{O}}} u(x + \xi) = M_2 u.$$

We shall be able to show similarly that this is the only solution of (8.100) in the same interval.□

Remark 8.9

The mapping S defined in the course of the proof of Theorem 8.9 is not concave. The problems of uniqueness and regularity are unsolved.□

9. COMMENTS

In formulating Q.V.I.'s it is as well not to commit the following error: the solution u of the Q.V.I. is not a solution of the V.I. (which may have a unique solution)

$$\left|\begin{array}{l} a(u,v - u) \geqslant (f,v - u), \\ \forall\, v \text{ such that } v \leqslant Mv,\ u \leqslant Mu. \end{array}\right.$$

In (1.13) we have defined a family of mappings T_λ, at least for λ large enough. The coercivity of the form $a(u,v) + \lambda(u,v)$ then allows us to define T_λ easily. In fact (at least if $f \quad L^\infty$) we can define T_λ for all $\lambda \geqslant 0$. From the point of view of the control interpretation this is the same T_0 which is the most interesting mapping.

Lemma 1.6 was inspired by J.L. MENALDI [2],[3].

The penalised problem introduced in Section 1.5 is natural if we remember the penalised problem associated with the V.I. of the obstacle. As we have said, the penalised problem obtained fits into the framework of Hamilton-Jacobi-Bellman equations studied in Chapter 3. Consequently it is interpreted as the optimal cost of a control problem for diffusion with jumps.

It would be interesting to give a probabilistic proof of the convergence.

The regularity problem for a Q.V.I. is very different from that of the regularity for a V.I.. For the obstacle problem ($u \leqslant \psi$), the regularity of u is, up to second order, virtually identical to that of ψ. This is easily understood, because u may be equal to ψ on a set of measure non-zero. If we consider the Q.V.I. as a V.I. with obstacle $\psi = M(u)$, the results that we can directly obtain thus are considerably reduced, since M(u) will simply be a continuous function, because u is a priori in $H^1(\mathcal{O}) \cap C^0(\bar{\mathcal{O}})$.

However, since the obstacle varies with u, that is not incompatible with the regularity of the solution u. The proof of regularity reduces to proving additional regularity properties for M(u). Let us notice that the problem of $W^{2,\infty}$-regularity remains open.

Numerous problems remain open for V.I.'s or Q.V.I.'s with quadratic Hamiltonian (Theorems 3.1 and 3.2).

The important questions are the uniqueness of the solution and the regularity. We may conjecture a positive answer.

For a purely probabilistic approach in simpler cases, cf. A. BENSOUSSAN - J.L. LIONS [1]. The difficulty in the matter of regularity arises from this (contrary to the case H = 0), that we do not know how to prove that $u \in W^{2,\infty}$ on the open sets whose closure is contained in the continuation set of C, and this is simply because $H(x,u,Du) \in L^\infty$ (and not $C^{0,\delta}$). It would be necessary to make more restrictive assumptions about the function $H(z,\lambda,p)$ (as in MOSCO [1]) to be able to obtain regularity.

In Section 4 we considered the mapping T_λ for λ large enough. We could consider T_λ for arbitrary λ. In Chapter 4 we shall need T_0.

The semi-group formulation of V.I.'s or Q.V.I.'s is interesting by virtue of the weakening of the concept of solution. It would be interesting to be able to associate effective numerical procedures with it, for after all the problem comes down to solving a linear program with an infinite number of constraints (in this matter we must indicate the works of R. GONZALES [1] and R. GONZALES - E. ROFMAN [1]). Here we have simply given an approximation by discretization of the time. It will also be necessary to consider an approximation of the semi-group $\Phi(t)$. In the case of discretization (simply in time) it would be interesting to obtain estimates of the error. The case $\alpha = 0$ (the ergodic case) remains largely open (cf., however, J.M. LASRY [1], M. ROBIN [4]).

In Section 6 we were able to consider systems of Q.V.I.'s consisting of operators M of the type introduced in Section 1. Quite generally, we can consider

$$u^m \leqslant \operatorname*{Inf}_{\mu \neq m} M^\mu u^\mu$$

where each M^μ is of type M.

It would be interesting to obtain the continuity of the maximum solution of (6.18) under less restrictive assumptions than those given in Section 6.5. In particular, can we obtain a uniform convergence of the approximations (6.16) when $k \to 0$, or of (6.21) when $h \to 0$.

The continuity methods described in Section 7 could be applied to the impulse control Q.V.I., provided that we work in $C^0(\bar{\mathcal{O}})$ and use the techniques of Section 3 to show that the solution of the V.I. stays in a bounded set of $C^\delta(\bar{\mathcal{O}})$ independent of the obstacle, in the interior of a certain class.

It would be very interesting to study more fundamentally the problem of Nash points considered in (8.30). In particular, is there a convenient approximation procedure? We can apply the technique of O. NAKOULIMA [1] to transform the double Q.V.I. (8.100) into a system of Q.V.I.'s.

Chapter 5
Evolutionary quasi-variational inequalities

INTRODUCTION

The object of study in this Chapter is evolutionary Q.V.I.'s. In practice we restrict ourselves to the impulse control Q.V.I, although we may consider the evolutionary Q.V.I.'s related to all those studied in Chapter 4.

In the elliptic case we meet an essential additional difficulty, namely the dependence with respect to t of the solution sought. This is the reason we are led to consider two concepts, that of weak Q.V.I., and that of strong Q.V.I.. In the former case the solution is only L^2 with respect to time, and the Q.V.I. is written in an integral form. In the latter, the solution is more regular with respect to time, and the Q.V.I. can be written for almost all t.

The study of weak and strong Q.V.I.'s is the object of Section 1. Section 2 is devoted to regularity. This time we look for a quite regular solution, unique furthermore. Sections 3 and 4 are devoted to the semi-group approach to evolutionary problems. Section 3 treats the object problem and Section 4 the problem of the implicit obstacle. This is the analogue of the stationary problems considered in Chapter 4, but in addition the evolutionary problems lead to non-linear semi-groups, which present interesting problems.

Using the same type of methods we could also study the analogue of various extensions carried out in Chapter 4: the case of a quadratic Hamiltonian, unbounded open sets, systems of Q.V.I.'s, "two-sided" Q.V.I.'s arising in games, etc. ... . This has not to be done, so as not to lengthen the book; however, we have briefly indicated (in Sections 5 and 6) a (partial) extension to second order hyperbolic Q.V.I.'s and to parabolic Q.V.I.'s of the second kind.

1. PARABOLIC Q.V.I.'s OF IMPULSE CONTROL

1.1 *Assumptions: Notations*

Let $\mathcal{O}$ be a bounded open set of R^n. Let us write $Q = \mathcal{O}\times\,]0,T[$, $\Gamma = \partial\mathcal{O}$, $\Sigma = \partial\mathcal{O}\times]0,T[$. Let there be given the functions

$$(1.1)\qquad \left|\begin{array}{l} a_{ij},\ a_j,\ a_o \in L^\infty(Q) \\ \Sigma a_{ij}\ \xi_i\ \xi_j \geq \alpha\ |\xi|^2 \quad \forall\ \xi \in R^n, \quad \alpha > 0\ . \end{array}\right.$$

For $u,v \in H^1(\mathcal{O})$, we define

$$(1.2)\qquad a(t;u,v) = \Sigma \int_{\mathcal{O}} a_{ij}(x,t)\ \frac{\partial u}{\partial x_j}\ \frac{\partial v}{\partial x_i}\, dx + \Sigma \int_{\mathcal{O}} a_j\ \frac{\partial u}{\partial x_j}\ v dx + \int_{\mathcal{O}} a_o u\ v dx\ .$$

We are also given

(1.3) $f \in L^2(Q)$, $f \geq 0$

(1.4) $\bar{u} \in H = L^2(\mathcal{O})$, $\bar{u} \geq 0$.

We shall also set

(1.5) $V = H_0^1(\mathcal{O})$ where $H^1(\mathcal{O})$.

When necessary we shall specify quantities.

Let us consider the operator M defined in (1.5), (1.6) of Chapter 4. We could make the data k and c_0 appearing in (1.5), (1.6) of Chapter 4 time-dependent. But to simplify matters we shall keep to the same operator M as in the stationary case.

We give the name of *"parabolic Q.V.I. of impulse control"* to the problem of finding u with

$$
(1.6)\quad \left|\begin{array}{l}
u \in L^2(0,T;V) \ , \quad u \leq Mu \\
\displaystyle\int_0^T [-(\frac{\partial v}{\partial t}, v-u) + a(t;u(t),v-u(t)) - (f(t),v-u(t))]dt + \\
\qquad\qquad + \frac{1}{2}|v(T)-\bar{u}|^2 \geq 0 \ , \\
\forall\, v \in \mathcal{K} = \{v \in L^2(0,T;V),\ \frac{\partial v}{\partial t} \in L^2(0,T;V'),\ v \leq Mu \quad \text{a.e.}\} \ .
\end{array}\right.
$$

Remark 1.1.

It is clear that (1.6) is a very weak formulation for an evolutionary problem. No regularity is demanded of $\frac{\partial u}{\partial t}$. Later we shall consider stronger forms for which $\frac{\partial u}{\partial t}$ has the same regularity.

Remark 1.2.

The function Mu only has a real meaning if u is bounded from below, which is implicitly assumed when we write Mu. In Chapter 4 we considered $u \geq 0$ (in fact $u \in L^\infty$, cf., Theorem 1.2 of Chapter 4 if $f \in L^p$, $p > \frac{n}{2}$), for we would like to consider the concept of minimal solution, which will not have a meaning here in the general case. We shall also have the condition $u \leq u^0$, at least in the general case. This is because of the absence of coercivity of the form a(u,v). This condition is automatically satisfied here (the evolutionary problem).

1.2 *Revision On Weak Parabolic V.I.'s*

We shall use a result on weak parabolic V.I.'s owed to F. MIGNOT and J.P. PUEL [1] (cf., also A. BENSOUSSAN, J.L. LIONS [1]).

Let an obstacle be given,

(1.7) $\psi \in L^2(Q)$.

Then we consider the problem

$$
(1.8)\quad \begin{array}{l}
u \in L^2(0,T;V) \ , \quad u \leq \psi \\
\displaystyle\int_0^T [-(\frac{\partial v}{\partial t}, v-u) + a(t;u(t),v-u(t)) - (f(t),v-u(t))]dt + \\
\qquad\qquad + \frac{1}{2}|v(T)-\bar{u}|^2 \geq 0 \\
\forall\, v \in \{v \in L^2(0,T;V),\ \frac{\partial v}{\partial t} \in L^2(0,T;V'),\ v \leq \psi\} \ .
\end{array}
$$

Then under the assumptions (1.1), (1.3), (1.4) (without the positivity hypothesis), (1.7), and assuming that the set of test functions v is non-empty, we can state that the set of functions u satisfying (1.8) is non-empty and has a *maximum* element. We shall call $\sigma(\psi)$ the maximum solution of (1.8).

This maximum solution is obtained as the decreasing limit of the penalised scheme

$$(1.9)\quad \begin{cases} -(\frac{\partial u^\varepsilon}{\partial t},v)+ a(t;u^\varepsilon,v) + \frac{1}{\varepsilon}((u^\varepsilon-\psi)^+,v) = (f,v) \\ u^\varepsilon(T) = \bar{u} \\ u^\varepsilon \in L^2(0,T;V), \quad \frac{\partial u^\varepsilon}{\partial t} \in L^2(0,T;V'). \end{cases}$$

Thus we have

$$(1.10)\quad u^\varepsilon \downarrow u \text{ when } \varepsilon \downarrow 0, \text{ and weak in } L^2(0,T;V).$$

The function u is the maximum solution of (1.6).

If we write $\sigma(f,\bar{u},\psi)$ in place of $\sigma(\psi)$, to indicate the dependence on $f,\bar{u}$, then σ is an increasing function of the arguments, in the sense that

$$(1.11)\quad \begin{cases} f \le \tilde{f}, \quad \bar{u} \le \tilde{\bar{u}}, \quad \psi \le \tilde{\psi} \quad \text{implies} \\ \sigma(f,\bar{u},\psi) \le \sigma(\tilde{f},\tilde{\bar{u}},\tilde{\psi}) . \end{cases}$$

Moreover,

$$(1.12)\quad \| \sigma(f,\bar{u},\psi)- \sigma(f,\tilde{\bar{u}},\tilde{\psi}) \|_{L^\infty} \le \| \psi - \tilde{\psi}\|_{L^\infty(Q)} + \| \bar{u}-\tilde{\bar{u}} \|_{L^\infty} .$$

We shall also introduce u^o, the solution of the parabolic equation

$$(1.13)\quad \begin{cases} - (\frac{\partial u^o}{\partial t},v) + a(t;u^o;v) = (f,v) \qquad \text{a.e. in } t \\ u^o(T) = \bar{u} \\ \forall v \in V . \end{cases}$$

1.3 *Existence of a Maximum Solution*

We can then state the following result for (1.6), owed to F. MIGNOT, J.P. PUEL [1]:

THEOREM 1.1: *Assume* (1.1), (1.3), (1.4), *and* (1.5) *of Chapter 4. Then the set of solutions of* (1.6) *is non-empty, and it has a maximum element, denoted* u. *Then:*

$$(1.14)\quad 0 \le u \le u^o.$$

Proof: First let us verify that every solution of (1.6) satisfies

$$(1.15)\quad u \le u^o.$$

In fact this is true at the level of the V.I. (1.8). We shall now prove that

$$(1.16)\quad u^\varepsilon \le u^o.$$

from which (1.15) will follow. Let us take $v = (u^\varepsilon - u^o)^+$ as a test function in (1.9) and (1.13), and consider their difference. We obtain

$$-(\frac{\partial}{\partial t}(u^\varepsilon-u^o),(u^\varepsilon-u^o)^+)+ a(t;u^\varepsilon-u^o,(u^\varepsilon-u^o)^+) + $$
$$+ \frac{1}{\varepsilon}((u^\varepsilon-\psi)^+,(u^\varepsilon-u^o)^+) = 0$$

whence

$$\frac{1}{2}|(u^\varepsilon-u^o)^+(t)|^2+\int_t^T a(t;(u^\varepsilon-u^o)^+,(u^\varepsilon-u^o)^+)dt \le 0$$

which certainly implies (1.16). Next we consider the iterative scheme

(1.17) $u^{n+1} = T(u^n)$ where $T = \sigma .\ M$.

By induction we verify that

(1.18) $u^o \ge u^1 \ge \dots \ge u^n \ge \dots \ge 0.$

We have just seen that $u^o \ge u^1$. Let us assume $u^{n-1} \ge u^n$, and let us show that $u^n \ge u^{n+1}$. We consider approximations $u^{n,\varepsilon}$ and $u^{n+1,\varepsilon}$, and are going to show that

(1.19) $u^{n,\varepsilon} \ge u^{n+1,\varepsilon}$

which implies the result desired. Therefore we have

$$-(\frac{\partial u^{n,\varepsilon}}{\partial t},v)+a(t;u^{n,\varepsilon},v)+\frac{1}{\varepsilon}((u^{n,\varepsilon}-Mu^{n-1})^+,v) = (f,v)$$
$$u^{n,\varepsilon}(T) = \bar{u}\ .$$

and

$$-(\frac{\partial u^{n+1,\varepsilon}}{\partial t},v)+a(t;u^{n+1,\varepsilon},v)+\frac{1}{\varepsilon}((u^{n+1,\varepsilon}-Mu^n)^+,v) = (f,v)$$
$$u^{n+1,\varepsilon}(T) = \bar{u}$$

Let us take

$$v = (u^{n+1,\varepsilon}-u^{n,\varepsilon})^+$$

and consider the difference. We obtain

$$-(\frac{\partial}{\partial t}(u^{n,\varepsilon}-u^{n+1,\varepsilon}),(u^{n+1,\varepsilon}-u^{n,\varepsilon})^+) +$$
$$+ a(t;u^{n,\varepsilon}-u^{n+1,\varepsilon},(u^{n+1,\varepsilon}-u^{n,\varepsilon})^+) +$$
$$+ \frac{1}{\varepsilon}((u^{n,\varepsilon}-Mu^{n-1})^+-(u^{n+1,\varepsilon}-Mu^n)^+,(u^{n+1,\varepsilon}-u^{n,\varepsilon})^+) = 0$$

Since $Mu^{n-1} \ge Mu^n$, the penalised term is negative or zero. From this it follows that

$$\frac{1}{2}|(u^{n+1,\varepsilon}-u^{n,\varepsilon})^+(t)|^2+\int_t^T a(s;(u^{n+1,\varepsilon}-u^{n,\varepsilon})^+,(u^{n+1,\varepsilon}-u^{n,\varepsilon}))ds \le 0$$

hence (1.19).

Furthermore, $u^o \ge 0$, therefore $Mu^o \ge 0$, hence $u^1 \ge 0$ etc. ... . So we obtain (1.18).

Let us also verify that if $\tilde{u}$ is a solution of (1.16), then

(1.20) $$u^n \geq \tilde{u}.$$

This is true for u^0. Let us assume (1.20) is satified for n-1, and show it holds for n. But we have

$$u^n = \sigma_0 M(u^{n-1})$$
$$u \leq \sigma_0 M(u)$$

Since $u^{n-1} \geq \tilde{u}$, we have

$$\sigma_0 M(u^{n-1}) \geq \sigma_0 M(\tilde{u})$$

thus

$$u^n \geq \tilde{u}.$$

Moreover, we can take v = 0 as a test function in the V.I. defining u^n. So we deduce

$$\int_0^T a(t;u^n,u^n)dt \leq C .$$

Since by (1.18) u^n stays in a bounded set of $L^2(Q)$, it follows that

(1.21) $$\| u^n \|_{L^2(0,T;V)} \leq C .$$

Therefore

(1.22) $$u^n \downarrow u \text{ and } u^n \rightarrow u \text{ weakly in } L^2(0,T;V).$$

But then

(1.23) $$Mu^n \downarrow Mu$$

and from

$$u^n \leq Mu^{n-1}$$

it follows that

(1.24) $$u \leq Mu.$$

Moreover, if $v \in L^2(0,T;V)$, $\frac{\partial v}{\partial t} \in L^2(0,T;V')$ and $v \leq Mu$, then $v \leq Mu^{n-1}$. therefore it is an admissable function for the problem u^n. From this we easily deduce that u is a solution of,(1.6). It is therefore the maximum solution, by (1.20).

1.4 *Rate of Convergence of the Iterative Process*

Let us now assume that

(1.25) $$f \in L^p(Q), \quad p > \frac{n}{2} + 1, \; \bar{u} \in C^0(\bar{\mathcal{O}}), \; \bar{u} \in V ,$$

(1.26) $\mathcal{O}$ a regular bounded open set.

We then have

$$u^o \in C^o(\bar{Q}). \tag{1.27}$$

From this it follows that the maximum solution of the Q.V.I. belongs to L^∞ also, as well as all the iterates u^n.

We are now going to prove the equivalent of the Hanouzet-Joly result (cf., Chapter 4, Section 1.3) in the parabolic case. For this we write

$$T = \sigma o M. \tag{1.28}$$

LEMMA 1.1: *The operator* T *maps* $L^\infty(Q)$ *into itself. It is a concave, increasing operator. If* $0 \le \mu \le \mathrm{Min}(1, \frac{k}{||u^o||})$ *and if* $z, \zeta \in L^\infty$, $0 \le z \le u^o$, $0 \le \zeta \le u^o$, *satisfy*

$$z - \zeta \le \zeta z \quad \text{a.e. } x,t, \text{ where } \gamma \in [0,1], \tag{1.29}$$

then

$$Tz - T\zeta \le \gamma(1-\mu)Tz . \tag{1.30}$$

Proof: The fact that T is concave and increasing follows from the properties of M,σ as in the elliptic case. Moreover, T maps the interval $[0,u^o]$ into itself. If we refer to the proof of Lemma 1.4 of Chapter 4, the property (1.30) will result from

$$T(0) \ge \mu u^o. \tag{1.31}$$

This is established with the aid of the penalised scheme. Let us write $u_1 = T(0)$ and u_1^ε as the solution of

$$\left| \begin{array}{l} -\frac{\partial}{\partial t}(u_1^\varepsilon, v) + a(t; u_1^\varepsilon, v) + \frac{1}{\varepsilon}((u_1^\varepsilon - k)^+, v) = (f,v) \\ u_1^\varepsilon(T) = \bar{u} \end{array} \right. \tag{1.32}$$

(note that $k = M(0)$); μu^o is a solution of

$$\left| \begin{array}{l} -\frac{\partial}{\partial t}(\mu u^o, v) + a(t, \mu u^o, v) = (\mu f, v) \\ \mu u^o(T) = \mu\bar{u} . \end{array} \right. \tag{1.33}$$

We then take $v = (\mu u^o - u_1^\varepsilon)^+$ as a test function in (1.32) and (1.33) and subtract. This gives

$$-\frac{\partial}{\partial t}((u_1^\varepsilon - \mu u^o), (\mu u^o - u_1^\varepsilon)^+) + a(t; u_1^\varepsilon - \mu u^o, (\mu u^o - u_1^\varepsilon)^+) + $$
$$+ \frac{1}{\varepsilon}((u_1^\varepsilon - k)^+, (\mu u^o - u_1^\varepsilon)^+) = 0$$

and $$(u_1^\varepsilon - \mu u^o)(T) = (1-\mu)\bar{u} \ge 0 .$$

But by the choice of μ

$$((u_1^\varepsilon - k)^+, (\mu u^o - u_1^\varepsilon)^+) = 0.$$

From this we easily deduce that $u_1^\varepsilon \geq \mu u^o$, and hence (1.31). □

COROLLARY 1.1: *Take the assumptions of Theorem 1.1 as well as* (1.25). *Then if* u *denotes the maximum solution of the Q.V.I. and* u^n *the sequence of functions obtained by the decreasing process, then*

$$\|u^n - u\|_{L^\infty} \leq C(1-\mu)^n . \tag{1.34}$$

Proof: We have

$$0 \leq u^o - u^1 \leq u^o.$$

So by Lemma 1.1,

$$u^1 - u^2 \leq (1 - \mu)u^1.$$

By induction we see that

$$0 \leq u^n - u^{n+1} \leq (1 - \mu)^n\, u^n \leq (1 - \mu)^n\, u^o.$$

and thus for $m \geq n$

$$0 \leq u^n - u^m \leq \frac{(1 - \mu)^n}{\mu}\, u^o$$

and when m tends to $+\infty$, we obtain (1.34). □

THEOREM 1.2: *Take the assumptions of Theorem 1.1, as well as* (1.25), (1.26) *and*

$$a_{ij} \in C^{\alpha, \frac{\alpha}{2}}(\bar{Q}), \quad \frac{\partial a_{ij}}{\partial x_k} \in L^\infty, \quad 0 < \alpha < 1. \tag{1.35}$$

In the case where $V = H^1$ *(Neumann's problem); we also assume* (1.39) *of Chapter 4, and*

$$\bar{u} \leq M\bar{u}, \ \bar{u} \in C^o(\bar{\mathcal{O}}), \ \bar{u} \in V. \tag{1.36}$$

Then the maximum solution of the Q.V.I. (1.6) *belongs to* $C^o(\bar{Q})$.

Proof: Thanks to the estimate (1.12) and (1.35), (1.36), we verify that $\psi \in C^o(\bar{Q})$ implies $\sigma(\psi) \in C^o(\bar{Q})$. It suffices to approximate ψ and $\bar{u}$ by regular functions in $C^o(\bar{Q})$, and to use the fact that if $\psi, \bar{u}$ are regular, and $f \in L^p(Q)$, $p \geq \frac{1}{2}n + 1$, then $\sigma(f,\bar{u},\psi)$ is regular. By Lemma 1.6 of Chapter 4, the operator M sends $C^o(Q)$ into itself if (1.39)(Chapter 4) is satisfied, and C^o_{+o} into C^o_+ without a particular assumption. Consequently, since u^o $C^o(\bar{Q})$, by induction we have

$$\sigma o M(u^n) = u^{n+1} \in C^o(\bar{Q}),$$

and thanks to Corollary 1.1 it follows that $u \in C^o(\bar{Q})$. □

1.5 *An Existence and Uniqueness Result*

We shall need the following result.

LEMMA 1.2: *Assume that* $a_{ij}(x)$ *are independent of* t *and satisfy* (1.1), (1.35). *Also assume that* a_j, a_o *are as in* (1.1) *and are independent of* t. *We consider* f *satisfying* (1.25) *and*

(1.37) $\quad f(x,t + h) \leq f(x,t)$ for all $h \geq 0$ a.e. $f \geq 0$.

We also consider $\psi \in C^o(\bar{Q})$, *satisfying*

(1.38) $\quad \psi(x,t+h) \leq \psi(x,t) \quad \forall h \geq 0, \quad \psi \geq 0$

and if $\quad V = H^1_0, \qquad \psi|_\Sigma \geq 0$,

$\bar{u} \in C^o(\bar{\mathcal{O}}), \quad \bar{u} \in V$ *and*

(1.39) $\quad 0 \leq \bar{u}(x) \leq \psi(x,T)$

$a(\bar{u},v) \leq (f(t),v)) \quad \forall v \in V,\ v \geq 0,$ a.e.

Let z *be the solutiom of the problem* (μ *a constant* ≥ 0).

(1.40)
$$\left| \begin{array}{l} -(\frac{\partial z}{\partial t},v) + a(z,v) + \mu((z-\psi)^+,v) = (f(t),v) \\ z(T) = \bar{u}, \qquad \forall v \in V, \text{ a.e.t.} \\ z \in L^2(0,T;V),\ \frac{\partial z}{\partial t} \in L^2(0,T;V'),\quad z \in C^o(\bar{Q}). \end{array} \right.$$

Then

(1.41)
$$\frac{\partial z}{\partial t} \leq 0,\ \frac{\partial z}{\partial t} \in L^2(Q),\ z \in L^\infty(0,T;V).$$

The norms of z *in* $L^\infty(0,T:V)$ *and* $\frac{\partial z}{\partial t}$ *in* $L^2(Q)$ *are bounded by constants independent of* μ *and* ψ *in the class* (1.38).

First Proof: We consider the discretisation $0,k,\ldots,Nk = T$ of $[0,T]$, and write

$$\psi_n(x) = \psi(x,nk)$$

$$f_n(x) = \frac{1}{k}\int_{nk}^{(n+1)k} f(x,t)dt .$$

Let us define a sequence $z_n \in V$ by the recurrence relations

(1.42)
$$\left| \begin{array}{l} z_o = \bar{u}, \quad z_n \in V \\ -(\frac{z_n - z_{n+1}}{k},v) + a(z_{n+1},v) + \\ + \mu((z_{n+1} - \psi_{N-n-1})^+,v) = (f_{N-n-1},v), \quad \forall v \in V . \end{array} \right.$$

Let us show by induction that

(1.43) $\quad z_{n+1} \geq z_n \geq 0.$

Notice that by the hypothesis (1.38), (1.37) we have

(1.44) $\quad \left| \ f_{N-n-1} \geq f_{N-n} \qquad \psi_{N-n-1} \geq \psi_{N-n} . \right.$

First we show that

$$(1.45)\qquad z_1 \geq \bar{u}.$$

By (1.42) we have

$$\left(\frac{z_1 - \bar{u}}{k}, v\right) + a(z_1, v) + \mu((z_1 - \psi_{N-1})^+, v) = (f_{N-1}, v)$$

so

$$(1.46)\qquad \left(\frac{z_1 - \bar{u}}{k}, v\right) + a(z_1 - \bar{u}, v) = -a(\bar{u}, v) + (f_{N-1}, v) - \mu((z_1 - \psi_{N-1})^+, v)\ .$$

Now take $v = (z_1 - \bar{u})^-$. By the second hypothesis (1.39) we have

$$(f_{N-1}, v) - a(\bar{u}, v) \geq 0 \text{ if } v \geq 0.$$

Moreover

$$(1.47)\qquad ((z_1 - \psi_{N-1})^+, (z_1 - \bar{u})^-) = 0.$$

for if $z_1 \geq \psi_{N-1}$ and $\bar{u} \geq z_1$, then

$$\psi_N \geq \bar{u} \geq z_1 \geq \psi_{N-1} \geq \psi_N,$$

hence (1.47). Therefore it follows from (1.46) that

$$\left(\frac{z_1 - \bar{u}}{k}, (z_1 - \bar{u})^-\right) + a(z_1 - \bar{u}, (z_1 - \bar{u})^-) \geq 0$$

from which (1.45) (assuming k small enough for the form $a(u,v) + \frac{1}{k}(u,v)$ to be coercive.

Now assume that $z_n \geq z_{n-1}$, and we shall now deduce (1.43) from it. By subtraction we have

$$(1.48)\qquad \begin{aligned} \left(\frac{z_{n+1} - z_n}{k}, v\right) + a(z_{n+1} - z_n, v) &= \left(\frac{z_n - z_{n-1}}{k}, v\right) + \\ &+ (f_{N-n-1} - f_{N-n}, v) + \\ &+ \mu((z_n - \psi_{N-n})^+ - (z_{n+1} - \psi_{N-n-1})^+, v)\ . \end{aligned}$$

Let us take $v = (z_{n+1} - z_n)^-$. We notice that

$$((z_n - \psi_{N-n})^+ - (z_{n+1} - \psi_{N-n-1})^+, (z_{n+1} - z_n)^-) \geq 0$$

for if

$$z_n \geq z_{n+1} \text{ and } z_{n+1} \geq \psi_{N-n-1},$$

then

$$z_n \geq \psi_{N-n-1} \geq \psi_{N-n}$$

and in this case

$$(z_n-\psi_{N-n})^+-(z_{n+1}-\psi_{N-n-1})^+ = z_n-z_{n+1}+\psi_{N-n-1}-\psi_{N-n} \geq 0.$$

Using the first hypothesis (1.44) also, and the recurrence, we obtain

$$\left(\frac{z_{n+1}-z_n}{k},(z_{n+1}-z_n)^-\right) + a(z_{n+1}-z_n,(z_{n+1}-z_n)^-) \geq 0$$

from which it follows that $(z_{n+1}-z_n)^- = 0$. Therefore (1.43). Positivity is easily verified by induction, taking $v = z_{n+1}^-$.

By virtue of the regularity of the coefficients a_{ij} we can always write

(1.49) $a(u,v) = a_o(u,v) + a_1(u,v)$

where $|a_1(u,v)| \leq C \|u\| |v|$

a_o *symetric* and

$a_o(u,v) \geq \beta \|u\|^2 - \lambda|u|^2$.

[In fact it suffices to note that

$$-\Sigma \frac{\partial}{\partial x_i}(a_{ij}\frac{\partial u}{\partial x_j}) = -\sum_{i,j}\frac{\partial}{\partial x_i}(\frac{1}{2}(a_{ij}+a_{ji})\frac{\partial u}{\partial x_j}) - -\frac{1}{2}\sum_{i,j}(\frac{\partial a_{ij}}{\partial x_i}-\frac{\partial a_{ji}}{\partial x_i})\frac{\partial u}{\partial x_j}].$$

Now take $v = z_{n+1}$ in (1.42). We obtain

$$(1.50)\qquad a(z_{n+1},z_{n+1}) - \left(\frac{z_n-z_{n+1}}{k},z_{n+1}\right) \leq (f_{N-n-1},z_{n+1}).$$

Let us define

$$(1.51)\qquad z_k(t) = z_{N-n-1},\ t \in\ ]nk,(n+1)k]$$

$$(1.52)\qquad \tilde{z}_k(t) = \frac{z_{N-n-1}(t-nk) + ((n+1)k-t)\,z_{N-n}}{k},\ t \in [nk,(n+1)k]$$

$$(1.53)\qquad f_k(t) = f_{n+1}\qquad t \in\ ]nk,(n+1)k]\ ;$$

now apply (1.50) with $n = N-1-j$, with j varying from 0 to $N-1$, and we obtain

$$a(z_{N-j},z_{N-j}) + \frac{1}{2k}(|z_{N-j}|^2 - |z_{N-j-1}|^2) \leq (f_j,z_{N-j})$$

and by adding over the j's lying between M and N - 1 this yields

$$(1.54)\qquad 2k\sum_{j=m}^{N-1} a(z_{N-j},z_{N-j}) + |z_{N-m}|^2 \leq |\bar{u}|^2 + 2k\sum_{j=m}^{N-1}(f_j,z_{N-j})$$

which, with the help of the definitions (1.51), (1.53)

$$2\int_{(m-1)k}^{(N-1)k} a(z_k(t),z_k(t))dt + |z_k(mk)|^2 \leq |\bar{u}|^2 + 2\int_{(m-1)k}^{(N-1)k} (f_k(t),z_k(t))dt .$$

Let us take m = [t/k] + 1, then we obtain

$$2\int_{[\frac{t}{k}]k}^{T-k} a(z_k(s),z_k(s))ds + |z_k(t)|^2 \leq |\bar{u}|^2 + 2\int_{[\frac{t}{k}]k}^{T-k} (f_k(s),z_k(s))ds .$$

Taking into account that $z_k(t) = \bar{u}$ for $t \in]T - k,T]$, and that

$$a(v,v) \geq \beta_1 \|v\|^2 - \lambda_1 |v|^2$$

this yields

$$\int_t^T \|z_k(s)\|^2 ds + |z_k(t)|^2 \leq C [1 + \int_{[\frac{t}{k}]k}^{T} + |z_k(s)|^2 ds] . \tag{1.55}$$

Notice that from (1.54) it follows that

$$|z_{N-m}|^2 \leq C [1 + k \sum_{j=m}^{N-1} |z_j|^2] \quad 0 \leq m \leq N-1$$

and from the discrete Gronwall Inequality it follows that $|z_{N-m}| \leq C$, which, when coupled with (1.55) shows that

$$\|z_k\|_{L^2(0,T;V)} \leq C , \quad \|z_k\|_{L^\infty(0,T;H)} \leq C . \tag{1.56}$$

In (1.42) let us now take $v = z_{n+1} - z_n \geq 0$. This yields

$$\frac{1}{k} |z_{n+1}-z_n|^2 + a(z_{n+1},z_{n+1}-z_n) \leq (f_{N-n-1},z_{n+1}-z_n) . \tag{1.57}$$

Using the decomposition of a(u,v) described above, then with (1.57) we can write

$$\begin{aligned} &\frac{1}{k} |z_{n+1}-z_n|^2 + \frac{1}{2} a_o(z_{n+1},z_{n+1}) - \frac{1}{2} a_o(z_n,z_n) + \\ &+ \frac{1}{2} a_o(z_{n+1}-z_n,z_{n+1}-z_n) \leq C \|z_{n+1}\| |z_{n+1}-z_n| + |f_{N-n-1}||z_{n+1}-z_n| \end{aligned} \tag{1.58}$$

or again

$$\begin{aligned} &\frac{1}{2k}|z_{n+1}-z_n|^2 + \frac{1}{2} a_o(z_{n+1},z_{n+1}) - \frac{1}{2} a_o(z_n,z_n) + \\ &+ \frac{1}{2} a_o(z_{n+1}-z_n,z_{n+1}-z_n) \leq C [k \|z_{n+1}\|^2 + k |f_{N-n-1}|^2] . \end{aligned} \tag{1.59}$$

From this relation and the definition of z_k, we easily deduce that

$$(1.60) \qquad \left\|\frac{\partial \tilde{z}_k}{\partial t}\right\|_{L^2(0,T;H)} \le C \quad , \|z_k\|_{L^\infty(0,T;V)} \le C \quad .$$

From (1.59) we also deduce that

$$\frac{|z_{n+1} - z_n|^2}{k} \le C$$

and therefore by the definition of z_k and $\tilde{z}_k$

$$(1.61) \qquad \|z_k - \tilde{z}_k\|_{L^\infty(0,T;H)} \le C\sqrt{k} \,.$$

From (1.42) applied to $n = N - 1 - j$, where $j = [\frac{t}{k}]k$, and from the definition of $\tilde{z}_k$ we deduce

$$(1.62) \qquad \begin{aligned} &- (\frac{\partial \tilde{z}_k}{\partial t}, v) + a(\tilde{z}_k([\tfrac{t}{k}]\, k), v) + \\ &+ \mu((\tilde{z}_k([\tfrac{t}{k}]\, k) - \psi([\tfrac{t}{k}]\, k))^+, v) = \\ &= (f_k([\tfrac{t}{k}]\, k), v) \ . \end{aligned}$$

Now extract a subsequence such that

$$\tilde{z}_k \to z \text{ weakly in } L^2(0,T;V)$$

$$\frac{\partial \tilde{z}_k}{\partial t} \to \frac{\partial z}{\partial t} \text{ weakly in } L^2(0,T;H).$$

thus

$$\tilde{z}_k \to z \text{ strongly in } L^2(0,T;H).$$

We notice that from (1.61)

$$\tilde{z}_k([\tfrac{t}{k}]k) - \tilde{z}_k(t) \to 0 \text{ in } L^\infty(0,T;H).$$

Since $\tilde{z}_k([\frac{t}{k}]k)$ remains bounded in $L^2(0,T;V)$ we also have

$$\tilde{z}_k([\tfrac{t}{k}]k) \to z \text{ weakly in } L^2(0,T;V) \text{ and}$$
$$\text{strongly in } L^2(0,T;H).$$

Multiply (1.62) by $\phi \in \mathcal{D}(]0,T[)$ and integrate over $(0,T)$. Using the preceding convergences we easily deduce that z satisfies equation (1.40). Moreover, $\tilde{z}_k(T) = z_0 = \bar{u}$.

If we write

$$W(0,T) = \{z \in L^2(0,T;V), \frac{\partial z}{\partial t} \in L^2(0,T;V')\}$$

the mapping $z \to z(T)$ is continous linear from $W(0,T)$ into H. Therefore $\tilde{z}_k(T)$

converges weakly to z(T) in H. Therefore z satisfies the second condition (1.40).

Now the solution of (1.40) is unique, and the solution obtained by the discretisation process satisfies (1.41), which completes the proof. □

Second Proof: Another proof consists of first assuming f and ψ to be regular, with (1.37),(1.38), then to pass the limit. In fact we can find a family of functions $f_\alpha, \psi_\alpha, \bar{u}_\alpha$ such that

$$f_\alpha, \psi_\alpha \in C^1(\bar{Q}), \quad \frac{\partial f_\alpha}{\partial t} \leq 0, \quad \frac{\partial \psi_\alpha}{\partial t} \leq 0,$$
$$f_\alpha \to f, \quad \psi_\alpha \to \psi \quad \text{in} \quad C^0(\bar{Q}),$$
$$\bar{u}_\alpha \in C^1(\bar{\mathcal{O}}) \cap V, \quad u_\alpha \to \bar{u} \quad \text{in} \quad C^0(\bar{\mathcal{O}}) \cap V,$$

with

$$0 \leq \bar{u}_\alpha \leq \psi(T) \text{ and } a(\bar{u}_\alpha, v) \leq (f_\alpha(T), v) \ \forall v \in V, \ v \geq 0.$$

If z_α denotes the corresponding solution of (1.40) it suffices to establish (1.41) for z_α. Consequently everything reduces to proving (1.41) holds, assuming the data to be regular, and with

$$\frac{\partial f}{\partial t} \leqslant 0, \quad \frac{\partial \psi}{\partial t} \leqslant 0 \text{ in } Q.$$

Differentiate (1.40) with respect to t, and write:

$$\frac{\partial z}{\partial t} = \zeta.$$

We obtain:

$$-(\frac{\partial \zeta}{\partial t}, v) + a(\zeta, v) + \mu(\frac{\partial}{\partial t}((z-\psi)^+), v) = (\frac{\partial f}{\partial t}, v).$$

Setting t = T in (1.40), and noting that $(\bar{u} - \psi)^+ = 0$, we obtain

$$-(\frac{\partial z}{\partial t}(T), v) + a(\bar{u}, v) = (f(T), v)$$

from which it follows that

$$\zeta(T) = \frac{\partial z(T)}{\partial t} \leqslant 0.$$

Then taking $v = \zeta^+$ in the equation for ζ, we obtain

$$-\frac{1}{2}\frac{d}{dt}|\zeta^+|^2 + a(\zeta^+) = (\frac{\partial f}{\partial t}, \zeta^+);$$

Integrating over (t,T), then since $\zeta^+(T) = 0$, we obtain

$$\frac{1}{2}|\zeta^+(t)|^2 + \int_t^T a(\zeta^+)\, d\sigma \leq 0$$

from which it follows that $\zeta^+ = 0$, and consequently $\frac{\partial z}{\partial t} \leqslant 0$.

We then take $v = -\frac{\partial z}{\partial t}$ in (1.40), and this yields

$$\left|\frac{\partial z}{\partial t}\right|^2 - \frac{1}{2}\frac{d}{dt} a_o(z(t)) - a_1(z,\frac{\partial z}{\partial t}) + \mu((z-\psi^+),-\frac{\partial z}{\partial t}) =$$
$$= (f, -\frac{\partial z}{\partial t}).$$

Therefore

$$\int_t^T \left|\frac{\partial z}{\partial t}(\sigma)\right|^2 d\sigma + \frac{1}{2} a_o(z(t)) \le \int_t^T |f|\left|\frac{\partial z}{\partial t}\right| d\sigma + c\int_t^T \|z\| \left|\frac{\partial z}{\partial t}\right| ds +$$
$$+ \frac{1}{2} a_o(z(T))$$

from which it follows that

$$\|z\|_{L^\infty(0,T;V)} + \left\|\frac{\partial z}{\partial t}\right\|_{L^2(0,T;H)} \le C ,$$

where c is independent of μ and ψ, and depends on the norm of f in $L^2(0,T;H)$ and on the norm of $\bar{u}$ in V. □

We next have:

THEOREM 1.3: *Taking the assumptions of* 1.2, *and*

(1.63) *the coefficients* a_{ij}, a_j, a_o *are independent of* t

(1.64) $f(x,t+h) \leqslant f(x,t) \quad \forall h \geqslant 0$

(1.65) $a(\bar{u},v) \leqslant (f(t),v) \quad \forall v \in V \geqslant 0 \quad$ a.e.t.

Then the maximum solution of the Q.V.I. (1.6) *satisfies*

(1.66) $\frac{\partial u}{\partial t} \le 0, \quad \frac{\partial u}{\partial t} \in L^2(0,T;H), \quad u \le Mu, \quad 0 \le u \le u^o$

$u \in L^2(0,T;V), \quad u \in C^o(\bar{Q})$

(1.67)
$$\left| \begin{array}{l} (-\frac{\partial u}{\partial t}, v(t)-u(t)) + a(u(t), v(t)-u(t)) - (f(t),v(t)-u(t)) \ge 0 \\ \text{a.e.t.}, \quad \forall v \in L^2(0,T;V), \quad v(t) \le Mu(t) \quad \text{a.e.} \\ u(T) = \bar{u}. \end{array}\right.$$

Furthermore, the solution of (1.66), (1.67) *is unique.*

Proof: Consider the following iterative method. Let $\mu(n)$ be an increasing sequence which tends to $+\infty$. Define $\tilde{u}^n$ by

(1.68) $\tilde{u}^o = u^o$

(1.69)
$$\left| \begin{array}{l} -(\frac{\partial \tilde{u}^n}{\partial t},v) + a(\tilde{u}^n,v) + \mu(n)\,((\tilde{u}^n - M\tilde{u}^{n-1})^+,v) = \\ \qquad = (f(t),v) \\ \tilde{u}^n(T) = \bar{u} \end{array}\right.$$

(1.70) $\tilde{u}^n \in L^2(0,T;V), \quad \frac{\partial \tilde{u}^n}{\partial t} \in L^2(0,T;H), \quad \frac{\partial \tilde{u}^n}{\partial t} \le 0 .$

By Lemma 1.2 we verify that (1.68),(1.69),(1.70) define one and only one sequence. In fact since the regularity theorems apply to $\tilde{u}^n \in C^0(\bar{Q})$, and since M is increasing, the obstacle $\psi = M(u^n)$ certainly satisfies (1.38), by induction.

Next we have

$$(1.71) \qquad \tilde{u}^0 \geq \tilde{u}^1 \ldots \geq \tilde{u}^n \ldots \geq 0 .$$

Let us verify simply that if $\tilde{u}^{n-1} \geqslant \tilde{u}^n$, then $\tilde{u}^n \geqslant \tilde{u}^{n+1}$. Take $v = (\tilde{u}^{n+1} - \tilde{u}^n)^+$ in (1.69) written with index n - 1, as well as in (1.69), and subtract the two equations. This yields:

$$(1.72) \qquad -(\frac{\partial}{\partial t}(\tilde{u}^{n+1} - \tilde{u}^n),(\tilde{u}^{n+1} - \tilde{u}^n)^+) + a(\tilde{u}^{n+1} - \tilde{u}^n,(\tilde{u}^{n+1} - \tilde{u}^n)^+) + + X = 0$$

where

$$\begin{aligned} X &= \mu(n+1)((\tilde{u}^{n+1}-M\tilde{u}^n)^+-(\tilde{u}^n-M\tilde{u}^{n-1})^+,(\tilde{u}^{n+1}-\tilde{u}^n)^+) + \\ &+ (\mu(n+1)-\mu(n))((\tilde{u}^n-M\tilde{u}^{n-1})^+,(\tilde{u}^{n+1}-\tilde{u}^n)^+) \\ &\geq \mu(n+1)\ ((\tilde{u}^{n+1} - M\tilde{u}^n)^+ - (\tilde{u}^n - M\tilde{u}^{n-1})^+,(\tilde{u}^{n+1}-\tilde{u}^n)^+) \\ &\geq 0 \end{aligned}$$

For if $\tilde{u}^{n+1} \geqslant \tilde{u}^n$, since $\tilde{u}^{n-1} \geqslant \tilde{u}^n$ implies $M\tilde{u}^{n-1} \geqslant M\tilde{u}^n$, we have $\tilde{u}^{n+1}-M\tilde{u}^n \geqslant \tilde{u}^n-M\tilde{u}^{n-1}$. But then (1.72) implies $(\tilde{u}^{n+1}-\tilde{u}^n)^+ = 0$. Applying Lemma 1.2 immediately gives

$$(1.73) \qquad \| \tilde{u}^n \|_{L^\infty(0,T;V)} \leq C \qquad \left\| \frac{\partial \tilde{u}^n}{\partial t} \right\|_{L^2(0,T;H)} \leq C .$$

From (1.71) and (1.73), we therefore have

$$(1.74) \qquad \tilde{u}^n \downarrow u$$

$$\tilde{u}^n \to u \text{ weakly in } L^2(0,T;V)$$

$$\frac{\partial \tilde{u}^n}{\partial t} \to \frac{\partial u}{\partial t} \text{ weakly in } L^2(0,T;V),$$

from which we deduce

$$(1.75) \qquad 0 \leqslant u \leqslant u^0, \ \frac{\partial u}{\partial t} \leqslant 0, \ u(T) = \bar{u}.$$

Furthermore, by taking $v = u^n$ we deduce from (1.69) that

$$\int_0^T((\tilde{u}^n - M\tilde{u}^{n-1})^+, \tilde{u}^n)dt \leq \frac{C}{\mu(n)} ;$$

now,

$$\tilde{u}^n \geqslant u,$$

whence

$$\int_0^T ((\tilde{u}^n - M\tilde{u}^{n-1})^+, u)\, dt \le \frac{C}{\mu(n)} .$$

Since $\tilde{u}^n \downarrow u$, $M\tilde{u}^n \downarrow Mu$, at the limit we therefore obtain

$$\int_0^T ((u - Mu)^+, u)\, dt = 0$$

and since $u \geqslant 0$, it follows that $u \leqslant Mu$.

From (1.69) we deduce that

$$-(\frac{\partial \tilde{u}^n}{\partial t}, v-\tilde{u}^n) + a(\tilde{u}^n, v-\tilde{u}^n) + \mu(n)((\tilde{u}^n - M\tilde{u}^{n-1})^+; v-\tilde{u}^n) = (f(t), v-\tilde{u}^n).$$

If $v \leqslant M\tilde{u}^{n-1}$, then

$$((\tilde{u}^n - M\tilde{u}^{n-1})^+, v-\tilde{u}^n) = ((\tilde{u}^n - M\tilde{u}^{n-1})^+, v - M\tilde{u}^{n-1} + M\tilde{u}^{n-1} - \tilde{u}^n) \le 0$$

and therefore

$$-(\frac{\partial \tilde{u}^n}{\partial t}, v-\tilde{u}^n) + a(\tilde{u}^n, v-\tilde{u}^n) \ge (f(t), v-\tilde{u}^n) . \tag{1.76}$$

Let

$$v \in L^2(0,T;V), \quad \frac{\partial v}{\partial t} \in L^2(0,T;V'), \quad v(t) \le M u(t) .$$

We can take $v = v(t)$ in (1.76), since $v(t) \leqslant M\tilde{u}^{n-1}(t)$, therefore

$$-(\frac{\partial v}{\partial t}, v-\tilde{u}^n) + (\frac{\partial}{\partial t}(v-\tilde{u}^n), v-\tilde{u}^n) + a(\tilde{u}^n, v-\tilde{u}^n) \ge (f(t), v-\tilde{u}^n) .$$

Integrating over $(0,T)$ we obtain

$$- \int_0^T (\frac{\partial v}{\partial t}, v-\tilde{u}^n) dt + \frac{1}{2} |v(T) - \bar{u}|^2 + \int_0^T a(\tilde{u}^n, v-\tilde{u}^n) dt \ge \int_0^T (f(t), v-\tilde{u}^n) dt . \tag{1.77}$$

Passing to limit in n, we then deduce

$$- \int_0^T (\frac{\partial v}{\partial t}, v-\tilde{u}) dt + \frac{1}{2} |v(T) - \bar{u}|^2 + \int_0^T a(u, v-u) dt \ge \int_0^T (f(t), v(t)-u(t)) dt . \tag{1.78}$$

Let

$$w \in L^2(0,T;V), \quad \frac{\partial w}{\partial t} \in L^2(0,T;V') \text{ and } w(t) \le Mu(t).$$

In (1.78) we take

$$v = \theta u + (1-\theta)w$$

thus

$$-\int_0^T (\theta\frac{\partial u}{\partial t} + (1-\theta)\frac{\partial w}{\partial t}, w-u)dt + \frac{1}{2}(1-\theta)\,|w(T) - \bar{u}|^2 + \\ + \int_0^T a(u, w-u)dt \geq \int_0^T (f(t), w-u)dt$$

and making $\theta \to 1$, it follows that

$$(1.79)\qquad -\int_0^T (\frac{\partial u}{\partial t}, w-u)dt + \int_0^T a(u,w-u)dt \geq \int_0^T (f(t), w-u)dt$$

Let t_0 be fixed and $\delta > 0$, $[t_0-\delta, t_0] \subset [0,T]$. Then consider a function $\theta(t)$, which satisfies, $\theta(t) = 0$ for $t \notin\,]t_0-\delta, t_0[$ and $0 \leqslant \theta(t) \leqslant 1$. Lastly, let $v \leqslant Mu(t_0)$. Then we take

$$w(t) = (1-\theta(t))u(t) + \theta(t)v$$

which is permissible, since for $t \in\,]t_0-\delta, t_0[$ we have

$$Mu(t) \geqslant Mu(t_0) \geqslant v, \text{ hence } w(t) \leqslant Mu(t).$$

Therefore we obtain

$$-\int_{t_0-\delta}^{t_0} (\frac{\partial u}{\partial t}, v-u)\,\theta(t)dt + \int_{t_0-\delta}^{t_0} a(u,v-u)\,\theta(t)dt \geq \\ \int_{t_0-\delta}^{t_0} (f(t), v-u)\theta(t)dt$$

Next we take a sequence of functions $\theta \to 1$ at every point of $]t_0-\delta, t_0[$, and

$$(1.80)\qquad \left|\begin{array}{l} -\displaystyle\int_{t_0-\delta}^{t_0} (\frac{\partial u}{\partial t}, v-u)dt + \int_{t_0-\delta}^{t_0} a(u,v-u)dt \geq \int_{t_0-\delta}^{t_0} (f(t), v-u(t))dt \\ \forall\ v \leq Mu(t_0). \end{array}\right.$$

Now let $w \in L^2(0,T;V)$ be such that $w(t) \leqslant Mu(t)$ a.e.. Let

$$w_k(t) = \frac{1}{k}\int_{nk}^{(n+1)k} w(s)ds = w_n \quad \text{for } t \in\,]nk,(n+1)k[\ .$$

For $s \in [nk, (n+1)k]$ we have

$$w(s) \leqslant Mu(s) \leqslant Mu(nk)$$

and therefore $w_n \leqslant Mu(nk)$. Applying (1.80) with $t_0 = nk$, $\delta = k$, $v = w_n$, we obtain

$$- \int_{(n-1)k}^{nk} (\frac{\partial u}{\partial t}, w_k(t+k)-u(t))dt + \int_{(n-1)k}^{nk} a(u,w_k(t+k)-u(t))dt \geq$$
$$\geq \int_{(n-1)k}^{nk} (f(t),w_k(t+k)-u(t))dt$$

and therefore by adding for $n = 1,\ldots, N-1$

$$- \int_0^{(N-1)k} (\frac{\partial u}{\partial t}, w_k(t+k)-u(t))dt + \int_0^{(N-1)k} a(u,w_k(t+k)-u(t))dt \geq$$
$$\geq \int_0^{(N-1)k} (f(t),w_k(t+k)-u(t))dt$$

so by making k tend to 0 we can easily deduce

$$(1.81) \qquad - \int_0^T (\frac{\partial u}{\partial t}, w(t)-u(t))dt + \int_0^T a(u,w-u)dt \geq \int_0^T (f(t),w(t)-u(t))dt$$
$$\forall\, w \in L^2(0,T;V), \quad w(t) \leq Mu(t) \quad \text{a.e.}$$

Now let $v \in L^2(0,T;V)$ be such that $v(t) \leqslant Mu(t)$ a.e.. Then for a fixed t_0 in $]0,T[$, we take $w(t) = v(t)$ on $]t_0-\eta, t_0+\eta[$ and $w(t) = u(t)$ otherwise, and obtain

$$- \int_{t_0-\eta}^{t_0+\eta} (\frac{\partial u}{\partial t}, v(t)-u(t))dt + \int_{t_0-\eta}^{t_0+\eta} a(u,v-u)dt \geq \int_{t_0-\eta}^{t_0+\eta} (f(t),v-u)dt \,.$$

Dividing by η and making η tend to 0, we deduce

$$- (\frac{\partial u}{\partial t}(t_0), v(t_0)-u(t_0)) + a(u(t_0),v(t_0)-u(t_0)) \geq (f(t_0),v(t_0)-u(t_0))$$

at every Lebesgue point of the function

$$- (\frac{\partial u}{\partial t}, v-u) + a(u,v-u) - (f,v-u)$$

hence the result (1.67).

In (1.78) we saw that u was a solution of the Q.V.I. (1.6). It is necessarily the maximum solution. In fact, since (1.69) is a penalised equation for the parabolic V.I. corresponding to the obstacle Mu^{n-1}, we have

$$\tilde{u}^n > \sigma 0 M(\tilde{u}^{n-1}).$$

Furthermore if u is a solution of the Q.V.I. (1.6), then

$$\tilde{u} > \sigma 0 M(\tilde{u}).$$

Therefore if we assume $\tilde{u} \leqslant \tilde{u}^{n-1}$, then $\tilde{u} \leqslant \tilde{u}^n$ necessarily. Now $\tilde{u} \leqslant u^0$, hence $\tilde{u} \leqslant \tilde{u}^n$ for all n, and therefore $\tilde{u} \leqslant u$.

It remains to establish the uniqueness of the solution of (1.66),(1.67).

We shall use Laetsch's argument (cf.(1.45) of Chapter 4). Let u and $\tilde{u}$ be two solutions of (1.66),(1.67), and let γ be the largest number lying between 0 and 1 such that

(1.82) $\gamma u \leqslant \tilde{u}.$ a.e.

We are going to show that $\gamma = 1$. Let us assume $0 \leqslant \gamma < 1$, and construct γ' such that

(1.83) $\gamma' u \leqslant \tilde{u}$ a.e. and $\gamma < \gamma' \leqslant 1.$

But $\gamma' u$ satisfies

$$(1.84)\quad \left| \begin{array}{l} (-\frac{\partial}{\partial t}(\gamma' u), v(t)-\gamma' u(t)) + a(\gamma' u(t), v(t)-\gamma' u(t)) - \\ \qquad -(\gamma' f(t), v(t)-\gamma' u(t)) \geq 0 \\ \forall v \in L^2(0,T;V), \quad v(t) \leq \gamma' Mu(t) \text{ a.e.} \\ \qquad \gamma' u(t) \leq \gamma' Mu(t) \end{array} \right.$$

and $\tilde{u}$ satisfies

$$(1.85)\quad \left| \begin{array}{l} (-\frac{\partial}{\partial t}\tilde{u}, v(t)-\tilde{u}) + a(\tilde{u}, v-\tilde{u}) - (f(t), v-\tilde{u}) \geq 0 \\ \forall v \in L^2(0,T;V), \quad v \leq M\tilde{u} \quad , \quad \tilde{u} \leq M\tilde{u} . \end{array} \right.$$

Let us take γ' so that

$$\gamma' Mu \leqslant M(\gamma u) \leqslant M\tilde{u} \text{ by (1.82)}.$$

Therefore it suffices to take γ' such that

$$(1.86)\quad \frac{1-\gamma'}{\gamma'-\gamma} = \frac{\|u\|}{k}$$

We can then take $\gamma' u - (\gamma' u - \tilde{u})^+$ as test function in (1.84) and $\tilde{u} + (\gamma' u - \tilde{u})^+$ in (1.85). By adding we obtain

$$-(\frac{\partial}{\partial t}(\tilde{u}-\gamma' u), (\gamma' u-\tilde{u})^+) + a(\tilde{u}-\gamma' u, (\gamma' u-\tilde{u})^+) \geq$$
$$\geq (1-\gamma')(f(t), (\gamma' u-\tilde{u})^+) .$$

Since $(\gamma' u - \tilde{u})^+(T) = 0$, we easily deduce (1.83), which completes the proof. □

Remark 1.3

The penalised scheme used in the course of the proof of Theorem 1.3, which is interesting in itself, is not indispensible. We could have used the decreasing iterative process of Theorem 1.1 if we had the equivalent of Lemma 1.2 for the V.I.'s, which is a consequence of Lemma 1.2 (independence of the estimates with respect to the parameter μ). □

Remark 1.4

It is convenient to use the following terminology: we shall speak of a *weak Q.V.I.* for (1.6), and of a *strong Q.V.I.* for (1.66),(1.67). □

Remark 1.5

It would be interesting to prove that the formulation (1.67) is equivalent to

$$(1.87)\quad \left| \begin{array}{l} -(\frac{\partial u}{\partial t}, v-u(t)) + a(u(t), v-u(t)) - (f(t), v-u(t)) \geq 0 \\ \qquad \text{a.e.t.} \quad \forall v \in V, \quad v \leq Mu(t). \end{array} \right.$$ □

1.6 *Penalised Problem*

With the Q.V.I. (1.6) we associate a penalised problem in the following way. Consider

(1.88)
$$\left|\begin{array}{l} -(\frac{\partial u^\varepsilon}{\partial t},v) + a(t;u^\varepsilon,v) + \frac{1}{\varepsilon}((u^\varepsilon - Mu^\varepsilon)^+,v) = (f(t),v) \\ \text{a.e.t.} \quad \forall v \in V, \\ u^\varepsilon(T) = \bar{u} \\ u^\varepsilon \in L^2(0,T;V), \quad \frac{\partial u^\varepsilon}{\partial t} \in L^2(0,T;V'). \end{array}\right.$$

LEMMA 1.3: *Under the assumption of 1.2. there exists one and only one solution of the problem* (1.88) *such that* $u^\varepsilon \in W^{2,1,p}(Q)$.

Proof: This results from the property that (1.88) is a Hamilton-Jacobi-Bellman equation (cf. Chapter 3, Section 3.4., in the case where $V = H_0^1$). In fact we can write

(1.89)
$$(u-Mu)^+ = -\inf_{\substack{0\le d\le 1\\ \xi\ge 0\\ x+\xi\in\bar{\mathcal{O}}}} d\,[u(x+\xi) + c_0(\xi)+k - u(x)]\ .$$

First of all let us notice that if we use the transformation

$$u_\varepsilon = \tilde{u}_\varepsilon e^{+\lambda_\varepsilon(T-t)}$$

then u is a solution of

(1.90)
$$-(\frac{\partial \tilde{u}_\varepsilon}{\partial t},v) + a(t;\tilde{u}_\varepsilon,v) + \lambda_\varepsilon(\tilde{u}_\varepsilon,v) + + \frac{1}{\varepsilon}((\tilde{u}_\varepsilon - \tilde{M}\tilde{u}_\varepsilon)^+,v) = (\tilde{f},v)$$

where

(1.91)
$$\tilde{M}\phi = k\,e^{-\lambda_\varepsilon(T-t)} + \inf_{\substack{\xi\ge 0\\ x+\xi\in\bar{\mathcal{O}}}} (c_0(\xi)e^{-\lambda_\varepsilon(T-t)} + \phi(x+\xi))$$

(1.92)
$$\tilde{f} = f\,e^{-\lambda_\varepsilon(T-t)}.$$

We choose λ_ε so that

(1.93)
$$\inf a_0 + \lambda_\varepsilon > \frac{1}{\varepsilon}\ .$$

The existence and uniqueness of the solution of (1.90) is equivalent to that of (1.88). Then let $z \in L^\infty(Q)$, and define ζ as the solution of

(1.94)
$$\left|\begin{array}{l} -(\frac{\partial \zeta}{\partial t},v) + a(t;\zeta,v) + \lambda_\varepsilon(\zeta,v) + \\ \qquad + \frac{1}{\varepsilon}((\zeta - \tilde{M}z)^+,v) = (\tilde{f},v) \\ \zeta \in W^{2,1,p}(Q)\ . \end{array}\right.$$

In particular this defines a mapping S of $L^\infty(Q)$ into itself. We are going to show it is a contraction. Let d(x) be a Borel function, with values in [0,1], such that

$$(\zeta - \tilde{M}z)^+ = d(\zeta - \tilde{M}z) \quad \text{a.e.} \tag{1.95}$$

Let us consider $z_1, z_2 \in L^\infty(Q)$ and ζ_1, ζ_2 solutions of (1.94). Let us take $v \geq 0$, by subtracting we obtain

$$-\left(\frac{\partial}{\partial t}(\zeta_1-\zeta_2),v\right)+ a(t;\zeta_1-\zeta_2,v) + \lambda_\varepsilon (\zeta_1-\zeta_2,v) + $$
$$+ \frac{1}{\varepsilon}(d_2(\zeta_1- \tilde{M}z_1)-d_2(\zeta_2 - \tilde{M}z_2),v) \leq 0$$

where d_2 is the function defined in (1.95) with respect to ζ_1. We take $v = (\zeta_1 - \zeta_2 - K)^+$, and obtain

$$-\left(\frac{\partial}{\partial t}(\zeta_1-\zeta_2-K),(\zeta_1-\zeta_2-K)^+\right)+ a(t;\zeta_1-\zeta_2-K,(\zeta_1-\zeta_2-K)^+)$$
$$+ K(a_0,(\zeta_1-\zeta_2-K)^+) +$$
$$+ \lambda_\varepsilon|(\zeta_1-\zeta_2-K)^+|^2 + K(\lambda_\varepsilon,(\zeta_1-\zeta_2-K)^+) +$$
$$+ \frac{1}{\varepsilon}(d_2(\zeta_1-\zeta_2-K)^+,(\zeta_1-\zeta_2-K)^+) + \frac{K}{\varepsilon}(d_2,(\zeta_1-\zeta_2-K)^+)$$
$$+ \frac{1}{\varepsilon}(d_2(\tilde{M}z_2- \tilde{M}z_1),(\zeta_1-\zeta_2-K)^+) \leq 0 . \tag{1.96}$$

But

$$\| \tilde{M}z_2- \tilde{M}z_1\|_{L^\infty} \leq \|z_1- z_2\|_{L^\infty}$$

therefore

$$K(a_0+ \lambda_\varepsilon+ \frac{d_2}{\varepsilon}) - \frac{d_2}{\varepsilon}(\tilde{M}z_2- \tilde{M}z_1) \geq K(\inf a_0+ \lambda_\varepsilon)- \frac{1}{\varepsilon}\| z_1-z_2\| = 0$$

if

$$K = \frac{\frac{1}{\varepsilon}\|z_1- z_2\|}{\inf a_0+ \lambda_\varepsilon} .$$

From (1.96) we then easily deduce that $(\zeta_1 - \zeta_2 - K)^+ = 0$, therefore $\zeta_1 - \zeta_2 \leq K$. By a symmetric argument we have in fact

$$\| \zeta_1 - \zeta_2\| \leq K.$$

By the choice of λ_ε ((1.93)), we immediately deduce that S is a contraction, and thus we have the desired result. □

THEOREM 1.4: *Take the assumptions of Theorem* 1.2.*; the sequence* u_ε *is decreasing when* $\varepsilon \downarrow 0$ *and* $u_\varepsilon \downarrow u$ *as well as in* $L^2(0,T;V)$, *where* u *is the maximum solution of the Q.V.I.* (1.6).

Proof: With (1.88) we associate a decreasing iterative scheme in the following way

$$u_0^\varepsilon = u^0 \tag{1.97}$$

then

$$(1.98)\qquad -(\frac{\partial u_n^\varepsilon}{\partial t},v)+a(t;u_n^\varepsilon,v)+\frac{1}{\varepsilon}((u_n^\varepsilon - Mu_{n-1}^\varepsilon)^+,v) = (f(t),v)$$

$$u_n^\varepsilon(T) = \bar{u} .$$

Now, when

$$(1.99)\qquad \begin{cases} u^0 \geq u_1^\varepsilon \dots \geq u_n^\varepsilon \dots \geq 0 \\ u_n^\varepsilon \downarrow u^\varepsilon \quad \text{when } n \uparrow +\infty. \end{cases}$$

This property is obtained exactly as for the decreasing scheme associated with the Q.V.I.. We also have

$$(1.100)\qquad u_n^\varepsilon \geqslant u^n \text{ and } u_n^\varepsilon \downarrow u^n \text{ when } \varepsilon \downarrow 0.$$

In fact, since (1.98) can be considered as a penalised equation relative to the parabolic V.I. whose obstacle is Mu_{n-1}^ε, we then have

$$u_n^\varepsilon \geqslant \sigma o M u_{n-1}^\varepsilon$$

Therefore, if by induction $u_{n-1}^\varepsilon \geqslant u^{n-1}$, we then have

$$u_n^\varepsilon \geqslant \sigma o M u^{n-1} = u^n$$

and so the first part of (1.100) (since $u_0^\varepsilon = u^0$).

Let us show the second part by induction. Let $\varepsilon_0 < \varepsilon_1$, and show that

$$(1.101)\qquad u_n^{\varepsilon_0} < u_n^{\varepsilon_1}$$

by assuming this property is satisfied up to order $n-1$. Now, we have

$$-(\frac{\partial}{\partial t}(u_n^{\varepsilon_0} - u_n^{\varepsilon_1}),(u_n^{\varepsilon_0} - u_n^{\varepsilon_1})^+)+a(t;u_n^{\varepsilon_0} - u_n^{\varepsilon_1},(u_n^{\varepsilon_0} - u_n^{\varepsilon_1})^+) + X = 0$$

where

$$X = \frac{1}{\varepsilon_0}((u_n^{\varepsilon_0} - Mu_{n-1}^{\varepsilon_0})^+ - (u_n^{\varepsilon_1} - Mu_{n-1}^{\varepsilon_1})^+,(u_n^{\varepsilon_0} - u_n^{\varepsilon_1})^+) +$$
$$+ (\frac{1}{\varepsilon_0} - \frac{1}{\varepsilon_1})((u_n^{\varepsilon_1} - Mu_{n-1}^{\varepsilon_1})^+,(u_n^{\varepsilon_0} - u_n^{\varepsilon_1})^+)$$
$$\geq \frac{1}{\varepsilon_0}((u_n^{\varepsilon_0} - Mu_{n-1}^{\varepsilon_0})^+ - (u_n^{\varepsilon_1} - Mu_{n-1}^{\varepsilon_1})^+,(u_n^{\varepsilon_0} - u_n^{\varepsilon_1})^+) .$$

But for $u_n^{\varepsilon_0} \geqslant u_n^{\varepsilon_1}$ we have

$$u_n^{\varepsilon_0} - Mu_{n-1}^{\varepsilon_0} \geqslant u_n^{\varepsilon_1} - Mu_{n-1}^{\varepsilon_1}$$

since by the induction hypothesis $Mu^{\varepsilon 1}_{n-1} \geqslant Mu^{\varepsilon 0}_{n-1}$. Therefore $X > 0$. (1.101) is easily deduced from this. However, if $u^{\varepsilon}_{n-1} \downarrow u^{n-1}$ then

$$Mu^{\varepsilon}_{n-1} \downarrow Mu^{n-1} .$$

Let

$$v \in L^2(0,T;V),\ \frac{\partial v}{\partial t} \in L^2(0,T;V')\ ,\ v \le Mu^{n-1},$$

then

$$-(\frac{\partial v}{\partial t},v-u^{\varepsilon}_n) - (\frac{\partial}{\partial t}(u^{\varepsilon}_n-v),v-u^{\varepsilon}_n)+ a(t;u^{\varepsilon}_n,v-u^{\varepsilon}_n)+$$
$$+ \frac{1}{\varepsilon}((u^{\varepsilon}_n-Mu^{\varepsilon}_{n-1})^+,v-u^{\varepsilon}_n) = (f(t),v-u^{\varepsilon}_n)$$

and

$$((u^{\varepsilon}_n- Mu^{\varepsilon}_{n-1})^+,v-u^{\varepsilon}_n) \le 0$$

hence

$$- \int_0^T(\frac{\partial v}{\partial t},v-u^{\varepsilon}_n)dt + \int_0^T a(t;u^{\varepsilon}_n,v-u^{\varepsilon}_n)dt + \frac{1}{2}|v(T)- \bar{u}|^2 \ge$$
$$\ge \int_0^T (f(t),v-u^{\varepsilon}_n)dt .$$

When $\varepsilon \downarrow 0$ we easily deduce from this that $u^{\varepsilon}_n \downarrow u_n$. Therefore we have proved (1.100). Starting from this we end the proof as in Theorem 1.4 of Chapter 4. □

2. REGULARITY

Here we give the parabolic analogue of the results presented in Section 2 of Chapter 4 for the elliptic case.

2.1 *The Levy-Stampacchia Inequality*

We give the Levy-Stampacchia Inequality for the parabolic case, which is owed to P. CHARRIER and G.M. TROIANIELLO [1]. We consider the parabolic V.I. (1.8) with obstacle (1.7). We shall assume that the obstacle satisfies

(2.1)
$$\left|\begin{array}{l} \psi \in L^2(0,T;V)\ (^1)\ ,\ \frac{\partial \psi}{\partial t} = \psi' \in L^2(0,T;V') \text{ and} \\ - <\psi',v> + a(t;\psi,v) = < L^+(t) - L^-(t),v > \\ \text{a.e.t},\quad \forall v \in V, \\ \text{where } L^+,L^- \in L^2(0,T;V') \text{ and are positive } (^2). \end{array}\right.$$

(1) If $V = H^1_0(\mathcal{O})$, we can replace "$\Psi \in L^2(0,T;V)$" by $\Psi \in L^2(0,T;H^1(\mathcal{O}))$ and $\Psi \geqslant 0$ on $\Gamma \times]0,T[$.
(2) i.e., a.e.t. $< L^+(t),v > \geqslant 0 \quad \forall v \geqslant 0, v \in V$.

Recall that $V = H^1(\mathcal{O})$ or $H^1_0(\mathcal{O})$. Since $\psi \in C(0,T;H)$, we can consider the value $\Psi(T)$. We also assume

(2.2) $\quad \bar{u} \leqslant \psi(T).$

Now consider the V.I. (1.8) in the strong form, say

$$(2.3)\qquad \left|\begin{array}{l} u \in L^2(0,T;V),\ u' \in L^2(0,T;V')\ ,\ u \leq \psi\ ,\ u(T) = \bar{u} \\ -\ (u'(t),v(t) - u(t)) + a(t;u(t),v(t)-u(t)) \geq (f(t),v(t)-u(t)) \\ \text{a.e.t.} \\ \forall v \in L^2(0,T;V) \quad \text{s.t.} \quad v(t) \in \psi(t) \quad \text{a.e.} \end{array}\right.$$

Our objective is to prove:

THEOROM 2.1: *Assume* $f \in L^2(Q)$, $\bar{u} \in H$ *as well as* (2.1),(2.2). *Then the problem* (2.3) *has one and only one solution. Furthermore, we have the following Levy-Stampacchia Inequality*

$$(2.4)\qquad \int_0^T (f,v)dt \geq \int_0^T [-(u',v) + a(u,v)]\, dt \geq$$

$$\geq \int_0^T (f,v)dt - \sup_{\substack{0 \leq w \leq v \\ w \in L^2(0,T;V)}} \{\int_0^T (f,w)dt - \int_0^T [-(\psi',w) + a(\psi,w)]\, dt\}$$

$$\forall v \in L^2(0,T;V), \quad v \geq 0\ . \ \square$$

First let us recall the relation between strong and weak V.I.'s.

LEMMA 2.1: *Every solution of* (2.3) *is a solution of* (1.8). *Conversely, if* $\psi \in L^2(0,T;V)$, $\psi' \in L^2(0,T;V')$ *then every solution of* (1.8) *such that* $u(T) = \bar{u}$ *and* $u' \in L^2(0,T;V')$ *is a solution of* (2.3).

Proof: Let u be a solution of (2.3) and v as in (1.8). Then we have

$$\int_0^T [-(v',v-u) + (v'-u',v-u) + a(u,v-u) - (f,v-u)]\, dt \geq 0$$

and thus (1.8), taking into account that $u(T) = \bar{u}$.

Conversely, let u be the solution of (1.8) such that $u' \in L^2(0,T;V')$, $u(T) = \bar{u}$. Take $v = \theta u + (1-\theta)w$ in (1.8), where $w \in L^2(0,T;V)$, $w' \in L^2(0,T;V)$ and $w \leqslant \psi$. Therefore

$$v - u = (1-\theta)(w-u) \text{ and } v(T) - \bar{u} = (1-\theta)(w(T) - u(T)).$$

From (1.8) we deduce

$$\int_0^T [-(\theta\, u' + (1-\theta)w',w-u) + a(u,w-u) - (f,w-u)]\, dt +$$
$$+ \frac{1}{2}\, |w(T)-u(T)|^2\, (1-\theta) \quad \geq 0$$

and on making θ tend to 1, we obtain

$$(2.5)\qquad \int_0^T [-(u',w-u) + a(u,w-u) - (f,w-u)]\, dt \geq 0$$
$$\forall w \in L^2(0,T;V), \quad w' \in L^2(0,T;V'), \quad w \leq \psi\ .$$

The relation (2.5) is in fact valid for $w \in L^2(0,T;V)$, $w \leqslant \psi$. In fact we consider a sequence $(\psi - w)_\varepsilon \to \psi - w$ in $L^2(0,T;V)$ and

$$(\psi - w)_\varepsilon \geqslant 0, \ (\psi - w)_\varepsilon \in L^2(0,T;V), \ (\psi - w)'_\varepsilon \in L^2(0,T;V')$$

(take

$$+ \varepsilon(\psi - w)'_\varepsilon + (\psi - w)_\varepsilon = \psi - w, \ (\psi - w)_\varepsilon(0) = 0).$$

Let us take

$$w_\varepsilon = \psi - (\psi - w)_\varepsilon$$

in (2.5). But $w_\varepsilon \to w$ in $L^2(0,T;V)$. Passing to the limit we obtain (2.5) with $w \in L^2(0,T;V)$, $w \leqslant \psi$. From this we easily deduce (2.3) by considering the Lebesgue points of the function $-(u',w-u) + a(u,w-u) - (f,w-u)$, and by applying (2.5) with $\tilde{w} = w$ on a small interval about a Lebesgue point, $\tilde{w} = u$ outside, then by passing to the limit. □

LEMMA 2.2: *Without loss of generality we can assume*

(2.6) $\quad f = 0, \ \bar{u} = 0.$

Proof: Let u^0 be the solution of the equation

$$-(u^{0'},v) + a(t;u^0,v) = (f,v) , \quad u^0(T) = \bar{u} \ ;$$

then $u - u^0$ is the solution of (2.3) with $f = 0$, $\bar{u} = 0$ and ψ replaced by $\psi - u^0$. Now $\tilde{\psi} = \psi - u^0 \in L^2(0,T;V)$, $\psi' \in L^2(0,T;V')$ and

$$-\langle \psi',v \rangle + a(t;\tilde{\psi},v) = \langle L^+(t) - L^-(t) - f,v \rangle$$
$$= \langle L^+(t) + f^+(t) - (L^-(t) + f^-(t)),v \rangle$$

since $f \in L^2(Q)$, therefore $f = f^+ - f^-$. Therefore $\tilde{\psi}$ satisfies (2.1). Now (2.4) is equivalent to

$$0 \geq \int_0^T [-(u-u^0)' + a(u-u^0,v)]dt \geq$$
$$\geq - \sup_{0 \leq w \leq v} \{ - \int_0^T [-((\psi-u^0)',w) + a(\psi-u^0,w)] dt \} \quad \square$$

In the case (2.6), we write (2.4) in the more convenient form

(2.7)
$$0 \geq \int_0^T [-(u',v) + a(u,v)] dt \geq \inf_{\substack{0 \leq w \leq v \\ w \in L^2(0,T,V)}} \int_0^T [-(\psi',w) + a(\psi,w)] dt ,$$
$$\forall v \in L^2(0,T;V), \quad v \geq 0 .$$

LEMMA 2.3: *Without loss of generality we can assume*

(2.8) $\quad a(t;u,u) \geqslant \beta \|u\|^2 \quad \forall u \in V.$

Proof: Take $f = 0$, $\bar{u} = 0$. If we perform the change of functions $u = \tilde{u}\exp(T-t)\lambda$, we are led to the same problem with a change into $a + \lambda(u,v)$, ψ changed into $\tilde{\psi} = \psi.\exp - (T-t)\lambda$. Moreover, (2.7) is equivalent to

$$0 \geq \int_0^T [-(\tilde{u}',v\exp(T-t)\lambda) + a(\tilde{u}, v\exp(T-t)\lambda) + \lambda(u,v\exp(T-t)\lambda)]dt \geq \inf_{0 \leq w \leq v} \int_0^T [-(\tilde{\psi}',w\exp(T-t)\lambda) + a(\tilde{\psi}, w\exp(T-t)\lambda) + (\tilde{\psi},w\exp(T-t)\lambda)]\,dt$$

and since we can change v into $v.\exp(T-t)\lambda$, w into $w.\exp(T-t)\lambda$, we can reduce to the same problem with $a + \lambda$, hence (2.8). □

Therefore in the remainder we can assume (2.6) and (2.8). The technique of proving Theorem 2.1. rests upon the elliptical regularisation method. We shall write

$$(2.9) \qquad \mathcal{V} = \{ v \in L^2(0,T;V) \quad , \quad v' \in L^2(0,T;H)\}$$

Then we begin by regularising the obstacle, and to do this we introduce the problem

$$(2.10) \qquad \int_0^T [\varepsilon(\psi'_\varepsilon, v') - (\psi'_\varepsilon, v) + a(\psi_\varepsilon, v)]dt + (\psi_\varepsilon(T), v(T)) = \int_0^T [-(\psi',v) + a(\psi,v)]dt + (\psi(T),v(T)$$
$$\forall v \in \mathcal{V}, \quad \psi_\varepsilon \in \mathcal{V}.$$

LEMMA 2.4: *The problem* (2.10) *has one and only one solution. Furthermore, when* $\varepsilon \to 0$

$$(2.11) \qquad \psi_\varepsilon \to \psi \text{ strongly in } L^2(0,T;V)$$

$$\psi'_\varepsilon \to \psi' \text{ weakly in } L^2(0,T;V')$$

$$\sqrt{\varepsilon}\psi'_\varepsilon \to 0 \text{ strongly in } L^2(0,T;H).$$

Proof: It will be convenient to write

$$(2.12) \qquad B_\varepsilon(u,v) = \int_0^T [\varepsilon(u',v') - (u',v) + a(u,v)]\,dt + (u(T),v(T))$$

$$(2.13) \qquad L(v) = \int_0^T [-(\psi',v) + a(\psi,v)]\,dt + (\psi(T),v(T)).$$

Then B_ε is a continuous bilinear form on $\mathcal{V} \times \mathcal{V}$ and

$$(2.14) \qquad B_\varepsilon(v,v) = \int_0^T \varepsilon|v'|^2 + \frac{1}{2}|v(T)|^2 + \frac{1}{2}|v(0)|^2 + \int_0^T a(v,v)dt,$$

therefore $B_\varepsilon(v,v)$ is coercive on $\mathcal{V}$. Furthermore, $L(v)$ is linear and continuous on $\mathcal{V}$, hence the existence and uniqueness of the solution ψ_ε.

We now establish a certain number of estimates. Now, we have

$$B_\varepsilon(\psi_\varepsilon,\psi_\varepsilon) = L(\psi_\varepsilon). \tag{2.15}$$

But

$$|L(\psi_\varepsilon)| \le C[\|\psi_\varepsilon\|_{L^2(V)} + |\psi_\varepsilon(T)|]$$

and so by (2.14) and the coercivity (2.8)

$$\beta\|\psi_\varepsilon\|^2_{L^2(V)} + \frac{1}{2}|\psi_\varepsilon(T)|^2 \le C[\|\psi_\varepsilon\|_{L^2(V)} + |\psi_\varepsilon(T)|]$$

from which it easily follows that

$$\|\psi_\varepsilon\|_{L^2(V)} \le C, \qquad |\psi_\varepsilon(T)|_H \le C. \tag{2.16}$$

Furthermore, by again making use of (2.15) together with (2.16), we also have

$$|\sqrt{\varepsilon}\psi'_\varepsilon|_{L^2(H)} \leqslant C. \tag{2.17}$$

We now establish an additional estimate. From (2.10) we immediately deduce (in the sense of distributions in t) that

$$\begin{aligned} -\varepsilon(\psi''_\varepsilon,v) - (\psi'_\varepsilon,v) + a(\psi_\varepsilon,v) &= \\ -(\psi',v) + a(\psi,v) & \\ \text{a.e.t.,} \quad \forall v \in V & \end{aligned} \tag{2.18}$$

and

$$\psi''_\varepsilon \in L^2(0,T;V'),$$

then replacing v by v(t), where $v \in \mathcal{V}$, and integrating over (0,T),

$$\begin{aligned} &-\varepsilon(\psi'_\varepsilon(T),v(T)) + \varepsilon(\psi'_\varepsilon(0),v(0)) + \\ &+ \int_0^T [\varepsilon(\psi'_\varepsilon,v') - (\psi'_\varepsilon,v) + a(\psi_\varepsilon,v)]\,dt = \\ &\qquad \int_0^T [-(\psi',v) + a(\psi,v)]\,dt \end{aligned}$$

which, after taking (2.10) into account, implies

$$\begin{aligned} -\varepsilon(\psi'_\varepsilon(T),v(T)) + \varepsilon(\psi'_\varepsilon(0),v(0)) &= (\psi_\varepsilon(T),v(T)) \\ &\quad - (\psi(T),v(T)) \\ \forall v \in \mathcal{V}. & \end{aligned}$$

Since we can choose v in $\mathcal{V}$ in such a way that v(T) and v(0) take arbitrary values, we then necessarily have

$$\left| \begin{aligned} \varepsilon\psi'_\varepsilon(T) + \psi_\varepsilon(T) &= \psi(T) \\ \psi'_\varepsilon(0) &= 0 \end{aligned} \right. \tag{2.19}$$

which constitutes the boundary conditions to be associated with (2.18). Using the estimate (2.16) we can write (2.18) in the form

$$\text{(2.20)} \qquad \varepsilon(\psi''_\varepsilon, v) + (\psi'_\varepsilon, v) = (g_\varepsilon, v)$$

where g remains in a bounded set of $L^2(0,T;V')$.

If $J : V \to V'$ is the canonical isomorphism of V in its dual, we can take $v = J^{-1}\psi'_\varepsilon$ in (2.20). From this, by the definition of the scalar product in V' it follows that

$$\frac{\varepsilon}{2}\frac{d}{dt}\|\psi'_\varepsilon\|^2_{V'} + \|\psi'_\varepsilon\|^2_{V'} = (g_\varepsilon, \psi'_\varepsilon)_{V'}$$

and therefore by taking the second condition of (2.19) we have

$$\int_0^T \|\psi'_\varepsilon\|^2_{V'}\, dt \le C\left(\int_0^T \|\psi'_\varepsilon\|^2_{V'}\, dt\right)^{1/2}$$

so

$$\|\psi'_\varepsilon\|_{L^2(0,T;V')} \leqslant C.$$

Now let us pick a subsequence, again denoted ψ_ε, such that

$$\text{(2.22)} \qquad \psi_\varepsilon \to \tilde\psi \text{ weakly in } L^2(0,T;V)$$

$$\psi_\varepsilon \to \tilde\psi \text{ weakly in } L^2(0,T;V')$$

$$\psi_\varepsilon(T) \to \tilde\psi(T) \text{ weakly in } H$$

$$\psi_\varepsilon(0) \to \tilde\psi(0) \text{ weakly in } H$$

$$\sqrt{\varepsilon}\psi'_\varepsilon \to \phi \text{ weakly in } L^2(0,T;H).$$

We can now pass to the limit in (2.10), and we obtain

$$\int_0^T [-(\tilde\psi', v) + a(\tilde\psi, v)]\, dt + (\tilde\psi(T), v(T)) =$$
$$= \int_0^T [-(\psi', v) + a(\psi, v)]dt + (\psi(T), v(T))$$
$$\forall v \in \mathcal{V}$$

which implies $\tilde\psi = \psi$, and the whole sequence converges.

It remains to prove the strong convergence. From (2.15) it again follows that

$$\int_0^T [\varepsilon|\psi'_\varepsilon|^2 + a(\psi_\varepsilon, \psi_\varepsilon)]\, dt + \frac{1}{2}|\psi_\varepsilon(T)|^2 + \frac{1}{2}|\psi_\varepsilon(0)|^2 =$$
$$= \int_0^T (-\psi', \psi_\varepsilon) + a(\psi, \psi_\varepsilon)\, dt + (\psi(T), \psi_\varepsilon(T))$$
$$= \int_0^T [-(\psi', \psi) + a(\psi, \psi)]\, dt + |\psi(T)|^2$$

$$= \frac{1}{2} \quad |\psi(T)|^2 + \frac{1}{2}|\psi(0)|^2 + \int_0^T a(\psi,\psi)\, dt .$$

But then

$$\int_0^T \varepsilon|\psi'_\varepsilon|^2\, dt + \int_0^T a(\psi_\varepsilon - \psi, \psi_\varepsilon - \psi) dt +$$

$$\frac{1}{2} |\psi_\varepsilon(T) - \psi(T)|^2 + \frac{1}{2} |\psi_\varepsilon(0) - \psi(0)|^2 \to 0$$

and therefore

(2.23) $\psi_\varepsilon \to \psi$ strongly in $L^2(0,T;V)$

$\sqrt{\varepsilon}\psi'_\varepsilon \to 0$ strongly in $L^2(0,T;H)$.

Now (2.20) means

$$\varepsilon\psi''_\varepsilon + \psi'_\varepsilon = g_\varepsilon \to \psi' \text{ strongly in } L^2(0,T;V')$$

from which, taking the initial condition (2.19) into account, we deduce

$$\overline{\lim} \int_0^T \| \psi'_\varepsilon \|^2_{V'} dt \leq \int_0^T \| \psi' \|^2_{V'} dt$$

which, when joined with the weak convergence of ψ'_ε to ψ', implies

(2.24) $\psi'_\varepsilon \to \psi'$ strongly in $L^2(0,T;V')$. □

Let us now consider the elliptic V.I. (regularised form of the parabolic V.I. (2.3))

(2.25)
$$\int_0^T [\varepsilon(u'_\varepsilon, v' - u'_\varepsilon) - (u'_\varepsilon, v - u_\varepsilon) + a(u_\varepsilon, v - u_\varepsilon)]\, dt +$$
$$+ (u_\varepsilon(T), v(T) - u_\varepsilon(T)) \geq 0$$
$$\forall v \in \mathcal{V}, \quad v \leq \psi_\varepsilon, \quad u \leq \psi_\varepsilon .$$

LEMMA 2.5: *The problem* (2.25) *has one and only one solution* u_ε. *Furthermore, we have the a priori estimates*

(2.26) $\|u_\varepsilon\|_{L^2(v)} \leqslant C$

(2.27) $|\sqrt{\varepsilon}u'_\varepsilon|_{L^2(H)} \leqslant C.$

Proof: The problem (2.25) may be written

(2.28)
$$\left| \begin{array}{l} B_\varepsilon(u_\varepsilon, v - u_\varepsilon) \geq 0 \\ \forall v \in \mathcal{V}, \; v \leq \psi_\varepsilon, \quad u_\varepsilon \in \mathcal{V}, \; u_\varepsilon \leq \psi_\varepsilon . \end{array} \right.$$

Since B_ε is a coercive continuous bilinear form, the general theory of V.I.'s applies, whence the existence and uniqueness of u^ε (note that $\psi_\varepsilon \in \mathcal{V}$). take $v = \psi_\varepsilon$, therefore (2.28) implies

$$B_\varepsilon(u_\varepsilon, u_\varepsilon) \leqslant B_\varepsilon(u_\varepsilon, \psi_\varepsilon)$$

so

$$\int_0^T [\varepsilon|u_\varepsilon'|^2 + a(u_\varepsilon, u_\varepsilon)]dt + \frac{1}{2}|u_\varepsilon(T)|^2 + \frac{1}{2}|u_\varepsilon(0)|^2 \le$$
$$\le \int_0^T [\varepsilon(u_\varepsilon', \psi_\varepsilon') + a(u_\varepsilon, \psi_\varepsilon) + (u_\varepsilon, \psi_\varepsilon')]\, dt + (u_\varepsilon(0), \psi_\varepsilon(0))$$

and by using the estimates for ψ_ε proved in Lemma 2.4., we can deduce (2.26), (2.27). □

The essential point is now to prove a Levy-Stampacchia inequality for the V.I. (2.25). We have:

LEMMA 2.6: *The following Levy-Stampacchia inequality is satisfied*

$$(2.29) \qquad 0 \ge B_\varepsilon(u_\varepsilon, v) \ge \inf_{\substack{0 \le w \le v \\ w \in \mathcal{V}}} B_\varepsilon(\psi_\varepsilon, w), \qquad \forall v \in \mathcal{V}, \quad v \ge 0 .$$

Proof: This is actually the general theorem that is a variant of Theorem 5.8., Chapter 2. For $v \in \mathcal{V}$ we write

$$(2.30) \qquad \mathcal{L}_\varepsilon(v) = \inf_{\substack{0 \le w \le v^+ \\ w \in \mathcal{V}}} B_\varepsilon(\psi_\varepsilon, w) - \inf_{\substack{0 \le w \le v^- \\ w \in \mathcal{V}}} B_\varepsilon(\psi_\varepsilon, w) .$$

Then $\mathcal{L}_\varepsilon$ is an element of $\mathcal{V}'$. In fact, first let $v \geqslant 0$, then

$$\mathcal{L}_\varepsilon(v) = \inf_{\substack{0 \le w \le v \\ w \in \mathcal{V}}} B_\varepsilon(\psi_\varepsilon, w) .$$

Linearity is shown in Theorem 5.8, Chapter 2. Let us show continuity. Now, by the definition of ψ_ε,

$$\mathcal{L}_\varepsilon(v) = \inf_{\substack{0 \le w \le v \\ w \in \mathcal{V}}} L(w)$$

and by assumption (2.1)

$$L(w) = \int_0^T \langle L^+(t) - L^-(t), w \rangle dt + (\psi(T), w(T))$$
$$\ge - \int_0^T \langle L^-(t), w \rangle dt$$

since $w \geqslant 0$, and $\psi(T) \geqslant 0$, therefore also, because $w \leqslant v$,

$$L(w) \ge - \int_0^T \langle L^-(t), v \rangle dt$$
$$\ge - C\|v\|_{L^2(V)}$$

and thus

$$0 \geq \mathcal{L}_\varepsilon(v) \geq - C \|v\|$$

whence the result. From this we deduce that $\mathcal{L}_\varepsilon$ is an element of $\mathcal{V}'$.

As in Theorem 5.8., Chapter 2, we next define the V.I.

(2.31) $$B_\varepsilon(z_\varepsilon, v-z_\varepsilon) \geq \mathcal{L}_\varepsilon(v-z_\varepsilon)$$
$$\forall v \geq u_\varepsilon \; , \; z_\varepsilon \geq u_\varepsilon, \quad v \in \mathcal{V} \; , \; z_\varepsilon \in \mathcal{V} \, .$$

Naturally, for $v \geqslant 0$, $v \in \mathcal{V}$, we have

$$B_\varepsilon(z_\varepsilon, v) \geqslant \mathcal{L}_\varepsilon(v)$$

and therefore the inequality on the right of (2.29) will result from

(2.32) $$z_\varepsilon = u_\varepsilon.$$

First we verify that

(2.33) $$z_\varepsilon \leqslant \psi_\varepsilon \, .$$

In fact

$$z_\varepsilon - (z_\varepsilon - \psi_\varepsilon)^+ \geqslant u_\varepsilon,$$

therefore it follows from (2.31) that

$$-B_\varepsilon(z_\varepsilon, (z_\varepsilon-\psi_\varepsilon)^+) \geq - \mathcal{L}_\varepsilon((z_\varepsilon-\psi_\varepsilon)^+)$$
$$\geq - B_\varepsilon(\psi_\varepsilon, (z_\varepsilon-\psi_\varepsilon)^+)$$

so

$$B_\varepsilon(z_\varepsilon - \psi_\varepsilon, (z_\varepsilon - \psi_\varepsilon)^+) \leqslant 0$$

and by (2.13)

$$B_\varepsilon((z_\varepsilon - \psi_\varepsilon)^+, (z_\varepsilon - \psi_\varepsilon)^+) \leqslant 0$$

which with coercivity implies $(z_\varepsilon - \psi_\varepsilon)^+ = 0$. But we can then take $v = z_\varepsilon$ in (2.28). This yields

(2.34) $$B_\varepsilon(u_\varepsilon, z_\varepsilon - u_\varepsilon) \geqslant 0$$

and by (2.31)

$$B_\varepsilon(z_\varepsilon, u_\varepsilon - z_\varepsilon) \geqslant \mathcal{L}_\varepsilon(u_\varepsilon - z_\varepsilon) \geqslant 0$$

which with (2.34) implies

$$B_\varepsilon(u_\varepsilon - z_\varepsilon, z_\varepsilon - u_\varepsilon) \geqslant 0$$

whence

$z_\varepsilon - u_\varepsilon = 0$. □

LEMMA 2.7: *We have the estimate*

(2.35) $$\|u'_\varepsilon\|_{L^2(V')} \leqslant C.$$

Proof: We have seen in Lemma 2.6. that $\forall v \in \mathcal{V}$

$$0 \geq B_\varepsilon(u_\varepsilon, v) \geq -C \|v\|_{L^2(V)} .$$

From this we deduce the estimate

(2.36) $$|B_\varepsilon(u_\varepsilon,v)| \geq C \|v\|_{L^2(V)} \qquad \forall v \in \mathcal{V} .$$

Consequently, we in fact have

(2.37) $$B_\varepsilon(u_\varepsilon,v) = \langle g_\varepsilon, v\rangle \qquad \forall v \in \mathcal{V}$$

where

(2.38) $$\|g_\varepsilon\|_{L^2(0,T;V')} \leqslant C.$$

From this we deduce

(2.39) $$-\varepsilon(u''_\varepsilon,v) - (u'_\varepsilon,v) + a(u_\varepsilon,v) = (g_\varepsilon,v)$$

$$\text{a.e.t.,} \quad \forall v \in V, \quad u''_\varepsilon \in L^2(0,T;V')$$

and the boundary conditions

(2.40) $$\left| \begin{array}{l} u'_\varepsilon(0) = 0 \\ \varepsilon u'_\varepsilon(T) + u_\varepsilon(T) = 0 . \end{array} \right.$$

We are therefore in a situation analogous to (2.19), (2.20), and (2.35) results. □

Proof of theorem 2.1

Consider a subsequence such that

$$u_\varepsilon \to u \text{ weakly in } L^2(0,T;V)$$

$$u'_\varepsilon \to u' \text{ weakly in } L^2(0,T;V')$$

$$\sqrt{\varepsilon}u'_\varepsilon \to \xi \text{ weakly in } L^2(0,T;H).$$

Let $v \in W(0,T) = \{v \in L^2(0,T;V),\ v' \in L^2(0,T;V')\}$ be such that $v \leqslant \psi$. Consider a sequence $(\psi - v)_k \geqslant 0$ belonging to $\mathcal{V}$ and $(\psi - v)_k \to \psi - v$ in $W(0,T)$. We take $v = \psi_\varepsilon - (\psi - v)_k$ in (2.25), and obtain

$$\int_0^T [\varepsilon\,(u'_\varepsilon, \psi'_\varepsilon - (\psi - v)'_k) - (u'_\varepsilon, \psi_\varepsilon - (\psi-v)_k) +$$

$$+ a(u_\varepsilon,\psi_\varepsilon-(\psi-v)_k)]dt + (u_\varepsilon(T),\psi_\varepsilon(T)-(\psi-v)_k(T)) \geq$$
$$\int_0^T a(u_\varepsilon,u_\varepsilon)dt + \frac{1}{2}|u_\varepsilon(T)|^2+\frac{1}{2}|u_\varepsilon(0)|^2 .$$

We can pass to the limit in ε by using Lemma 2.4. and weak l.s.c., obtaining

$$\int_0^T [-(u', \psi-(\psi-v)_k)+ a(u,\psi -(\psi-v)_k)]dt +$$
$$+ (u(T),\psi(T)-(\psi-v)_k(T)) \geq \int_0^T a(u,u)dt + \frac{1}{2}|u(T)|^2+\frac{1}{2}|u(0)|^2 .$$

Next, we make k tend to $+\infty$, from which it follows that

$$\int_0^T [-(u',v) + a(u,v)]dt +$$
$$+(u(T),v(T)) \geq \int_0^T a(u,u)dt + \frac{1}{2}|u(T)|^2+\frac{1}{2}|u(0)|^2$$

so again

$$\text{(2.41)} \quad \int_0^T [-(u',v-u)+ a(u,v-u)]dt + (u(T),v(T)-u(T)) \geq 0$$
$$\forall v \in W(0,T), \quad v \leq \psi , \quad u \leq \psi .$$

From (2.41) we deduce

$$\int_0^T [-(u',v)+ a(u,v)]dt + (u(T),v(T)) \leq 0$$
$$\forall v \in W(0,T), \quad v \geq 0 .$$

Now, we can take $v = \psi - u$, whence

$$\int_0^T [-(u',\psi-u)+ a(u,\psi-u)]dt + (u(T),\psi(T)-u(T)) \leq 0$$

but on taking $v = \psi$ in (2.41) we obtain a reverse inequality, hence

$$\text{(2.42)} \quad \int_0^T [-(u',\psi-u) + a(u,\psi-u)]dt + (u(T),\psi(T)-u(T)) = 0 .$$

Now, an argument to the level of that for (2.42) allows us to state

$$\text{(2.43)} \quad \int_0^T [\varepsilon(u_\varepsilon',\psi_\varepsilon'-u_\varepsilon')-(u_\varepsilon',\psi_\varepsilon-u)+ a(u_\varepsilon,\psi_\varepsilon-u_\varepsilon)]dt +$$
$$+ (u_\varepsilon(T),\psi_\varepsilon(T)- u_\varepsilon(T)) = 0$$

Thanks to the convergence properties of ψ_ε, we deduce that

$$\text{(2.44)} \quad \int_0^T |u_\varepsilon'|^2 dt + \int_0^T a(u_\varepsilon,u_\varepsilon)dt + \frac{1}{2}|u_\varepsilon(T)|^2+\frac{1}{2}|u_\varepsilon(0)|^2$$
$$\to \int_0^T [-(u',\psi)+ a(u,\psi)]dt + (u(T),\psi(T))$$

and by (2.42)

$$= \int_0^T a(u,u)dt + \frac{1}{2}\,|u(T)|^2 + \frac{1}{2}|u(0)|^2 .$$

From this it follows that

(2.45) $u_\varepsilon \to u$ strongly in $L^2(0,T;V)$

$\sqrt{\varepsilon}u'_\varepsilon \to 0$ strongly in $L^2(0,T;H)$

$u_\varepsilon(T) \to u(T)$ strongly in H

$u_\varepsilon(0) \to u(0)$ strongly in H.

Let $v \in V$, and set $\phi_\varepsilon(t) = (u'_\varepsilon(t),v)$. Then starting from (2.39) and (2.40) we can write

$$\varepsilon\,\phi'_\varepsilon(t) + \phi_\varepsilon(t) = h_\varepsilon(t)$$
$$\phi_\varepsilon(0) = 0$$

where

$$h_\varepsilon(t) = a(u_\varepsilon(t),v) - \langle g_\varepsilon(t),v\rangle \in L^2(0,T)$$

and h_ε then stays in a bounded set $L^2(0,T)$. But then

$$\frac{1}{2}\,\varepsilon\,\frac{d}{dt}\,|\phi_\varepsilon(t)|^2 + \phi_\varepsilon(t)^2 = h_\varepsilon(t)\quad\phi_\varepsilon(t)$$

so

$$\varepsilon\,|\phi_\varepsilon(T)|^2 + \int_0^T \phi_\varepsilon(t)^2 dt = \int_0^T h_\varepsilon(t)\phi_\varepsilon(t)dt$$

from which it follows (and this is the interesting thing) that $\sqrt{\varepsilon}\phi_\varepsilon(T)$ is bounded, so

$\sqrt{\varepsilon}(u'_\varepsilon(T),v)$ is bounded $\forall v \in V$,

therefore

(2.46) $\sqrt{\varepsilon}u'_\varepsilon(T)$ stays bounded in V'.

But the second relation (2.40) then implies

(2.47) $u_\varepsilon(T) \to 0$ in V'

which with (2.45), implies

(2.48) $u(T) = 0$.

Therefore, starting from (2.41), we have, in fact,

$$(2.49)\qquad \int_0^T [-(u',v-u)+ a(u,v-u)]dt \geq 0$$
$$\forall v \in W(0,T), \quad v \leq \psi \quad , \quad u \leq \psi \;.$$

But then, as for (2.5), the inequality (2.48) is indeed valid for all $v \in L^2(0,T;V)$, $v \leqslant \psi$.

We have therefore proved the existence of a solution of (2.3). This solution is necessarily unique, since if u_1 and u_2 are two solutions, then

$$-(u_1'(t) - u_2'(t),u_2(t) - u_1(t)) + a(t;u_1 - u_2,u_2 - u_1) \geqslant 0$$

so, since $u_1(T) = u_2(T)$,

$$\frac{1}{2}|u_1(0)-u_2(0)|^2 + \int_0^T a(u_1-u_2,u_1-u_2)dt \leq 0$$

hence $u_1 = u_2$.

It remains to verify (2.7). But by the property (2.28) and the definition of ψ_ε, (taking account of $\psi(T) \geqslant 0$ and $u(T) = 0$) we have

$$0 \geq \int_0^T [\, \varepsilon\, (u_\varepsilon',v')-(u_\varepsilon',v)+ a(u_\varepsilon,v)\,]dt \geq$$
$$\geq \inf_{\substack{0 \leq w \leq v \\ v \in \mathcal{V}}} \int_0^T [-(\psi',w) + a(\psi,w)]dt \;.$$

Passing to the limit, we obtain (2.7) for all $v \in \mathcal{V}$. Since this relation may be written

$$0 \geq M(v) \geq N(v) \qquad \forall v \in \mathcal{V}, \qquad v \geq 0$$

where M,N are linear and continuous on $L^2(0,T;V)$, in fact the inequality holds for all $v \in L^2(0,T;V)$. This completes the proof of Theorem 2.1. □

Remark 2.1

In BL [1] another proof of Theorem 2.1 will be found for the case of the Dirichlet problem, but which requires more assumptions about the coefficients. □

COROLLARY 2.1: ([1]) *Take the conditions of Theorem* 2.1, *together with* $V = H_0^1(\mathcal{O})$ *and* $\psi \in L^2(0,T;H^1(\mathcal{O}))$, $\psi \geqslant 0$ *on* Γ, *and*

$$(2.50)\qquad -\frac{\partial\psi}{\partial t} + A(t)\psi + a_0\psi \geq g^* \in L^2(Q) \quad ([2]) \;.$$

Then the solution of the V.I. (2.3) *satisfies*

$$-\frac{\partial u}{\partial t} + A(t)u \in L^2(Q)$$

and

$$(2.51)\qquad f \geqslant -\frac{\partial u}{\partial t} + A(t)u + a_0 u \geqslant f \wedge g^*.$$

---

(1) A direct proof of Corollary 2.1 is given afterwards.

(2) $A(t) = -\sum_{i,j} \frac{\partial}{\partial x_i} a_{ij} \frac{\partial}{\partial x_j} + \sum_i a_i \frac{\partial}{\partial x_i}$

Proof: Now,

$$-(\psi',w)+a(\psi,w)\geqslant(g^*,w).$$

Therefore the sup in the last term of (2.4) may be written

$$\sup_{\substack{0\le w\le v\\ w\in L^2(V)}}\left\{\int_0^T (f,w)\,dt-\int_0^T[-(\psi',w)+a(\psi,w)]\,dt\right\}\le \sup_{0\le w\le v}\left\{\int_0^T (f-g^*,w)\,dt\right\}$$

$$=\int_0^T ((f-g^*)^+,v)\,dt$$

and therefore (2.4) is written as

$$\int_0^T (f,v)\,dt\ge\int_0^T[(-u'+Au+a_0u,v)]\,dt$$

$$\ge\int_0^T (f\wedge g^*,v)\,dt$$

whence we easily deduce (2.50).□

The principles of another proof of Theorem 2.1 and of Corollary 2.1

We are now going to give the principles of another proof of Theorem 2.1 which are based on homographic penalisation (used in Chapter 2 in the case of stationary Q.V.I.'s). This proof is shorter and more intuitive than the preceding one, but assumes additional specifications of the structure of L^+ and L^- in (2.1) for the case $V=H^1(\mathcal{O})$. We next give some indications for the case $V=H_0^1(\mathcal{O})$.□

Let us assume, therefore, that $V=H^1(\mathcal{O})$, and let us say that an element $F\in L^2(0,T;V')$ has *natural structure* if

$$\text{(2.52)}\qquad\left|\begin{array}{l}\langle F,v\rangle=\displaystyle\int_{\mathcal{O}} F_{\mathcal{O}}(x,t)v(x)\,dx+\int_\Gamma F_\Gamma v\,d\Gamma\;,\\ \text{with}\\ F_{\mathcal{O}}\in L^2(\mathcal{O}\times]0,T[)\;,\quad F_\Gamma\in L^2(0,T;H^{-\frac12}(\Gamma))\;,\end{array}\right.$$

and we shall say that F is "positive with natural structure" if in the representations above we have

$$\text{(2.53)}\qquad F_{\mathcal{O}}\geqslant 0\quad\text{and}\quad F_\Gamma\geqslant 0.$$

We shall then prove Theorem 2.1 under the assumptions of Theorem 2.1, and with the addition of

(2.54) L^+ and L^- are positive with natural structure, in (2.1),

Consider the function $g\in L^2(0,T;V')$ defined by

$$\text{(2.55)}\qquad\langle g,v\rangle=\int_{\mathcal{O}} g_{\mathcal{O}}v\,dx+\int_\Gamma g_\Gamma v\,d\Gamma\quad,$$

where $g_{\mathcal{O}}$ and g_Γ are $\geqslant 0$ and *bounded*, and are defined further below.

For $\varepsilon>0$ we define u_ε by

(2.56)
$$\left|\begin{array}{l} -(\frac{\partial u_\varepsilon}{\partial t},v)+ a(u_\varepsilon,v) + \langle g \dfrac{u_\varepsilon - \psi}{|u_\varepsilon-\psi| + \varepsilon},v\rangle = \langle f-g,v\rangle \\ \forall v \in V , \end{array}\right.$$

(2.57) $u_\varepsilon(T) = 0$ (e.g., we then use Lemma 2.2).

and we show that u_ε exists, satisfying

(2.58)
$$u_\varepsilon \in L^2(0,T;V), \quad \frac{\partial u_\varepsilon}{\partial t} \in L^2(0,T;V') .$$

The essential point is the following: *for a suitable choice of* g (made explicit below)

(2.59) $u_\varepsilon \leqslant \psi.$

In fact we re-write (2.54) in the form

(2.60)
$$\begin{aligned} - (\frac{\partial(u_\varepsilon-\psi)}{\partial t},v)+ a(u_\varepsilon-\psi,v) + \langle g \frac{u_\varepsilon- \psi}{|u_\varepsilon-\psi|+\varepsilon} ,v\rangle &= \\ = \langle f-g + \frac{\partial\psi}{\partial t} ,v\rangle-a(\psi,v) &= \\ = \langle f-g - (L^+- L^-),v\rangle &, \end{aligned}$$

and in (2.60) we take $v = (u_\varepsilon - \psi)^+$, which is permissible, from which we deduce that

(2.61)
$$\begin{aligned} - \frac{1}{2} \frac{d}{dt} |(u_\varepsilon-\psi)^+|^2 + a((u_\varepsilon-\psi)^+)+ \langle g \frac{(u_\varepsilon-\psi)^+}{|u_\varepsilon-\psi|+\varepsilon}, (u_\varepsilon-\psi)^+\rangle &= \\ = \langle f-g -(L^+-L^-),(u_\varepsilon-\psi)^+\rangle . \end{aligned}$$

Here we have assumed that

(2.62)
$$\left|\begin{array}{l} \langle L^\pm,v\rangle = \int_{\mathcal{O}} L^\pm_{\mathcal{O}} \; v \, dx + \int_\Gamma L^\pm_\Gamma \; v \, d\Gamma , \\ L^\pm_{\mathcal{O}} , \; L^\pm_\Gamma \geq 0 . \end{array}\right.$$

By possibly making an extra passage to the limit, we can assume that $f,L^\pm,L^\pm_\Gamma$ are *bounded*. The second member of (2.61) is therefore negatino (and therefore $(u_\varepsilon - \psi)^+ = 0$, whence we get (2.59)) if we choose $g_{\mathcal{O}}$ and g_Γ with

$$g_{\mathcal{O}}, g_\Gamma \geqslant 0$$

and

$$g - (g_{\mathcal{O}} + L^+_{\mathcal{O}} - L^-_{\mathcal{O}}))^+ ,\leqslant 0, \; g_\Gamma = (L^+_\Gamma - L^-_\Gamma)^- .$$

We can therefore take (what is the best choice possible here)

(2.63)
$$g_{\mathcal{O}} = (f - (L^+_{\mathcal{O}} - L^-_{\mathcal{O}}))^+, \quad g_\Gamma = (L^+_\Gamma - L^-_\Gamma)^- .$$

Thus we have (2.59), and we can rewrite (2.56) in the form

$$-(\frac{\partial u_\varepsilon}{\partial t},v)+a(u_\varepsilon,v)-(f,v) = -<g\ \frac{\varepsilon}{\varepsilon+\psi-u_\varepsilon},v> . \tag{2.64}$$

If we take $v \leqslant 0$, we therefore have

$$\int_0^T[(-\frac{\partial u_\varepsilon}{\partial t},v)+a(u_\varepsilon,v)-(f,v)]dt \le 0 . \tag{2.65}$$

Furthermore, since $\frac{\varepsilon}{\varepsilon+\psi-u_\varepsilon} \leqslant 1$, we also have

$$\int_0^T[-(\frac{\partial u_\varepsilon}{\partial t},v)+a(u_\varepsilon,v)-(f,v)]dt \ge -\int_0^T[\int_{\mathcal{O}} g\ v\ dx + \int_\Gamma g_\Gamma v d\Gamma]dt . \tag{2.66}$$

But

$$\begin{aligned}\sup_{0\le w\le v}&\int_0^T\{(f,w)-[-(\frac{\partial\psi}{\partial t},w)+a(\psi,w)]\}\,dt = \\ &= \sup_{0\le w\le v}\int_{\mathcal{O}\times(0,T)}[f-(L^+-L^-)]w\ dx\ dt - \int_{\Gamma\times(0,T)}(L_\Gamma^+-L_\Gamma^-)w\ d\Gamma\ dt] \\ &= \int_{\mathcal{O}\times(0,T)} g\ v\ dx\ dt + \int_{\Gamma\times(0,T)} g_\Gamma w\ d\Gamma\ dt\end{aligned}$$

where $g_{\mathcal{O}}$ and g_Γ are given by (2.63). Therefore (2.66) becomes

$$\begin{aligned}\int_0^T[-(\frac{\partial u_\varepsilon}{\partial t},v)+a(u_\varepsilon,v)-(f,v)]dt &\ge \\ \ge -\sup_{0\le w\le v}\int_0^T\{(f,w)-[(-\frac{\partial\psi}{\partial t},w)+a(\psi,w)]\}\,dt .\end{aligned}$$

Looking at this another way, it follows that, when $\varepsilon \to 0$, u_ε stays in a bounded open set of $L^2(0,T;V)$ and $\frac{\partial u_\varepsilon}{\partial t}$ stays in a bounded open set of $L^2(0,T;V')$. Hence we can pick a subsequence, again denoted u_ε, such that

$$\left|\begin{aligned}&u_\varepsilon \to \tilde{u} \quad \text{weakly in } L^2(0,T;V), \\ &\frac{\partial u_\varepsilon}{\partial t} \to \frac{\partial u}{\partial t} \quad \text{weakly in } L^2(0,T;V').\end{aligned}\right. \tag{2.67}$$

But we deduce from (2.64) that for $v \leqslant \psi$:

$$\left|\begin{aligned}&\int_0^T[-(\frac{\partial u_\varepsilon}{\partial t},v-u_\varepsilon)+a(u_\varepsilon,v-u_\varepsilon)-(f,v-u_\varepsilon)]\,dt = \\ &= \int_0^T -<\frac{\varepsilon\ g}{\varepsilon+\psi-u_\varepsilon},v-\psi>dt + X_\varepsilon , \\ &X_\varepsilon = -\varepsilon\int_0^T <g,\frac{\psi-u_\varepsilon}{\varepsilon+\psi-u_\varepsilon}> \ dt .\end{aligned}\right. \tag{2.68}$$

The first term of the second member of (2.68) is $\geqslant 0$, and $|X_\varepsilon| \leqslant C$. Therefore we deduce from (2.66) that $\tilde{u}$ is the solution of the V.I. we want, and hence $\tilde{u} = u$. We can pass to the limit in (2.65),(2.67), hence the Theorem is proved.□

The case $V = H_0^1(\mathcal{O})$. *Proof of Corollary 2.1 by homographic penalisation*

We can now, therefore, take $V = H_0^1(\mathcal{O})$ and ψ satisfying

$$(2.69) \quad \left| \begin{array}{l} \psi \in L^2(0,T;H^1(\mathcal{O})), \quad \psi \geq 0 \quad \text{on } \Gamma \times]0,T[\quad , \\ \dfrac{\partial\psi}{\partial t} \in L^2(0,T;H^1(\mathcal{O})) \quad \text{and} \quad (2.52). \end{array} \right.$$

We can approximate ψ by a sequence ψ_α of regular functions (in the sense that $\psi_\alpha \to \psi$ in $L^2(0,T;H^1(\mathcal{O}))$ and $\dfrac{\partial\psi_\alpha}{\partial t} \to \dfrac{\partial\psi}{\partial t}$ in $L^2(0,T;H^{-1}(\mathcal{O}))$, with the ψ_α's satisfying the analogue of (2.52). We then work as in the preceding proof, but with $V = H_0^1(\mathcal{O})$, (1)

$$\langle g,v\rangle = \int_{\mathcal{O}} g_{\mathcal{O}}\, v \, dx \ ,$$

and where $g_{\mathcal{O}}$ is given by

$$(2.70) \qquad g_{\mathcal{O}} = (f - g^*)^+ .$$

We must verify that g so defined satisfies

$$g_{\mathcal{O}} \geq f - \left(-\frac{\partial\psi}{\partial t} + A\psi + a_0\,\psi\right)$$

which is satisfied if $g_{\mathcal{O}} \geqslant f - g^*$.

Starting from (2.66) we then easily obtain that

$$-\frac{\partial u}{\partial t} + Au + a_0 u - f \geq -(f-g^*)^+$$

thus

$$-\frac{\partial u}{\partial t} + Au + a_0 u \geq \inf(f,g^*)$$

whence (2.53).□

2.2 *Regularity of the Q.V.I. Solution*

We take the conditions of Theorem 1.2, with $V = H_0^1(\mathcal{O})$. To simplify matters we shall take

$$\bar{u} = 0.$$

---

(1) In the approximation by homographic penalisation we can take $v = (u_\varepsilon - \psi)^+$, thanks to the fact $\psi \geqslant 0$ on Γ, and therefore this choice of v is also in $H_0^1(\mathcal{O})$.

The approach presented in the elliptic case consists of obtaining additional properties of the implicit obstacle Mu, and then deducing from them the regularity of the solution. To obtain this information about the obstacle we use the fact that in the continuation set[1] we have the equation

$$(A + a_0)u = f.$$

In the present case, in the weak form of the Q.V.I. (1.6), we cannot state that

$$-\frac{\partial u}{\partial t} + A(t)u + a_0 u = f$$

holds in the continuation set. Therefore there is a difficulty in adapting the method used in the elliptic case. We are going to proceed differently, *using the penalised equation.*

We write the penalised problem in the form (cf. (1.88))

(2.71) $$\left| \begin{array}{l} -\frac{\partial u^\varepsilon}{\partial t} + (A(t) + a_0)\, u^\varepsilon + \frac{1}{\varepsilon}(u^\varepsilon - Mu^\varepsilon)^+ = f \\ u^\varepsilon|_\Sigma = 0 \ , \quad u^\varepsilon(x,T) = 0 \\ u^\varepsilon \in W^{2,1,p}(Q) \ . \end{array} \right.$$

However, by Theorem 1.4 the sequence u^ε is decreasing, and it converges to u, the maximum solution of the Q.V.I. (1.6).

Since $u \in C^0(\bar{Q})$, Dini's Theorem allows us to state that

(2.72) $$u^\varepsilon \to u \quad \text{in } C^0(\bar{Q}).$$

We introduce

(2.73) $$C = \{x \in \bar{\mathcal{O}} \mid \min_{t \in [0,T]} (Mu(x,t) - u(x,t)) > 0\} .$$

The set C is an open set of $\bar{\mathcal{O}}$. Next consider

(2.74) $$C' = \{x \in \mathcal{O} \mid \min_{t \in [0,T]} (Mu(x,t) - u(x,t)) > \frac{k}{2}\}$$

which is not empty since $\Gamma \subset C'$, because

$$Mu|_\Gamma = k \quad \text{and} \quad u|_\Gamma = 0.$$

Let ε_0 be such that for $\varepsilon \leqslant \varepsilon_0$ we have

(2.75) $$\|(u^\varepsilon - Mu^\varepsilon) - (u - Mu)\|_{C^0(\bar{Q})} \leq \frac{k}{4}$$

therefore

(2.76) $$x \in \bar{C}', \quad t \in [0,T] \quad \text{implies}$$

$$u^\varepsilon(x,t) - Mu^\varepsilon(x,t) \leq -\frac{k}{2} + \frac{k}{4} < 0 .$$

(1) The set $\{x \mid u < Mu\}$.

Consequently (2.71) implies

(2.77) $$\left| \begin{array}{l} -\frac{\partial u^\varepsilon}{\partial t} + A(t)u^\varepsilon + a_0 u^\varepsilon = f \quad \text{in} \quad \mathcal{O} \cap C' \times (0,T) \\ u^\varepsilon|_\Sigma = 0, \quad u^\varepsilon(x,T) = 0. \end{array} \right.$$

But then, by O.A. LADYZHENSKAYA, V.A. SOLONNIKOV and N.N. URALT'SEVA [1], we can state that

(2.78) $$\left| \begin{array}{l} \forall \ D \text{ an open subset of } \mathcal{O} \cap C' \text{ such that} \\ \bar{D} \subset C', \text{ we have} \\ \| u^\varepsilon \|_{W^{1,2,p}(D \times (0,T))} \le C [\|f\|_{L^p(Q)} + \|u^\varepsilon\|_{L^p(Q)}] \end{array} \right.$$

where the constant depends on D,C' and the coefficients of the operator, but not upon f, nor on ε. Since u^ε is bounded, from this we deduce:

LEMMA 2.8: *The maximum solution* u *of the Q.V.I.* (1.6) *satisfies*

(2.79) $$u \in W^{2,1,p}(D \times (0,T)) \quad \forall \ D \quad \textit{as in } (2.78).$$

Proof: This follows immediately from the estimate (2.78).□

The estimate (2.79) will be insufficient, in fact. In order to obtain more results on regularity we must assume

(2.80) $$a_{ij} \in C^{1+\alpha, \frac{\alpha}{2}}(\bar{Q}) \ a_i, a_0 \in C^{\alpha, \frac{\alpha}{2}}(\bar{Q}) ,$$

(2.81) $$f \in C^{\alpha, \frac{\alpha}{2}}(\bar{Q}).$$

Under these conditions we obtain a result stronger than (2.78);

(2.82) $$\left| \begin{array}{l} \forall \ D \text{ an open subset of } \mathcal{O} \cap C' \text{ such that} \\ \bar{D} \subset C', \text{ we have} \\ \| u^\varepsilon \|_{C^{2+\alpha, 1+\frac{\alpha}{2}}(\bar{D} \times [0,T])} \le C[|f|_{C^{0,\alpha}(\bar{Q})} + |u^\varepsilon|_{C^{0}(\bar{Q})}]. \end{array} \right.$$

Since u is bounded, from this we deduce:

LEMMA 2.9: *The maximum solution of the Q.V.I.* (1.6) *satisfies*

(2.83) $$u \in C^{2+\alpha, 1+\frac{\alpha}{2}}(\bar{D} \times [0,T]), \quad \forall \ D \quad \textit{as in } (2.82) \ \square$$

We then have:

LEMMA 2.10: *Under the assumptions of Theorem 1.2, together with* (2.71),(2.80), (2.81) *and* $V = H_0^1(\mathcal{O})$, *the obstacle* Mu *satisfies the additional property*

(2.84) $$Mu \in W^{1,\infty}(\bar{Q}).$$

Proof: We assume that u si continued by 0 outside $\mathcal{O}$. Define a Borel mapping of $R^n \times [0,T] \to (R^n)^+$ such that

$$(2.85)\quad \left|\begin{array}{l} Mu(x,t) = k + c_0(\xi(x,t)) + u(x+\xi(x,t),t) \quad \forall\, x,t \\ \xi(x,t) = 0 \quad \forall\, x \notin \mathcal{O}\ ,\ x+\xi(x,t) \in \bar{\mathcal{O}} \text{ si } x \in \bar{\mathcal{O}}\ ,\ t \in [0,T]\ . \end{array}\right.$$

Then, as in Chapter 4, (2.6), we have

$$(2.86)\quad u(x+\xi(x,t),t) \leq Mu\,(x+\xi(x,t),t) - k\ .$$

Let

$$(2.87)\quad D = \{\, x \in \mathcal{O} \mid \underset{t\in[0,T]}{\text{Min}} \quad (Mu(x,t)-u(x,t)) > \frac{3k}{4} \,\}$$

therefore D satisfies the condition (2.82). By (2.86)

$$x + \xi(x,t) \in \bar{D}\ ,\ \forall\, x \in \bar{\mathcal{O}}\ ,\ t \in [0,T]\ .$$

By Lemma 2.9 we have (in particular)

$$(2.88)\quad \left|\begin{array}{llll} |u(y,t) - u(y',t)| & \leq C_D|y-y'| & \forall\, y,y' \in \bar{D}, & t \in [0,T] \\ |u(y,t) - u(y,t')| & \leq C_D|t-t'| & \forall\, y \in \bar{D}\ , & t,t' \in [0,T]\ . \end{array}\right.$$

Now let ρ_0 be such that

$$(2.89)\quad \left|\begin{array}{lll} |y-y'| \leq \rho_0 & \text{implies} & |Mu(y',t)-u(y',t)-(Mu(y,t)-u(y,t)| \leq \\ y,y' \in \bar{\mathcal{O}} & & \leq \dfrac{k}{8}\ . \\ t \in [0,T] & & \end{array}\right.$$

Consider a unit vector χ of R^n, θ real, and x_0 fixed in $\bar{\mathcal{O}}$, $t \in [0,T]$. For $|\theta| \leqslant \rho_0$ let us set

$$\begin{array}{l} y = x_0 + \xi(x_0,t) \\ y' = y + \theta\chi \quad ,\quad y'' = \text{projection of } y' \text{ onto } \bar{\mathcal{O}} \end{array}$$

then by (2.89), using (2.86), and from $|y'' - y| \leqslant |y' - y| \leqslant \rho_0$,

$$Mu(y'',t)-u(y'',t) \geq Mu(y,t)-u(y,t) - \frac{k}{8} > k - \frac{k}{8} = \frac{7k}{8}$$

which implies $y'' \in \bar{D}$, therefore by (2.88)

$$|u(y'',t)-u(y,t)| \leq C_D|y-y''| \leq C_D|y-y'|\ .$$

But

$$\begin{array}{ll} Mu(x_0,t) & = k + c_0(\xi_0(x,t)) + u(y,t) \\ Mu(x_0+\theta\chi\ ,t) & \leq k + c_0(\xi_0(x,t)) + u(y';t) \\ & = k + c_0(\xi_0(x,t)) + u(y'',t) \end{array}$$

(since u is constant and equal to 0 outside of $\mathcal{O}$), therefore

$$Mu(x_0+\theta\chi,t) - Mu(x_0,t) \le C_D|\theta| \quad , \forall\ |\theta|<\rho_0 ,$$

χ a unit vector of R^n, $x_0 \in \bar{\mathcal{O}}$.

Consequently

$$(2.90)\qquad |Mu(x,t)-Mu(x',t)| \le K\ |x-x'| \quad \forall x,\ x' \in \bar{\mathcal{O}},$$

$t \in [0,T]$, where K is a constant.

Moreover, for $t,t' \in [0,T]$

$$\begin{aligned} Mu(x,t')-Mu(x,t) &\le u(x+\xi(x,t),t') \\ &\quad - u(x+\xi(x,t),t) \\ &\le C_D|t-t'| \end{aligned}$$

and therefore

$$(2.91)\qquad |Mu(x,t')-Mu(x,t)| \le C_D|t-t'| \qquad \forall x \in \bar{\mathcal{O}}\ ,\ t,t' \in [0,T]$$

which completes the proof of the property (2.84).□

Consider the function $\gamma(x)$ already introduced Section 2.2 of Chapter 4; let

$$(2.92)\qquad \gamma(x) = k + \inf_{\substack{\xi \ge 0 \\ x+\xi\in\Gamma}} c_0(\xi)$$

and assume

$$(2.93)\qquad \gamma \in W^{2,\infty}(\mathcal{O}).$$

We then have:

LEMMA 2.11: *Under the assumptions of Lemma* (2.10), *together with* (2.93), *we have*

$$(2.94)\qquad A(t) \geqslant -C, \quad t \text{ a.e. } in\ H^{-1}(\mathcal{O}).$$

Proof: By Lemma (2.10) AMu belongs to $L^\infty(0,T;H^{-1}(\mathcal{O}))$. Therefore it suffices to prove (2.94) in the sense of $H^{-1}(Q)$, and more generally in the sense of distributions in Q. Now,

$$Mu(x,t) \le \gamma(x) \qquad \forall (x,t) \in Q .$$

If $Mu(x,t) = \gamma(x)$ on Q, then (2.94) immediately follows from (2.93). Otherwise there exists $x_0,t_0 \in Q$ such that $Mu(x_0,t_0) < \gamma(x_0)$.

We then set

$$Q_\varepsilon^1 = \{x,t \in Q| \quad Mu(x,t) < \gamma(x) - \tfrac{\varepsilon}{2}\}$$

which is non-empty for ε small enough. Let

$$Q_\varepsilon^2 = \{x,t \in Q \mid Mu(x,t) > \gamma(x) - \varepsilon\}$$

then

$$Q = Q_\varepsilon^1 \cup Q_\varepsilon^2.$$

Set

(2.95) $$\psi_\varepsilon(x,t) = Mu(x,t) \wedge (\gamma(x)-\varepsilon) .$$

By (2.92) everything reduces to proving that

(2.96) $A\psi_\varepsilon \geqslant -c$ in the sense of distributions in Q_ε^1

where c does not depend on ε. Define

$$\mathcal{E}(x,t) = \{ \xi \geq 0 \mid c_0(\xi) + u(x+\xi,t) = \inf_{\eta\geq 0} (c_0(\eta)+u(x+\eta,t)) \quad x + \mathcal{E}(x,t) \in \bar{\mathcal{O}}, \ \forall x \in \mathcal{O}, \ t\in[0,T]\}.$$

Therefore $\mathcal{E}(x,t)$ is compact for all fixed x,t in $\bar{Q}$. Set

(2.97) $$d^\varepsilon = \inf_{\{x,t\in \bar{Q}_\varepsilon^1\}} d(x+\mathcal{E}(x,t),\Gamma) .$$

Then, as in Lemma 2.3 of Chapter 4,

(2.98) $$d^\varepsilon > 0.$$

Let χ be a unit vector of R^n, and θ be real such that

(2.99) $$|\theta| \leq \frac{d^\varepsilon}{2} \wedge \rho_0$$

where ρ_0 has been defined in (2.89). Let $(x,t) \in Q_\varepsilon^1$, then $d(x + \xi(x,t),\Gamma) \geqslant d^\varepsilon$ and by (2.86),(2.87) $x + \xi(x,t) \in \bar{D}$ necessarily, because $x + \xi(x,t)$ cannot be on the boundary,

(2.100) $$x + \xi(x,t) \in D.$$

Also, for $0 \leqslant |\lambda| \leqslant 1$, $x + \lambda\theta\chi + \xi(x,t)$ satisfies

$$d(x+ \lambda\theta\chi+\xi(x,t),\Gamma) \geq \frac{d^\varepsilon}{2} ,$$

thanks to the choice of θ. Furthermore, with the notations of Lemma 2.10,

$$y = x + \xi(x,t), \qquad y' = y'' = y + \lambda\theta\chi,$$

and also necessarily

(2.101) $$x + \lambda\theta\chi + \xi(x,t) \in D.$$

As in Lemma 2.3, Chapter 4, we deduce from this that

(2.102) $$\frac{1}{\theta^2} [Mu(x+\theta\chi,t)+ Mu(x-\theta\chi,t)- 2Mu(x,t)] \leq C_D$$

where

$$C_D = \sup_{\substack{y,y'\in\bar{D}\\ y\neq y'\\ t\in[0,T]}} \frac{|Du(y,t)-Du(y',t)|}{|y-y'|},$$

The inequality (2.102) holds for all $x,t \in Q_\varepsilon^1$. From this we also deduae that

(2.103) $$\frac{1}{\theta^2}[\psi_\varepsilon(x+\theta\chi,t) + \psi_\varepsilon(x-\theta\chi,t) - 2\psi_\varepsilon(x,t)] \leq C \,. \quad \forall x,t \in Q_\varepsilon^1.$$

Next we consider $\beta_\varepsilon \in \mathcal{D}(Q_\varepsilon^1)$ and we work as in the proof of Lemma 2.3 of Chapter 4; t is a simple parameter. We obtain (2.96) and also that

$$A\psi_\varepsilon \geqslant -c \quad \text{in} \quad H^{-1}(Q).$$

By making ε tend to 0 we obtain the result desired.□

We are then able to give the regularity result:

THEOREM 2.2: *Take the assumptions of Theorem 1.2, together with* $\bar{u} = 0$, (2.80), (2.81), (2.93) *and* $V = H_0^1$. *The the maximum solution of the Q.V.I.* (1.6) *satisfies*

(2.104) $$u \in W^{2,1,p}(Q), \quad \forall p;\ p \geqslant 2,\ p < \infty,$$

(2.105) $$-\frac{\partial u}{\partial t} + A(t)u \in L^\infty(Q).$$

Furthermore, the solution with the regularity (2.104), (2.105) *is unique.*

Proof: Let us consider

(2.106) $$\left| \begin{array}{l} -\dfrac{\partial \tilde{u}^\varepsilon}{\partial t} + (A(t)+a_0)\tilde{u}^\varepsilon + \dfrac{1}{\varepsilon}(\tilde{u}^\varepsilon - Mu)^+ = f \\ \tilde{u}^\varepsilon|_\Sigma = 0, \quad \tilde{u}^\varepsilon(x,T) = 0, \quad \tilde{u}^\varepsilon \in W^{2,1,p}(Q) \end{array} \right.$$

and

$$\tilde{v}^\varepsilon = \tilde{u}^\varepsilon - Mu$$

therefore

$$-\frac{\partial \tilde{v}^\varepsilon}{\partial t} + (A(t)+a_0)\tilde{v}^\varepsilon + \frac{1}{\varepsilon}\tilde{v}^{\varepsilon+} = f - \left(-\frac{\partial}{\partial t}Mu + (A+a_0)Mu\right)$$
$$\leqslant \text{(using Lemmas 2.10 and 2.11)} \leqslant f + c \leqslant c_1$$
$$\tilde{v}^\varepsilon|_\Sigma = -k, \quad \tilde{v}^\varepsilon(x,T) = -k,$$

so again

$$-\frac{\partial \tilde{v}^\varepsilon}{\partial t} + (A(t)+a_0)\tilde{v}^\varepsilon + \frac{1}{\varepsilon}\tilde{v}^\varepsilon \leq c_1$$

and therefore by the Maximum Principle $\tilde{v}^\varepsilon \leqslant \varepsilon c_1$, so again $\frac{\tilde{v}^{\varepsilon+}}{\varepsilon} \leqslant c_1$, so that $-\frac{\partial \tilde{v}^\varepsilon}{\partial t} + (A(t) + a_0)\tilde{v}^\varepsilon$ is bounded in L^∞, and $\tilde{v}^\varepsilon$ also stays in a bounded set of $\mathbb{W}^{2,1,p}(Q)$. But then $\tilde{u}^\varepsilon$ stays in a bounded set of $W^{1,\infty}(\bar{Q})$.

Using this estimate we show that $\tilde{u}^\varepsilon \downarrow u$ and weakly in $W^{1,\infty}$, and is a (strong) solution of the V.I.

$$
(2.107)\quad \left|\begin{array}{l}
-(\frac{\partial u}{\partial t}, v(t)-\tilde{u}(t)) + a(t;\tilde{u}(t), v(t)-\tilde{u}(t)) \geq (f(t), v(t)-\tilde{u}(t)) \\
\forall\, v \in L^2(0,T;V) \quad \text{s.t.} \quad v(t) \leq Mu(t) \quad \text{a.e.} \\
\tilde{u} \in L^2(0,T;V), \quad \frac{\partial \tilde{u}}{\partial t} \in L^2(Q), \quad \tilde{u}(t) \leq Mu(t).
\end{array}\right.
$$

But if we compare equations (2.89) and (2.106) we have

$$
\|\tilde{u}^\varepsilon - u^\varepsilon\|_{L^\infty} \leq \|Mu - Mu^\varepsilon\|_{L^\infty} \to 0
$$

therefore

$$
\tilde{u} = u,
$$

so that we can state that

$$
(2.108)\quad \left|\begin{array}{l}
u \in W^{1,\infty}(Q), \quad u|_\Sigma = 0, \; u(x,T) = 0 \\
-(\frac{\partial u}{\partial t}, v(t)-u(t)) + a(t;u(t), v(t)-u(t)) \geq (f(t), v(t)-u(t)) \\
\forall v \in L^2(0,T;V) \quad \text{s.t.} \quad v(t) \leq Mu(t) \quad \text{a.e.}
\end{array}\right.
$$

We can then apply the Levy-Stampacchia inequality (Corollary 2.1) to (2.108), where the obstacle is $\psi = Mu$, therefore

$$
(2.109)\quad f \geq -\frac{\partial u}{\partial t} + Au + a_0 u \geq \inf(f, -\frac{\partial Mu}{\partial t} + AMu + a_0 Mu).
$$

It follows from (2.109) and from Lemmas 2.10 and 2.11 that

$$
f \geq -\frac{\partial u}{\partial t} + Au + a_0 u \geq f \wedge (-c_2)
$$

and consequently

$$
-\frac{\partial u}{\partial t} + Au \in L^\infty(Q),
$$

which certainly implies that $u \in \mathbb{W}^{2,1,p}(Q)$ for all p.

The method of T. LAETSCH usd in Theorem 1.3 shows that the bounded positive solution of (2.108) is unique. Since every regular solution of the Q.V.I. is a solution of (2.108), the proof of the result is complete.□

2.3 *Iterative Method*

We are able to state:

THEOREM 2.3: *Take the assumptions of Theorem 2.2. Let* $0 \leqslant u_0 \leqslant u^0$. *Then the sequence starting from* u_0,

$$u_{n+1} = Tu_n$$

converges in L^∞ *to the regular solution of the Q.V.I.* (1.6) *(i.e. that satisfies* (2.104),(2.105)*).*

Proof: The operator T has been defined in (1.28). The result rests upon Lemma 1.1. Now,

$$u^0 - u_0 \leqslant u^0$$

and therefore by Lemma 1.1,

$$u^1 - u_1 \leqslant (1 - \mu)u^1,$$

and quite generally,

$$0 \leqslant u^n - u_n \leqslant (1 - \mu)^n u^n \leqslant (1 - \mu)^n u_0,$$

and therefore

$$u^n - u_n \to 0 \quad \text{in} \quad L^\infty.$$

But we know that $u^n \downarrow u$ is the maximum solution and the unique regular solution of the Q.V.I. (1.6). Since the limit is in $C^0(\bar{Q})$, Dini's Theorem implies that $u^n \to u$ in L^∞. Therefore $u_n \to u$ in L^∞ also.□

3. APPROXIMATE SEMI-GROUP FOR THE INEQUALITY

We again take the spaces B and C introduced in Section 5 of Chapter 4. Consider the semi-group $\Phi(t)$ satisfying

$$\text{(3.1)} \qquad \begin{aligned} &\Phi(t) : B \to B \ , \quad \Phi(0) = I \\ &\Phi(t+s) = \Phi(t)\ \Phi(s) \\ &\Phi(t)\gamma \geq 0 \quad \text{if} \quad \gamma \geq 0 \\ &\|\Phi(t)\| \leq 1 \end{aligned}$$

$$\text{(3.2)} \qquad \begin{aligned} &\Phi(t) : C \to C \\ &\sup_{0 \leq s \leq T} \|\Phi(t)f(s)-f(s)\|_C \to 0 \qquad t \downarrow 0,\ \forall f \in C(0,T;C). \end{aligned}$$

Let there be given

(3.3) $L \in B$, $t \to \Phi(t)L(x)$ is (Lebesgue) measureable for all $x \in E$,

$$\int_0^\infty e^{-\alpha t}\Phi(t)L\,dt \in C, \quad \alpha \geqslant 0.$$

We write $-\mathcal{A}$ for the infinitesimal generator of $\Phi(t)$ in C, and $D(\mathcal{A})$ for its domain (we recall that $D(\mathcal{A})$ is dense in C, cf. Chapter 4, Lemma 5.2).

Let there also be given

(3.4) $\bar{u} \in C.$

3.1 *Equation*

We consider the following problem:

(3.5)
$$\left| \begin{array}{l} u(.) \in C([0,T];\, C)\,, \quad u(0) = \bar{u} \\ u(t) = \displaystyle\int_0^{t-s} e^{-\alpha\sigma}\, \Phi(\sigma)L\, d\sigma + e^{-\alpha(t-s)}\Phi(t-s)u(s) \\ \forall\ s \le t \in [0,T] \end{array} \right.$$

LEMMA 3.1: *The problem* (3.5) *has one and only one solution.*

Proof: Uniqueness is clear, since by taking $s = 0$ in (3.5) we have

$$u(t) = \int_0^t e^{-\alpha\sigma}\, \Phi(\sigma)L d\sigma + e^{-\alpha t}\, \Phi(t)\, \bar{u}$$

Also, if we consider this as a definition of $u(t)$, then by assumptions (3.2) and (3.3), $u(t) \in C([0,T]);C)$, $u(0) = \bar{u}$, and

$$u(t) = \int_0^{t-s} e^{-\alpha\sigma}\, \Phi(\sigma)L d\sigma + \int_{t-s}^{t} e^{-\alpha\sigma}\Phi(\sigma)L d\sigma + e^{-\alpha t}\, \Phi(t)\bar{u}$$
$$= \int_0^{t-s} e^{-\alpha\sigma}\Phi(\sigma)L d\sigma + e^{-\alpha(t-s)}\, \Phi(t-s)u(s)$$

whence (3.5).□

Remark 3.1

We can write equation (3.5) in the form

(3.6) $$u(t) = \int_s^t e^{-\alpha(t-\lambda)}\, \Phi(t-\lambda)L d\lambda + e^{-\alpha(t-s)}\, \Phi(t-s)u(s)\,. \quad \square$$

Set

$$u(t) = T(t)\bar{u}.$$

LEMMA 3.2: $T(t)$ *is a contraction (non-linear) semi-group on* C.

Proof: By definition,

(3.8) $T(t):C \to C, \quad T(0) = I.$

Let us set

$$v(t) = T(t)T(\theta)\bar{u}, \quad t \geqslant 0.$$

Therefore

$$v(t) = \int_0^t e^{-\alpha\sigma} \Phi(\sigma)Ld\sigma + e^{-\alpha t} \Phi(t)\, u(\theta) .$$

But by (3.5)

$$u(t+\theta) = \int_0^t e^{-\alpha\sigma} \Phi(\sigma)Ld\sigma + e^{-\alpha t} \Phi(t)u(\theta) \cdot$$

whence

$$u(t + \theta) = v(t),$$

so

(3.9) $$T(t + \theta) = T(t)T(\theta).$$

However, by (3.6)

$$T(t)\bar{u}_1 - T(t)\bar{u}_2 = e^{-\alpha t} \Phi(t)(\bar{u}_1 - \bar{u}_2)$$

therefore

(3.10) $$\|T(t)\bar{u}_1 - T(t)\bar{u}_2\| \le \|\bar{u}_1 - \bar{u}_2\| .$$

Lastly, by (3.6) it is clear that

(3.11) $T(t)\bar{u} \to \bar{u}$ in C as $t \to 0$.□

We now prove a formula of Trotter for T(t) (cf. M.G. CRANDALL - T.M. LIGGETT [1]). For $\lambda > 0$ consider

(3.12) $$R_\lambda(\bar{u}) = \int_0^\infty e^{-(\frac{1}{\lambda}+\alpha)t} \Phi(t)(\frac{\bar{u}}{\lambda} + L)\, dt .$$

Our objective is to prove:

THEOREM 3.1: *Assume* (3.1),(3.2),(3.3). *Then*

(3.13) *for all* $t > 0$, $R^n_{t/n}(u) \to T(t)u$ *in* C *as* $n \to \infty$.□

The proof of Theorem 3.1 requires some preliminary lemmas.

LEMMA 3.3: *We have*

(3.14) $$\|R^k_\lambda(\bar{u}_1) - R^k_\lambda(\bar{u}_2)\| \le \|\bar{u}_1 - \bar{u}_2\| \quad \forall\lambda > 0,\ k \ge 1.$$

Proof: It suffices to prove (3.14) for k = 1, which immediately follows from (3.12).□

LEMMA 3.4: *We have*

(3.15) $R_\lambda \circ T(\mu) = T(\mu)\circ R_\lambda$ *for all* $\lambda,\mu > 0$.

Proof: By definition

$$(3.16)\qquad R_\lambda \circ T(\mu)\bar{u} = \int_0^\infty e^{-(\frac{1}{\lambda}+\alpha)t}\Phi(t)[\frac{T(\mu)\bar{u}}{\lambda} + L]\,dt$$
$$= \int_0^\infty e^{-(\frac{1}{\lambda}+\alpha)t}\Phi(t)[L + \frac{e^{-\alpha\mu}\Phi(\mu)\bar{u}}{\lambda} + \frac{\int_0^\mu e^{-\alpha\sigma}\Phi(\sigma)L\,d\sigma}{\lambda}]dt\ ,$$

Then

$$(3.17)\qquad T(\mu)\circ R_\lambda\,\bar{u} = \int_0^\mu e^{-\alpha\sigma}\Phi(\sigma)L\,d\sigma + e^{-\alpha\mu}\Phi(\mu)[\int_0^\infty e^{-(\frac{1}{\lambda}+\alpha)t}\Phi(t)(\frac{\bar{u}}{\lambda} + L)dt\,]\ .$$

If we compare (3.16) and (3.17) we establish that the terms in $\bar{u}$ are equal. It therefore remains to prove that

$$(3.18)\qquad \int_0^\infty e^{-(\frac{1}{\lambda}+\alpha)t}\Phi(t)\,[L + \frac{\int_0^\mu e^{-\alpha\sigma}\Phi(\sigma)L\,d\sigma}{\lambda}]dt =$$
$$= \int_0^\mu e^{-\alpha\sigma}\Phi(\sigma)L\,d\sigma + e^{-\alpha\mu}\Phi(\mu)\int_0^\infty e^{-(\frac{1}{\lambda}+\alpha)t}\Phi(t)L\,dt.$$

Let X be the left hand member of (3.18). Then

$$X = \int_0^\infty e^{-(\frac{1}{\lambda}+\alpha)t}\Phi(t)L\,dt + \int_0^\infty \frac{e^{-\frac{t}{\lambda}}}{\lambda}(\int_0^\mu e^{-\alpha(\sigma+t)}\Phi(\sigma+t)L\,d\sigma)dt$$
$$= \int_0^\infty e^{-(\frac{1}{\lambda}+\alpha)t}\Phi(t)L\,dt + \int_0^\infty \frac{e^{-\frac{t}{\lambda}}}{\lambda}(\int_t^{t+\mu} e^{-\alpha\sigma}\Phi(\sigma)L\,d\sigma)dt$$
$$= \int_0^\infty e^{-(\frac{1}{\lambda}+\alpha)t}\Phi(t)L\,dt + \int_0^\infty -\,de^{-\frac{t}{\lambda}}\ (\int_t^{t+\mu} e^{-\alpha\sigma}\Phi(\sigma)\ L\,d\sigma)$$
$$= \int_0^\mu e^{-\alpha\sigma}\Phi(\sigma)L\,d\sigma + \int_0^\infty e^{-\frac{t}{\lambda}}\,e^{-\alpha(t+\mu)}\,\Phi(t+\mu)\ L\ dt$$

which is the right hand member of (3.18).□

Proof of Theorem 3.1: Let us set

$$u^0 = \int_0^\infty e^{-\alpha t}\,\Phi(t)L\ dt$$

therefore

$$u^0 = \int_0^\infty e^{-(\frac{1}{\lambda}+\alpha)t}\Phi(t)(L + \frac{u^0}{\lambda})dt$$

also, and by (3.12)

$$R_\lambda(\bar{u}) = \int_0^\infty e^{-(\frac{1}{\lambda}+\alpha)t}\,\Phi(t)\,\frac{(\bar{u}-u^0)}{\lambda}\,dt + u^0 . \tag{3.19}$$

Since we also have

$$u^0 = \int_0^t e^{-\alpha\sigma}\,\Phi(\sigma)L\,d\sigma + e^{-\alpha t}\,\Phi(t)\,u^0$$

it follows from (3.6) that

$$T(t)\bar{u} = e^{-\alpha t}\,\Phi(t)(\bar{u} - u^0) + u^0 . \tag{3.20}$$

Let us now show that it suffices to prove the result (3.13) under the additional assumption

$$\bar{u} - u^0 \in D(\mathcal{A}). \tag{3.21}$$

In fact, let $v_k \in D(\mathcal{A})$ and $v_k \to \bar{u} - u^0$ in C. Let us set

$$u_k = v_k + u^0 \to \bar{u} \quad \text{in C.}$$

Then

$$R^n_{t/n}(\bar{u}) - T(t)\bar{u} = R^n_{t/n}(\bar{u}) - R^n_{t/n}(u_k) + $$
$$+ R^n_{t/n}(u_k) - T(t)u_k + T(t)u_k - T(t)\bar{u}$$

and therefore by Lemma 3.3 and (3.10)

$$\|R^n_{t/n}(\bar{u}) - T(t)\bar{u}\| \le 2\,\|u_k - \bar{u}\| +$$
$$+ \|R^n_{t/n}(u_k) - T(t)u_k\|$$

and thus it suffices to have (3.13) with u_k instead of u. Now u_k satisfies (3.21). Therefore we assume (3.21). This implies that there exists $U \in C$ such that

$$\bar{u} - u^0 = \int_0^\infty e^{-\beta t}\,\Phi(t)(U + \beta(\bar{u} - u^0))\,dt \qquad \forall\, \beta > 0 \tag{3.22}$$

therefore

$$\bar{u} - u^0 = \int_0^\infty e^{-(\frac{1}{\lambda}+\alpha)t}\,\Phi(t)(U + (\tfrac{1}{\lambda}+\alpha)(\bar{u}-u^0))\,dt$$

and by (3.19)

$$-R_\lambda(\bar{u}) + \bar{u} = \int_0^\infty e^{-(\frac{1}{\lambda}+\alpha)t}\,\Phi(t)(U + \alpha(\bar{u}-u^0))dt$$

$$= \left(\int_0^\infty e^{-\sigma}\,\Phi(\sigma \frac{\lambda}{1+\lambda\alpha})(U + \alpha(\bar{u}-u^0))d\sigma\right) \frac{\lambda}{1+\lambda\alpha}$$

so

$$\frac{\bar{u} - R_\lambda(\bar{u})}{\lambda} - (U + \alpha(\bar{u}-u^0)) = \int_0^\infty e^{-\sigma}\left(\frac{\Phi(\frac{\sigma\lambda}{1+\lambda\alpha})}{1+\lambda\alpha} - I\right)(U + \alpha(\bar{u}-u^0))d\sigma .$$

Since $U + \alpha(\bar{u} - u^0) \in C$ it follows from the second assumption (3.2) that

$$\left\| \frac{\bar{u} - R_\lambda(\bar{u})}{\lambda} - (U + \alpha(\bar{u}-u^0)) \right\| \to 0 \quad \text{when } \lambda \to 0. \tag{3.23}$$

Further, by (3.22)

$$\bar{u}-u^0 = \int_0^\lambda e^{-\beta t}\Phi(t)(U+\beta(\bar{u}-u^0))dt + e^{-\beta\lambda}\Phi(\lambda)(\bar{u}-u^0)$$

and for fixed λ this relation has a meaning with $\beta = \alpha$, therefore

$$\bar{u}-u^0 = \int_0^\lambda e^{-\alpha t}\Phi(t)(U+\alpha(\bar{u}-u^0))dt + e^{-\alpha\lambda}\,\Phi(\lambda)(\bar{u}-u^0) \tag{3.24}$$

and thus by (3.20)

$$\bar{u} - T(\lambda)\bar{u} = \int_0^\lambda e^{-\alpha t}\Phi(t)(U + \alpha(\bar{u}-u^0))dt$$

hence again

$$\left\| \frac{\bar{u} - T(\lambda)\bar{u}}{\lambda} - (U + \alpha(\bar{u}-u^0)) \right\| \to 0 \quad \text{when } \lambda \to 0. \tag{3.25}$$

Next, since $T(t)$ is a semi-group,

$$R^n_{\frac{t}{n}}\,\bar{u} - T(t)\bar{u} = R^n_{\frac{t}{n}}\,\bar{u} - T(\tfrac{t}{n})^n\,\bar{u}$$

$$= \sum_{k=0}^{n-1} \left(R^{n-k}_{\frac{t}{n}}\;T(\tfrac{t}{n})^k\,\bar{u} - R^{n-k-1}_{\frac{t}{n}}\;T(\tfrac{t}{n})^{k+1}\,\bar{u}\right).$$

Since $R_{t/n}$ is a contraction,

$$\left\| R^{n-k}_{\frac{t}{n}}\;T(\tfrac{t}{n})^k\,\bar{u} - R^{n-k-1}_{\frac{t}{n}}\;T(\tfrac{t}{n})^{k+1}\,\bar{u} \right\| \le$$

$$\le \left\| R_{\frac{t}{n}}\;T(\tfrac{t}{n})^k\,\bar{u} - T(\tfrac{t}{n})^{k+1}\,\bar{u} \right\|$$

and by Lemma 3.4,

$$= \| T(\tfrac{t}{n})^k R_{\frac{t}{n}} \bar{u} - T(\tfrac{t}{n})^{k+1} \bar{u} \|$$

and since T(t/n) is a contraction

$$\leq \| R_{\frac{t}{n}} \bar{u} - T(\tfrac{t}{n}) \bar{u} \|.$$

Therefore

$$\| R^n_{\frac{t}{n}} (\bar{u}) - T(t)(\bar{u}) \| \leq n \| R_{\frac{t}{n}} \bar{u} - T(\tfrac{t}{n}) \bar{u} \|$$

$$= t \left\| \frac{R_{t/n} \bar{u} - \bar{u}}{t/n} - \frac{T(\frac{t}{n}) \bar{u} - \bar{u}}{t/n} \right\|$$

$$\leq \left\| t \frac{\bar{u} - R_{t/n} \bar{u}}{t/n} - (U+\alpha(\bar{u}-u^0)) \right\| +$$

$$+ t \left\| \frac{\bar{u} - T(\frac{t}{n})\bar{u}}{\frac{t}{n}} - (U+\alpha(\bar{u}-u^0)) \right\| \to 0 \quad \text{when } n \to \infty$$

by (3.23) and (3.25). The proof of (3.13) is complete.□

3.2 *Evolutionary Inequalities*

Let there be given a function $\Psi(t)$ such that

$$(3.26) \quad \left| \begin{array}{l} \Psi \in C([0,T];\, C) \\ \bar{u} \leq \Psi(0)\ . \end{array} \right.$$

We introduce the following problem

$$(3.27) \quad \left| \begin{array}{l} u(.) \in C([0,T];\, C)\ , \quad u(0) = \bar{u}\ , \\ u(t) \leq \Psi(t) \qquad \forall t \in [0,T]\ ; \\ u(t) \leq \displaystyle\int_0^{t-s} e^{-\alpha\sigma}\Phi(\sigma)L\,d\sigma + e^{-\alpha(t-s)}\Phi(t-s)u(s), \\ \forall\, s \leq t \in [0,T]\ . \end{array} \right.$$

Our objective is to prove the following result:

THEOREM 3.2: *Assume* (3.1),(3.2),(3.3) *and* (3.26). *Then the set of functions satisfying* (3.27) *is non-empty and has a maximum element.*□

We are going to study (3.27) by the penalised problem technique. Consider the problem

(3.28)
$$u_\varepsilon \in C([0,T];\ C)\ ,\quad u_\varepsilon(0) = \bar{u}$$
$$u_\varepsilon(t) = e^{-\alpha(t-s)}\Phi(t-s)u_\varepsilon(s) + $$
$$+ \int_s^t e^{-\alpha(t-\lambda)}\Phi(t-\lambda)\ [L - \frac{1}{\varepsilon}(u_\varepsilon(\lambda)-\Psi(\lambda))^+]\ d\lambda\ ,$$
$$\forall s \le t \le T\ .$$

LEMMA 3.5: *The problem* (3.28) *is equivalent to*

(3.29)
$$\left|\begin{array}{l} u_\varepsilon \in C([0,T]\ ;\ C) \\ u_\varepsilon(t) = e^{-\alpha t}\Phi(t)\bar{u} + \int_0^t e^{-\alpha(t-\lambda)}\Phi(t-\lambda)[L - \frac{1}{\varepsilon}(u_\varepsilon(\lambda)-\Psi(\lambda))^+]\ d\lambda\ . \end{array}\right.$$

Proof: It is clear that if u_ε is the solution of (3.28), then on taking $s = 0$ we see that u_ε is the solution of (3.29). Conversely, if u_ε is the solution of (3.29), then $u_\varepsilon(0) = \bar{u}$, then if we set

(3.30)
$$g(\lambda) = L - \frac{1}{\varepsilon}\ (u_\varepsilon(\lambda) - \Psi(\lambda))^+$$

then

$$u_\varepsilon(t) = e^{-\alpha t}\Phi(t)\bar{u} + \int_0^t e^{-\alpha(t-\lambda)}\Phi(t-\lambda)g(\lambda)\ d\lambda$$
$$= \int_s^t e^{-\alpha(t-\lambda)}\Phi(t-\lambda)g(\lambda)d\lambda + \int_0^s e^{-\alpha(t-\lambda)}\Phi(t-\lambda)g(\lambda)d\lambda + $$
$$+ e^{-\alpha t}\ \Phi(t)\ \bar{u}$$
$$= \int_s^t e^{-\alpha(t-\lambda)}\Phi(t-\lambda)g(\lambda)d\lambda + e^{-\alpha(t-s)}\Phi(t-s)[\Phi(s)\bar{u} + $$
$$+ \int_0^s e^{-\alpha(s-\lambda)}\Phi(s-\lambda)g(\lambda)d\lambda]$$

and therefore u_ε is the solution of (3.28).□

LEMMA 3.6: *If* u_ε *is the solution of* (3.28), *then if* $\beta \geqslant 0$, u_ε *is also the solution of*

(3.31)
$$u_\varepsilon(t) = e^{-\alpha t}\Phi(t)\bar{u} + \int_0^t e^{-\beta(t-\lambda)}\Phi(t-\lambda)[L+(\beta-\alpha)u_\varepsilon(\lambda) - $$
$$- \frac{1}{\varepsilon}\ (u_\varepsilon(\lambda) - \Psi(\lambda))^+\ ]d\lambda\ .$$

Proof: Now,

$$u_\varepsilon(t) = e^{-\alpha(t-s)}\Phi(t-s)u_\varepsilon(s) + $$
$$+ \int_s^t e^{-\alpha(t-\lambda)}\Phi(t-\lambda)g(\lambda)d\lambda\ .$$

Multiplying by $e^{-(\beta-\alpha)(t-s)}$ we deduce from this that

$$e^{-(\beta-\alpha)(t-s)} u_\varepsilon(t) = e^{-\beta(t-s)}\Phi(t-s)\, u_\varepsilon(s) + $$
$$+ e^{-(\beta-\alpha)(t-s)} \int_s^t e^{-\alpha(t-\lambda)}\Phi(t-\lambda)g(\lambda)d\lambda .$$

Now integrate this relation with respect to s over the interval [0,t], and we obtain

$$(3.32) \qquad (1-e^{-(\beta-\alpha)t})u_\varepsilon(t) = \int_0^t e^{-\beta(t-s)}\Phi(t-s)u_\varepsilon(s)(\beta-\alpha)ds + $$
$$+ \int_0^t ds\; e^{-(\beta-\alpha)(t-s)} \int_s^t e^{-\alpha(t-\lambda)}\Phi(t-\lambda)g(\lambda)d\lambda .$$

Now, if we integrate by parts,

$$\int_0^t ds\; e^{-(\beta-\alpha)(t-s)} \int_s^t e^{-\alpha(t-\lambda)}\Phi(t-\lambda)g(\lambda)d\lambda =$$
$$- e^{-(\beta-\alpha)t} \int_0^t e^{-\alpha(t-\lambda)}\Phi(t-\lambda)g(\lambda)\, d\lambda$$
$$+ \int_0^t e^{-\beta(t-s)}\Phi(t-s)g(s)ds$$

and since

$$u_\varepsilon(t) - \int_0^t e^{-\alpha(t-\lambda)}\Phi(t-\lambda)g(\lambda)d\lambda = e^{-\alpha t}\,\Phi(t)\bar{u}$$

we easily deduce from (3.32) that (3.31) is satisfied.□

LEMMA 3.7: *The problem* (3.29) *has one and only one solution.*

Proof: By Lemma 3.6 equation (3.29) is equivalent to

$$(3.33) \qquad u_\varepsilon(t) = e^{-(\alpha+\frac{1}{\varepsilon})t}\Phi(t)\bar{u} + \int_0^t e^{-(\alpha+\frac{1}{\varepsilon})(t-\lambda)}\Phi(t-\lambda)[\, L + $$
$$+ \frac{1}{\varepsilon}\, u_\varepsilon(\lambda) - \frac{1}{\varepsilon}(u_\varepsilon(\lambda)-\psi(\lambda))^+]\; d\lambda$$
$$= e^{-(\alpha+\frac{1}{\varepsilon})t}\Phi(t)\bar{u} + \int_0^t e^{-(\alpha+\frac{1}{\varepsilon})(t-\lambda)}\Phi(t-\lambda)[L+ \frac{1}{\varepsilon}\, u_\varepsilon(\lambda)\wedge \psi(\lambda)]\; d\lambda .$$

We define a mapping T_ε of C([0,T];C) into itself by the formula

$$(3.34) \qquad T_\varepsilon z(t) = e^{-(\alpha+\frac{1}{\varepsilon})t}\Phi(t)\bar{u} + \int_0^t e^{-(\alpha+\frac{1}{\varepsilon})(t-\lambda)}\Phi(t-\lambda)[\, L + $$
$$+ \frac{1}{\varepsilon}\, z(\lambda) \wedge\psi(\lambda)]\; d\lambda ,$$
$$T_\varepsilon z_1(t) - T_\varepsilon z_2(t) = \frac{1}{\varepsilon} \int_0^t e^{-(\alpha+\frac{1}{\varepsilon})(t-\lambda)}\Phi(t-\lambda)\,[z_1(\lambda)\wedge\psi(\lambda) -$$
$$- z_2(\lambda)\wedge \psi(\lambda)]\; d\lambda .$$

We see that

$$\|T_\varepsilon z_1(t) - T_\varepsilon z_2(t)\| \le \frac{1}{\varepsilon}\int_0^t e^{-(\alpha+\frac{1}{\varepsilon})(t-s)} \|z_1(s)-z_2(s)\| \, ds$$

whence

$$(3.35) \qquad \|T_\varepsilon z_1 - T_\varepsilon z_2\|_{C(0,T;C)} \le \frac{1}{1+\alpha\varepsilon}\|z_1 - z_2\|_{C(0,T;C)} .$$

Therefore if $\alpha > 0$, T is a contraction with ratio < 1, whence the existence and uniqueness of the fixed point. Otherwise, we set

$$u_{n+1} = T_\varepsilon u_n, \quad u_0 \text{ arbitrary}$$

whence

$$\|u_{n+1}(t) - u_n(t)\| \le \frac{1}{\varepsilon}\int_0^t \|u_n(s) - u_{n-1}(s)\| \, ds$$

and by iteration

$$\|u_{n+1}(t) - u_m(t)\| \le \frac{1}{\varepsilon^n}\int_0^t \frac{(t-s)^{n-1}}{(n-1)!} \|u_1(s) - u_0(s)\| \, ds$$

therefore

$$\|u_{n+1} - u_n\|_{C(0,T;C)} \le \frac{1}{\varepsilon^n} \frac{T^n}{n!} \|u_1 - u_0\|_{C(0,T;C)} .$$

From this it follows that u_n is a Cauchy sequence in $C(0,T;C)$, whence the existence of a fixed point. This fixed point is unique, for if $u,\tilde{u}$ are two fixed points, then

$$\|u(t) - \tilde{u}(t)\| \le \frac{1}{\varepsilon}\int_0^t \|u(s) - \tilde{u}(s)\| \, ds$$

and by Gronwall's Lemma $u - \tilde{u} = 0$.□

LEMMA 3.8: *We have*

$$(3.36) \qquad u_\varepsilon \leqslant u_{\varepsilon'} \quad \textit{if } \varepsilon \leqslant \varepsilon$$

and

$$(3.37) \qquad \|u_\varepsilon\| \leqslant K.$$

Proof: Now,

$$T_\varepsilon u_{\varepsilon'}(t) = e^{-(\alpha+\frac{1}{\varepsilon})t}\Phi(t)\bar{u} + \int_0^t e^{-(\alpha+\frac{1}{\varepsilon})(t-\lambda)}\Phi(t-\lambda)\,[L + + \frac{1}{\varepsilon}u_\varepsilon'(\lambda)\wedge\psi(\lambda)\,]\,d\lambda$$

and

$$u'_\varepsilon(t) = e^{-(\alpha+\frac{1}{\varepsilon})t}\Phi(t)\bar{u} + \int_0^t e^{-(\alpha+\frac{1}{\varepsilon})(t-\lambda)}\Phi(t-\lambda)\,[L +$$

$$+ (\frac{1}{\varepsilon} - \frac{1}{\varepsilon'})u_{\varepsilon'}(\lambda) + \frac{1}{\varepsilon'}\, u_{\varepsilon'}(\lambda) - \frac{1}{\varepsilon'}(u_{\varepsilon'}(\lambda) - \psi(\lambda))^+\,]d\lambda$$

$$= e^{-(\alpha+\frac{1}{\varepsilon})t}\Phi(t)\bar{u} + \int_0^t e^{-(\alpha+\frac{1}{\varepsilon})(t-\lambda)}\Phi(t-\lambda)\,[L +$$

$$+ \frac{1}{\varepsilon}\, u_{\varepsilon'}(\lambda) \wedge \psi(\lambda) + (\frac{1}{\varepsilon} - \frac{1}{\varepsilon'})(u_{\varepsilon'} - \psi)^+(\lambda)\,]\, d\lambda$$

therefore

$$T_\varepsilon u_{\varepsilon'}(t) - u_{\varepsilon'}(t) = \int_0^t e^{-(\alpha+\frac{1}{\varepsilon})(t-\lambda)}\Phi(t-\lambda)(\frac{1}{\varepsilon'} - \frac{1}{\varepsilon})(u_{\varepsilon'} - \psi)^+(\lambda)\, d\lambda$$

$$\leqslant 0 \quad \text{if } \varepsilon \leqslant \varepsilon'.$$

Since T_ε is non-decreasing, from this we deduce that

$$T^n_\varepsilon u'_\varepsilon \leqslant u_{\varepsilon'},$$

and when $n \to \infty$, we certainly obtain (3.36). Furthermore, by (3.29)

$$u_\varepsilon(t) \leq e^{-\alpha t}\Phi(t)\,\bar{u} + \int_0^t e^{-\alpha(t-\lambda)}\Phi(t-\lambda)\, L\, d\lambda$$

$$\leq e^{-\alpha t} C_\alpha + \alpha \int_0^t e^{-\alpha(t-\lambda)} C_\alpha\, d\lambda$$

where

$$C_\alpha = \text{Max}\,(\|\bar{u}\|\,,\ \frac{\|L\|}{\alpha})$$

therefore we again have

$$u_\varepsilon(t) \leqslant C_\alpha \quad \text{if } \alpha > 0.$$

If $\alpha = 0$, then

$$u_\varepsilon(t) \leq \|\bar{u}\| + T\,\|L\| = C_0.$$

Next we prove an inequality in the opposite direction. Now,

$$T_\varepsilon 0 = e^{-(\alpha+\frac{1}{\varepsilon})t}\bar{u} + \int_0^t e^{-(\alpha+\frac{1}{\varepsilon})(t-\lambda)}\Phi(t-\lambda)\,[L - \frac{1}{\varepsilon}\,\psi^-\,]d\lambda$$

$$\geq -e^{-(\alpha+\frac{1}{\varepsilon})t}\bar{u}^- - \int_0^t e^{-(\alpha+\frac{1}{\varepsilon})(t-\lambda)}\Phi(t-\lambda)[L^- + \frac{1}{\varepsilon}\,\Psi^-]\, d\lambda$$

$$\geq -\ \text{Max}\,(\|\bar{u}^-\|,\ \frac{\varepsilon\,\|L^-\| + \|\Psi^-\|}{\varepsilon\alpha+1})$$

$$\geq -\ \text{Max}\,(\|\bar{u}^-\|\,,\ \|\Psi^-\| + \frac{\varepsilon\|L^-\|}{\varepsilon\alpha+1}) = -\,k_\varepsilon\,.$$

If $\alpha > 0$, then

$$k_\varepsilon \le k = \text{Max}\,(\|\bar{u}^-\|, \quad \|\Psi^-\| + \frac{\|L^-\|}{\alpha}).$$

If $z \geqslant -k$, then

$$T_\varepsilon z(t) \ge -\text{Max}\,(\|\bar{u}^-\| , \quad \frac{\varepsilon\,\|L^-\| + k}{\alpha\varepsilon + 1}) \quad \ge -k .$$

Therefore by iteration

$$u_\varepsilon \geqslant -k.$$

If $\alpha = 0$, by (3.29) we write

$$u_\varepsilon(t) = \Phi(t)\bar{u} + \int_0^t \Phi(t-\lambda)\,[L + \frac{1}{\varepsilon} v(\lambda)(\psi(\lambda) - u_\varepsilon(\lambda))]\,d\lambda$$

where

$$v(\lambda) = \begin{cases} 1 & \text{if} \quad u_\varepsilon(\lambda) > \psi(\lambda) \\ 0 & \text{if} \quad u_\varepsilon(\lambda) \le \psi(\lambda) \end{cases}$$

therefore (cf. Lemma 3.9 below)

$$(3.38) \qquad u_\varepsilon(t) = (\exp - \frac{1}{\varepsilon}\int_0^t v(s)ds)\,\Phi(t)\bar{u} + \int_0^t (\exp - \frac{1}{\varepsilon}\int_s^t v(\lambda)d\lambda\,)\,\Phi(t-s)\,[L + \frac{v(s)}{\varepsilon}\,u_\varepsilon(s) + \frac{1}{\varepsilon}v(s)\,(\psi(s) - u_\varepsilon(s))]\,ds$$

therefore

$$u_\varepsilon(t) \ge -\|\bar{u}^-\| - T\,\|L^-\| - \|\psi^-\| .$$

And from this we certainly deduce the property (3.37).

LEMMA 3.9: *Equation* (3.38) *holds.*

Proof: This result generalises Lemma 3.6. In fact, set

$$g(\lambda) = L + \frac{1}{\varepsilon}v(\lambda)\;(\psi(\lambda) - u_\varepsilon(\lambda))$$

and then

$$u_\varepsilon(t) = \Phi(t-s)\,u_\varepsilon(s) + \int_s^t \Phi(t-\lambda)g(\lambda)d\lambda .$$

Now multiply by

$$\frac{1}{\varepsilon} v(s) \exp - \frac{1}{\varepsilon} \int_s^t v(\lambda) d\lambda .$$

From this we deduce

$$\frac{1}{\varepsilon} v(s) u_\varepsilon(t) \exp - \frac{1}{\varepsilon} \int_s^t v(\lambda) d\lambda = \frac{v(s)}{\varepsilon}(\exp - \frac{1}{\varepsilon} \int_s^t v(\lambda) d\lambda) \Phi(t-s) u_\varepsilon(s) +$$
$$+ \frac{v(s)}{\varepsilon}(\exp - \frac{1}{\varepsilon} \int_s^t v(\lambda) d\lambda) \int_s^t \Phi(t-\lambda) g(\lambda) d\lambda.$$

We integrate this relation with respect to s over the interval [0,t] and deduce that

$$u_\varepsilon(t) (1 - \exp - \frac{1}{\varepsilon} \int_0^t v(\lambda) d\lambda) =$$
$$= \int_0^t \frac{v(s)}{\varepsilon}(\exp - \frac{1}{\varepsilon} \int_s^t v(\lambda) d\lambda) \Phi(t-s) u_\varepsilon(s) ds -$$
$$-(\exp - \frac{1}{\varepsilon} \int_0^t v(\lambda) d\lambda) \int_0^t \Phi(t-\lambda) g(\lambda) d\lambda +$$
$$+ \int_0^t \Phi(t-s) g(s) (\exp - \frac{1}{\varepsilon} \int_s^t v(\lambda) d\lambda) ds$$

from whence (3.38) easily results.□

LEMMA 3.10: *Let* u *be a solution of* (3.27), *then*

(3.39) $\quad u \leqslant u^\varepsilon.$

Proof: Now,

$$u(t) \leq e^{-\alpha t} \Phi(t) \bar{u} + \int_0^t e^{-\alpha s} \Phi(s) L \, ds$$

and therefore by Lemma 3.6,

$$u(t) \leq e^{-(\alpha + \frac{1}{\varepsilon}) t} \Phi(t) \bar{u} + \int_0^t e^{-(\alpha + \frac{1}{\varepsilon})(t-\lambda)} \Phi(t-\lambda) [L +$$
$$+ \frac{1}{\varepsilon} u(\lambda)] \, d\lambda$$

and since $u(t) \leqslant \psi(t)$,

$$T_\varepsilon u(t) = e^{-(\alpha + \frac{1}{\varepsilon}) t} \Phi(t) \bar{u} + \int_0^t e^{-(\alpha + \frac{1}{\varepsilon})(t-\lambda)} \Phi(t-\lambda) [L +$$
$$+ \frac{1}{\varepsilon} u(\lambda)] \, d\lambda$$

whence

$$T_\varepsilon u \geqslant u,$$

and by iteration, followed by passing to the limit, we obtain (3.39).□

LEMMA 3.11: *Let* $\tilde{\bar{u}},\tilde{\psi}$ *satisfy the same conditions as* $\bar{u},\psi$ *(i.e.* (3.4),(3.26)*). Let* $\tilde{u}_\varepsilon$ *be the corresponding solution of* (3.29). *Then*

$$\|u_\varepsilon - \tilde{u}_\varepsilon\|_{C(0,T;C)} \le \text{Max}\{\|\bar{u} - \tilde{\bar{u}}\|_C , \|\psi - \tilde{\psi}\|_{C(0,T;C)}\}. \tag{3.40}$$

Proof: Let M = the right hand side of (3.40). Then

$$T_\varepsilon 0 - \tilde{T}_\varepsilon 0 = e^{-(\alpha+\frac{1}{\varepsilon})t}\Phi(t)(\bar{u} - \tilde{\bar{u}}) + \int_0^t e^{-(\alpha+\frac{1}{\varepsilon})(t-\lambda)}\Phi(t-\lambda)\left[\frac{1}{\varepsilon}\psi(\lambda) - \frac{1}{\varepsilon}\tilde{\psi}(\lambda)\right]d\lambda$$

therefore

$$\|T_\varepsilon 0 - \tilde{T}_\varepsilon 0\|_{C(0,T;C)} \le M.$$

If $z,\tilde{z} \in C(0,T;C)$ satisfy

$$\|z - \tilde{z}\| \leqslant M$$

then

$$T_\varepsilon z - \tilde{T}_\varepsilon \tilde{z} = e^{-(\alpha+\frac{1}{\varepsilon})t}\Phi(t)(\bar{u} - \bar{u}) + \int_0^t d\lambda\, e^{-(\alpha+\frac{1}{\varepsilon})(t-\lambda)}\frac{\Phi(t-\lambda)}{\varepsilon}[z(\lambda)\wedge\psi(\lambda) - \tilde{z}(\lambda)\wedge\tilde{\psi}(\lambda)]$$

and therefore, since

$$z(\lambda)\wedge\psi(\lambda) - \tilde{z}(\lambda)\wedge\tilde{\psi}(\lambda) \le M$$

we have

$$T_\varepsilon z(t) - \tilde{T}_\varepsilon \tilde{z}(t) \le Me^{-(\alpha+\frac{1}{\varepsilon})t} + \frac{M(1-e^{-(\alpha+\frac{1}{\varepsilon})t})}{1+\alpha\varepsilon} \le M$$

and by symmetry

$$\|T_\varepsilon z - \tilde{T}_\varepsilon \tilde{z}\|_{C(0,T;C)} \le M.$$

From this we deduce by recurrence that

$$\|T_\varepsilon^n 0 - \tilde{T}_\varepsilon^n 0\|_{C(0,T;C)} \le M$$

and by making n tend to +∞ we obtain (3.40).□

LEMMA 3.12: *Let* $\psi \in C(0,T;C)$. *Define*

$$\psi_n(t) = e^{-nt}\,\Phi(t)\psi(0) + \int_0^t e^{-n(t-s)}\,\Phi(t-s)n\psi(s)ds \,. \tag{3.41}$$

Then

$$\psi_n \to \psi \quad \text{in } C(0,T;C). \tag{3.42}$$

Proof: Now,

$$\psi_n(t) = e^{-nt}\Phi(t)\psi(0) + \int_0^{nt} e^{-\sigma}\,\Phi(\tfrac{\sigma}{n})\;\psi(t-\tfrac{\sigma}{n})\;d\sigma$$

and

$$\psi(t) = e^{-nt}\;\;\psi(t) + \int_0^{nt} e^{-\sigma}\,\psi(t)\;d\sigma$$

whence

$$\begin{aligned}\psi_n(t)-\psi(t) &= e^{-nt}(\Phi(t)\psi(0)-\psi(t)) + \\ &+ \int_0^{nt} e^{-\sigma}(\Phi\;(\tfrac{\sigma}{n}\;)\;\psi(t-\tfrac{\sigma}{n}) - \psi(t))\;d\sigma\,.\end{aligned} \tag{3.43}$$

Let $\delta > 0$, and let us consider $t \leqslant \delta$. Then

$$\|\psi(t) - \bar{\psi}(0)\|_C \le \rho_1(\delta)$$

$$\|\Phi(t)\quad\psi(0) - \psi(0)\|_C \le \rho_2(\delta)$$

$$\|\psi(t-\tfrac{\sigma}{n}) - \psi(0)\|_C \le \rho_1(\delta) \quad \text{for} \quad \sigma \le nt\,.$$

Therefore

$$\begin{aligned}\sup_{0\le t\le\delta} \|\psi_n(t) - \psi(t)\| &\le 3\,\rho_1(\delta) + \rho_2(\delta) + \\ &+ \int_0^\infty e^{-\sigma}\|\Phi(\tfrac{\sigma}{n})\;\psi(0) - \psi(0)\|\;d\sigma\end{aligned} \tag{3.44}$$

$$\begin{aligned}\sup_{\delta\le t\le T} \|\psi_n(t)-\psi(t)\| &\le C\,e^{-n\delta} + \\ &+ \int_0^\infty e^{-\sigma} \sup_{0\le t\le T} \|\Phi(\tfrac{\sigma}{n})\psi(t)-\psi(t)\|\;d\sigma + \\ &+ \int_0^\infty e^{-\sigma}\,\rho_1(\tfrac{\sigma}{n})\;d\sigma\end{aligned} \tag{3.45}$$

where

$$\rho_1(\delta) = \sup_{\substack{t_1,t_2\in[0,T]\\ |t_1-t_2|\le\delta}} \|\psi(t_1) - \psi(t_2)\|$$

$$\rho_2(\delta) = \sup_{0 \le t \le \delta} \| \Phi(t)\psi(0) - \psi(0) \|$$

and $\rho_1(\delta), \rho_2(\delta) \to 0$ as $\delta \to 0$. Moreover, ρ_1, ρ_2 are bounded.

By assumption (3.2),

$$\sup_{0 \le t \le T} \| \Phi(\tfrac{\sigma}{n}) \psi(t) - \psi(t) \| \to 0 \quad \text{as } n \to \infty, \ \forall \sigma > 0,$$

and stays bounded, therefore

$$\int_0^\infty e^{-\sigma} \sup_{0 \le t \le T} \| \Phi(\tfrac{\sigma}{n})\psi(t) - \psi(t) \| \, d\sigma \to 0 \quad \text{as } n \to \infty.$$

Then from (3.44),(3.45) we deduce that

$$\| \psi_n - \psi \|_{C(0,T;C)} \le 3\rho_1(\delta) + \rho_2(\delta) + Ce^{-n\delta} + O(n)$$

where $O(n) \to 0$ as $n \to \infty$. From this we easily deduce that

$$\| \psi_n - \psi \|_{C(0,T;C)} \to 0$$

whence the result desired.□

We can rewrite (3.41) in the form

$$\begin{aligned} \text{(3.46)} \qquad \psi_n(t) &= e^{-\alpha t} \Phi(t) \psi(0) + \\ &+ \int_0^t e^{-\alpha(t-s)} \Phi(t-s)(n\psi(s) + (\alpha - n)\psi_n(s))ds \\ &= e^{-\alpha t}\Phi(t)\psi(0) + \int_0^t e^{-\alpha(t-s)} \Phi(t-s) \Lambda_n(s) ds \end{aligned}$$

where $\Lambda_n \in C(0,T;C)$.

LEMMA 3.13: *The sequence* $u_\varepsilon \downarrow u$ *and*

(3.47) $\quad u_\varepsilon \to u$ *in* $C(0,T;C)$.

Proof: By Lemma 3.8 $u_\varepsilon \downarrow u$ pointwise. We are going to establish (3.47) under the assumption

(3.48) $$\psi(t) = e^{-\alpha t}\Phi(t)\psi(0) + \int_0^t e^{-\alpha(t-s)} \Phi(t-s)\Lambda(s)ds$$

where

$$\Lambda \in C(0,T;C).$$

Everything can be reduced to this case without any loss of generality.

In fact we consider $\psi_n \to \psi$ in $C(0,T;C)$ defined in (3.46). Let $u_{\varepsilon n}$ be relative to ψ_n. By Lemma 3.11 we have

$$\| u_{\varepsilon n} - u_\varepsilon \| \le \| \psi_n - \psi \|$$

therefore by pointwise convergence we also have

$$\| u_n - u \| \le \| \psi_n - \psi \| .$$

From this we deduce that if we show that

$$\| u_{\varepsilon n} - u_n \|_{C(0,T;C)} \to 0 \quad \text{for all fixed } n$$

then (3.47) holds. Therefore assume (3.48). By (3.29) we then have

$$u_\varepsilon(t) - \psi(t) = e^{-\alpha t}\Phi(t)(\bar{u}-\psi(0)) + \\ + \int_0^t e^{-\alpha(t-\lambda)}\Phi(t-\lambda)[\tilde{L}(\lambda) - \tfrac{1}{\varepsilon}(u_\varepsilon(\lambda)-\psi(\lambda))^+]\, d\lambda$$

where $\tilde{L} = L - \Lambda$. Let us set

$$\tilde{u}_\varepsilon(t) = u_\varepsilon(t) - \psi(t),$$

then

$$\begin{aligned}
(3.49)\qquad \tilde{u}_\varepsilon(t) &= e^{-\alpha t}\Phi(t)(\bar{u}-\psi(0)) + \\
&\quad + \int_0^t e^{-\alpha(t-\lambda)}\Phi(t-\lambda)[\tilde{L}(\lambda) - \frac{1}{\varepsilon}\tilde{u}_\varepsilon^+(\lambda)]\, d\lambda \\
&= e^{-(\alpha+\frac{1}{\varepsilon})t}\Phi(t)(\bar{u}-\psi(0)) + \\
&\quad + \int_0^t e^{-(\alpha+\frac{1}{\varepsilon})(t-\lambda)}\Phi(t-\lambda)[\tilde{L}(\lambda) + \frac{1}{\varepsilon}\tilde{u}_\varepsilon - \frac{1}{\varepsilon}\tilde{u}_\varepsilon^+(\lambda)]\, d\lambda \\
&\le \int_0^t e^{-(\alpha+\frac{1}{\varepsilon})(t-\lambda)}\Phi(t-\lambda)\,\tilde{L}(\lambda)\, d\lambda \\
&\le \frac{\varepsilon}{1+\alpha\varepsilon}\,\| \tilde{L}^+ \|
\end{aligned}$$

whence

$$\frac{\tilde{u}_\varepsilon(t)}{\varepsilon} \le \| \tilde{L}^+ \|$$

and therefore

$$(3.50)\qquad \frac{\tilde{u}_\varepsilon^+(t)}{\varepsilon} \le \| \tilde{L}^+ \| \qquad \forall\, t \in [0,T] .$$

We have seen in Lemma 3.8 that

$$T_\varepsilon u_{\varepsilon'}(t) - u_{\varepsilon'}(t) = \int_0^t e^{-(\alpha+\frac{1}{\varepsilon})(t-\lambda)}\Phi(t-\lambda)(\tfrac{1}{\varepsilon'} - \tfrac{1}{\varepsilon})\tilde{u}_{\varepsilon'}^+(\lambda)\, d\lambda$$

and therefore by (3.50) we have

$$\geq - \int_0^t e^{-(\alpha+\frac{1}{\varepsilon})(t-\lambda)} \frac{\varepsilon'}{\varepsilon} \|\tilde{L}^+\| \, d\lambda$$

(3.51) $$T_\varepsilon u_{\varepsilon'}(t) - u_{\varepsilon'}(t) \geq - \varepsilon' \|\tilde{L}^+\| .$$

Next we have

(3.52) $$\begin{aligned} T_\varepsilon^2 u_{\varepsilon'}(t) - \psi(t) &= e^{-(\alpha+\frac{1}{\varepsilon})t} \Phi(t)(\bar{u}-\psi(0)) + \\ &+ \int_0^t e^{-(\alpha+\frac{1}{\varepsilon})(t-\lambda)} \Phi(t-\lambda)[\tilde{L}(\lambda) - \frac{1}{\varepsilon}\psi(\lambda) + \frac{1}{\varepsilon} T_\varepsilon u_{\varepsilon'}(\lambda) \wedge \psi(\lambda)] \, d\lambda \\ &= e^{-(\alpha+\frac{1}{\varepsilon})(t-\lambda)} \Phi(t-\lambda) \, [\tilde{L}(\lambda) - \frac{1}{\varepsilon} (T_\varepsilon u_{\varepsilon'}(\lambda) - \psi(\lambda))^-] d\lambda . \end{aligned}$$

By (3.51)

$$T_\varepsilon u_{\varepsilon'}(t)\psi(t) \geq \tilde{u}_{\varepsilon'}(t) - \varepsilon' \|\tilde{L}^+\|$$

which is negative by (3.50), therefore

$$-(T_\varepsilon u_{\varepsilon'}(t) - \psi(t))^- \geq \tilde{u}_{\varepsilon'}(t) - \varepsilon' \|\tilde{L}^+\|$$

whence, by (3.52),

$$\begin{aligned} T_\varepsilon^2 u_{\varepsilon'}(t) - \psi(t) &\geq e^{-(\alpha+\frac{1}{\varepsilon})t} \Phi(t)(\bar{u}-\psi(0)) + \\ &+ \int_0^t e^{-(\alpha+\frac{1}{\varepsilon})(t-\lambda)} \Phi(t-\lambda)[\tilde{L}(\lambda) + \frac{\tilde{u}_{\varepsilon'}(\lambda)}{\varepsilon} - \frac{\varepsilon'}{\varepsilon} \|L^+\|] . \end{aligned}$$

But by (3.49)

$$\begin{aligned} \tilde{u}_{\varepsilon'}(t) &= e^{-(\alpha+\frac{1}{\varepsilon})t} \Phi(t)(\bar{u}-\psi(0)) + \\ &+ \int_0^t e^{-(\alpha+\frac{1}{\varepsilon})(t-\lambda)} \Phi(t-\lambda)[L(\lambda) + \frac{1}{\varepsilon} \tilde{u}_{\varepsilon'}(\lambda) - \frac{1}{\varepsilon'} \tilde{u}^+_{\varepsilon'}(\lambda)] \, d\lambda \\ &\leq e^{-(\alpha+\frac{1}{\varepsilon})t} \Phi(t)(\bar{u}-\psi(0)) + \\ &\quad + \int_0^t e^{-(\alpha+\frac{1}{\varepsilon})(t-\lambda)} \Phi(t-\lambda)[\tilde{L}(\lambda) + \frac{1}{\varepsilon} \tilde{u}_{\varepsilon'}(\lambda)] \, d\lambda \end{aligned}$$

therefore

$$\begin{aligned} T_\varepsilon^2 u_{\varepsilon'}(t) - \psi(t) &\geq \tilde{u}_{\varepsilon'}(t) - \int_0^t e^{-(\alpha+\frac{1}{\varepsilon})(t-\lambda)} \Phi(t-\lambda) \frac{\varepsilon'}{\varepsilon} \|\tilde{L}^+\| \, d\lambda \\ &\geq \tilde{u}_{\varepsilon'}(t) - \varepsilon' \|\tilde{L}^+\| \end{aligned}$$

so again

$$T_\varepsilon^2\, u_{\varepsilon'}(t) - u_{\varepsilon'}(t) \geq -\varepsilon' \;\|\tilde{L}^+\|$$

and quite generally,

$$T_\varepsilon^k\, u_{\varepsilon'}(t) - u_{\varepsilon'}(t) \geq -\varepsilon' \;\|\tilde{L}^+\|$$

whence we finally have

$$u_\varepsilon(t) - u_{\varepsilon'}(t) \geq -\varepsilon' \;\|\tilde{L}^+\| .$$

By making ε tend to 0, we deduce from this that

$$u(t) - u_{\varepsilon'}(t) \geq -\varepsilon' \;\|\tilde{L}^+\|$$

and consequently we have proved

$$\sup_t \| u_\varepsilon(t) - u(t)\|_B \leq \varepsilon \;\|\tilde{L}^+\| \tag{3.53}$$

if ψ is of the form (3.48). This certainly proves (3.47).□

Proof of Theorem 3.2: Now, $u_\varepsilon \downarrow u$ and $u_\varepsilon \to u$ in $C(0,T;C)$. Multiplying (3.28) by ε and making ε tend to 0, from this we deduce that

$$\int_s^t e^{-\alpha(t-\lambda)}\Phi(t-\lambda)(u(\lambda)-\psi(\lambda))^+ d\lambda = 0 \qquad \forall t > s \;.$$

We now fix $t > 0$ and $h < t$, $h \downarrow 0$, therefore

$$\frac{1}{h} \int_0^h e^{-\alpha\sigma}\Phi(\sigma)(u(t-\sigma)-\psi(t-\sigma))^+ \,d\sigma = 0 \;.$$

Now,

$$\frac{1}{h} \int_0^h e^{-\alpha\sigma}\Phi(\sigma)(u(t-\sigma)-\psi(t-\sigma))^+ d\sigma \;\to\; (u(t)-\psi(t))^+$$

in C when $(u(t) - \psi(t))^+ \in C(0,T;C)$. Therefore

$$u(t) \leqslant \psi(t) \quad \forall\, t > 0$$

and since $u(0) = \bar{u} \leqslant \psi(0)$, we have

$$u(t) \leqslant \psi(t) \quad \forall\, t \in [0,T].$$

Also, we always have by (3.28) that

$$u_\varepsilon(t) \leq e^{-\alpha(t-s)}\Phi(t-s)u_\varepsilon(s) + \int_s^t e^{-\alpha(t-\lambda)}\Phi(t-\lambda)\, L\, d\lambda$$

and therefore

$$u(t) \leq \int_0^{t-s} e^{-\alpha\sigma}\Phi(\sigma)L d\sigma + e^{-\alpha(t-s)}\Phi(t-s)u(s)$$
$$\forall \quad s \leq t \in [0,T]$$

thus u satisfies (3.27). By Lemma 3.10 u is the maximum solution of the set (3.27).□

3.3 *Non-Linear Semi-Group*

3.3.1 *Semi-group of the penalised problem*

Here we consider the penalised problem (3.29) with the assumption

(3.54) $\psi(t) = \psi$.

Then we set

(3.55) $u_\varepsilon(t) = S_\varepsilon(t)\bar{u}$.

We then have:

THEOREM 3.3: *Assume* (3.1),(3.2),(3.3) *and* $\bar{u},\psi \in C$. *Then the operator* $S_\varepsilon(t)$ *of* C *into* C *defined by* (3.55) *is a non-linear contraction semi-group, and*

(3.56) $S_\varepsilon(t)\bar{u} \to \bar{u}$ *in* C *when* $t \to 0$.

Proof: Let us set

$$v_\varepsilon(t) = S_\varepsilon(t)S_\varepsilon(\theta)\bar{u}$$

therefore

$$v_\varepsilon(t) = e^{-\alpha(t-s)}\Phi(t-s)v_\varepsilon(s) + \int_s^t e^{-\alpha(t-\lambda)}\Phi(t-\lambda)[L - \frac{1}{\varepsilon}(v_\varepsilon(\lambda) - \psi)^+]d\lambda$$
$$v_\varepsilon(0) = S_\varepsilon(\theta)\bar{u}.$$

Now, by (3.28) we also have

$$u_\varepsilon(t+\theta) = e^{-\alpha(t-s)}\Phi(t-s)u_\varepsilon(s+\theta) + \int_{s+\theta}^{t+\theta} e^{-\alpha(t+\theta-\lambda)}\Phi(t+\theta-\lambda)[L - \frac{1}{\varepsilon}(u_\varepsilon(\lambda)-\psi)^+]d\lambda$$
$$= e^{-\alpha(t-s)}\Phi(t-s)u_\varepsilon(s+\theta) + \int_s^t e^{-\alpha(t-\lambda)}\Phi(t-\lambda)[L - \frac{1}{\varepsilon}(u_\varepsilon(\lambda+\theta)-\psi)^+]d\lambda$$

which, coupled with

$$u_\varepsilon(\theta) = S_\varepsilon(\theta)\bar{u}$$

shows that

$$v_\varepsilon(t) = u_\varepsilon(t + \theta) = S_\varepsilon(t + \theta)\bar{u}$$

and therefore S_ε is certainly a semi-group. By Lemma 3.11 we also have that

$$\| S_\varepsilon(t)\bar{u} - S_\varepsilon(t)\,\tilde{\bar{u}} \| \le \|\bar{u} - \tilde{\bar{u}}\| .$$

The property (3.56) immediately results from the continuity of $\Phi(t)$.□

We now want to prove Trotter's formula for the semi-group $S_\varepsilon(t)$. Consider

$$R_{\lambda\varepsilon}(\bar{u}) = z_\varepsilon \tag{3.57}$$

defined by the equation

$$z_\varepsilon = \int_0^\infty e^{-(\alpha+\frac{1}{\lambda})t}\,\Phi(t)\left(L + \frac{\bar{u}}{\lambda} - \frac{1}{\varepsilon}(z_\varepsilon - \psi)^+\right)dt \tag{3.58}$$

which defines $z_\varepsilon \in C$ uniquely (cf. Lemma 5.1.3 of Chapter 4). Our objective is to prove:

THEOREM 3.4: *Take the assumptions of Theorem 3.3. Then*

$$\left|\begin{array}{l} \forall t > 0 \qquad R^n_{\frac{t}{n},\varepsilon}(\bar{u}) \to S_\varepsilon(t)\bar{u} \quad \textit{in} \quad C, \\ \textit{when} \quad n \to \infty . \end{array}\right. \tag{3.59}$$

We shall use some preliminary results.

LEMMA 3.14:

$$\| R^k_{\lambda\varepsilon}(\bar{u}_1) - R^k_{\lambda\varepsilon}(\bar{u}_2) \| \le \|\bar{u}_1 - \bar{u}_2\| , \quad \forall\, \lambda > 0,\ k \ge 1 . \tag{3.60}$$

Proof: It suffices to prove (3.60) with k = 1. We recall that

$$z_\varepsilon = \int_0^\infty e^{-(\alpha+\frac{1}{\lambda}+\frac{1}{\varepsilon})t}\,\Phi(t)\left(L + \frac{\bar{u}}{\lambda} + \frac{1}{\varepsilon} z_\varepsilon \wedge \psi\right)dt$$
$$= T_\varepsilon z_\varepsilon$$

where

$$T_\varepsilon z = \int_0^\infty e^{-(\alpha+\frac{1}{\lambda}+\frac{1}{\varepsilon})t}\,\Phi(t)\left(L + \frac{\bar{u}}{\lambda} + \frac{1}{\varepsilon} z \wedge \psi\right) dt .$$

Let us assume that

$$\|z_1 - z_2\| \le \|\bar{u}_1 - \bar{u}_2\|$$

then we immediately verify that

$$\| T_\varepsilon z_1 - T_\varepsilon z_2 \| \le \|\bar{u}_1 - \bar{u}_2\|$$

and therefore

$$\|z_\varepsilon^1 - z_\varepsilon^2\| \le \|\bar{u}_1 - \bar{u}_2\|$$

from which we deduce (3.60).□

LEMMA 3.15: *Assume that* $\bar{u} - u^0 \in D(\mathcal{A})$, *so*

(3.61) $$\bar{u} - u^0 = \int_0^\infty e^{-\beta t}\Phi(t)(U + \beta(\bar{u} - u^0))dt \qquad \forall \beta > 0,$$

$U \in C$.

Then

(3.62) $$\frac{\bar{u} - R_{\lambda\varepsilon}(\bar{u})}{\lambda} \to U + \alpha(\bar{u} - u^0) + \frac{1}{\varepsilon}(\bar{u} - \psi)^+$$

in C *when* $\lambda \to 0$.

Proof: Now,

$$\bar{u} - u^0 = \int_0^\infty e^{-(\frac{1}{\lambda}+\alpha)t}\Phi(t)(U + (\frac{1}{\lambda}+\alpha)(\bar{u} - u^0))dt$$

(3.63) $$\bar{u} - R_{\lambda\varepsilon}(\bar{u}) = \bar{u} - z_{\varepsilon\lambda}$$
$$= \int_0^\infty e^{-(\alpha+\frac{1}{\lambda})t}\Phi(t)(U + \alpha(\bar{u}-u^0) + \frac{1}{\varepsilon}(z_{\varepsilon\lambda}-\psi)^+)dt .$$

However, by Lemma 5.4 of Chapter 4 we have

$$z_{\varepsilon\lambda} \le \frac{\| L + \frac{\bar{u}}{\lambda} \|}{\alpha + \frac{1}{\lambda}}$$

$$z_{\varepsilon\lambda} \ge -\left(\|\psi^-\| + \frac{\| (L + \frac{\bar{u}}{\lambda})^- \|}{\alpha + \frac{1}{\lambda}} \right)$$

therefore

(3.64) $$\|z_{\varepsilon\lambda}\| \le \|\psi\| + \|L\| + \|\bar{u}\|$$

for $\lambda < 1$, for all ε.

It follows from (3.63) that

(3.65) $$\|\bar{u} - z_{\varepsilon\lambda}\| \le \frac{C}{\varepsilon}\lambda$$

where C is a constant independent of λ and of ε. We then deduce from (3.63) that

$$\frac{\bar{u} - R_{\lambda\varepsilon}(\bar{u})}{\lambda} - (U+\alpha(\bar{u}-u^0) + \frac{1}{\varepsilon}(\bar{u}-\psi)^+) =$$

$$= \int_0^\infty e^{-\sigma}\left(\frac{\Phi(\frac{\sigma\lambda}{1+\lambda\alpha})}{1+\lambda\alpha} - I\right)(U+\alpha\ (\bar{u}-u^0)+ \frac{1}{\varepsilon}(\bar{u}-\psi)^+)d\sigma$$

$$+ \frac{1}{\lambda}\int_0^\infty e^{-(\alpha+\frac{1}{\lambda})t}\Phi(t)[\frac{1}{\varepsilon}(z_{\varepsilon\lambda}-\psi)^+ - \frac{1}{\varepsilon}(\bar{u}-\psi)^+]\ dt\ .$$

Using (3.65) and assumption (3.2), from this we easily deduce the result (3.62).□

LEMMA 3.16: *Again assume* (3.61). *Then*

(3.66) $$\left|\ \begin{aligned} &\frac{\bar{u} - S_\varepsilon(\lambda)\bar{u}}{\lambda} - (U+\alpha(\bar{u}-u^0) + \frac{1}{\varepsilon}(\bar{u}-\psi)^+) \\ &\to\ 0\ \ \textit{in}\ \ C\ ,\ \ \textit{when}\ \ \lambda \to 0. \end{aligned}\right.$$

Proof: We have (cf. (3.24))

$$\bar{u}-u^0 = \int_0^\lambda e^{-\alpha t}\Phi(t)(U+\alpha(\bar{u}-u^0))dt + e^{-\alpha\lambda}\Phi(\lambda)(\bar{u}-u^0)$$

and therefore by (3.29)

$$\bar{u} - S_\varepsilon(\lambda)\bar{u} = \int_0^\lambda e^{-\alpha t}\Phi(t)(U+\alpha(\bar{u}-u^0))dt$$

$$+ \frac{1}{\varepsilon}\int_0^\lambda e^{-\alpha(\lambda-s)}\Phi(\lambda-s)(u_\varepsilon(s)-\psi)^+\ ds$$

$$= \int_0^\lambda e^{-\alpha t}\Phi(t)(U+\alpha(\bar{u}-u^0)+ \frac{1}{\varepsilon}(u_\varepsilon(\lambda-t)-\psi)^+)dt$$

whence

$$\frac{\bar{u} - S_\varepsilon(\lambda)\bar{u}}{\lambda} - (U+\alpha(\bar{u}-u^0)+ \frac{1}{\varepsilon}(\bar{u}-\psi)^+)$$

$$= \int_0^1 (e^{-\lambda s}\Phi(\lambda s)-I)(U+\alpha(\bar{u}-u^0)+ \frac{1}{\varepsilon}(\bar{u}-\psi)^+)ds\ +$$

$$+ \frac{1}{\lambda}\int_0^\lambda e^{-\alpha t}\Phi(t)[\frac{1}{\varepsilon}(u_\varepsilon(\lambda-t)-\psi)^+ - \frac{1}{\varepsilon}(\bar{u}-\psi)^+]\ dt$$

$$= I + II$$

and

$$II = \int_0^1 e^{-\alpha\lambda s}\Phi(\lambda s)[\frac{1}{\varepsilon}(u_\varepsilon(\lambda(1-s))-\psi)^+ - \frac{1}{\varepsilon}(\bar{u}-\psi)^+\ ]ds\ .$$

From this we easily deduce that I and II converge to 0 in C as $\lambda \to 0$, whence the result (3.66).□

We now draw our inspiration from M.G. CRANDALL - T.M. LIGGETT [1], Section 3. We define

(3.67) $$R_{\lambda,\varepsilon}(t)\bar{u} = \left(I + \lambda\frac{I - S_\varepsilon(t)}{t}\right)^{-1} \bar{u} .$$

LEMMA 3.17: *We have*

(3.68) $$R_{\lambda,\varepsilon}(t)\bar{u} \to R_{\lambda,\varepsilon}\,\bar{u} \quad \textit{when} \quad t \to 0 , \quad \forall\, \bar{u} \in C , \quad \lambda > 0.$$

Proof: We shall not write ε, which is fixed. We notice that

(3.69) $$\| R_\lambda(t)\bar{u}_1 - R_\lambda(t)\bar{u}_2 \| \le \| \bar{u}_1 - \bar{u}_2 \| .$$

In fact, let

$$z_1 = R_\lambda(t)\bar{u}_1 , \qquad z_2 = R_\lambda(t)\bar{u}_2$$

hence, by (3.67)

$$z_1 + \lambda \frac{(z_1 - S(t)z_1)}{t} = \bar{u}_1$$

and therefore

$$(z_1 - z_2)(t+\lambda) = t(\bar{u}_1 - \bar{u}_2) + \lambda(S(t)z_1 - S(t)z_2)$$

whence

$$(t+\lambda)\, \|z_1 - z_2\| \le t\, \|\bar{u}_1 - \bar{u}_2\| + \lambda \|z_1 - z_2\|$$

which implies (3.69) for all $t > 0$.

Next we have

(3.70) $$\begin{aligned} \| R_\lambda(t)\bar{u} - R_\lambda\bar{u} \| &= \| R_\lambda(t)\bar{u} - R_\lambda(t)\, R_\lambda(t)^{-1} R_\lambda\bar{u} \| \\ &\le \| \bar{u} - (I + \lambda\frac{(I - S(t))}{t}) R_\lambda\bar{u} \| . \end{aligned}$$

By (3.58) and the definition of u^0 we have

$$\begin{aligned} R_\lambda\bar{u} - u^0 &= \int_0^\infty e^{-(\alpha+\frac{1}{\lambda})t}\, \Phi(t) \left(\frac{\bar{u}-u^0}{\lambda} - \frac{1}{\varepsilon}(R_\lambda\bar{u}-\psi)^+\right) dt \\ &= \int_0^\infty e^{-\beta t}\Phi(t)\left(\frac{\bar{u}-u^0}{\lambda} - (\alpha+\frac{1}{\lambda})(R_\lambda\bar{u}-u^0) - \right. \\ &\qquad \left. - \frac{1}{\varepsilon}(R_\lambda\bar{u}-\psi)^+ + \beta\,(R_\lambda\bar{u}-u^0)\right)dt \quad \forall\, \beta > 0 . \end{aligned}$$

therefore $R_\lambda\bar{u} - u^0 \in D(\mathcal{A})$, and if we compare with (3.61) there corresponds to it a function U given by

$$U = \frac{\bar{u}-u^0}{\lambda} - (\alpha+\frac{1}{\lambda})(R_\lambda\bar{u}-u^0) - \frac{1}{\varepsilon}(R_\lambda\bar{u}-\psi)^+ .$$

Applying Lemma 3.16, we deduce from this that

$$\frac{R_\lambda \bar{u} - S(t)R_\lambda \bar{u}}{t} \to \frac{1}{\lambda}(\bar{u} - R_\lambda \bar{u}) \quad \text{in C when } t \to 0.$$

and therefore the right hand side of (3.70) tends to 0 when $t \to 0$, whence the result desired.□

Next consider the differential equation in C

(3.71)
$$\left|\begin{array}{l} \dfrac{dz}{dt} + \dfrac{z - S_\varepsilon(\lambda)z}{\lambda} = 0 \\ z(0) = \bar{u} \in C \end{array}\right.$$

which defines a semi-group depending on the parameter λ,

(3.72) $$z(t) = S_{\lambda,\varepsilon}(t)\bar{u}.$$

LEMMA 3.18: *We have*

(3.73) $$S_{\lambda,\varepsilon}(t)\bar{u} = \lim R^n_{\frac{t}{n},\varepsilon}(\lambda)\, \bar{u} \quad \textit{in } C.$$

Proof: We omit writing the index ε. First of all let us show that

(3.74) $$\|S_\lambda(t)\bar{u}_1 - S_\lambda(t)\bar{u}_2\| \leq \|\bar{u}_1 - \bar{u}_2\| .$$

In fact, by (3.71) we have

$$\frac{d}{dt}(e^{\frac{t}{\lambda}} z) = \frac{e^{\frac{t}{\lambda}}}{\lambda} S(\lambda)z$$

therefore if

$$z_1(t) = S_\lambda(t)\bar{u}_1, \quad z_2(t) = S_\lambda(t)\bar{u}_2 ,$$

we have

$$\frac{d}{dt}(e^{\frac{t}{\lambda}}(z_1 - z_2)) = \frac{e^{\frac{t}{\lambda}}}{\lambda} (S(\lambda)z_1 - S(\lambda)z_2)$$

whence

$$e^{\frac{t}{\lambda}}(z_1(t) - z_2(t)) \leq \|\bar{u}_1 - \bar{u}_2\| + \\ + \int_0^t \frac{e^{\frac{s}{\lambda}}}{\lambda} \|z_1(s) - z_2(s)\|\, ds .$$

Interchanging the roles of z_1, z_2, from this we deduce

$$e^{\frac{t}{\lambda}}\| z_1(t) - z_2(t) \| \leq \|\bar{u}_1 - \bar{u}_2\| + \int_0^t \frac{e^{\frac{s}{\lambda}}}{\lambda} \| z_1(s) - z_2(s) \|\, ds$$

and by Gronwall's inequality

$$e^{\frac{t}{\lambda}} \|z_1(t) - z_2(t)\| \le e^{\frac{t}{\lambda}} \|\bar{u}_1 - \bar{u}_2\|$$

i.e. (3.74). Also, it is clear that $S_\lambda(t)$ defines a non-linear semi-group on C. Its infinitesimal generator is $\frac{S(\lambda) - I}{\lambda}$, the domain of which is the C and the resolvant $R_\beta(\lambda)$.

Next we have

$$R^n_{\frac{t}{n}}(\lambda)\bar{u} - S_\lambda(t)\bar{u} = R^n_{\frac{t}{n}}(\lambda)\bar{u} - S_\lambda(\tfrac{t}{n})^n \bar{u}$$
$$= \sum_{k=0}^{n-1} (R^{n-k}_{\frac{t}{n}}(\lambda) S_\lambda(\tfrac{t}{n})^k \bar{u} - R^{n-k-1}_{\frac{t}{n}}(\lambda) S_\lambda(\tfrac{t}{n})^{k+1} \bar{u})$$

whence

$$\|R^n_{\frac{t}{n}}(\lambda)\bar{u} - S_\lambda(t)\bar{u}\| \le \sum_{k=0}^{n-1} \|R_{\frac{t}{n}}(\lambda) S_\lambda(\tfrac{kt}{n})\bar{u} - S_\lambda(\tfrac{t}{n}) S_\lambda(\tfrac{kt}{n})\bar{u}\| \tag{3.75}$$
$$= \sum_{k=0}^{n-1} \|R_{\frac{t}{n}}(\lambda)\, z(\tfrac{kt}{n}) - S_\lambda(\tfrac{t}{n}) z(\tfrac{kt}{n})\| \quad .$$

Now,

$$\frac{S_\lambda(\frac{t}{n}) z(\frac{kt}{n}) - z(\frac{kt}{n})}{\frac{t}{n}} = \frac{z((k+1)\frac{t}{n}) - z(\frac{kt}{n})}{\frac{t}{n}}$$
$$= \frac{1}{\frac{t}{n}} \int_{\frac{kt}{n}}^{(k+1)\frac{t}{n}} \frac{S(\lambda)z(\sigma) - z(\sigma)}{\lambda}\, d\sigma$$
$$= \frac{S(\lambda)z(\frac{kt}{n}) - z(\frac{kt}{n})}{\lambda} +$$
$$+ \frac{1}{\frac{t}{n}} \int_{\frac{kt}{n}}^{(k+1)\frac{t}{n}} \frac{S(\lambda)z(\sigma) - S(\lambda)z(\frac{kt}{n}) + z(\frac{kt}{n}) - z(\sigma)}{\lambda}\, d\sigma \ .$$

Now

$$\sup_{\frac{kt}{n} \le \sigma \le (k+1)\frac{t}{n}} \|z(\sigma) - z(\tfrac{kt}{n})\| \le \frac{t}{n} \sup_{0 \le s \le t} \left\|\frac{z(s) - S(\lambda)z(s)}{\lambda}\right\|$$
$$= \frac{t}{n}\rho$$

therefore, collecting the results together, we have

$$(3.76)\qquad \|S_\lambda(\tfrac{t}{n})z(\tfrac{kt}{n})-z(\tfrac{kt}{n})-\frac{t}{n}\frac{(S(\lambda)z(\frac{kt}{n})-z(\frac{kt}{n}))}{\lambda}\| \le \frac{2}{\lambda}(\tfrac{t}{n})^2\rho \qquad \forall\ k=0,..,n-1$$

Furthermore,

$$R_{\frac{t}{n}}(\lambda)z(\tfrac{kt}{n})+\frac{t}{n}\Big[\frac{R_{\frac{t}{n}}(\lambda)\ z(\frac{kt}{n})-S(\lambda)R_{\frac{t}{n}}(\lambda)z(\frac{kt}{n})}{\lambda}\Big]=z(\tfrac{kt}{n})$$

therefore

$$R_{\frac{t}{n}}(\lambda)\ z(\tfrac{kt}{n})-z(\tfrac{kt}{n})-\frac{t}{n}\Big(\frac{S(\lambda)z(\frac{kt}{n})-z(\frac{kt}{n})}{\lambda}\Big)=$$

$$\frac{t}{n}\Big[\frac{S(\lambda)R_{\frac{t}{n}}(\lambda)z(\frac{kt}{n})-S(\lambda)z(\frac{kt}{n})}{\lambda}-\Big(\frac{R_{\frac{t}{n}}(\lambda)z(\frac{kt}{n})-z(\frac{kt}{n})}{\lambda}\Big)\Big].$$

From (3.77) we deduce that

$$\|R_{\frac{t}{n}}(\lambda)z(\tfrac{kt}{n})-z(\tfrac{kt}{n})\|\le\frac{t}{n}\rho\ +$$
$$+\frac{2t}{n}\|R_{\frac{t}{n}}(\lambda)z(\tfrac{kt}{n})-z(\tfrac{kt}{n})\|$$

therefore

$$(3.78)\qquad \|R_{\frac{t}{n}}(\lambda)z(\tfrac{kt}{n})-z(\tfrac{kt}{n})-\frac{t}{n}\frac{S(\lambda)z(\frac{kt}{n})-z(\frac{kt}{n})}{\lambda}\|\le\frac{2}{\lambda}(\tfrac{t}{n})^2\frac{\rho}{1-\frac{2}{\lambda}\frac{t}{n}}.$$

From (3.76) and (3.78) it follows that

$$\|R_{\frac{t}{n}}(\lambda)z(\tfrac{kt}{n})-S_\lambda(\tfrac{t}{n})z(\tfrac{kt}{n})\|\le\frac{2}{\lambda}\rho\ (\tfrac{t}{n})^2\ \Big[1+\frac{1}{1-\frac{2}{\lambda}\frac{t}{n}}\Big]$$

which, together with (3.75), shows that

$$\|R^n_{\frac{t}{n}}(\lambda)\bar{u}-S_\lambda(t)\bar{u}\|\le\frac{2}{\lambda}\rho\ \frac{t}{n}\,t\ \Big[1+\frac{1}{1-\frac{2}{\lambda}\frac{t}{n}}\Big]$$

whence (3.73).□

LEMMA 3.19: *We have*

$$(3.79)\qquad \|S_{\lambda,\varepsilon}(t)\bar{u}-R^n_{\frac{t}{n},\varepsilon}(\lambda)\bar{u}\|\le\frac{2t}{\sqrt{n}}\ \frac{\|\bar{u}-S_\varepsilon(\lambda)\bar{u}\|}{\lambda}.$$

Lemma 3.19 gives an estimate of the convergence that is very much more interesting than that of Lemma 3.18. This estimate will play a fundamental role in the proof of Theorem 3.4. However, Lemma 3.19 does not replace Lemma 3.18, because in the proof we shall need to know a priori that the result (3.73) holds. It follows from a sharp estimate of M.G. CRANDALL - T.M. LIGGETT [1]. We shall only give the broad outline of the proof. We omit writing ε.

Set

$$J_\beta = R_\beta(\lambda) = (I + \beta \ B)^{-1}$$

where

$$B = \frac{I - S(\lambda)}{\lambda} .$$

We have

$$(3.80) \qquad \begin{aligned} \|J_\beta u - u\| &\le \|J_\beta u - J_\beta \ J_\beta^{-1} u\| \\ &\le \| u - J_\beta^{-1} u\| = \| u - (I+\beta \ B)u\| \\ &\le \beta \ \| Bu \| . \end{aligned}$$

Now,

$$(3.81) \qquad \begin{aligned} J_\beta u &= J_\gamma \ (J_\beta u + \gamma \ B \ J_\beta \ u) \\ &= J_\gamma \ (J_\beta u + \frac{\gamma(u - J_\beta u)}{\beta}) \\ &= J_\gamma(\frac{\gamma}{\beta} u + \frac{\beta-\gamma}{\beta} J_\beta \ u) . \end{aligned}$$

For $0 \leqslant j \leqslant n$, $0 \leqslant k \leqslant m$ we set

$$\begin{aligned} a_{kj} &= \| J_\beta^j \ u - J_\gamma^k \ u\| \\ &= \| J_\beta^j \ u - J_\beta(\frac{\beta}{\gamma} \ J^{k-1} u + \frac{\gamma-\beta}{\gamma} J^k \ u) \| \\ &\le \| J_\beta^{j-1} \ u - \frac{\beta}{\gamma} J_\gamma^{k-1} u - \frac{\gamma-\beta}{\gamma} J_\gamma^k \ u\| \\ &\le \frac{\beta}{\gamma} \| J_\beta^{j-1} u - J_\gamma^{k-1} u\| + \frac{\gamma-\beta}{\gamma} \|J_\beta^{j-1} u - J_\gamma^k u \| \end{aligned}$$

if

$$j,k > 0 \quad \text{and} \quad \gamma \leqslant \beta > 0,$$

therefore

$$a_{kj} \le \frac{\beta}{\gamma} \ a_{k-1,j-1} + \frac{\gamma-\beta}{\gamma} \ a_{k,j-1} .$$

From this we deduce that

$$(3.82) \qquad \| J_\beta^n \ u - J_\gamma^m \ u\| \le \sum_{j=0}^{m-1} (\frac{\beta}{\gamma})^j (\frac{\gamma-\beta}{\gamma})^{n-j} \ C_n^j \| \ J^{m-j} \ u - u \| + \qquad \text{(Contd)}$$

(Contd) $$+ \sum_{j=m}^{n} \left(\frac{\beta}{\gamma}\right)^m \left(\frac{\gamma-\beta}{\gamma}\right)^{j-m} C_{j-1}^{m-1} \| J^{n-j} u - u\|$$

for $n \geqslant m$

and

$$\leq \sum_{j=0}^{n} \left(\frac{\beta}{\gamma}\right)^j \left(\frac{\gamma-\beta}{\gamma}\right)^{n-j} C_n^j \| J^{m-j} u - u\|$$

if $n \leqslant m$.

Next we use the following identities with

$$n \geqslant m \quad \text{and } p,q \geqslant 0, \quad p + q = 1,$$

(3.83) $$\sum_{j=0}^{m} C_n^j p^j q^{n-j}(m-j) \leq \sqrt{(np-m)^2 + npq}$$

and

(3.84) $$\sum_{j=m}^{n} C_{j-1}^{m-1} p^m q^{j-m}(n-j) \leq \sqrt{\frac{mq}{p^2} + \left(\frac{mq}{p} + m - n\right)^2} .$$

Let us set

$$p = \frac{\beta}{\gamma}, \quad q = \frac{\gamma - \beta}{\gamma} .$$

We have

$$\| J_\gamma^{m-j} u - u\| \leq (m-j) \| Bu\| \gamma$$

therefore by (3.82) and the estimates (3.83),(3.84), for $n \geqslant m$ and $\gamma \geqslant \beta$ we obtain

$$\begin{aligned} \|J_\beta^n - J_\gamma^m u\| &\leq \sqrt{(np-m)^2 + npq}\ \gamma \|Bu\| + \\ &\quad + \sqrt{\frac{mq}{p^2} + \left(\frac{mq}{p} + m - n\right)^2}\ \beta \|Bu\| \\ &= \left\{ \sqrt{(n\beta - m\gamma)^2 + n\beta(\gamma-\beta)} + \right. \\ &\quad \left. \sqrt{m(\gamma-\beta)\gamma + (m\gamma - n\beta)^2} \right\} \|Bu\| . \end{aligned}$$

Next we take

$$\beta = \frac{t}{n}, \quad \gamma = \frac{t}{m} \quad (n \geqslant m),$$

and we obtain

$$\| J_{\frac{t}{n}}^n u - J_{\frac{t}{m}}^m u\| \leq 2t\left(\frac{1}{m} - \frac{1}{n}\right)^{1/2} \| Bu\| .$$

We then make n tend to ∞ and use Lemma 3.18 to obtain (3.79).□

LEMMA 3.20: *For* $n\lambda \leqslant t < (n+1)\lambda$ *we have*

(3.85) $$\| S_{\lambda,\varepsilon}(t)\bar{u} - S_\varepsilon(t)\bar{u}\| \leq (\sqrt{n} + e) \sup_{0\leq \sigma\leq \lambda} \|S_\varepsilon(\sigma)\bar{u} - \bar{u}\| .$$

Proof: If we set $z_\lambda(t) = z(\lambda t)$, we deduce from (3.71) that

$$\frac{dz_\lambda}{dt} = S(\lambda)z_\lambda(t) - z_\lambda(t)$$
$$z_\lambda(0) = \bar{u}$$

and since $S(\lambda)$ is a contraction, by a result of H. BREZIS - A. PAZY [1] we have

$$\| z_\lambda(n) - S(\lambda)^n \bar{u}\| \leq \sqrt{n} \;\; \|S(\lambda)\bar{u} - \bar{u} \|$$

therefore

(3.86) $$\| S_\lambda(\lambda n)\bar{u} - S(n\lambda)\bar{u}\| \leq \sqrt{n} \;\; \|S(\lambda)\bar{u} - \bar{u}\| .$$

Also, let us take $t \leqslant \lambda$ and in (3.71) let us set

$$\zeta(t) = z(t) - \bar{u},$$

therefore

$$\frac{d\zeta(t)}{dt} = \frac{S(\lambda)(\zeta + \bar{u}) - (\zeta + \bar{u})}{\lambda}$$
$$\zeta(0) = 0$$

so

$$\frac{d}{dt} e^{\frac{t}{\lambda}} (t) = e^{\frac{t}{\lambda}} \left[\frac{S(\lambda)(\zeta(t) + \bar{u}) - \bar{u}}{\lambda}\right]$$
$$= e^{\frac{t}{\lambda}} \left[\frac{S(\lambda)(\zeta(t) + \bar{u}) - S(\lambda)\bar{u} + S(\lambda)\bar{u} - \bar{u}}{\lambda}\right]$$
$$\leq e^{\frac{t}{\lambda}} \left\|\frac{\zeta(t)}{\lambda}\right\| + e^{\frac{t}{\lambda}} \frac{\|S(\lambda)\bar{u} - \bar{u}\|}{\lambda}$$
$$\leq e^{\frac{t}{\lambda}} \frac{\|\zeta(t)\|}{\lambda} + e^{\frac{t}{\lambda}} \frac{\|S(\lambda)\bar{u} - \bar{u}\|}{\lambda} .$$

From this we easily deduce for $t \leqslant \lambda$

$$e^{\frac{t}{\lambda}} \|\zeta(t)\| \leq \int_0^t e^{\frac{s}{\lambda}} \frac{\|\zeta(s)\|}{\lambda} ds + (e-1) \|S(\lambda)\bar{u} - \bar{u}\|$$

and by Gronwall's inequality

(3.87) $$\|S_\lambda(t)\bar{u} - \bar{u}\| \leq (e-1) \|S(\lambda)\bar{u} - \bar{u}\| \quad \text{for } 0 \leqslant t \leqslant \lambda.$$

Then let

$$n\lambda \leqslant t < (n + 1)\lambda,$$

and we have

$$\begin{aligned} \| S_\lambda(t)\bar{u} - S(t)\bar{u} \| &\le \| S_\lambda(n\lambda)\bar{u} - S(n\lambda)\bar{u} \| \\ &\quad + \| S_\lambda(t)\bar{u} - S_\lambda(n\lambda)\bar{u} \| + \\ &\quad + \| S(t)\bar{u} - S(n\lambda)\bar{u} \| \\ &\le \sqrt{n} \| S(\lambda)\bar{u} - \bar{u} \| \\ &\quad + \| S_\lambda(t-n\lambda)\bar{u} - \bar{u} \| \\ &\quad + \| S(t-n\lambda)\bar{u} - \bar{u} \| \end{aligned}$$

from which we easily deduce (3.85).□

Proof of Theorem 3.4: Since $R^n_{t/n,\varepsilon}$ and $S_\varepsilon(t)$ are contractions, it suffices to prove the result (3.59) with $u - u^0 \in D(\mathcal{A})$, which is dense in C. We omit ε.

By the expression for $S(\lambda)\bar{u} - \bar{u}$ given in Lemma 3.16, for $\lambda \leqslant 1$ we have

$$\sup_{0 \le \sigma \le \lambda} \| S(\sigma)\bar{u} - \bar{u} \| \le \lambda K$$

where K is a constant independent of λ, $(\lambda \leqslant 1)$. Therefore by Lemma 3.20 we have

$$\| S_\lambda(t)\bar{u} - S(t)\bar{u} \| \le \left(\sqrt{\left[\frac{t}{\lambda}\right]} + e\right) K\lambda \le K\left(\sqrt{\lambda t} + e\lambda\right). \tag{3.88}$$

By Lemma 3.19 we also have

$$\| S_\lambda(t)\bar{u} - R^n_{\frac{t}{n}}(\lambda)\,\bar{u} \| \le \frac{2Kt}{\sqrt{n}}. \tag{3.89}$$

Next we have

$$\begin{aligned} \| R^n_{\frac{t}{n}}\bar{u} - S(t)\bar{u} \| &\le \| S(t)\bar{u} - S_\lambda(t)\bar{u} \| + \\ &\quad + \| S_\lambda(t)\bar{u} - R^n_{\frac{t}{n}}(\lambda)\bar{u} \| + \\ &\quad + \| R^n_{\frac{t}{n}}(\lambda)\bar{u} - R^n_{\frac{t}{n}}\bar{u} \| \\ &< K(\sqrt{\lambda t} + e\lambda) + 2K\frac{t}{\sqrt{n}} + \\ &\quad + \| R^n_{\frac{t}{n}}(\lambda)\bar{u} - R^n_{\frac{t}{n}}\bar{u} \| . \end{aligned}$$

With n fixed we make λ tend to 0, then by using Lemma 3.17 and the fact that $R_{t/n}(\lambda)$ is a contraction, from the preceding inequality we easily deduce the estimate

(3.90) $$\|R^n_{\frac{t}{n}}\bar{u} - S(t)\bar{u}\| \leq 2K\frac{t}{\sqrt{n}}$$

which completes the proof of the Theorem.□

3.3.2 Semi-group of the evolutionary inequality

We consider the problem (3.27) with

(3.91) $$\psi(t) = \psi, \quad \bar{u} \in C, \quad \bar{u} \leqslant \psi.$$

We write

(3.92) $$u(t) = S(t)\bar{u}.$$

We then have:

THEOREM 3.5: *Assume* (3.1),(3.2),(3.3) *and* (3.91). *Then* S(t) *defines a contraction semi-group on*

$$\mathcal{C} = \{\bar{u} \in C \mid \bar{u} \leqslant \psi\}$$

and, furthermore,

$$S(t)\bar{u} \to \bar{u} \quad \textit{in } C \textit{ when } t \to 0.$$

Proof: Let us set

$$v(t) = S(t)S(\theta)\bar{u},$$

therefore

$$v(t) \leq \int_0^{t-s} e^{-\alpha\sigma}\Phi(\sigma)L\,d\sigma + e^{-\alpha(t-s)}\Phi(t-s)v(s)$$

$$v(t) \leq \psi$$

$$v(0) = S(\theta)\bar{u}.$$

Now $u(t + \theta)$ satisfies the same relations as v, which is the maximum solution, therefore

$$u(t + \theta) \leqslant v(t).$$

Moreover, let us set

$$\tilde{u}(t) = \begin{vmatrix} v(t - \theta) & \text{for } t \geqslant \theta, \\ u(t) & \text{for } t \geqslant \theta. \end{vmatrix}$$

We see that

(3.93) $$\tilde{u}(t) \leq \int_0^{t-s} e^{-\alpha\sigma}\Phi(\sigma)L\,d\sigma + e^{-\alpha(t-s)}\Phi(t-s)\tilde{u}(s)$$

This is clear if $s \leqslant t \leqslant \theta$. If $\theta \leqslant s \leqslant t$, then

$$v(t-\theta) \leq \int_0^{t-s} e^{-\alpha\sigma}\Phi(\sigma)L d\sigma + e^{-\alpha(t-s)}\Phi(t-s)v(s-\theta)$$

whence (3.93) again. If $s \leqslant \theta \leqslant t$, we have

$$v(t-\theta) \leq \int_0^{t-\theta} e^{-\alpha\sigma}\Phi(\sigma)L d\sigma + e^{-\sigma(t-\theta)}\Phi\,(t-\theta)u(\theta)$$

and

$$u(\theta) \leq \int_0^{\theta-s} e^{-\alpha\sigma}\Phi(\sigma)L d\,\sigma + e^{-\alpha(\theta-s)}\Phi(\theta-s)u(s)$$

whence

$$v(t-\theta) \leq \int_0^{t-s} e^{-\alpha\sigma}\Phi(\sigma)L d\sigma + e^{-\alpha(t-s)}\Phi(t-s)u(s)$$

so we have (3.93) again. Since

$$\tilde{u}(t) \leqslant \psi \quad \text{and} \quad \tilde{u}(0) = \bar{u}$$

we have

$$\tilde{u}(t) \leqslant u(t),$$

whence

$$v(t-\theta) \leqslant u(t) \quad \text{if} \quad t \geqslant \theta,$$

so

$$v(t) \leqslant u(t+\theta),$$

and therefore, in fact,

$$v(t) = u(t+\theta),$$

so

$$S(t)S(\theta)\bar{u} = S(t+\theta)\bar{u}.$$

Therefore $S(t)$ is a semi-group on $\mathcal{C}$. Since

$$u_\varepsilon(\cdot) \to u(\cdot) \quad \text{in } C(0,T;C)$$

and using the property (3.40), we see that

$$\| S(t)\bar{u}_1 - S(t)\bar{u}_2 \| \leq \| \bar{u}_1 - \bar{u}_2 \|$$

and therefore $S(t)$ is a contraction. Since $u(\cdot) \in C(0,T;C)$ we certainly have

$$S(t)\bar{u} \to \bar{u} \text{ in } C, \text{ as } t \to 0$$

which completes the proof.

We are now going to prove Trotter's formula. We define

$$(3.94) \qquad T_\lambda(\bar{u}) = z_\lambda$$

by the problem

$$(3.95)\qquad z \leqslant \psi, \quad z \in C,$$

$$z \leq \int_0^t e^{-(\alpha+\frac{1}{\lambda})t}\,\Phi(t)(L + \frac{\bar{u}}{\lambda})ds + e^{-(\alpha+\frac{1}{\lambda})t}\,\Phi(t)z$$

and z_λ is the maximum element satisfying (3.95).

Our objective is to prove:

THEOREM 3.6: *Take the assumptions of Theorem 3.5 and* $\Phi(t) = 1$. *Then*

$$(3.96)\qquad \forall\, t > 0 \qquad R^n_{\frac{t}{n}}(\bar{u}) \to S(t)\bar{u} \quad \textit{in}\ C$$

$$\forall\, \bar{u} \in \mathcal{C}, \quad \textit{when} \quad n \to \infty .$$

Proof: By Lemma 5.7 of Chapter 4 we have

$$(3.97)\qquad R_{\lambda,\varepsilon}(\bar{u}) \to R_\lambda(\bar{u}) \quad \text{in } C, \text{ when } \varepsilon \to 0,\ \forall\, \lambda > 0,\ \bar{u} \in C.$$

From this it immediately follows that $R_\lambda(\bar{u})$ is a contraction. Furthermore, by Lemma 3.13 we also have

$$(3.98)\qquad S_\varepsilon(t)\bar{u} \to S(t)\bar{u} \quad \text{in } C \text{ when } \varepsilon \to 0,\ \forall\, \bar{u} \in C.$$

On the other hand, recalling that $z_\lambda = R_{\lambda,\varepsilon}(\bar{u})$ is the solution of

$$z_\varepsilon = \int_0^\infty e^{-(\alpha+\frac{1}{\lambda}+\frac{1}{\varepsilon})t}\,\Phi(t)(L + \frac{\bar{u}}{\lambda} + \frac{1}{\varepsilon} z_\varepsilon \wedge \psi)dt$$

then

$$\|R_{\lambda,\varepsilon}(\bar{u}_1,\psi_1) - R_{\lambda,\varepsilon}(\bar{u}_2,\psi_2)\| \leq \operatorname{Max}(\|\bar{u}_1-\bar{u}_2\|, \|\psi_1-\psi_2\|)$$

therefore we also have

$$(3.99)\qquad \|R_\lambda(\bar{u}_1,\psi_1) - R_\lambda(\bar{u}_2,\psi_2)\| \leq \operatorname{Max}(\|\bar{u}_1-\bar{u}_2\|, \|\psi_1-\psi_2\|) .$$

Also

$$(3.100)\qquad \|S(t)(\bar{u}_1,\psi_1) - S(t)(\bar{u}_2,\psi_2)\| \leq \operatorname{Max}(\|\bar{u}_1-\bar{u}_2\|, \|\psi_1-\psi_2\|).$$

We define

$$R^n_\lambda(\bar{u}_1,\psi_1) = R_\lambda(R^{n-1}_\lambda(\bar{u}_1,\psi_1),\psi_1)$$

and easily verify, using (3.99), that

$$(3.101)\qquad \|R^n_\lambda(\bar{u}_1,\psi_1) - R^n_\lambda(\bar{u}_2,\psi_2)\| \leq \operatorname{Max}(\|\bar{u}_1-\bar{u}_2\|, \|\psi_1-\psi_2\|).$$

Let us write $(\bar{u} - u^0)_k, (\psi - u^0)_k$ for two sequences belonging to $D(\mathcal{A})$ converging respectively to $\bar{u} - u^0$ and $\psi - u^0$ in C. Let ε_k be a sequence of numbers $\geqslant 0$ tending to 0 such that

$$\varepsilon_k \geq (u-u^0)_k - (u-u^0) + (\psi-u^0) - (\psi-u^0)_k .$$

We write

$$\psi_k = u^0 + (\psi-u^0)_k + \varepsilon_k$$
$$\bar{u}_k = u^0 + (\bar{u}-u^0)_k$$

and

$$\psi_k \to \psi, \quad \bar{u}_k \to \bar{u} \quad \text{in C}, \quad \psi_k \geqslant \bar{u}_k .$$

By (3.100) and (3.101) it suffices us to prove the desired result (3.96) under the additional assumption

(3.102) $\quad \bar{u} - u^0 \in D(\mathcal{A}), \quad \psi - u^0 \in D(\mathcal{A}),$

since

$$(\psi - u^0)_k + \varepsilon_k \in D(\mathcal{A}) \quad \text{(because } \Phi(t)1 = 1).$$

Then (cf. Lemma 3.16)

(3.103) $\quad \bar{u}- u_\varepsilon(t) = \int_0^t e^{-\alpha s}\Phi(s)[U +\alpha(\bar{u}-u^0))ds + \frac{1}{\varepsilon}(u_\varepsilon(t-s)-\psi)^+] \, ds$

But, thanks to (3.102),

$$u_\varepsilon(t)- u_0 = e^{-(\alpha+ \frac{1}{\varepsilon})t}\Phi(t)(\bar{u}-u_0) + $$
$$+ \int_0^t e^{-(\alpha+ \frac{1}{\varepsilon})(t-s)}\Phi(t-s)[\tfrac{1}{\varepsilon}(u_\varepsilon(s)-u_0)- \tfrac{1}{\varepsilon}(u_\varepsilon(s)-\psi)^+] \, ds$$
$$\psi- u_0 = e^{-(\alpha+ \frac{1}{\varepsilon})t}\Phi(t)(\psi- u_0) +$$
$$+ \int_0^t e^{-(\alpha+ \frac{1}{\varepsilon})(t-s)}\Phi(t-s)[V+ (\alpha+ \tfrac{1}{\varepsilon})(\psi-u_0)] \, ds$$

therefore

$$u_\varepsilon(t)-\psi = e^{-(\alpha+ \frac{1}{\varepsilon})t}\Phi(t)(\bar{u}-\psi) +$$
$$+ \int_0^t e^{-(\alpha+ \frac{1}{\varepsilon})(t-s)}\Phi(t-s)[-V- \alpha(\psi-u_0) +$$
$$+ \tfrac{1}{\varepsilon}(u_\varepsilon(s)-\psi) - \tfrac{1}{\varepsilon}(u_\varepsilon(s)-\psi)^+] \, ds \quad \leqslant \qquad \text{(Contd)}$$

(Contd) $$\leq \int_0^t e^{(\alpha+\frac{1}{\varepsilon})(t-s)} \Phi(t-s)[(-V-\alpha(\psi-u_0))^+]ds$$

and therefore

$$(u_\varepsilon(t) - \psi)^+ \leq \varepsilon \; \|(V+\alpha(\psi-u_0))^-\|$$

and then substituting in (3.103) we deduce that

$$\|\bar{u}-u_\varepsilon(t)\| \leq t\,[\|U + \alpha(\bar{u}-u^0)\| + \|(V+\alpha(\psi-u_0))^-\|$$

so

(3.104) $$\|S_\varepsilon(t)\bar{u} - \bar{u}\| \leqslant Kt$$

where K is a constant independent of ε and t.

However, if we then refer to the estimate (3.90), we can write

(3.105) $$\|R^n_{\frac{t}{n},\varepsilon}\bar{u} - S_\varepsilon(t)\bar{u}\| \leq 2K\frac{t}{\sqrt{n}} .$$

We then have

$$\|R^n_{\frac{t}{n}}(\bar{u}) - S(t)\bar{u}\| \leq \|R^n_{\frac{t}{n}}(\bar{u}) - R^n_{\frac{t}{n},\varepsilon}(\bar{u})\|$$
$$+ \|R^n_{\frac{t}{n},\varepsilon}(\bar{u}) - S_\varepsilon(t)\bar{u}\| + \|S_\varepsilon(t)\bar{u} - S(t)\bar{u}\|$$
$$\leq \|R^n_{\frac{t}{n}}(\bar{u}) - R^n_{\frac{t}{n},\varepsilon}(\bar{u})\| + 2K\frac{t}{\sqrt{n}} +$$
$$+\|S_\varepsilon(t) - S(t)\bar{u}\| .$$

Making ε tend to 0 with n fixed, and using (3.97),(3.98), from this we deduce that

(3.106) $$\|R^n_{\frac{t}{n}}(\bar{u}) - S(t)\bar{u}\| \leq 2K\frac{t}{\sqrt{n}}$$

whence the result desired.□

3.4 *Asymptotic Behaviour*

For $\alpha > 0$ we consider the maximum solution u of

(3.107) $$\left| \begin{array}{l} u \in C, \quad u \leq \psi \\ u \leq e^{-\alpha t}\Phi(t)u + \int_0^t e^{-\alpha\sigma}\Phi(\sigma)L\,d\sigma \end{array} \right. .$$

Our objective is to prove:

THEOREM 3.7: *Take the assumptions of Theorem 3.5 and* $\Phi(t)1 = 1$, $\alpha > 0$. *Then if* $S(t)\bar{u}$ *is defined by* (3.92), *we have*

$$S(t)\bar{u} \to u \quad \text{in } C \text{ when } t \to \infty. \tag{3.108}$$

Proof: We shall set $u = S(\psi)$ as the maximum solution of (3.107), and

$$u(t) = S(t)\bar{u} = S(t;\bar{u},\psi).$$

Since

$$\|S(\psi_1) - S(\psi_2)\| \le \|\psi_1 - \psi_2\|$$

$$\|S(t;\bar{u}_1,\psi_1) - S(t;\bar{u}_2,\psi_2)\| \le \text{Max } \|\psi_1-\psi_2\| \ , \ \|\bar{u}_1-\bar{u}_2\|)$$

it suffices to establish (3.108) by further assuming that

$$\bar{u} - u^0 \in D(\mathcal{A}), \qquad \psi - u^0 \in D(\mathcal{A})$$

(cf. (3.102)).

Write u_ε and $u_\varepsilon(t)$ for the solutions of the penalised problems corresponding to (3.107) and $S(t)\bar{u}$ (cf. (3.29)). We can consider $u_\varepsilon - u^0$ as the solution of a penalised problem with $L = 0$ and ψ changed into $\psi - u^0$. With the notations of Lemma 5.7 we have

$$\tilde{L} = -(V + \alpha(\psi - u^0))$$

(where $V \in C$ was introduced in Theorem 3.6), therefore

$$\|u_\varepsilon - u\| \le \varepsilon \quad \|(V + \alpha(\psi - u^0))^-\|. \tag{3.109}$$

By Lemma 3.13 the same estimate holds for $u_\varepsilon(t)$, i.e.,

$$\|u_\varepsilon(t) - u(t)\| \le \varepsilon \ \|(V+\alpha(\psi-u^0))^-\| \tag{3.110}$$

and therefore it suffices to show that for the penalised problem

$$u_\varepsilon(t) \to u_\varepsilon \quad \text{in } C \text{ when } t \to \infty, \ \varepsilon \text{ fixed.} \tag{3.111}$$

We then consider the operators $T_\varepsilon : C \to C$ and $T_c(\cdot) : C(0,T;C) \to C(0,T;C)$ whose fixed points are u_c and $u_\varepsilon(\cdot)$ respectively, and we set

$$u_n^\varepsilon = T_\varepsilon^n \, 0 \qquad n \geqslant 1,$$

$$u_n^\varepsilon(t) = T_\varepsilon^n(t) \ 0 \quad n \geqslant 1.$$

Then (cf. Lemma 5.3, Chapter 4)

$$\|T_\varepsilon z_1 - T_\varepsilon z_2\| \le \frac{1}{1+\alpha\varepsilon} \ \|z_1 - z_2\|$$

and (cf. (3.55))

$$\|T_\varepsilon z_1(.) - T_\varepsilon z_2(.)\|_{C(0,T;C)} \le \frac{1}{1+\alpha\varepsilon}\|z_1 - z_2\|_{C(0,T;C)}$$

whence, by setting $k_\varepsilon = \dfrac{1}{1+\alpha\varepsilon}$, we deduce that

$$\text{(3.112)} \qquad \|u - u_n^\varepsilon\| \le \frac{k_\varepsilon^{n+1}}{1-k_\varepsilon}\|u_1^\varepsilon\|$$

$$\|u(.) - u_n^\varepsilon(.)\| \le \frac{k_\varepsilon^{n+1}}{1-k_\varepsilon}\|u_1^\varepsilon(.)\| .$$

However, since $\alpha > 0$, by Lemma 3.8 $\|u_1^\varepsilon(t)\| \leqslant K$, a constant independent of ε and t, therefore

$$\text{(3.113)} \qquad \|u(t) - u_n^\varepsilon(t)\| \le K \frac{k_\varepsilon^{n+1}}{1-k_\varepsilon} .$$

The estimates (3.112) and (3.113) show that is suffices to prove that for *fixed* ε and n we have

$$\text{(3.114)} \qquad u_n^\varepsilon(t) \to u_n^\varepsilon \quad \text{when } t \to \infty \text{ in } C.$$

It then suffices to proceed by induction using the following result:

(3.115) if $z(\cdot) \in C(0,T;C)$ for all finite t, and

$z(t) \to z_\infty$ in C when $t \to \infty$, then

$T_\varepsilon z(t) \to T_\varepsilon z_\infty$ in C when $t \to \infty$.

Let us therefore prove (3.115). Now,

$$T_\varepsilon z(t) - T_\varepsilon z_\infty = e^{-(\alpha+\frac{1}{\varepsilon})t}\Phi(t)\,\bar{u} + \\ + \int_0^t e^{-(\alpha+\frac{1}{\varepsilon})}\Phi(\sigma)[L + \frac{1}{\varepsilon}\psi \wedge z(t-\sigma)]d\sigma . \\ - \int_0^\infty e^{-(\alpha+\frac{1}{\varepsilon})}\Phi(\sigma)[L + \frac{1}{\varepsilon}\psi \wedge z_\infty]\,d\sigma$$

It is clear that it suffices to show that

$$\int_0^t e^{-(\alpha+\frac{1}{\varepsilon})}\Phi(\sigma)(\psi \wedge z(t-\sigma))d\sigma \to \int_0^\infty e^{-(\alpha+\frac{1}{\varepsilon})}\Phi(\sigma)\psi \wedge z_\infty\, d\sigma$$

in C when $t \to \infty$, or again

$$X_t = \int_0^t e^{-(\alpha+\frac{1}{\varepsilon})}\Phi(\sigma)(\psi \wedge z(t-\sigma) - \psi \wedge z_\infty)d\sigma \to 0 \quad \text{in } C \text{ when } t \to \infty.$$

Now,

$$X_t = X_t^1 + X_t^2,$$

with

$$X_t^1 = \int_0^{\frac{t}{2}} e^{-(\alpha+\frac{1}{\varepsilon})} \Phi(\sigma)(\psi \wedge z(t-\sigma) - \psi \wedge z_\infty)d\sigma$$

$$X_t^2 = \int_{\frac{t}{2}}^{t} e^{-(\alpha+\frac{1}{\varepsilon})} \Phi(\sigma)(\psi \wedge z(t-\sigma) - \psi \wedge z_\infty)d\sigma$$

and

$$X_t^2 \to 0 \quad \text{in } C \text{ when } t \to \infty, \text{ and}$$

$$\| X_t^1 \| \leq \sup_{s \geq \frac{t}{2}} \|z(s) - z_\infty\| \frac{1}{\alpha + \frac{1}{\varepsilon}} \to 0$$

when $t \to \infty$. This completes the proof of the result (3.108).□

4. APPROXIMATE SEMI-GROUP FOR THE INEQUALITY WITH IMPLICIT OBSTACLE

4.1 *Non-Linear Semi-Group*

Here we consider a situation similar to that of Section 5.2 in Chapter 4. We are given the semi-group $\Phi(t)$ satisfying (3.1),(3.2). We are given L,α as in (3.3), and we further assume

(4.1) $\quad L \geqslant 0.$

At last we consider M such that

(4.2)
$$\begin{cases} M : C \to C \quad , \quad \|M(\phi_1) - M(\phi_2)\| \leq \|\phi_1 - \phi_2\| \quad , \\ M\phi_1 \leq M\phi_2 \quad \text{si} \quad \phi_1 \leq \phi_2 \\ M \text{ concave}, \quad M(0) \geq k > 0 \,. \end{cases}$$

Lastly we are given

(4.3) $\quad \bar{u} \in C, \quad \bar{u} \geqslant 0, \quad \bar{u} \leqslant M\bar{u}.$

Consider the following problem:

(4.4)
$$\begin{cases} u(.) \in C([0,T]; C) \quad , \quad u(0) = \bar{u} \\ u(t) \leq Mu(t) \qquad \forall\, t \in [0,T] \\ u(t) \leq \int_0^{t-s} e^{-\alpha\sigma}\Phi(\sigma)Ld\sigma + e^{-\alpha(t-s)}\,\Phi(t-s)u(s) \\ \qquad\qquad \forall \quad s \leq t \in [0,T]\,. \end{cases}$$

We are going to prove:

THEOREM 4.1: *Assume* (3.1),(3.2),(3.3),(4.1),(4.2),(4.3). *Then the set of functions* u *satisfying* (4.4) *is non-empty and has a maximum element. If write this maximum element as* u(t), *then*

$$u(t) = S(t)\bar{u}$$

defines a (non-linear) contraction semi-group on

$$\mathcal{C} = \{\bar{u} \quad C | \bar{u} \geqslant 0, \bar{u} \leqslant M\bar{u}\}. \square$$

We proceed with the proof in several steps.

Let z satisfy

$$z \in C(0,T;C), \quad z(0) = \bar{u}. \tag{4.5}$$

We consider the maximum element of (3.27) corresponding to the obstacle

$$\psi(t) = Mz(t)$$

which satisfies the condition (3.26). We thus define an operator T on the subset of C(0,T;C) defined in (4.5) that takes it into itself. We begin with a result similar to that of Lemma 1.1. There is a small difficulty owed to the fact of 0 not belonging to the subset (4.5).

We shall write

$$u^0(t) = \int_0^t e^{-\alpha\sigma}\Phi(\sigma)Ld\sigma + e^{-\alpha t}\Phi(t)\bar{u} \tag{4.6}$$

and thanks to assumptions (4.1),(4.3), $u^0(t) \geqslant 0$. Furthermore, $u^0(\cdot)$ belongs to the set (4.5).

LEMMA 4.1: *Let* $0 \leqslant \mu \leqslant k/\|Mu^0\|$. *If* z,ζ *satisfy* (4.5) *and* $0 \leqslant z, \zeta \leqslant u^0$ *as well as*

$$z(t) - \zeta(t) \leqslant \gamma z(t) \quad \text{for all } t \in [0,T], \ \gamma \in [0,1], \tag{4.7}$$

then

$$Tz(t) - T\zeta(t) \leqslant \gamma(1-\mu)Tz(t) \quad \text{for all } t \in [0,T]. \tag{4.8}$$

Proof: Let us write

$$\theta = 1 - \gamma(1-\mu), \quad u = Tz, \quad \tilde{u} = T\zeta.$$

It is then a matter of showing that

$$\theta u \leqslant \tilde{u}.$$

By the definition of u we have

$$\left| \begin{aligned} &\theta u(t) \leq \theta\, Mz(t) \\ &\theta u(t) \leq \theta \int_0^{t-s} e^{-\alpha\sigma}\Phi(\sigma)Ld\sigma + \theta\, e^{-\alpha(t-s)}\Phi(t-s)u(s) \\ &\quad \leq \int_0^{t-s} e^{-\alpha\sigma}\Phi(\sigma)Ld\sigma + e^{-\alpha(t-s)}\Phi(t-s)(\theta u(s)) \\ &\theta u(0) = \theta\bar{u} \leq \bar{u}\,. \end{aligned} \right. \tag{4.9}$$

We are going to show that

(4.10) $\theta Mz \leqslant M\zeta$.

By (4.7) it suffices to show that

$$\theta Mz \leqslant M(z(1 - \gamma)),$$

but by the concaveness of M it again suffices to show that

$$\theta Mz \leqslant (1 - \gamma)M(z) + \gamma M(0),$$

and by the definition of θ,

(4.11) $\mu Mz \leqslant M(0)$.

Since $z \leqslant u^0$, $Mz \leqslant Mu^0$, and $M(0) \geqslant k$, and therefore by the choice of μ, we have $k \geqslant Mu^0$, hence (4.11) holds, and consequently (4.10).

Consequently, by (4.9),

$$\theta u \leqslant \sigma(M\zeta, \theta\bar{u})$$

where $\sigma(\psi,\bar{u})$ denotes the maximum element of (3.27) (making the dependence upon $\psi,\bar{u}$ explicit). Next we use the fact that $\sigma(\psi,\bar{u})$ is increasing with respect to $\psi,\bar{u}$, therefore

$$\theta u \leqslant \sigma(M\zeta,\bar{u}) = \tilde{u},$$

whence the desired result.□

Remark 4.1

The fact that σ is an increasing function of its arguments is easily shown from the formula (3.34) giving T_ε, and therefore by the same property for $T_\varepsilon(\psi,\bar{u}) = u_\varepsilon$.

Remark 4.2

Since $u^0 \geqslant 0$, we have $Mu^0 \geqslant M_0 \geqslant k$, and therefore $\mu \leqslant 1$.□

Proof of Theorem 4.1: Consider the recurrence procedure

$$u^{n+1} = Tu^n, \quad u^0 \text{ defined by (4.6)}.$$

By Lemma 4.1 we easily deduce (cf. Corollary 1.1) that $u^n \to u \in C(0,T;C)$, $u(0) = \bar{u}$, u belonging to the set (4.4). It necessarily is a the maximum element, for if $\tilde{u}$ belonged to the set (4.4), then $\tilde{u} \leqslant u^0$ and T would be increasing, by recurrence $\tilde{u} \leqslant u^n$, therefore $\tilde{u} \leqslant u$.

The fact the S(t) is a semi-group on $\mathcal{C}$ is proved as in Theorem 3.5. The fact that S(t) is contracting is shown by induction. In fact, let $\bar{u}_1,\bar{u}_2 \in \mathcal{C}$ and $u_1^n(t),u_2^n(t)$ be sequences defined by the iteration procedure above. Lastly, let $u_1(t),u_2(t)$ be such that

$$u_i(t) = S(t)\bar{u}_i, \quad i = 1,2.$$

Now,

$$\|u_1^0 - u_2^0\| \leqslant \|\bar{u}_1 - \bar{u}_2\|.$$

Then, thanks to (3.40) and the analogous property for the inequality, we have

$$\| u_1^1 - u_2^1 \| \le \text{Max} \left(\|u_1 - u_2\| , \| M u_1^0 - M u_2^0 \| \right) \le \| \bar{u}_1 - \bar{u}_2 \| .$$

Quite generally we show that

$$\| u_1^n - u_2^n \| \le \|\bar{u}_1 - \bar{u}_2\|$$

whence, in the limit,

$$\|u_1(t) - u_2(t)\| \le \| \bar{u}_1 - \bar{u}_2 \|$$

which is the result desired.□

4.2 *Trotter's Formula*

We define $R_\lambda(u) : C \to C$ by the problem

(4.12) $\quad z \leqslant Mz, \quad z \in C,$

$$z \le \int_0^t e^{-(\alpha + \frac{1}{\lambda})s} \Phi(s)(L + \frac{\bar{u}}{\lambda}) ds + e^{-(\alpha + \frac{1}{\lambda})t} \Phi(t) z$$

and $z_\lambda = R_\lambda(\bar{u})$ is the maximum solution of (4.12).

Our objective is to prove:

THEOREM 4.2: *Take the assumptions of Theorem 4.1. Then*

(4.13) $\quad \forall\, t > 0 \ , \quad R^n_{\frac{t}{n}}(\bar{u}) \to S(t)\bar{u} \ , \quad n \to +\infty$

(4.14) $\quad \left| \begin{array}{l} \forall\, \bar{u} \in \mathcal{C} \quad \textit{such that} \\ \bar{u} \le \int_0^t e^{-\alpha\sigma} \Phi(\sigma) L d\sigma + e^{-\alpha t} \Phi(t)\, \bar{u} \ , \quad \forall\, t \ge 0 . \ \square \end{array} \right.$

Remark 4.3

If $\bar{u}$ satisfies the condition (4.14), it is clear that u satisfies the conditions (4.4), therefore

(4.15) $\quad \bar{u} \leqslant u(t) \quad$ for all t.□

We shall need the penalised problem associated with (4.4), namely,

(4.16) $$u_\varepsilon(t) = e^{-\alpha t} \Phi(t) \bar{u} + \int_0^t e^{-\alpha(t-s)} \Phi(t-s)[L - \frac{1}{\varepsilon}(u_\varepsilon(s) - Mu_\varepsilon(s))^+] \, ds .$$

To study the problem (4.16) we consider an iteration procedure analogous to that used in the proof of Theorem 4.1, say,

$$(4.17)\quad \left|\begin{array}{l} u_\varepsilon^0(t) = u^0(t) \qquad \text{(cf. (4.6))}. \\ u_\varepsilon^k(t) = e^{-\alpha t}\Phi(t)\bar{u} + \int_0^t e^{-\alpha(t-s)}\Phi(t-s)[L - \frac{1}{\varepsilon}(u_\varepsilon^k(s) - Mu_\varepsilon^{k-1}(s))^+]ds \\ u_\varepsilon^k \in C(0,T;C)\ . \end{array}\right.$$

The sequence u_ε^k is well defined (cf. (3.28)). It is clear that $u_\varepsilon^k \leqslant u^0$. Also, by induction we show that

$$(4.18)\qquad 0 \leqslant u_\varepsilon^k \leqslant u^0.$$

In fact, we have (cf. Lemma 3.6)

$$u_\varepsilon^k(t) = e^{-(\alpha+\frac{1}{\varepsilon})t}\Phi(t)\bar{u} + \int_0^t e^{-(\alpha+\frac{1}{\varepsilon})(t-s)}\Phi(t-s)[L + + \frac{1}{\varepsilon}u_\varepsilon^k(s) - \frac{1}{\varepsilon}(u_\varepsilon^k(s) - Mu_\varepsilon^{k-1}(s))^+]ds$$

and since

$$Mu_\varepsilon^{k-1} \geqslant 0,$$

we have

$$u_\varepsilon^k \leqslant 0 \qquad \text{(cf. Lemma 3.7)}.$$

We now define

$$T^\varepsilon z = \zeta^\varepsilon$$

to be the solution of

$$(4.19)\qquad \zeta^\varepsilon(t) = e^{-(\alpha+\frac{1}{\varepsilon})t}\Phi(t)\bar{u} + \int_0^t e^{-(\alpha+\frac{1}{\varepsilon})(t-s)}\Phi(t-s)[L+\frac{1}{\varepsilon}\zeta^\varepsilon(s)\wedge Mz(s)]ds.$$

LEMMA 4.2: *Let* $0 \leqslant \mu \leqslant k/\|Mu^0\|$. *If* z_1, z_2 $C(0,T;C)$ *satisfy* $0 \leqslant z_1, z_2 \leqslant u^0$ *and*

$$(4.20)\qquad z_1(t) - z_2(t) \leqslant \gamma z_1(t) \quad \text{for all } t \in [0,T],\ \gamma \in [0,1],$$

then

$$(4.21)\qquad T^\varepsilon z_1(t) - T^\varepsilon z_2(t) \leqslant \gamma(1-\mu)T^\varepsilon z_1(t) \quad \text{for all } t \in [0,T].$$

Proof: Let us set

$$u = T^\varepsilon z_1, \quad \tilde{u} = T^\varepsilon z_2,$$

Therefore we have

$$u(t) = e^{-(\alpha+\frac{1}{\varepsilon})t}\Phi(t)\bar{u} + \int_0^t e^{-(\alpha+\frac{1}{\varepsilon}\, t-s)}\Phi(t-s)[L + \frac{1}{\varepsilon}u \wedge Mz_1]\, ds$$

therefore on setting $\theta = 1 - \gamma(1 - \mu)$ we have

$$\theta u(t) \leq e^{-(\alpha+\frac{1}{\varepsilon})t}\Phi(t)\bar{u} + \int_0^t e^{-(\alpha+\frac{1}{\varepsilon})(t-s)}\Phi(t-s)\,[L + \frac{1}{\varepsilon}\theta u \wedge \theta M z_1]ds. \tag{4.22}$$

Now, as in Lemma 4.1, by the choice of μ we have

$$\theta Mz_1 \leqslant Mz_2. \tag{4.23}$$

It easily follows from (4.22) and (4.23) that

$$\theta u(t) \leqslant \tilde{u}(t),$$

whence (4.21).□

LEMMA 4.3: *There exists one and only one solution of* (4.16) *in* C(0,T;C). *Moreover, we then have the estimate*

$$\|u_\varepsilon - u_\varepsilon^k\| \leq \frac{(1-\mu)^k}{\mu}\|u^0\| = c\,(1-\mu)^k . \tag{4.24}$$

Proof: As in Corollary 1.1, T^ε has an unique fixed point u_ε, and (4.24) is satisfied.□

LEMMA 4.4: *Let* $\bar{u} \in \mathcal{C}$, *then*

$$u_\varepsilon \to S(\cdot)\bar{u} \quad \textit{in } C(0,T;C) \textit{ when } \varepsilon \to 0. \tag{4.25}$$

Proof: If we consider the iteration procedure defined in the course of the proof of Theorem 4.1, we also have

$$\|u^k - u\| \leqslant c(1 - \mu)^k$$

and therefore, thanks to (4.24), everything reduces to showing that

$$u_\varepsilon^k \to u^k \quad \text{in } C(0,T;C) \text{ for all fixed } k, \text{ when } \varepsilon \to 0. \tag{4.26}$$

The property is obvious for k = 0. Let us assume that it holds for k - 1. Define $\tilde{u}_\varepsilon^k$ by

$$\tilde{u}_\varepsilon^k(t) = e^{-\alpha t}\Phi(t)\bar{u} + \int_0^t e^{-\alpha(t-s)}\Phi(t-s)[L - \frac{1}{\varepsilon}(\tilde{u}_\varepsilon^k(s) - Mu^{k-1}(s))^+]ds.$$

By Lemma 3.11 we have

$$\begin{aligned} \|\tilde{u}_\varepsilon^k - u_\varepsilon^k\| &\leq \|M_\varepsilon u^{k-1} - M\, u^{k-1}\| \\ &\leq \| u_\varepsilon^{k-1} - u^{k-1}\| . \end{aligned} \tag{4.27}$$

Further, by Lemma 3.13 we have

$$\|u_\varepsilon^k - u^k\| \to 0 \quad \text{when } \varepsilon \to 0,$$

which when coupled with (4.27) and the induction hypothesis certainly implies (4.26), whence the result desired.□

Set

$$u_\varepsilon(t) = S_\varepsilon(t)\bar{u} \tag{4.28}$$

and we then easily show that S_ε defines a (non-linear) contraction semi-group on C. Therefore

$$\| S_\varepsilon(.)\bar{u} - S(.)\bar{u} \|_{C(0,T;C)} \to 0 \qquad \forall \bar{u} \in \mathcal{C} . \tag{4.29}$$

Next we define the penalised operator $R_{\lambda,\varepsilon}(\bar{u})$ of R_λ, so

$$z_\varepsilon = R_{\lambda,\varepsilon}(\bar{u})$$

is the solution of

$$z_\varepsilon = \int_0^\infty e^{-(\alpha+\frac{1}{\lambda}+\frac{1}{\varepsilon})t}\Phi(t)(L + \frac{\bar{u}}{\lambda} + \frac{1}{\varepsilon} z_\varepsilon \wedge Mz_\varepsilon)\, dt . \tag{4.30}$$

We solve (4.30) by considering a recurrence method

$$z_\varepsilon^k = \int_0^\infty e^{-(\alpha+\frac{1}{\lambda}+\frac{1}{\varepsilon})t}\Phi(t)(L + \frac{\bar{u}}{\lambda} + \frac{1}{\varepsilon} z_\varepsilon^k \wedge M z_\varepsilon^{k-1})dt . \tag{4.31}$$

Define the mapping $T_\lambda^\varepsilon z = \zeta$ by

$$\zeta = \int_0^\infty e^{-(\alpha+\frac{1}{\lambda}+\frac{1}{\varepsilon})t}\Phi(t)(L + \frac{\bar{u}}{\lambda} + \frac{1}{\varepsilon}\zeta \wedge Mz)dt . \tag{4.32}$$

Set

$$u_{0\lambda} = \int_0^\infty e^{-(\alpha+\frac{1}{\lambda})t}\Phi(t)(L + \frac{\bar{u}}{\lambda})dt \tag{4.33}$$

therefore

$$\| u_{0\lambda} \| \le \| \int_0^\infty e^{-\alpha t}\Phi(t)L\, dt\| + \|\bar{u}\| = c_0 . \tag{4.34}$$

Since

$$\zeta = \int_0^\infty e^{-(\alpha+\frac{1}{\lambda})t}\Phi(t)(L + \frac{\bar{u}}{\lambda} - \frac{1}{\varepsilon}(\zeta - Mz)^+)\, dt \leqslant u_{\lambda 0},$$

we see that T_λ^ε maps the interval $[0,u_{0\lambda}]$ into itself.

LEMMA 4.4: *Let* $0 \leqslant \mu \leqslant k/Mc_0$. *If* $z_1, z_2 \in C$ *belong to* $[0,u_{0\lambda}]$ *and*

(4.35) $$z_1 - z_2 \leqslant \gamma z_1, \quad \gamma \in [0,1],$$

then

(4.36) $$T_\lambda^\varepsilon z_1 - T_\lambda^\varepsilon z_2 \leqslant \gamma(1-\mu)T_\lambda^\varepsilon z_1.$$

Proof: This is very similar to Lemma 4.2. It suffices to show that

$$\mu M u_{0\lambda} \leqslant k,$$

which is satisfied under the conditions stated.□

LEMMA 4.5: *There exists one and only one solution of* (4.30) *in* C. *Moreover,*

(4.37) $$\| z_\varepsilon - z_\varepsilon^k \| \leq \frac{(1-\mu)^k}{\mu} c_0 .$$

Proof: This is identical to that of Lemma 4.3. Let us notice that in Lemma 4.3 we can also take $\mu \leqslant \frac{k}{Mc_0}$, and that we have the same estimate as (4.37) for $\|u_\varepsilon - u_\varepsilon^k\|$.□

LEMMA 4.6: *Let* $\bar{u} \in C$. *Then*

(4.38) $$R_{\lambda,\varepsilon}(\bar{u}) \to R_\lambda(\bar{u}) \text{ in C when } \varepsilon \to 0, \text{ for all fixed } \lambda > 0.$$

Proof: Analogous to that of Lemma 4.4, using as intermediary the sequence z_ε^k and and the estimate of Lemma 4.5.

LEMMA 4.7: *For all* $\bar{u} \in C$ *we have the estimate*

(4.39) $$\| S_\varepsilon(t)\bar{u} - R^n_{\frac{t}{n},\varepsilon} \bar{u}\| \leq \frac{2t}{\sqrt{n}} \sup_\lambda \frac{\|S_\varepsilon(\lambda)\bar{u} - \bar{u}\|}{\lambda}.$$

Proof: We take our inspiration from proofs already carried out. We recall that we wrote

$$u^0 = \int_0^\infty e^{-\alpha t} \Phi(t)L\, dt .$$

First of all, if $u - u^0 \in D(\mathcal{A})$, i.e. $\bar{u} - u^0$ satisfies (3.61), then

(4.40) $$\frac{\bar{u} - S_\varepsilon(\lambda)\bar{u}}{\lambda} \to U + \alpha(\bar{u} - u^0) + \frac{1}{\varepsilon}(\bar{u} - M\bar{u})^+ \text{ in C when } \lambda \to 0.$$

Similarly, we have

$$(4.41)\quad \left|\begin{array}{l} \dfrac{\bar{u} - S_\varepsilon(\lambda)\bar{u}}{\lambda} \to U + \alpha(\bar{u} - u^0) + \dfrac{1}{\varepsilon}(\bar{u} - M\bar{u})^+ \\ \text{in } C, \text{ when } \lambda \to 0. \end{array}\right.$$

Next, as in (3.67), we introduce

$$(4.42)\quad R_{\lambda,\varepsilon}(t)\,\bar{u} = \left(I + \lambda\,\frac{I - S_\varepsilon(t)}{t}\right)^{-1} \bar{u}\ .$$

We establish that

$$(4.43)\quad \left|\begin{array}{l} R_{\lambda,\varepsilon}(t)\bar{u} \to R_{\lambda,\varepsilon}\bar{u} \text{ in } C \text{ when } t \to 0, \\ \forall\, u \in C, \quad \lambda,\varepsilon > 0. \end{array}\right.$$

Then we consider the semi-group $S_{\lambda,\varepsilon}(t)$ (cf. (3.72)), and show that

$$(4.44)\quad R^n_{\frac{t}{n},\varepsilon}(\lambda)\,\bar{u} \to S_{\lambda,\varepsilon}(t)\,\bar{u} \quad \text{in } C \text{ when } n \to \infty.$$

The essential step consists of then proving the estimate

$$(4.45)\quad \|S_{\lambda,\varepsilon}(t)\,\bar{u} - R^n_{\frac{t}{n},\varepsilon}(\lambda)\,\bar{u}\| \le \frac{2t}{\sqrt{n}}\ \frac{\|\bar{u} - S_\varepsilon(\lambda)\,\bar{u}\|}{\lambda}\ .$$

We also have

$$(4.46)\quad \|S_{\lambda,\varepsilon}(t)\,\bar{u} - S_\varepsilon(t)\,\bar{u}\| \le (\sqrt{n} + e) \sup_{0 \le \sigma \le \lambda} \|S_\varepsilon(\sigma)\bar{u} - \bar{u}\|\ .$$

From this we deduce

$$\begin{aligned} \|R^n_{\frac{t}{n},\varepsilon}\bar{u} - S_\varepsilon(t)\,\bar{u}\| &\le \|S_\varepsilon(t)\,\bar{u} - S_{\lambda,\varepsilon}(t)\,\bar{u}\| + \\ &+ \|S_{\lambda,\varepsilon}(t)\,\bar{u} - R^n_{\frac{t}{n},\varepsilon}(\lambda)\,\bar{u}\| + \\ &+ \|R^n_{\frac{t}{n},\varepsilon}(\lambda)\,\bar{u} - R^n_{\frac{t}{n},\varepsilon}\bar{u}\| \\ &\le \frac{2t}{\sqrt{n}} \sup_\theta \frac{\|\bar{u} - S_\varepsilon(\theta)\,\bar{u}\|}{\theta} + \\ &+ \|S_\varepsilon(t)\,\bar{u} - S_{\lambda,\varepsilon}(t)\,\bar{u}\| \\ &+ \|R^n_{\frac{t}{n},\varepsilon}(\lambda)\bar{u} - R^n_{\frac{t}{n},\varepsilon}\bar{u}\| \end{aligned}$$

and with n fixed we make λ tend to 0. Using (4.46) and (4.43), from this we deduce the estimate (4.39), which is only of interest if the sup on the right hand side is finite.□

Proof of Theorem 4.2: The assumptions about $\bar{u}$ (cf. (4.14)) are going to allow us to bound the quantity $\sup \dfrac{\|S_\varepsilon(\lambda)\bar{u} - \bar{u}\|}{\lambda}$ by a constant independent of ε.

Since R_λ and $S(\lambda)$ are contractions, it suffices to prove the result (4.13) under the additional assumption $u - u^0 \in D(\mathcal{A})$. Under these conditions we have

$$\bar{u} - u_\varepsilon(t) = \int_0^t e^{-\alpha s}\Phi(s)[L + \alpha(\bar{u} - u^0) + \frac{1}{\varepsilon}(u_\varepsilon(t-s) - Mu_\varepsilon(t-s))^+]ds$$

$$\geq \int_0^t e^{-\alpha s}\Phi(s)[L + \alpha(\bar{u} - u^0)]ds$$

$$\geq -Kt.$$

But by assumption (4.14) we have (4.15), and since

$$u(t) \leqslant u_\varepsilon(t)$$

we have

$$\bar{u} \leqslant u\varepsilon(t),$$

whence

$$\|S_\varepsilon(t)\bar{u} - \bar{u}\| \leqslant Kt.$$

Using Lemmas 4.4 and 4.6, from this, as in the proof of Theorem 3.6, we deduce that

$$(4.47) \qquad ||R^n_{\frac{t}{n}}(\bar{u}) - S(t)\,\bar{u}\,|| \leq 2K\frac{t}{\sqrt{n}}$$

which completes the proof of the desired result.□

Remark 4.4

We refer to L. BARTHELEMY [1] and L. BARTHELEMY - F. CATTE [1] for other aspects related to Trotter's formula for Q.V.I.'s.□

Remark 4.5

The assumption (4.14) has already arisen in Theorem 1.3.□

5. HYPERBOLIC Q.V.I.'s

5.1 *Method of Approach*

In this Section we are going to give an example of a hyperbolic Q.V.I., i.e., more precisely, a Q.V.I. for a *second order hyperbolic system*.

This example, introduced in B.L. [1], and which we round off a little here, has no other motivation than a taste of 'symmetry' with respect to the parabolic theory that has gone before. Some other examples demonstrating the same flavour are to be found in the following Section.

5.2 *Statement of the Problem: Statement of the Principal Results*

In the open set $Q = \mathcal{O} \times]0,T[$ we look for a pair of functions u_1 and u_2 such that

$$(5.1) \qquad \left| \begin{array}{l} \dfrac{\partial^2 u_1}{\partial t^2} - \Delta u_1 - f_1 \leq 0 \;, \quad \dfrac{\partial u_1}{\partial t} - \dfrac{\partial u_2}{\partial t} - k_2 \leq 0 \;, \\ (\dfrac{\partial^2 u_1}{\partial t^2} - \Delta u_1 - f_1)(\dfrac{\partial u_1}{\partial t} - \dfrac{\partial u_2}{\partial t} - k_2) = 0 \end{array} \right.$$

(5.2)
$$\left|\begin{array}{l} \dfrac{\partial^2 u_2}{\partial t^2} - \Delta u_2 - f_2 \le 0 \quad , \quad \dfrac{\partial u_2}{\partial t} - \dfrac{\partial u_1}{\partial t} - k_1 \le 0 \quad , \\ \left(\dfrac{\partial^2 u_2}{\partial t^2} - \Delta u_2 - f_2\right)\left(\dfrac{\partial u_2}{\partial t} - \dfrac{\partial u_1}{\partial t} - k_1\right) = 0 \end{array}\right.$$

with the boundary conditions

(5.3) $$\frac{\partial u_i}{\partial \nu} = 0, \quad i = 1,2, \quad \text{on } \Sigma = \Gamma \times]0,T[,$$

and with initial conditions

(5.4) $$u_i(x,0) = u_i^0(x), \quad \frac{\partial u_i}{\partial t}(x,0) = u_i^1(x) \quad , \quad i = 1,2 \; , \; x \in \mathcal{O} .$$

The constraints

$$\frac{\partial u_1}{\partial t} - \frac{\partial u_2}{\partial t} - k_2 \le 0 \; , \quad \frac{\partial u_2}{\partial t} - \frac{\partial u_1}{\partial t} - k_1 \le 0 \; ,$$

where the k_i *are constants*, are compatible if and only if $k_1 + k_2 \geqslant 0$; *we shall assume that*

(5.5) $$k_1 + k_2 > 0.$$

The initial data will be assumed to be:

(5.6) $$u_i^0 \in H^2(\mathcal{O}), \quad \frac{\partial u_i^0}{\partial \nu} = 0, \quad u_i^1 \in H^1(\mathcal{O}) \; , \quad i = 1,2 \quad ,$$

and we shall make the *compatibility assumption*

(5.7) $$u_1^1 - u_2^1 - k_2 \leqslant 0, \quad u_2^1 - u_1^1 - k_1 \leqslant 0 \quad \text{in } \mathcal{O}.$$

The *minima* assumptions for f_1 and f_2 will be

(5.8) $$f_i, \; \frac{\partial f_i}{\partial x_j} \in L^2(Q) \; , \; \forall \, j, \quad i = 1, 2 \; .$$

The first *existence theorem* is:

THEOREM 5.1: *Assume that* (5.5),...,(5.8) *hold. Then there exist* u_1 *and* u_2 *such that*

(5.9)
$$\left|\begin{array}{l} u_i \in L^\infty(0,T;H^1(\mathcal{O})), \quad \dfrac{\partial u_i}{\partial t} \in L^\infty(0,T;L^2(\mathcal{O})) \; , \\ \dfrac{\partial^2 u_i}{\partial t^2} - \Delta u_i \in L^2(Q) \quad , \end{array}\right.$$

(5.10)
$$\left|\begin{array}{l} u_1 - u_2 \in L^\infty(0,T;H^2(\mathcal{O})) \quad , \quad \dfrac{\partial(u_1-u_2)}{\partial t} \in L^\infty(0,T;H^1(\mathcal{O})) \; , \\ \dfrac{\partial^2 u_1}{\partial t^2} - \dfrac{\partial^2 u_2}{\partial t^2} \in L^2(0,T;L^2(\mathcal{O})) \end{array}\right.$$

and they satisfy (5.1),...,(5.4).

We can obtain a (*slight!*) improvement in regularity by adding an extra assumption about f:

THEOREM 5.2: *Take the conditions of Theorem 5.1, and assume, in addition, that*

(5.11) $$\frac{\partial f_i}{\partial t} \in L^2(Q), \quad i = 1,2.$$

We can then find solutions u_1, u_2 *of* (5.1),...,(5.4) *with* (5.9),(5.10) *and, in addition,*

(5.12) $$\frac{\partial^2 u_1}{\partial t^2} - \frac{\partial^2 u_2}{\partial t^2} \in L^\infty(0,T;L^2(\mathcal{O})) \; .$$

The *uniqueness* problem is still unsolved; this uniqueness does seem very likely; a result which adds some comfort to this conjecture is:

THEOREM 5.3: *Take the conditions of Theorem 5.1. Let* u_1, u_2 *and* $\hat{u}_1, \hat{u}_2$ *be two pairs of solutions of* (5.1),...,(5.4) *with* (5.9),(5.10). *Then*

(5.13) $$u_1 - \hat{u}_1 = u_2 - \hat{u}_2 \quad \textit{in } Q.$$

Remark 5.1

In the preceding statements of theorems we can replace $-\Delta$ by A, given by

(5.14) $$A\phi = - \sum \frac{\partial}{\partial x_i}\left(a_{ij} \frac{\partial \phi}{\partial x_j}\right)$$

where

(5.15)
$$\left|\begin{array}{l} a_{ij} = a_{ji} \in W^{1,\infty}(\mathcal{O}) \; , \\ \displaystyle\sum_{i,j=1}^{n} a_{ij}(x)\, \xi_i \xi_j \geq \alpha \sum_{i=1}^{n} \xi_i^2 \; , \quad x \in \mathcal{O}, \quad \alpha > 0 \; . \end{array}\right.$$

In what follows we are going to write A instead of $-\Delta$; as usual, we shall set

(5.16) $$a(u,v) = \sum_{i,j=1}^{n} \int_{\mathcal{O}} a_{ij} \frac{\partial u}{\partial x_j} \frac{\partial v}{\partial x_i}\, dx, \quad a(v) = a(v,v) \; . \; \square$$

5.3 *Proof of the Existence*

We shall set

$$\frac{\partial u_i}{\partial t} = u_i' \;, \qquad \frac{\partial^2 u_i}{\partial t^2} = u_i'' \quad \text{and} \tag{5.17}$$

$$\beta_{1\varepsilon} = u_{1\varepsilon}' - u_{2\varepsilon}' - k_2 \;, \qquad \beta_{2\varepsilon} = u_{2\varepsilon}' - u_{1\varepsilon}' - k_1 \;.$$

Consider the *penalised system* (for $\varepsilon > 0$)

$$\left| \begin{aligned} & u_{1\varepsilon}'' + Au_{1\varepsilon} + \frac{1}{\varepsilon}\beta_{1\varepsilon}^+ = f_1 \;, \\ & u_{2\varepsilon}'' + Au_{2\varepsilon} + \frac{1}{\varepsilon}\beta_{2\varepsilon}^+ = f_2 \quad \text{in} \quad Q \;, \end{aligned} \right. \tag{5.18}$$

$$\frac{\partial u_{i\varepsilon}}{\partial \nu} = 0 \quad \text{on} \quad \Sigma \;, \quad i = 1, 2 \;, \tag{5.19}$$

$$u_{i\varepsilon}(0) = u_i^0 \;, \quad \frac{\partial u_{i\varepsilon}}{\partial t}(0) = u_i^1 \;, \quad i = 1,2 \;. \tag{5.20}$$

We are going to establish some a priori estimates which will show that the problem (5.18),(5.19),(5.20) has an unique solution (the uniqueness will be shown by standard procedures not expounded here); the a priori estimates will next allow us to pass to the limit when $\varepsilon \to 0$.□

A priori estimates (I)

When no ambiguity arises we shall suppress the index "ε", in order to simplify the calculations. Hence we shall set:

$$w = u_1 - u_2, \quad u_i = u_{i\varepsilon}, \quad \beta_i = \beta_{i\varepsilon}. \tag{5.21}$$

Now,

$$w'' + Aw + \frac{1}{\varepsilon}(\beta_1^+ - \beta_2^+) = f_1 - f_2 \;. \tag{5.22}$$

Now take the scalar product of (5.22) with Aw'. Notice that

$$Aw' = A\beta_1 = - A\beta_2. \tag{5.23}$$

Therefore we obtain

$$\begin{aligned} (w'',Aw') + (Aw,Aw') + \frac{1}{\varepsilon}(\beta_1^+, A\beta_1) + \frac{1}{\varepsilon}(\beta_2^+, A\beta_2) = \\ = (f_1 - f_2, Aw') \;. \end{aligned} \tag{5.24}$$

Consequently,

$$\frac{1}{2}\frac{d}{dt}[a(w') + |Aw|^2] + \frac{1}{\varepsilon}[a(\beta_1^+) + a(\beta_2^+)] = a(f_1 - f_2, w') \tag{5.25}$$

from which we deduce that as $\varepsilon \to 0$,

(5.26) $\left|\begin{array}{l} u_{1\varepsilon} - u_{2\varepsilon} \text{ stays in a bounded set of } L^\infty(0,T;H^2(\mathcal{O})), \\ u'_{1\varepsilon} - u'_{2\varepsilon} \text{ stays in a bounded set of } L^\infty(0,T;H^1(\mathcal{O})). \end{array}\right.$

Remark 5.2

To obtain (5.26) we must assume that the boundary Γ of $\mathcal{O}$ is regular enough. Let us further note that we have used only the property

(5.27) $$f_1 - f_2 \in L^2(0,T;H^1(\mathcal{O})),$$

which allows us to generalise (5.8) a little.□

A priori estimate (II)

Let us now take the scalar product of (5.22) with

(5.28) $$w'' = \beta'_1 = -\beta'_2.$$

This yields

$$|w''|^2 + \frac{1}{\varepsilon}(\beta_1^+, \beta_1') + \frac{1}{\varepsilon}(\beta_2^+, \beta_2') = (f_1 - f_2 - Aw, w'')$$

or again,

(5.29) $$|w''(t)|^2 + \frac{1}{2}\frac{d}{dt}[|\beta_1^+(t)|^2 + |\beta_2^+(t)|^2] = (f_1 - f_2 - Aw, w'').$$

But by the compatibility relations (5.7) we have:

(5.30) $$\beta_i^+(0) = 0, \; i = 1,2$$

and we therefore deduce from (5.29) that as $\varepsilon \to 0$

(5.31) $u''_{1\varepsilon} - u''_{2\varepsilon}$ stays in a bounded set of $L^2(0,T;L^2(\mathcal{O}))$. □

A priori estimate (III)

It follows from (5.26),(5.31) and equation (5.22) that

(5.32) $$\left\| \frac{1}{\varepsilon}(\beta_1^+ - \beta_2^+) \right\|_{L^2(Q)} \leq C.$$

But using (5.5) we see that

(5.33) $$(\beta_1^+, \beta_2^+)_{L^2(Q)} = 0$$

so that when $\varepsilon \to 0$,

(5.34) $\frac{1}{\varepsilon}\beta_{i\varepsilon}^+$ stays in a bounded set of $L^2(Q)$.□

A priori estimate (IV)

From (5.18) and (5.34) it follows that

(5.35) $u''_{i\varepsilon} + Au_{i\varepsilon}$ stays in a bounded set of $L^2(Q)$.□

Passage to the limit

By the preceding a priori estimates we see that we can pick out a subsequence, again denoted u_i, such that

$$(5.36)\quad \left| \begin{array}{ll} u_{i\varepsilon} \to u_i & \text{weak star in } L^\infty(0,T;H^1(\mathcal{O})), \\ \dfrac{\partial u_{i\varepsilon}}{\partial t} \to \dfrac{\partial u_i}{\partial t} & \text{weak star in } L^\infty(0,T;L^2(\mathcal{O})). \end{array} \right.$$

and

$$(5.37)\quad \left| \begin{array}{ll} u_{1\varepsilon} - u_{2\varepsilon} \to u_1 - u_2 & \text{weak star in } L^\infty(0,T;H^2(\mathcal{O})), \\ u'_{1\varepsilon} - u'_{2\varepsilon} \to u'_1 - u'_2 & \text{weak star in } L^\infty(0,T;H^1(\mathcal{O})), \\ u''_{1\varepsilon} - u''_{2\varepsilon} \to u''_1 - u''_2 & \text{weakly in } L^2(Q). \end{array} \right.$$

Consequently we have, in particular,

(5.38) $u'_{1\varepsilon} - u'_{2\varepsilon} \to u'_1 - u'_2$ strongly in $L^2(Q)$.

By (5.34) we have:

$$\beta^+_{i\varepsilon} \to 0 \quad \text{in } L^2(Q),$$

which, when coupled with (5.38), shows that

(5.39) $u'_1 - u'_2 - k_2 \leq 0, \quad u'_2 - u'_1 - k_1 \leq 0$.

From $(5.18)_1$ we deduce that

(5.40) $u''_{i\varepsilon} + Au_{i\varepsilon} \to u''_i + Au_i$ *weakly* in $L^2(Q)$.

In (5.40),

$$(u'_{1\varepsilon} - u'_{2\varepsilon} - k_2)^- \to (u'_1 - u'_2 - k_2)^- \quad \text{strongly in } L^2(Q),$$

and by (5.39),

$$(u'_1 - u'_2 - k_2)^- = -(u'_1 - u'_2 - k_2)$$

so that in the limit (5.40) gives

$$(u''_1 + Au_1 - f_1)(u'_1 - u'_2 - k_2) = 0 .$$

We obtain a result analogous with $(5.18)_2$, which completes the proof of Theorem 5.1.□

Proof of Theorem 5.2: We now assume that (5.11) holds(1). We differentiate (5.22) with respect to t:

$$w''' + Aw' + \frac{1}{\varepsilon}\frac{\partial}{\partial t}(\beta_1^+ - \beta_2^+) = f_1' - f_2' . \tag{5.41}$$

From (5.22) (and from $\beta_i^+(0) = 0$) we deduce that

$$w''(0) = (f_1-f_2)(0)-A(u_1'-u_2') \in L^2(\mathcal{O}) . \tag{5.42}$$

Take the scalar multiplication of (5.41) with $w'' = \beta_1' - \beta_2'$, and we then deduce

$$\frac{1}{2}\frac{d}{dt}[|w''(t)|^2 + a(w'(t))] + \frac{1}{\varepsilon}\left[\left|\frac{\partial(\beta_1^+)}{\partial t}\right|^2 + \left|\frac{\partial(\beta_2^+)}{\partial t}\right|^2\right] = (f_1' - f_2', w'') \tag{5.43}$$

whence we deduce that

(5.44) $u_1'' - u_2''$ stays in a bounded set of $L^\infty(0,T;L^2(\mathcal{O}))$,

from which the result follows.□

5.4 *The Uniqueness Problem*

We take the conditions of Theorem 5.1. Notice that we can write (5.1),(5.2) in the form

$$\left|\begin{array}{l} (u_1'' + Au_1 - f_1, v_1 - u_1') \geq 0 \quad ,(u_2'' + Au_2 - f_2, v_2 - u_2') \geq 0 \\ \forall\, v_1, v_2 \in H^1(\mathcal{O}) \times H^1(\mathcal{O}), \text{ with} \\ v_1 \leq u_2' + k_2 \quad , \quad v_2 \leq u_1' + k_1 \\ \text{and with} \\ u_1' \leq u_2' + k_2 \quad , \quad u_2' \leq u_1' + k_1 \end{array}\right. \tag{5.45}$$

In the first inequality of (5.45), *for example*, the scalar product means more precisely:

$$(u_1''(t) + Au_1(t) - f_1(t), v_1 - u_1'(t)) \geq 0 \quad \text{a.e. in } t$$

The proof of (5.45) is immediate.

Then let $\hat{u}_1, \hat{u}_2$ be a (possible) second pair of solutions of (5.45), which we shall say is a solution of $(5.45)^\wedge$.

In (5.45) we then take v_1 and v_2 as:

$$\left|\begin{array}{l} v_1 = u_1' + (\hat{u}_1' - \hat{u}_2' - k_2) - (u_1' - u_2' - k_2) , \\ v_2 = u_2' + (\hat{u}_2' - k_1 - \hat{u}_1') - (u_2' - u_1' - k_1) . \end{array}\right. \tag{5.46}$$

(1) It will suffice that $\partial(f_1 - f_2)/\partial t \in L^2(Q)$.

These choices are permissible; for example,

$$v_1 = u_2' + k_2 + (\hat{u}_1' - \hat{u}_2' - k_2)$$

and since

$$\hat{u}_1' - \hat{u}_2' - k_2 \le 0,$$

so we have

$$v_1 \leqslant u_2' + k_2.$$

In (5.45)^ we take $\hat{v}_1$ and $\hat{v}_2$ by

$$\left| \begin{aligned} \hat{v}_1 &= \hat{u}_1' + (u_1' - u_2' - k_2) - (\hat{u}_1' - \hat{u}_2' - k_2) , \\ \hat{v}_2 &= \hat{u}_2' + (u_2' - u_1' - k_1) - (\hat{u}_2' - \hat{u}_1' - k_1) \end{aligned} \right. \tag{5.47}$$

a choice which is also permissible.

Now add the corresponding results; setting

$$\phi_i = u_i - \hat{u}_i \tag{5.48}$$

this yields

$$\left| \begin{aligned} &-(\phi_1'' + A\phi_1, \phi_1' - \phi_2') \ge 0 , \\ &(\phi_2'' + A\phi_2, \phi_1' - \phi_2') \ge 0 \end{aligned} \right. \tag{5.49}$$

from which we deduce that

$$(\phi_1'' - \phi_2'' + A(\phi_1 - \phi_2), \phi_1' - \phi_2') \le 0$$

i.e.,

$$\frac{d}{dt} [|\phi_1' - \phi_2'|^2 + a(\phi_1 - \phi_2)] \le 0 .$$

Since $\phi_1(0) = \phi_2(0) = (0)$, from which we deduce that

$$\phi_1 - \phi_2 = 0, \tag{5.50}$$

which proves Theorem 5.3.□

Remark 5.3

From (5.13) we deduce that the sets

$$E_1 = \{x,t \mid x,t \in Q, \ u_1' = u_2' + k_2 \} \tag{5.51}$$

and

$$E_2 = \{x,t \mid x,t \in Q, \ u_2' = u_1' + k_1 \} \tag{5.52}$$

are defined uniquely. Now,

(5.53) $$u_i'' + Au_i = f_i \quad \text{in} \quad Q\setminus E_i \ , \quad i = 1,2$$

and the identical equations for $\hat{u}_i$ in the same set $Q\setminus E_i$. But it is not obvious that u_i and $\hat{u}_i$ satisfy the same boundary conditions on ∂E_i nor, furthermore, that it has a meaning *for want of a regularity result on the free boundary* ∂E_i.□

6. PARABOLIC Q.V.I.'s OF THE SECOND KIND AND REMARKS

6.1 *Parabolic Q.V.I.'s of the Second Kind*

We keep the notations of Section 5, with A given by (5.14) with (5.15). We seek u_1 and u_2 such that

(6.1) $$\left| \begin{array}{l} u_1' + Au_1 - f_1 \le 0 \ , \quad u_1' - u_2' - k_2 \le 0 \ , \\ (u_1' + Au_1 - f_1)(u_1' - u_2' - k_2) = 0 \quad \text{in} \quad Q \ , \end{array} \right.$$

(6.2) $$\left| \begin{array}{l} u_2' + Au_2 - f_2 \le 0 \ , \quad u_2' - u_1' - k_1 \le 0 \ , \\ (u_2' + Au_2 - f_2)(u_2' - u_1' - k_1) = 0 \quad \text{in} \quad Q \end{array} \right.$$

with the *boundary conditions*

(6.3) $$\frac{\partial u_i}{\partial \nu_A} = 0 \quad \text{on} \quad \Sigma \ , \quad i = 1,2$$

and with the *initial conditions*

(6.4) $$u_i(0) = u_i^0 \quad \text{on} \quad \Omega \ , \quad i = 1,2 \ .$$

We are going to prove:

THEOREM 6.1: *Assume that*

(6.5) $$f_i \in L^2(0,T;H^1(\mathcal{O})), \ \frac{\partial f_i}{\partial t} \in L^2(Q) \ , \quad i = 1,2$$

(6.6) $$u_i^0 \in H^2(\mathcal{O}) \ , \ i = 1,2, \text{ with } f_1(0)-f_2(0)-A(u_1^0-u_2^0) \in H^1(\mathcal{O}) \ .$$

Then there exist functions u_1 *and* u_2 *such that*

(6.7) $$u_i \in L^2(0,T;H^2(\mathcal{O})), \ \frac{\partial u_i}{\partial t} \in L^2(Q) \ , \quad i = 1,2$$

(6.8) $$\left| \begin{array}{l} u_1 - u_2 \in L^\infty(0,T;H^2(\mathcal{O})) \ , \\ u_1' - u_2' \in L^\infty(0,T;H^1(\mathcal{O})) \ , \ u_1'' - u_2'' \in L^2(Q) \ , \end{array} \right.$$

satisfying (6.1),...,(6.4) *also.*

Proof: Consider the penalised system

(6.9) $$u_{1\varepsilon}' + Au_{1\varepsilon} + \frac{1}{\varepsilon}\beta_1^+ = f_1 \ , \quad u_{2\varepsilon}' + Au_{2\varepsilon} + \frac{1}{\varepsilon}\beta_2^+ = f_2 \quad \text{in} \quad Q$$

where, as in Section 5,

$$(6.10) \qquad \beta_{1\varepsilon} = u'_{1\varepsilon} - u'_{2\varepsilon} - k_2 \, , \quad \beta_{2\varepsilon} = u'_{2\varepsilon} - u'_{1\varepsilon} - k_1$$

and where the boundary and initial conditions are analogous to (6.3) and (6.4). When no ambiguity arises we leave out the index ε. We establish a priori estimates which prove the existence (and the uniqueness) of a solution of (6.9),(6.3),(6.4) for $\varepsilon > 0$, and which allow us to pass to the limit.

Setting

ww $$w = u_{1\varepsilon} - u_{2\varepsilon},$$

we obtain

$$(6.11) \qquad w' + Aw + \frac{1}{\varepsilon}(\beta_1^+ - \beta_2^+) = f_1 - f_2$$

Taking the scalar product with $Aw' = A\beta_1 = - A\beta_2$, this yields

$$(6.12) \qquad a(w'(t)) + \frac{1}{2}\frac{d}{dt}|Aw(t)|^2 + \frac{1}{\varepsilon}\, a(\beta_1^+) + a(\beta_2^+) = a(f_1 - f_2, w')$$

from which we deduce that as $\varepsilon \to 0$,

$$(6.13) \qquad \left| \begin{array}{ll} A(u_{1\varepsilon} - u_{2\varepsilon}) & \text{stays in a bounded set of } L^\infty(0,T;L^2(\mathcal{O})), \\ u'_{1\varepsilon} - u'_{2\varepsilon} & \text{stays in a bounded set of } L^2(0,T;H^1(\mathcal{O})). \end{array} \right.$$

From (6.13) and (6.11) we deduce that

$$\frac{1}{\varepsilon}(\beta_1^+ - \beta_2^+) = f_1 - f_2 - w' - Aw \quad \text{stays in a bounded set of } L^2(Q),$$

and consequently

$$(6.14) \qquad \frac{1}{\varepsilon}\beta_i^+ \quad \text{stays in a bounded set of } L^2(Q), \ i = 1,2$$

Then (6.9) and (6.14) (and the initial and boundary conditions for $u_{i\varepsilon}$) imply that

$$(6.15) \qquad \left| \begin{array}{ll} u_{i\varepsilon} & \text{stays in a bounded set of } L^2(0,T;H^2(\mathcal{O})), \\ \dfrac{\partial u_{i\varepsilon}}{\partial t} & \text{stays in a bounded set of } L^2(Q), \ i = 1,2 \end{array} \right.$$

It is not obvious that these estimates are sufficient for passing to the limit in ε. But we can obtain additional estimates by differentiating (6.11) with respect to t. This yields:

$$(6.16) \qquad w'' + Aw'' + \frac{1}{\varepsilon}\frac{\partial}{\partial t}(\beta_1^+ - \beta_2^+) = f'_1 - f'_2 \, .$$

From (6.11) we deduce that

$$(6.17) \qquad w'(0) + \frac{1}{\varepsilon}[(w'(0) - k_2)^+ - (-w'(0) - k_1)^+] = g$$

where

$$g = (f_1 - f_2)(0) - A(u_1^0 - u_2^0) \in H^1(\mathcal{O}) \tag{6.18}$$

by (6.6). The solution of (6.17) gives:

$$w'(x,0) = g(x) \quad \text{if} \quad -k_1 \le g(x) \le k_2 ,$$

$$w'(x,0) = \frac{k_2 + \varepsilon g(x)}{1+\varepsilon} \quad \text{if} \quad g(x) \ge k_2 ,$$

$$w'(x,0) = \frac{-k_1 + \varepsilon g(x)}{1+\varepsilon} \quad \text{if} \quad g(x) \le -k_1 ,$$

so that

(6.19) $w'(0)$ stays in a bounded set of $H^1(\mathcal{O})$ when $\varepsilon \to 0$.

Taking the scalar product of (6.16) with w'' yields:

$$|w''(t)|^2 + \frac{1}{2}\frac{d}{dt} a(w'(t)) + \frac{1}{\varepsilon}\left[\left|\frac{\partial(\beta_1^+)}{\partial t}\right|^2 + \left|\frac{\partial(\beta_2^+)}{\partial t}\right|^2\right] = (f_1' - f_2', w'') \tag{6.20}$$

from which, using (6.19), we deduce that

(6.21)
$$\begin{cases} u_{1\varepsilon}' - u_{2\varepsilon}' & \text{stays in a bounded set of } L^\infty(0,T;H^1(\mathcal{O})), \\ u_{1\varepsilon}'' - u_{2\varepsilon}'' & \text{stays in a bounded set of } L^2(Q). \end{cases}$$

We can now pick out a subsequence from $u_{i\varepsilon}$, again denoted $u_{i\varepsilon}$, such that

(6.22)
$$\begin{cases} u_{i\varepsilon} \to u_{i\varepsilon} & \text{weakly in } L^\infty(0,T;H^2(\mathcal{O})), \\ u_{i\varepsilon}' \to u_{i\varepsilon}' & \text{weakly in } L^2(Q),\ i = 1,2, \end{cases}$$

(6.23)
$$\begin{cases} u_{1\varepsilon} - u_{2\varepsilon} \to u_1 - u_2 & \text{weak star in } L^\infty(0,T;H^2(\mathcal{O})) \\ u_{1\varepsilon}' - u_{2\varepsilon}' \to u_1' - u_2' & \text{weak star in } L^\infty(0,T;H^1(\mathcal{O})) \\ u_{1\varepsilon}'' - u_{2\varepsilon}'' \to u_1'' - u_2'' & \text{weakly in } L^2(Q). \end{cases}$$

In particular it follows from (6.23) that

(6.24) $u_{1\varepsilon}' - u_{2\varepsilon}' \to u_1' - u_2'$ *strongly* in $L^2(Q)$,

and we can then pass to the limit, as in Theorem 5.1, which proves the Theorem.□

Remark 6.1

Uniqueness is an unsolved problem. Exactly as in Theorem 5.3, we shall show that:

(6.25) | if u_1, u_2 and $\hat{u}_1, \hat{u}_2$ are two pairs of solutions of (6.1),...,(6.4) with the properties of Theorem 6.1, then $u_1 - \hat{u}_1 = u_2 - \hat{u}_2$.□

Remark 6.2

The problem under consideration can be written in the Q.V.I. form:

$$(6.26)\quad \left|\begin{array}{l} (u_1'+ Au_1- f_1,\ v_1-u_1') \geq 0,\ (u_2'+ Au_2- f_2,\ v_2-u_2') \geq 0\ , \\ \forall\ v_1,v_2 \text{ with } v_1 \leq u_2'+ k_2,\ v_2 \leq u_1'+ k_1, \text{ and with} \\ u_1' \leq u_2'+ k_2,\ u_2' \leq u_1'+ k_1,\ v_i \in H^1(\mathcal{O})\ , \end{array}\right.$$

and the conditions (6.4).

This is a Q.V.I. for a *parabolic* operator, but in the *'constraints' some derivatives with respect to* t *appear*. In the case of V.I.'s, this situation has been called a "V.I. of the Second Kind" in G. DUVAUT - J.L. LIONS [1]. This is why we propose the terminology "Q.V.I. of the Second Kind" for Q.V.I.'s of the type (6.26).□

6.2 *Various Remarks*

Remark 6.3

Using the same notations as before, we can consider the hyperbolic Q.V.I.

$$(6.27)\quad \left|\begin{array}{l} (\gamma u_1'' + u_1'+ Au_1- f_1,\ v_1-u_1') \geq 0\ , \\ (\gamma u_2'' + u_2'+ Au_2- f_2,\ v_2-u_2') \geq 0 \\ \forall\ v_1,\ v_2,\ v_1 \leq u_2' + k_2,\ v_2 \leq u_1'+ k_1, \text{ with } u_1' \leq u_2'+ k_2\ , \\ u_2' \leq u_1'+ k_1\ ,\ \text{où}\ \gamma > 0\ ; \end{array}\right.$$

and the initial conditions are as in Section 5.

We can then find a solution $u_{1\gamma},u_{2\gamma}$ of (6.27) such that (under the assumptions of Theorem 6.1) $u_{1\gamma},u_{2\gamma} \to u_1,u_2$ when $\gamma \to 0$, for the weak (or weak star) topology corresponding to the properties (6.7),(6.8), where u_1,u_2 is a solution of the problem (6.1),...,(6.4).□

Remark 6.4

Here is an example (constructed ad hoc, to be sure!) of a hyperbolic Q.V.I. where we can prove existence *and* uniqueness: we introduce

$$(6.28)\quad M(v) = \lambda\int_{\mathcal{O}} vdz + \mu,\quad \mu \text{ and } \lambda \text{ constants} > 0;$$

here we assume that $\mathcal{O}$ *is bounded* (which was not essential in the preceding results). Then if f is given, with

$$(6.29)\quad f,\ \frac{\partial f}{\partial t} \in L^2(Q),$$

and if we assume that

$$(6.30)\quad \lambda \text{meas}(\mathcal{O}) < 1,$$

we can prove that *there exists one and only one function* u *such that*

$$(6.31)\quad \left|\begin{array}{l} u \in L^\infty(0,T;H^2(\mathcal{O})),\quad u' \in L^\infty(0,T;H^1(\mathcal{O}))\ , \\ u'' \in L^2(Q) \end{array}\right.$$

with

(6.32) $$\left|\begin{array}{l} u''-\Delta u-f \le 0\ ,\quad u'-M(u') \le 0\ , \\ (u''-\Delta u-f)(u'-M(u')) = 0 \quad \text{in} \quad Q \end{array}\right.$$

with the *initial conditions*

(6.33) $$u(0) = u'(0) = 0 \quad \text{in } \Omega,$$

and the *boundary conditions*

(6.34) $$\left|\begin{array}{l} \frac{\partial u}{\partial \nu} \le 0\ ,\quad \frac{\partial u}{\partial t} - M\,(\frac{\partial u}{\partial t}) \le 0\ , \\ (\frac{\partial u}{\partial \nu})(\frac{\partial u}{\partial t} - M(\frac{\partial u}{\partial t})) = 0 \quad \text{on} \quad \Sigma\ . \end{array}\right.$$

The Q.V.I. formulation is then

(6.35) $$\left|\begin{array}{l} (u'',v-u')+a(u,v-u') \ge (f,v-u') \quad \forall\, v \in H^1(\mathcal{O}) \quad \text{with} \\ v \le M(u')\ ,\quad u' \le M(u') \end{array}\right.$$

and initial conditions (6.33).□

Remark 6.5

The following set of inequalities:

(6.36) $$\left|\begin{array}{l} u_1''-\Delta u_1-f_1 \le 0,\quad u_1-k_2-u_2 \le 0\ , \\ (u_1''-\Delta u_1-f_1)(u_1-k_2-u_2) = 0 \quad \text{in} \quad Q\ , \end{array}\right.$$

(6.37) $$\left|\begin{array}{l} u_2''-\Delta u_2-f_2 \le 0,\quad u_2-k_1-u_1 \le 0\ , \\ (u_2''-\Delta u_2-f_2)(u_2-k_1-u_1) = 0 \quad \text{in} \quad Q \end{array}\right.$$

with the same boundary and initial conditions as in the preceding case, seems to be an unsolved problem. [As to the problem in Section 5, the constraints of the type $u_1' - k_2 - u_2' \leqslant 0$ have been replaced by $u_1 - k_2 - u_2 \leqslant 0$.]□

Remark 6.6

Another unsolved problem is the following: consider the 'model' operator M for impulse control:

(6.38) $$Mu(x) = k + \inf u(x+\xi), \qquad \xi \geqslant 0;$$

we then look for the solution u (using the notation (5.16)) of

(6.39) $$\left|\begin{array}{l} (u'',v-u')+a(u,v-u') \ge (f,v-u') \quad \forall\, v \in H^1(\mathcal{O})\ , \\ v \le Mu'\ ,\quad u' \le M(u')\ , \end{array}\right.$$

with u(0) and u'(0) given.

We shall ignore the matter of whether this problem has a solution (even the question of replacing M(u') by M(u) and $u' \leqslant M(u')$ by $u \leqslant M(u)$ in (6.39)). It seems very likely, at least in one-dimensional space, that this is the case. In, fact, if, *in dimension one* we consider the *penalised equation*

(6.40) $$u''_\varepsilon - \frac{\partial^2 u_\varepsilon}{\partial x^2} + \frac{1}{\varepsilon}(u'_\varepsilon - M(u'_\varepsilon))^+ = f$$

(M being given by (6.38)), with

(6.41) $$\frac{\partial u_\varepsilon}{\partial x} = 0 \quad \text{on the boundary } \partial\mathcal{O},$$

with $u_\varepsilon(0)$ and $u'_\varepsilon(0)$ being given, we then obtain an a priori estimate by taking the scalar product of (6.40) with

$$Au' = -\frac{\partial^2 u'_\varepsilon}{\partial x^2} ;$$

in fact this yields

(6.42) $$\left| \begin{aligned} &\frac{1}{2}\frac{d}{dt}[a(u'_\varepsilon(t)) + |Au'_\varepsilon(t)|^2] + \frac{1}{\varepsilon}((u'_\varepsilon - M(u'_\varepsilon))^+, Au'_\varepsilon) = \\ &\qquad = (f, Au'_\varepsilon) = a(f, u'_\varepsilon) \end{aligned} \right.$$

(assuming that $f \in L^2(0,T;H^1(\mathcal{O}))$). But if $\frac{\partial v}{\partial \nu} = 0$:

$$\begin{aligned} ((v-M(v))^+, -\Delta v) &= (\frac{d}{dx}(v-M(v))^+, \frac{dv}{dx}) = \\ &= (\frac{d}{dx}(v-M(v))^+, \frac{d(v-M(v))}{dx}) + (\frac{d}{dx}(v-M(v))^+, \frac{dM(v)}{dx}) \end{aligned}$$

and if $v(x_0) > M(v)(x_0)$, then $M(v)$ = constant in the neighbourhood of x_0, so that

(6.43) $$((v-M(v))^+, -\Delta v) = |\frac{d}{dx}(v-M(v))^+|^2 .$$

From (6.42),(6.43) it follows that when $\varepsilon \to 0$,

(6.44) $$\left| \begin{aligned} &u_\varepsilon \quad \text{stays in a bounded set of } L^\infty(0,T;H^2(\mathcal{O})), \\ &u'_\varepsilon \quad \text{stays in a bounded set of } L^\infty(0,T;H^1(\mathcal{O})), \end{aligned} \right.$$

(6.45) $$\frac{1}{\sqrt{\varepsilon}}(u'_\varepsilon - M(u'_\varepsilon))^+ \quad \text{stays in a bounded set of } L^2(Q).$$

We also have

$$|u'_\varepsilon(T)| \leqslant C,$$

so that *if $\mathcal{O}$ is bounded*, from (6.40) we deduce that

$$\frac{1}{\varepsilon}\int_Q (u'_\varepsilon - M(u'_\varepsilon))^+ dx\, dt = \int_Q f\, dx\, dt - \int_{\mathcal{O}} u'_\varepsilon(T) dx + \int_{\mathcal{O}} u'(x) dx$$

whence

$$\frac{1}{\varepsilon}(u'_\varepsilon - M(u_\varepsilon))^+ \text{ stays in a bounded set of } L^1(Q). \tag{6.46}$$

These a priori estimates do not seem to be sufficient for proving the existence of a solution, but they do render the existence very plausible.□

7. COMMENTS

The profitable concept of a weak solution of Q.V.I.'s is that of the maximum solution. Here there is not, as there is in the elliptic case, a concept of minimum solution. For example, in the case of bounded data the weak Q.V.I. can have several solutions, the good solution is that which is the maximum, although the stationary Q.V.I. has an unique solution. The uniqueness is obtained whenever we impose conditions of time-regularity.

The Levy-Stampacchia has been proved for the spaces H^1 and H_0^1. In P. CHARRIER and G.M. TROIANIELLO [1], there will be found the equivalent for an intermediate space V (essentially functions vanishing on a piece of the boundary). The proof of regularity of the maximum weak solution is more delicate than in the elliptic case, as explained in Section 2.2. M. MATZEU - M.A. VIVALDI [1] have proved the regularity of the strong solution. The employment of the penalised problem for proving regularity apparently is a new technique.

The case where the operator is perturbed by a quadratic Hamiltonian remains a problem that is still largely unsolved in the evolutionary case (cf. Section 3, Chapter 4).

The results of Sections 3 and 4 are presented for the first time here. The non-linear semi-groups of Sections 3 and 4, and especially the proof of Trotter's formula, were motivated by the the thesis of L. BARTHELEMY [1], which, by starting from accretive properties and the general theory of non-linear contraction semi-groups, defines a contraction semi-group by Trotter's formula. This is a semi-group in L^∞. This semi-group give a concept of a solution for a parabolic Q.V.I.. This concept coincides, in fact, with that of the strong solution of a Q.V.I.. This result is compared with that indicated in Theorem 4.2, where we have shown that Trotter's formula is valid for subsolutions of the stationary problem.

It will also be noted that the existence of a strong solution is proved in Theorem 1.3 under similar assumptions.

It would be very interesting to study the "infinitesimal generator" of the non-linear semi-groups introduced in this Chapter.

Chapter 6
Impulse control and the interpretation of quasi-variational inequalities

INTRODUCTION

In this Chapter we give the connection between Q.V.I.'s and the stochastic control problems that motivated their introduction. So as not to burden the reader we shall limit ourselves to the essentials. Whilst assimilating the methods of the present Chapter the reader could easily interpret other cases. Section 1 is devoted to the interpretation of the impulse control Q.V.I. in the elliptic case and under Dirichlet conditions. The evolutionary case does not pose particular problems (at the level of the interpretation), but, of course, we interpret the 'good solution' of the parabolic Q.V.I., namely the maximum solution. Section 2 is devoted to the interpretation of increasing and decreasing iterative methods. They both correspond to very natural impulse control problems.

Section 3 is devoted to the case of an unbounded open set. We take R^n, for simplicity. The new and interesting case to consider corresponds to the case where the data are not bounded (when they are, the case is the standard one, i.e., very similar to that of Section 1). The difficulty arises from having several solutions of the Q.V.I. in an adequate interval of functions. In particular, there is one and only one maximum solution.

We then give the probabilistic interpretation of *two* (the maximum and minimum) solutions. However, we show that the more interesting solution is the *minimum* solution. We must not confuse this case with that of the weak *parabolic* Q.V.I., where the 'good' solution was the *maximum* solution.

Section 4 is devoted to the probabilistic interpretation of the semi-group formulations of problems with an (explicit or implicit) obstacle. Thus we have a framework which regroups not only the problem considered in Section 1 (diffusion processes stopped on the boundary of an open set), but also reflected diffusion processes, and the diffusion processes with jumps of Chapter 2 and 3.

The generality is paid for by a probabilistic model that is a bit weighty, but which can be dispensed with in particular, in the case of continuous processes, or where we have a trajectorial representation.

1. THE PROBLEM OF STATIONARY IMPULSE CONTROL WITH STOP AT THE EXIT FROM AN OPEN SET

General Approach

In this section our objective is to interpret the Q.V.I. (1.7) of Chapter 4 in terms of the assumptions of Theorem 1.2, Chapter 4, with $V = H_0^1$.

1.1 *Assumptions: Notations: The Problem*

To make the reading easier, we restate the assumptions and notations of theorem 1.2, Chapter 4:

$$
(1.1)\quad \left|\begin{array}{l} a_{ij} = a_{ji} \in W^{1,\infty}(R^n) \quad , \; a_i,\, a_0 \in L^{\infty}(R^n) \\ \Sigma\; a_{ij}\, \xi_i\, \xi_j \geq \alpha\, |\xi|^2 \quad , \quad \forall\, \xi \in R^n \,, \; \alpha > 0 \,, \\ a_0(x) \geq \beta > 0 , \end{array}\right.
$$

$$
(1.2)\quad \left|\begin{array}{l} f \in L^p(\mathcal{O}),\; f \geq 0,\; p > \frac{n}{2} \quad ; \quad \mathcal{O} \text{ a regular bounded open set} \\ \text{of } R^n \,, \end{array}\right.
$$

$$
(1.3)\quad \left|\begin{array}{l} k > 0, \quad c_0 : (R^n)^+ \to R^+, \text{ continuous, } c_0(0) = 0 \,, \\ \text{sub-linear, non-decreasing.} \end{array}\right.
$$

After having defined

$$
(1.4)\quad a(u,v) = \sum_{i,j} \int_{\mathcal{O}} a_{ij}(x) \frac{\partial u}{\partial x_j} \frac{\partial v}{\partial x_i}\, dx + \sum_i \int_{\mathcal{O}} a_i \frac{\partial u}{\partial x_i} v \; dx + \int_{\mathcal{O}} a_0 u\, v\, dx \,, \quad u,v \in H^1(\mathcal{O})
$$

$$
(1.5)\quad M\phi(x) = k + \inf_{\substack{\xi \geq 0 \\ x+\xi \in \bar{\mathcal{O}}}} [\phi(x+\xi) + c_0(\xi)]
$$

we proved in Theorem 1.2 that there exists one and only one solution of the problem

$$
(1.6)\quad \left|\begin{array}{l} a(u,v-u) \geq (f,v-u) \quad , \quad u \in H_0^1 \,, \; u \geq 0, \\ u \in C^0(\bar{\mathcal{O}}) \,, \quad u \leq Mu \,, \\ \forall\, v \in H_0^1 \,, \quad v \leq Mu \quad \text{p.p.} \end{array}\right.
$$

We shall now define an *impulse control* problem. To do this we shall write

$$
(1.7)\quad \sigma \in W^{1,\infty} \qquad \frac{\sigma^2}{2} = a \quad (\text{matrix } a_{ij})
$$

$$
(1.8)\quad b_i = - a_i + \sum_j \frac{\partial a_{ij}}{\partial x_j}
$$

so that

$$
\begin{aligned} A &= - \sum_{i,j} \frac{\partial}{\partial x_i} a_{ij} \frac{\partial}{\partial x_j} + \sum_i a_i \frac{\partial}{\partial x_i} \\ &= - \sum_{i,j} a_{ij} \frac{\partial^2}{\partial x_i \partial x_j} - \sum_i b_i \frac{\partial}{\partial x_i} \,. \end{aligned}
$$

Let us introduce

$$(1.9)\quad \left|\begin{array}{l} \Omega_0 = C([0,\infty);R^n) \;,\quad x(s,\omega) \equiv \omega(s) \;, \\ \mathcal{M}_0 = \sigma(x(s), s \geq 0) \quad , \quad \mathcal{M}_s^t = \sigma(x(\lambda),\ s \leq \lambda \leq t), \\ \mathcal{M}_s = \mathcal{M}_s^\infty \;,\quad \mathcal{M}^s = \mathcal{M}_0^s \;. \end{array}\right.$$

So we may have

(1.10) P probability on $(\Omega_0, \mathcal{M}_0)$

$w_0(t)$ n-dimensional standard Wiener process.

Since σ is Lipschitz continuous we can solve the Ito equation

$$(1.11)\quad dx = \sigma(x(t))dw_0(t), \quad x(0) = x.$$

We give the name of impulse control to the following set

$$(1.12)\quad W = \begin{array}{l} \theta^1, \theta^2, \ldots, \theta^n, \ldots \\ \xi^1, \xi^2, \ldots, \xi^n, \ldots \end{array}$$

θ^n stopping time of $\mathcal{M}^t$

ξ^n R.V. $\mathcal{M}^{\theta^n}$ measurable, $(R^n)^+$ - valued

$$\left|\begin{array}{l} \theta^n \leq \theta^{n+1} \\ \theta^n \to +\infty \quad \text{p.s.} \quad (\theta^n = +\infty \text{ is possible}). \end{array}\right.$$

For a given impulse control W we can solve the equation

$$(1.13)\quad \left|\begin{array}{l} dx = \sigma(x(t))dw_0(t) + \Sigma\, \delta(t-\theta^i)\xi^i \\ x(0) = x \;. \end{array}\right.$$

We actually interpret (1.13) in the following way. First we solve a set of Ito equations with random initial conditions, say

$$(1.14)\quad \left|\begin{array}{l} dx^n = \sigma(x^n)dw_0 \\ x^n(\theta^n) = x^{n-1}(\theta^n) + \xi^n \;, \quad n \geq 1 \\ dx^0 = \sigma(x^0)dw_0 \\ x^0(0) = x \;. \end{array}\right.$$

Then

$$(1.15)\quad \begin{array}{l} \text{if } \theta^n < \theta^{n+1} \\ x(t) = x^n(t), \quad \forall\, t \in [\theta^n, \theta^{n+1}[\;; \\ \text{if } \theta^n = \theta^{n+1}, \text{ and (to fix our ideas)} \end{array}$$

$$\theta^{n-1} < \theta^n = \theta^{n+1} < \theta^{n+2}$$

$$x(\theta^{n+1}-0) = x^{n-1}(\theta^{n+1})$$

$$x(\theta^{n+1}) = x^{n+1}(\theta^{n+1})$$

$$= x(\theta^{n+1}-0) + \xi^n + \xi^{n+1} .$$

Analogous definitions are to be applied when more than 2 impulse instants coincide. Let us note that the last condition of (1.12) implies that at most a finite number of impulses may coincide within a finite distance.

We have thus defined an adapted (to $\mathcal{M}^s$) cadlag process that is the solution of

$$\text{(1.16)} \qquad x(t) = x + \int_0^t \sigma(x(s))dw_0(s) + \xi^1 + \dots\xi^{\nu_t}$$
$$\nu_t = \max\{ n \mid \theta^n \le t \} ,$$

which is the integral formulation of (1.13). The uniqueness is easy, because the impulses do not depend upon x(.).

We now want to complete (1.16) with the help of a drift term. For this we use Girsanov's transformation. Now define P^x_W by

$$\text{(1.17)} \qquad \left.\frac{d\,P^x_W}{dP}\right|_{\mathcal{M}^t} = \exp\{ \int_0^t \sigma^{-1}(x(s))b(x(s))dw_0(s) - \frac{1}{2}\int_0^t |\sigma^{-1}b(x(s))|^2\, ds\}$$

and set

$$\text{(1.18)} \qquad w(t) = w^0(t) - \int_0^t \sigma^{-1}b(x(s))ds$$

and for the system $(\Omega_0, \mathcal{M}_0, \mathcal{M}^t, P^x_W, w)$ the process x is the solution of the equation

$$\text{(1.19)} \qquad x(t) = x + \int_0^t b(x(s))ds + \int_0^t \sigma(x(s))dw(s) + \xi^1 + \dots + \xi^{\nu_t}$$
$$\text{where } \nu_t = \max\{ n \mid \theta^n \le t \} .$$

Notice that

$$\text{(1.20)} \qquad \theta^n \to \infty,\ P^x_W \text{ a.s.}$$

In fact

$$(\theta^n \uparrow \infty) = \bigcup_p (\theta^N \leqslant p, \forall n)$$

and

$$P(\theta^n \leqslant p, \forall n) = 0$$

implies

$$P_W^x(\theta^n \leqslant p, \forall n) = 0$$

whence the result (1.20).

Finally we define a cost functional by the formula

$$(1.21)\quad \begin{aligned} J^x(W) = E_W^x \Big[&\int_0^\tau f(x(t))(\exp - \int_0^t a_0(x(s))ds)dt + \\ &+ \sum_n (k + c_0(\xi^n))(\exp - \int_0^{\theta^n} a_0(x(s))ds)\, \chi_{\theta^n<\infty} \Big] \end{aligned}$$

where τ is defined by

$$(1.22)\quad \tau = \inf\{t \mid x(t^-) \text{ where } x(t) \notin \mathcal{O}\},$$

in fact τ depends on x (the initial condition) and $W, \tau \equiv \tau_W^x$. Since the process x is right continuous and $\mathcal{O}$ is a bounded open set, τ is an $\mathcal{M}^t$ stopping time (cf., E. DYNKIN [2], p. 188).

1.2 *Statement of Results*

THEOREM 1.1: *Assume* (1.1), (1.2), (1.3). *The solution* u *of* (1.6) *is then explicitly given by*

$$(1.23)\quad u(x) = \inf_W J_x(W).$$

Furthermore, there exists an optimal impulse control. □

Here, let us give the definition of the optimal control $\hat{W}$. First of all we define a function $\hat{\xi}(x)$ such that

$$(1.24)\quad \left| \begin{array}{l} \hat{\xi}(x) \text{ is positive Borel, } x + \hat{\xi}(x) \in \bar{\mathcal{O}}, \\ \forall x \in \bar{\mathcal{O}}, \\ Mu(x) = k + c_0(\hat{\xi}(x)) + u(x + \hat{\xi}(x)), \quad \forall x \in \bar{\mathcal{O}}. \end{array} \right.$$

Such a mapping exists because u, c_0 are continuous, and $\bar{\mathcal{O}}$ is bounded. Next we define

$$\left| \begin{array}{l} dx^0 = \sigma(x^0)\, dw_0 \\ x^0(0) = x \end{array} \right.$$

$$\left| \begin{array}{l} S^0 = \text{exit time of } x^0 \text{ from } \mathcal{O} \\ T^0 = \inf\{t \mid u(x^0(t)) = Mu(x^0(t))\} \end{array} \right.$$

$$\left| \begin{array}{l} \hat{\theta}^1 = T^0, \text{ if } T^0 < S^0, +\infty \text{ otherwise} \\ \hat{\xi}^1 = \hat{\xi}(x^0(\hat{\theta}^1)) \text{ if } \hat{\theta}^1 < \infty, \text{ arbitrary if } \hat{\theta}^1 = +\infty. \end{array} \right.$$

Since $Mu|_\Gamma = k$, the point $x^0(\hat{\theta}^1) \in \mathcal{O}$, if $\hat{\theta}^1 \leqslant \infty$. Having defined the process x^{n-1} as well as $\hat{\theta}^n$, $\hat{\xi}^n$, we define x^n, $\hat{\theta}^{n+1}$, ξ^{n+1} as follows

$$(1.25)\quad \Big| \; dx^n = \sigma(x^n)dw^0 \qquad x^n(\hat{\theta}^n) = x^{n-1}(\hat{\theta}^n) + \hat{\xi}^n$$

and

(1.26)
$$\left|\begin{array}{l} S^n = \text{first exit time of } x^n \text{ from } \mathcal{O} \\ \text{after } \hat{\theta}^n \\ T^n = \inf\{\, t \geq \hat{\theta}^n \mid u(x^n(t)) = Mu(x^n(t))\} \\ \hat{\theta}^{n+1} = T^n \quad \text{si} \quad T^n < S^n, \quad + \infty \text{ otherwise} \\ \hat{\xi}^{n+1} = \hat{\xi}(x^n(\hat{\theta}^{n+1})), \text{ if } \hat{\theta}^{n+1} < \infty \text{ , arbitrary} \\ \qquad\qquad \text{if } \hat{\theta}^{n+1} = + \infty . \end{array}\right.$$

Then the control

(1.27)
$$\hat{W} = (\hat{\theta}^1, \hat{\xi}^1; \ldots ; \hat{\theta}^n, \hat{\xi}^n; \ldots)$$

is optimal.

1.3 *Additional Results on Optimal Stopping Time Problems*

We are going to need some additional results on the problems of optimal stopping times.

Consider a "usual" system $\Omega, \mathcal{Q}, P, \mathfrak{F}^t, w(t)$. Let there be two stopping times θ, θ' such that $\theta \leqslant \theta'$ a.s., and let there be a R.V. η that is $\mathfrak{F}^\theta$ measurable and is R^n-valued.

We consider a process $y(t)$ such that

(1.28)
$$\left|\begin{array}{l} y(t) \text{ is continuous} \\ (1-\chi_\theta(t))y(t) \quad \text{is adapted} \\ y(t) = \eta + \int_0^t (1-\chi_\theta(s))\beta(s)ds + \int_0^t (1-\chi_\theta(s))\sigma(y(s))dw(s) \end{array}\right.$$

where β is an adapted process that is bounded (by a deterministic constant). It is clear that we have

(1.29)
$$y(t) = \eta \text{ for } t \leqslant \theta.$$

We write τ_θ for the first exit time after θ of the process $y(t)$ from the open set $\mathcal{O}$. τ_θ is then a stopping time. In fact, let $x_0 \in \mathcal{O}$ be arbitrary, and let us define

$$z(t) = x_0\, \chi_\theta(t) + (1-\chi_\theta(t))y(t) .$$

Then $z(t)$ is an adapted process that is cadlag, whose unique point of discontinuity is θ. Write σ_s for the first exit time from $\mathcal{O}$ of the process $z(t)$ after s. Then we have

$$\sigma_\theta = \tau_\theta \text{ since } z(t) = y(t) \text{ for } t \geqslant \theta.$$

Also let

$$\hat{\sigma}_s = \inf\{\, t \geqslant s \mid z(t^-) \text{ where } z(t) \notin \mathcal{O}\}$$

which Dynkin calls the first exit time from the *interior* of $\mathcal{O}$. Because of the special form of the process $z(t)$ we easily verify that

$$\hat{\sigma}_s = \sigma_s .$$

But as $\mathcal{O}$ is a bounded open set and z is cadlag, by E. DYNKIN [2], p. 188, $\hat{\sigma}_s$ is an $\mathfrak{F}^t$ stopping time.

Now since $x_0 \in \mathcal{O}$, $\sigma_\theta = \sigma_0$, and thus σ_θ is certainly an $\mathfrak{F}^t$ stopping time.

We shall begin with some preliminary results.

LEMMA 1.1: *Let ϕ be an adapted bounded set, θ a stopping time of $\mathfrak{F}^t$. Then*

$$\chi_{\theta<\infty}\int_{0}^{\theta\wedge T} \phi(t)dw(t)$$

converges, when $T \to +\infty$, a.s. to a R.V. which by definition is

$$\chi_{\theta<\infty}\int_0^{\theta} \phi(t)dw(t).$$

Proof: Let $\Omega_N = \{\omega \mid \theta < N\}$, then $\Omega_N \subset \Omega_{N+1}$ and

$$\Omega = \bigcup_N \Omega_N \cup (\theta = +\infty).$$

Let us set

$$X_N = \chi_{\theta<\infty} \int_0^{\theta\wedge N} \phi(t)dw(t).$$

For $\omega \in \Omega_{N_0}$ and $N \geq N_0$, $X_N(\omega) = X_{N_0}$, therefore X_N converges as on $\bigcup_N \Omega_N$ and on $(\theta = +\infty)$, $X_N = 0$, thus X_N converges a.s. □

We now assume that

(1.30) $\beta(t) = b(y(t))$ for $t \in [\theta, \theta' \wedge \tau_\theta[$ a.s.

LEMMA 1.2: *Let $f \geqslant 0$, then we have the estimate*

(1.31) $$E\,[(\int_\theta^{\theta'\wedge\tau_\theta} f(y(t))dt)\chi_{\theta<\infty}] \leq C\,|f|_{L^p} \quad , \quad p > \frac{n}{2}$$

where C does not depend on f, nor on θ,θ', but does depend on $\mathcal{O}$ and on the bounds on b,σ,σ^{-1}.

Proof: We can always assume $f \in \mathcal{D}(\mathcal{O})$, $f \geq 0$. For fixed T we carry out an inverse Girsanov transformation. Let us set

$$\tilde{w}(t) = w(t) + \int_0^t (1-\chi_\theta(s))\sigma^{-1}\,\beta(s)ds$$

and

$$\frac{d\tilde{P}^T}{dP}\Big|_{\mathcal{F}^T} = \exp\left[\int_0^T (1-\chi_\theta(t))\sigma^{-1}\beta\, dw - \frac{1}{2}\int_0^T (1-\chi_\theta(t))|\sigma^{-1}\beta|^2\, dt\right]$$

for Ω, $\mathcal{F}^T$, $\tilde{P}^T$, $\mathcal{F}^t$, $\tilde{w}(t)$ is for $t \leqslant T$, an n-dimensional standard Wiener process, and the process $y(t)$ satisfies

$$y(t) = \eta + \int_0^t (1-\chi_\theta)(t)\sigma(y(t))\, d\tilde{w}(t) .$$

Moreover, we have the inversion formula

$$\frac{dP}{d\tilde{P}^T}\Big|_{\mathcal{F}^T} = \exp\left[-\int_0^T (1-\chi_\theta)(t)\, \sigma^{-1}\beta\, d\tilde{w}(t) - \frac{1}{2}\int_0^T (1-\chi_\theta)(t)|\sigma^{-1}\beta(t)|^2\, dt\right].$$

Let us set

$$\begin{aligned} X_T &= E\int_{\theta\wedge T}^{\theta'\wedge T\wedge\tau_\theta} f(u(t))dt \\ &= \tilde{E}^T\left[\left(\int_{\theta\wedge T}^{\theta'\wedge T\wedge\tau_\theta} f(y(t))dt\right)\right]\frac{dP}{d\tilde{P}^T} \\ &\leq C_T\sqrt{\tilde{E}^T\int_{\theta\wedge T}^{\theta'\wedge T\wedge\tau_\theta} f^2(y(t))dt} \end{aligned} .$$

Now write $h = f^2$ and consider the problem with limits

$$(1.32)\qquad \left|\begin{aligned} &-\frac{\partial u}{\partial t} - \sum_{i,j} a_{ij}(x)\frac{\partial^2 u}{\partial x_i\partial x_j} = h \\ &u|_\Sigma = 0, \qquad u(x,T) = 0. \end{aligned}\right.$$

Since $a_{ij} \in W^{1,\infty}$, h is regular as well as $\mathcal{O}$, and it follows from O.A.LADYZHENSKAYA V.A.SOLONNIKOV-N.N.URAL'TSEVA [1] that (1.32) has one and only one solution in $C^{2+\alpha,1+\frac{\alpha}{2}}(\bar{Q})$ where $Q = \mathcal{O}\times]0,T[$. By considering a function belonging to $C^{2,1}(R^n\times[0,T])$ which coincides with u on $\bar{Q}$, using Ito's formula we obtain

$$(1.33)\qquad \begin{aligned} \tilde{E}^T u(y(\theta\wedge T),\theta\wedge T) &= \tilde{E}^T u(y(\theta'\wedge T\wedge\tau_\theta),\theta'\wedge T\wedge\hat{\tau}_\theta) + \\ &+ \tilde{E}^T\int_{\theta\wedge T}^{\theta'\wedge T\wedge\tau_\theta} h(y(s))ds. \end{aligned}$$

Also the solution of (1.32) satisfies the estimate

$$\|u\|_{W^{2,1,p}(Q)} \leq C\, |h|_{L^p(Q)}$$

and therefore for $p > \frac{n}{2} + 1$,

$$\|u\|_{C^\alpha(\bar{Q})} \leq C\,|h|_{L^p(Q)}$$

which, on taking (1.33) into account, implies

$$\tilde{E}^T \int_{\theta\wedge T}^{\theta'\wedge T\wedge\tau_\theta} h(y(s))ds \leq C_T\,|h|_{L^p} \quad , \quad p > \frac{n}{2} + 1 .$$

Since $h = f^2$, we obtain

$$(1.34) \qquad E \int_{\theta\wedge T}^{\theta'\wedge T\wedge\tau_\theta} f(y(t))dt \leq C_T\,|f|_{L^p} \quad , \quad p > n + 2 ,$$

where C_T does not depend on f. Now by considering the Dirichlet problem

$$Au = f \ , \quad u \in W^{2,p}(\mathcal{O}) \ , \quad u|_\Gamma = 0 \ ,$$

and first assuming $p > n + 2$, we deduce from (1.34) and (1.30) by passing to the limit, after regularisation and using Ito's formula, that

$$Eu(y(\theta\wedge T)) = Eu(y(\theta'\wedge T\wedge\tau_\theta)) + \\ + E\int_{\theta\wedge T}^{\theta'\wedge T\wedge\tau_\theta} f(y(t))dt .$$

But we have the estimate

$$\|u\|_{C^0(\bar{\mathcal{O}})} \leq C|f|_{L^p} \quad , \quad p > \frac{n}{2}$$

which allows us to obtain a better estimate than (1.34), namely

$$E\int_{\theta\wedge T}^{\theta'\wedge T\wedge\tau_\theta} f(y(t))dt \leq C\,|f|_{L^p} \quad , \quad p > \frac{n}{2}$$

where C does not depend on T. Using Fatou's Lemma and making T tend to $+\infty$, we deduce the desired result. □

COROLLARY 1.1: *We have*

$$(1.35) \qquad \chi_{\theta'\wedge\tau_\theta < \infty} = \chi_{\theta<\infty} \qquad \text{a.s.}$$

and if g *is adapted and bounded,*

$$(1.36) \qquad E[\,\chi_{\theta<\infty} \int_\theta^{\theta'\wedge\tau_\theta} g(y(t)).dw(t) \mid \mathfrak{F}^\theta\,] = 0 .$$

Proof: If we take f = 1 in (1.31), we obtain

$$E\ \chi_{\theta<\infty}\,(\theta'\wedge\tau_\theta - \theta) < \infty$$

and so if $\theta \leqslant \infty$, $\theta'\wedge\tau_\theta \leqslant \infty$ a.s.. This immediately implies (1.35). By Lemma 1.1

and (1.35) we have a.s.

$$(1.37) \qquad \chi_{\theta<\infty} \int_{\theta}^{\theta'\wedge\tau_\theta} g(y(t)).dw(t) = \lim_{T\to\infty} \chi_{\theta'\wedge\tau_\theta<\infty} \int_{\theta\wedge T}^{\theta'\wedge\tau_\theta\wedge T} g(y(t)).dw(t)$$

$$= \lim_{T\to\infty} \chi_{\theta<\infty} \int_{\theta\wedge T}^{\theta'\wedge\tau_\theta} g(y(t).dw(t) .$$

By Lemma 1.2

$$E\, \chi_{\theta<\infty} \int_{\theta\wedge T}^{\theta'\wedge\tau_\theta\wedge T} |g(y(t))|^2 dt \to E\, \chi_{\theta<\infty} \int_{\theta}^{\theta'\wedge\tau_\theta} |g(y(t))|^2\, dt < \infty$$

when $T \to \infty$

and therefore the limit (1.37) also holds in the sense of L^2. But then since the conditional expectation is a projector,

$$E\, [\chi_{\theta<\infty} \int_{\theta}^{\theta'\wedge\tau_\theta} g(y(t)).dw(t) \mid \mathcal{F}^\theta] = \lim_{T\to\infty} E[\chi_{\theta<\infty} \int_{\theta\wedge T}^{\theta'\wedge\tau_\theta\wedge T} g(y(t)).dw(t) \mid \mathcal{F}^\theta]$$

in the sense of L^2,

and so (1.36) easily follows. □

We now consider the I.V.

$$(1.38) \qquad \left| \begin{array}{l} a(\phi, v-\phi) \geq (f, v-\phi) \\ \quad \forall\, v \in H_0^1,\ v \leq \psi\ ,\ \phi \in H_0^1,\ \phi \leq \psi\ . \end{array} \right.$$

where ψ is given and is such that

$$(1.39) \qquad \psi \in C^o(\bar{\mathcal{O}})\ , \quad \psi|_\Gamma \geq 0\ .$$

We know (cf., for example A. BENSOUSSAN-J.L. LIONS [1]) that $\phi \in C^o(\bar{\mathcal{O}})$. We can define

$$(1.40) \qquad \hat{\theta} = \inf\, \{t \geqslant \theta \mid \phi(y(t)) = \psi(y(t))\}$$

which is the first *exit* time after θ of the process $y(t)$, from the continuation set

$$C = \{x \mid \phi(x) < \psi(x)\}.$$

ϕ can be continued by 0 outside $\mathcal{O}$, and ψ can be continued by a strictly positive function, at least outside a neighbourhood of $\mathcal{O}$. From this it follows that C is an open set of R^n whose complement is compact, and therefore that $\hat{\theta}$ is a stopping time. In fact the situation is identical to that of τ_θ, except that we are no longer considering a bounded open set of R^n but an open set of R^n whose complement is compact. The argument carried out for τ_θ above remains valid, by E. DYNKIN [2], p. 188. We certainly deduce from this that $\hat{\theta}$ is a stopping time. We then have:

THEOREM 1.2: *Assume* (1.1), (1.2), (1.39). *Let* θ,θ' *be as in* (1.28). *Assume that* (1.30) *holds. Then*

$$(1.41) \qquad \phi(y(\theta))\, \chi_{\theta<\infty} \leq E\, [\chi_{\theta<\infty} \int_{\theta}^{\hat{\theta}'\wedge\tau_\theta} f(y(s))(\exp - \int_{\theta}^{s} a_0(y(\lambda))d\lambda)ds\ +$$

$$+ \chi_{\theta<\infty} \phi(y(\theta'\wedge\tau_\theta))\exp - \int_\theta^{\theta'\wedge\tau_\theta} a_0(y(\lambda))d\lambda \mid \mathfrak{F}^\theta] \qquad \text{a.s.}$$

If $\theta' = \hat{\theta}$ *(hence assume that* (1.30) *holds with* $\theta' = \hat{\theta}$*), then*

$$(1.42)\quad \begin{aligned} \phi(y(\theta))\, \chi_{\theta<\infty} &= E[\chi_{\theta<\infty} \int_\theta^{\hat{\theta}\wedge\tau_\theta} f(y(s))(\exp- \int_\theta^s a_0(y(\lambda))d\lambda)\, ds + \\ &+ \chi_{\theta<\infty} \phi(y(\hat{\theta}\wedge\tau_\theta))\exp - \int_\theta^{\hat{\theta}\wedge\tau_\theta} a_0(y(\lambda))d\lambda \mid \mathfrak{F}^\theta] \qquad \text{a.s.} \end{aligned}$$

Proof: First of all let us prove (1.41). We first assume ψ is regular, and that $\psi \in \mathcal{D}(\bar{\mathcal{O}})$. Then the solution ϕ of the I.V. (1.38) belongs to $W^{2,p}(\mathcal{O})$ and

$$(1.43)\quad \left| \begin{aligned} & A\phi + a_0\phi \le f \quad \text{a.e.} \\ & \phi \le \psi \\ & (A\phi + a_0\phi - f)(\phi - \psi) = 0 \quad \text{a.e.} \\ & \phi|_\Gamma = 0 . \end{aligned} \right.$$

Consider a sequence $\phi_n \in C^2(R^n)$,

$$\phi_n \to \phi \text{ in } W^{2,p}(\mathcal{O}) .$$

By Ito's formula, and thanks to (1.30),

$$\begin{aligned} & \phi_n(y(\theta'\wedge\tau_\theta \wedge T))\exp - \int_0^{\theta'\wedge\tau_\theta\wedge T} a_0(y)ds = \\ & = \phi_n(y(\theta\wedge T))\exp - \int_0^{\theta\wedge T} a_0(y(s))ds + \\ & + \int_{\theta\wedge T}^{\theta'\wedge\tau_\theta\wedge T} (-A\phi_n - a_0\phi_n)(y(s))(\exp - \int_0^s a_0(y)d\lambda)ds + \\ & \qquad + \int_{\theta\wedge T}^{\theta'\wedge\tau_\theta} D\phi_n.\sigma(y(s))(\exp - \int_0^s a_0 d\lambda)ds . \end{aligned}$$

Multiply by $\chi_{\theta<\infty}$, and make T tend to $+\infty$. Using Lemma 1.1 and Corollary 1.1 we can go to the limit a.s.. Next take the conditional expectation with respect to $\mathfrak{F}^\theta$. Again thanks to Corollary 1.1, the contribution of the stochastic integral vanishes. Thus we obtain

$$(1.44)\quad \begin{aligned} \phi_n(y(\theta))\chi_{\theta<\infty} &= E[\int_\theta^{\theta'\wedge\tau_\theta} \chi_{\theta<\infty}(A\phi_n + a_0\phi_n)(y(s))\exp - \int_\theta^s a_0(y)d\lambda + \\ &+ \chi_{\theta<\infty}\, \phi_n(y(\theta'\wedge\tau_\theta))\exp - \int_\theta^{\theta'\wedge\tau_\theta} a_0(y)ds \mid \mathfrak{F}^\theta] \qquad \text{a.s.} \end{aligned}$$

Now

$$A\phi_n + a_0\phi_n \to A\phi + a_0\phi \text{ in } L^p(\mathcal{O}) .$$

From Lemma 1.2 it follows that

$$E\, \chi_{\theta<\infty} \int_{\theta}^{\theta'\wedge\tau_\theta} |(A\phi_n + a_0\phi_n - A\phi - a_0\phi)(y(s))|ds \to 0 \quad .$$

Using the property that the conditional expectation is a contraction in L^1, we deduce that at least for a subsequence

$$E[\chi_{\theta<\infty} \int_{\theta}^{\theta'\wedge\tau_\theta} (A\phi_n + a_0\phi_n)(y(s))(\exp -\int_{\theta}^{s} a_0(y)d\lambda)ds|\ \mathcal{F}^\theta] \to$$
$$E[\chi_{\theta<\infty} \int_{\theta}^{\theta'\wedge\tau_\theta} (A\phi + a_0\phi)(y(s))(\exp -\int_{\theta}^{s} a_0(y)d\lambda)ds|\ \mathcal{F}^\theta] \quad \text{a.s.}$$

Since $\phi = 0$ outside $\mathcal{O}$, we can always assume that

$\phi_n \to \phi$ uniformly in the whole space.

We can then go to the limit a.s. in (1.44) at least for a subsequence. From this we deduce

$$\phi(y(\theta))\ \chi_{\theta<\infty} = E[\int_{\theta}^{\theta'\wedge\tau_\theta} \chi_{\theta<\infty} (A\phi + a_0\phi)(y(s))\exp -\int_{\theta}^{s} a_0(y)d\lambda \; +$$

$$(1.45) \qquad + \chi_{\theta<\infty}\, \phi(y(\theta'\wedge\tau_\theta))\exp -\int_{\theta}^{\theta'\wedge\tau_\theta} a_0(y)ds|\ \mathcal{F}^\theta] \quad \text{a.s.}$$

(1.41) is deduced from this by using the inequality (1.43). Then by applying (1.45) with $\theta' = \hat{\theta}$, and using

$$(A\phi + a_0\phi - f)\ \chi_C = 0$$

we easily deduce (1.42). We have therefore proved the result we wanted, at least in the case of regular obstacles.

In the general case, for (1.41) we can proceed by regularising the obstacle. Consider a sequence

$$\psi_k \in \mathcal{D}(\bar{\mathcal{O}}) \qquad \psi_k|_\Gamma \geq 0 \text{ and } \psi_k \to \psi \quad \text{in } C^0(\bar{\mathcal{O}}) \ .$$

We know that

$$\phi_k \to \phi \text{ in } C^0(\bar{\mathcal{O}})$$

and therefore converges uniformly in all the space. Starting from this, we can write (1.41) for ϕ_k and go to the limit.

To prove (1.42) in the general case we cannot proceed as easily by regularisation, because we then have a stopping time $\hat{\theta}_k$ which varies with k. Technically, it is easier to use the penalised approximation, so

$$(1.46) \qquad \left|\begin{array}{l} A\phi^\varepsilon + a_0\,\phi^\varepsilon + \frac{1}{\varepsilon}(\phi^\varepsilon - \psi)^+ = f \\ \phi^\varepsilon|_\Gamma = 0 \ , \qquad \phi^\varepsilon \in W^{2,p}(\mathcal{O}) \ . \end{array}\right.$$

We set

(1.47) $\hat{\theta}^{\varepsilon} = \inf\{ t \geq \theta \mid \phi^{\varepsilon}(y(t)) \geq \psi(y(t))\}$

which is also a stopping time of $\mathfrak{F}^t$, for reasons similar to $\hat{\theta}$. We know that

(1.48) $\phi^{\varepsilon} \to \phi$ in $C^0(\bar{\mathcal{O}})$ and $\phi^{\varepsilon} \geq \phi$.

Let us first verify that

(1.49) $\hat{\theta}^{\varepsilon} \to \hat{\theta}$.

such an inequality is obvious if $\hat{\theta} = \infty$. When $\hat{\theta} < \infty$, by the definition of $\hat{\theta}$ (cf., (1.40)) we have

$$\phi(y(\hat{\theta})) = \psi(y(\hat{\theta}))$$

and therefore by the second part of (1.48)

$$\phi^{\varepsilon}(y(\hat{\theta})) \geq \psi(y(\hat{\theta}))$$

which certainly implies (1.49). Let us now show that

(1.50) $\hat{\theta}^{\varepsilon} \to \hat{\theta}$ a.s..

For ω such that $\hat{\theta} = \theta$, it immediately follows from (1.49) that $\hat{\theta}^{\varepsilon} = \theta = \hat{\theta}$, and therefore (1.50) is clear for such ω's.

We can consider the values of ω such that $\theta < \hat{\theta}$. So let δ_{ω} be such that

$$\hat{\theta} > \theta + \delta_{\omega} .$$

For

$$\theta \leq s \leq \hat{\theta} - \delta_{\omega}$$

we have

$$\phi(y(s)) < \psi(y(s)).$$

Also, by the first part of (1.48), we can find $\varepsilon_0(\omega)$ such that $\varepsilon \leq \varepsilon_0(\omega)$

$$\|\phi^{\varepsilon} - \phi\| < \min_{\theta \leq s \leq \hat{\theta} - \delta_{\omega}} [\psi(y(s)) - \varphi(y(s))]$$

hence for $\varepsilon \leq \varepsilon_0(\omega)$ and $\theta \leq s \leq \hat{\theta} - \delta_{\omega}$, we have

$$\phi^{\varepsilon}(y(s)) < \phi(y(s)) + \psi(y(s)) - \phi(y(s))$$
$$= \psi(y(s))$$

which implies

$$\hat{\theta}^{\varepsilon} \geq \hat{\theta} - \delta_{\omega}.$$

In fact for $\delta < \delta_{\omega}$, a similar argument shows that there exists $\varepsilon_0(\delta,\omega)$ such that for $\varepsilon \leq \varepsilon_0$ we have

$$\hat{\theta}^{\varepsilon} \geq \hat{\theta} - \delta$$

which certainly proves (1.50).

By (1.49), and if (1.30) holds for $\theta' = \hat{\theta}$, we deduce that (1.30) still holds with $\theta' = \hat{\theta}^{\varepsilon}$. But in the same way we proved (1.42), we verify that in the case of a regular obstacle, we have

$$\begin{aligned}(1.51)\quad \phi^{\varepsilon}(y(\theta))\chi_{\theta<\infty} = [E\ \chi_{\theta<\infty}\int_{\theta}^{\hat{\theta}^{\varepsilon}\wedge\tau_{\theta}} f(y(t))(\exp-\int_{\theta}^{t} a_0(y(s))ds)dt \\ + \chi_{\theta<\infty}\ \phi^{\varepsilon}(y(\hat{\theta}^{\varepsilon}\wedge\tau_{\theta}))\exp-\int_{\theta}^{\hat{\theta}^{\varepsilon}\wedge\tau_{\theta}} a_0(y(t))dt \mid \mathcal{F}^{\theta}] \quad \text{a.s.}\end{aligned}$$

Using (1.50) and the uniform convergence of ϕ^{ε} to ϕ, we can pass to the limit in ε in (1.51), which concludes the proof of (1.42). □

1.4 *Proof of The Principal Result*

We shall now go onto the proof of Theorem 1.1.

We may consider the Q.V.I. (1.6) as a V.I. for the obstacle $\psi = Mu$ which satisfies the conditions (1.39). With the notations of Section 1.1, and for the system $(\Omega_0, \mathcal{M}_0, \mathcal{M}^t, P_W^x)$, where W is an arbitrary impulse control, the process x^n defined in (1.14) satisfies the equation

$$\begin{aligned}(1.52)\quad x^n(t) = x^n(\theta^n) + \int_0^t (1-\chi_{\theta^n}(s))b(x(s))ds + \\ + \int_0^t (1-\chi_{\theta^n}(s))\sigma(x^n(s))dw(s).\end{aligned}$$

Using the notations of (1.3) we take $y = x^n$, $\theta = \theta^n$, $\tau^n = \tau_{\theta^n}$, and $\theta' = \theta^{n+1}$. By (1.15), the assumption (1.30) is satisfied. We then apply (1.41) with $\phi = u$, and we obtain

$$\begin{aligned}(1.53)\quad u(x^n(\theta^n))\chi_{\theta^n<\infty} \le E_W^x [\chi_{\theta^n<\infty}\int_{\theta^n}^{\theta^{n+1}\wedge\tau^n} f(x^n(s))(\exp-\int_{\theta^n}^{s} a_0(x^n(\lambda))d\lambda)ds \\ + u(x^n(\theta^{n+1}\wedge\tau^n))\chi_{\theta^n<\infty}\exp-\int_{\theta^n}^{\theta^{n+1}\wedge\tau^n} a_0(x^n(s))ds \mid \mathcal{M}^{\theta^n}] \quad \text{a.s..}\end{aligned}$$

We now notice that

$$(1.54)\quad \theta^n < \tau \quad \text{implies} \quad \theta^{n+1}\wedge\tau^n = \theta^{n+1}\wedge\tau .$$

The property is clear if $\theta^n = \theta^{n+1}$. Therefore let us assume $\theta^n < \theta^{n+1}$. Then

$$\begin{aligned}(1.55)\quad & x^n(s) = x(s) \text{ for } s\in[\theta^n, \theta^{n+1}[\\ & x^n(s) = x(s^-) \text{ for } s\in]\theta^n, \theta^{n+1}].\end{aligned}$$

Since $\theta^n < \tau$, we have

$$x(\theta^n), x(\theta^n-)\in\mathcal{O}.$$

If $\theta^{n+1} \leqslant \tau^n$, then

$$x^n(s)\in \mathcal{O} \text{ for } s\in[\theta^n,\theta^{n+1}[,$$

therefore by (1.55),

$$x(s) \in \mathcal{O} \text{ for } s \in [\theta^n, \theta^{n+1}[,$$

and

$$x(s^-) \in \mathcal{O} \text{ for } s \in [\theta^n, \theta^{n+1}[,$$

therefore $\theta^{n+1} \leqslant \tau$; we also have

$\theta^n < \tau^n$, because if

$\theta^n = \tau^n < \theta^{n+1}$, then $x(\tau^n) \notin \mathcal{O}$, and therefore $\tau^n \geqslant \tau$, which is impossible if $\theta^n < \tau$.

Let us now assume that $\theta^{n+1} > \tau^n$, therefore $\theta^n < \tau^n < \theta^{n+1}$, and then

$$x(\tau^n) = x(\tau^{n-}) = x^n(\tau^n) \notin \mathcal{O},$$

therefore $\tau^n = \tau$. Therefore we have proved (1.54). Now multiply (1.53) by $\chi_{\theta^n < \tau}$ which is $\mathcal{M}^{\theta^n}$ measurable. Using (1.54) we obtain

$$(1.56) \qquad u(x^n(\theta^n))\chi_{\theta^n < \tau} \leq E_W^x[\chi_{\theta^n<\tau} \int_{\theta^n \wedge \tau}^{\theta^{n+1} \wedge \tau} f(x(s))(\exp -\int_{\theta^n \wedge \tau}^{s} a_0 d\lambda) ds$$
$$+ u(x^n(\theta^{n+1} \wedge \tau^n))\chi_{\theta^n<\tau} \exp - \int_{\theta^n \wedge \tau}^{\theta^{n+1} \wedge \tau} a_0 ds \mid \mathcal{M}^{\theta^n}] \qquad \text{a.s.}$$

Next we use the property that $u \leqslant Mu$, and since $u \geqslant 0$, and $u|_\Gamma = 0$, by Lemma 1.6 of Chapter 4

$$Mu(x) = k + \inf_{\xi \geq 0} \{c_0(\xi) + u(x+\xi)\} ,$$

Consequently

$$u(x) \leq k + c_0(\xi) + u(x+\xi) \qquad \forall \xi \geq 0 .$$

This inequality is applied with

$$x = x^n(\theta^{n+1}), \text{ if } \theta^{n+1} < \infty , \quad \xi = \xi^{n+1} .$$

Therefore

$$(1.57) \qquad u(x^n(\theta^{n+1})) \leq k + c_0(\xi^{n+1}) + u(x^{n+1}(\theta^{n+1}))$$
$$\text{if } \theta^{n+1} < \infty$$

where we have used the definition (1.14) of $x^{n+1}(\theta^{n+1})$. By Corollary 1.1,

$$\theta^{n+1} \wedge \tau^n < \infty \text{ if } \theta^n < \infty, \text{ a.s. } P_W^x ,$$

therefore

$$(1.58)\qquad u(x^n(\theta^{n+1}\wedge\tau^n))\chi_{\theta^n<\infty} = u(x^n(\theta^{n+1}))\chi_{\theta^{n+1}<\tau^n}\,\chi_{\theta^n<\infty}$$

for

$$u(x^n(\tau^n))\,\chi_{\tau^n\le\theta^{n+1}}\,\chi_{\theta^n<\infty} = 0\ .$$

Therefore we also have

$$(1.59)\qquad u(x^n(\theta^{n+1}\wedge\tau^n))\,\chi_{\theta^n<\tau} = u(x^n(\theta^{n+1}))\,\chi_{\theta^{n+1}<\tau^n}\,\chi_{\theta^n<\tau}$$

and by (1.57)

$$\begin{aligned}&\le (k+c_0(\xi^{n+1}))\chi_{\theta^{n+1}<\tau^n}\,\chi_{\theta^n<\tau} + \\ &\quad + u(x^{n+1}(\theta^{n+1}))\,\chi_{\theta^{n+1}<\tau^n}\,\chi_{\theta^n<\tau} \\ &\le (k+c_0(\xi^{n+1}))\,\chi_{\theta^{n+1}<\infty} + \\ &\qquad + u(x^{n+1}(\theta^{n+1}))\,\chi_{\theta^{n+1}<\tau}\ .\end{aligned}$$

Substituting in (1.56), we obtain

$$(1.60)\qquad \begin{aligned}&u(x^n(\theta^n))\chi_{\theta^n<\tau}\exp - \int_0^{\theta^n} a_0(x(s))ds \le \\ &\le E_W^x\Big[\int_{\theta^n\wedge\tau}^{\theta^{n+1}\wedge\tau} f(x(s))(\exp - \int_0^s a_0(x(\lambda))d\lambda)ds + \\ &+ ((k+c_0(\xi^{n+1}))\,\chi_{\theta^{n+1}<\infty} + u(x^{n+1}(\theta^{n+1}))\chi_{\theta^{n+1}<\tau})\exp-\int_0^{\theta^{n+1}} a_0 ds\,|\,\mathcal{M}^{\theta^n}\Big]\ .\end{aligned}$$

To this we add the relations (1.60) when n varies from 0 to N-1, after taking the mathematical expectation, and assuming that $\theta^0 = 0$. Assuming $x\in\mathcal{O}$, we obtain

$$(1.61)\qquad \begin{aligned}&u(x) \le E_W^x\Big[\int_0^{\theta^N\wedge\tau} f(x(s))(\exp-\int_0^s a_0(x(\lambda))d\lambda)ds + \\ &+ \sum_{n=1}^N (k+c_0(\xi^n))\,\chi_{\theta^n<\infty}\exp - \int_0^{\theta^n} a_0(x(\lambda))d\lambda + \\ &\qquad + u(x^N(\theta^N))\,\chi_{\theta^N<\tau}\exp - \int_0^{\theta^N} a_0(x(\lambda))d\lambda\Big]\ .\end{aligned}$$

Let us consider the sequence

$$\chi_{\theta^n<\tau}\exp - \int_0^{\theta^n} a_0(x(s))ds$$

which is a decreasing sequence. It converges a.s. to a variable H. If H = 0 a.s., then from (1.61) and from $\theta^N\to\infty$ a.s., it follows that

$$u(x) \leq E_W^x[\int_0^\tau f(x(s))\exp - \int_0^s a_0(x(\lambda))d\lambda\, ds + $$
$$+ \sum_{n=1}^{\infty} (k+ c_0(\xi^n)) \chi_{\theta^n < \infty} \exp - \int_0^{\theta^n} a_0(x(s))ds] .$$

Consequently, we have shown that

$$u(x) \leqslant J^x(W), \text{ for W such that } H = 0 \text{ a.s..}$$

Now let us assume that W is such that $H > 0$ on a set of positive probability. We then necessarily have, $J^x(W) = + \infty$, since

$$J^x(W) \geq k\, E^x \sum_{n=1}^{N} \chi_{\theta^n<\tau} \exp - \int_0^{\theta^n} a_0(x(s))ds \to \infty .$$

Thus we have proved that

(1.62) $\quad u(x) \leqslant J^x(W) \qquad \forall W,$

at least if $x \in \mathcal{O}$. If $x \notin \mathcal{O}$, then $u(x) = 0$, therefore (1.62) is obviously satisfied.

It is now a question of showing the control $\hat{W}$ defined in (1.27) is optimal. let us first notice that if $x \notin \mathcal{O}$, then $\hat{\theta}^1 = + \infty$ and $\tau = 0$. Hence $u(x) = 0$ and $J^x(\hat{W}) = 0$. Therefore we can assume that $x \in \mathcal{O}$. First we must establish that $\hat{W}$ is an admissable control. The sequence $\hat{\theta}^n$ certainly is a sequence of stopping times. So we are going to show that

(1.63) $\quad \hat{\theta}^{n+1} > \hat{\theta}^n$ a.s. if $\hat{\theta}^n < \infty$. (1)

In fact let us assume $\hat{\theta}^n < \infty$, then we have

$$\begin{aligned} u(x^n(\hat{\theta}^n)) &= u(x^{n-1}(\hat{\theta}^n) + \hat{\xi}^n) \\ &= u(x^{n-1}(\hat{\theta}^n) + \hat{\xi}(x^{n-1}(\hat{\theta}^n))) \\ &= u(x^{n-1}(\hat{\theta}^n)) - k - c_0(\hat{\xi}^n) \\ &\leq k + u(x^{n-1}(\hat{\theta}^n) + \xi + \hat{\xi}^n)) + c_0(\xi + \hat{\xi}^n) \\ &\quad - k - c_0(\hat{\xi}^n) \end{aligned}$$

and thanks to the sub-linearity of c_0, we see that

$$\begin{aligned} u(x^n(\hat{\theta}^n)) &\leq u(x^{n-1}(\hat{\theta}^n) + \hat{\xi}^n + \xi) + c_0(\xi) \\ &= u(x^n(\hat{\theta}^n) + \xi) + c_0(\xi) + k - k , \quad \forall\, \xi \geq 0 . \end{aligned}$$

But then

$$u(x^n(\hat{\theta}^n)) \leqslant Mu(x^n(\hat{\theta}^n)) - k < Mu(x^n(\hat{\theta}^n))$$

which certainly implies $\hat{\theta}^n < \hat{\theta}^{n+1}$. Hence we have (1.63).

---

(1) This condition is not indispensible for $\hat{W}$ to be admissible.

We now have to show that

(1.64) $\theta^n \to \infty$ a.s..

Let us set

$$\rho(\delta) = \sup_{\substack{|x-y| \le \delta \\ x,y \in \bar{\mathcal{O}}}} |u(x)-u(y)| .$$

By the continuity of u on $\bar{\mathcal{O}}$, ρ is an increasing function that tends to 0 when $\delta \to 0$. Now, by (1.24) we have

$$|u(x+ \hat{\xi}(x)) - u(x)| \ge k$$

therefore

$$\rho(|\hat{\xi}(x)|) \ge k$$

which implies

(1.65) $|\hat{\xi}(x)| \ge L > 0 \quad \forall x \in \bar{\mathcal{O}}$.

Also by (1.25) we have

(1.66) $$x^N(\hat{\theta}^N) \chi_{\hat{\theta}^N < \infty} = [x + \sum_{j=1}^{N-1} \int_{\hat{\theta}^j}^{\hat{\theta}^{j+1}} \sigma(x^j(t))dw_0(t) + $$
$$+ \xi^1 + \dots + \xi^N] \chi_{\hat{\theta}^N < \infty} .$$

If $\hat{\theta}N < \infty$, we know that $x^N(\hat{\theta}N) \in \bar{\mathcal{O}}$, and by (1.65) and by the positivity of the components of the vector $\hat{\xi}$, we have

(1.67) $|\xi^1 + \dots + \xi^N| \ge L \sqrt{N} \quad$ if $\quad \hat{\theta}^N < \infty$.

Let us now consider the set

$$\tilde{\Omega}_0 = \{ \lim_N \hat{\theta}^N = \Lambda < \infty \} .$$

Let us also define the process

(1.68) $$\zeta(t) = \begin{vmatrix} 0 & \text{if} & t \ge \Lambda \\ \sigma(x^j(t)) & \text{if} & \hat{\theta}^j \le t < \hat{\theta}^{j+1} \end{vmatrix}$$

then (1.66) may be written

$$x^N(\hat{\theta}^N) \chi_{\hat{\theta}^N < \infty} = [x + \int_0^{\hat{\theta}^N} \zeta(s)dw_0(s) + \xi^1 + \dots + \xi^N] \chi_{\hat{\theta}^N < \infty}$$

therefore we also have

$$x^N(\hat{\theta}^N) \chi_{\Lambda < \infty} - \chi_{\Lambda < \infty} [x + \int_0^{\hat{\theta}^N} \zeta(s)dw_0(s)] = \chi_{\Lambda < \infty}(\xi^1 + \dots + \xi^N)$$

and on $\tilde{\Omega}_0$, for fixed ω, the left hand side remains bounded when $N \to \infty$, and then by (1.67) the right hand side tends to $+\infty$. Therefore we have $P(\Omega_0) = 0$, which certainly implies (1.64). We notice that

$$\hat{\theta}^{n+1} \wedge \tau_{\hat{\theta}^n} = \hat{\theta}^{n+1} \wedge S^n = T^n \wedge S^n ,$$

therefore (1.30) is satified. We can then apply (1.42) with $y = x^n$, $\hat{\theta}^n$, $\hat{\theta} = \hat{\theta}^{n+1}$ and $\phi = u$. After multiplication by $\chi_{\hat{\theta}^n < \tau}$, we obtain

$$(1.69) \qquad u(x^n(\hat{\theta}^n))\, \chi_{\hat{\theta}^n<\tau} = E^x_{\hat{W}}[\chi_{\hat{\theta}^n<\tau} \int_{\hat{\theta}^n}^{\hat{\theta}^{n+1}\wedge\tau^n} f(x^n)(\exp - \int_{\hat{\theta}^n}^{s} a_0) ds + u(x^n(\hat{\theta}^{n+1}\wedge\tau^n))\, \chi_{\hat{\theta}^n<\tau} \exp - \int_{\hat{\theta}^n}^{\hat{\theta}^{n+1}\wedge\tau^n} a_0\, ds \mid \mathcal{M}^{\hat{\theta}^n}] \quad \text{a.s.}$$

However, we have

$$(1.70) \qquad u(x^n(\hat{\theta}^{n+1}\wedge\tau^n))\, \chi_{\hat{\theta}^n<\tau} = [k + c_0(\hat{\xi}^{n+1}) + u(x^{n+1}(\hat{\theta}^{n+1}))] \times \chi_{\hat{\theta}^{n+1}<\tau}\, \chi_{\hat{\theta}^n<\tau} .$$

Now if $\hat{\theta}^{n+1} < \infty$, it follows from (1.26) that

$$\hat{\theta}^{n+1} < S^n = \tau^n,$$

therefore

$$(1.71) \qquad \chi_{\hat{\theta}^{n+1}<\infty} = \chi_{\hat{\theta}^{n+1}<\tau^n} .$$

We also have $x^n(\hat{\theta}^n), x^n(\hat{\theta}^{n+1}) \in \mathcal{O}$, therefore $x^{n-1}(\hat{\theta}^n) \in \mathcal{O}$ also. Consequently, $\hat{\theta}^{n+1} \leqslant \tau$, and $\hat{\theta}^n < \tau$, because $\hat{\theta}^n < \hat{\theta}^{n+1}$. Therefore

$$(1.71') \qquad \chi_{\hat{\theta}^{n+1}<\infty} = \chi_{\hat{\theta}^{n+1}<\tau}\, \chi_{\hat{\theta}^n<\tau} .$$

Also

$$(1.72) \qquad u(x^{n+1}(\hat{\theta}^{n+1}))\chi_{\hat{\theta}^{n+1}<\tau}\, \chi_{\hat{\theta}^n<\tau} = u(x^{n+1}(\hat{\theta}^{n+1}))\, \chi_{\hat{\theta}^{n+1}<\infty} = u(x^{n+1}(\hat{\theta}^{n+1}))\, \chi_{\hat{\theta}^{n+1}<\tau} + u(x^{n+1}(\hat{\theta}^{n+1}))\, \chi_{\hat{\theta}^{n+1}=\tau}\, \chi_{\hat{\theta}^{n+1}<\infty} .$$

Now if $\hat{\theta}^{n+1} < \infty$, then from (1.26),

$$\hat{\theta}^{n+1} = T^n, x^n(T^n) \in \mathcal{O} .$$

Since $\hat{\theta}^n < \hat{\theta}^{n+1} < \hat{\theta}^{n+2}$, we have

$$x(\hat{\theta}^{n+1}) = x^{n+1}(\hat{\theta}^{n+1}) \qquad x(\hat{\theta}^{n+1} - 0) = x^n(\hat{\theta}^{n+1}) \in \mathcal{O} .$$

But then if $\hat{\theta}^{n+1} = \tau$, then necessarily $x(\tau) \notin \mathcal{O}$, and therefore

$$u(x^{n+1}(\hat{\theta}^{n+1}))\ \chi_{\hat{\theta}^{n+1}=\tau}\ \chi_{\hat{\theta}^{n+1}<\infty} = 0 .$$

Using (1.71),(1.72) we deduce from (1.70) that

$$(1.73)\qquad u(x^n(\hat{\theta}^{n+1}\wedge\tau^n))\ \chi_{\hat{\theta}^n<\tau} = (k + c_0(\hat{\xi}^{n+1}))\ \chi_{\hat{\theta}^{n+1}<\infty} + $$
$$+\ u(x^{n+1}(\hat{\theta}^{n+1}))\ \chi_{\hat{\theta}^{n+1}<\tau} .$$

We deduce from (1.70) and (1.69) that (1.60) holds, with equality instead of inequality. In addition, we finally see that

$$u(x) = E^x_{\hat{W}}\Big[\int_0^{\hat{\theta}^N\wedge\tau} f(x(s))\exp - \int_0^s a_0(x(\lambda))d\lambda\ +$$
$$+ \sum_{n=1}^N (k+ c_0(\hat{\xi}^n))\ \chi_{\hat{\theta}^n<\infty}\ \exp - \int_0^{\hat{\theta}^n} a_0(x(\lambda))d\lambda\ +$$
$$+\ u(x^N(\hat{\theta}^N))\ \chi_{\hat{\theta}^n<\tau}\ \exp - \int_0^{\hat{\theta}^N} a_0(x(s))ds\Big]$$

and by the positivity of u

$$\geq E^x_{\hat{W}}\Big[\int_0^{\hat{\theta}^N\wedge\tau} f(x(s))\ \exp - \int_0^s a_0 d\lambda\ +$$
$$+ \sum_{n=1}^N (k+ c_0(\hat{\xi}^n))\ \chi_{\hat{\theta}^n<\infty}\ \exp - \int_0^{\hat{\theta}^n} a_0 d\lambda\Big]$$

and by making N tend to $+\infty$ we obtain

$$u(x) \geqslant J^x(\hat{W})$$

which, taking (1.62) into account, shows that $\hat{W}$ is an optimal impulse control. □

2. INTERPRETATION OF ITERATIVE METHODS

2.1 *Decreasing Method*

Consider the sequence of functions, $u^n(x)$ defined by

$$(2.1)\qquad a(u^0,v) = (f,v) \qquad \forall\ v \in H^1_0\ ,\quad u^0 \in H^1_0$$

$$(2.2)\qquad \left|\begin{array}{l} a(u^{n+1},v-u^{n+1}) \geq (f,v-u^{n+1}) \\ u^{n+1} \in H^1_0\ ,\quad u^{n+1} \leq Mu^n \\ \forall\ v \in H^1_0\ ,\quad v \leq Mu^n . \end{array}\right.$$

Then by induction we establish that

(2.3) $u^n \in C^0(\bar{\mathcal{O}})$, $u^n \geq 0$.

Let μ be such that $\mu < 1$ and

(2.4) $\mu \| u^0 \|_{L^\infty} \leqslant k$

then (cf., Chapter 4, by a variant of Lemma 1.4) we have

(2.5) $0 \leq u(x) \leq u^n(x) \leq u(x) + \dfrac{(1-\mu)^n}{\mu} \| u^0 \|$.

Our objective is to give the probabilistic interpretation of $u^n(x)$. Let us introduce

(2.6) $\mathcal{W}^n = \{W \mid \theta^p = +\infty \text{ for } p \geqslant n\}$,

and then have:

THEOREM 2.1: *Using the assumption of Theorem 1.1, the sequence* u^n *defined by* (2.1), (2.2) *satisfies*

(2.7) $u^{n-1}(x) = \inf\limits_{W \in \mathcal{W}^n} J^x(W), n \geqslant 1.$

Furthermore, there exists an optimal control.

Proof: For $n = 1$, $\theta^1 = +\infty$, therefore $x(t) = x^0(t)$, and $J^x(W)$ does not depend on W for $W \in \mathcal{W}^1$, and (2.7) is none other than the probabilistic interpretation of the solution u^0 of the equation (2.1). We now assume $n > 1$ and use a method similar to the proof of Theorem 1.1 by employing the fact that u^{n+1} is a solution of V.I. whose obstacle is Mu^n.

Therefore let $W \in \mathcal{W}^{n+1}$, and consider $0 \leqslant j \leqslant n-1$. We now establish a relation similar to (1.56), namely

(2.8)
$$u^{n-j}(x^j(\theta^j))\chi_{\theta^j<\tau} \leq E^x_W[\chi_{\theta^j<\tau} \int_{\theta^j\wedge\tau}^{\theta^{j+1}\wedge\tau} f(x(s))(\exp -\int_0^s a_0 d\lambda)d\lambda + u^{n-j}(x^j(\theta^{j+1}\wedge\tau^j))\, \chi_{\theta^j<\tau} \exp - \int_{\theta^j\wedge\tau}^{\theta^{j+1}\wedge\tau} a_0 ds \mid \mathcal{M}^{\theta^j}].$$

Next we use the property

$$u^{n-j} \leqslant Mu^{n-1-j}$$

Therefore (cf., (1.57)

(2.9) $u^{n-j}(x^j(\theta^{j+1})) \leqslant k + c_0(\xi^{j+1}) + u^{n-j-1}(x^{j+1}(\theta^{j+1}))$

if $\theta^{j+1} < \infty$, a.s.

We then argue as for (1.61) to obtain

(2.10) $$u^{n-j}(x^j(\theta^j))\,\chi_{\theta^j<\tau}\exp -\int_0^{\theta^j} a_0(x(s))ds \le$$
$$\le E_W^x\Big[\int_{\theta^j\wedge\tau}^{\theta^{j+1}\wedge\tau} f(x(s))(\exp -\int_0^s a_0 d\lambda)ds + ((k+c_0(\xi^{j+1}))\,\chi_{\theta^{j+1}<\infty} +$$
$$+ u^{n-(j+1)}(x^{j+1}(\theta^{j+1}))\,\chi_{\theta^{j+1}<\tau})\exp -\int_0^{\theta^{j+1}} a_0(x(s))ds \,|\, \mathcal{M}^{\theta^j}\Big] .$$

Now take the mathematical expectation and add the relations (2.10) for j varying between 0 and n-1. We obtain

$$u^n(x) \le E_W^x\Big[\int_0^{\theta^n\wedge\tau} f(x(s))(\exp -\int_0^s a_0 d\lambda)ds +$$
$$+ \sum_{j=1}^{n} (k+c_0(\xi^j))\,\chi_{\theta^j<\infty}\exp -\int_0^{\theta^j} a_0 ds +$$
$$u^0(x^n(\theta^n))\,\chi_{\theta^n<\tau}\exp -\int_0^{\theta^n} a_0(x(s))ds\Big] .$$

Since $\theta^{n+1} = +\infty$, we obtain

(2.11) $$u^n(x) \leqslant J^x(W),\ \forall W \in \mathcal{W}^{n+1} .$$

Next we prove the existence of an optimal control. Define $\hat{\xi}^n(x)$ as the positive Borel function such that

(2.12) $$u^n(x) = k + c_0(\hat{\xi}^n(x)) + u^{n-1}(x+\hat{\xi}^n(x)) \quad \forall\, x \in \bar{\mathcal{O}} ,$$
$$n \ge 1 .$$

Next define

$$\left|\begin{array}{l} dx^0 = \sigma(x^0)\,dw_0 \\ x^0(0) = x , \end{array}\right.$$
$$T_n^0 = \inf\{\, t \mid u^n(x^0(t)) = Mu^{n-1}(x^0(t))\}$$
$$S^0 = \text{exit time of } x^0 \text{ from } \mathcal{O}$$

and

(2.13) $$\left|\begin{array}{l} \hat{\theta}_n^1 = T_n^0 \quad \text{if} \quad T_n^0 < S^0 , \quad +\infty \quad \text{otherwise}, \\ \hat{\xi}_n^1 = \hat{\xi}^n(x^0(\hat{\theta}_n^1)) \quad \text{if} \quad \hat{\theta}_n^1 < \infty , \text{ otherwise arbitrary}, \end{array}\right.$$

We again have $x^0(\hat{\theta}_n^1) \in \mathcal{O}$ if $\hat{\theta}_n^1 < \infty$, for $Mu^{n-1}|_\Gamma = k$. Having defined the process x_n^{j-1} as well as $\hat{\theta}_n^j, \hat{\xi}_n^j$ for $j = 1,\ldots,n-2$, we define $x_n^j, \hat{\theta}_n^{j+1}, \hat{\xi}_n^{j+1}$ as follows

(2.14) $$\left|\begin{array}{l} dx_n^j = \sigma(x_n^j)\,dw_0 \\ x_n^j(\hat{\theta}_n^j) = x_n^{j-1}(\hat{\theta}_n^1) + \hat{\xi}_n^j \end{array}\right.$$ (Contd)

$$T_n^j = \inf \{t \geqslant \hat{\theta}_n^j \mid u^{n-j}(x_n^j(t)) = Mu^{n-j-1}(x_n^j(t))\}$$

$$S_n^j = \text{exit time of } x_n^j \text{ from } \mathcal{O}, \text{ after } \hat{\theta}_n^j$$

then

$$(2.15) \quad \left| \begin{array}{l} \hat{\theta}_n^{j+1} = T_n^j \quad \text{if} \quad T_n^j < S_n^j \;, \quad +\infty \quad \text{otherwise}, \\ \hat{\xi}_n^{j+1} = \hat{\xi}^{n-j}(x_n^j(\hat{\theta}_n^{j+1})) \quad \text{if} \quad \hat{\theta}_n^{j+1} < \infty \;, \text{ otherwise arbitrary}, \end{array} \right.$$

and naturally we set

$$(2.16) \quad \left| \begin{array}{l} \hat{\theta}_n^j = +\infty \quad \text{if} \quad j \geq n+1 \\ \xi_n^j \quad \text{arbitrary}. \end{array} \right.$$

We are now going to verify that

$$(2.17) \quad \hat{\theta}_n^j < \hat{\theta}_n^{j+1} \text{ if } \hat{\theta}_n^j < \infty.$$

But

$$u^{n-j}(x_n^j(\hat{\theta}_n^j)) + k + c_0(\hat{\xi}_n^j) = u^{n-j+1}(x_n^{j-1}(\hat{\theta}_n^j))$$

and since

$$u^{n+1-j} \leq Mu^{n-j}$$

$$u^{n-j}(x_n^j(\hat{\theta}_n^j)) \leq u^{n-j}(x_n^{j-1}(\hat{\theta}_n^j) + \hat{\xi}_n^j + \xi) +$$

$$+ c_0(\hat{\xi}_n^j + \xi) + k - k - c_0(\hat{\xi}_n^j)$$

and by the sub-linearity of c_0

$$\leq u^{n-j}(x_n^j(\hat{\theta}_n^j) + \xi) + c_0(\xi) + k - k$$

therefore

$$u^{n-j}(x_n^j(\hat{\theta}_n^j)) \leqslant Mu^{n-j}(x_n^j(\hat{\theta}_n^j)) - k$$

and since $u^{n-j} \leqslant u^{n-j-1}$, we have, finally,

$$u^{n-j}(x_n^j(\hat{\theta}_n^j)) \leqslant Mu^{n-j-1}(x_n^j(\hat{\theta}_n^j)) - k$$

which, by the definition of $\hat{\theta}_n^{j+1}$ (cf., (2.15)), proves (2.17). We thus have defined a control $\hat{W}_n \in \mathcal{W}_{n+1}$ and we easily verify that

$$u^n(x) = J^x(\hat{W}_n)$$

which completes the proof of the result desired. □

Starting from the result of Theorem 2.1, it is possible to show that u^n converges in $C^0(\bar{\mathcal{O}})$ without reference to the result (2.5) of Hanouzet-Joly. We obtain, however, an arithmetic, and not geometric convergence. The (probabilistic) method that we give is owed to J.L. MENALDI [1]. However, we shall need the additional hypothesis (1)

(2.18) $\quad b(x)$ is Lipschitz continuous.

In this case we can solve (1.19) on a space with fixed probability $(\Omega_0, \mathcal{M}_0, \mathcal{M}^t, P, w)$, i.e., where P and w do not depend on the impulse control chosen. We then have

$$J^x(W) = E\Big[\int_0^\tau f(x(t))\Big(\exp - \int_0^t a_0 ds\Big)dt + \sum_n (k + c_0(\xi^n))\Big(\exp - \int_0^{\theta^n} a_0(x(s))ds\Big)\chi_{\theta^n < \infty}\Big]. \tag{2.19}$$

We shall write

$$u^*(x) = \operatorname*{Inf}_W J^x(W) \tag{2.20}$$

which defines $u^*(x)$ (we shall not use the result of Theorem 1.1). It is clear that

$$u^*(x) \leqslant u^0(x)$$

(take $\theta^1 = +\infty$), consequently, without loss of generality, in the evaluation of (2.20) we can restrict ourselves to the controls W such that

$$\left|\begin{array}{l} \displaystyle\int_0^\tau f(x(t))\Big(\exp - \int_0^t a_0 ds\Big)dt < \infty \quad \text{a.s. } P. \\ \theta^n < \theta^{n+1} \quad \text{a.s.} \end{array}\right. \tag{2.21}$$

Let W be an arbitrary impulse control satisfying (2.21). To it we can associate a control denoted $\tilde{W}_n \in \mathcal{W}_n$. In fact it suffices to write

$$\tilde{\theta}^j_n = \theta^j \text{ for } j \leqslant n-1$$

$$\tilde{\theta}^j_n = +\infty \text{ for } j \geqslant n.$$

We denote as $\tilde{x}(t)$ the trajectory associated with $\tilde{W}_n$. It is clear that

$$x(t) = \tilde{x}(t) \text{ for } t < \theta^n.$$

Let $\tilde{\tau}$ be the exit time of $\dot{\mathcal{O}}$ (cf., (1.21)). We have

$$\left|\begin{array}{l} \tilde{\tau} = \tau \text{ if } \tau < \theta^n \text{ or } \theta^n = +\infty \\ \tilde{\tau} = \tau_{\theta^{n-1}} = \tau^{n-1} \text{ if } \tau \geqslant \theta^n \text{ and } \theta^n < \infty. \end{array}\right. \tag{2.22}$$

(1) For the proof. This allows us to avoid some technical difficulties.

By noting that the cost of the control W is greater than or equal to that of the control $\tilde{W}_n$, we have

(2.23)
$$J^x(W) - J^x(\tilde{W}_n) \geq E[\int_0^\tau f(x(t))(\exp - \int_0^t a_0 ds)dt - \\ - \int_0^{\tilde{\tau}} f(\tilde{x}(t))(\exp - \int_0^t a_0(\tilde{x}(s))ds)dt]$$

and by (2.22)

$$= E\,[(\int_0^\tau f(x(t))(\exp - \int_0^t a_0 ds)dt - \int_0^{\tilde{\tau}} f(x(t))(\exp - \int_0^t a_0(\tilde{x}(s))ds)dt) \times \\ \times \chi_{\theta^n < \infty}\ \chi_{\tau \geq \theta^n}\ ]$$

$$= E\,[(\int_{\theta^{n-1}}^\tau f(\exp - \int_0^t a_0 ds)dt - \int_{\theta^{n-1}}^{\tau\,\theta^{n-1}} f(\tilde{x})(\exp - \int_0^t a_0(\tilde{x})ds)dt)\chi_{\theta^n<\infty}\ \chi_{\tau\geq\theta^n}]$$

$$\geq - E\ [\int_{\theta^{n-1}}^{\tau\,\theta^{n-1}} f(\tilde{x}(t))(\exp - \int_0^t a_0(\tilde{x}(s))ds)dt\ ]\ \chi_{\theta^{n-1}<\infty}$$

$$= - E\, u^0(x(\theta^{n-1}))(\exp - \int_0^{\theta^{n-1}} a_0(x(s))ds)\ \chi_{\theta^{n-1}<\infty}\ .$$

$$\geq - \|u^0\|_{L^\infty}\ \dot{E}\ (\exp - \int_0^{\theta^{n+1}} a_0(x(s))ds)\ \chi_{\theta^{n-1}<\infty}\,.$$

But, furthermore, we can always assume, as we saw above

$$J^x(W) \leqslant u^0(x) \leqslant \|u^0\|$$

therefore

$$E \sum_j (k + c_0(\xi^j))(\exp - \int_0^{\theta^j} a_0(x(s))ds)\ \chi_{\theta^j<\infty} \leq \|u^0\|$$

but the left member of the preceeding inequality is greater than or equal to

$$k(n-1)E(\exp - \int_0^{\theta^{n-1}} a_0(x(s))ds)\ \chi_{\theta^{n-1}<\infty}$$

thus

$$E(\exp - \int_0^{\theta^{n-1}} a_0(x(s))ds)\ \chi_{\theta^{n-1}<\infty} \leq \frac{\|u^0\|}{k(n-1)}$$

which with (2.23) implies

$$J^x(W) - J^x(\tilde{W}_n) \geq - \frac{\|u^0\|^2}{k(n-1)}\ .$$

By Theorem 2.1, we have

$$J^x(\tilde{W}_n) \geq u^{n-1}(x)$$

and therefore finally

$$u^*(x) \geq u^{n-1}(x) - \frac{\|u^0\|^2}{k(n-1)} .$$

Now,

$$u^n(x) \geq u^*(x)$$

whence

$$\| u^n - u^* \|_{L^\infty} \leq \frac{\|u^0\|^2}{kn} . \tag{2.24}$$

2.2 *Increasing Method*

The increasing method is defined in the following way. Starting from

$$u_0 = 0 \tag{2.25}$$

and knowing u_n, we solve

$$\left| \begin{array}{l} a(u_{n+1}, v-u_{n+1}) \geq (f, v-u_{n+1}) \\ u_{n+1} \in H_0^1 , \quad u_{n+1} \leq Mu_n \\ \forall\, v \in H_0^1 , \quad v \leq Mu_n . \end{array} \right. \tag{2.26}$$

We then have the properties

$$\left| \begin{array}{l} u_n \in C^0(\bar{\mathcal{O}}), \quad u_n \geq 0 \\ 0 \leq u_n(x) \leq u(x) \leq u_n(x) + \frac{(1-\mu)^n}{\mu} \| u^0 \| . \end{array} \right. \tag{2.27}$$

Our objective is to give the probabilistic interpretation of $u_n(x)$. For every impulse control W we define

$$\left| \begin{array}{l} J_x^n(W) = E_W^x \left[\int_0^{\tau\wedge\theta^n} f(x(t)) (\exp - \int_0^t a_0(x(s))ds)dt + \right. \\ \left. + \sum_{j=1}^n (k + c_0(\xi^j)) (\exp - \int_0^{\theta^j} a_0(x(s))ds)\, \chi_{\theta^j < \infty} \right] . \end{array} \right. \tag{2.28}$$

We then have:

THEOREM 2.2: *Take the assumptions of Theorem 1.1. Then the sequence* $u_n(x)$ *defined by* (2.25),(2.26) *satisfies*

$$u_n(x) = \operatorname*{Inf}_W J_x^n(W) \tag{2.29}$$

Furthermore, there exists an optimal control.

Proof: Assume $n \geqslant 1$. Arguing as in Theorem 2.1 we again obtain (2.10), with u_{n-j} and u_{n-j-1} instead of u^{n-j} and u^{n-j-1}, respectively. We then take the mathematical expectation and add the relation obtained for j varying between 0 and n-1. Using that $u_0 = 0$, we immediately obtain

$$u_n(x) \geqslant J^{x,n}(W) \tag{2.30}$$

Next we prove the existence of an optimal control. Define $\hat{\xi}_n(x)$ as a positive Borel function such that

$$u_n(x) = k + c_0(\hat{\xi}_n(x)) + u_{n-1}(x+ \hat{\xi}_n(x)) \quad \forall\, x \in \bar{\mathcal{O}} . \quad n \geq 1 \tag{2.31}$$

Next define

$$\left| \begin{array}{l} dx^0 = \sigma(x^0)\, dw_0 \\ x^0(0) = x \end{array} \right.$$

$$T^{0,n} = \inf \{t \mid u_n(x^0(t)) = Mu_{n-1}(x^0(t))\}$$

$$S^0 = \text{exit time of } x^0 \text{ from } \mathcal{O}$$

and

$$\left| \begin{array}{lll} \hat{\theta}^{1,n} = T^{0,n} & \text{if } T^{0,n} < S^0 , & + \infty \quad \text{otherwise} \\ \hat{\xi}^{1,n} = \hat{\xi}_n(x^0(\hat{\theta}^{1,n})) & \text{if } \hat{\theta}^{1,n} < \infty & \text{otherwise arbitrary.} \end{array} \right. \tag{2.32}$$

Having defined the process $x^{j-1,n}$ as well as $\hat{\theta}^{j,n}, \hat{\xi}^{j,n}$ for $j = 1,\ldots,n-1$, we define $x^{j,n}, \hat{\theta}^{j+1,n}, \hat{\xi}^{j+1,n}$ by the formula

$$\left| \begin{array}{l} dx^{j,n} = \sigma(x^{j,n})\, dw_0 \\ x^{j,n}(\hat{\theta}^{j,n}) = x^{j-1,n}(\hat{\theta}^{j,n}) + \hat{\xi}^{j,n} \end{array} \right. \tag{2.33}$$

$$T^{j,n} = \inf\{ t \geq \hat{\theta}^{j,n} \mid u_{n-j}(x^{j,n}(t)) = Mu_{n-j-1}(x^{j,n}(t))\}$$

$$S^{j,n} = \text{exit time of } x^{j,n} \text{ from } \mathcal{O} \text{, after } \hat{\theta}^{j,n} ,$$

then

$$\left| \begin{array}{lll} \hat{\theta}^{j+1,n} = T^{j,n} & \text{if } T^{j,n} < S^{j,n} , & + \infty \quad \text{otherwise} \\ \hat{\xi}^{j+1,n} = \hat{\xi}_{n-j}(x^{j,n}(\hat{\theta}^{j+1,n})) & \text{if } \hat{\theta}^{j+1,n} < \infty & \text{otherwise arbitrary.} \end{array} \right.$$

and

$$\left| \begin{array}{l} \hat{\theta}^{j,n} = +\infty \ \text{ if } j \geqslant n+1 \\ \xi^{j,n} \text{ arbitrary.} \end{array} \right.$$

Let us show that the control $\hat{W}^n$ thus defined is optimal. We shall write $x^j, \hat{\theta}^j$ for $x^{j,n}, \hat{\theta}^{j,n}$. By applying (1.42) we have

(2.34) $$u_{n-j}(x^j(\hat{\theta}^j))\chi_{\hat{\theta}^j<\infty} = E_{\hat{W}}^x [\chi_{\hat{\theta}^j<\infty} \int_{\hat{\theta}^j}^{\hat{\theta}^{j+1}\wedge\tau^j} f(x(s))(\exp-\int_{\hat{\theta}^j}^s a_0 d\lambda)ds + + u_{n-j}(x^j(\hat{\theta}^{j+1}\wedge\tau^j))\, \chi_{\hat{\theta}^j<\infty} \exp - \int_{\hat{\theta}^j}^{\hat{\theta}^{j+1}\wedge\tau^j} a_0 ds \mid \mathcal{M}^{\hat{\theta}^j}].$$

if $\hat{\theta}^{j+1} < \infty$, it follows from (2.33) that $\hat{\theta}^{j+1} < S^j = \tau^j$

therefore

(2.35) $$\chi_{\hat{\theta}^{j+1}<\infty} = \chi_{\hat{\theta}^{j+1}<\tau^j}.$$

Also

(2.36) $$\hat{\theta}^{j+1} \leqslant \infty \text{ implies } \hat{\theta}^{j+1} < \tau.$$

In fact if $\hat{\theta}^{j+1} < \infty$, then $x^j(t) \in \mathcal{O}$ for $t \in [\hat{\theta}^j\ \hat{\theta}^{j+1}]$, as well as $x^{j-1}(t)$ for $t \in [\hat{\theta}^{j-1}, \hat{\theta}^j]$ etc. ... From the construction of $x(t)$, we deduce

$$x(t) \in \mathcal{O} \quad , \quad \forall\, t < \hat{\theta}^{j+1} \text{ and } \quad x(\hat{\theta}^{j+1} - 0) \in \mathcal{O}$$

hence (2.36), and $x(\tau) \notin \mathcal{O}$ if $\tau = \hat{\theta}^{j+1}$. We also have

(2.37) $$\hat{\theta}^j < \infty \text{ implies } \hat{\theta}^{j+1} \wedge \tau^j = \hat{\theta}^{j+1} \wedge \tau.$$

In fact if $\hat{\theta}^{j+1} < \infty$, then $\hat{\theta}^{j+1} \wedge \tau = \hat{\theta}^{j+1}$ and $\hat{\theta}^{j+1} \wedge \tau^j = \hat{\theta}^{j+1}$. If $\hat{\theta}^{j+1} = \infty$, then

$$x(t) = x^j(t) \quad \forall t \geqslant \hat{\theta}^j$$

and $x(\hat{\theta}^j - 0) \in \mathcal{O}$, therefore $\tau = \tau^j$ necessarily. Also

(2.38) $$u_{n-j}(x^j(\hat{\theta}^{j+1}\wedge\tau^j))\, \chi_{\hat{\theta}^j<\infty} = u_{n-j}(x^j(\hat{\theta}^{j+1}))\, \chi_{\hat{\theta}^{j+1}<\tau^j}$$
$$= u_{n-j}(x^j(\hat{\theta}^{j+1}))\, \chi_{\hat{\theta}^{j+1}<\infty}$$
$$= (k + c_0(\hat{\xi}^{j+1}))\, \chi_{\hat{\theta}^{j+1}<\infty} + u_{n-j-1}(x^{j+1}(\hat{\theta}^{j+1}))\chi_{\hat{\theta}^{j+1}<\infty}.$$

Therefore, starting from (2.34) and from (2.35), (2.36), (2.37), (2.38), we have

(2.39) $$E_{\hat{W}}^x\, u_{n-j}(x^j(\hat{\theta}^j))(\exp - \int_0^{\hat{\theta}^j} a_0(x(s))ds)\, \chi_{\hat{\theta}^j<\infty} =$$
$$= E_{\hat{W}}^x\, \chi_{\hat{\theta}^j<\infty} \int_{\hat{\theta}^j\wedge\tau}^{\hat{\theta}^{j+1}} f(x(s))(\exp-\int_0^s a_0(x(\lambda))d\lambda)ds +$$
$$+ E_{\hat{W}}^x\, u_{n-j-1}(x^{j+1}(\hat{\theta}^{j+1}))\, (\exp - \int_0^{\hat{\theta}^{j+1}} a_0(x(s))ds)\, \chi_{\hat{\theta}^{j+1}<\infty} +$$

(Contd)

$$+ E^x_{\hat{W}} (k + c_0(\hat{\xi}^{j+1})) \chi_{\hat{\theta}^{j+1} < \infty} \exp - \int_0^{\hat{\theta}^{j+1}} a_0(x(s))ds .$$

Let $\nu \leqslant n$ be the largest (random) integer such that

$$\hat{\theta}^j < \infty \text{ if } j \leqslant \nu$$

then by (2.35),

$$\hat{\theta}^j \leqslant \tau \text{ if } j \leqslant \nu.$$

So we have

$$\int_0^\tau f(\exp- \int_0^s a_0)ds = \sum_{j=0}^{\nu} \int_{\hat{\theta}^j \wedge \tau}^{\hat{\theta}^{j+1}} f(\exp - \int_0^s a_0)ds$$
$$= \sum_{j=0}^{n} \chi_{\hat{\theta}^j < \infty} \int_{\hat{\theta}^j \wedge \tau}^{\hat{\theta}^{j+1} \wedge \tau} f(\exp - \int_0^s a_0)ds .$$

Summing the relations (2.39) for $j = 0,\ldots,n-1$, and using $u_0 = 0$, we deduce

(2.40) $$u_n(x) = J^{x,n}(\hat{W}^n)$$

which completes the proof of the desired result. □

Remark 2.1:

The proof of (2.40) is slightly different from those carried out in Theorems 1.1 and 2.1. This is because we can use the property $\hat{\theta}^{j,n} < \hat{\theta}^{j+1,n}$, which is not necessarily true. However, the proof carried out for Theorem 2.2 could have been used for Theorem 1.1 and 2.1. □

We can prove an estimate similar to (2.24). If (provisionally) we define

$$u^*(x) = \operatorname*{Inf}_W J^x(W)$$

Then since

$$J^n_x(W) \leqslant J^x(W)$$

It is clear that

$$u_n(x) \leqslant u^*(x).$$

Also

$$u^*(x) - u_n(x) \leq J^x(\hat{W}^n) - J^n_x(\hat{W}^n)$$
$$= E^x_{\hat{W}^n} [\int_{\hat{\theta}^n}^\tau f(x(t))(\exp-\int_0^t a_0(x(s))ds)dt \; \chi_{\hat{\theta}^n < \infty}$$
$$= E^x_{\hat{W}^n} \; u^0(x(\hat{\theta}^n))(\exp - \int_0^{\hat{\theta}^n} a_0(x(s))ds) \; \chi_{\hat{\theta}^n < \infty} \qquad \text{(Contd)}$$

$$\leq \|u^0\| \; E^x_{\hat{W}^n}\left(\exp - \int_0^{\hat{\theta}^n} a_0(x(s))ds\right) \chi_{\hat{\theta}^n < \infty} \; .$$

Now

$$k\, n\, E^x_{\hat{W}^n}\left(\exp - \int_0^{\hat{\theta}^n} a_0(x(s))ds\right) \chi_{\hat{\theta}^n < \infty}$$

$$\leq J^n_x(\hat{W}^n) = u_n(x) \leqslant \|u^0\|$$

and therefore

$$0 \leq u^*(x) - u_n(x) \leq \frac{\|u^0\|^2}{kn} \; . \quad \square$$

3. THE UNBOUNDED CASE

Method of Approach

We consider the impulse control problem in the case of unbounded open sets and unbounded cost functions (cf., Chapter 4, Section 4). To simplify matters a little we shall work in R^n with drift terms and regular diffusion terms. We shall take the adaptation coefficient as constant.

3.1 *The Problem: Assumptions: Notations*

We are given

(3.1)
$$\begin{cases} a_{ij} = a_{ji} \in W^{1,\infty}(R^n) \\ \Sigma\, a_{ij}\, \xi_i\, \xi_j \geq \alpha|\xi|^2 \quad \forall\, \xi \in R^n, \; \alpha > 0 \; , \end{cases}$$

(3.2)
$$\begin{cases} a_i \in L^\infty \text{ such that} \\ b_i = - a_i + \sum_j \dfrac{\partial a_{ij}}{\partial x_j} \in W^{1,\infty}(R^n) \; , \end{cases}$$

(3.3) $\quad f$ measurable $0 \leqslant f(x) \leqslant f_0(1 + |x|^s), s \geqslant 0,$ (1)

(3.4)
$$\begin{cases} k > 0, \quad c_0 : (R^n)^+ \to R^+, \text{ continuous, } c_0(0) = 0 \; , \\ \text{sub-linear, non-decreasing, } c_0(\xi) \to +\infty \text{ if} \\ |\xi| \to \infty \quad , \; c_0(\xi) \leq d|\xi|^\gamma \quad . \end{cases}$$

We write

(3.5) $\quad \sigma \in W^{1,\infty}$ such that $\dfrac{\sigma^2}{2} = a,$

---

(1) Or even an exponential, so that (3.15) remains satisfied.

(3.6)
$$\begin{aligned} A &= -\sum_{i,j} \frac{\partial}{\partial x_i} a_{ij} \frac{\partial}{\partial x_j} + \sum_i \frac{\partial}{\partial x_i} \\ &= -\sum_{i,j} a_{ij} \frac{\partial^2}{\partial x_i \partial x_j} - \sum_i b_i \frac{\partial}{\partial x_i} . \end{aligned}$$

We now consider a probability space $\Omega, \mathcal{A}, P, \mathcal{F}^t$, and an R^n-valued (standard) Wiener process with respect to $\mathcal{F}^t$. An impulse control is a set

(3.7)
$$\begin{aligned} W = \ & \theta^1, \theta^2, \ldots, \theta^n, \ldots \\ & \xi^1, \xi^2, \ldots, \xi^n, \ldots \end{aligned}$$

θ^n stopping times of $\mathcal{F}^t$

ξ^n R^n-valued $\mathcal{F}^{\theta^n}$-measurable R.V.

$\theta^n \le \theta^{n+1}$,

$\theta^n \uparrow +\infty$ a.s. ($\theta^n = +\infty$ is possible).

For a given impulse control we can solve (in the strong sense) the equation

(3.8)
$$\begin{aligned} & dy = b(y(t))dt + \sigma(y(t))dw(t) + \Sigma\delta(t-\theta^i)\xi^i \\ & y(0) = x . \end{aligned}$$

We shall write

$$y(t) = y_x(t),$$

and define the functional

(3.9)
$$J^x(W) = E\left[\int_0^\infty e^{-\beta t} f(y_x(t))dt + \sum_n (k + c_0(\xi^n))e^{-\beta\theta^n} \chi_{\theta^n < \infty} \right].$$

Our objective is to characterize the lower bound of $J^x(W)$.

3.2 *Study of the Equation*

As in Section 4 of Chapter 4 we consider

$$\beta_\mu(x) = \exp - \mu(|x|^2 + 1)^{1/2}$$

and the spaces $L^{p,\mu}$, $W^{1,p,\mu}$, $W^{2,p,\mu}$ (notice that $\mathcal{O} = R^n$). We write $H^\mu = L^{2,\mu}$, $V^\mu = W^{1,2,\mu}$, and $(\ ,\ )_\mu$, $((\ ,\ ))_\mu$ for the scalar products in H^μ and V^μ respectively. Now define

(3.10)
$$a(u,v) = \Sigma \int_{R^n} a_{ij} \frac{\partial u}{\partial x_j} \frac{\partial v}{\partial x_i} \beta_\mu^2 dx + \Sigma \int_{R^n} \left(a_i - \frac{\sum_j a_{ji} x_j \mu}{(1+|x|^2)^{1/2}}\right) \frac{\partial u}{\partial x_i} v \beta_\mu^2 dx$$

therefore

$$(3.11)\quad \begin{cases} a_\mu(u,v) = \int Au\ v\ \beta_\mu^2\ dx \\ \text{if } u \in W^{2,p,\mu} \text{ and } v \in V^\mu . \end{cases}$$

There exists (for fixed μ) a positive λ such that

$$(3.12)\quad a_\mu(v,v) + \lambda|v|_\mu^2 \geqslant \gamma\,||v||_\mu^2 \quad \forall v \in V^\mu .$$

Furthermore, $f \in H^\mu, \forall\mu > 0$.

Here we are interested in the equation

$$(3.13)\quad a_\mu(u,v) + \beta(u,v)_\mu = (f,v)_\mu \quad \forall v \in V^\mu .$$

Now write (the notation is changed to avoid some confusion. cf., Chapter 4 (4.7))

$$(3.14)\quad \tilde{u}(x) = F^0\beta_{-\mu_0}(x) = F^0 \exp \mu_0(|x|^2 + 1)^{1/2}.$$

by what we saw in Lemma 4.1, Chapter 4, we have

$$(3.15)\quad a_\mu(\tilde{u},v) + \beta(\tilde{u},v)_\mu \geqslant (f,v)_\mu \quad \forall v \in V^\mu,\ v \geqslant 0$$

at least for a suitable choice of constants F^0 and $\mu_0 > 0$ and $\mu > \mu_0$.

LEMMA 3.1: *Equation* (3.13) *has a minimum solution and a maximum solution in the interval*

$$(3.16)\quad K = \{u \in H^\mu \mid 0 \leqslant u \leqslant \tilde{u}\}.$$

Proof: Consider the iterative methods

$$(3.17)\quad \begin{cases} a_\mu(u^{n+1},v) + (\lambda+\beta)(u^{n+1},v)_\mu = (f,v)_\mu + \lambda(u^n,v)_\mu \\ u^0 = \tilde{u} \end{cases}$$

and

$$(3.18)\quad \begin{cases} a_\mu(u_{n+1},v) + (\lambda+\beta)(u_{n+1},v)_\mu = (f,v)_\mu + \lambda(u_n,v)_\mu \\ u_0 = 0 . \end{cases}$$

By (3.12) we define the sequences u^n and u_n uniquely in V^μ. We easily verify that u_n is increasing, u^n is decreasing, and that

$$0 \leqslant u_n \leqslant u^n \leqslant \tilde{u}.$$

Since u_n, u^n stay bounded in H^μ, it follows from (3.17),(3.18) that u^n, u_n are bonded in V^μ. We easily go to the limit to conclude that

$$u^n \downarrow \bar{u},\ u_n \uparrow \underline{u} \text{ are solutions of (3.13).}$$

We easily verify also that $\bar{u}$ is a maximum solution and $\underline{u}$ is a minimum solution. □

In the remainder we shall consider only the *minimum* solution of (3.13) which we shall denote by u (and by u^0 later on). It is this solution which will have a probabilistic interpretation. First of all we define

$$(3.19) \qquad f_M(x) = f(x) \wedge M$$

and u_M a solution of

$$(3.20) \qquad a_\mu(u_M,v) + \beta(u_M,v)_\mu = (f_M,v)_\mu, \quad u_M \in V^\mu.$$

In fact the solution of (3.20) in L^∞ is unique and

$$(3.21) \qquad \| u_M \|_{L^\infty} \leqslant \frac{1}{\beta} \| f_M \|_{L^\infty} .$$

LEMMA 3.2: *The sequence* $u_M \uparrow u$ *is a minimum solution of* (3.13).

Proof: Define an increasing process $u_{M,n}$ relative to u_M. Since $f_M \leqslant f$, we easily verify by induction that

$$u_{M,n} \leqslant u_n$$

and therefore

$$(3.22) \qquad u_M \leqslant u.$$

Also u_M is increasing, and therefore converges to u^*. Since by (3.20)

$$a_\mu(u_M,v) + (\beta + \lambda)(u_M,v)_\mu = (f_M + \lambda u_M,v)_\mu$$

it follows from (3.12) that u_M is bounded in V^μ. By passing to the limit, we deduce from this that u^* is a solution of (3.13), and since $u^* \leqslant u$, then $u^* = u$ necessarily, hence the result. □

Notice that there may be an infinity of solutions in K defined in (3.16). In fact let us take $\mathcal{O} = R$, and for (3.13) the equation

$$-\frac{\partial^2 u}{\partial x^2} + \beta u = f$$

where f is a constant > 0 and where β is > 1.

If we write

$$\tilde{u} = F_0 e^{\sqrt{\beta}(x^2 + 1)^{1/2}}$$

(thus $\mu_0 = \sqrt{\beta}$) we find

$$-\frac{\partial^2 \tilde{u}}{\partial x^2} + \beta \tilde{u} = F_0 \sqrt{\beta}\, e^{\sqrt{\beta}(x^2+1)^{1/2}} (x^2+1)^{-1} [\sqrt{\beta} - (x^2+1)^{-1/2}]$$

Since β > 1, we have:

$$\sqrt{\beta} - (x^2 + 1)^{-1/2} \geqslant \gamma > 0$$

and therefore

$$-\frac{\partial^2\tilde{u}}{\partial x^2} + \beta\tilde{u} \geq f \quad \text{if}$$

$$\gamma\, F_0 \sqrt{\beta} \inf_x [e^{\sqrt{\beta}(x^2+1)^{1/2}} (x^2+1)^{-1}] \geq f$$

i.e.,

$$F_0 \geq \frac{4f\, e^{-2}}{\gamma\beta^{3/2}} .$$

The relation is a fortiori satisfied if we increase F_0, and therefore we can always assume $F^0 \geqslant \frac{2f}{\beta}$. If we then consider

$$u = \frac{f}{\beta} + ce^{x\sqrt{\beta}} , \quad 0 \leq c \leq f/_{\beta}$$

we see that u is a solution with $0 \leqslant u \leqslant \tilde{u}$; we now only have to verify that $u \leqslant \tilde{u}$; now since $F_0 \geqslant \frac{2f}{\beta}$, it suffices to verify that

$$\frac{f}{\beta} + ce^{\sqrt{\beta}\, x} \leq \frac{2f}{\beta}\, e^{\sqrt{\beta}(x^2+1)^{1/2}} \qquad \text{if} \quad 0 \leq c \leq f/_{\beta}$$

which follows immediately.

We can then prove:

THEOREM 3.1: *Assume* (3.1), (3.2), (3.3). *Then the minimum solution of* (3.13) *is continuous and given explicitly by*

$$u(x) = E \int_0^{\infty} e^{-\beta t} f(y_x(t))dt . \tag{3.23}$$

Proof: Continuity follows from the local regularity, since u satisfies

$$Au + \beta u = f.$$

Also we have

$$u_M(x) = E \int_0^{\infty} e^{-\beta t} f_M(y_x(t))dt$$

hence it easily follows by the monotone convergence Theorem that

$$u_M(x) \uparrow E \int_0^{\infty} e^{-\beta t} f(y_x(t))dt$$

whence the result, using Lemma 3.2. □

Remark 3.1

We have thus proved the estimate

$$E\int_0^\infty e^{-\beta t} f(y_x(t))dt \le F^0 \exp \mu_0(|x|^2+1)^{1/2}$$

We could also obtain an estimate starting from $E|y_x(t)|^s$ by using Ito's formula. □

3.3 *Study of the Variational Inequality*

In the remainder we write

$$u^0(x) = E\int_0^\infty e^{-\beta t} f(y_x(t))dt \tag{3.24}$$

which is therefore continuous in V^μ and is a minimum solution of (3.13) in the interval $[0,\tilde{u}]$.

We are now given an obstacle ψ satisfying

$$\left|\begin{array}{l}\psi \quad \text{continuous} \\ 0 \le \psi \quad ; \quad \psi \in H^\mu .\end{array}\right. \tag{3.25}$$

and are interested in the V.I.

$$\left|\begin{array}{l} a_\mu(u,v-u) + \beta(u,v-u)_\mu \ge (f,v-u)_\mu \\ \qquad \forall v \in V^\mu \quad , \quad v \le \psi \\ \qquad u \in V^\mu \quad , \quad u \le \psi . \end{array}\right. \tag{3.26}$$

We begin with:

LEMMA 3.3: *Assume* (3.1), (3.2), (3.3) *and* (3.25). *Then the V.I.* (3.26) *has a minimum solution and a maximum solution in the interval* $[0,u^0]$. □

Proof: Consider increasing and decreasing methods:

$$\left|\begin{array}{l} a_\mu(u_{n+1},v-u_{n+1}) + (\beta+\lambda)(u_{n+1},v-u_{n+1})_\mu \ge (f,v-u_{n+1})_\mu + \\ \qquad\qquad\qquad\qquad +\lambda(u_n,v-u_{n+1})_\mu \\ \qquad \forall v \in V^\mu \quad , \quad v \le \psi \\ \qquad u \in V^\mu \quad , \quad u \le \psi \\ u_0 = 0 \end{array}\right. \tag{3.27}$$

and u^n is a decreasing method, beginning at u^0. We easily verify that $u_n \uparrow \underline{u}$ is a minimum solution (in $[0,u^0]$) of (3.26) and $u^n \downarrow \bar{u}$ a maximum solution (in $[0,u^0]$) of (3.26). □

LEMMA 3.4: *The minimum solution of* (3.26) *is continuous. The maximum solution* $\bar{u}$ *is u.s.c.*

Proof: We begin by noticing that if f is bounded, then u^0 is bounded, and the solution of (3.26) is unique, moreover it is continuous. The uniqueness results from the mapping $T_\lambda z = \zeta$ defined by

$$a_\mu(\zeta,v-\zeta) + (\beta+\lambda)(\zeta,v-\zeta)_\mu \geqslant (f+\lambda z,v-\zeta)_\mu$$

maps L^∞ into itself and is a contraction.

To prove continuity we first notice that the penalised problem

$$a_\mu(u^\varepsilon,v) + \beta(u^\varepsilon,v)_\mu + \frac{1}{\varepsilon}((u^\varepsilon-\psi)^+,v)_\mu = (f,v)_\mu$$

which has a bounded and *continuous* unique solution, satisfies

$$u^\varepsilon \downarrow u \text{ when } \varepsilon \downarrow 0$$

thus u is u.s.c. Let us also consider $\psi_R(x) = \psi\theta_R(x)$, where

$$\theta_R(x) = \theta(\frac{x}{R})$$

with $\theta = 1$ for $|x| \leqslant 1$, $\theta = 0$ for $|x| \geqslant 2$, $0 \leqslant \theta \leqslant 1$, θ regular.

Let u_R be the solution of the V.I. corresponding to the obstacle ψ_R. Then u_R is continuous, for ψ_R is uniformly continuous and bounded, therefore it can be approximated in the 'sup' norm by a sequence of regular and bounded functions, denoted $\psi_{R,k}$. Now,

$$\| u_R - u_{R,k}\|_{L^\infty} \leqslant \| \psi_R - \psi_{R,k}\|_{L^\infty}$$

where $u_{R,k}$ is the solution of the V.I. corresponding to the obstacle $\psi_{R,k}$. For $u_{R,k}$ local regularity is satisfied. From this we certainly deduce that u_R is continuous. Now we set $\psi_q(x) = \psi_{2q}$

$u_q(x) = u_{2q}$. Then we have

$$\psi_q \leqslant \psi_{q+1} \text{ (as } \psi_q \leqslant \theta_{q+1})$$

therefore

$$u_q \leqslant u_{q+1}. \text{ in fact } u_q \uparrow u.$$

In fact we have

$u_q \leqslant u$ (since $\psi_q \leqslant \psi$) and

$$(3.28)\quad \left|\begin{array}{l} a_\mu(u_q,v-u_q) + \beta\,(u_q,v-u_q)_\mu \geq (f,v-u_q)_\mu \\ \forall\, v \leq \psi_q, \quad u_q \leq \psi_q\,. \end{array}\right.$$

On taking $v = 0$, we see that u_q remains bounded in V^μ.

If we write $u^* = \lim \uparrow u_q$, then $u_q \to u^*$ in V^μ weakly. Now let $v \in V^\mu$, $v \leqslant \psi$ then $v_q = v\theta_q \leqslant \psi_q$ and $v_q \to v$ in V^μ strongly. On taking $v = v_q$ in (3.28) and passing to the limit we verify that u^* is a solutiom of the V.I., therefore $u^* = u$. Since u_q is continuous it follows that u is l.s.c. Since it is already u.s.c. we see that u is continuous.

Let us now consider the general case with f not bounded. Take $f_M = f \wedge M$, and write u_M for the solution of the V.I. corresponding to f_M instead of f. Then u_M is continuous and bounded. Let us verify that

(3.29) $\quad u_M \uparrow \underline{u}$ when $M \uparrow +\infty$.

Consider the increasing iterates $u_{M,n}$ and u_n. Since $f_M \leqslant f$, by induction we have

$$u_{M,n} \leqslant u_n$$

therefore

(3.30) $\quad u_M \leqslant \underline{u}$.

Also $u_M \uparrow u^*$, and it is easy to see that u^* is a solution of the V.I., hence (3.29) From (3.29) we deduce that $\underline{u}$ is l.s.c..

We now write as u_M^0 the solution of the equation corresponding to f_M instead of f, and $\tilde{u}_M$ for the solution of the V.I..

$$\text{(3.31)}\quad \begin{aligned} &a_\mu(\tilde{u}_M, v-\tilde{u}_M) + \beta(\tilde{u}_M, v-\tilde{u}_M) \geq 0 \\ &\forall\, v \in V^\mu,\quad v \leq \psi - u_M^0,\quad \tilde{u}_M \leq \psi - u_M^0 \end{aligned}$$

$\tilde{u}_M$ is then the unique solution of (3.31) in the interval $[-u_M^0, 0]$. From what we saw above, $\tilde{u}_M$ is continuous and bounded. Furthermore,

(3.32) $\quad \tilde{u}_M = u_M - u_M^0$.

Since the obstacle $\psi - u_M^0$ decreases when M increases, we know that $\tilde{u}_M$ decreases when M increases. Since $u_M \uparrow \underline{u}$ and $u_M^0 \uparrow u^0$, we have

$$\tilde{u}_M \downarrow \underline{u} - u^0.$$

Since $\tilde{u}_M$ is continuous, we see that $\underline{u} - u^0$ is u.s.c., therefore since u^0 is continuous, $\underline{u}$ is u.s.c.. Since $\underline{u}$ is already l.s.c. we have proved that $\underline{u}$ is continuous. To prove that the maximum solution $\bar{u}$ is u.s.c. we consider the penalised problem

(3.33) $\quad a_\mu(u^\varepsilon, v) + \beta(u^\varepsilon, v)_\mu + (\frac{1}{\varepsilon}(u^\varepsilon - \psi)^+, v)_\mu = (f,v)_\mu$.

More precisely, we denote by u^ε the *maximum* solution of (3.33) in the interval $[0,u^0]$. It is obtained by starting from the decreasing process

$$\begin{aligned} &a_\mu(u^{\varepsilon,n+1}, v) + (\beta+\lambda)(u^{\varepsilon,n+1}, v)_\mu + \\ &+ (\tfrac{1}{\varepsilon}(u^{\varepsilon,n+1} - \psi)^+, v)_\mu = (f,v)_\mu + \lambda(u^{\varepsilon,n}, v)_\mu \\ &u^{\varepsilon,0} = u^0 . \end{aligned}$$

Now $u^{\varepsilon,n} \downarrow u^\varepsilon$ if $n \uparrow \infty$ and $u^{\varepsilon,n} \downarrow u^n$ when $\varepsilon \downarrow 0$. From this we deduce, as in Chapter 4, Theorem 1.4, that $u^\varepsilon \downarrow \bar{u}$ when $\varepsilon \downarrow 0$. But u^ε was the solution of an equation and it is continuous by local regularity. Hence $\bar{u}$ is certainly u.s.c.. □

Now consider a stopping time θ of $\mathcal{F}^t$ (cf., Section 3.1) and the functional

(3.34) $\quad J_x(\theta) = E\left[\int_0^\theta e^{-\beta t} f(y_x(t))\,dt + \psi(y_x(\theta)) e^{-\beta\theta} \chi_{\theta<\infty}\right]$.

We are going to prove:

LEMMA 3.5: *We have*

$$\underline{u}(x) = \inf J_x(\theta)$$

Furthermore, there exists an optimal stopping time $\hat{\theta}$ *defined by*

$$\hat{\theta} = \inf \{t \geqslant 0 \mid \underline{u}(y_x(t)) = \psi(y_x(t))\}.$$

Proof: Since $\underline{u}$ is continuous, the continuation set

$$C = \{x \mid \underline{u}(x) < \psi(x)\}$$

is an open set of R^n, therefore $\hat{\theta}$ is a stopping time. Also, we know that $u_M \uparrow \underline{u}$. Furthermore, for u_M, we have the usual probabilistic interpretation, so

$$u_M(x) = \operatorname*{Inf}_{\theta} J_x^M(\theta) \tag{3.35}$$

with

$$J_x^M(\theta) = E\,[\int_0^\theta e^{-\beta t} f_M(y_x(t))dt + \psi(y_x(\theta))e^{-\beta\theta}\chi_{\theta<\infty}].$$

But

$$J_x^M(\theta) \leqslant J_x(\theta)$$

therefore

$$\underline{u}(x) \leqslant \operatorname*{Inf}_{\theta} J_x(\theta) \tag{3.36}$$

Further, let $\hat{\theta}^M$ be the optimal stopping time for (3.35), so

$$\hat{\theta}^M = \inf \{t \geqslant 0 \mid u_M(y_x(t)) = \psi(y_x(t))\}.$$

Since $u_M \leqslant \underline{u}$, it is clear that $\hat{\theta}^M \geqslant \hat{\theta}$. But then (cf., Theorem 1.2, (1.42)) we can write

$$u_M(x) = E[\int_0^{\hat{\theta}} e^{-\beta t} f_M(y_x(t))dt + e^{-\beta\hat{\theta}} u_M(y_x(\hat{\theta}))\chi_{\hat{\theta}<\infty}]$$

from which we deduce by Fatou's Lemma that

$$E[\int_0^{\hat{\theta}} e^{-\beta t} f(y_x(t))dt + e^{-\beta\hat{\theta}}\underline{u}(y_x(\hat{\theta}))\chi_{\hat{\theta}<\infty}] \leq \underline{u}(x). \tag{3.37}$$

But by the definition of $\hat{\theta}$

$$\underline{u}(y_x(\hat{\theta}))\chi_{\hat{\theta}<\infty} = \psi(y_x(\hat{\theta}))\chi_{\hat{\theta}<\infty}$$

and so (3.37) implies

$$\underline{u}(x) \geq J_x(\hat{\theta})$$

which with (3.36) proves that

$$\underline{u}(x) = J_x(\hat{\theta}) = \inf_{\theta} J_x(\theta). \quad \square$$

We can then prove:

THEOREM 3.2: *Assume* (3.1), (3.2), (3.3) *and* (3.25). *Then there exists one and only one solution of* (3.26) *in the interval* $[0,u^0]$. *This solution is continuous, and*

$$u(x) = \operatorname{Inf}_{\theta} J_x(\theta) = J_x(\hat{\theta}) \tag{3.38}$$

where

$$\hat{\theta} = \inf \{t \geq 0 \mid u(y_x(t)) = \psi(y_x(t))\}. \tag{3.39}$$

Proof: By (3.33) the penalised problem implies

$$Au^{\varepsilon} + \beta u^{\varepsilon} + \frac{1}{\varepsilon}(u^{\varepsilon} - \psi)^{+} = f. \tag{3.40}$$

If θ is a stopping time and τ_R the exit time of $y_x(t)$ from the ball of radius R, it follows from the local regularity of u^{ε} and from equation (3.40) that

$$\begin{aligned} u^{\varepsilon}(x) &\leq E\Big[\int_0^{\theta\wedge\tau_R} e^{-\beta t} f(y_x(t))dt + \\ &\qquad + e^{-\beta\theta\wedge\tau_R} u^{\varepsilon}(y_x(\theta\wedge\tau_R))\Big] \\ &\leq E\Big[\int_0^{\theta\wedge\tau_R} e^{-\beta t} f(y_x(t))dt + \\ &\qquad + e^{-\beta\theta\wedge\tau_R} u^0(y_x(\theta\wedge\tau_R))\Big] = u^0(x) . \end{aligned} \tag{3.41}$$

Since $u^{\varepsilon} \downarrow \bar{u}$ when $\varepsilon \downarrow 0$, by using Lebesgue's Theorem we can pass to the limit in (3.41), and we thus obtain

$$\bar{u}(x) \leq E\Big[\int_0^{\theta\wedge\tau_R} e^{-\beta t} f(y_x(t))dt + e^{-\beta\theta\wedge\tau_R} \bar{u}(y_x(\theta\wedge\tau_R))\Big] . \tag{3.42}$$

We know that

$$\tau_R \to +\infty, \text{ if } R \to \infty;$$

but

$$E\, e^{-\beta\,\tau_R} \bar{u}(y_x(\tau_R)) \leq E\, e^{-\beta\tau_R} u^0(y_x(\tau_R)) \to 0 \tag{3.43}$$

when $R \to \infty$, for

$$u^0(x) = E[\int_0^{\tau_R} e^{-\beta t} f(y_x(t))dt + e^{-\beta\tau_R} u^0(y_x(\tau_R))]$$
$$= E \int_0^{\infty} e^{-\beta t} f(y_x(t))dt$$

hence the second part of (3.43).

Now,

$$E[\int_0^{\theta\wedge\tau_R} e^{-\beta t}f(y_x(t))dt + e^{-\beta\theta\wedge\tau_R} \bar{u}(y_x(\theta \wedge\tau_R))] \leq$$
$$\leq E[\int_0^{\theta} e^{-\beta t}f(y_x(t))dt + e^{-\beta\theta} \bar{u}(y_x(\theta))\chi_{\theta<\infty}] +$$
$$+ E\, e^{-\beta\,\tau_R} u^0(y_x(\tau_R))$$

and hence by (3.43) we deduce from (3.42) that

$$\bar{u}(x) \leqslant J_x(\theta)$$

and since θ is arbitrary

$$\bar{u}(x) \leqslant \inf_\theta J_x(\theta)$$

which, with Lemma 3.5, proves the result desired. □

We now prove the analogue of Theorem 1.2 for the V.I. (3.26).

LEMMA 3.6: *Let θ,θ' be two stopping times with $\theta \leqslant \theta'$, then*

$$(3.44)\qquad e^{-\beta\theta}u(y_x(\theta))\, \chi_{\theta<\infty} \leq E\,[(\int_\theta^{\theta'} f(y(s))e^{-\beta s}ds)\, \chi_{\theta<\infty} + e^{-\beta\theta'} u(y_x(\theta'))\, \chi_{\theta'<\infty} \mid \mathfrak{F}^{\theta}] \quad \text{a.s.}$$

If

$$\hat{\theta} = \inf\,\{t > \theta \mid u(y_x(t)) = \psi(y_x(t))\}$$

then

$$(3.45)\qquad e^{-\beta\hat{\theta}}u(y_x(\hat{\theta}))\chi_{\hat{\theta}<\infty} = E\,[(\int_\theta^{\hat{\theta}} f(y(s))e^{-\beta s})\, \chi_{\hat{\theta}<\infty} + e^{-\beta\hat{\theta}} u(y_x(\hat{\theta}))\, \chi_{\hat{\theta}<\infty} \mid \mathfrak{F}^{\hat{\theta}}] \quad \text{a.s.}$$

Proof: Let us first remark that

$$(3.46)\qquad u^0(y_x(\theta))\, e^{-\beta\theta}\, \chi_{\theta<\infty} = E[\,(\int_\theta^{\theta'} f(y(s))e^{-\beta s}ds)\chi_{\theta<\infty} + u^0(y_x(\theta'))e^{-\beta\theta'}\chi_{\theta'<\infty} \mid \mathfrak{F}^{\theta}\,] \quad \text{a.s.}$$

In fact let us first assume f is bounded and regular, then u^0 is bounded and C^2. Ito's formula gives us

$$u^0(y_x(\theta\wedge\tau_R))e^{-\beta\theta\wedge\tau_R} = E[\int_{\theta\wedge\tau_R}^{\theta'\wedge\tau_R} f(y(s))e^{-\beta s}\,ds + \tag{3.47}$$
$$+ u^0(y_x(\theta'\wedge\tau_R))e^{-\beta\theta'\wedge\tau_R} \mid \mathfrak{F}^\theta] \quad \text{a.s.}$$

where τ_R denotes the exit time from the ball of radius R, then we easily verify that (3.47) remains valid for f as in (3.3). Furthermore

$$u^0(y(\theta\wedge\tau_R))\, e^{-\beta\theta\wedge\tau_R} \to u^0(y(\theta))e^{-\beta\theta}\,\chi_{\theta<\infty} \quad \text{a.s.}$$
$$\text{when } R\uparrow\infty$$

$$\int_{\theta\wedge\tau_R}^{\theta'\wedge\tau_R} f(y(s))e^{-\beta s}\,ds \uparrow (\int_\theta^{\theta'} f(y(s))e^{-\beta s}ds)\chi_{\theta<\infty} \quad \text{a.s.}$$
$$E[\, u^0(y_x(\theta'))e^{-\beta\theta'}\,\chi_{\theta'<\tau_R} \mid \mathfrak{F}^\theta\,]$$
$$+ E\,[u^0(y_x(\tau_R))e^{-\beta\tau_R}\,\chi_{\theta'\geq\tau_R} \mid \mathfrak{F}^\theta\,]$$

and

$$E\,[u^0(y_x(\tau_R))\, e^{-\beta\tau_R}\,\chi_{\theta'\geq\tau_R} \mid \mathfrak{F}^\theta\,] \leq$$
$$E\,[u^0(y_x(\tau_R))\, e^{-\beta\tau_R} \mid \mathfrak{F}^\theta] \to 0 \quad \text{in } L^1$$

by (3.43). Moreover,

$$E\,[u^0(y_x(\theta'))e^{-\beta\theta'}\chi_{\theta'<\tau_R} \mid \mathfrak{F}^\theta] \uparrow E[u^0(y_x(\theta'))e^{-\beta\theta'}\chi_{\theta'<\infty} \mid \mathfrak{F}^\theta].$$

From the preceding convergences it follows that we can pass to the limit in (3.47) when $R\uparrow+\infty$ in the L^1 sence. (3.46) is deduced from this.

For the penalised problem similar considerations allow us to write

$$e^{-\beta\theta}u^\varepsilon(y(\theta))\,\chi_{\theta<\infty} = E[(\int_\theta^{\theta'} f(y(s))e^{-\beta s}ds)\,\chi_{\theta<\infty} \tag{3.48}$$
$$-\frac{1}{\varepsilon}(\int_\theta^{\theta'}(u^\varepsilon-\psi)^+(y(s))e^{-\beta s}ds)\chi_{\theta<\infty} +$$
$$+ e^{-\beta\theta'}\,u^\varepsilon(y(\theta'))\chi_{\theta'<\infty} \mid \mathfrak{F}^\theta\,] \quad \text{a.s.}$$

from which, since $u \leq u^\varepsilon$, we deduce

$$e^{-\beta\theta}\,u(y(\theta))\,\chi_{\theta<\infty} \leq E\,[(\int_\theta^{\theta'} f(y(s))e^{-\beta s}ds)\,\chi_{\theta<\infty}] +$$
$$+ E\,[e^{-\beta\theta'}\,u^\varepsilon(y(\theta'))\chi_{\theta'<\infty} \mid \mathfrak{F}^\theta] \quad \text{a.s.}$$

and making ε tend to 0 (taking into account $u^\varepsilon \leqslant u^0$) it follows from Lebesgue's Theorem that the relation (3.44) holds. Next set

$$\hat\theta^\varepsilon = \inf \{t \geqslant \theta \mid u^\varepsilon(y_x(t)) \geqslant \psi(y_x(t))\}$$

and applying (3.48) with $\theta' = \hat\theta$, we obtain

$$(3.49) \qquad e^{-\beta\theta}\, u^\varepsilon(y(\theta))\chi_{\theta<\infty} = E[\, (\int_\theta^{\hat\theta^\varepsilon} f(y(s))e^{-\beta s}ds)\chi_{\theta<\infty} + \\ + e^{-\beta\hat\theta^\varepsilon}\, u(y(\hat\theta^\varepsilon))\chi_{\hat\theta^\varepsilon<\infty} \mid \mathcal{F}^\theta\,] .$$

Let us show that

$$(3.50) \qquad \hat\theta^\varepsilon \to \hat\theta \text{ a.s..}$$

We already have

$$\hat\theta^\varepsilon \leqslant \hat\theta.$$

It suffices to consider ω such that $\theta < \hat\theta$. Then let $0 < \delta < \hat\theta - \theta$. For

$$\theta \leqslant s \leqslant \hat\theta - \delta$$

We have

$$u(y(s)) < \psi(y(s)).$$

Since $u^\varepsilon \downarrow u$, and since u is continuous, it follows from Dini's Theorem that $u^\varepsilon \to u$ uniformly on every compact set. Therefore we can find $\varepsilon^0(\delta,\omega)$ such that for $\varepsilon \leqslant \varepsilon^0(\delta,\omega)$ we have

$$\sup_{\theta\leq s \leq \hat\theta-\delta} |u^\varepsilon(y(s))-u(y(s))| < \operatorname{Min}_{\theta\leq s\leq \hat\theta-\delta} [\psi(y(s)) - \\ - u(y(s))]$$

hence for $\varepsilon \leqslant \varepsilon^0(\delta,\omega)$ and $s \in [\theta,\hat\theta - \delta]$ we have

$$u^\varepsilon(y(s)) < u(y(s)) + \psi(y(s)) - u(y(s)) \\ = \psi(y(s))$$

which implies $\hat\theta^\varepsilon \geqslant \hat\theta - \delta$. Therefore (3.50) holds. From this we deduce (noticing that $\hat\theta < \infty$ implies $\hat\theta^\varepsilon < \infty$)

$$(3.51) \qquad e^{-\beta\hat\theta^\varepsilon}\, u^\varepsilon(y(\hat\theta^\varepsilon))\chi_{\hat\theta<\infty} \to e^{-\beta\hat\theta}\, u(y(\hat\theta))\chi_{\hat\theta<\infty} \qquad \text{a.s.}$$

but (3.49) implies

$$e^{-\beta\theta}\, u^\varepsilon(y(\theta))\, \chi_{\theta<\infty} \geq E\,[(\int_\theta^{\hat\theta^\varepsilon} f(y(s))e^{-\beta s}ds)\, \chi_{\theta<\infty} + \\ + e^{-\beta\hat\theta^\varepsilon}\, u^\varepsilon(y(\hat\theta^\varepsilon))\, \chi_{\hat\theta<\infty} \mid \mathcal{F}^\theta\,]$$

and by applying Fatou's Theorem, using (3.51), when $\varepsilon \to 0$ we obtain

$$e^{-\beta\theta}\, u(y(\theta))\, \chi_{\theta<\infty} \geq E\left[\left(\int_{\theta}^{\hat\theta} f(y(s))e^{-\beta s}ds\right)\chi_{\theta<\infty} + \right.$$

$$\left. + e^{-\beta\hat\theta}\, u(y(\hat\theta))\, \chi_{\hat\theta<\infty} \,\middle|\, \mathfrak{F}^{\theta}\right]$$

which, with (3.44), implies (3.45). □

3.4 *Study of the Quasi-Variational Inequality*

For $\phi \geqslant 0$ we write

(3.52) $$M\phi(x) = k + \inf_{\xi\geq 0} (c_0(\xi) + \phi(x+\xi)).$$

We are interested in the Q.V.I.

(3.53) $$\left|\begin{array}{l} a_{\mu}(u,v-u) + \beta(u,v-u)_{\mu} \geq (f,v-u)_{\mu} \\ \forall\, v \in V^{\mu} \;,\; v \leq Mu \;,\; u \leq Mu \;,\; u \in V^{\mu} \end{array}\right.$$

and more particularly in the solutions of (3.53) such that

(3.54) $$0 \leqslant u \leqslant u^0.$$

Our first objective is to prove:

THEOREM 3.3: *Assume* (3.1), (3.2), (3.3), (3.4). *Then in the interval* (3.54) *the Q.V.I.* (3.53) *has a maximum solution* $\bar{u}$ *and a minimum solution* $\underline{u}$. *The solution* $\underline{u}$ *is continuous, and is the limit of the increasing process. The solution* $\bar{u}$ *is continuous, and is the limit of the decreasing process.* □

We use variants of what was done in Section 4 of Chapter 4. For $z \in [0,u^0]$ we define $Tz = \zeta$ to be the solution of the V.I. (3.26), with $\psi = Mz$. In Theorem 3.2 we saw that ζ was defined uniquely and ζ was continuous, $\zeta \in V^{\mu}$. We write $\mathcal{O}_H$ for the ball with centre 0 and radius H. (1)

We are going to prove

LEMMA 3.7: $\forall H > 0$, *there exists* C_H *and* $\delta_H \in]0,1[$ *such that*

(3.55) $$\|Tz\|_{C^{0,\delta_H}(\bar{\mathcal{O}}_H)} \leq C_H \qquad \forall\, z \text{ continuous}$$

such that $\quad 0 \leq z \leq u^0$.

Proof: We easily verify that Mz is continuous (thanks, in particular, to the condition $c_0(\xi) \to +\infty$ if $|\xi| \to \infty$). We shall set $\psi = Mz$. Also we have (cf., (3.4))

(3.56) $$\psi(x) \leqslant \psi(z) + d\,|x - z|^{\gamma} \qquad \forall x \leqslant z.$$

Consider a ball $Q_H \supset\supset \mathcal{O}_H$ and for $x_0 \in \mathcal{O}_H$, the Green's function $G = G_H^{x_0}$ defined in (4.12) of Chapter 4. We take $n \geqslant 3$ for simplification. Next define $G\rho = G_{\rho,H}^{x_0}$ by

(1) The notation $B_R(x_0)$ will also be used (which explains why we shall not follow the notation of Section 4 of Chapter 4).

(4.14) of Chapter 4. Now take $\mathcal{O}_H \in \mathcal{D}(Q_H)$ and $\mathcal{O}_H = 1$ on $\bar{\mathcal{O}}_H$, $0 \le \mathcal{O}_H \le 1$. Next take $B_R(x_0)$, $(R \le 1)$, $\tau_R = \tau$, $\tau_R = 1$ on B_R, $\tau_R = 0$ on $R^n - B_{2R}$, $0 \le \tau_R \le 1$, τ_R regular, and

$$(3.57) \qquad |\nabla \tau_R| \le \frac{C}{R} \quad , \quad \| D^2 \tau_R \| \le \frac{C}{R^2} \ .$$

We define

$$T_R(x_0) = \{x \mid x_i \le x_{0i} - 2R\} \cap B_{4R}$$

$$\zeta_R = \frac{1}{|T_R(x_0)|} \int_{T_R} \zeta(x)\,dx \qquad (\zeta = Tz)$$

$$\xi_R = \zeta_R - d(6R)^\gamma \ .$$

The proof afterwards is identical to that of Proposition 4.1 of Chapter 4. □

Proof of Theorem 3.3: Consider the increasing process $u_{n+1} = Tu_n$, $u_0 = 0$. We verify that $u_n \uparrow \underline{u}$, and thanks to Lemma 3.7,

$$\|u_n - \underline{u}\|_{C^0(\bar{\mathcal{O}}_H)} \to 0$$

$$\|Mu_n - M\underline{u}\|_{C^0(\bar{\mathcal{O}}_H)} \to 0$$

for all fixed H, when $n \to \infty$. We then take

$$\tau_H = \tau\left(\frac{x}{H}\right) \text{ where } \tau = 1 \text{ on } \mathcal{O}_1,\ 0 \text{ on } R^n - \mathcal{O}_2,\ 0 \leqslant \tau \leqslant 1 \text{ and } \tau \text{ regular,}$$

If $v \in V^\mu$, then $\tau_H v \to v$ in V^μ. If $v \in V^\mu$, $v \leqslant M\underline{u}$, then

$$(v-\varepsilon)\tau_H \le (M\underline{u}-\varepsilon)\tau_H$$
$$= M\, u_n\, \tau_H + (M\underline{u} - Mu_n - \varepsilon)\tau_H$$

and for $n \geqslant N(\varepsilon,H)$

$$\leqslant Mu_n\, \tau_H \leqslant Mu_n .$$

As in the proof of Theorem 4.2, we easily deduce that $\underline{u}$ is a minimum solution. Similar considerations with the increasing process allow us to finish. □

We now solve the impulse control problem for (3.9). Let u be a *continuous* relation of (3.53) (thus the case for $\underline{u}$ and $\bar{u}$). To it we will associate a control $\hat{W}$ defined in the following way. Let $\hat{\xi}(x)$ be the bond mapping from R^n into R^{n+1} such that

$$(3.58) \qquad Mu(x) = k + c_0(\hat{\xi}(x)) + u(x + \hat{\xi}(x)) \ \forall x.$$

We define x^0 by

$$(3.59) \qquad \left| \begin{array}{l} dx^0 = b(x^0)dt + \sigma(x^0)dw \\ x^0(0) = x \end{array} \right.$$

and

(3.60)
$$\begin{cases} \hat{\theta}^1 = \inf\{ t \mid u(x^0(t)) = Mu(x^0(t))\} \\ \hat{\xi}^1 = \hat{\xi}(x^0(\hat{\theta}^1)) \quad \text{si} \quad \hat{\theta}^1 < \infty , \text{ arbitrary if } \hat{\theta}^1 = +\infty . \end{cases}$$

Having defined the process x^{n-1} as well as $\hat{\theta}^n, \hat{\xi}^n$, we define x^n, $\hat{\theta}^{n+1}, \hat{\xi}^{n+1}$ as follows:

(3.61)
$$\begin{cases} dx^n = b(x^n)dt + \sigma(x^n)dw \\ x^n(\hat{\theta}^n) = x^{n-1}(\hat{\theta}^n) + \hat{\xi}^n , \end{cases}$$

(3.62)
$$\begin{cases} \hat{\theta}^{n+1} = \inf \{t \geq \hat{\theta}^n \mid u(x^n(t)) = Mu(x^n(t))\} \\ \hat{\xi}^{n+1} = \hat{\xi}(x^n(\hat{\theta}^{n+1})) \quad \text{if} \quad \hat{\theta}^{n+1} < \infty \quad , \text{ arbitrary if } \hat{\theta}^{n+1} = \infty . \end{cases}$$

We write

(3.63) $$\hat{W} = (\hat{\theta}^1, \hat{\xi}^1; \ldots ; \hat{\theta}^n, \hat{\xi}^n; \ldots)$$

As for (1.63) we verify that

(3.64) $$\hat{\theta}^n < \hat{\theta}^{n+1} \qquad \text{a.s. if } \hat{\theta}^n < \infty.$$

Let us set

$$y(t) = x^n(t) \text{ for } t \in [\hat{\theta}^n, \hat{\theta}^{n+1}[.$$

We shall now see that $\hat{W}$ is admissible (it remains to establish that $\hat{\theta}^n \to \infty$ a.s.). In this case y(t) is the trajectory associated with the impulse control $\hat{W}$. We have:

LEMMA 3.8: *For all* N,

(3.65)
$$u(x) = E \Big[\int_0^{\hat{\theta}^N} f(y(s))e^{-\beta s} ds + \sum_{n=1}^{N} (k + c_0(\hat{\xi}^n)) e^{-\beta\hat{\theta}^n} \chi_{\hat{\theta}^n < \infty} \Big] + u(y(\hat{\theta}^N)) \chi_{\hat{\theta}^N < \infty} e^{-\beta\hat{\theta}^N} .$$

Proof: Consider u as the (unique) solution of a V.I., corresponding to the obstacle $\psi = Mu$, which is continuous because u is assumed continuous. By a variant of Lemma 3.6 we have

$$e^{-\beta\hat{\theta}^n} u(x^n(\hat{\theta}^n)) \chi_{\hat{\theta}^n<\infty} = E\Big[\int_{\hat{\theta}^n}^{\hat{\theta}^{n+1}} f(x^n(s))e^{-\beta s} \chi_{\hat{\theta}^n<\infty} + e^{-\beta\hat{\theta}^{n+1}} u(x^n(\hat{\theta}^{n+1})) \chi_{\hat{\theta}^{n+1}<\infty} \;\Big|\; \mathcal{F}^{\hat{\theta}^n} \Big] .$$

Now

$$u(x^n(\hat{\theta}^{n+1})) \chi_{\hat{\theta}^{n+1}<\infty} = (k + c_0(\hat{\xi}^{n+1})) \chi_{\hat{\theta}^{n+1}<\infty}$$ (Contd)

(Contd) $+ u(x^{n+1}(\hat{\theta}^{n+1}))\, \chi_{\hat{\theta}^{n+1}<\infty}$

and therefore

$$E\, e^{-\beta\hat{\theta}^n} u(y(\hat{\theta}^n))\, \chi_{\hat{\theta}^n<\infty} = E\, [(\int_{\hat{\theta}^n}^{\hat{\theta}^{n+1}} f(y(s))e^{-\beta s})\, \chi_{\hat{\theta}^n<\infty} + (k + c_0(\hat{\xi}^{n+1})\, e^{-\beta\hat{\theta}^{n+1}} \chi_{\hat{\theta}^{n+1}<\infty}\,] + E\, e^{-\beta\hat{\theta}^{n+1}} u(y(\hat{\theta}^{n+1}))\, \chi_{\hat{\theta}^{n+1}<\infty}\,. \tag{3.66}$$

By induction we deduce from this relation that

$$E\, e^{-\beta\hat{\theta}^n} u(y(\hat{\theta}^n))\, \chi_{\hat{\theta}^n<\infty} < \infty \qquad \forall\, n\,. \tag{3.67}$$

Consequently we can add the relations (3.66) when n varies between 0 and N-1 (by convention $\hat{\theta}^0 = 0$, $y(\hat{\theta}^0) = x$). (1). We obtain

$$u(x) = E\, [\sum_{n=0}^{N-1} \chi_{\hat{\theta}^n<\infty} \int_{\hat{\theta}^n}^{\hat{\theta}^{n+1}} f(y(s))e^{-\beta s}\, ds + \sum_{n=1}^{N} (k + c_0(\hat{\xi}^n))e^{-\beta\, \hat{\theta}^n} \chi_{\hat{\theta}^n<\infty}\,] + E\, e^{-\beta\hat{\theta}^N} u(y(\hat{\theta}^N))\, \chi_{\hat{\theta}^N<\infty}\,.$$

Now,

$$\sum_{n=0}^{N-1} \chi_{\hat{\theta}^n<\infty} \int_{\hat{\theta}^n}^{\hat{\theta}^{n+1}} f(y(s))\, e^{-\beta s} ds = \int_0^{\hat{\theta}^N} f(y(s))e^{-\beta s} ds$$

hence the result (3.65). □

LEMMA 3.9: *The control $\hat{W}$ is admissible.*

Proof: It suffices to show that

$$\hat{\theta}^n \uparrow +\infty \text{ a.s. when } n \uparrow \infty\,. \tag{3.68}$$

But by (3.65) we have

$$\sum_{n-1}^{\infty} e^{-\beta\cdot\hat{\theta}^n} \chi_{\hat{\theta}^n<\infty} < \infty \qquad \text{a.s.}$$

which immediately implies (3.68). □

We now pass on to the interpretation of the decreasing and increasing methods.

---

(1) Naturally, if u = Mu we have $\hat{\theta}^1 = 0$, therefore the rule (3.64) only applies for $n \geqslant 1$.

Therefore we consider u^n defined by

$$u^{n+1} = Tu^n,\ u^0 \text{ defined by (3.24).}$$

Now consider

(3.69) $$\mathcal{W}^n = \{W \mid \theta^p = +\infty \text{ for } p \geq n\}.$$

We have:

LEMMA 3.10: *The sequence* u^n *satisfies*

(3.70) $$u^{n-1}(x) = \underset{W \in \mathcal{W}}{\text{Min}} \ J^x(W).$$

Proof: For $W \in \mathcal{W}^{n+1}$, $0 \leq j \leq n-1$, we have

$$e^{-\beta\theta^j} u^{n-j}(x^j(\theta^j))\chi_{\theta^j<\infty} \leq E\Big[\chi_{\theta^j<\infty}\int_{\theta^j}^{\theta^{j+1}} f(x^j(s))e^{-\beta s}\,ds + e^{-\beta\theta^{j+1}} u^{n-j}(x^j(\theta^{j+1}))\chi_{\theta^{j+1}<\infty} \ \Big|\ \mathcal{F}^{\theta^j}\Big]$$

from which we easily deduce (cf., Theorem 2.1) that

$$\begin{aligned} &Eu^{n-j}(x^j(\theta^j))\chi_{\theta^j<\infty} e^{-\beta\theta^j} \leq \\ &\leq E\Big[\chi_{\theta^j<\infty}\int_{\theta^j}^{\theta^{j+1}} f(y(s))e^{-\beta s}\,ds + \\ &+ (k+c_0(\xi^{j+1}))\,e^{-\beta\theta^{j+1}}\chi_{\theta^{j+1}<\infty} + \\ &+ u^{n-(j+1)}(x^{j+1}(\theta^{j+1}))\,e^{-\beta\theta^{j+1}}\chi_{\theta^{j+1}<\infty}\Big] \end{aligned}$$

and by adding

$$\begin{aligned} u^n(x) \leq E\Big[&\int_0^{\theta^n} f(y(s))\,e^{-\beta s}ds + \\ &+ \sum_{j=1}^{n}(k+c_0(\xi^j))\,e^{-\beta\theta^j}\chi_{\theta^j<\infty} + \\ &+ u^0(y(\theta^n))e^{-\beta\theta^n}\chi_{\theta^n<\infty}\Big] \end{aligned}$$

but

$$E\,u^0(y(\theta^n))e^{-\beta\theta^n}\chi_{\theta^n<\infty} = E\Big(\int_{\theta^n}^{+\infty} f(y(s))e^{-\beta s}\,ds\Big)\chi_{\theta^n<\infty}$$

and therefore

$$u^n(x) \leq J^x(W) \qquad \forall\ W \in \mathcal{W}^{n+1}.$$

We also verify that

$$u^n(x) = J^x(\hat{W}_n) \text{ where } \hat{W}_n \in \mathcal{W}_{n+1}$$

hence the desired result. □

We now interpret the increasing method. For this, for every control W we define W

$$(3.71) \qquad J_x^n(W) = E \left[\int_0^{\theta^n} f(y(t))e^{-\beta t}\, dt + \right.$$
$$\left. + \sum_{j=1}^{n} (k + c_0(\xi^j))e^{-\beta\theta^j} \chi_{\theta^j < \infty} \right].$$

LEMMA 3.11: *There holds*

$$(3.72) \qquad u_n(x) = \text{Min } J_x^n(W).$$

Proof: This is analogous to that of Lemma 3.10 (cf., Theorem 2.2 also). □

We can then prove:

THEOREM 3.4: *Assume* (3.1), (3.2), (3.3), (3.4). *Then the minimum solution* $\underline{u}(x)$ *of the Q.V.I.* (3.53) *has the following probabilistic interpretation:*

$$(3.73) \qquad u(x) = \underset{W}{\text{Min}}\ J^x(W).$$

The maximum solution is interpreted by

$$(3.74) \qquad \bar{u}(x) = \underset{W \in \mathcal{W}}{\text{Inf}}\ J^x(W)$$

where we have set

$$(3.75) \qquad \mathcal{W} = \{ W \mid E\, e^{-\beta\theta^n}\, \bar{u}(y(\theta^n))\chi_{\theta^n < \infty} \to 0 \text{ when } n \to \infty\} .$$

Proof: By Lemma (3.11) we have

$$u_n(x) \leqslant J_x^n(W) \leqslant J_x(W)$$

and therefore

$$(3.76) \qquad \underline{u}(x) \leqslant \underset{W}{\text{Inf}}\ J_x(W).$$

But by Lemma 3.8

$$\underline{u}(x) \leqslant J_x^N(\hat{W}) \qquad \forall N$$

and therefore

$$\underline{u}(x) \geqslant J_x(\hat{W})$$

which with (3.76) shows that

(3.77) $$\underline{u}(x) = J_x(\hat{W}) = \operatorname*{Inf}_{W} J_x(W).$$

Furthermore, $\mathcal{W}^n \subset \mathcal{W}$ for all n, therefore

$$u^n(x) \geq \operatorname*{Inf}_{W\in\mathcal{W}} J^x(W)$$

which implies

(3.78) $$\bar{u}(x) \geq \operatorname*{Inf}_{W\in\mathcal{W}} J^x(W)$$

But we verify, similarly to Lemma 3.8, that for every control W we have

$$\bar{u}(x) \leq E\,[\int_0^{\theta^N} f(y(s))e^{-\beta s}\,ds +$$
$$+ \sum_{n=1}^{N} (k+ c_0(\xi^n))\, e^{-\beta\theta^n} \chi_{\theta^n<\infty}] +$$
$$+ E\,\bar{u}(y(\theta^N))\, e^{-\beta\theta^N} \chi_{\theta^N<\infty}$$

and thus if $W \in \mathcal{W}$, we obtain

$$\bar{u}(x) \leq J^x(W)$$

which, with (3.78), completes the desired result. □

4. PROBABILISTIC INTERPRETATION OF SEMI-GROUP FORMULATIONS

4.1 *The Obstacle Problem*

Let E be locally compact, separable, complete, metric space, and $\mathcal{E}$ the Borel σ-algebra. We define the spaces B and C (cf., Chapter 4, Section 5), as

B ≡ B(E) the bounded Borel functions on E,

C ≡ C(E) the bounded and uniformly continuous functions on E.

We are given a Markov semi-group $\Phi(t) \in \mathcal{L}(B;B)$ satisfying

(4.1) $$\Phi(t+s) = \Phi(t)\,\Phi(s)$$

(4.2) $$\Phi(0) = I$$

(4.3) $$\Phi(t)1 = 1$$

(4.4) $$\Phi(t)f \geq 0 \quad \forall\, f \in B,\ f \geq 0 .$$

Since E is not necessarily compact, we shall need the space

(4.5) $$\hat{C} = \{f \in C \mid \forall\varepsilon, \exists K_\varepsilon \text{ compact such that } |f(x)| < \varepsilon \text{ for } x \notin K_\varepsilon\}.$$

We then assume that

(4.6) $\Phi(t) : C \to C$

(4.7) $\Phi(t) : \hat{C} \to \hat{C}$

(4.8) $\Phi(t)f \to f$ in C, when $t \downarrow 0$, $\forall f \in C$.

For all $\Gamma \in \mathcal{E}$, we set $(x \in E, t \geq 0)$

(4.9) $P(x,t;\Gamma) = \Phi(t) \quad \chi_\Gamma(x)$.

With $\Phi(t)$ we can associate a Markov process whose transition probability function is given by (4.9).

Consider the space

$$\tilde{\Omega} = D([0,\infty);E)$$

of E-valued cadlag process. We write an element of $\tilde{\Omega}$ as $\tilde{\omega} \equiv \tilde{\omega}(.)$, and the canonical process as $x(t;\tilde{\omega}) = \tilde{\omega}(t)$. We define

$$\tilde{\mathcal{M}} = \sigma(x(t), t \geq 0), \qquad \tilde{\mathcal{M}}_s^t = \sigma(x(\lambda),\ s \leq \lambda \leq t) \ .$$

By E. DYNKIN [1],[2], for all fixed $x \in E$ there exists one and only one probability Q_x on $\tilde{\Omega}, \tilde{\mathcal{M}}$ such that $\tilde{\Omega}$, $\tilde{\mathcal{M}}$, Q_x, $\tilde{\mathcal{M}}^t$, $x(t)$ is *a quasi-left continuous, right continuous, strong Markov process* satisfying

(4.10) $Q_x(x(t) \in \Gamma) = P(x,t;\Gamma) \quad \forall x,t,\Gamma$.

We write

$$\tilde{\mathcal{N}} = \tilde{\mathcal{M}} \text{ completion for } Q_x, \forall x$$

$$\tilde{\mathcal{N}}^t = \tilde{\mathcal{M}}_0^{t+0} \text{ completion for } Q_x, \forall x$$

then the process keeps the same properties when $\tilde{\mathcal{M}}$ is replaced by $\tilde{\mathcal{N}}$ and $\tilde{\mathcal{M}}_0^t$ by $\tilde{\mathcal{N}}^t$. The process $x(t)$ is then *standard* (cf., E. DYNKIN [2], I.I. GIKHMAN-A.V. SKOROKHOD [1]).

Given

(4.11) $\psi \in C$

(4.12) $L \in B$, $t \to \Phi(t)L$ is measurable from $[0, \infty)$ into C.

we consider the set of functions u satisfying

(4.13)
$$\left| \begin{array}{l} u \in C, \qquad u \leq \psi \\ u \leq \int_0^t e^{-\alpha s} \Phi(s)L\, ds + e^{-\alpha t} \Phi(t)u, \quad \forall t \geq 0 . \end{array} \right.$$

By Theorem 5.1 of Chapter 4 the set of u's satisfying (4.13) is non-empty and has a maximum element. Our objective is to give the probabilistic interpretation of the maximum element, also written as u.

Let θ be a stopping time of $\tilde{\mathcal{N}}^t$, and define the functional

(4.14) $$J^x(\theta) = E_x\left[\int_0^\theta e^{-\alpha s} L(x(s))ds + e^{-\alpha\theta} \psi(x(\theta))\chi_{\theta < \infty}\right]$$

where E_x is the expectation with respect to Q_x.
We have:

THEOREM 4.1: *Assume* (4.1), (4.2), (4.3), (4.4), (4.6), (4.7), (4.8), (4.11), (4.12). *Then the maximum solution of* (4.13) *is given explicitly by*

(4.15) $u(x) = \text{Inf } J^x(\theta)$.

Furthermore, there exists an optimal stopping time $\hat{\theta}$ *defined by*

(4.16) $\hat{\theta} = \inf \{t \mid u(x(t)) = \psi(x(t))\}$

($\hat{\theta}$ may take the value $+\infty$).

Proof: Write β_t for the translation operator of $\tilde{\Omega}$ into itself. Let ξ be a random variable belonging to L^1. By the Markov property we have

$$E_x [\beta_t \cdot \xi | \tilde{\eta}^t] = E_{x(t)} \xi .$$

Apply this relation to

$$\xi = u(x(s_2-s_1)) e^{-\alpha(s_2-s_1)} + \int_0^{s_2-s_1} L(x(s)) e^{-\alpha s} ds$$

then

$$\beta_{s_1} \xi = u(x(s_2))e^{-\alpha(s_2-s_1)} + \int_{s_1}^{s_2} L(x(s))e^{-\alpha(s-s_1)} ds$$

and thus we obtain

$$E_x [u(x(s_2))e^{-\alpha s_2} + \int_{s_1}^{s_2} L(x(s))e^{-\alpha s} ds | \tilde{\eta}^{s_1}] = \\ = E_{x(s_1)} \xi e^{-\alpha s_1} ,$$

and by (4.13) we deduce

$$\geq u(x(s_1))e^{-\alpha s_1} .$$

Consequently the process

$$u(x(s))e^{-\alpha s} + \int_0^s L(x(\lambda))e^{-\alpha\lambda} d\lambda$$

is an $\tilde{\eta}^s$, Q_x sub-martingale. By Doob's Theorem we deduce

$$u(x) \leq E_x [u(x(\theta\wedge T))e^{-\alpha\theta\wedge T} + \int_0^{\theta\wedge T} L(x(\lambda))e^{-\alpha\lambda} d\lambda] ,$$

$\forall\theta$ a stopping time and T fixed,

and since $u \leq \psi$, after making T tend to $+\infty$ we obtain

$$u(x) \leq J^x(\theta), \ \forall\theta. \tag{4.17}$$

Next we consider the penalised problem (cf., Chapter 4, (5.17)). We may write

$$u_\varepsilon = \int_0^t e^{-\alpha s} \Phi(s)(L - \frac{1}{\varepsilon}(u - \psi)^+)ds + e^{-\alpha t} \Phi(t)u_\varepsilon . \tag{4.18}$$

Arguing as above, we see that the process

$$u_\varepsilon(x(s))e^{-\alpha s} + \int_0^s [L(x(\lambda)) - \frac{1}{\varepsilon}(u_\varepsilon - \psi)^+(x(\lambda))] e^{-\alpha\lambda} d\lambda$$

is an $\tilde{\eta}^s$, Q_x-martingale, and therefore by using Doob's Theorem again we obtain

$$u_\varepsilon(x) = E_x [u_\varepsilon(x(\theta))e^{-\alpha\theta}\chi_{\theta<\infty} + \int_0^\theta (L(x(s)) - \frac{1}{\varepsilon}(u_\varepsilon - \psi)^+(x(s)))e^{-\alpha s} ds] \tag{4.19}$$

for every stopping time θ. Let us write

$$\hat{\theta}^\varepsilon = \inf \{t \mid u_\varepsilon(x(t)) \geq \psi(x(t))\}.$$

As the set $\{x \mid u_\varepsilon(x) \geq \psi(x)\}$ is closed, and the process $x(t)$ is standard, by E. DYNKIN [2] $\hat{\theta}^\varepsilon$ is a stopping time of $\tilde{\eta}^t$. Let us notice that by the definition of $\hat{\theta}^\varepsilon$ we have

$$t < \hat{\theta}^\varepsilon \text{ implies } (u_\varepsilon - \psi)^+(x(t)) = 0. \tag{4.20}$$

Also, $\hat{\theta}$ is a stopping time of $\tilde{\eta}^t$ for the same reason as $\hat{\theta}^\varepsilon$.

We are going to show that

$$u(x) = J^x(\hat{\theta})$$

which, with (4.17), will complete the proof of the desired result.

If x satisfies $u(x) = \psi(x)$, then $\hat{\theta} = 0$, and

$$J^x(\hat{\theta}) = \psi(x),$$

and hence (4.23). We can assume

$$u(x) < \psi(x).$$

Let δ_0 be such that $u(x) < \psi(x) - \delta_0$. Now take $\delta < \delta_0$, therefore $u(x) < \psi(x) - \delta$. Next let

$$\theta^\delta = \inf \{ t \mid u(x(t)) \geq \psi(x(t)) - \frac{\delta}{2}\}$$

which is also an $\tilde{\eta}^t$ stopping time. Since $u_\varepsilon \to u$ in C, there exists ε_δ such that for $\varepsilon \leq \varepsilon_\delta$ we have

$$\| u_\varepsilon - u \| \leq \frac{\delta}{4} .$$

Then for $\varepsilon \leqslant \varepsilon_\delta$ and $t < \theta^\delta$, we have

$$u_\varepsilon(x(t)) \leq u(x(t)) + \frac{\delta}{4} < \psi(x(t)) - \frac{\delta}{2} + \delta/_4$$
$$< \psi(x(t)) - \delta/_4$$

which implies $\hat{\theta}^\varepsilon \geqslant \theta^\delta$. We then apply (4.19) with $\theta = \theta^\delta$, taking (4.20) into account, and obtain

$$u_\varepsilon(x) = E_x[u_\varepsilon(x(\theta^\delta))e^{-\alpha\theta^\delta}\chi_{\theta^\delta<\infty} \int_0^{\theta^\delta} L(x(s))e^{-\alpha s}\, ds]$$

and since $u_\varepsilon \to u$ in C,

(4.21) $$u(x) = E_x\, [u(x(\theta^\delta))e^{-\alpha\theta^\delta}\chi_{\theta^\delta<\infty} + \\ + \int_0^{\theta^\delta} L(x(s))e^{-\alpha s}\, ds] .$$

Let us show that

(4.22) $\theta^\delta \uparrow \hat{\theta}$ a.s. when $\delta \downarrow 0$.

Let us assume $\hat{\theta} \leqslant \infty$, then θ^δ is increasing when $\delta \downarrow 0$ and $\theta^\delta < \hat{\theta}$. Therefore $\theta^\delta \uparrow \Lambda$. Since the process is right continuous

$$u(x(\theta^\delta)) \geqslant \psi(x(\theta^\delta)) - \frac{\delta}{2} .$$

Thanks to the process's left continuity we also have

$$x(\theta^\delta) \to x(\Lambda) \text{ a.s. on the set } \hat{\theta} < \infty$$

and since u,ψ are continuous

$$u(x(\Lambda)) \geqslant \psi(x(\Lambda))$$

which implies $\Lambda \geqslant \hat{\theta}$, therefore $\Lambda = \hat{\theta}$ necessarily. Let us now assume that $\hat{\theta} = +\infty$, then

(4.23) $u(x(t)) < \psi(x(t))\ \forall t$.

Let us assume that $\theta^\delta \uparrow \Lambda < \infty$, then we verify, as above, that

$$u(x(\Lambda)) \geqslant \psi(x(\Lambda))$$

which contradicts (4.23). Therefore we have (4.22).

By the left quasi-continuity we have

$$x(\theta^\delta) \to x(\hat{\theta}) \text{ a.s. on } \{\hat{\theta} < \infty\}$$

and therefore

$$u(x(\theta^\delta))e^{-\alpha\theta^\delta}\chi_{\theta^\delta<\infty} \to u(x(\hat{\theta}))e^{-\alpha\hat{\theta}}\chi_{\hat{\theta}<\infty} .$$

Since u is bounded we can pass to the limit in (4.27) using Lebesgue's Theorem. we obtain

(4.24) $$u(x) = E_x \left[u(x(\hat{\theta}))e^{-\alpha\hat{\theta}} \chi_{\theta<\infty} + \int_0^{\hat{\theta}} L(x(s))e^{-\alpha s}ds\right].$$

But if $\hat{\theta} < \infty$, it follows from the process's right continuity that

$$u(x(\hat{\theta})) = \psi(x(\hat{\theta}))$$

and so we have (4.23). □

4.2 *Implicit Obstacle*

We now consider the case where $\psi = Mu$ (cf., Chapter 4, Section 5.2). Assume

(4.25) $L \geq 0.$

(4.26) M non-linear operator from $C \to C$ such that
$\|M\phi_1 - M\phi_2\| \leq \|\phi_1 - \phi_2\|$
M concave, increasing
$M(0) \geq k$.

By Theorem 5.2 of Chapter 4 the set of functions satisfying

(4.27) $$\begin{cases} u \in C, \quad u \leq Mu \\ u \leq \int_0^t e^{-\alpha s}\, \Phi(s)\, L\, ds + e^{-\alpha t}\, \Phi(t)u \end{cases}$$

is non-empty and has a maximum element denoted u.

As our objective is to interpret u as the lower bound of an impulse control problem, we must specify the operator M. We consider a situation analogous to that of M. ROBIN [3].

We are given

(4.28) E_{ad} compact $\subset E$

(4.29) $c(x,\xi): E \times E_{ad} \to R^+$,
continuous and bounded, $c(x,\xi) \geq k > 0$,
uniformly continuous at x, and uniformly with respect to ξ. (1).

For $\phi \in C$ we set

(4.30) $$M\phi(x) = \operatorname*{Inf}_{\xi \in E_{ad}} [c(x,\xi) + \phi(x+\xi)] \quad .$$

It is easy to verify that the operator M defined by (4.30) satisfies the properties (4.26).

We are now going to define an impulse control problem. First let us consider the σ-algebra

(1) Therefore $\forall\varepsilon\ \exists\ \delta$ such that $d(x,x') \leq \varepsilon$ implies $|c(x,\xi) - c(x',\xi)| \leq \varepsilon, \forall \xi$.

$\tilde{\mathcal{M}}_t = \tilde{\mathcal{M}}_t^{\infty}$ ($\tilde{\mathcal{M}}_t = \beta_t\ \tilde{\mathcal{M}}$, i.e., every element Δ_t of $\tilde{\mathcal{M}}_t$ may be written in the form $\Delta_t = \beta_t\ \Delta$, with $\Delta \in \tilde{\mathcal{M}}$).

We define a measure Q_{xt} on $\tilde{\Omega}$, $\tilde{\mathcal{M}}_t$ by writing for $\Delta_t \in \tilde{\mathcal{M}}_t$ ($\Delta_t = \beta_t\ \Delta$)

(4.31) $$Q_{xt}(\Delta_t) = Q_x(\Delta).$$

For $t \leqslant s$ and $\Gamma \in \mathcal{E}$, we have

(4.32) $$\begin{aligned} E_{xt}(x(s) \in \Gamma) &= E_x(x(s-t) \in \Gamma) \\ &= P(x,x-t;\ \Gamma) \\ &= \Phi(s-t)\ \chi_\Gamma(x)\ . \\ &= P(x,t,s;\Gamma) \end{aligned}$$

Let $x_0 \in E$ be (arbitrarily) fixed. We define Q°_{xt} on $\tilde{\Omega}, \tilde{\mathcal{M}}$ by

(4.33) $$Q^\circ_{xt} = \left| \begin{array}{ll} \delta_{\tilde{x}_0} \times Q_{xt}\ , & \text{if } t < \infty \\ \delta_{\tilde{x}_0} & \text{if } t = \infty \end{array} \right.$$

In other words for $\Delta_t \in \tilde{\mathcal{M}}_t$ and $\Delta^t \in \tilde{\mathcal{M}}_0^{t-}$, if we write

$\tilde{x}_0 \in \tilde{\Omega}$, defined by $\tilde{x}_0(s) = x_0 \forall s$

then we have

(4.34) $$Q^\circ_{xt}(\Delta^t \cap \Delta_t) = \chi_{\Delta^t}(\tilde{x}_0) Q_{xt}(\Delta_t).$$

It is understood that Q_{xt} restricted to $\tilde{\mathcal{M}}_t$ coincides with Q_{xt}, so that $\tilde{\Omega}, \tilde{\mathcal{M}}_t, Q_{xt}$, $\tilde{\mathcal{M}}_t^s$, $x(s)$ is a strong (stationary) Markov process that is right continuous and left quasi-continuous.

We shall set

(4.35) $$\left| \begin{array}{lll} \tilde{\mathcal{N}}_t = \tilde{\mathcal{M}}_t & \text{completion for } Q^\circ_{xu} & \forall\ u \le t,\ \forall\ x \\ \tilde{\mathcal{N}}^s_t = \tilde{\mathcal{M}}^{s+0}_t & \text{completion for } Q^\circ_{xu} & \forall\ u \le t,\ \forall\ x\ , \end{array} \right.$$

and

$$\tilde{\mathcal{N}} = \tilde{\mathcal{N}}_0 = \tilde{\mathcal{M}} \quad \text{completion for } Q_x = Q^\circ_{x0} \quad \forall\ x.$$

We next write

(4.36) $$\begin{aligned} \Omega &= \tilde{\Omega}^N \quad , \qquad N = \{0,1,2,\ldots\} \\ \mathcal{N} &= \tilde{\mathcal{N}}^{\otimes N}\ . \end{aligned}$$

If $\omega \in \Omega$, then $\omega = (\omega_0,\ldots,\omega_n,\ldots)$ with $\omega_n \in \tilde{\tilde{\Omega}}$.

Let us write $\mathcal{N}_n$ for the σ-algebra of $\mathcal{N}$ generated by the events of the form $\tilde{\Omega} \times \ldots \times \tilde{\Omega} \times A \times \tilde{\Omega} \ldots . \tilde{\Omega} \ldots .$ (the A at the $(n-1)$-th component belongs to $\tilde{\mathcal{N}}$), and

$\mathcal{N}_n$ = σ-algebra generated by the events of the form $A_0 \times \ldots \times A_n \times \tilde{\Omega} \times \ldots$ where $A_0,\ldots,A_n$ varies in $\tilde{\mathcal{N}}$.

We have

$$\mathcal{N}_{n+1} = \sigma(\mathcal{N}_n \cup \tilde{\mathcal{N}}_{n+1}) \quad , \mathcal{N}_0 = \tilde{\mathcal{N}}_0 .$$

Define the canonical process

$$x_n(t;\omega) = \omega_n(t) = x(t;\omega_n), \ n = 0,1,\dots$$

and the σ-algebras ($t \leq s$)

$$\mathcal{N}^s_{t,n+1} = \sigma(\mathcal{N}^s_{t,n} \cup \tilde{\mathcal{N}}^s_{t,n+1})$$

where $\tilde{\mathcal{N}}^s_{t,n}$ = sub-σ-algebra of $\tilde{\mathcal{N}}_n$ (where $A \in \tilde{\mathcal{N}}^s_t$, cf., above)

$$\mathcal{N}^s_{t,0} = \tilde{\mathcal{N}}^s_t .$$

The process $x_n(t;\omega)$ is adapted to $\mathcal{N}^t_{0,n}$.

We now define the concept of impulse control. Given

$$W = (\theta^1, \xi^1; \theta^2, \xi^2, \dots; \theta^n, \xi^n; \dots)$$

where the θ^n are stopping times of $\mathcal{N}^t_{0,n-1}$, $n \geq 1$, such that

$$(4.37) \quad \begin{cases} \theta^n \leq \theta^{n+1} ; \\ \xi^n \quad E_{ad}\text{-valued R.V., which is } \mathcal{N}^{\theta^n}_{0,n-1} \text{ measurable.} \end{cases}$$

We are going to associate a family $P^n_{x,W}$ of probabilities on $\Omega, \mathcal{N}_n$, $n = 0,\dots$ with the impulse control W and $x \in E$. To do this we begin by defining a probability $\mathcal{M}^n_{\omega,W}$ on $(\Omega, \tilde{\mathcal{N}}_{n+1})$ depending on a parameter $\omega \in \Omega$. We write

$$(4.38) \quad \mathcal{M}^n_{\omega,W} = \overset{\circ}{Q}_{x_{n-1}(\theta^n) + \xi^n, \theta^n} \quad , \quad n \geq 1 .$$

Next, by recurrence we define

$$(4.39) \quad \begin{cases} P^{n+1}_{x,W} = P^n_{x,W} \otimes \mathcal{M}^{n+1}_{\omega,W} & n \geq 0 \\ P^0_{x,W} = Q_x . \end{cases}$$

To be exact, if we consider an event of $\mathcal{N}_{n+1}$ of the form $\Delta \cap \Lambda$, where $\Delta \in \mathcal{N}_n$ and $\Lambda \in \tilde{\mathcal{N}}_{n+1}$ (these events generate $\mathcal{N}_{n+1}$), then we have

$$(4.40) \quad P^{n+1}_{x,W}(\Delta \cap \Lambda) = E^{P^n_{x,W}}[\chi_\Delta(\omega)\, \mu^{n+1}_{\omega,W}(\Lambda)] .$$

In particular, if $\Delta \in \mathcal{N}_n$,

$$P^{n+1}_{x,W}(\Delta) = P^n_{x,W}(\Delta)$$

so that the family defined thus is compatible.

Since the space $\tilde{\Omega}$ is Polish it follows from Kolmogorov's Theorem that the family $P^n_{x,w}$ can be uniquely extended to the σ-algebra $\mathcal{N}$. We denote the probability defined thus on $(\tilde{\Omega}, \mathcal{N})$ by $P_{x,w}$.

We shall say that the control W is admissable if

(4.41) $\theta^n \to +\infty$, if $n \uparrow \infty$, $P_{x,W}$ a.s. $\forall x$

(4.42) $\theta^{n+1} = \tau_{n+1}(\theta^n, \omega_n)$ where

$\tau_{n+1}(s,\tilde{\omega})$ is defined on $R^+ \times \tilde{\Omega} \to R^+$ and Borel,

$\tau_{n+1}(s,\tilde{\omega}) \geq s$

and $\forall s$ $\tau_{n+1}(s,\omega_n)$ is an $\tilde{\mathcal{N}}^t_{s,n}$ stopping time;

$\theta^0 = 0.$

Notice that θ^{n+1} is an $\mathcal{N}^t_{0,n}$-stopping time.

We then define a functional associated with W and $x \in E$ by the formula

$$J^x(W) = E_{x,W}\left[\int^{\infty} e^{-\alpha t} L(x(t))dt + \sum_{n=1}^{\infty} c(x_{n-1}(\theta^n), \xi^n) e^{-\alpha\theta^n} \chi_{\theta^n < \infty}\right] \tag{4.43}$$

where we have set

$$x(t;\omega) = x_n(t;\omega) \text{ for } t \in [\theta^n, \theta^{n+1}[\tag{4.44}$$

(with $\theta^0 = 0$ by convention) (1).

Let us notice that we have

$$\int_0^{\infty} e^{-\alpha t} L(x(t))dt = \sum_{n=0}^{\infty} \chi_{\theta^n < \infty} \int_{\theta^n}^{\theta^{n+1}} e^{-\alpha t} L(x_n(t))dt . \tag{4.45}$$

We now define a particular control which will be denoted $\hat{W}$. First of all, by the hypothesis (4.28), (4.29) there exists a Borel mapping $\hat{\xi}(x)$ from $E \to E_{ad}$ such that

$$Mu(x) = c(x,\hat{\xi}(x)) + u(x + \hat{\xi}(x)) \ \forall x. \tag{4.46}$$

Let us define

$$\left|\begin{array}{l} \hat{\theta}^1 = \inf\{t \geq 0 \mid u(x_0(t)) = Mu(x_0(t))\} \\ \hat{\xi}^1 = \hat{\xi}(x_0(\hat{\theta}^1)), \text{ si } \hat{\theta}^1 < \infty \quad , \text{ arbitrary if } \hat{\theta}^1 = \infty \end{array}\right. . \tag{4.47}$$

---

(1) We must not confuse $x(t;\omega)$ with $x(t;\omega_n) = \omega_n(t)$.

Quite generally we may write

(4.48) $$\hat{\tau}(s;\tilde{\omega}) = \inf \{t \geq s \mid u(x(t;\tilde{\omega})) = Mu(x(t;\tilde{\omega}))\}$$

and

(4.49) $$\left| \begin{array}{l} \tau_{n+1}(s,\omega_n) = \hat{\tau}(s;\omega_n) \\ \hat{\theta}^{n+1} = \tau_{n+1}(\hat{\theta}^n,\omega_n) \qquad n = 0,1,\ldots \end{array} \right.$$

(4.50) $$\hat{\xi}^{n+1} = \hat{\xi}(x_n(\hat{\theta}^{n+1})) .$$

The sequence $\hat{\theta}^1,\ldots,\hat{\theta}^n,\ldots$ certainly satisfies (4.42). Thus we define an impulse control $\hat{W}$ to which we associate a probability $P_{x,\hat{W}}$. We shall see later that $\hat{W}$ is admissable (i.e., satisfies (4.41)).

Our objective is to prove:

THEOREM 4.2: *Assume* (4.1),...,(4.4) and (4.6), (4.7), (4.8). *Also assume* (4.12), (4.25), (4.28), (4.29). *Then the maximum solution of* (4.27) *is given by*

(4.51) $$u(x) = \operatorname*{Inf}_{W} J^x(W) = J^x(\hat{W}),$$

where $\hat{W}$ *is defined by the formula*

Proof: Let W be an arbitrary impulse control. First let us notice that if s is fixed, $\tau(s;\tilde{\omega})$ is an $\tilde{\mathcal{H}}_s^t$ -stopping time, then we have

(4.52) $$\chi_{s<\infty}\, u(x)e^{-\alpha s} \leq E^{Q_{xs}}[(\int_s^{\tau} e^{-\alpha\lambda} L(x(\lambda;\tilde{\omega}))d\lambda)\, \chi_{s<\infty} + \chi_{\tau<\infty}\, u(x(\tau))\, e^{-\alpha\tau}] .$$

This results from the second inequality of (4.27) and from the definition of Q_{xs}, which implies that

$$\int_s^t e^{-\alpha\lambda} L(x(\lambda))d\lambda + u(x(t))e^{-\alpha t}$$

is a Q_{xs}, $\mathcal{H}_s^t$ -sub-martingale. In (4.52) we can replace $E^{Q_{xs}}$ by $E^{\mathring{Q}_{xs}}$, and we can apply (4.52) with $\tilde{\omega} = \omega_n$, $s = \theta^n$, $\tau = \theta^{n+1}$, $x = x_{n-1}(\theta^n) + \xi^n$ (the values of s and x so chosen do not depend upon ω_n). Therefore

$$\chi_{\theta^n<\infty}\, u(x_{n-1}(\theta^n)+ \xi^n)\, e^{-\alpha\theta^n} \leq$$

$$\leq E^{\mathring{Q}_{x_{n-1}(\theta^n) + \xi^n,\theta^n}}[(\int_{\theta^n}^{\theta^{n+1}} e^{-\alpha\lambda} L(x_n(\lambda))d\lambda)\, \chi_{\theta^n<\infty} + \chi_{\theta^{n+1}<\infty}\, u(x_n(\theta^{n+1}))\, e^{-\alpha\theta^{n+1}}]$$

so

$$E^{\mathring{Q}_{x_{n-1}(\theta^n)+\xi^n,\theta^n}} \chi_{\theta^n<\infty}\, u(x_n(\theta^n))\, e^{-\alpha\theta^n} \leq$$

$$\leq E^{\overset{\circ}{Q}_{x_{n-1}(\theta^n)+\xi^n,\theta^n}} [(\int_{\theta^n}^{\theta^{n+1}} e^{-\alpha\lambda} L(x_n(\lambda))d\lambda)\chi_{\theta^n<\infty} + \\ + \chi_{\theta^{n+1}<\infty} u(x_n(\theta^{n+1})) e^{-\alpha\theta^{n+1}}]$$

and by taking the expectation with respect to $E^{P^{n-1}_{x,W}}$ by using (4.39) we obtain

$$(4.53) \quad E^{P^n_{x,W}} \chi_{\theta^n<\infty} u(x_n(\theta^n)) e^{-\alpha\theta^n} \leq \\ \leq E^{P^n_{x,W}} [\chi_{\theta^n<\infty} \int_{\theta^n}^{\theta^{n+1}} e^{-\alpha\lambda} L(x_n(\lambda))d\lambda] + \\ + E^{P^n_{x,W}} \chi_{\theta^{n+1}<\infty} u(x_n(\theta^{n+1})) e^{-\alpha\theta^{n+1}} .$$

Notice that

$$(4.54) \quad E^{P^n_{x,W}} [\chi_{\theta^n<\infty} \int_{\theta^n}^{\theta^{n+1}} e^{-\alpha\lambda} L(x_n(\lambda))d\lambda] = \\ = E^{P_{x,W}} [\chi_{\theta^n<\infty} \int_{\theta^n}^{\theta^{n+1}} e^{-\alpha\lambda} L(x_n(\lambda))d\lambda \ ]$$

and since

$$u(x) \leq c(x,\xi) + u(x+\xi) \qquad \forall\, \xi \in E_{ad}$$

$$u(x_n(\theta^{n+1})) \leq c(x_n(\theta^{n+1}),\xi^{n+1}) + u(x_n(\theta^{n+1})+\xi^{n+1}) \quad \text{if } \theta^{n+1}<\infty .$$

Furthermore

$$E^{P_{x,W}} \chi_{\theta^{n+1}<\infty} e^{-\alpha\theta^{n+1}} u(x_n(\theta^{n+1}) + \xi^{n+1}) \\ = E^{P^{n+1}_{x,W}} \chi_{\theta^{n+1}<\infty} e^{-\alpha\theta^{n+1}} u(x_{n+1}(\theta^{n+1})) .$$

Collecting the results we obtain

$$(4.55) \quad E^{P_{x,W}} \chi_{\theta^n<\infty} u(x_n(\theta^n)) e^{-\alpha\theta^n} \leq \\ E^{P_{x,W}} [\chi_{\theta^n<\infty} \int_{\theta^n}^{\theta^{n+1}} e^{-\alpha\lambda} L(x_n(\lambda))d\lambda + \\ + \chi_{\theta^{n+1}<\infty} e^{-\alpha\theta^{n+1}} c(x_n(\theta^{n+1}),\xi^{n+1})] + \\ + E^{P^{n+1}_{x,W}} \chi_{\theta^{n+1}<\infty} e^{-\alpha\theta^{n+1}} u(x_{n+1}(\theta^{n+1})) .$$

Adding the relations (4.55) for n varying between 0 and N-1, we obtain

$$u(x) \leq E_{x,W} \Big[\sum_{n=0}^{N-1} \chi_{\theta^n < \infty} \int_{\theta^n}^{\theta^{n+1}} e^{-\alpha\lambda} L(x_n(\lambda))d\lambda$$
$$+ \sum_{n=1}^{N} c(x_{n-1}(\theta^n),\xi^n)\, e^{-\alpha\theta^n} \chi_{\theta^n < \infty} \Big] +$$
$$+ E_{x,W}\, \chi_{\theta^N < \infty}\, e^{-\alpha\theta^N} u(x_N(\theta^N)) .$$

Taking account of (4.41) we easily deduce from this that

(4.56) $\quad u(x) \leqslant J_x(W)\ \forall W.$

Considering (4.27) as an obstacle problem with $\psi = Mu$ (cf., (4.24)), we can also write

(4.57)
$$\chi_{s<\infty}\, u(x) e^{-\alpha s} = E^{Q_{xs}} \Big[Mu(x(\hat{\tau}))\, \chi_{\hat{\tau}<\infty}\, e^{-\alpha\hat{\tau}} +$$
$$+ \int_s^{\hat{\tau}} e^{-\alpha\lambda} L(x(\lambda;\tilde{\omega}))d\lambda \Big]$$
$$= E^{Q_{xs}} \Big[u(x(\hat{\tau}) + \hat{\xi}(x(\hat{\tau})))\, \chi_{\hat{\tau}<\infty}\, e^{-\alpha\hat{\tau}} +$$
$$+ c(x(\hat{\tau}),\ \hat{\xi}(x(\hat{\tau})))\, \chi_{\hat{\tau}<\infty}\, e^{-\alpha\hat{\tau}} +$$
$$+ \int_s^{\hat{\tau}} e^{-\alpha\lambda} L(x(\lambda;\tilde{\omega}))d\lambda \Big] .$$

Arguing as above, and using the definition of $\hat{\theta}^n, \hat{\xi}^n$, we then obtain

(4.58)
$$u(x) = E_{x,\hat{W}} \Big[\sum_{n=0}^{N-1} \chi_{\hat{\theta}^n < \infty} \int_{\hat{\theta}^n}^{\hat{\theta}^{n+1}} e^{-\alpha\lambda} L(x_n(\lambda))d\lambda +$$
$$+ \sum_{n=1}^{N} c(x_{n-1}(\theta^n),\ \hat{\xi}^n)\, e^{-\alpha\hat{\theta}^n} \chi_{\hat{\theta}^n < \infty} \Big] +$$
$$+ E_{x,W}\, \chi_{\hat{\theta}^N < \infty}\, e^{-\alpha\hat{\theta}^N} u(x_N(\hat{\theta}^N)) , \quad \forall N .$$

Since $c \geqslant k$, and since $L,\ u \geqslant 0$, we deduce from this that

$$\sum_{n=1}^{\infty} e^{-\alpha\hat{\theta}^n} \chi_{\hat{\theta}^n < \infty} < \infty \qquad P_{x,\hat{W}} \quad \text{a.s.}$$

Therefore we necessarily have $\hat{\theta}^n \to \infty$ $P_{x,\hat{W}}$ a.s. when $n \to \infty$. It then follows from (4.58) that

$$u(x) = J_x(\hat{W}) .$$

which completes the proof of the desired result. □

5. COMMENTS

We have given the interpretation of the Q.V.I. (1.6) in $H_0^1 \cap C^0$; to do this we have no further need of regularity.

The results of the unbounded case are given here for the first time. They are quite consistent with what we obtain from dynamic programming in discrete time (cf., A. BENSOUSSAN [2]).

It would be interesting to give a purely analytical proof of the uniquness result for the V.I. (3.26).

We note the role of the (local) estimates of Hölder type in proving the convergence of increasing and decreasing processes, since the technique of B. HANOUZET-J.L. JOLY does not apply in the unbounded case. We could, however, give a probabilistic proof of these results by using the explicit formula for the functions u_n and u^n, but at the price of additional assumptions about the function f.

In the evolutionary case this would probably be the better technique, since the Hölder estimates, although very likely, are not yet established in the literature.

In Section 4 we allowed a small confusion in notation. A priori $\tilde{Q}_{xt}$ or $\mathring{Q}_{xt}$ is a probabliity on $\tilde{\Omega}, \tilde{\eta}$. We wrote $P^{0}_{x,W} = Q_x$ as the probability on $\Omega, \tilde{\eta}_0$, and therefore assimilated $(\Omega, \tilde{\eta}_0)$ into a space of the type $(\tilde{\Omega}, \tilde{\eta})$. This is legitimate, since for the events of $\tilde{\eta}_0$ only the first component ω_0 of $(\omega_0, \omega_1, \ldots) = \tilde{\omega}$ counts. Similarly $(\tilde{\Omega}, \tilde{\eta}_n)$ for all n is assimilated into a space of the type $(\tilde{\Omega}, \tilde{\eta})$.

The limitation (4.42) on the admissable sequence of impulse times is not actually necessary (for the optimal control it is taken as satisfied). However, this assumption allows us to simplify the proof of Theorem 4.2. It allows us to avoid using conditional expectations. Thanks to this assumption we simply manipulate expectations possibly depending upon parameters. Thus we have a proof that is essentially algebraic.

For the probabilistic interpretation of discretised Q.V.I.'s we refer to A. BENSOUSSAN-M. ROBIN [1] and A. BENSOUSSAN [3].

For purely probabilistic aspects of impulse control we refer to J.P. LEPELTIER-B. MARCHAL [2], N. EL KAROUI [2], J.M. BISMUT [1], H. NAGAI [1].

Bibliography

S. AGMON, A. DOUGLIS, L. NIRENBERG [1]: 'Estimates near the Boundary for Solutions of Elliptic Partial Differential Equations', *Comm. Pure Appl. Math.*, **XII**, (1959), 623-727.

R.F. ANDERSON and A. FRIEDMAN [1]: 'A Quality Control Problem and Q.V.I.', *A.R.M.A.*, **63**, (1977), 205-252.

[2]: 'Multidimensional Quality Control Problem and Q.V.I.', *Trans. A.M.S.*, (1978).

R.F. ANDERSON and S. OREY [1]: *Small Random Perturbations of Dynamical Systems with Reflecting Boundary.*

J.P. AUBIN [1]: *Mathematical Methods of Game and Economic Theory*, (North Holland, Amsterdam), (1979).

C. BAIOCCHI [1]: 'Problemi di frontiera libera e disequazioni quasi-variazionali', *Atti. Giorn. Analisi Convessa e Applicazioni, Roma*, (1974), 11-20.

[2]: 'Free Boundary Problems in the Theory of Fluid Flow through Porous Media', *Proc. Int. Congress Math., Vancouver, Vol. II*, (1974), 237-243.

[3]: 'Studio di un problema quasi-variazionale convesso a problemi di frontiera libera', *Boll. U.M.I.*, (4), (1975), 589-613.

C. BAIOCCHI, A. CAPELO [1]: *Disequazioni variazionali e quasivariazionali. Applicazioni a problemi di frontiera libera, Vol.s 1,2*, (Pitagora Editrice, Bologna), (1978).

L. BARTHELEMY [1]: *Application de la théorie des semi-groupes non-linéaires dans L^∞ à l'étude d'une classe d'inéquations quasi-variationnelles*, (Thèse 3e cycle), (Université de Franche-Comté, Besançon , (1980); ('Application of Non-Linear Semi-Group Theory in L^∞ to the Study of a Class of Quasi-Variational Inequalities').

L. BARTHELEMY, F. CATTE 1 : *Application de la théorie des semi-groupes non-linéaires dans L^∞ a l'étude d'une classe d'inéquations quasi-variationnelles*, (to appear); ('Application of Non-Linear Semi-Group Theory in L^∞ to the Study of a Class of Quasi-Variational Inequalities').

J.A. BATHER [1]: 'A Control Chart Model and a Generalized Stopping Problem for Brownian Motion' *Math. Oper. Res.*, **1**, (1976), 209-224.

R. BELLMAN [1]: *Dynamic Programming*, (Princeton University Press), (1957).

A. BENSOUSSAN [1]: 'On the Semi-Group Approach to Variational and Quasi-Variational Inequalities', *Proceedings of the 1st Franco-South East Asian Conference on Mathematical Sciences.*

[2]: 'Stochastic Control in Discrete Time and Applications to the Theory of Production, *Math. Programming*.

[3]: *Stochastic Control by Functional Analysis Methods*, (North Holland, Amsterdam), (1982).

A. BENSOUSSAN, H. BREZIS and A. FRIEDMAN [1]: 'Estimates on the Free Boundary for Q.V.I., *Comm. P.D.E.*, **2**, (1977), 297-321.

A. BENSOUSSAN, M. CROUHY and J.M. PROTH [1]: *Mathematical Theory of Production Planning*, (North Holland), (1983).

A. BENSOUSSAN, J. FREHSE, U. MOSCO [1]: *A Stochastic Impulse Control Problem with Quadratic Growth Hamiltonian and the Corresponding Quasi-Variational Inequality*, *J. Reine Angew. Math.*.

A. BENSOUSSAN and A. FRIEDMAN [1]: 'Non-Zero Sum Stochastic Differential Games with Stopping Times and Free Boundary Problems', *Trans. A.M.S.*, **231**, (1977), 275-327.

A. BENSOUSSAN, M. GOURSAT et J.L. LIONS [1]: 'Contrôle impulsionnel et I.Q.V. stationnaires', *C.R. Acad. Sci. Paris*, **276**, (1973), 1279-1284; (Impulse Control and Stationary Q.V.I.'s).

A. BENSOUSSAN, G. HURST, B. NÅSLÜND [1]: *Management Applications of Modern Control Theory*, (North Holland), (1974).

A. BENSOUSSAN, R. KLEINDORFER, C.S. TAPIERO [1] (ed.s): *Applied Stochastic Control in Econometrics and Management Science*, (North Holland, Amsterdam), (1980).

A. BENSOUSSAN and J. LESOURNE [1]: 'Optimal Growth of a Self-Financing Firm in an Uncertain Environment', in *Applied Stochastic Control in Econometrics and Management Science*, (A. Bensoussan, R. Kleindorfer, C.S. Tapiero, ed.s), (North Holland, Amsterdam), (1980).

A. BENSOUSSAN et J.L. LIONS [1]: *Applications des Inéquations Variationnelles en Contrôle Stochastique*, (Dunod, Paris), (1978); (*Applications of Variational Inequalities in Stochastic Control*, (North Holland, Amsterdam), (1982)).

[2]: 'Nouvelle formulation de problèmes de contrôle impulsionnel et applications', *C.R. Acad. Sci. Paris*, **276**, (1973), 1189-1192; ('New Formulation of Impulse Control Problems and Applications').

[3]: 'Contrôle Impulsionnel et I.Q.V. d'évolution', *C.R. Acad. Sci. Paris*, **276**, (1973), 1333-1338; ('Impulse Control and Evolutionary Q.V.I.'s').

[4]: 'Contrôle impulsionnel et contrôle continu. Méthode des I.Q.V. non-linéaires', *C.R. Acad. Sci. Paris*, **278**, (1974), 675-679; (Impulse Control and Continuous Control. The Method of Non-Linear Q.V.I.'s').

[5]: 'Contrôle impulsionnel et systèmes d'I.Q.V.', *C.R. Acad. Sci. Paris*, **278**, (1974), 747-751; ('Impulse Control and Systems of Q.V.I.'s').

[6]: 'Sur de nouveaux problèmes aux limites pour des opérateurs hyperboliques', *C.R. Acad. Sci. Paris*, **278**, (1974), 1345-1349; ('On New Boundary Problems for Hyperbolic Operators').

[7]: 'Sur le contrôle impulsionnel et les I.Q.V. d'évolution', *C.R. Acad. Sci. Paris*, **280**, (1975), 1049-1053; ('On Impulse Control and Evolutionary Q.V.I.'s').

[8]: 'Nouvelles méthodes en contrôle impulsionnel', *J.A.M.O.*, **1**, (1975), 289-312; ('New Methods in Impulse Control').

[9]: 'Sur l'approximation numerique d'I.Q.V. stationnaires', *Proc. Int. Symp. Computing Methods in Applices Sciences and Engineering, Versailles 1973, Paris, II*, 325-338, *Lecture Notes in Computer Science*, **11**, (Springer-Verlag, Heidelberg), (1974); ('On the Numerical Approximation of Stationary Q.V.I.'s').

[10]: 'Propriétés des I.Q.V. décroissantes', in *Analyse Convexe et ses applications*, *Lecture Notes in Econ. Math. Systems*, **102**, (Springer-Verlag, Heidelberg), (1974), 66-84; ('Properties of Decreasing Q.V.I.'s', in *Convex Analysis and Its Applications*).

[11]: 'I.Q.V. dépendant d'un parametre', *Ann. S.N.S., Pisa*, **4**, (1977), 231-255; ('A Q.V.I. depending on a Parameter').

[12]: *Diffusion Processes in Bounded Domains and Singular Perturbation Problems for V.I.'s with Neumann Boundary Conditions. Probabilistic Methods in Differential Equations, Lecture Notes in Mathematics*, **451**, (Springer-Verlag, Heidelberg), (1974).

[13]: 'A Remark on an Equation of Dynamic Programming', *I.E.E.E. Automatic Control*, **AC-26**, No.5, (October, 1981), (Special Issue dedicated to R. Bellman).

A. BENSOUSSAN, J.L. MENALDI [1]: *Stochastic Control of Diffusion with Jumps in a Bounded Domain*, (to appear).

A. BENSOUSSAN, M. ROBIN [1]: *On the Convergence of the Discrete Time Dynamic Programming Equation for General Semi-Groups*, (*S.I.A.M. Control*).

P. BILLINGSLEY [1]: *Convergence of Probability Measures*, (Wiley), (1968).
G. BIRKHOFF [1]: *Lattice Theory*, A.M.S. Colloq. Publ., **XXV**, (A.M.S., New York), (1948).
M. BIROLI [1]: 'Sur les I.Q.V. paraboliques avec contraintes sur le bord', *C.R. Acad. Sci. Paris*, **283**, (1976), 705-708; ('On Parabolic Q.V.I.'s with Constraints at the Boundary').
[2]: 'Sur la G-convergence pour les I.Q.V.', *C.R. Acad. Sci. Paris*, **284**, (1977), 947-950; ('On G-Convergence for Q.V.I.'s').
[3]: 'Sur la G-convergence pour les I.Q.V.', *Boll. U.M.I.*, (5), (1977), 540-550; ('On G-Convergence for Q.V.I.'s').
J.M. BISMUT [1]: 'Contrôle de processus alternants et applications', *Z. Wahrscheinlichkeitstheorie*, **47**, (1979), 241-288; ('Control of Alternating Processes and Applications').
L. BOCCARDO et I. CAPUZZO DOLCETTA [1]: 'Disequazioni quasi variazionali con funzione d'ostacolo quasi-limitata, esistenza di soluzioni e G-convergenza', *Boll. U.M.I.*, (5), 15.B, (1978), 370-385.
C.M. BRAUNER and B. NICOLAENKO [1]: *Homographic Approximation of Free Boundary Problems Characterized by Elliptic Variational Inequalities*, (to appear).
A. BRETON et C. LEGUAY [1]: *Application du contrôle stochastique à la gestion des centrales thermiques et hydrauliques*, *Lecture Notes Econ. Math. Systems*, **107**, (Springer-Verlag, Heidelberg), (1975), 728-744; (Application of Stochastic Control to the Management of Thermal and Hydro-Electric Power Stations).
H. BREZIS, A. PAZY [1]: 'Accretive Sets and Differential Equations in Banach Spaces', *Israel J. Math.*, **9**, (1970), 367-383.
H. BUHLMANN [1]: *Mathematical Methods in Risk Theory*, (Springer-Verlag, Berlin), (1970).
L.A. CAFFARELLI, A. FRIEDMAN [1]: 'Regularity of the Solution of the Quasi-Variational Inequality for the Impulse Control Problem, I,II', *Communications in P.D.E.*, **3**, (1978), 745-753.
M. CHALEYAT-MAUREL, N. EL KAROUI et B. MARCHAL [1]: *Réflexion discontinue et systèmes stochastiques*, (to appear).
P. CHARRIER, G.M. TROIANIELLO [1]: 'On Strong Solutions to Parabolic Unilateral Problems with Obstacles Dependent on Time', *J. Math. Anal. Applic.*, **64**, (1978).
P. CHARRIER et M.A. VIVALDI [1]: 'Existence d'une solution reguliere d'une I.Q.V. d'evolution avec conditions de Dirichlet', *Boll. U.M.I.*, (5), (1977), 579-589; ('Existence of a Regular Solution of an Evolutionary Q.V.I. with Dirichlet Boundary Conditions').
P. COLLETER, F. DELEBECQUE, F. FALGARONE, J.P. QUADRAT [1]: 'Application of Stochastic Control Methods to the Management of Energy Production in New Caledonia', in *Applied Stochastic Control in Econometrics and Management Science*, (A. Bensoussan, P. Kleindorfer, C.S. Tapiero, ed.s), (North Holland), (1980).
V. COMINCIOLI [1]: 'On Some Oblique Derivative Problems Arising in the Fluid Flow in Porous Media', *J.A.M.O.*, **1**, (1975), 313-336.
M.G. CRANDALL, T.M. LIGGETT [1]: 'Generation of Semi-Groups of Nonlinear Transformations in General Banach Spaces', *Amer. J. Math.*, **93**, (1977), 265-298.
F. DELEBECQUE, J.P. QUADRAT [1]: 'Applications of Stochastic Control Methods in the Management of Hydropower Production', in *1st International Conference on Mathematical Modeling, September 1977, St Louis, (Missouri)*.
[2]: *Contribution of Stochastic Control Singular Perturbation Averaging and Team Theories to an Example of Large-Scale Systems: Management of Hydropower Production*.
[3]: 'Application de l'identification et du contrôle stochastique à la gestion des réservoirs, *Colloque sur la Theorie des Systèmes et la gestion des Services Publics*, (Presses Université, Montréal), (1975); ('Application of Identification and Stochastic Control to the Management of Reservoirs').
I.C. DOLCETTA, M.A. VIVALDI [1]: 'Existence d'une solution reguliere d'une I.Q.V. elliptique sur un domaine non borne', *C.R. Acad. Sci. Paris*, **284**, (1977), 1033-1036; ('Existence of a Regular Solution of an Elliptic Q.V.I. on an Unbounded Domain').
C. DOLEANS-DADE [1]: 'Quelques applications de la formule de changement de variables pour les semi-martingales', *Z. Wahrscheinlichkeitstheorie*, (1970), 181-194; ('Some Applications of the Change of Variables Formula for Semi-Martingales').

C. DOLEANS-DADE, P.A. MEYER [1]: *Intégrales stochastiques par rapport aux martingales locales, Séminaire de Probabilités, IV, Lecture Notes in Mathematics*, (Springer-Verlag, Heidelberg); (*Stochastic Integrals with Respect to Local Margingales, Probability Seminar, IV*).
G. DUVAUT et J.L. LIONS [1]: *Les Inéquations en Mécanique et en Physique*, (Dunod Editeur, Paris), (1972); (*Inequalities in Mechanics and Physics*, (Springer), (1976)).
E.B. DYNKIN [1]: *Theory of Markov Processes*, (Pergamon Press, New York), (1960).
[2]: *Markov Processes*, (Springer-Verlag, Berlin), (1965).
I. EKELAND, R. TEMAM [1]: *Analyse Convexe et Problèmes Variationnels*, (Dunod, Paris), (1974); (*Convex Analysis and Variational Problems*, (North Holland)).
N. EL KAROUI [1]: *Processus de réflexion sur* $\mathbb{R}^n$*, Séminaire de Probabilités, IX, Lecture Notes in Mathematics*, **465**, (Springer-Verlag, Heidelberg), (1975); (*Reflection Processes on* $\mathbb{R}^n$*, Probability Seminar, IX*).
[2]: *Méthodes probabilistes en contrôle stochastique*, (to appear); (*Probability Methods in Stochastic Control*).
L.C. EVANS, J.L. MENALDI [1]: *Gradient Bounds for Solutions of Degenerate Variational Inequalities*, (to appear).
G. FAYOLLE and M. ROBIN [1]: 'Optimum Queuing Policies for Multi-Processor Computers', in *Modelling and Performance Evaluation of Computer Systems*, (North Holland), (1976).
W. FLEMING [1]: 'Optimal Control of Markov Processes', *International Congress of Mathematicians*, (Warsaw), (1983).
W. FLEMING and R. RISHEL [1]: *Deterministic and Stochastic Optimal Control*, (Springer-Verlag, Berlin), (1975).
J. FREHSE [1]: 'On the Signorini Problem and Variational Problems with Thin Obstacles', *Ann. S.N.S. Pisa, IV*, **2**, (1977), 343-362.
[2]: 'On the Smoothness of Solutions of V.I.'s with Obstacles', *Proc. Semester on P.D.E., Banach Center, Warsaw, 1978.*
[3]: Personal communication.
J. FREHSE et U. MOSCO [1]: 'Sur la régularité des solutions faibles de certaines inéquations variationnelles et quasi-variationnelles non-linéaires du contrôle stochastique', *C.R. Acad. Sci. Paris*, (1979); ('On the Regularity of Weak Solutions of Certain Non-Linear Variational and Quasi-Variational Inequalities of Stochastic Control').
[2]: 'Variational Inequalities with One-Sided Irregular Obstacles', *Manuscripta Math.*, **28**, (1979), 219-233.
[3]: *Irregular Obstacles and Quasi-Variational Inequalities of Stochastic Impulse Control*, (to appear).
A. FRIEDMAN [1]: 'A Class of Parabolic Q.V.I.'s, II.', *J. Diff. Eq.*, **22**, (1976), 379-401.
[2]: 'Optimal Stopping Problems in Stochastic Control', *S.I.A.M. Review*, **21**, (1979), 71-80.
[3]: *Stochastic Differential Equations, Vol.s I, II*, (Academic Press, New York), (1976).
A. FRIEDMAN and R. JENSEN [1]: 'A Parabolic Q.V.I. Arising in Hydraulics', *Ann. S.N.S. Pisa*, (4), (1975), 421-468.
[2]: 'Elliptic Q.V.I. and Applications to a Non-Stationary Problem in Hydraulics', *Ann. S.N.S. Pisa*, (4), (1976), 47-88.
A. FRIEDMAN, D. KINDERLEHRER [1]: 'A Class of Parabolic Q.V.I.'s', *J. Diff. Eq.*, **21**, (1976), 395-416.
A. FRIEDMAN and M. ROBIN [1]: 'The Free Boundary for V.I.'s with Non-Local Operators', *S.I.A.M. J. Control and Optimization*, **16**, (1978), 347-372.
E. GAGLIARDO [1]: 'Ulteriori proprieta di alcune classi di funzioni in piu variabili', *Ricerche Mat.*, **8**, (1959), 24-51.
M.G. GARRONI, B. HANOUZET, J.L. JOLY [1]: *Régularité pour la solution d'un système d'I.Q.V.*, (Université de Bordeaux 1).
M.G. GARRONI, G.M. TROIANIELLO [1]: 'Some Regularity Results and A Priori Estimates for Solutions of Variational and Quasi-Variational Inequalities', in *Proc. Conf. on Recent Methods in Non-Linear Analysis and Applications, Roma, 1978*, (E. de Giorgi, E. Magenes, U. Mosco, ed.s), (Pitagora Editori, Bologna), (1979).

F.W. GEHRING [1]: 'The L^p-Integrability of the Partial Derivatives of a Quasi-Conformal Mapping', *Acta Math.*, **130**, (1973), 265-277.
M. GIAQUINTA, G. MODICA [1]: 'Regularity Results for Some Classes of Higher Order Non-Linear Elliptic Systems', *J. Reine Angew. Math.*.
I.I. GIKHMAN, A.V. SKOROKHOD [1]: *The Theory of Stochastic Processes, Vol.s I, II, III*, (Springer-Verlag, Berlin), (1975).
G. GILARDI [1]: 'Studio di una disequazione quasi-variazionale relativa ad un problema di filtrazione in tre dimensioni', *Ann. M.P.A.*, (4), (1977), 1-17.
R. GONZALES [1]: (Thèse de 3e cycle), (Université Paris Dauphine), (1979).
R. GONZALES, E. ROFMAN [1]: Paper presented to the *Colloquium on Systems, Tel Aviv, December 1980*, (to appear).
M. GOURSAT et G. MAAREK [1]: 'Nouvelle approche des problèmes de gestion de stocks. Comparaison avec les méthodes classiques', *Rapport Laboria*, No.148, (1976); ('New Approach to Stock Management Problems. Comparison with Classical Methods').
M. GOURSAT et S. MAURIN [1]: *Méthodes de résolution numérique des I.Q.V., Proc. International Symposium at Rocquencourt, Lecture Notes Econ. Math. Systems*, **107**, (1975), 585-609; (*Numerical Methods of Solving Q.V.I.'s*).
M. GOURSAT et J.P. QUADRAT [1]: 'Analyse numerique d'I.Q.V. elliptiques associees a des problemes de controle impulsionnel', *Rapport Laboria*, (1976); ('Numerical Analysis for Elliptic Q.V.I.'s Associated with Impulse Control Problems').
P. GRISVARD [1]: *Boundary Value Problems in Non-Smooth Domains*, (North Holland), (1982).
B. HANOUZET et J.L. JOLY [1]: 'Un résultat de régularité pour une I.Q.V. du type de Neumann intervenant dans un problème de contrôle impulsionnel', *J.M.P.A.*, **9**, (1977), 327-337; ('A Regularity Result for an I.Q.V. of Neumann Type Arising in a an Impulse Control Problem).
[2]: 'Méthode d'ordre dans l'interprétation de certaines inéquations variationnelles et applications', *C.R. Acad. Sci. Paris, Série A*, **281**, (1975), 373; ('Ordering Method in the Interpretation of Certain Variational Inequalities, and Applications').
[3]: 'Convergence uniforme des itérés définissant la solution d'une inéquation quasi-variationnelle', *C.R. Acad. Sci. Paris, Série A*, **286**, (1978); ('Uniform Convergence of Iterates Defining the Solution of a Quasi-Variational Inequality').
[4]: 'Résultats de régularité pour certaines I.Q.V.', *U.E.R. Math. et Informatique*, (Université de Bordeaux 1), (1978); ('Regularity Results for Certain Q.V.I.'s').
C. HOLLAND [1]: *A New Energy Characterization of the Smallest Eigenvalue of the Schrödinger Equation*, (Preprint).
J.L. JOLY [1]: 'Résultats d'existence et de régularité pour certaines inéquations quasi-variationnelles', in *Lectures at the School on Recenti sviluppi ed applicazioni della teoria delle disequazioni variazionali, Erice 2-12 marzo 1975*.
J.L. JOLY et U. MOSCO [1]: 'Sur les I.Q.V.', *C.R. Acad. Sci. Paris*, **279**, (1974), 499-502; ('On Q.V.I.'s').
[2]: 'Remarques sur les I.Q.V.', in *Proc. International Symposium at Rocquencourt, Lecture Notes Econ. Math. Systems*, **107**, (Springer-Verlag, Heidelberg), (1975), 625-642; ('Remarks on Q.V.I.'s').
J.L. JOLY, U. MOSCO and G. TROIANIELLO [1]: 'On the Regular Solution of a Quasi-Variational Inequality Connected with a Problem of Stochastic Impulse Control', *J.M.A.A.*, **61**, (1979), 2.
T. KOMATSU [1]: 'Markov Processes Associated with Certain Integro-Differential Operators', *Osaka J. Math.*, **10**, (1973), 271-303.
N. KRYLOV [1]: *Controlled Diffusion Processes*, (Springer-Verlag, Berlin), (1980).
P. LABORDE [1]: 'Problèmes Quasi-Variationnels en viscoplasticité avec écrouissage', *C.R. Acad. Sci. Paris*, **283**, (1976), 393-396; ('Quasi-Variational Problems in Viscoplasticity with Hammer-Hardening').
O.A. LADYZHENSKAYA, V.A. SOLONNIKOV, N.N. URAL'TSEVA [1]: *Linear and Quasi-Linear Equations of Parabolic Type*, Vol. 23, (American Mathematical Socity, Providence), (1968).
O.A. LADYZHENSKAYA, N.N. URAL'TSEVA [1]: *Linear and Quasi-Linear Elliptic Equations*, (Academic Press, New York), (1979).

Th. LAETSCH [1]: 'A Uniqueness Theorem for Elliptic Q.V.I.'s', *J. Func. Analysis*, **18**, (1975), 286-287.
J.M. LASRY [1]: *Contrôle stochastique ergodique*, (Thèse d'Etat), (Université Paris Dauphine), (1974).
C. LEGUAY [1]: 'Application of Stochastic Control to the Problem of Optimal Energy Management', in *Applied Stochastic Control in Econometrics and Management Science*, (A. Bensoussan, P. Kleindorfer, C.S. Tapiero ed.s), (North Holland), (1980).
S. LENHART: *Integro-Differential Operators Associated with Diffusion Processes with Jumps*, (to appear).
J.P. LEPELTIER, B. MARCHAL [1]: 'Problème des martingales et équations différentielles stochastiques associées à un opérateur integro-différentiel', *Ann. Inst. Henri Poincaré*, **XII**, No.1, (1976), 43-103; ('Martingale Problems and Stochastic Differential Equations Associated with an Integro-Differential Operator').
[2]: *Théorie générale du contrôle impulsionnel*, (to appear).
J.L. LIONS [1]: 'Sur la théorie du contrôle', *Proc. International Congress of Mathematicians, Vancouver 1974, Vol. 1*, 139-154; ('On the Theory of Control').
[2]: 'On the Numerical Approximation of Problems of Impulse Controls', *Proc. Optimisation Techniques, Lecture Notes in Computer Science*, **27**, (1975), 232-251, (Springer-Verlag).
J.L. LIONS, E. MAGENES [1]: *Problemes aux limites non-homogenes et applications, Vol.s 1, 2*, (1968), *Vol. 3*, (1970), (Dunod, Paris); (*Non-Homogeneous Boundary Problems and Applications*, (Springer-Verlag), (1972)).
J.L. LIONS et B. MALGRANGE [1]: 'Sur l'unicité rétrograde dans les problèmes mixtes paraboliques', *Math. Scand.*, **8**, (190), 277-286; ('On Backward Uniqueness in Parabolic Mixed Problems').
J.L. LIONS, G. STAMPACCHIA [1]: 'Variational Inequalities', *Comm. Pure Applied Math.*, **XX**, (1967), 493-519.
P.L. LIONS [1]: *Generalized Solutions of Hamilton-Jacobi Equations*, (Pitman), (1982).
[2]: 'Hamilton-Jacobi-Bellman Equations and the Optimal Control of Stochastic Systems, *International Congress of Mathematicians*, (Warsaw), (1983).
P.L. LIONS et J.L. MENALDI [1]: 'Optimal Control of Stochastic Integrals and Hamilton-Jacobi-Bellman Equations', *S.I.A.M. Control and Opt.*, **20**, (1982), 58-81, 82-95.
P.L. LIONS, J.L. MENALDI and A.S. SNITSMAN [1]: 'Construction de processus de diffusion réfléchis par pénalisation du domaine', *C.R. Acad. Sci. Paris*, **1**, **292**, (1981), 559-562; ('Construction of Reflected Diffusion Processes by Penalisation of the Domain').
E. MAGENES [1]: 'Topics in Parabolic Equations: Some Typical Free Boundary Problems', in *Boundary Value Problems for Linear Evolution Partial Differential Equations*, (D. Reidel, Dordrecht), (1977), 239-312.
M. MATZEU, M.A. VIVALDI [1]: 'On the Regular Solution of a Non-Linear Parabolic Quasi-Variational Inequality Related to a Stochastic Control Problem', *Commun. P.D.E.*.
J. MENALDI [1]: 'Sur le problème de contrôle impulsionnel et l'I.Q.V. associée', *C.R. Acad. Sci. Paris*, **284**, (1977), 1499-1502; ('On the Impulse Control Problem and the Associated Q.V.I.').
[2]: Thesis, (Paris), (1980).
[3]: 'On the Optimal Impulse Control Problem for Degenerate Diffusions', *S.I.A.M. Control and Opt.*, **18**, No.6, (1980).
[4]: 'Stochastic Variational Inequality for Reflected Diffusion', *Indiana Univ. Math. J.*, **32**, (1983), No. 4.

J.L. MENALDI and M. ROBIN [1]: Book on *Quasi-Variational Inequalities and Impulse Control*, (to be published).
P.A. MEYER [1]: *Probabilités et Potentiel*, (Hermann, Paris), (1966); (*Probability and Potential*).
N.G. MEYERS, A. ELCRAT [1]: 'Some Results on Regularity for Solutions of Non-Linear Elliptic Systems and Quasi-Regular Functions', *Duke Math. J.*, **42**, (1975), 121-136.
J.C. MIELLOU [1]: 'A Mixed Relaxation Algorithm Applied to Q.V.I.'s', *Lecture Notes in Computer Science*, **41**, (1976), 192-199, (Springer-Verlag).

F. MIGNOT [1]: 'Contrôle dans les I.V. elliptiques', *J. Func. Anal.*, **22**, (1976), 130-185; ('Control in Elliptic V.I.'s').
F. MIGNOT et J.P. PUEL [1]: 'I.V. Paraboliques avec convexes dépendant du temps. Applications aux I.Q.V. d'évolution', *A.R.M.A.*, **64**, (1977), 59-91; ('Parabolic V.I.'s with Time-Dependent Convex Sets. Applications to Evolutionary Q.V.I.'s').
[2]: 'I.V. et I.Q.V. hyperboliques du 1er ordre', *J.M.P.A.*, (9), (1976), 353-378; ('Hyperbolic V.I.'s and Q.V.I.'s of the First Order').
U. MOSCO [1]: 'On Some Non-Linear Quasi-Variational Inequalities and Implicit Complementarity Problems in Stochastic Control Theory', in *Proc. Internat. School on V.I.'s and Applications, Erice, 1978*, (W. Cottle, Giannessi, and J.L. Lions ed.'s), (Wiley), (1980).
[2]: 'Non-Linear Q.V.I.'s and Stochastic Impulse Control Theory', *Proc. Conf. on Differential Equations, IV, Prague, 1977*, Lecture Note Series, (Springer-Verlag).
[3]: 'Implicit Variational Problems and Quasi-Variational Inequalities', in *Summer School on 'Non-Linear Operators and the Calculus of Variations', Brussels, 1975.*
C.B. MORREY [1]: *Multiple Integrals in the Calculus of Variations*, Die Grundlagen der Mathematischen Wissenschaften, **130**, (Springer-Verlag, Berlin-Heidelberg-New York), (1966).
J. MOSSINO [1]: 'Sur certains I.Q.V. apparaissant en physique', *C.R. Acad. Sci. Paris*, **282**, (1976), 187-190; ('On Certain Q.V.I.'s Appearing in Physics').
H. NAGAI [1]: 'On an Impulse Control of Additive Processes', *Z. Wahrscheinlichkeitstheorie*, 53, (1980), 1-16.
O. NAKOULIMA [1]: *Etude d'une I.V. bilaterale et d'un système d'I.Q.V. unilatérales associées*, (Thesis), (Bordeaux), (1977); (*Study of a Two-Sided V.I. and of a System of Associated One-Sided Q.V.I.'s*).
J. NEVEU [1]: *Calcul des Probabilités*, (Masson, Paris), (1964); (*Probability Calculus*).
L. NIRENBERG [1]: 'An Extended Interpolation Inequality', *Ann. Scuola Norm. Sup. Pisa*, **20**, (1966), 733-737.
M. NISIO [1]: 'On a Non-Linear Semi-Group Attached to Stochastic Optimal Control', *R.I.M.S. Kyoto Univ.*, **13**, (1976), 513-537.
K.R. PARTHASARATHY [1]: *Probability Measures on Metric Spaces*, (Academic Press, New York), (1967).
M. PIERRE [1]: *Problèmes d'évolution avec constraintes unilatérales et potentiels paraboliques*, (to appear); (*Evolutionary Problems with One-Sided Constraints and Parabolic Potentials*).
J.P. QUADRAT [1]: Thesis, (Paris), (1981).
M. ROBIN [1]: 'Contrôle impulsionnel avec retard pour les processus de diffusion', *C.R. Acad. Sci. Paris*, **282**, (1976), 463-466; ('Impulse Control with Delay for Diffusion Processes').
[2]: 'Contrôle optimal de files d'attente', *Rapport Laboria*, No.117, (IRIA-Laboria), (1975).
[3]: Thesis, (Paris), (1978).
[4]: 'On Some Impulse Control Problems with Long Run Average Cost', *S.I.A.M. Control Opt.*, (1980).
S.P. SETHI, G. THOMPSON [1]: Book on *Production Theory*, (to be published).
M. SHIMA [1]: 'Analytical Construction of the Free Boundary for Non-Linear Impulse Control Problems and Its Application to the Optimal Design of Aerator Systems', *Rapport Laboria*, No.194, (IRIA-Laboria), (1976).
A.V. SKOROKHOD [1]: *Studies in the Theory of Random Processes*, (Addison-Wesley), (1965).
D. STROOCK [1]: 'Diffusion Processes Associated with Levy Generators', *Z. Wahrscheinlichkeitstheorie*, **32**, (1975), 209-244.
D. STROOCK and S.R.S. VARADHAN [1]: 'Diffusion Processes with Boundary Conditions', *Comm. Pure Appl. Math.*, **24**, (1971), 147-225.
[2]: *Multi-Dimensional Diffusion Processes*, (Springer-Verlag, Berlin), (1979).
L. TARTAR [1]: 'I.Q.V. abstraites', *C.R. Acad. Sci. Paris*, **278**, (1974), 1193-1196; ('Abstract Q.V.I.'s').

[2]: 'I.Q.V.', *Atti Giornate Analisi convessa e applicazioni, Roma*, (1974), 94-100; ('Q.V.I.'s').
R.T. VESCAN [1]: 'Quasi-Variational Inequalities Solved by a Non-Void Intersection Property', *J.M.A.A.*, (1982), (to appear).
S. WATANABE [1]: 'On Stochastic Differential Equations for Multi-Dimensional Diffusion Processes with Boundary Conditions', *J. Math. Kyoto Univ.*, **11**, No.1, (1971), **11**, No.3, (1971).
J. ZABCZYK [1]: *Semi-Group Methods in Stochastic Control*, (University of Montreal), (CRM), (working paper).
[2]: Address to the *International Congress of Mathematicians*, (Warsaw), (1983).

Achevé d'imprimer par l'Imprimerie Dumas 42100 Saint-Étienne (Loire)
Dépôt légal : juin 1984 — N° Imprimeur : 26598
Imprimé en France